Condensed Matter Theories

Volume 22

Condensed Matter Theories

Volume 22

Heidi Reinholz
Institut für Physik
Universität Rostock, Germany

Gerd Röpke
Institut für Physik
Universität Rostock, Germany

Manuel de Llano
Instituto de Investigaciones en Materiales
Universidad Nacional Autonoma de Mexico, Mexico

Editors

World Scientific

NEW JERSEY • LONDON • SINGAPORE • BEIJING • SHANGHAI • HONG KONG • TAIPEI • CHENNAI

Published by

World Scientific Publishing Co. Pte. Ltd.
5 Toh Tuck Link, Singapore 596224
USA office: 27 Warren Street, Suite 401-402, Hackensack, NJ 07601
UK office: 57 Shelton Street, Covent Garden, London WC2H 9HE

British Library Cataloguing-in-Publication Data
A catalogue record for this book is available from the British Library.

CONDENSED MATTER THEORIES — Volume 22

ISBN-13 978-981-270-958-5
ISBN-10 981-270-958-4

PREFACE

The *Thirtieth International Workshop on Condensed Matter Theories* (CMT30) was held at the Max-Planck-Insitut für Physik Komplexer Systeme (MPIPKS) in Dresden, Germany on 5–10 June, 2006. It provided a forum in which to discuss advanced theoretical research in the area of quantum many-body theory. The Workshop was highly interdisciplinary, with the emphasis placed on the common concerns of theorists applying many-particle methods in such diverse areas as solid-state, low-temperature, nuclear, particle, statistical, and biological physics, as well as in quantum field theory, quantum information and the theory of complex systems. Additionally, overview lectures on current experimental research with respect to strongly-correlated electronic systems were included in the program.

The series of *International Workshop on Condensed Matter Theories* was initiated in 1977 in São Paulo, Brazil as a small two-week-long gathering of about 20 physicists. Since then, it has been held annually and uninterruptedly. Although the original meetings in the series were designed to bring together US and Latin American physicists working in similar fields, later workshops have become truly international meetings which have been held at various locations around the world. The Brazil meeting was followed by later workshops in Trieste, Italy (1978); Buenos Aires, Argentina (1979); Caracas, Venezuela (1980); Mexico City (1981); St. Louis, USA (1982); Altenberg, Germany (1983); Granada, Spain (1984); San Francisco, USA (1985); Argonne, USA (1986); Oulu, Finland (1987); Taxco, Mexico (1988); Campos do Jordão, Brazil (1989); Elba, Italy (1990); Mar del Plata, Argentina (1991); San Juan, Puerto Rico (1992); Nathiagali, Pakistan (1993); Valencia, Spain (1994); Caracas, Venezuela (1995); Pune, India (1996); Luso, Portugal (1997); Nashville, USA (1998); Ithaca, Greece (1999); Buenos Aires, Argentina (2000); Canberra, Australia (2001); Luso, Portugal (2002); Toulouse, France (2003); St. Louis, USA (2004); and Kizu, Kyoto, Japan (2005). The MPIPKS in Dresden was eminently worthy of the Workshop's 30th anniversary. It served in particular as a place where the physics of complex systems was at the center of attention.

Since 1985, workshop invited talks have been published annually in a *Condensed Matter Theories* proceedings series, thus making the main results of these Workshops available to a larger community. The present volume of *Condensed Matter Theories* contains the contributions of most of the invited talks delivered at the Dresden workshop as well as a selection from the poster presentations. Points of view exchanged between participants from distinct scientific branches in modern techniques used in condensed many-body theories gave rise to a highly stimulating atmosphere. The main areas covered at this workshop were: Fermi liquids, Bose liquids, strongly-correlated electronic systems, magnetism, superconductivity, nuclear systems, density-functional theory, molecular-dynamics simulations, biophysics, quantum information, and new formalisms.

We would like to express our deep gratitude to Professor Peter Fulde and the host institute for its excellent workshop facilities. We gratefully thank the Max Planck Society for financial support. Also, we thankfully acknowledge the financial support of the US Army Research Office in support of US participants.

Special thanks go to Marita Schneider for providing the organizational infrastructure, and also to John Adams for taking on the technical management. All of the participants enjoyed the opportunity to spend some time in the beautifully restored city of Dresden and its surroundings, such as the "Swiss Saxonian region" and the "Festung Königstein" that most of us visited during an excursion.

On behalf of the Program Committee and the Local Organizing Committee, we thank all the members of the International Advisory Committee for their constant support. We wish the organizers of future Workshops the very best in fostering the fine tradition of these *International Workshops on Condensed Matter Theories*, which will continue to flourish.

Editors

H. Reinholz, Rostock, Germany
G. Röpke, Rostock, Germany
M. de Llano, Mexico City, Mexico
28 February 2007

Local Organizing Committee

for the XXXth International Workshop on Condensed Matter Theories

P. Fulde	– Max-Planck-Institut für Physik Komplexer Systeme, Dresden, Germany
H. Reinholz (Chair)	– Universität Rostock, Germany
G. Röpke (Vice Chair)	– Universität Rostock, Germany
M. Schneider	– Max-Planck-Institut für Physik Komplexer Systeme, Dresden, Germany

Program Committee

for the XXXth International Workshop on Condensed Matter Theories

R. F. Bishop	– UMIST, Manchester, UK
M. P. Das	– The Australian National University, Canberra, Australia
P. Fulde	– Max-Planck-Institut für Physik Komplexer Systeme, Dresden, Germany
G. Ortiz	– Indiana University, USA
H. Reinholz (Chair)	– Universität Rostock, Germany
G. Röpke (Vice Chair)	– Universität Rostock, Germany

Participants

of the XXXth International Workshop on Condensed Matter Theories

John Adams	Rostock, Germany
Sasha Alexandrov	Loughborough, UK
Gerasimos Anagnostatos	Attiki, Greece
Frithjof Anders	Bremen, Germany
George Baker Jr.	Los Alamos, USA
Christian D. Batista	Los Alamos, USA
Raymond F. Bishop	Manchester, UK
Lesser Blum	Piscataway, USA
Helga Böhm	Linz, Austria
Henrik Bohr	Lyngby, Denmark
Jordi Boronat	Barcelona, Spain
Wolfram Brenig	Braunschweig, Germany
John Briggs	Freiburg, Germany
Mark S. Byrd	Illinois,USA
Charles Campbell	Minneapolis, USA
Andrey V. Chubukov	Madison, USA
John W. Clark	St. Louis, USA
Frank Crowne	Adelphi, USA
Mukunda P. Das	Canberra, Australia
Manuel de Llano	Mexico City, Mexico
Willem H. Dickhoff	St. Louis, USA
Helmut Eschrig	Dresden, Germany
Carsten Fortmann	Rostock,Germany
Shigeji Fujita	Buffalo, USA
Peter Fulde	Dresden, Germany
Paola Gori-Giorgi	Paris, France
Hardy Gross	Berlin, Germany
Susana Hernández	Buenos Aires, Argentina
Naoum Karchev	Sofia, Bulgaria
Victor Khodel	St. Louis, USA
Dmitri Khveshchenko	Chapel Hill, USA
Marcus Kollar	Augsburg, Germany
Eckhard Krotscheck	Linz, Austria
Karl Kürten	Wien, Austria
Abdollah Langari	Dresden, Germany
Howard Lee	Athens, USA

David E. Logan	Oxford, UK
Kavita Louis	Dresden, Germany
Eduardo Ludeña	Caracas, Venezela
Fazley Bary Malik	Illinois, USA
Tofik Mamedov	Ankara, Turkey
Michael D. Miller	Washington State, USA
Lubos Mitas	Raleigh, USA
Klaus Morawetz	Chemnitz, Germany
David Neilson	Camerino, Italy
Satoshi Nishimoto	Dresden, Germany
Banaz Omar	Rostock, Germany
Beate Paulus	Dresden, Germany
Axel Pelster	Essen, Germany
Araceli Noemi Proto	Buenos Aires, Argentina
João da Providência	Coimbra, Portugal
Michael Pustilnik	Atlanta, USA
Khandker F. Quader	Ohio, USA
Thomas Raitza	Rostock, Germany
Julius Ranninger	Grenoble, France
Ronald Redmer	Rostock, Germany
Heidi Reinholz	Rostock, Germany
Johannes Richter	Magdeburg, Germany
Manfred L. Ristig	Köln, Germany
Gerd Röpke	Rostock, Germany
Erich Runge	Ilmenau, Germany
Mikko J. Saarela	Oulu, Finnland
Virulh Sa-yakanit	Bangkog, Thailand
Kevin Schmidt	Tempe, USA
Urich Schollwöck	Aachen, Germany
Peter Schuck	Orsay, France
Andrea Sengebusch	Rostock, Germany
Heinrich Stolz	Rostock, Germany
Eric Suraud	Toulouse, France
Bilal Tanatar	Ankara, Turkey
Joseph Tiggesbäumker	Rostock, Germany
Arkadiusz Wojs	Wroclaw, Poland
Paul Ziesche	Dresden, Germany
Roland Zimmermann	Berlin, Germany
Gertrud Zwicknagl	Braunschweig, Germany

CONTENTS

Part A Fermi Liquids

Part J New Formalisms

Part A

Fermi Liquids

PRESSURE COMPARISON BETWEEN THE SPHERICAL CELLULAR MODEL AND THE THOMAS-FERMI MODEL

GEORGE A. BAKER, JR

Los Alamos National Laboratory, Theoretical Division,
Los Alamos, NM 87545, USA
gbj@viking.lanl.gov

The widely used Thomas Fermi model always produces pressure which is less than or equal to that of the ideal Fermi gas. On the other hand the spherical cellular model, in certain regions of the temperature-density plane, can produce pressures which are greater than that of the ideal gas. This phenomenon is investigated.

Keywords: Thomas-Fermi model; spherical cellular model.

1. Introduction

I have found an interesting feature of the pressure as produced by the spherical cellular model. Except for hydrogen, for low to moderate temperatures and densities, the pressure can exceed that of the ideal Fermi gas very significantly. It has been shown previously[1] that for certain types of attractive potentials that this result is possible. This effect is seen in the spherical cellular in part of the one-phase region. There is a physical explanation for this pressure phenomenon. As the electrons descend into the nuclear potential well, they move faster and faster which motion results in larger and larger momentum transfer at the surface of the sphere. That means that the pressure can be quite large.

On the other hand as I show, the pressure for the widely used Thomas-Fermi model. (See Figure 1.) is always less than or equal to that for the ideal gas. This disparity (Compare Figure 2 with Figure 1.) in theoretical results may have very significant effects in some practical applications. It is, I think, of considerable interest to resolve whether in actual physical systems this feature occurs.

The system under consideration is a partially ionized gas of nuclei of charge Ze where $-e$ is the electronic charge. I will compare two models of this system where each is confined to a finite sized sphere.

2. Thomas-Fermi Model

First the Thomas-Fermi theory[2] is given by

$$\rho = \int_0^\infty \frac{2 \times 4\pi p^2/h^3}{\exp[(p^2/2m - eV)/kT + \eta] + 1} dp, \tag{1}$$

where eV is the potential energy, It is convenient to define the auxiliary functions

$$I_n(\eta) = \int_0^\infty \frac{y^n}{\exp(y - \eta) + 1} dy. \tag{2}$$

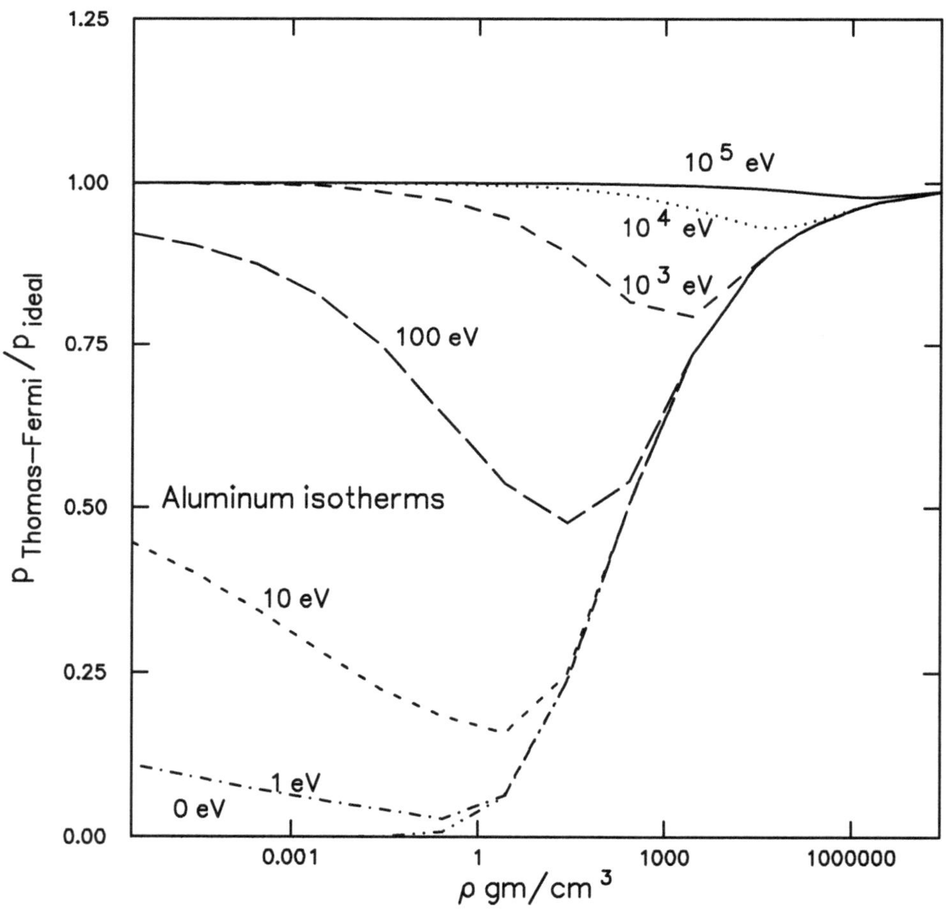

Fig. 1. The Thomas-Fermi model electron pressure divided by the ideal gas pressure vs. density for the case of Aluminum ($Z = 13$).

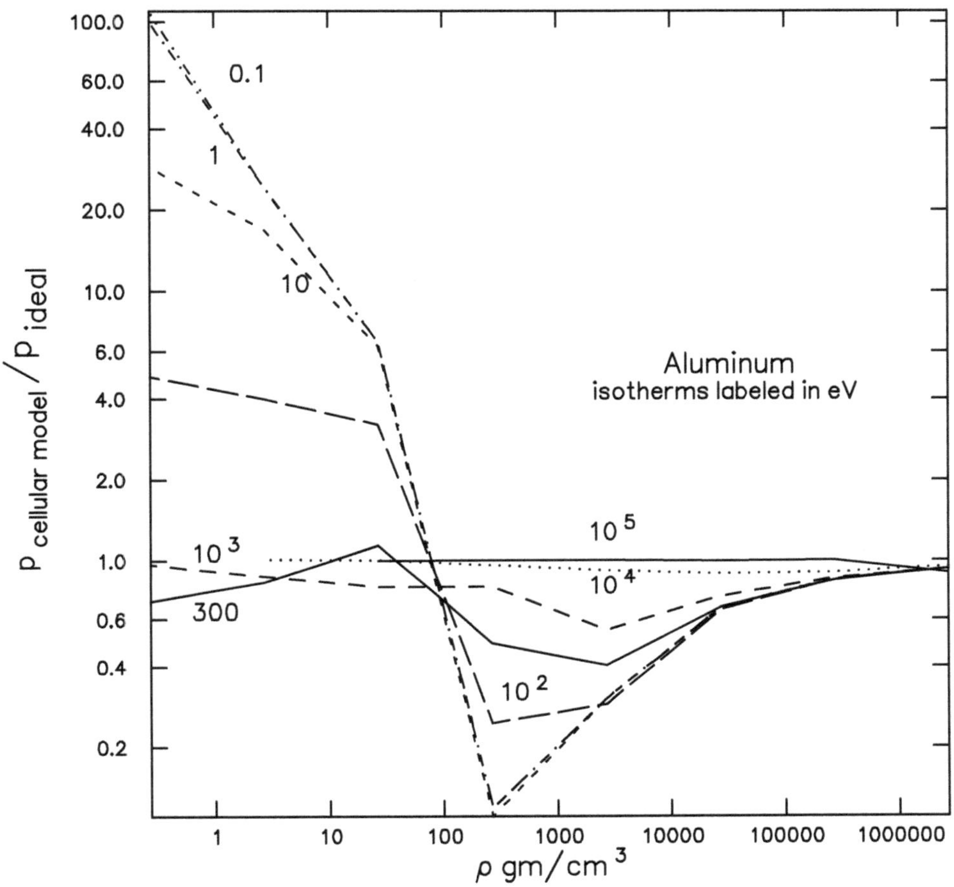

Fig. 2. The spherical cellular model electron pressure divided by the ideal gas pressure vs. density for the case of Aluminum ($Z = 13$). Notice that in Figure 1, the pressure is always less than the ideal pressure. Notice in Figure 2, for low densities and $T \leq 300$ eV the pressure can exceed the ideal.

The Poisson equation is then used to determine $V(r)$. It yields

$$\frac{1}{r}\frac{d^2}{dr^2}[rV(r)] = \frac{16\pi^2}{h^3}e(2mkT)^{1/2}I_{1/2}(eV(r)/kT - \eta). \tag{3}$$

The boundary conditions are $\lim_{r\to 0} rV(r) = Ze$ and the Z equals the integral of the density $\rho(r)$ over a sphere of radius r_b where $4\pi r_b^3 N/3 = \Omega$ where N is the number of atoms and Ω is the volume. The latter condition determines $\eta(T, r_b)$ The pressure is given by

$$\frac{p\Omega}{N} = \frac{32\pi^2 kT}{9}\left(\frac{2mkT}{h^2}\right)^{3/2} I_{3/2}(eV(r_b)/kT - \eta). \tag{4}$$

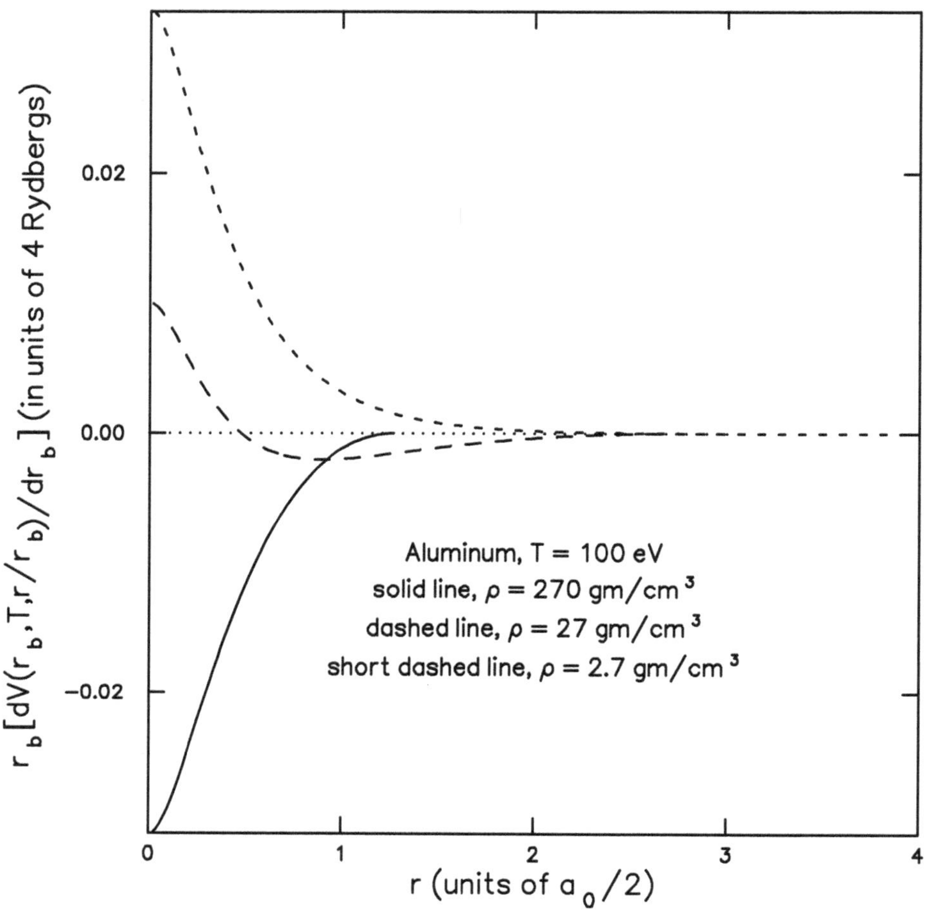

Fig. 3. $r_b\, \partial V(r_b, T, r/r_b)/\partial r_b|_T$ for the spherical cellular model as a function of r for the case of Aluminum ($Z = 13$).

Feynman, Metropolis and Teller[3] have shown that the solution of these equations for any Z can be obtained from that for any other by means of similarity transformations.

3. Spherical Cellular Model

Next I describe, briefly, the spherical cellular model.[4–6] First I remark for the ideal gas that, if space is divided into identical cubes, and if one sums over the eigenvalues in the cubes and integrates over the first Brillouin zone, then the exact result for the ideal gas is obtained. It is an approximation of a few percent to go to spherical cells instead.

The quantum statistical mechanics formulæ on which the usual formulation is

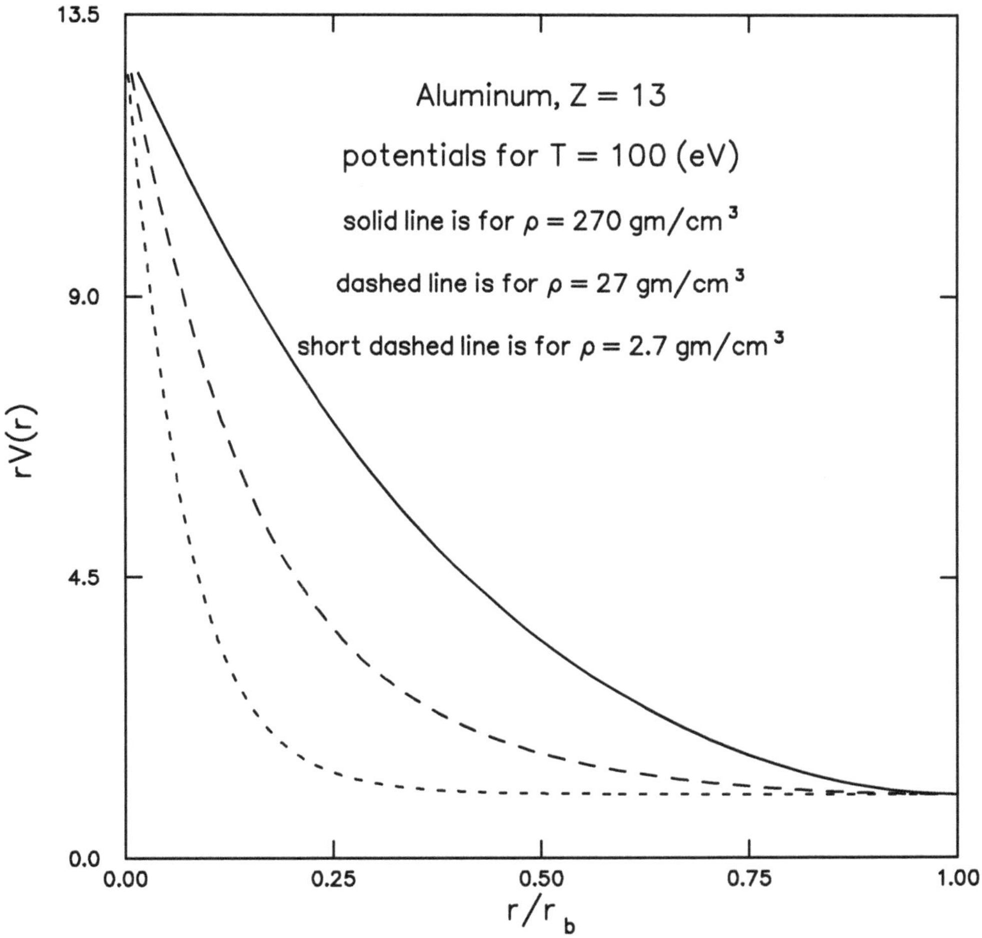

Fig. 4. The potentials $V(r_b, T, r/r_b)$ for the spherical cellular model as a function of r/r_b for the case of Aluminum ($Z = 13$).

based are given for Fermi-Dirac statistics.[7] The grand canonical partition function is

$$\begin{aligned}\mathcal{Q}(\Omega, T) &= \textstyle\sum_{N=0} \exp[N\mu(\Omega, T)/(kT)]Q_N(\Omega, T) \\ &= \textstyle\prod_j \{1 + \exp[(\mu(\Omega, T) - \epsilon_j)/(kT)]\}.\end{aligned} \tag{5}$$

where the ϵ_j's are the state energies. Differentiating by μ (the chemical potential), determines μ as the solution of

$$N = \sum_j \frac{1}{\exp[(\epsilon_j - \mu)/(kT)] + 1}, \tag{6}$$

where N is the average number of occupied states. The Helmholtz free energy is

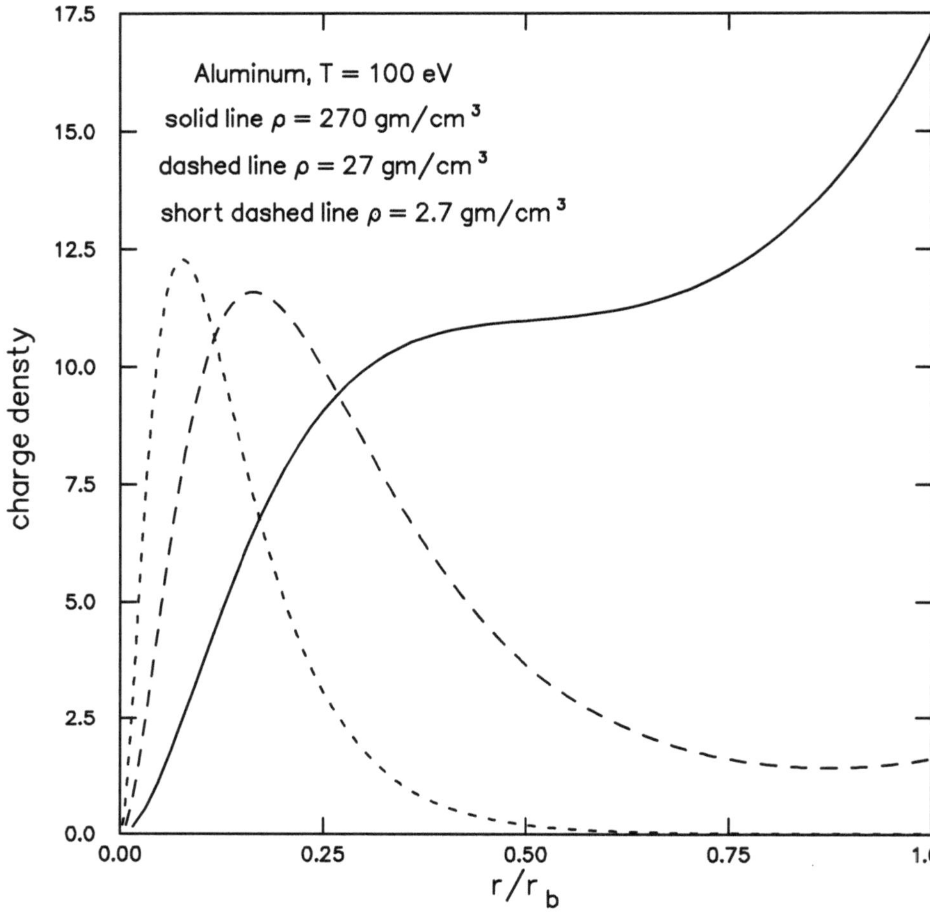

Fig. 5. The charge densities for several densities for the spherical cellular model as a function of r/r_b for the case of Aluminum ($Z = 13$).

then given as

$$A(\Omega, T) = \mathcal{N}\mu(\Omega, T) - kT \sum_j \ln\{1 + \exp[(\mu(\Omega, T) - \epsilon_j)/(kT)]\}. \tag{7}$$

From it, one can deduce the pressure as

$$p\Omega = -\frac{1}{3} r_b \left.\frac{\partial A}{\partial r_b}\right|_T = -\frac{1}{3} \sum_j \frac{r_b \left.\dfrac{\partial \epsilon_j}{\partial r e r_b}\right|_T}{\exp[(\epsilon_j - \mu)/(kT)] + 1}. \tag{8}$$

In order to make the calculations numerically feasible, I have introduced the independent electron approximation inside each atomic sphere.[5] The Schrodinger equa-

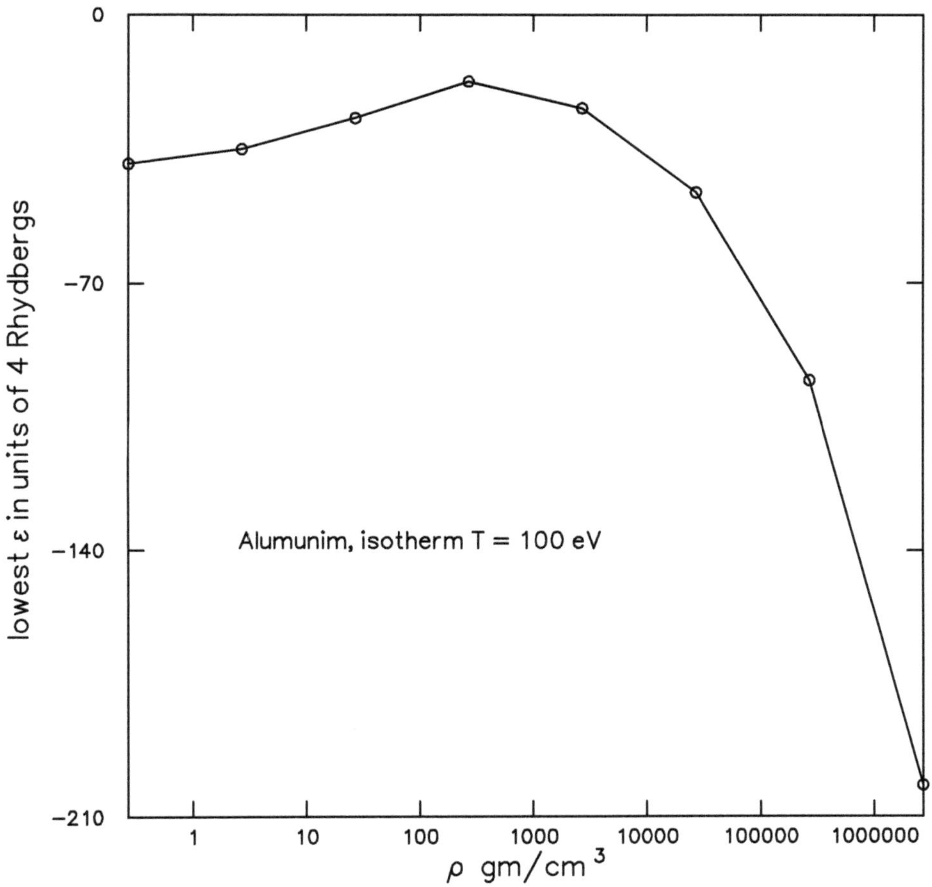

Fig. 6. The lowest $\epsilon(\vec{k}=\vec{0})$ vs. the density ρ.

tion to be solved is

$$\left\{\frac{\hbar^2}{2m^*}\left[k^2-2i\vec{k}\cdot\vec{\nabla}-\nabla^2\right]-\frac{e^2Z^{-\frac{1}{3}}v(r_b,T,r/r_b)}{r}\right.$$
$$\left.-\frac{e^2Z^{-\frac{1}{3}}}{2r_b}\left(\frac{r}{r_b}\right)^2F\left(\frac{y^2}{Z}\right)\right\}\phi_{l,\lambda}(\vec{r})=\mathcal{E}_{l,\lambda}(\vec{k})\phi_{l,\lambda}(\vec{r}), \tag{9}$$

with the boundary conditions

$$\vec{n}\cdot\vec{\nabla}\phi_{\text{even}}(Z^{\frac{1}{3}}\vec{\hat{r}}_b)=0,\quad \phi_{\text{odd}}(Z^{\frac{1}{3}}\vec{\hat{r}}_b)=0, \tag{10}$$

where $\vec{n}$ is the unit vector normal to the sphere, and $y^2 = Ze^2/(r_bkT)$ is a dimensionless strength of the Coulomb interaction. $\hat{r}_b$ is the radius of the spherical cell r_b over $Z^{1/3}$. The effective mass m^* is defined elsewhere.[5] In the spherical cell approximation the normalization condition which determines the chemical potential

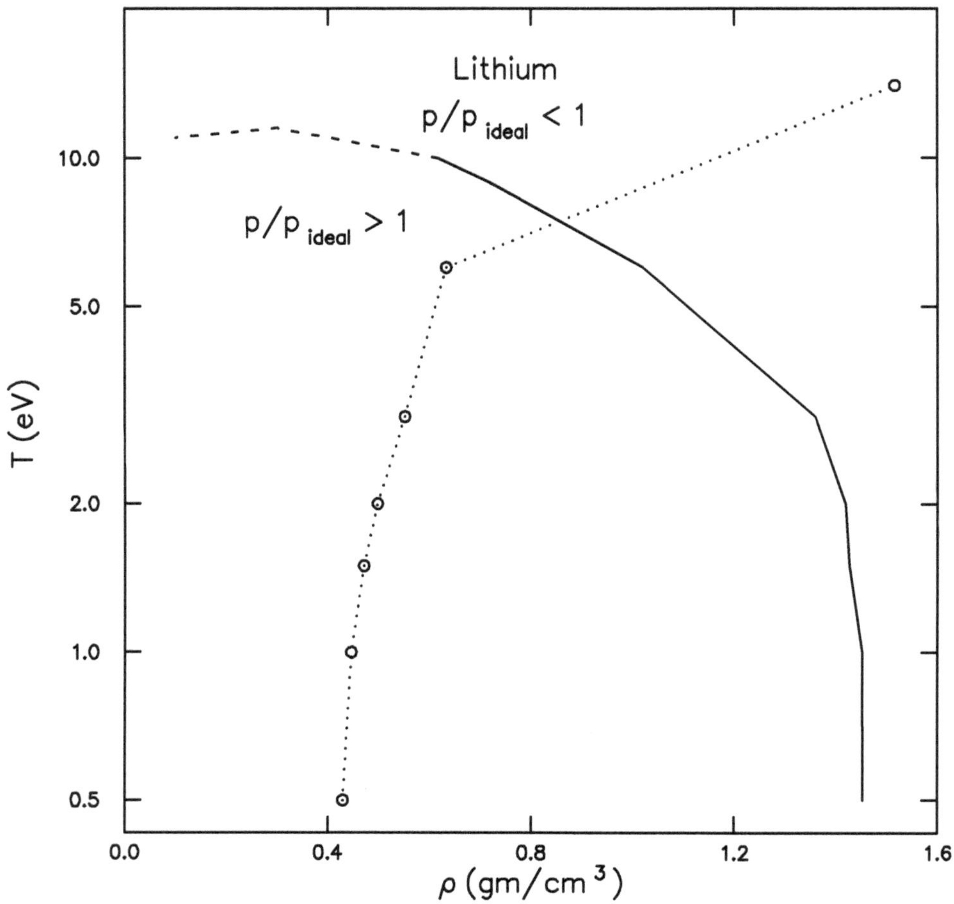

Fig. 7. The region where the electron pressure exceeds that of the ideal gas for the case of lithium ($Z = 3$). The dashed boundary is interpolated between two adjacent temperatures. The solid lines connect the data points. The dotted lines are (the straight lines connecting the data points on) the phase boundary. The open circles are those data points.

μ is

$$1 = 3\sum_{l=0}^{\infty}(2l+1)\sum_{n=0}^{\infty}\int_0^1 d\hat{\kappa}\,\hat{\kappa}^2 \tag{11}$$

$$\times\left\{\frac{1}{1+\exp\left\{(1.5\sqrt{\pi}\zeta)^{\frac{2}{3}}\left[e_{l,n}+\frac{1}{2}\left(1+\frac{m^*}{m}\right)\left(\hat{\kappa}^2+\hat{\kappa}\hat{\Delta}_{n+[(l+1)/2]}\right)\right]-\mu/kT\right\}}\right.$$

$$+\frac{1}{1+\exp\left\{(1.5\sqrt{\pi}\zeta)^{\frac{2}{3}}\left[e_{l,n}+\frac{1}{2}\left(1+\frac{m^*}{m}\right)\left(\hat{\kappa}^2-\hat{\kappa}\hat{\Delta}_{n+[(l+1)/2]}\right)\right]-\mu/kT\right\}}\Bigg\},$$

where now the dimensionless form of the eigenvalue is, $e_{l,n} = 2m\epsilon_{l,n}(\vec{0})/(\hbar^2 k_B^2)$, where $k_B = (9\pi/2)^{1/3}/r_B$. The $\hat{\Delta}_{n+[(l+1)/2]}$ are defined elsewhere.[5] For the corresponding ϵ,

$$\begin{aligned}
&\epsilon_{l,\lambda}(\vec{k}) = \\
&\frac{1}{2}\left(1+\frac{m^*}{m}\right)\left[\mathcal{E}_{l,\lambda}(\vec{k})-\frac{1}{2}\left\langle\phi_{l,\lambda}(\vec{r})\left|\left(\frac{Z^{\frac{2}{3}}e^2}{r}-\frac{e^2v(r_b,T,r/r_b)Z^{-\frac{1}{3}}}{r}\right)\right|\phi_{l,\lambda}(\vec{r})\right\rangle\right] \\
&\quad-\frac{1}{2}\left(1-\frac{m^*}{m}\right)\left\langle\phi_{l,\lambda}(\vec{r})\left|\frac{v(r_b,T,r/r_b)e^2Z^{-\frac{1}{3}}}{r}\right|\phi_{l,\lambda}(\vec{r})\right\rangle \\
&\quad+\frac{m^*e^2Z^{-\frac{1}{3}}}{4mr_b}F\left(\frac{y^2}{Z}\right)\left\langle\phi_{l,\lambda}(\vec{r})\left|\frac{r^2}{r_b^2}\right|\phi_{l,\lambda}(\vec{r})\right\rangle+\Delta\epsilon,
\end{aligned} \tag{12}$$

where the state independent part is

$$\begin{aligned}
&\Delta\epsilon_{||} = \\
&Z^{-\frac{1}{3}}\left\{\frac{3e^2}{10r_b}g(y^2)+\frac{3Ze^2}{5r_b}F(y^2Z)+\frac{e^2}{r_b}F\left(\frac{y^2}{Z}\right)\left[\frac{3}{4}-\left(\frac{3Z}{\pi\zeta}\right)^{\frac{1}{3}}f_{\frac{1}{2}}(z(\zeta))\right]\right\}, \\
&\Delta\epsilon_{\text{anti}-||} = Z^{-\frac{1}{3}}\left\{\frac{3e^2}{10r_b}g(y^2)+\frac{3Ze^2}{5r_b}F(y^2Z)+\frac{3e^2}{4r_b}F\left(\frac{y^2}{Z}\right)\right\}.
\end{aligned} \tag{13}$$

For the anti-parallel case $m^*/m = 1$ and I drop the $f_{\frac{1}{2}}$ term. m^* is the effective mass in the parallel spin case. Account has been taken of the double counting of the electron-electron interactions. The potential is determined self-consistently by the use of Poisson's equation which leads to

$$v(r_b,T,r/r_b) = 1+\left(\frac{Z-1}{Z}\right)r_b\int_{(r/r_b)}^{1}\left(1-\frac{r}{\beta r_b}\right)\mathcal{D}(r_b,T,\beta)\,d\beta. \tag{14}$$

$\mathcal{D}(r_b,T,\beta)$ is the electron density. For the case of a uniform density $\mathcal{D}(r_b,T,r/r_b) = 3Zr^2/r_b^3$, and $v(r_b,T,r/r_b) = Z+\frac{1}{2}(Z-1)[(r/r_b)^3-3(r/r_b)]$. I use this value to start the iteration procedure to find the solution of these equations.

4. Pressure Comparison

Notice that in Figure 1, the pressure is always less than the ideal pressure. Notice in Figure 2, for low densities and $T \leq 300$ eV the pressure can exceed the ideal. Under current investigation is the point that the region where p/p_{ideal} is greater than 5 to 10 is either thermodynamically forbidden or in the two phase region. In either case, one would not expect reliable results from the spherical cellular model. The quantity plotted in Figure 3 is in the pressure equation and for positive values can increase the pressure above the ideal. Comparing Figures 4 and Figure 5 we see that as the density decreases the peak of the electron density moves to progressively stronger values of the potential. Consequently the electrons move faster and faster. This motion is related to an increase in the pressure. For low temperatures the integral over the Brillouin zone gives the ideal gas pressure and the positive slope of ϵ as seen in Figure 6 adds to that pressure. Figure 7 shows the region where the pressure exceeds that of the ideal gas for the case of lithium ($Z = 3$).

The cellular model incorporates the following low temperature physics. Note that the pressures depend on the partial derivative of the state energies with respect to volume. At very high densities the lowest eigenvalue is very negative so that partial derivative is negative. As one reaches intermediate densities an atom like state is seen. Here the lowest eigenvalue decreases as the volume restriction is removed. The integration over the first Brillouin zone acts like the ideal-gas, kinetic energy and the change in the eigenvalue gives extra pressure. The physics here is complicated and the simple spherical cellular model may have over estimated this effect. Part, but not all, of the region in which the overpressure is seen in the spherical cellular model is thermally unstable as the internal energy decreases as the temperature increases. In aluminum for temperatures greater than 100 to 150 eV depending on the density, or less than a few eV or for a density greater than 27 gm/cm^3 the results are thermodynamically stable. At extreme dilutions, the entropy causes there to be full ionization so this limit for the spherical cellular model is again the ideal Fermi gas. The Thomas-Fermi model shows no such "overpressure."

DOES THIS "OVERPRESSURE" OCCUR IN REAL SYSTEMS?

There are some experimental results[8] on aluminum for $\rho = 0.1, 0.3$ gm/cm^3 at temperatures of a few eV. They indicate a system of very little ionization with the pressure less than that of an ideal gas of electrons and ions.

This work was carried out under the auspices of the National Nuclear Security Administration of the U. S. Department of Energy at the Los Alamos National Laboratory under Contract No. DE-AC52-06NA25396. The author is pleased to acknowledge a partial travel support from the U.S. Army Research Office. The views, opinions, and/or findings contained in this article, are those of the author and should not be construed as an official department of the army position, policy or decision, unless so designated by other documentation.

References

1. G. A. Baker, Jr., *J. Stat. Phys.* **110**, 971 (2003).
2. S. Eliezer, A. Ghatak, and H. Hora, *Fundamentals of Equations of State* (World Scientific, Singapore, 2002).
3. 3. R. P. Feynman, N. Metropolis, and E. Teller, *Phys. Rev.* **75**, 1561 (1949).
4. 4. G. A. Baker, Jr., *Phys. Rev. E* **56**, 5216 (1997).
5. 5. G. A. Baker, Jr., *Phys. Rev. E* **68**, 056112 (2003).
6. 6. G. A. Baker, Jr., to be published *Phys. Rev. E.*
7. 7. K. Huang, *Statistical Mechanics* (Wiley, New York, 1963).
8. 8. P. Renaudin, C. Blancard, J Clérouin, G. Faussurier, P. Noiret, and V. Recoules, *Phys. Rev. Let.* **91**, 075002 (2003).

PAIR EXCITATIONS AND VERTEX CORRECTIONS IN FERMI FLUIDS AND THE DYNAMIC STRUCTURE FUNCTION OF TWO-DIMENSIONAL ^{3}He

H. M. BÖHM[1*], H. GODFRIN[2], E. KROTSCHECK[1], H. J. LAUTER[3], M. MESCHKE[2] and M. PANHOLZER[1]

[1] *Inst. f. Theoretische Physik, Johannes Kepler Universität, A-4040 Linz, Austria*
**E-mail: helga.boehm@jku.at*

[2] *Centre de Recherches sur les Tres Basses Temperatures, C.N.R.S., B.P. 166X, 38042 Grenoble, France*

[3] *Institute Laue Langevin, 38042 Grenoble Cedex, France*

We use the equations–of–motion approach for time–dependent *pair* correlations in strongly interacting Fermi liquids to develop a theory of the excitation spectrum and the single–particle self energy in such systems. We present here the fully correlated equations and their approximate solutions for ^{3}He. Our theory has the following properties: It reduces to both, i) the "correlated" random–phase approximation (RPA) for strongly interacting fermions if the two–particle–two–hole correlations are ignored, and, ii) to the correlated Brillouin–Wigner perturbation theory for boson quantum fluids in the appropriate limit. iii) It preserves the two first energy–weighted sum rules, and systematically improves upon higher ones. iv) A familiar problem of the standard RPA is that it predicts a roton energy that lies more than a factor of two higher than what is found in experiments. A popular cure for this is to introduce an effective mass in the Lindhard function. No such *ad–hoc* assumption is invoked in our work. We demonstrate that the inclusion of correlated pair–excitations improves the dispersion relation significantly. Finally, a novel form of the density response function is derived that arises from vertex corrections in the proper polarization.

Keywords: Fermi fluids; density response function; multi–pair excitations; ^{3}He; correlated basis functions.

1. Introduction

The method of choice for studying the strongly interacting quantum liquids ^{3}He and ^{4}He is the Jastrow–Feenberg variational method and its extension in the form of correlated basis functions (CBF) theory.[1] The theory builds the *correlations* of the system into the the basis of the Hilbert space in a natural and straightforward manner. While ground state properties are accessible in many cases via simulation studies, the above approach remains the most powerful tool for a quantitatively reliable prediction of dynamic properties. In the last decade the excitation spectrum of the Bose system has been successfully investigated for a variety of configurations, ranging form the basic works of Refs. 2,3 to recent studies of the influence of phonon decay on the transport current.[4] In all these cases the inclusion of dynamic, *i.e.* time–dependent *pair* correlations has proved necessary. It is not only plausible, but has been demonstrated convincingly in several cases, that short-ranged fluctuations become relevant for excitations with wavelengths that are comparable to the inter-particle distance. A special case of such pair excitations was already introduced by Feynman[5] in the form of "backflow" effects.

The theory of pair fluctuations is very well understood for Bose systems, but is still in its beginnings for fermions. We expect that also in the the case of Fermi systems, especially in ^{3}He whose properties are dominated by the strong, short-ranged repulsion, the dispersion of phonons is visibly influenced by short-ranged fluctuations. In addition, accounting for these processes is indispensable for a description of the damping of long-wavelength phonons. This work reports progress in our development of a theory for bulk ^{3}He with the ultimate goal of reaching a level of accuracy as well as consistency comparable to the theory for bosons.

Central to our investigations is the density-density response function $\chi(q,\omega)$ connected to the dynamic structure function by

$$S(q,\omega) = -\frac{1}{\pi}\Im m \chi(q,\omega)\,. \tag{1}$$

In Fermi liquids $\chi(q,\omega)$ is often taken to be of the "RPA" form

$$\chi^{\mathrm{RPA}}(q,\omega) \;=\; \frac{\chi_0(q,\omega)}{1-\widetilde{V}_{\text{p-h}}(q)\,\chi_0(q,\omega)}\,, \tag{2}$$

where χ_0 is the Lindhard function and $\widetilde{V}_{\text{p-h}}\,(q)$ a suitable effective interaction (also termed, "pseudo-potential[6]" and "local field corrected potential[7]"). One can determine $\widetilde{V}_{\text{p-h}}\,(q)$ through the ω^0 and ω^1 sum rules, but this leads to a collective mode that is energetically much higher than what is found in experiments. An *ad-hoc* repair of this problem is to introduce an effective mass[8] m^* in χ_0. The effective mass in both three– and two–dimensional ^{3}He is far from constant,[9,10] it exhibits a strong peak around the Fermi momentum k_F, and another one at approximately $2.5\,k_F$. Introducing an m^* is unsatisfactory mainly from two points of view: First, we want to develop a manifestly microscopic description of the low-temperature properties of ^{3}He, and second, it introduces inconsistencies because the important sum rules are violated, which makes $\widetilde{V}_{\text{p-h}}\,(q)$ undetermined. The above example of a successful description of the phonon–roton spectrum in ^{4}He systems – which does not even introduce the concept of an "effective mass" – leads one to expect that the same physical mechanisms, *viz.* dynamic pair correlations, lower the phonon-roton spectrum not only in ^{4}He but also in ^{3}He, without violating the two above sum rules.

An even more compelling argument against introducing an effective mass is provided by experimental data on the dynamic structure function of two-dimensional ^{3}He. An example of the experimental data is shown in Fig. 1. Here the collective mode is found *inside* the particle-hole continuum.[11] It would be quite difficult to explain the feature that the maximum strength of a possible collective mode is found inside the continuum with the form (2) of the density density response function, in particular this feature *cannot* be reproduced by changing m^* in the RPA formula (2).

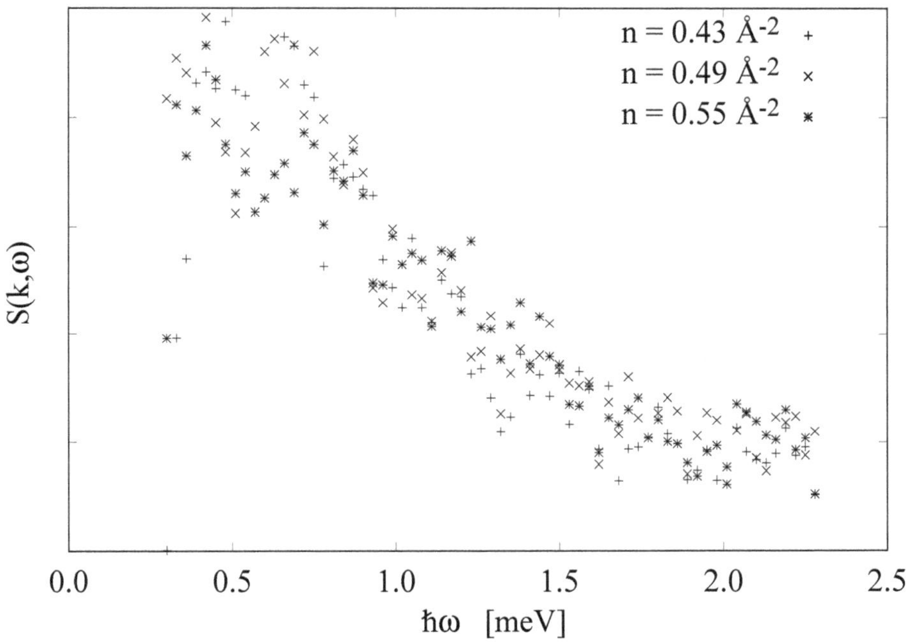

Fig. 1. The figure shows experimental data of the dynamic structure function $S(k,\omega)$ in two-dimensional ^{3}He for wave numbers $1.1\,\text{Å}^{-1} \leq k \leq 1.2\,\text{Å}^{-1}$. These data were obtained by three of us (H. G., H.-J. L and M. M.) at the IN14 instrument at the Institute Laue-Langevin in Grenoble.

2. Equations of Motion

2.1. *Correlated equations of motion*

We describe the excited state wave function of strongly interacting fermions by a generalized time-dependent Hartree-Fock form

$$\left|\Psi(t)\right\rangle = \frac{1}{\sqrt{\mathcal{N}}} e^{-itH_{00}/\hbar} F\, e^{U(t)} \left|\Phi_0\right\rangle , \tag{3}$$

where $\mathcal{N}$ denotes the normalization integral, $\left|\Phi_0\right\rangle$ the ground state of a model system (normally a Slater determinant) and H_{00} the energy of the correlated ground state $F\left|\Phi_0\right\rangle$. The most general excitation operator U contains $n-$particle amplitudes

$u_{(n)}(t)$

$$U = \sum_n \sum_{\substack{\mathbf{p}_1 \cdots \mathbf{p}_n \\ \mathbf{h}_1 \ldots \mathbf{h}_n}} u_{\mathbf{p}_1 \ldots \mathbf{p}_n \mathbf{h}_1 \ldots \mathbf{h}_n}(t)\, a^\dagger_{\mathbf{p}_1} \ldots a^\dagger_{\mathbf{p}_n} a_{\mathbf{h}_n} \ldots a_{\mathbf{h}_1}$$
$$\equiv \sum_n \sum_{(ph)^n} u_{(n)}(t)\, \widehat{X}_n \,. \tag{4}$$

These amplitudes are determined by minimizing the action integral $\mathcal{S} = \int dt\, \mathcal{L}$ corresponding to the time–dependent Schrödinger equation:

$$\Big[\frac{\delta}{\delta u^*_{(n)}} - \frac{d}{dt}\frac{\delta}{\delta \dot{u}^*_{(n)}}\Big]\mathcal{L} = 0 \qquad \text{with} \qquad \mathcal{L} = \langle \Psi \big| H + H^{\rm ext}(t) + \frac{\hbar}{i}\frac{\partial}{\partial t} \big| \Psi \rangle \,. \tag{5}$$

Here, $H^{\rm ext}(t) \equiv \int d^3r\, h^{\rm ext}(\mathbf{r};t)\, \hat{\rho}(\mathbf{r})$ describes a small and real external perturbation where $\hat{\rho}(\mathbf{r})$ is the density operator. Considering $H^{\rm ext}$ as a perturbation, linear equations of motion (EOMs) are derived by expanding $\mathcal{L}$ to order $(U^\dagger)^2$. Using the abbreviation $H' \equiv H{-}H_{00}$ and denoting expectation values taken in the free ground state $\langle \Phi_0 | \ldots | \Phi_0 \rangle$ by $\langle \ldots \rangle_0$ these EOMs read

$$-\frac{\hbar}{i}\Big(\big\langle \widehat{X}^\dagger F^\dagger F \dot{U} \big\rangle_0 - \big\langle \widehat{X}^\dagger F^\dagger F \big\rangle_0 \big\langle F^\dagger F \dot{U} \big\rangle_0 \Big) = \big\langle \widehat{X}^\dagger\, F^\dagger H^{\rm ext} F \big\rangle_0 - \big\langle\, F^\dagger H^{\rm ext} F \big\rangle_0$$
$$+ \big\langle \widehat{X}^\dagger F^\dagger H' F U \big\rangle_0 + \big\langle \widehat{X}^\dagger U^\dagger F^\dagger H' F \big\rangle_0 - \big\langle U^\dagger F^\dagger H' F \big\rangle_0 \big\langle \widehat{X}^\dagger F^\dagger F \big\rangle_0 \,. \tag{6}$$

In the standard notation of correlated basis functions (CBF) formalism[12]

$$\frac{\langle \widehat{X}^\dagger_n F^\dagger H' F \widehat{X}'_{n'} \rangle_0}{\langle F^\dagger F \rangle_0} = H'_{X_n, X'_{n'}}\, z_{X_n}\, z_{X'_{n'}} \; :=$$
$$\frac{\langle F \widehat{X}_n \Phi_0 \,| H' |\, F \widehat{X}'_{n'} \Phi_0 \rangle}{\sqrt{\langle \widehat{X}^\dagger_n F^\dagger F \widehat{X}_n \rangle_0 \, \langle \widehat{X}'^\dagger_{n'} F^\dagger F \widehat{X}'_{n'} \rangle_0}} \; \frac{\sqrt{\langle \widehat{X}^\dagger_n F^\dagger F \widehat{X}_n \rangle_0}}{\sqrt{\langle F^\dagger F \rangle_0}} \; \frac{\sqrt{\langle \widehat{X}'^\dagger_{n'} F^\dagger F \widehat{X}'_{n'} \rangle_0}}{\sqrt{\langle F^\dagger F \rangle_0}} \tag{7}$$

and the analogous convention for the off-diagonal normalization integrals

$$\frac{\langle \widehat{X}^\dagger_n F^\dagger F \widehat{X}'_{n'} \rangle_0}{\langle F^\dagger F \rangle_0} \equiv N_{X_n, X'_{n'}}\, z_{X_n}\, z_{X'_{n'}} \tag{8}$$

the EOMs take the form

$$\sum_{n'} \sum_{(p'h')^{n'}} \Big\{ -\frac{\hbar}{i}\, \dot{u}_{(n')} \Big(z_{X'_{n'}} N_{X_n, X'_{n'}} - z_{X'_{n'}} N_{X_n, 0} N_{0, X'_{n'}} \Big) \Big\} = H^{\rm ext}_{X_n, 0} +$$
$$\textstyle\sum_{n'} \sum_{(p'h')^{n'}} \Big\{ u_{(n')}\, z_{X'_{n'}} H'_{X_n, X'_{n'}} + u^*_{(n')} \Big(z_{X'_{n'}} H'_{X_n X'_{n'}, 0} - z_{X'_{n'}} N_{X_n, 0} H'_{X'_{n'}, 0} \Big) \Big\} \tag{9}$$

(for a uniform system, we have $z_{X_n X'_{n'}} {=} z_{X_n} z_{X'_{n'}} + \mathcal{O}(N^{-1})$). Technically, we restrict ourselves to single particle and pair fluctuations. We spell out these equations in detail for single-particle and pair fluctuations in the appendix; here we adopt a more pragmatic strategy: In essence, the CBF theory provides a means[13,14] to map the strong, bare interaction of the system onto weak, *effective* interactions. This mapping is uniquely determined by the diagrammatic analysis of off-diagonal correlated matrix elements,[15] generally one can define non-local effective interactions from overlap integrals of the Hamiltonian between correlated states (as examples we mention $H_{ph,p'h'}$ and $H_{hh'pp',hh'p''p''}$; which, in turn, defines a pair operator $\mathcal{H}$ via the anti-symmetrized expectation values $\langle ph'|\mathcal{H}|hp'\rangle_{\rm a}$, and $\langle pp'|\mathcal{H}|p''p'''\rangle_{\rm a}$). Once this mapping is performed, the originally strongly interacting system can be regarded as only moderately interacting in these new effective interactions.

2.2. *The Hartree Fock (HF) limit*

To see the content of our theory of excitations at a somewhat simpler level, we treat the problem in this section in the Hartree-Fock approximation, *i.e.* the correlation operator F is replaced by unity: $F\,,F^\dagger\to 1$. Consequently, all $z-$factors become 1 and the normalization overlap integrals are antisymmetrized delta−functions.

In the spirit of the ideas outlined at the end of Sec. 2.1 the interaction in this HF approximation should clearly be interpreted as an *effective* interaction which will ultimately be obtained from the correlated theory. We also should allow for different effective interactions in different channels. Denoting empty (occupied) states of the Slater determinant by p_i ="particles" (h_i="holes"), the "particle–hole interaction" $\widetilde{V}_{\rm p\text{-}h}$ arises from matrix elements of the form $\langle h_1p_2|\mathcal{H}|p_1h_2\rangle$, whereas we would have a "particle–particle" interaction $\widetilde{V}_{\rm p-p}(r)$ arising from matrix elements of the type $\langle p_1p_2|\mathcal{H}|p_1'p_2'\rangle$ (and analogously for "hole–hole" terms). Finally, another effective interaction W arises from all other (=mixed) channels. Anticipating such effective interactions (all of which are normally non-local), we start with inspecting the EOMs for a *weakly* interacting system.

Eqs. (9) simplify to

$$\frac{-\hbar}{i}\,\mathcal{A}\dot{u}_{(n)} \;=\; H^{\rm ext}_{X_n,0} + \sum_{n'}\sum_{(p'h')^{n'}}\left\{ u_{(n')}\,H'_{X_n,X'_{n'}} + u^*_{(n')}\,H'_{X_nX'_{n'},0}\right\}, \qquad (10)$$

where $\mathcal{A}$ denotes antisymmetrisation in all indices. In our case the fluctuation operator $U(t)$ in Eq. (4) consists of 1p-1h and 2p-2h fluctuations

$$\begin{array}{ll} \widehat{X}_1 = a^\dagger_{\mathbf{p}}a_{\mathbf{h}} & u_{(1)}(t) =: c_{ph} \\ \widehat{X}_2 = a^\dagger_{\mathbf{p}}a^\dagger_{\mathbf{p}'}a_{\mathbf{h}'}a_{\mathbf{h}} & u_{(2)}(t) =: \frac{1}{2}d_{pp',hh'}\;. \end{array} \qquad (11)$$

The pair correlations are symmetric under a pairwise interchange of indices, antisymmetry needs to be accounted for the hole (or particle) indices alone:

$$d_{pp',hh'} = d_{p'p,h'h}\,; \qquad d_{pp',(hh')_a} \equiv d_{pp',hh'} - d_{pp',h'h}\;. \qquad (12)$$

Abbreviating the particle-hole excitation energies with $e_{ph}\equiv e_p - e_h$ the EOMs read explicitly

$$\left[i\hbar\frac{\partial}{\partial t} - e_{ph}\right] c_{ph}(t) = h^{\rm ext}_{ph}(t) + \langle ph'|\widetilde{V}_{\text{p-h}}\,|hp'\rangle_a\, c_{p'h'} + \tag{13}$$
$$\langle pp'|\widetilde{V}_{\text{p-h}}\,|hh'\rangle_a\, c^*_{p'h'} + \langle ph'|W|p''p'\rangle_a\, d_{p''p',hh'} - \langle h''h'|W|hp'\rangle_a\, d_{pp',h''h'}\,,$$

and

$$\left[i\hbar\,\frac{\partial}{\partial t} - e_{ph} - e_{p'h'}\right] d_{pp',hh'} =$$
$$\big\langle pp'\big|W\big|hp''\big\rangle\, c_{p''h'} + \big\langle pp'\big|W\big|p''h'\big\rangle\, c_{p''h} - \big\langle ph''\big|W\big|hh'\big\rangle\, c_{p'h''} - \big\langle h''p'\big|W\big|hh'\big\rangle\, c_{ph''}$$
$$+\left[\big\langle h''p'\big|\widetilde{V}_{\text{p-h}}\,\big|p''h'\big\rangle_a\, d_{pp'',hh''} - \big\langle h''p'\big|\widetilde{V}_{\text{p-h}}\,\big|hp''\big\rangle_a\, d_{pp'',h''h'} + \left(\begin{smallmatrix} p\leftrightarrow p' \\ h\leftrightarrow h'\end{smallmatrix}\right)\right]$$
$$+\,\big\langle pp'\big|\widetilde{V}_{\text{p-p}}\big|p''p'''\big\rangle\, d_{p''p''',hh'} + \big\langle h''h'''\big|\widetilde{V}_{\text{h-h}}\big|hh'\big\rangle\, d_{pp',h''h'''}\,. \tag{14}$$

3. Solution Strategy

Without loss of generality we assume harmonic time dependence for the perturbation

$$h_{\rm ext}(\mathbf{r},t) = h_{\rm ext}(\mathbf{r},\omega)\ \left(e^{i\omega t} + e^{-i\omega t}\right) \tag{15}$$

and consequently a similar form for the excitation amplitudes.

In this first implementation of the theory we aim towards a form that is, as far as possible, analytic and preserves the analytic structure and the physical content of CBF theory for Bose systems. We therefore neglect exchange effects and also contributions of the "ladder" type, and we restrict the set of two–pair amplitudes to functions that are matrix elements of a *local* operator

$$\sum_{pp'hh'} d_{ph,p'h'}\ a^\dagger_p a^\dagger_{p'} a_{h'} a_h \longrightarrow \sum_{qq'} d(\mathbf{q},\mathbf{q}')\,[\hat\rho_{\mathbf{q}}\hat\rho_{\mathbf{q}'} - \hat\rho_{\mathbf{q}+\mathbf{q}'}]\,, \tag{16}$$

where $\hat\rho_{\mathbf{q}}$ is the density operator in momentum space. Comparing our theory to descriptions using Feynman diagrams, our approach corresponds to accounting for the proper polarization graphs shown in Fig. 2. Clearly, the graphs no. 3 and 5–7 represent vertex corrections, which are beyond the RPA (*i.e.* these terms are not properly accounted for in response functions of the form of Eq. (2)).

We further simplify the *c*-*d* coupling by replacing the amplitudes $c_{ph}(t)$ in Eq. (14) by their Fermi-sea averages. With these approximations the EOMs can be solved for c_{ph} and $d_{ph,p'h'}$ which, in turn, yield the linear response wave function and thus the induced density $\delta\rho(\mathbf{r},t)$.

Introducing

$$\chi_0^{(\pm)}(q,\omega) = \pm\frac{1}{N}\sum_h \frac{n_h\,\overline{n}_{\mathbf{h}+\mathbf{q}}}{\hbar\omega \mp e_{h+q,h} + i0^+}\,, \tag{17}$$

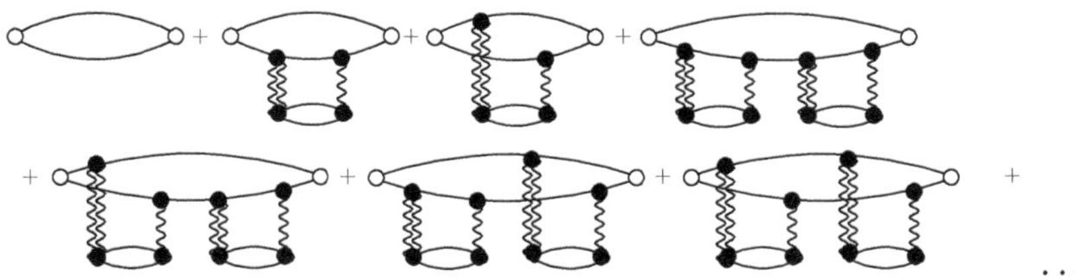

Fig. 2. Typical proper polarization Feynman diagrams taken into account in our approach. (The single and double wavy lines denote the bare and screened interactions, respectively. The lines denote free single particle propagators).

(where $n_k = \theta(k_{\rm F} - k) \equiv 1 - \overline{n}_k$), the final result for the density-density response function can be expressed as follows:

$$\chi^{\rm Pair}(q,\omega) = \frac{\kappa(q,\omega)}{1 - \kappa(q,\omega)\widetilde{V}_{\text{p-h}}(q)} \tag{18}$$

$$\kappa(q,\omega) = \frac{\chi_0^{(+)}}{1-\chi_0^{(+)}\mathcal{W}_+} + \frac{\chi_0^{(-)}}{1-\chi_0^{(-)}\mathcal{W}_-}\,, \tag{19}$$

$$\mathcal{W}_\pm = \mathcal{W}(q,\mp\omega) = -\frac{1}{2}\sum_{\mathbf{q}'} \frac{\overline{\mathcal{W}_3}(\mathbf{q}-\mathbf{q}',\mathbf{q}')\,\overline{\mathcal{W}_3}(\mathbf{q}-\mathbf{q}',\mathbf{q}')}{\pm\hbar\omega - \varepsilon(\mathbf{q}-\mathbf{q}') - \varepsilon(q')}\,. \tag{20}$$

Here $\overline{\mathcal{W}_3}$ is a three-"phonon" vertex

$$\overline{\mathcal{W}_3}(\mathbf{q},\mathbf{q}') = \frac{\widetilde{V}_{\text{p-h}}(q)\ S_F(q) + \widetilde{V}_{\text{p-h}}(q')\ S_F(q')}{\sqrt{S_{\rm F}(q)\,S_{\rm F}(q')}}\,\frac{S_{\rm F}^{(3)}(\mathbf{q},\mathbf{q}',\mathbf{q}+\mathbf{q}')}{S_{\rm F}(\mathbf{q}+\mathbf{q}')}\,. \tag{21}$$

with $S_{\rm F}$ nd $S_{\rm F}^{(3)}$ denoting the static structure factors of the non-interacting Fermi gas. In $\varepsilon(q)$ we recover the weakly interacting limit of the Bogoljubov dispersion relation

$$\varepsilon(q) = \frac{\hbar^2 q^2}{2mS_{\rm F}(q)} 1 + S_{\rm F}(q)\,\widetilde{V}_{\text{p-h}}(q) \tag{22}$$

$$\approx \frac{\hbar^2 q^2}{2mS_{\rm F}(q)}\sqrt{1 + 4mS_{\rm F}^2(q)\,\widetilde{V}_{\text{p-h}}(q)\,/\hbar^2 q^2} \approx \frac{\hbar^2 q^2}{2m\,S(q)}\,. \tag{23}$$

This dispersion relation (23) can be obtained, for example, by a "collective approximation" of the RPA equation (2).

There are two important differences between the density response function from the EOMs for the strongly interacting case and its weakly interacting counterpart. First, $\varepsilon(q)$ in Eq.(20) becomes the *full* Bogoljubov spectrum corresponding to the exact interacting structure factor $S(q)$, and, second, a multi–phonon contribution $\widetilde{V}_{\rm mp}(q,\omega)$ must be added in the denominator of Eq. (18). This latter term is of no immediately transparent structure and numerical investigations show that its inclusion has only a marginal effect on the numerical results.[16]

We end this section by noting that it is easily proved that $\chi^{\rm Pair}(q,\omega)$ obeys the ω^0 and ω^1 sum rules, *independently* of the specifics of $\mathcal{W}(q,\omega)$. This means that $\widetilde{V}_{\text{p-h}}(q)$ is *uniquely* defined by the static structure function through the first two energy weighted sumrules.

4. Applications to Two-Dimensional ^{3}He

We show in Fig. 3 results for the dynamic structure factor $S(k,\omega)$ of two–dimensional ^{3}He for an areal density of $n = 0.049\,\text{Å}^{-2}$ and 4 different wave numbers in the vicinity of the experimental data. The effective interaction was calculated through the ω^0 sum rule from the static $S(k)$ obtained by diffusion Monte Carlo calculations in Ref. 17. In the spirit of the above discussion, the multi–phonon contribution has been neglected for the time being.

Results are shown for the RPA as well as for the CBF theory in Fig. 3. The three figures for $k = 0.85\,\text{Å}^{-1}$, $k = 1.01\,\text{Å}^{-1}$ and $k = 1.15\,\text{Å}^{-1}$ all show a collective mode and the familiar particle-hole continuum. Note that the boundaries of the single-particle-hole continuum are identical in RPA and CBF. (This is to be contrasted with a theory invoking a m^*, as the latter changes χ^0 and thus the particle-hole continuum). It is clearly seen from the figure that the collective mode in CBF theory is shifted appreciably to lower energies and is about to merge into the continuum at $k = 1.15\,\text{Å}^{-1}$. Evidently, pair fluctuations are quite able to move the location of the zero sound mode without effecting the particle-hole continuum. Therefore, an adjustment of the effective mass is unnecessary to lower the collective mode, this was also the case[18] for 3d ^{3}He.

While the RPA still yields a well separated collective mode at somewhat higher wave numbers, the CBF phonon has merged into the continuum and $S(k,\omega)$ displays a form similar to the RPA curve in the last part of Fig. 3. The case of the wave number $k = 1.34\,\text{Å}^{-1}$ is most interesting: In both RPA and CBF, the collective mode has merged with the single-particle-hole continuum. But whereas the RPA spectrum still has a strong peak close to the upper boundary of the continuum, the overall form of the CBF corrected spectrum is now very close to the experimentally observed one.

From these results we conclude that that the essential mechanism determining the shape of the dynamic structure function in ^{3}He is that of fluctuating pair correlations. Our study must, of course, be considered preliminary. A number of physical mechanisms are still unaccounted for:

- We have found that the collective mode can be lowered without introducing an *ad-hoc* effective mass. Nevertheless a consistent view of multi–pair and single (quasi–)particle properties appears important. In previous work[9,10] we have taken working formulas from perturbation theory and have combined these with effective interactions generated from CBF theory to compute m^*. A rigorous justification for this procedure is still lacking.
- In ^{4}He, the same approximations used here lead to a significant lowering

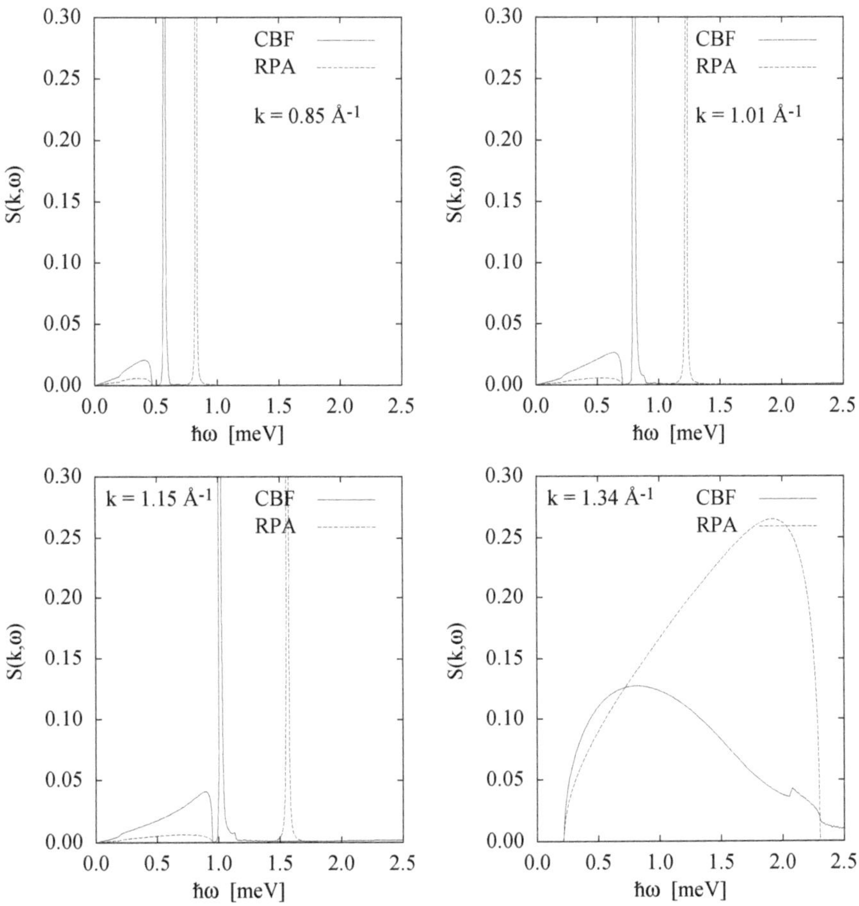

Fig. 3. Dynamic structure factor of a ^{3}He layer at an areal density $n = 0.049\,\text{Å}^{-2}$ for four different wave numbers as shown in the plots. Full lines: CBF-theory results Dashed lines: RPA results as defined by Eq. (2).

of the phonon-roton spectrum, but still do *not* reproduce the experiments. A theory with fewer approximations is needed to lower the energy of the collective mode even further.[3] The same is expected here, note that our collective mode at the experimental wave number is still too high.

- At the experimental wave number $1.1\,\text{Å}^{-1} \leq k \leq 1.2\text{Å}^{-1}$, the particle-hole continuum terminates at about 1 meV, whereas the data still show significant strength up to about twice that energy. To explain this feature with single-particle excitations would require an effective mass that is significantly less than 1, which does not seem realistic.[19–21] Therefore, a part of the continuum seen in Fig. 1 must be due to multi-pair-excitations. These have been treated in this work as "collective modes" (note that the energy denominator in Eq. (20) contains only collective excitations); clearly this approximation must sooner or later be eliminated.

Our first priority is to investigate the effect of pair-fluctuations on quasi–particle excitations in the system within this same formalism, and to combine this aspect of the theory with the multi-pair-theory of the dynamic structure function. In previous work,[9,10] we have used the so-called G0W approximation without rigorous justification. Such a justification be achieved by studying the time–dependent wave function corresponding to a definite single particle-hole amplitude, *i.e.* by omitting the sum over ph (only) in the $c_{ph}-$ contribution in Eq. 4. Then one can proceed in exactly the same way as above, leading to a consistent way of studying collective excitations and quasi–particle properties. Work in this direction is in progress.

In summary, the introduction of dynamic multi-pair correlations, so successful in ^{4}He, also proves necessary for obtaining the correct phonon dispersion in ^{3}He. A promising extension is the inclusion of spin fluctuations, which holds the potential of understanding the large effective mass in ^{3}He. As a matter of course the theory can also be applied to other strongly interacting Fermi fluids and to electrons.

Appendix

We include henceforth the factors $z_{X'_{n'}}$ in the definition of the correlation amplitudes $u_{(n)}$ and conveniently split off the diagonal parts contained in the correlated EOMs

$$\begin{aligned} N_{X_n, X'_{n'}} &= \quad \delta_{n,n'}\, \mathcal{A}\delta_{X_n, X'_{n'}} + \widetilde{N}_{X_n, X'_{n'}} \\ H'_{X_n, X'_{n'}} &= \delta_{n,n'}\, e_{X_n}\, \mathcal{A}\delta_{X_n, X'_{n'}} + \widetilde{H}'_{X_n, X'_{n'}} \;. \end{aligned} \tag{24}$$

The $n-$particle excitation energies in this definition are essentially the differences of single–particle and –hole energies

$$e_{X_n} \;=\; H_{X_n, X_n} - H_{00} \;=\; \sum_i (e_{p_i} - e_{h_i}) + \mathcal{O}(\tfrac{1}{N}) \tag{25}$$

(N is the particle Number). Defining the correlation amplitude vector as

$$\mathbf{u} \equiv \left(u_{(1)}, \ldots, u_{(n)}, u^*_{(1)}, \ldots, u^*_{(n)}\right) \equiv \left(c_{p''h''},\, d_{p''p''',h''h'''},\, c^*_{p''h''},\, d^*_{p''p''',h''h'''}\right) \tag{26}$$

the correlated EOMs can be cast into matrix form

$$\left[-\begin{pmatrix} \underline{\underline{\mathbf{M}}} & \underline{\underline{\mathbf{O}}} \\ \underline{\underline{\mathbf{O}}} & -\underline{\underline{\mathbf{M}}} \end{pmatrix} \frac{\hbar}{i}\frac{\partial}{\partial t} - \begin{pmatrix} \underline{\underline{\Omega_0}} & \underline{\underline{\mathbf{O}}} \\ \underline{\underline{\mathbf{O}}} & \underline{\underline{\Omega_0}} \end{pmatrix} \right] \cdot \mathbf{u} = \begin{pmatrix} \underline{\underline{\mathbf{A}}} & \underline{\underline{\mathbf{B}}} \\ \underline{\underline{\mathbf{B}^*}} & \underline{\underline{\mathbf{A}^*}} \end{pmatrix} \cdot \mathbf{u}\,, \tag{27}$$

where $\underline{\underline{\mathbf{O}}}$ is the 2×2 zero matrix, $\underline{\underline{\Omega_0}}$ contains the excitation energies

$$\underline{\underline{\Omega_0}} := \begin{pmatrix} (e_p - e_h)\, \delta_{\substack{pp'' \\ hh''}} & 0 \\ 0 & (e_p - e_h + e_{p'} - e_{h'})\, \mathcal{A}\delta_{\substack{pp'' \\ hh''}}\, \delta_{\substack{p'p''' \\ h'h'''}} \end{pmatrix} \tag{28}$$

with

$$\mathcal{A}\delta_{\substack{pp'' \\ hh''}}\, \delta_{\substack{p'p''' \\ h'h'''}} := \delta_{\substack{pp'' \\ hh''}}\, \delta_{\substack{p'p''' \\ h'h'''}} - \delta_{\substack{pp''' \\ hh''}}\, \delta_{\substack{p'p'' \\ h'h'''}} - \delta_{\substack{pp'' \\ hh'''}}\, \delta_{\substack{p'p''' \\ h'h''}} + \delta_{\substack{pp''' \\ hh'''}}\, \delta_{\substack{p'p'' \\ h'h''}} \;. \tag{29}$$

The time derivative terms have factors arising from the normalization overlap integrals contained in $\underline{\underline{\mathbf{M}}}$

$$\underline{\underline{\mathbf{M}}} = \begin{pmatrix} \delta_{\substack{pp'' \\ hh''}} & 0 \\ 0 & \mathcal{A}\delta_{\substack{pp'' \\ hh''}}\delta_{\substack{p'p''' \\ h'h'''}} \end{pmatrix}$$

$$+ \begin{pmatrix} \widetilde{N}_{ph,p''h''} & \frac{1}{2}N_{ph,p''h''p'''h'''} \\ N_{php'h',p''h''} & \frac{1}{2}\left[\widetilde{N}_{php'h',p''h''p'''h'''} - N_{0,php'h'}N_{0,p''h''p'''h'''}\right] \end{pmatrix}, \tag{30}$$

(in the bulk system $N_{ph,0} = 0$). Finally, the dynamics is governed by the Hamiltonian contributions

$$\underline{\underline{\mathbf{A}}} := \begin{pmatrix} \widetilde{H}'_{ph,p''h''} & \frac{1}{2}H_{ph,p''h''p'''h'''} \\ H'_{php'h',p''h''} & \frac{1}{2}\widetilde{H}'_{php'h',p''h''p'''h'''} \end{pmatrix} \tag{31}$$

$$\underline{\underline{\mathbf{B}}} := \begin{pmatrix} H'_{php''h'',0} & \frac{1}{2}H_{php''h''p'''h''',0} \\ \overline{H}'_{php'h'p''h'',0} & \frac{1}{2}\overline{H}'_{php'h'p''h''p'''h''',0} \end{pmatrix} \tag{32}$$

$$\overline{H}'_{php'h'p''h'',0} \equiv \left[H'_{php'h'p''h'',0} - N_{php'h',0}H'_{p''h'',0}\right]$$

$$\overline{H}'_{php'h'p''h''p'''h''',0} \equiv \left[H'_{php'h'p''h''p'''h''',0} - N_{php'h',0}H'_{p''h''p'''h''',0}\right]$$

Obviously, Eq. (27) is similar to the corresponding dynamic CBF equations containing single–particle excitations only,[12] the matrices $\underline{\underline{\mathbf{A}}}$ and in particular $\underline{\underline{\mathbf{B}}}$, however, being much more intricate.

Acknowledgments

This work was supported by the Austrian Science Fund (FWF) project P18134-N08. We thank V. Grau, J. Boronat and J. Casulleras for providing the Monte Carlo data[17] of structure functions in two-dimensional ^{3}He.

References

1. E. Feenberg, *Theory of Quantum Fluids* (Academic, New York, 1969).
2. B. E. Clements, E. Krotscheck and C. J. Tymczak, *Phys. Rev. B* **53**, 12253 (1996).
3. V. Apaja and M. Saarela, *Phys. Rev. B* **57**, p. 5358 (1998).
4. V. Apaja and E. Krotscheck, A microscopic view of confined quantum liquids, in *Microscopic Approaches to Quantum Liquids in Confined Geometries*, eds. E. Krotscheck and J. Navarro (World Scientific, Singapore, 2002) pp. 205–268.
5. R. P. Feynman and M. Cohen, *Phys. Rev.* **102**, 1189 (1956).
6. N. Iwamoto and D. Pines, *Phys. Rev. B* **29**, 3924 (1984).
7. K. S. Singwi and M. P. Tosi, *Solid State Phys.* **36**, 177 (1981).
8. H. R. Glyde, B. Fåk, N. H. van Dijk, H. Godfrin, K. Guckelsberger and R. Scherm, *Phys. Rev. B* **61**, 1421 (2000).
9. B. L. Friman and E. Krotscheck, *Phys. Rev. Lett.* **49**, 1705 (1982).

10. E. Krotscheck and J. Springer, *J. Low Temp. Phys.* **132**, p. 281 (2003).
11. H. Godfrin, M. Meschke and H. J. Lauter, private communication (2003).
12. E. Krotscheck, Theory of correlated basis functions, in *Introduction to Modern Methods of Quantum Many–Body Theory and their Applications*, eds. A. Fabrocini, S. Fantoni and E. Krotscheck, Advances in Quantum Many–Body Theory, Vol. 7 (World Scientific, Singapore, 2002) pp. 267–330.
13. J. M. C. Chen, J. W. Clark and D. G. Sandler, *Z. Physik A* **305**, 223 (1982).
14. E. Krotscheck, *Phys. Rev. A* **26**, 3536 (1982).
15. E. Krotscheck and J. W. Clark, *Nucl. Phys. A* **328**, 73 (1979).
16. K. E. Schörkhuber, Multi–particle–hole excitations in many–body fermi systems, PhD thesis, Institut für Theoretische Physik, Johannes Kepler University, Linz, Austria 2002.
17. V. Grau, J. Boronat and J. Casulleras, *Phys. Rev. Lett.* **89**, 045301 (2002).
18. H. M. Böhm, E. Krotscheck, K. Schörkhuber and J. Springer, Pair excitations in fermi fluids, in *Proceedings of the XXIVth International Conference on Low Temperature Physics*, ed. Y. Takano (AIP, New York, 2006).
19. D. S. Greywall, *Phys. Rev. B* **41**, 1842 (1990).
20. K.-D. Morhard, C. Bäuerle, J. Bossy, Y. Bunkov, S. N. Fisher and H. Godfrin, *Phys. Rev. B* **53**, 2658 (1996).
21. A. Casey, H. Patel, J. Nyéki, B. P. Cowan and J. Saunders, *Phys. Rev. Lett.* **90**, 115301 (2003).

CONDENSATION OF HELIUM IN WEDGES

E. S. HERNÁNDEZ

Departamento de Física, Facultad de Ciencias Exactas y Naturales, Universidad de Buenos Aires and Consejo Nacional de Investigaciones Científicas y Técnicas, 1428 Buenos Aires, Argentina

F. ANCILOTTO

INFM-DEMOCRITOS and Dipartimento di Fisica "G. Galilei", Università di Padova, via Marzolo 8, I-35131 Padova, Italy

M. BARRANCO, R. MAYOL and M. PI

Departament ECM, Facultat de Física and IN^{2}UB, Universitat de Barcelona, 08028 Barcelona, Spain

We investigate the adsorption of helium in alkali wedges at zero temperature within a fully unrestricted density–functional theory. We have found a sequence of shapes that includes one filling transition on the nonwettable Cs and one emptying transition for every alkali, as well as a threshold for metastable configurations where helium does not populate the wedge apex.

Keywords: Wedge geometry; wetting; filling and emptying transitions.

1. Introduction

A variety of theoretical and experimental studies of the growth of adsorbed films have demonstrated that the geometry of a substrate exposed to a vapor phase plays a significant role in the wetting behavior of the fluid. While on a planar surface the growth is governed by the interactions between adatoms and between adsorbate and substrate, on real surfaces exhibiting steps, corners, stripes and other manifestations of disorder that may give rise to roughness, randomness and even fractality, the liquid-substrate interface adjusts to the local imperfections to keep the surface free energy at a global minimum.[1–5] Experiments on X-ray scattering at grazing incidence can resolve such interfacial structures.[6] Current manufacturing techniques are in use in nanofluidics, where very small amounts of fluid are stored and may flow in narrow channels,[7] and various experimental methods permit the creation of structured or imprinted surfaces within the micrometer scale.[5] Moreover, using chemical patterning one can control fabrication at the nanoscale[8] and for some applications involving structured semiconductor surfaces, highly regular, although nonhomogeneous surface structures can be produced (see, *i.e.*, Ref. 3). It is then of interest to investigate adsorption of liquid films on structured substrates in the nanoscopic, mesoscopic and macroscopic domains.

Since the early 90's, it has been pointed out that wedge wetting may constitute a paradigm for nontrivial interfacial profiles, and it has been recently suggested that fluids adsorbed in a planar wedge can be regarded as a generic class of structured inhomogeneous fluids.[9] In a series of papers, Parry and coworkers (see 10,11 and refs. cited therein) have shown that the wedge geometry introduces classes of interfacial phase transitions, giving rise to a family of interfacial critical phenomena controlled by the geometric parameters that define the structured surface. As a speculation,[12] it has been also suggested that wedge-like geometries might offer the possibility of observing effects of prewetting. From the experimental viewpoint, adsorption of Ar on patterned Si[4] shows that the crossover between a planar and a geometrically dominated regime is similar to that found numerically by Rascón and Parry.[11] More recently, Checco *et al* employ chemically nanopattered surfaces as templates for the condensation of nanometer-sized liquid drops of ethanol.[8] The experimental results show clearly that condensed nanoliquids can be controlled through the vapor phase and that their shape can be measured in situ using noninvasive scanning probe techniques.

As put forward in Ref. 12, wedge wetting represents an interesting physical problem in itself, in addition to serving as a prelude to the study of wetting of corrugated surfaces. In this philosophy, in the present work we study the adsorption of helium in alkali wedges at zero temperature within a fully unrestricted density–functional theory (DFT) as described in the following sections.

2. Classical Theories of Adsorption in Corners

In this section we review the main approaches applied to the description of the various meniscus shapes of a liquid adsorbed in a linear wedge with aperture on the (x, z) plane and in the thermodynamic limit in the y direction, $L_y \to \infty$, along with the possible shape transitions that may take place as one changes the wedge aperture, the vapor presure or the equilibrium temperature.

Within classical theories of wetting phenomena on planar surfaces, simple thermodynamics arguments show that by changing the state of a liquidlike layer on a substrate in a specifically chosen way, it is possible to provoke unlimited growth of the thickness l, giving rise to a wetting phase transition. If the growth takes place a) by approaching the liquid-vapor coexistence curve from the vapor side it is called complete wetting, b) by changing the state of the fluid along the coexistence line towards the wetting temperature T_w is it known as critical wetting if the growth is continuous, or first-order wetting if the height l of the film changes discontinuously to a macroscopic value. At given temperature T, Young's equation gives the contact angle $\theta(T)$ at the triple line of a macroscopic drop

$$\cos\theta(T) = \frac{\sigma_{sv} - \sigma_{sl}}{\sigma_{lv}} \tag{1}$$

with v, l, s respectively denoting the vapor, liquid and solid phases and with $\sigma_{ij}(T)$ the surface tension at the ij interface. The transition is first order or continuous depending on the manner on which $\cos\theta$ approaches unity as $T \to T_w$; if $\partial \cos\theta/\partial \mathrm{T} \to$ const > 0 the transition is first order, and continuous if the constant vanishes.

Now if one considers a linear wedge with aperture 2φ exposed to vapor below saturation, i.e., at a chemical potential $\mu = \mu_0 + \Delta\mu < \mu_0$, it has been shown by macroscopic, thermodynamic reasoning,[12] regarding both equilibrium and stability conditions, that the resulting configurations are unique and stable *iff* $\theta + \varphi < \pi/2$. If $\theta + \varphi > \pi/2$, no macroscopically stable adsorbate is possible. With varying T at fixed $\Delta\mu$, a macroscopic adsorbate appears in the well when

$$\theta(T, \Delta\mu) = \pi/2 - \varphi \tag{2}$$

then in a wedge, the wetting transition at coexistence is shifted to $T_\varphi < T_w$ such that

$$\theta(T_\varphi, 0) = \pi/2 - \varphi \tag{3}$$

Such a discontinuous change in behavior as $|\theta - \pi/2| = \varphi$ has been corroborated in microgravity experiments.[13]

Note that Eq. (2) is not restricted to coexistence. For $\varphi < \pi/2$, $T = T_\varphi(\Delta\mu)$ is the locus of a continuous wetting transition for $\Delta\mu < 0$, that becomes very similar to capillary condensation in the narrow wedge limit $\varphi \to 0$. An important limitation is neglect of gravity, which is a valid approximation provided that all relevant lengths are small compared to the capillary length $a_c = (\sigma_{lv}/m\rho_0 g)^{1/2}$ at bulk density ρ_0.

More formally, the shape of equilibrium film-vapor interface configurations can be determined by minimizing the total thermodynamic potential Ω that includes the cost in free energy to distort a flat interface into a curved meniscus. For a finite fluid system, the grandpotential functional decomposes into a bulk Ω_b, a surface Ω_s and a line Ω_l contribution, which depend on different characteristics of the density configuration as follows:

a) The bulk contribution depends only on the bulk densities of liquid and vapor, whose equilibrium values $\bar{\rho}_l, \bar{\rho}_v$ minimize Ω_b.

b) The surface contribution is identical to the surface free energy of the corresponding flat substrate and depends on the asymptotic layer thickness l and –parametrically– on the liquid and vapor densities $\bar{\rho}_l, \bar{\rho}_v$ at equilibrium. The actual width $\bar{l}$ minimizes Ω_s, thus T_w and other wetting properties of the system are fixed by the surface free energy.

c) Only Ω_l depends on the interfacial profile $z(x)$ and –parametrically– on l and the densities. The condition $\delta\Omega_l/\delta z(x) = 0$ at the equilibrium values of thickness and densities gives the equilibrium profile.

Usually, one searches to minimize the difference $\Delta\Omega = \Omega - \Omega_0$ between surface–plus–line grandpotentials (per unit length) of the wedge filled with liquid with respect to that filled with vapor, with the general form.

$$\Delta\Omega = \int_0^L dx\, \varepsilon[x, z(x)] - (\mu_l - \mu_v)\, n \tag{4}$$

The various descriptions offered in the literature consider either a fixed number of fluid particles per unit length $n = N/L$

$$n = \rho_0 \int_0^L dx\, [z(x) - |x| \operatorname{cotg} \varphi] \tag{5}$$

at saturation pressure[1] or a fixed volume V.[3] Models also differ in the specific choice of the energy density kernel in Eq. (4); the latter always contains the curvature energy ε_c,

$$\varepsilon_c = \sigma_{lv}\, \left(1 + z_x^2\right)^{1/2} \tag{6}$$

but while the more phenomenological approaches resort to the empirical surface tensions to introduce the surface contribution to the functional, the so-called effective Hamiltonian model includes a substrate-adatom interaction to represent more acurately the influence of the external confining field.[3] This viewpoint is similar to the earlier proposal for the liquid profile of a wetting liquid in a corner put forward in Ref. 1. In the first case, the equilibrium shape of a nonwetting liquid can be analytically computed and shown to correspond to a circle with radius $R = \sigma_{lg}/\left(\rho_0\, \mu_l\right)$ that intersects the walls at θ, located at a finite height $(0, z_c)$. If $\frac{\pi}{2} + \varphi \geq \theta > \frac{\pi}{2} - \varphi$, the meniscus is convex, and concave for $\theta < \frac{\pi}{2} - \varphi$. A planar meniscus indicative of a wedge filling transition appears when $\theta = \frac{\pi}{2} - \varphi$. Moreover, a spherical bridge that intersects the walls at the contact angle θ can exist *if* $z_c > R$, implying $\theta > \pi/2 + \varphi$. This rigorous geometrical outcome strictly coincides with the more phenomenological description attained by means of thermodynamical arguments.[12]

3. Density Functional Description of Wedge Wetting

Recently, we have presented the first complete calculation of the structure of liquid ^{4}He confined to a concave nanoscopic wedge with alkaline walls, as a function of the opening angle of the walls, within a finite-range density functional formalism.[14] For a translationally invariant system along the y–direction, the equilibrium density profile $\rho(x, z)$ of the helium atoms depends on the (x, z) coordinates only. We compute the density profile that minimizes the zero temperature grandpotential $\Omega = E - \mu\, N$, with N the grandcanonical ensemble average of the particle number operator,

$$N = \int d^3\mathbf{r}\, \rho(\mathbf{r}) \equiv L \int\!\!\int dx\, dz\ \ \rho(x, z) \tag{7}$$

The variation $\delta\Omega/\delta\rho$ gives rise to the integrodifferential Euler-Lagrange equation

$$\left[-\frac{\hbar^2}{2m}\nabla^2 + V(\rho) + V_s(x, z)\right] \sqrt{\rho(x, z)} = \mu \sqrt{\rho(x, z)} \tag{8}$$

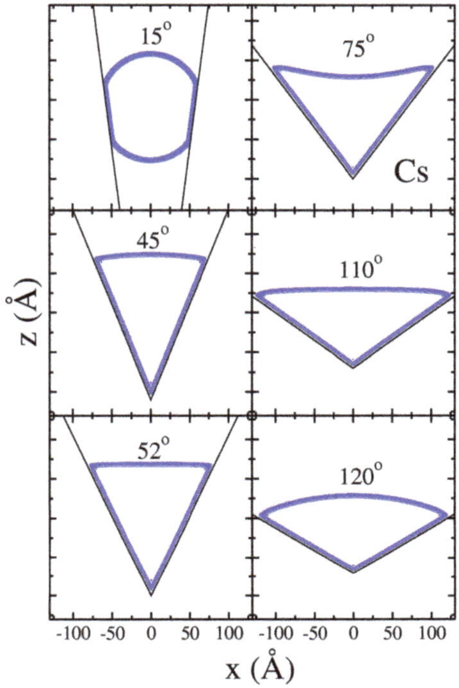

Fig. 1. Density profiles of helium at $n = 280$ Å^{-1} in a Cs corner for several angular openings. The distance between major tick marks on the z axis is 50 Å. The center of the 15° configuration is about 400 Å above the apex of the wedge. Equidensity lines are drawn from $\rho = 0.9\rho_0$ to $0.1\rho_0$ in steps of $0.1\rho_0$.

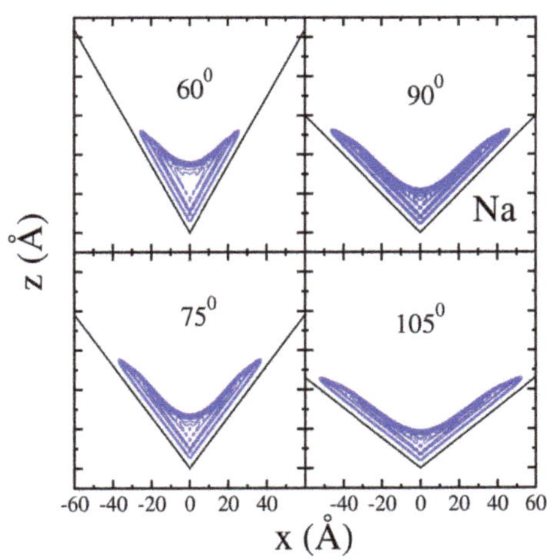

Fig. 2. Density profiles of helium at $n = 15$ Å^{-1} in a Na corner for several angular openings. The distance between major tick marks on the z axis is 20 Å. Equidensity lines as in Fig. 1.

with $V(\rho)$ the effective potential constructed as the first functional derivative of the potential energy density per unit length, and with $V_s(x, z)$ the confining field. In

Ref. 14, we have selected the full Orsay-Trento (OT) DF.[15] We have also chosen to approximate $V_s(x, z)$ by the summation of two planar potentials, each of the form given in Ref. 16, at an angle 2φ. Double-counting effects are unimportant for values of 2φ below 120°, far from the wedge apex and for the fairly high linear densities here considered. Equation (8) is solved by an imaginary time method and employing a Fast Fourier Transform algorithm to generate the convolution terms in $V(\rho)$ from the atom density $\rho(x, z)$.

We illustrate the main results of our calculation for the cases of cesium and sodium, which are respectively not wetted and wetted by ^{4}He at zero temperature. In either case, we first compute a sequence of pancake-shaped systems on the corresponding flat surface characterized by the linear density $n = N/L$. The chemical potential of the helium particles as a function of n asymptotically approaches the bulk figure $\mu_0 = -7.15$ K in the case of Cs, and the prewetting chemical potential $\mu_{pw} = -8.75$ K for Na. Up to linear densities of several hundreds Å^{-1}, the density profiles of the bars confirm the nanoscopic spatial scale of the systems.

Typical shape profiles for helium at various apertures 2φ are shown in Fig. 1 for a Cs wedge with a linear ^{4}He density of 280 Å^{-1}, and in Fig. 2 for Na walls with n = 15 Å^{-1}. Starting from a convex pancake on a flat Cs surface and decreasing the opening, one can generate a sequence of changes with the following landmarks: a) a filling transition where the meniscus curvature vanishes takes place for an opening near 110°, b) a concave meniscus appears for smaller angles, that persists within a large angular interval; c) an 'emptying' transition happens, near 52°, where the meniscus becomes planar; d) the convex shape reappears for even smaller angles; e) a bridge is formed, that will be expelled from the wedge for sufficiently narrow apertures. It should be kept in mind that in our philosophy, the term 'transition' does not refer to a thermodynamic phase transition, but to a crossover between two different regimes characterized by geometry and energetics. The bridge profiles are perfectly circular, in agreement with classical theories of the preceding Section. By contrast, in Fig. 2 we appreciate the spread of a wetting liquid in a strongly attractive Na corner, with a rather deep concave meniscus at all angles above $2\varphi_F \approx 30^o$. An emptying transition takes place at $2\varphi_F$, after which, the sample evolves into a nanoscopic bridge.

The above shape sequences are in strict correspondence with energetics as a function of the wedge opening and of linear density. Figure 3 shows the chemical potential of helium atoms in a Cs wedge parametrized by $n = 70$, 140 and 280 Å^{-1}. We see that the three curves cross at the bulk value μ_0 for the filling angle $2\varphi_F = 110^o$ and for the emptying one $2\varphi_E = 52^o$. Between these openings the chemical potential μ lies below the bulk value μ_0 with positive derivative $d\mu/dn$, thus the corresponding wedge configurations are stable. Instead, both above the filling angle φ_F and below the emptying one φ_E, μ lies above its thermodynamic limit with negative $d\mu/dn$, thus corresponding to metastable systems. These characteristics are confirmed by the behavior of the grandpotential per unit length $\Omega/L = E/L - \mu n$, where E/L is the energy per unit length; at φ_F, Ω/L is independent of n and

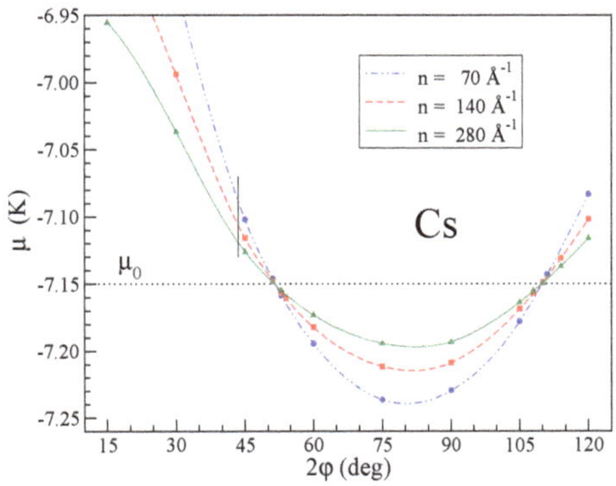

Fig. 3. Chemical potential of helium atoms in a Cs wedge as a function of angular opening 2φ for linear densities 70, 140 and 280 Å^{-1}. Wedge configurations to the left of the vertical segment near 43° have higher energies than a flat pancake with the same linear density.

vanishes for these large nanoscopic amounts of fluid. Accordingly, these helium samples undergo the kind of filling transition derived for macroscopic systems.[3] In Ref. 14 we have propossed an effective way of determining the contact angle θ of ^{4}He on Cs by means of DF calculations, looking for the value of $2\varphi_F \sim 110^o$ where the grandpotential is independent of n. This procedure gives a contact angle $\theta = \pi/2 - \varphi_F = 35^o$ for ^{4}He on Cs, in agreement with experimental evidence and earlier theoretical, DF-based, estimates.[17,18]

For the stable configurations in the range $[\varphi_E, \varphi_F]$, the disappearance of the solid-vacuum interface at the meniscus contact line, leaving only a liquid-vacuum interface at the free surface, is the source of the energy gain. Instead, as φ becomes smaller than φ_E , the system shifts to a nonwetting regime; in this case, it is energetically favorable to hold vacuum, rather than liquid, near the apex. Consequently, an inner convex meniscus appears away from the corner, forming a bridge with increased energy.

It is interesting to note that within the interval $[\varphi_E, \varphi_F]$, two stable profiles exist for each coverage; in particular, this is the case for the flat meniscii configurations at $\mu = \mu_0$. This is a genune finite size effect intrinsic to the current nanoscopic samples, since in macroscopic systems only a filling transition is expected, according to the classical theories in Section 2. In fact, in the truly macroscopic regime, all equilibrium configurations take the bulk values, *i.e.*, chemical potential and saturation density, regardless the wedge opening.

The mechanism of condensation is better understood examining the adsorption isotherms $\mu(n)$ in a wedge with fixed aperture. No condensation can take place in the wedge for openings below φ_E, since the ^{4}He configurations in the wedge are always metastable with respect to the bulk. For all angles in the stable domain, condensation takes place with a jump in the linear density from $n = 0$ to a finite value n_F. By using the Maxwell construction (or, equivalently, by identifying the crossing between the $\mu(n)$ and E/N curves) one can identify the chemical potential

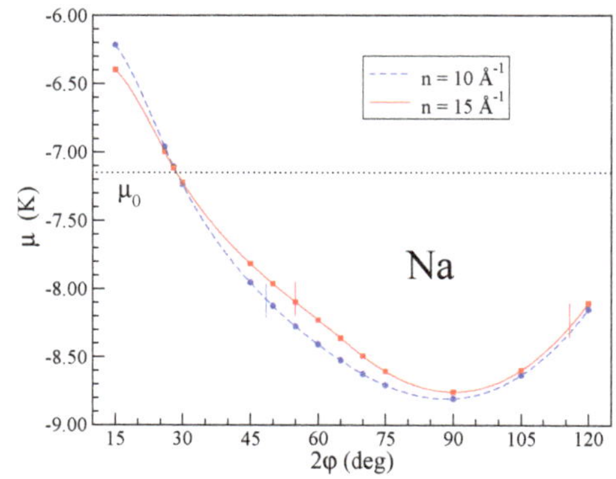

Fig. 4. Chemical potential of helium atoms in a Na wedge as a function of angular opening 2φ for linear densities $n = 10$ and 15 $\mathrm{\AA}^{-1}$. For $n = 15$ $\mathrm{\AA}^{-1}$, wedge configurations within the vertical segments are stable with respect to the infinite pancake configuration. For $n = 10$ $\mathrm{\AA}^{-1}$, wedge configurations to the right of the corresponding vertical segment are stable with respect to the infinite pancake configuration.

and density at the corresponding jump. This is indicated in Fig. 5, where we show the size of the "pre-filling/pre-emptying" jump in density as a function of angle for a Cs wedge (circles) together with the corresponding chemical potential (squares). We realize that n_F diverges at the filling and emptying transitions and encounters a shallow minimum around 2.4 $\mathrm{\AA}^{-1}$ at $2\varphi = 90^o$. Correspondingly, the chemical potential at the jump displays a minimum of about -7.5 K. In the strong Na adsorber, only the divergence at the emptying opening shows up; starting from large angles, the jump density n_F decreases monotonically from the planar prewetting value near 0.14 $\mathrm{\AA}^{-1}$ (per unit perpendicular length) to a minimum around 0.11 $\mathrm{\AA}^{-1}$. In either case, we have found that for the linear densities under consideration helium bridges are metastable (i.e. have higher energies) with respect to helium pancakes deposited on planar surfaces. These limits are shown in Figs. 3 and 4 by vertical segments. Similarly, Cs wedges corresponding to openings $2\varphi > 120^o$ are metastable with respect to pancakes with the same number of atoms. In these metastability regions the pancake, which is a symmetry breaking configuration, is energetically favored with respect to the axially symmetric configuration.

We stress that one of the most important, newest features of the results here presented is precisely the occurrence of an 'emptying' transition and the formation of metastable bridges. This is not expected on the basis of macroscopic thermodynamic arguments, and even seems to contradict the expectation that a narrow wedge should behave as a small pore or slit and be filled by capillary condensation (CC) below SVP. In fact, the latter is the behavior to be expected for sufficiently large helium samples. For the nanoscopic amounts of material here considered, bridges formed between two parallel plates destabilize as the distance between walls overcomes some critical value.[19] This is largely a nontrivial effect, in view of recent experimental evidence on the formation of helium-void micropuddles on rough Cs surfaces.[20]

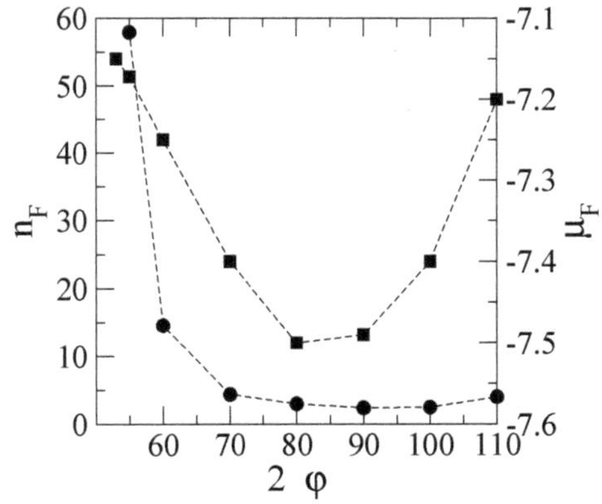

Fig. 5. Prefilling jump n_F in linear density (circles) and corresponding chemical potential μ_F (squares) as obtained from the adsorption isotherms of helium on Cs wedges as functions of wedge opening (in degrees). The lines are drawn to guide the eye. The divergence at an angle slightly above 110° is not shown.

4. Summary

In the first detailed, fully unrestricted, DFT calculation of the structure and energetics of nanoscopic samples of ^{4}He in alkali linear wedges at zero temperature, we have found a sequence of shapes that includes one filling transition on the non-wettable Cs, and a threshold for metastable configurations where helium does not populate the wedge apex. Condensation at negative grandpotential, accompanied by pre-filling jumps in the linear density, takes place for angles above threshold –and below the filling aperture in the case of Cs– in any alkaline wedge. When the helium fluid is near saturation, the angle for the filling transition is stable against changes in the linear density and can be identified with the contact angle on a planar substrate. We have found that metastable configurations appear for finite average numbers of ^{4}He atoms per unit length; it is important to keep in mind that a droplet with finite atom number N on Cs is metastable, not only with respect to bulk, but also with respect to a thin film at finite coverage N/A. Inspired by recent experiments on physisorption on nanostructured substrates, we suggest that deposition of *finite*, nanoscopic amounts of helium on alkali wedges, could lead to actual confirmation of our current predictions.

Acknowledgments

This work was supported by grants PIP2391/00 from CONICET and PICT03-08450 from ANPCYT, Argentina, FIS2005-01414 from DGI, Spain (FEDER), and 2005SGR-00343 from Generalitat of Catalunya. F. A. acknowledges funding from CESCA-CEPBA, Barcelona, through the program HPC-Europa Transnational Access, and Project num. CPDA033545 of Padova University.

References

1. E. Cheng and M. W. Cole, *Phys. Rev. B* **41**, 9650 (1990).
2. M. Napiórkowski, W. Koch and S. Dietrich, *Phys. Rev. B* **45**, 5760 (1992).
3. K. Rejmer, S. Dietrich and M. Napiórkowski *Phys. Rev. E* **60**, 4027 (1999).
4. L. Bruschi, A. Carlin and G. Mistura, *Phys. Rev. Lett.* **89**, 166101 (2002).
5. P. Lenz and R. Lipowsky, *Phys. Rev. Lett.* **80**, 1920 (1998).
6. A. Hanke, M. Krech, F. Schlesener and S. Dietrich, *Phys. Rev. E* **60**, 5163 (1999).
7. M. Cieplak, J. Koplik and J. R. Banavar, *Phys. Rev. Lett.* **96**, 114502 (2006).
8. A. Checco, O. Gang and B. M. Ocko, *Phys. Rev. Lett.* **96**, 056104 (2006).
9. J. R. Henderson, *Phys. Rev. E* **69**, 061613 (2004).
10. A. O. Parry, C. Rascón and A. J. Wood, *Phys. Rev. Lett.* **83**, 5535 (1999).
11. C. Rascón and A. O. Parry, *Nature* **407**, 986 (2000).
12. E. H. Hauge, *Phys. Rev. A* **46**, 4994 (1992).
13. P. Concus and R. Finn, *Phys. Fluids* **10**, 39 (1996).
14. E. S. Hernández, F. Ancilotto, M. Barranco, R. Mayol and M. Pi, *Phys. Rev. B* **73**, 245406 (2006).
15. F. Dalfovo, A. Lastri, L. Pricaupenko, S. Stringari and J. Treiner, *Phys. Rev. B* **52**, 1193 (1995).
16. A. Chizmeshya, M. W. Cole and E. Zaremba, *J. Low Temp. Phys.* **110**, 677 (1998).
17. F. Ancilotto, A. Sartori and F. Toigo, *Phys. Rev. B* **58**, 5085 (1998); F. Ancilotto, F. Faccin and F. Toigo, *Phys. Rev. B* **62**, 17035 (2000).
18. E. Rolley and C. Guthmann, *J. Low Temp. Phys* ***108**, 1 (1997)*; D. Ross, J.E. Rutledge, and P. Taborek, *Science* **278**, 664 (1997); J. E. Rutledge, D. Ross, and P. Taborek, *J. Low Temp. Phys.* **113**, 811 (1998).
19. R. Mayol, F. Ancilotto, M. Barranco, E. S. Hernández and M. Pi, J. Low Temp. Phys, in print (2007).
20. J. Klier, P. Leiderer, D. Reinelt and A. F. G. Wyatt, *Phys. Rev. B* **72**, 245410 (2005).

NON-FERMI-LIQUID BEHAVIOR FROM THE FERMI-LIQUID APPROACH

V. A. KHODEL*, J. W. CLARK† and H. Li#

McDonnell Center for the Space Sciences
and Department of Physics, Washington University,
St. Louis, MO 63130, USA
**E-mail: vak@wuphys.wustl.edu*
†E-mail: jwc@wuphys.wustl.edu
#E-mail: hli@artsci.wustl.edu

V. M. YAKOVENKO

Condensed Matter Theory Center
and Center for Superconductivity Research,
Department of Physics, University of Maryland,
College Park, Maryland 20742-4111, USA
E-mail: yakovenk@mu.edu

M. V. ZVEREV

**Russian Research Centre Kurchatov Institute,*
Moscow, 123182, Russia
E-mail: zverev@mbslab.kiae.ru

Non-Fermi liquid behavior of strongly correlated Fermi systems is derived within the Landau approach. We attribute this behavior to a phase transition associated with a rearrangement of the Landau state that leads to flattening of a portion of the single-particle spectrum $\epsilon(\mathbf{p})$ in the vicinity of the Fermi surface. We demonstrate that the quasiparticle subsystem responsible for the flat spectrum possesses the same thermodynamic properties as a gas of localized spins. Theoretical results compare favorably with available experimental data. While departing radically from prevalent views on the origin of non-Fermi-liquid behavior, the theory advanced here is nevertheless a conservative one of in continuing to operate within the general framework of Landau theory.

Keywords: Strongly correlated Fermi systems.

1. Introduction

This year is the jubilee of Fermi liquid (FL) theory. Fifty years ago L. D. Landau published an article[1] devoted to evaluation of thermodynamic properties of three-dimensional (3D) liquid ^{3}He, in which he postulated a one-to-one correspondence between the totality of real, decaying single-particle (sp) excitations of the Fermi liquid and a system of immortal interacting quasiparticles. Two distinguished features specify this system: (i) its quasiparticle number is equal to the particle number (the so-called Landau-Luttinger theorem), and (ii) its entropy S, given by the ideal-Fermi-gas combinatorial expression, coincides with the entropy of the actual helium system. Treating the ground-state energy E_0 as a functional of the quasiparticle momentum distribution $n(p)$, Landau derived a formula

$$n(p,T) = \left[1 + e^{\epsilon(p)/T}\right]^{-1} \tag{1}$$

for the quasiparticle momentum distribution $n(p,T)$ that resembles the corresponding Fermi-Dirac formula for the ideal Fermi gas. However, the quasiparticle energy $\epsilon(p) = \delta\Omega/\delta n(p)$ is a variational derivative of the thermodynamic potential $\Omega = E - \mu N$, where E is the ground-state energy and μ, the chemical potential, with respect to the momentum distribution $n(p)$, rather than the bare sp energy ϵ_p^0 entering the respective Fermi-Dirac formula. As we shall see, in strongly correlated Fermi systems, this noncoincidence is crucial.

Landau's quasiparticle pattern of low-temperature phenomena in Fermi liquids is universally recognized as a cornerstone of condensed-matter theory. However, a qwirk of fate is that although originally FL theory was created for the explanation of properties of 3D liquid ^{3}He, discrepancies between theoretical predictions and experimental data on the low-T spin susceptibility $\chi(T)$ and the ratio $C(T)/T$ of the specific heat $C(T)$ to temperature T (both these must be constant in FL theory) first came to right in this liquid. The deviations, rather small at extremely low T, increase with temperature. In their book, Nozières and Pines[2] attribute the departures to the damping of sp excitations. During more than thirty years this claim was regarded as a terminal diagnosis of FL theory. However, the situation was turned upside down about ten years ago when 2D liquid ^{3}He was studied quantitatively in a region of densities ρ where this system gradually evolves from a weakly correlated gas to a strongly correlated liquid.[3,4] The experimental data turned out to be even more challenging to interpret than those for the 3D counterpart. Indeed, in the density region where strong correlations set in, departures from predictions of FL theory exhibit themselves more strongly, the *lower* the temperature. This definitely rules out damping of sp excitations as the cause for failure of FL theory, since damping vanishes as $T \to 0$.

The departures in question occur in the density region where the FL effective mass $M^*(\rho)$, extracted from the 2D liquid ^{3}He specific heat data[4] using the FL formula $C_{\rm FL}(T)/T = p_F M^*/3$, is found to be enhanced, with the enhancement rapidly increasing as ρ approaches a critical value $\rho_s \simeq 0.072$ Å^{-2}, beyond which 2D liquid ^{3}He solidifies. Significantly, the behavior of $\chi(T)$ changes drastically well before ρ reaches ρ_s: the product $\chi(T)T$ ceases to vanish at $T \to 0$, in agreement with the Curie law and in contrast to predictions of standard FL theory. In doing so, the value of this product *gradually increases* as $\rho \to \rho_s$. Analogous behavior has been observed for other strongly correlated Fermi liquids (see e.g. Ref. 5).

Various theories have been suggested in attempts to identify the source of non-Fermi-liquid (NFL) behavior in strongly correlated Fermi systems. Many of them are based on a theory advanced in Refs. 6,7, in which the NFL behavior is attributed to spin fluctuations. However, the spin-fluctuation scenario fails to explain experimental results for thermodynamic properties of heavy-fermion metals in the vicinity of the so-called quantum critical point, especially in the presence of external magnetic fields.[8–11]

The gradually emerging Curie behavior of $\chi(T)$ in 2D liquid ^{3}He is sometimes attributed to a localization phase transition. Unfortunately for this explanation, a homogeneous system without impurities has no more chance to be slightly localized than a women has to be slightly pregnant. The Curie-Weiss behavior of $\chi(T)$ in heavy-fermion metals is ordinarily described within the Anderson $s-d$ model, in which localized d states acquire a nonzero spin as a result of interaction with delocalized s electrons. However, within this scenario it is hard to reconcile the localization of electrons with Cooper pairing, which is found to be unexpectedly strong in some cases.[12]

We shall argue that the occurrence of the Curie-like term in low-temperature behavior of the magnetic susceptibility $\chi(T)$ of a homogeneous Fermi liquid, which *gradually evolves under change of input parameters*, is a signature of a phase transition known by fermion condensation. This phase transition is associated with a rearrangement of *single-particle degrees of freedom*, rather than collective ones.

2. Rearrangement of sp Degrees of Freedom in Finite Fermi Systems

It is instructive to begin studying such a rearrangement in finite Fermi systems, in which damping of single-particle excitations does not occur. This closes off one favorite escape route of critical readers and referees. It is the conventional wisdom of textbooks that under variation of input parameters, two sp levels may repel or cross one other. However, as we shall demonstrate, the familiar dichotomy misses a further alternative: levels can in fact merge.[13] This phenomenon is made possible by the variation of sp energies with level occupation numbers – a property central to Landau theory. A primary condition for merging to occur is that the Landau-Migdal interaction function f is repulsive in coordinate space,[13] which holds for the effective nn and pp interactions in the nuclear interior[14] and for the electron-electron interaction in atoms.

Consider a schematic model involving three equidistant neutron levels, separated by an energy distance D in an open shell of a spherical nucleus. The levels are denoted $-$, 0, and $+$, in order of increasing energy. The sp energies ϵ_λ and wave functions $\varphi_\lambda(\mathbf{r}) = R_{nl}(r)\Phi_{jlm}(\mathbf{n})$ are solutions of

$$[p^2/2M + \Sigma(\mathbf{r},\mathbf{p})]\varphi_\lambda(\mathbf{r}) = \epsilon_\lambda\varphi_\lambda(\mathbf{r}), \tag{2}$$

where Σ stands for the self-energy. In even-even spherical nuclei, which in their ground states have total angular momentum $J = 0$ due to pairing correlations, the energies ϵ_λ are independent of the magnetic quantum number m associated with the total sp angular momentum j. We suppose that the level $-$ is filled, the level

+ is empty, and N neutrons are added to the level 0, changing the density $\rho(r)$ by $\delta\rho(r) = NR^2_{n_0 l_0}(r)/4\pi$.

In what follows, we shall retain only a major, spin- and momentum-independent part V of the self-energy Σ and a primary, $\delta(r)$-like portion of the Landau-Migdal interaction function f. Accordingly, the FL relation between Σ and ρ responsible for the variation of $\epsilon_\lambda(n)$ with n reduces to[14]

$$\delta V(r) = f[\rho(r)]\delta\rho(r) \ . \tag{3}$$

When particles are added to the system, all energy levels are shifted somewhat, but the level that receives the particles is affected more strongly than the others. For the sake of simplicity, the diagonal and nondiagonal matrix elements of f are assigned the respective values

$$u = \int R^2_{nl}(r) f\left[\rho(r)\right] R^2_{nl}(r) r^2 dr/4\pi \,, \quad w = \int R^2_{nl}(r) f\left[\rho(r)\right] R^2_{n_1 l_1}(r) r^2 dr/4\pi \,, \tag{4}$$

independently of the quantum numbers nl, $n_1 l_1$.

Based on these assumptions and results, the dimensionless shifts $\xi_k(N) = \left[\epsilon_k(N) - \epsilon_k(0)\right]/D$ for $k = 0, +, -$ are given by

$$\xi_0(N) = n_0 U \ , \quad \xi_+(N) = \xi_-(N) = n_0 W \ , \tag{5}$$

where $n_k = N_k/(2j_k+1)$ is the occupation number of the level k, $U = u(2j_0+1)/D$, and $W = w(2j_0+1)/D$. It is readily verified that if $fp_F M/\pi^2 \sim 1$, where $p_F = \sqrt{2M\epsilon_F}$ and ϵ_F is the Fermi energy, then the first of the integrals (4) has a value $u \simeq \epsilon_F/A$ and therefore $U \sim 1$, since $D \sim \epsilon_F/A^{2/3}$ in spherical nuclei.

According to Eqs. (5) at $(U - W) > 1$, the difference $d(N) = 1 + \xi_+(N) - \xi_0(N)$ changes sign at $n_{0c} = 1/(U - W)$, before filling of the level + is complete. At $n_0 > n_{0c}$, in the standard scenario provided by Hartree-Fock theory, all added quasiparticles must resettle into the empty sp level +. However, not all of the quasiparticles can take part in the migration process, since the situation would then be reversed, and the roles of the levels interchanged: the formerly empty level, lying above the formerly occupied one, would have the maximum positive energy shift, rendering migration impossible. Thus, the standard Fermi-liquid filling scenario, which prescribes that *one and only one* sp level lying exactly at the Fermi surface can remain unfilled, while all others must be completely occupied or empty, encounters a catastrophe.

This catastrophe is resolved as follows.[13,15] Migration occurs until the sp energies of the two levels in play coincide. As a result, both of the levels, 0 and +, become *partially* occupied – an impossible situation for the standard Landau state. Solution of the problem reduces to finding the minimum of the relevant energy functional

$$E_0 = \epsilon_0(0)N_0 + \epsilon_+(0)N_+ + \frac{1}{2}\left[u(N_0^2 + N_+^2) + 2wN_0N_+\right] \tag{6}$$

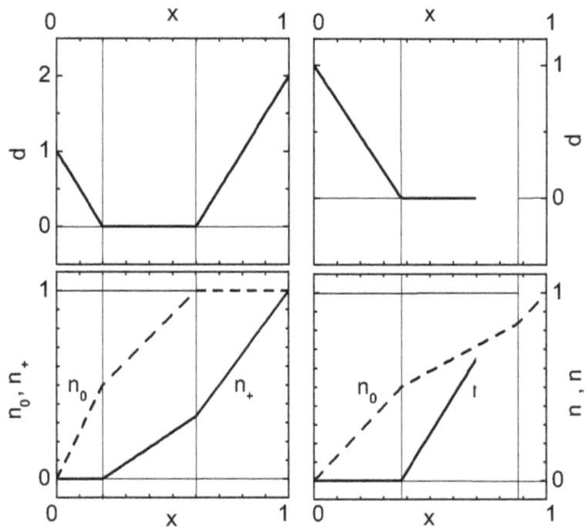

Fig. 1. Top panels: Dimensionless distance $d = (\epsilon_+ - \epsilon_0)/D$ between levels + and 0 as a function of the ratio $x = N/(2j_0+2j_++2)$. Lower panels: Occupation numbers n_k for levels 0 and +. Input parameters: $U = V = 3, W = 1$. For the left column, the ratio $r \equiv (2j_0+1)/(2j_++1) = 2/3$; for the right, $r = 3$.

with $N_k = \sum_m n_{km}$, through a variational condition

$$\frac{\delta E_0}{\delta n_{0m}} = \frac{\delta E}{\delta n_{+m_1}} = \mu\,, \qquad \forall m, m_1\,, \tag{7}$$

where μ is the chemical potential. Such a condition first appeared in Ref. 15, where homogeneous Fermi systems were addressed without attention to the degeneracy of sp levels. Eqs. (7) are conveniently rewritten as conditions

$$\epsilon_0(N) = \epsilon_0(0) + N_0 u + N_+ w = \mu\,, \quad \epsilon_+(N) = \epsilon_+(0) + N_0 w + N_+ u = \mu \tag{8}$$

for coincidence of the sp energies ϵ_0 and ϵ_+, which, at $N > N_c = (2j_0+1)/(U-W)$, yield $N_0 = \frac{1}{2}(N + N_c)$ and $N_+ = \frac{1}{2}(N - N_c)$.

Results from numerical calculations are plotted in Fig. 1, which consists of two columns, each made up of two plots. The upper panels show the dimensionless ratio $d(x) = [\epsilon_+(x) - \epsilon_0(x)]\,/D$ versus $x = N/(2j_0+2j_++2) \in [0,1]$. The lower panels give the occupation numbers $n_+(x)$ and $n_0(x)$. We observe that there are three different regimes: in two of them $d \neq 0$ and there exist well-defined sp excitations, and in the third, the energies of the levels 0 and + coincide at zero. Passage through the three regimes can be regarded as a second-order phase transition, with the occupation number n_+ treated as an order parameter.

The sp levels remain merged until one of them is completely filled. If the level 0 fills first, as in the left column of Fig. 1, then under further increase of N, quasiparticles fill the level +, signaling that the distance $d(N)$ again becomes positive. This behavior resembles the repulsion of two levels of the *same symmetry* in quantum mechanics, although here one deals with the sp levels of *different symmetry.* In a case where level + becomes fully occupied before the level 0 does, as in the right column, the distance $d(N)$ becomes negative, and the two levels just cross each other at this point.

In systems without pairing correlations, for example in atoms, a pair of particles added to any sp level with $l \neq 0$ always has total angular momentum $J \neq 0$ (Hund's rule), in principle destroying spherical symmetry and lifting the m-degeneracy of the sp energies ϵ_{km}, thereby complicating the analysis of merging. Here we only sketch the final results (for details, see Ref. 13). The energy functional

$$E = \sum \epsilon_k(0) n_{km} + \frac{1}{2} \sum f_{km,k_1m_1} n_{km} n_{k_1m_1} \tag{9}$$

replaces the functional (6), and the interaction matrix f_{km,k_1m_1} replaces the matrix elements (4). The variational equations

$$\mu = \epsilon_k(0) + \sum f_{km,k_1m_1} n_{k_1m_1} \,, \tag{10}$$

in which the sum runs over some states of the last unfilled shell that undergo merging, are to be solved numerically. Merging provides a qualitative explanation of the fact that the chemical properties of rare-earth elements differ little, in spite of marked variation in atomic numbers. Such an explanation is to some extent complementary to the textbook argument that the collapse of the electron $4f$-orbital is responsible for the remarkable similarity of the chemical properties of the rare-earth elements.

It is worth noting that if the number of equations (10) to be solved becomes large, reasonable results can be obtained with the replacement of summation by integration to yield[15]

$$\mu = \epsilon_k^0 + 2 \int f(\mathbf{k}, \mathbf{k}_1)\, n(\mathbf{k}_1)\, d^3k_1/(2\pi)^3 \,. \tag{11}$$

As will now be demonstrated, the analysis of merging of sp levels in finite systems helps us understand what is going on in infinite matter. Let us consider a model of a heavy-fermion metal in which the sp spectrum, evaluated usually in local-density approximation, is exhausted by (i) a wide band that disperses through the Fermi surface, and (ii) a narrow band situated below the Fermi surface at a distance D_n. To facilitate the analysis, we assume that only the diagonal matrix element f_{nn} of the interaction function f referring to the narrow band is significant, the others being negligible. The shift $\delta\epsilon_n$ in the location of the narrow band due to switching on the intraband interactions is given by a formula $\delta\epsilon_n = f_{nn}\rho_n$ analogous to Eq. (5), where ρ_n is the density of the narrow band. Suppose now that the correction $\delta\epsilon_n$ exceeds the distance D_n. In this case, the HF scenario calls for the narrow band to be completely emptied; but then the shift $\delta\epsilon_n$ must vanish. To eliminate this inconsistency, it must be the case that only a fraction of the particles leave the narrow band, in just the right proportion to equalize the chemical potentials of the two bands. The feedback mechanism we have described positions the narrow band exactly at the Fermi surface, resolving a long-standing problem with the LDA scheme.

3. Quantum Critical Point in a Homogeneous Fermi Liquid

In this section, we investigate properties of Fermi liquids in the vicinity of the FL quantum critical point (QCP), i.e. in a region close to the critical density ρ_∞ where the effective mass M^* diverges. Standard FL theory, which knows nothing about the QCP, tells us that properties of Fermi liquids are similar to those of an ideal Fermi gas, with differences merely involving a numerical factor, the ratio M^*/M. However, at the critical density ρ_∞ where M^* diverges, the FL spectrum $\epsilon_{\rm FL}(p;\rho)$ vanishes identically, signalling that the standard FL theory requires a cure. Remarkably, the Landau approach itself contains a medicine to treat the disease. The cure lies in the relation[1]

$$\frac{\partial\epsilon(p)}{\partial\mathbf{p}} = \frac{\mathbf{p}}{M} + 2\int f(\mathbf{p},\mathbf{p}_1)\frac{\partial n(\epsilon(p_1))}{\partial\mathbf{p}_1}d^3p_1/(2\pi)^3\,, \tag{12}$$

which connects the quasiparticle spectrum $\epsilon(p)$ and the momentum distribution $n(\epsilon)$ through the Landau interaction function $f(\mathbf{p},\mathbf{p}_1)$. Referring back to the preceding section in which we considered the merging of sp levels in finite Fermi systems, this relation has its conceptual counterpart in Eq. (3). From Eq. (12) one finds

$$M/M^*(\rho,T=0) = 1 - F_1^0(\rho)/3\,, \tag{13}$$

having introduced the dimensionless first harmonic $F_1^0(\rho) = f_1(p_F,p_F)N_0$, where $N_0 = p_F M/\pi^2$ and $f_1(p_F,p_F)$ is the first harmonic of $f(\mathbf{p}_1,\mathbf{p}_2)$. From Eq. (13) we infer that realization of the divergence of M^* hinges on the presence of sufficiently large velocity-dependent components in f, since one must have $F_1^0(\rho_\infty) = 3$.

Quantitative studies of the spectrum $\epsilon(p)$ first appeared in connection with the problem of fermion condensation (see below) some forty years after the creation of FL theory. Close to the critical density where $M^*(\rho)$ diverges, the group velocity $d\epsilon(p;\rho\to\rho_\infty)/dp$ ceases to be constant, acquiring a parabola-like shape.[16,17] This is seen from Fig. 2, where the microscopically evaluated[17] sp spectrum and group velocity of a 2D electron gas are shown.

The structure of the sp spectrum $\epsilon(p,\rho)$ in the vicinity of the Fermi-liquid QCP can be elucidated without resorting to the help of a computer. Instead one expands the relevant quantities on both sides of Eq. (12) in Taylor series, thereby obtaining

$$d\epsilon/dp \simeq p_F/M^*(\rho,T) + v_2(p-p_F)^2/Mp_F\ . \tag{14}$$

Inserting this expression into Eq. (12) and performing rather lengthy algebra, one finds[18]

$$M/M^*(T,\rho_\infty) \propto T^{2/3}\ . \tag{15}$$

Thus, within the Landau approach, the density of states diverges as $T^{-2/3}$ when the density ρ approaches ρ_∞. The corresponding asymptotic behaviors of the spin susceptibility and the ratio $C(T)/T$ are $\chi(T,\rho_\infty) \propto C(T,\rho)/T \propto T^{-2/3}$, while the entropy $S(T,\rho_\infty)$ and the thermal expansion $\beta(T,\rho_\infty)$ behave as $T^{1/3}$.

Imposition of a static external magnetic field H brings into play a new dimensionless parameter $R = \mu_B H/T$ and opens another arena for testing the

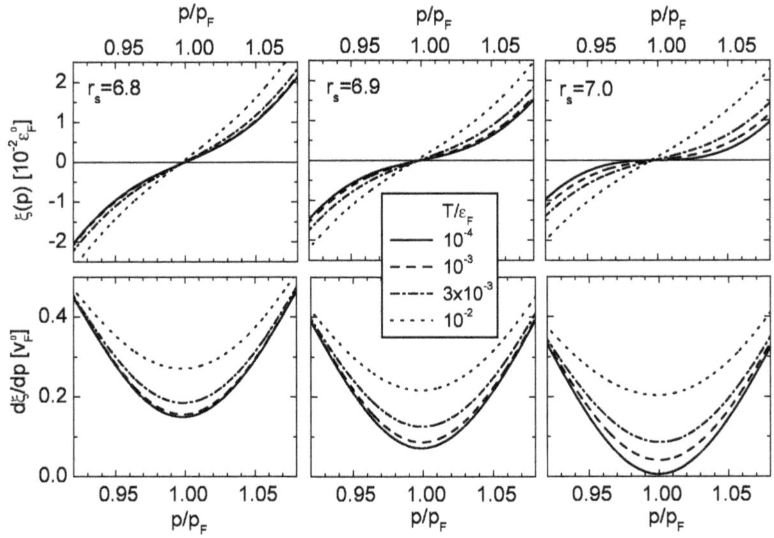

Fig. 2. Single-particle spectrum $\epsilon(p)$ in units of $10^{-2}\,\epsilon_F^0$ (top panels) and its derivative $d\epsilon(p)/dp$ in units of $v_F^0 = p_F/M$ (bottom panels) for the 2D electron gas at $r_s = 6.8$ (left column), $r_s = 6.9$ (middle column), and $r_s = 7.0$ (right column). Spectra and derivatives are shown as functions of p/p_F at four values of temperature in units of $\epsilon_F^0 = p_F^2/2M$.

Landau approach. The function $n(\epsilon(p))$ entering Eq. (12) is then replaced by $[n(\epsilon_+(p)) + n(\epsilon_-(p))]/2$, where $n(\epsilon_\pm(p)) = [1 + \exp(\epsilon(p)/T \pm R/2))]^{-1}$. Proceeding as before, one uncovers a scaling behavior

$$M/M^*(T, H, \rho_\infty) \propto T^{2/3} a(R) \tag{16}$$

of the effective mass, where $a(R)$ has been evaluated in a closed form in Ref. 18. In the limit $T \to 0$ or equivalently $R \to \infty$, this behavior simplifies, yielding the analytic form[18,19] $M^*(T = 0, H, \rho_\infty) \propto H^{2/3}$ for the effective mass. Thus, on the metallic side of the phase transition associated with the divergence of the effective mass, imposition of a static magnetic field satisfying $\mu_B H > T$ renders the effective mass $M^*(T, H, \rho_\infty)$ finite, promoting the recovery of Landau FL theory. This behavior, consistent with the experiment, remains elusive in any approach involving spin fluctuations as a basic ingredient.

Along the same lines, one finds that close to the QCP, the magnetic moment and AC spin susceptibility display a scaling behavior. Following Ref. 11, Fig. 3 presents the results of numerical calculations of these quantities as functions of the normalized temperatures T/T_P and T/T_M. We see that the model developed here reproduces the experimental scaling behaviors of both the spin susceptibility[11] of the heavy-fermion metal $CeRu_2Si_2$ and the specific heat[20] of the heavy-fermion compound $YbRh(Si_{0.95}Ge_{0.05})_2$, *without any adjustable parameters.*

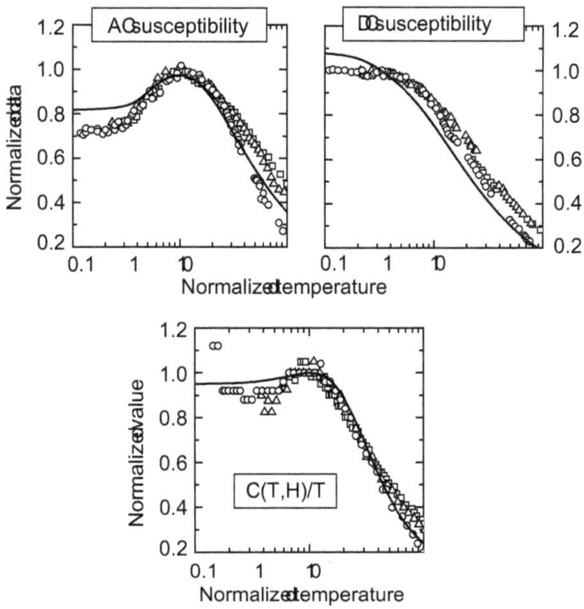

Fig. 3. Top panels: Normalized magnetic susceptibility $\chi(T,H)/\chi(T_P)$ (top-left panel) and normalized magnetization $\mathcal{M}(T,H)/\mathcal{M}(T_P)$ (top-right panel) for $CeRu_2Si_2$ in magnetic fields 0.20 mT (squares), 0.39 mT (triangles), and 0.94 mT (circles), plotted against normalized temperature T/T_P (Ref. 11), where T_P is the temperature at peak susceptibility. The solid curves trace the universal behavior predicted by the present theory. Bottom panel: The normalized ratio $C(T,H)T_M/C(T_M)T$ for $YbRh(Si_{0.95}Ge_{0.05})_2$ in magnetic fields 0.05 T (squares), 0.1 T (triangles), and 0.2 T (circles), versus the normalized temperature T/T_M (Ref. 20), where T_M is the temperature at maximum ratio $C(T,H)/T$. The solid curve shows the prediction of our theory.

4. Going Beyond the Point of Fermion Condensation

For decades, there has been virtually universal acceptance of Landau's hypothesis that in homogeneous systems of fermions there exists a one-to-one T-independent correspondence between the sp spectrum $\epsilon(p)$ and the momentum p. This postulate (see the text accompanying formula (4) in Ref. 1), implying that the quasiparticle group velocity $d\epsilon/dp$ at the Fermi surface is practically *independent of* T, has been regarded as a cornerstone of FL theory. However, numerical investigations of Eq. (12) performed in recent years[16] have demonstrated that this hypothesis is incorrect in the vicinity of the QCP. This flaw was in fact uncovered already in 1990's in connection with an analysis of the necessary condition for stability of the ground state with respect to the rearrangement of the ground-state momentum distribution $n(p)$. The pertinent stability condition reads

$$\delta E_0(n) = 2\int \epsilon(p)\delta n(p)d^3p/(2\pi)^3 \geq 0\ , \tag{17}$$

thus requiring nonnegativity of the variation of the ground-state energy E_0 under any admissible variation of $n(p)$. In the case of the Landau state where $n(p)$ is simply the Fermi step $n_F(p)$, this implies that the sign of $\epsilon(p)$ must coincide with the sign of the difference $p - p_F$. However, it turns out that there exists a class of NFL solutions having a completely flat portion $\epsilon(p) = 0$ in a region embracing the Fermi surface,[15,21–23] which in principle provides a different minimum of the functional $E_0(n)$. Flattening of the sp spectrum in this way is inescapably accompanied by another peculiarity: occupation numbers $n_*(p)$ in the flat region differ from 0 and 1. As a result, the Fermi surface swells from a surface to a volume in 3D and from a line to a surface in 2D. The phenomenon of swelling of the Fermi surface, first documented in Ref. 15 and called fermion condensation, is its distinctive signature. We have already encountered swelling of the Fermi surface in Sec. 2, where the merging of sp levels was studied. The set of states with $\epsilon(\mathbf{p}) = 0$ is called the *fermion condensate* (FC) in transparent analogy with the boson condensate that forms at $T = 0$ in a Bose liquid, and in which macroscopically many particles have energy coincident with the chemical potential μ.

The existence of ground states with occupation numbers $0 < n_*(p) < 1$ follows directly from the fundamental idea of the FL approach that E_0 is a functional of $n(p)$. To see this more clearly, it is instructive to invoke a mathematical correspondence with the functional $E_0(\rho)$ of statistical physics. If the interaction is small, the latter attains its minimum for solutions describing *gases*. However, if the interaction between the particles is sufficiently strong, then nontrivial solutions of the variational condition $\delta\Omega/\delta\rho = 0$ on the thermodynamic potential $\Omega = E_0 - \mu N$ describe *liquids*. From a mathematical point of view, the functional $\Omega(n)$ must have two types of solutions. Assuredly, in weakly interacting systems, its minimum is attained at $n_F(p) = \theta(p - p_F)$, which minimizes the kinetic energy alone. On the other hand, if the potential energy becomes strong, there arise nontrivial solutions of the variational condition

$$\frac{\delta\Omega}{\delta n(\mathbf{p})} = 0\,, \quad \mathbf{p} \in C\,, \tag{18}$$

over a *finite domain C adjacent to the Fermi surface.*[15,21–23] In Landau's formulation of Fermi-liquid theory, at $T = 0$ the derivative $\delta E/\delta n(\mathbf{p})$ is just the sp energy, while $\delta\Omega/\delta n(\mathbf{p})$ is the sp energy $\epsilon(\mathbf{p})$ measured from the chemical potential μ. Thus we arrive at the relation

$$\epsilon(\mathbf{p}) = 0\,, \quad \mathbf{p} \in C\,, \tag{19}$$

characterizing the fermion condensate.

In contrast to other second-order phase transitions, no symmetry is violated in the FC phase transition, and therefore the choice of the associated order parameter is not immediate. The FC density $n_*(p)$ can be treated as such an order parameter. Upon rewriting Eq. (1) as $\epsilon(\mathbf{p}, T) = T\ln[(1 - n(\mathbf{p}))/n(\mathbf{p})]$ and inserting into this expression of the FC solution $n_*(\mathbf{p})$, one finds[22]

$$\epsilon(\mathbf{p}, T \to 0) = T \ln \frac{1 - n_*(\mathbf{p})}{n_*(\mathbf{p})}\,, \quad \mathbf{p} \in C\,. \tag{20}$$

Thus at low T, the energy degeneracy of FC is lifted, the FC plateau in $\epsilon(\mathbf{p})$ being inclined with a slope proportional to temperature.[22]

This salient feature of the FC spectrum exhibits itself in the magnetic susceptibility $\chi(T)$, providing Curie-Weiss behavior $\chi(T) \propto 1/T$ in normal states of systems with a FC. Indeed, upon inserting the function $n_*(\mathbf{p})$ into the standard FL relation

$$\chi_0 = -2\mu_e^2 \int \frac{dn(\epsilon_{\mathbf{p}})}{d\epsilon_{\mathbf{p}}}\, d^3p/(2\pi)^3 \equiv \frac{2\mu_e^2}{T} \int n(\mathbf{p})\,[1 - n(\mathbf{p})]\, d^3p/(2\pi)^3 \,, \tag{21}$$

we find a Curie contribution to the spin susceptibility[24] given by

$$\chi_*(T) = \frac{\kappa\,\mu_e^2}{T} \,, \quad \kappa = 2 \int n_*(\mathbf{p})\,[1 - n_*(\mathbf{p})]\, d^3p/(2\pi)^3 \,. \tag{22}$$

The effective Curie constant in Eq. (22) *is reduced by the dimensionless parameter* κ *relative to the standard Curie law* $\chi_0 = \mu_e^2/T$. Accounting for the spin interaction amplitude g_0 generates the Curie-Weiss law $\chi(T) = \mu_e^2\kappa/(T - \Theta_W)$ with a Weiss temperature $\Theta_W = g_0\kappa\mu_e^2$. Such a behavior of $\chi(T)$ is observed in ^{3}He films, where the low-T Curie constant is about 4 times smaller than that for high T, as shown in Fig. 1 of Ref. 3, which gives $\kappa \approx 0.25$ in this case.

Another important feature inherent in systems having a FC stems from the fact that the variational condition (18) holds at finite temperatures,[22] provided the thermodynamic potential is redefined according to the conventional relation $\Omega = E - \mu N - TS$, where the entropy S is given by the Landau formula

$$S = -2 \int [n(\mathbf{p}) \ln n(\mathbf{p}) + (1 - n(\mathbf{p})) \ln(1 - n(\mathbf{p}))]\, d^3p/(2\pi)^3 \,. \tag{23}$$

Upon inserting here the distribution $n_*(p)$, we infer that systems with a FC possess *a finite, independent of* T *entropy* S_*.[15,22,23] The entropy-excess value, proportional to the total FC density, changes gradually under variation of input parameters. Although it does not contribute to the specific heat, it produces an enormous enhancement of the thermal expansion coefficient $\beta = \partial V/\partial T \equiv -\partial S/\partial P$ and the Grüneisen ratio $\Gamma = \beta/C$.[25] Experiment[26] shows that in normal states of several heavy-fermion metals, β is in fact temperature-independent at low T and exceeds typical values for ordinary metals by a factor 10^3–10^4. With $\beta \to$ const and $C(T) \to 0$, the Grüneisen ratio $\Gamma = \beta/C$ diverges at low T, as is observed experimentally.[27]

It should be emphasized that the existence of the residual entropy S_* at zero temperature contradicts the third law of thermodynamics (the Nernst theorem). To ensure that $S = 0$ at $T = 0$, localized spins are known to order magnetically due to spin-spin interactions. Similarly, a system with a FC must experience some sort of low-temperature phase transition eliminating the excess entropy S_*, e.g. the second-order phase transition to a superconducting state.[15] The presence of the FC in the ground state exhibits itself in a jump $\Delta C \approx 4.7\,\kappa$ of the specific heat at T_c, governed by the FC parameter κ in the Curie law[28] (22). Thus, the ratio $\Delta C/C_n$ can be very high when T_c is low, because $C_n \to 0$ as $T \to 0$ while ΔC remains finite. Such a situation is encountered for example in the heavy-fermion metal $CeCoIn_5$, with $T_c = 2.3K$ and $\Delta C/C_n \approx 4.5$, over three times higher than the BCS value.[29]

5. Discussion

In any conventional homogeneous Fermi liquid, e.g. liquid ^{3}He, the momentum $\mathbf{p}$ of an added particle can be associated with a certain quasiparticle. Similarly, in most spherical odd nuclei, the total angular momentum J in the ground state is carried by an odd quasiparticle. In atomic physics, the electronic configuration of ions of elements belonging to the principal groups of the periodic table repeats that of preceding atoms. By contrast, beyond the point of fermion condensation in homogeneous matter or the point of merging of sp levels in finite systems, the ground state features a multitude of quasiparticle terms and therefore exhibits a different, more complicated character, as in the comparison of a chorus with a dominant soloist.

Notwithstanding evident commonalities, there is a considerable difference between conditions for the "level-mergence" phenomenon in homogeneous Fermi liquids and in finite Fermi systems with degenerate sp levels. In the former, the presence of a significant velocity-dependent component in the interaction function f is needed to promote fermion condensation, while in the latter, sp levels can merge even if f is momentum-independent. The reason for this difference is simple: in the homogeneous case, the matrix elements u and w of Eq. (4) are equal to each other, implying zero energy gain due to the rearrangement when velocity-dependent forces are absent.

The salient feature of the fermion condensation phenomenon discussed in this paper is the flattening of the quasiparticle dispersion $\epsilon(\mathbf{p})$ at the Fermi level on the ordered side of the driving phase transition. Prime consequencies of this flattening are: (i) the magnetic susceptibility follows the Curie-Weiss law with an effective Curie constant proportional to the FC parameter κ, (ii) the entropy has a temperature-independent term S_* that greatly increases the thermal expansion coefficient $\beta = -\partial S/\partial P$ at low T, (iii) the excess entropy S_* released below the superconducting transition temperature T_c dramatically reduces β, enhancing the specific-heat jump $\Delta C/C_n$. All these features go unexplained in any other microscopic approach.

On the disordered side of the transition, close to the QCP, the proposed scenario adequately explains the low-T data on the spin susceptibility, predicting $\chi^{-1}(T) \sim T^\alpha$ in the critical density region with a critical exponent $\alpha \simeq 2/3$. The spin-fluctuation model (SFM) fails to produce $\alpha < 1$. Further, our scenario explains the scaling behavior $\chi^{-1} \sim T^\alpha F(H/T)$ of the spin susceptibility in static magnetic fields, whereas the SFM fails to do so. Finally, within the scenario advanced here, FL behavior is recovered close to the QCP by imposing a tiny magnetic field satisfying $\mu_B H > T$. In the SFM there is no such provision for reinstating FL theory.

6. Conclusion

The publication fifteen years ago of the first article[15] on the theory of fermion condensation, in which NFL behavior was deduced *from the Landau approach itself,*

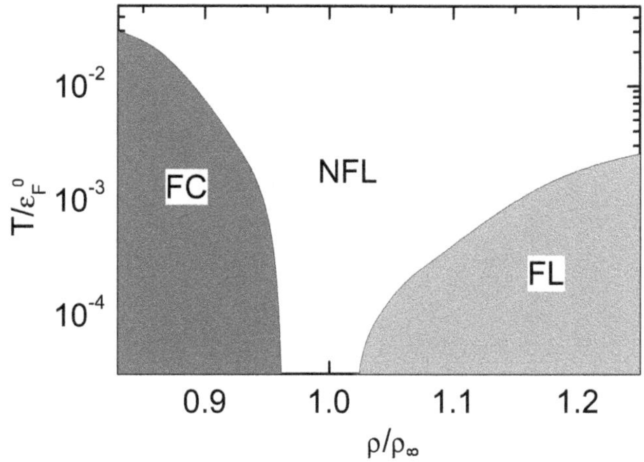

Fig. 4. Phase diagram of 2D electron gas in $(T, \rho/\rho_\infty)$ variables. Curves show crossovers between usual Fermi liquid (FL), fermion condensate (FC) and non-Fermi liquid (NFL) phase with the critical index $\alpha \simeq 0.6$.

triggered a wave of the criticism and disbelief. The judgment, "This theory is an artifact of the Hartee-Fock method" was typical. By now the debates on this subject have become pointless: numerical solution of the Landau equation (12) is the best way to calm down the critics. Due to lack of space, we present here only a phase diagram of the two-dimensional homogeneous electron gas calculated from Eq. (12). This phase diagram, as plotted in Fig. 4, essentially coincides with predictions of the theory of fermion condensation. Distinctions occur only at extremely low temperatures in the region adjacent to the QCP (for details see Ref. 16). Thus, it is fair to say that the theory of fermion condensation is just another chapter of Fermi-liquid theory. Accordingly, we abandon the conventional attitude of standard FL theory that properties of Fermi liquids are always similar to those of an ideal Fermi gas, since as we have seen, there exists a whole region of densities where this is not the case. Still to be answered is the important question: is the Fermi-liquid approach[1] relevant to the experimental situation in strongly correlated Fermi systems close to the QCP and beyond it? With the passage of time, a positive answer seems to be more and more evident.

We thank A. Chubukov, P. Coleman, P. Fulde, H. Li, K. Kikoin, G. Lonzarich, J. Thompson, J. Saunders, F. Steglich, and G. Stewart for fruitful discussions. This research was supported by Grant No. NS-8756.2006.2 (VAK and MVZ) from the Russian Ministry of Education and Science and the McDonnell Center for the Space Sciences (VAK).

References

1. L. D. Landau, JETP **30**, 1058, (1956); Sov. Phys. JETP **3**, 920, (1957).
2. D. Pines and P. Nozières, *Theory of quantum liquids*, v. 1, (W. A. Benjamin, New York-Amsterdam, 1966).

3. C. Bäuerle, Yu. M. Bun'kov, A. S. Chen, S. N. Fisher, and H. Godfrin, J. Low Temp. **110**, 333 (1998).
4. A. Casey, H. Patel, J. Nyeki, B. P. Cowan, and J. Saunders, Phys. Rev. Lett. **90**, 115301 (2003).
5. O. Prus, Y. Yaish, M. Reznikov, U. Srivan, and V. M. Pudalov, Phys. Rev. B **67**, 205407 (2003).
6. J. A. Hertz, Phys. Rev. B **14**, 1165 (1976).
7. A. J. Millis, Phys. Rev. B **48**, 7183 (1993).
8. P. Coleman, C. Pepin, Q. Si, and R. Ramazashvili, J. Phys.: Condens. Matter **13**, R723 (2001).
9. R. K"chler, N. Geischler, P. Gegenwart, T. Cichorek, K. Neumaier. O. Tegas, C. Geibel, J. A. Mydosh, F. Steglich, L. Zhu, and Q. Si, Phys. Rev. Lett. **91**, 066405 (2003).
10. P. Gegenwart, J. Custers, T. Tayama, K. Tenya, C. Geibel, O. Trovarelli, F. Steglich, and K. Neumaier, Acta Phys. Pol. B **34**, 323 (2003).
11. D. Takahashi, S. Abe, H. Mizuno, D. A. Tayurski, K. Matsumoto, H. Suzuki, and Y. Onuki, Phys. Rev. B **67**, 180407 (2003).
12. N. J. Curro, T. Caldwell, E. O. Bauer, L. A. Morales, M. J. Graf, Y. Bang, A. V. Balatsky, J. D. Thompson, and J. L. Sarrao, Nature **434**, 622 (2005).
13. V. A. Khodel, J. W. Clark, Haochen Li and M. V. Zverev, cond-mat 0609583.
14. A. B. Migdal, *Theory of Finite Fermi Systems and Applications to Atomic Nuclei* (Wiley, New York, 1967).
15. V. A. Khodel and V. R. Shaginyan, JETP Letters **51**, 553 (1990); Condensed Matter Theories **12**, 221 (1997).
16. M. V. Zverev and M. Baldo, JETP **87**, 1129 (1998); J. Phys.: Condens. Matter **11**, 2059 (1999).
17. V. V. Borisov and M. V. Zverev, JETP Letters, **81**, 503 (2005).
18. J. W. Clark, V. A. Khodel, and M. V. Zverev, Phys. Rev. **71**, 012401 (2004).
19. V. R. Shaginyan, JETP Letters **77**, 99 (2003); **79**, 344 (2004).
20. J. Custers, P. Gegenwart, H. Wilhelm, K. Neumaier, Y. Tokiwa, O. Trovarelli, F. Steglich, C. Pepin, and P. Coleman, Nature **424**, 524 (2003).
21. G. E. Volovik, JETP Letters **53**, 222 (1991).
22. P. Nozières, J. Phys. I France **2**, 443 (1992).
23. V. A. Khodel, V. V. Khodel, and V. R. Shaginyan, Phys. Rep. **249**, 1 (1994).
24. M. V. Zverev and V. A. Khodel, JETP Lett. **79**, 635 (2004).
25. M. V. Zverev, V. A. Khodel, V. R. Shaginyan and M. Baldo, JETP Lett. **65**, 863 (1997).
26. N. Oeschler, P. Gegenwart, M. Lang, R. Movshovich, J. L. Sarrao, J. D. Thompson, and F. Steglich, Phys. Rev. Lett. **91**, 076402 (2003).
27. R. Küchler, N. Oeschler, P. Gegenwart, T. Cichorek, K. Neumaier, O. Tegus, C. Geibel, J. A. Mydosh, F. Steglich, L. Zhu, and Q. Si, Phys. Rev. Lett. **91**, 066405 (2003); **93**, 096402 (2004).
28. V. A. Khodel, M. V. Zverev, and V. M. Yakovenko, Phys. Rev. Lett. **95**, 236402 (2005).
29. C. Petrovic, P. G. Pagliuso, M. F. Hundley, R. Movshovich, J. L. Sarrao, J. D. Thompson, Z. Fisk, and P. Monthoux, J. Phys.: Condens. Matter **13**, L337 (2001).

THEORY OF THIRD SOUND AND STABILITY OF THIN ^{3}He –^{4}He SUPERFLUID FILMS

M. D. MILLER[†,*] and E. KROTSCHECK[†]

[†] *Institut für Theoretische Physik, Johannes-Kepler-Universität, A-4040 Linz, Austria*
E-mail: eckhard.krotscheck@jku.at

[*] *Department of Physics and Astronomy, Washington State University, Pullman, WA 99164-2814, USA*
E-mail: mdm@wsu.edu

In this paper, we summarize the results of recent studies of third sound in thin, superfluid ^{3}He -^{4}He mixture films and the relation of the third sound spectrum to the question of the films' thermodynamic stability. We have considered films on several representative substrates: Nuclepore, glass, Li and Na. Our approach utilizes the variational, hypernetted chain/Euler-Lagrange (HNC–EL) theory as applied to inhomogeneous boson systems to calculate chemical potentials for both the ^{4}He superfluid film and the physisorbed ^{3}He. Numerical density derivatives of the chemical potentials lead to the sought-after third sound speeds. On all substrates, the third sound speeds show a series of oscillations as a function of film coverage that is driven by the layered structure of the ^{4}He film. We find that the effect on the third sound response of adding a small amount of ^{3}He to the ^{4}He film can depend sensitively on the particular ^{4}He film coverage. The third sound speed can either increase or decrease. In fact, in some regimes, the added ^{3}He destabilizes the film and can drive "layering transitions" leading to quite complicated geometric structures of the film in which the outermost layer consists of phase–separated regimes of ^{3}He and ^{4}He. Finally, we examine the range of applicability of the usual film–averaged hydrodynamic description. We find that at least up to film thicknesses of six liquid layers, there is no regime in which this hydrodynamic description is applicable.

Keywords: ^{3}He –^{4}He Thin Films; Third Sound; Stability.

1. Introduction

As first pointed out by Atkins, low–temperature ^{4}He films can support propagating hydrodynamic modes despite being much thinner than the viscous penetration depth.[1] These modes, termed third sound, are surface–type, density-fluctuations which are neither purely longitudinal nor purely transverse in character. In the low-temperature limit, the third sound speed, in a model double–layer mixture film, can be written as[2,3]

$$\left(\frac{c_3^2}{c_{30}^2}\right) = 1 - \frac{\rho_3}{\rho_4}\left[1 - \frac{1}{(1 + h_3/h_4^0)^4}\right], \qquad (1)$$

where ρ_4 and ρ_3 are ^{4}He and ^{3}He number densities, respectively, h_4 is the thickness of the mobile layer, h_3 is the thickness of the ^{3}He Andreev state, and

$$m_4 c_{30}^2 = (h_4^0 - h_0)\frac{3\alpha_s}{(h_4^0)^4} \tag{2}$$

is the third sound speed for pure ^{4}He. In Eq. (2), α_s is the substrate-helium van der Waals constant, and h_0 is the thickness of the immobile ^{4}He layer next to the substrate.

This hydrodynamic approach is valid for films that can be characterized by a single thickness parameter: for example, h_4, or h_3. This is the case for thick films which can be imagined as simply "a slice of the bulk"; that is, uniform films whose dynamic response is governed by long ranged van der Waals type interactions.

The basic results to be discussed below were published in a pair of publications. In the first paper,[4] we investigated third sound on two model substrates: Nuclepore and glass. These may be considered "strongly binding" since the first one or two layers of helium near the solid substrate are solid and immobile. We showed that the third sound speed developed oscillations in parallel with the layered structure of the mobile ^{4}He film, these oscillations can become strong enough to destabilize the film, leading to "layering transitions". This behavior was obtained first by Clements, Krotscheck and Lauter.[5] In the second paper[6] we extended these investigations to the weakly binding substrates, Li and Na.

We are aware of three third sound experiments for ^{3}He -^{4}He mixture thin films. Two experiments were reported by Hallock and co-workers. The first, by Valles, Heinrichs, and Hallock,[7] was on a borosilicate glass substrate. Data was taken for two values of the ^{4}He film thickness, 3.578 layers and 5.274 layers, where one "conventional" ^{4}He layer $\ell_4 = 3.6$ Å. The second, by Sheldon and Hallock,[8] was on a Nuclepore substrate. Here, the film thickness was 3.65 ℓ_4. The substrate, Nuclepore, is a polycarbonate material threaded with pores of nominal diameter ≈ 2000 Å. There is at least one other mixture third sound experiment in the literature. Noiray, Sornette, Romagnan, and Laheurte[9] reported results for third sound on both Nuclepore and glass. These authors found a transition (they called a *stratification transition*) as a function of ^{4}He film thickness in which the ^{3}He was separated from the ^{4}He for large film thicknesses and mixes into the film at small film thickness. In Ref. 4, we pointed out that this behavior evolves smoothly out of the submonolayer mixture limit and can be easily understood in terms of ^{3}He energetics.

The square of the third sound speed in the mixture film as a function of ^{3}He coverage is linear for small coverages. That is, in the small ^{3}He coverage limit, we can write

$$m_4 c_3^2 \backsimeq m_4 c_{30}^2 \left(1 - \Delta\theta_3\right), \tag{3}$$

where $\theta_3 = (\sigma_3/\sigma_{3\ell})$ is the ^{3}He coverage in units of monolayers, and $\sigma_{3\ell} \equiv 1/\ell_3^2 = 0.065$ Å^{-2} is the areal density at "conventional" monolayer completion. By inspection, the experimental slope in the Nuclepore case,[8] is $\Delta_{np} \backsimeq \frac{1}{2}$. The linear regime in Nuclepore extends out to 0.6 ℓ_3. For glass, the results qualitatively resemble older thick ^{3}He film results of Ref. 2: *little or no linear regime at low coverages.* If we assume that the very first few data points determine a line then we find that the slope is smaller for the thicker film, and approximately $\Delta_{gl}(3.578) \simeq 0.8$ and $\Delta_{gl}(5.274) \simeq 0.6$.

One focus of this research is a comprehensive examination of the quantity Δ as a function of the surface coverage and substrate potential, and the consequences of its dependence on film thickness and substrate potential.

The values of the van der Waals constants for Li, Na, Nuclepore, and glass substrates can be found tabulated in Refs. 4 and 6. ^{4}He interacts strongly with Nuclepore and glass, and weakly with Li and Na. For Nuclepore and glass there are one or two immobile layers which effectively screen the strong, short–ranged part of the substrate potential. We have modeled these, as described in Ref. 10, by one or two *static* layers. There are, on the other hand, no immobile layers on Li and Na, in fact, there are no physically stable monolayer configurations on a Na substrate.

On the screened Nuclepore and glass substrates, and, similarly, on Li,[11] the short–ranged repulsion of the bare ^{4}He-^{4}He interaction causes a well–defined layered structure in the film. The attraction of ^{4}He to a Na substrate is much weaker, and the layered structure is much less pronounced.

One of our primary goals is to examine the extent to which our results depend on the layered structure of the films as seen in the strongly–binding substrates. It is, for example, conceivable that hydrodynamic asymptotic behavior sets in earlier in films that display a less pronounced layered structure. As far as we know, no experimental work exists for the weakly–binding substrates and so those results, to be discussed below, will all be predictions. In Sec. 2, we briefly review the microscopic hypernetted chain/Euler-Lagrange (HNC/EL) theory. In Sec. 3, we discuss the results of the microscopic calculations for the Nuclepore and glass model substrates. Sec. 4 is the Conclusion.

2. Variational Theory

In this section, we shall very briefly sketch the calculational apparatus. Detailed descriptions of the hypernetted chain Euler-Lagrange (HNC–EL) theory as applied to thin helium films can be found in Refs. 4 and 6. A very detailed discussion of the variational approach to the inhomogeneous boson system with numerous further references can be found in the review article in Ref. 12. In the first subsection, we summarize the equations needed to describe the ^{4}He film in a planar geometry. In the following two subsections, we introduce a ^{3}He adatom and indicate how one calculates the physisorption single-particle states and chemical potential.

2.1. ^{4}*He films*

The system consists of N ^{4}He atoms adsorbed to a solid substrate of area $\mathcal{A}$. The substrate-helium interaction is modeled by an appropriate external field $U_{\rm ext}(\mathbf{r}_i)$.

The ^{4}He film is described by the N-body Hamiltonian

$$H_N = \sum_{i=1}^{N} \left[-\frac{\hbar^2}{2m_4} \nabla_i^2 + U_{\rm ext}(\mathbf{r}_i) \right] + \sum_{i<j}^{N} V(|\, \mathbf{r}_i - \mathbf{r}_j \,|) \,, \tag{4}$$

and a Feenberg variational wave function

$$\Psi_N(\mathbf{r}_1, \ldots, \mathbf{r}_N) = \exp \frac{1}{2} \left[\sum_i u_1^{(4)}(\mathbf{r}_i) + \sum_{i<j} u_2^{(4,4)}(\mathbf{r}_i, \mathbf{r}_j) + \sum_{i<j<k} u_3^{(4,4,4)}(\mathbf{r}_i, \mathbf{r}_j, \mathbf{r}_k) \right], \tag{5}$$

that allows for the breaking of translational invariance and isotropy as demanded by the external field. The ground-state of the system is determined by functionally minimizing the expectation value of the Hamiltonian in the space of all trial functions permitted in the *ansatz* (5).

The first task is to manipulate the expectation value of the Hamiltonian in a way that makes it accessible to numerical treatment. Symmetry breaking requires non–trivial one–body correlations as embodied in $u_1^{(4)}(\mathbf{r}_1)$. The general strategy is to use the Born-Green-Yvon (BGY) and the HNC equations for inhomogeneous systems to eliminate the variational functions $u_n^{(4,\ldots,4)}(\mathbf{r}_1, \ldots, \mathbf{r}_n)$ in favor of the physically observable density, $\rho_4(\mathbf{r})$, the pair distribution function, $g^{(4,4)}(\mathbf{r}, \mathbf{r}')$, and higher order distribution functions as needed.

The variational energy expectation value is then a functional of these distribution functions, given by

$$E_N[\rho_4(\mathbf{r}), g^{(4,4)}(\mathbf{r}, \mathbf{r}'), \ldots] = \int d^3r_1 \ldots d^3r_N \Psi_N(\mathbf{r}_1, \ldots, \mathbf{r}_N) H \Psi_N(\mathbf{r}_1, \ldots, \mathbf{r}_N) \,. \tag{6}$$

If triplet correlations are neglected for the time being, one can, express E_N entirely in terms of the one body density $\rho_4(\mathbf{r}_1)$ and the pair distribution function $g^{(4,4)}(\mathbf{r}_1, \mathbf{r}_2)$.

The functions $\rho_4(\mathbf{r}_1)$ and $g^{(4,4)}(\mathbf{r}_1, \mathbf{r}_2)$ are determined by minimization of the total energy.

$$\frac{\delta E_N}{\delta \rho_4(\mathbf{r})} = 0, \qquad \frac{\delta E_c}{\delta g^{(4,4)}(\mathbf{r}, \mathbf{r}')} = 0 \,, \tag{7}$$

subject to the constraint of fixed particle number

$$N = \int d^3r_1 \rho_4(\mathbf{r}_1) \ . \tag{8}$$

The chemical potential μ_4 enters the theory as a Lagrange parameter to enforce the constraint (8).

The central quantities of interest in this work are the ^{3}He and ^{4}He coverage dependences of the third sound speed, c_3. The third sound speed can be calculated in two different ways, namely as the hydrodynamic derivative

$$m_4 c_3^2 = \sigma_4 \frac{d\,\mu_4(\sigma_4)}{d\,\sigma_4} \,, \tag{9}$$

where σ_4 is the ^{4}He *coverage*,

$$\sigma_4 = \int_{-\infty}^{\infty} dz_1 \rho_4(\mathbf{r}_1), \tag{10}$$

and from the long–wavelength limit of the low–lying excitations.

In the former case, it can be shown[4] that

$$\sigma_4 \frac{d\,\mu_4(\sigma_4)}{d\,\sigma_4} = \frac{\sigma_4}{2}\left(\sqrt{\rho_4}\middle|\left[H_1^{(4)} + 2\tilde{V}_{\rm p-h}^{(44)}\right]^{-1}\middle|\sqrt{\rho_4}\right)^{-1}, \tag{11}$$

in an obvious inner product type notation, and we define the tilde notation $\tilde{V}_{p-h}^{(44)} \equiv \sqrt{\rho_4(\mathbf{r})}\, V_{\rm p-h}^{(44)}(\mathbf{r},\mathbf{r}')\, \sqrt{\rho_4(\mathbf{r}')}$. The quantity $V_{\rm p-h}^{(44)}(\mathbf{r},\mathbf{r}')$ is known as the *particle–hole interaction* and is defined by

$$V_{\rm p-h}^{(ij)}(\mathbf{r},\mathbf{r}') \equiv \frac{\delta V_{\rm H}^{(i)}(\mathbf{r})}{\delta \rho_j(\mathbf{r}')} = \frac{\delta^2 E_{\rm c}}{\delta\rho_i(\mathbf{r})\delta\rho_j(\mathbf{r}')}, \tag{12}$$

where $i, j = 3, 4$, $E_{\rm c}$ is the *correlation energy*, and $V_{\rm H}^{(i)}(\mathbf{r})$ is a generalized Hartree potential that depends implicitly on the density and the pair distribution function. The precise forms for $H_1^{(4)}$, the so-called single particle Hamiltonian, and $E_{\rm c}$, the correlation energy, can be found in Ref. 12. (We generalize the definition of the particle-hole interaction to different species in (12) as this will be needed below.)

An alternative derivation is to look at the long–wavelength limit of the low–lying excitations, which is, in our case, third sound. One then arrives[13] at exactly the same expression (11); however, in that case the particle–hole interaction $V_{\rm p-h}^{(44)}(\mathbf{r},\mathbf{r}')$ is given by the HNC approximation, which agrees with the definition (12) only in an exact theory. Thus, comparing the speed of third sound as obtained from the density derivative of the chemical potential with the one obtained from long–wavelength excitations gives some estimate for the accuracy of the HNC approximation for $V_{\rm p-h}^{(44)}(\mathbf{r},\mathbf{r}')$.

2.2. ^{3}*He single particle states*

The Hamiltonian of the $N+1$ particle system consisting of N ^{4}He atoms and one ^{3}He is

$$H_{N+1} = -\frac{\hbar^2}{2m_3}\nabla_0^2 + U_{\rm ext}(\mathbf{r}_0) + \sum_{i=1}^{N} V(|\mathbf{r}_0 - \mathbf{r}_i|) + H_N\,. \tag{13}$$

We adopt the convention that the coordinate $\mathbf{r}_0$ refers to the ^{3}He particle and coordinates $\mathbf{r}_i$, with $i = 1\ldots N$, to the ^{4}He film particles. The generalization of the wave function (5) to an inhomogeneous N-particle ^{4}He system with a single ^{3}He atom is

$$\begin{aligned}\Psi_{N+1}&(\mathbf{r}_0,\mathbf{r}_1,\ldots,\mathbf{r}_N) \qquad (14)\\ &= \exp\frac{1}{2}\left[u_1^{(3)}(\mathbf{r}_0) + \sum_{1\le i\le N} u_2^{(3,4)}(\mathbf{r}_0,\mathbf{r}_i) + \sum_{1\le i<j\le N} u_3^{(3,4,4)}(\mathbf{r}_0,\mathbf{r}_i,\mathbf{r}_j)\right]\Psi_N(\mathbf{r}_1,\ldots,\mathbf{r}_N)\,.\end{aligned}$$

The energy change due to adding one ^{3}He atom into the system is the ^{3}He chemical potential $\mu_3 \equiv E_{N+1} - E_N$, where E_{N+1} is the energy of the system containing one ^{3}He and N ^{4}He atoms, and E_N is the energy of the unperturbed ^{4}He film, (6). Here E_{N+1} is to be understood as the energy expectation value of the Hamiltonian (13) with respect to the wave function (14). The ^{3}He chemical potential μ_3 appears as the Lagrange multiplier to ensure the normalization $\int d^3r_0\rho_3(\mathbf{r}_0) = 1$. The two-body Euler equation is derived by variation of the ^{3}He chemical potential with respect to $g^{(34)}(\mathbf{r}_0, \mathbf{r}_1)$.

In a manner similar to the derivation of the third sound speed, we can calculate the σ_4 derivative of μ_3:

$$\sigma_3 \frac{d\mu_3}{d\sigma_4} = \int d^3r\rho_3(\mathbf{r}) \frac{dV_{\mathrm{H}}^{(3)}(\mathbf{r})}{d\sigma_4} = 2\int d^3rd^3r' \sqrt{\rho_3(\mathbf{r})}\tilde{V}_{\mathrm{p-h}}^{(34)}(\mathbf{r}, \mathbf{r}') \frac{d\sqrt{\rho_4(\mathbf{r}')}}{d\sigma_4}, \tag{15}$$

which is the desired relationship.

2.3. ^{3}*He dynamics*

The natural generalization of the variational approach to excited states is to allow for a time-dependence of the variational functions $u_n(\mathbf{r}_0, \ldots, \mathbf{r}_n; t)$. We outline only the basic steps, the reader is referred to Ref. 14 for details. We begin by separating the *kinematic* from the *dynamic* correlations and write the time–dependent variational wave function in the form

$$\phi(t) = \frac{1}{\sqrt{\langle \psi_{N+1} \mid \psi_{N+1}\rangle}} e^{-iE_{N+1}t/\hbar}\psi_{N+1}(\mathbf{r}_0, \mathbf{r}_1, \ldots \mathbf{r}_N; t) , \tag{16}$$

where $\psi_{N+1}(\mathbf{r}_0, \mathbf{r}_1, \ldots \mathbf{r}_N; t)$ contains the dynamic, time-dependent correlations. Consistent with the general strategy of variational methods, we include the time dependence in the one-particle *and* two-particle ^{3}He -^{4}He correlations, *i.e.* we write

$$\psi_{N+1}(\mathbf{r}_0, \mathbf{r}_1, \ldots \mathbf{r}_N; t) = \exp \frac{1}{2}\Big[\delta u_1^{(3)}(\mathbf{r}_0; t) + \sum_{1 \le i \le N} \delta u_2^{(3,4)}(\mathbf{r}_0, \mathbf{r}_i; t)\Big] \Psi_{N+1}(\mathbf{r}_0, \mathbf{r}_1, \ldots, \mathbf{r}_N) . \tag{17}$$

The time independent part remains the same as defined in Eq. (14). The time-dependent correlations are determined by searching for a stationary state of the action integral

$$\begin{aligned} \mathcal{S} &= \int_{t_0}^{t} \mathcal{L}(t)dt , \\ \mathcal{L}(t) &= \langle \phi(t)| H_{N+1} - i\hbar\frac{\partial}{\partial t} |\phi(t)\rangle , \end{aligned} \tag{18}$$

where H_{N+1} is the Hamiltonian (13) of the ^{3}He plus ^{4}He –film system.

We find the following expression for the ^{3}He self-energy

$$\Sigma_{ij}(\omega) = -\sum_{mk} \frac{W_{mk}^{*(i)} W_{mk}^{(j)}}{\hbar\omega_m + t_k - \hbar\omega}, \tag{19}$$

where the $\hbar\omega_m$ are the collective excitations of the background liquid. The ^{3}He – ^{3}He -ripplon vertex function $W_{mq}^{(p)}$ occurring in the self-energy (19) has the form of an effective interaction within the theory of correlated basis functions,[15–17] its specific form is irrelevant for our further discussions. In Sec. 3 below, the excited state energies that are reported are those corrected by time-dependent pair correlations.

3. Results

In the small ^{3}He concentration regime, we can write

$$\mu_4(N_3, N_4) \backsimeq \mu_{40} + \left(\frac{\partial \mu_4}{\partial \sigma_3}\right)_0 \sigma_3 \,, \tag{20}$$

where σ_3 is the ^{3}He areal density. We can use this to obtain a third sound speed, in terms of the incompressibility $(\equiv mc_3^2)$, in the small ^{3}He concentration limit. By taking the derivative with respect to σ_4, and putting the result in the form of the low-coverage hydrodynamic equations of Sec. 1, we find

$$m_4 c_3^2 = m_4 c_{30}^2 \left(1 - \Delta_{\text{thermo}}\theta_3\right) , \tag{21}$$

where θ_3 is the coverage in units of ^{3}He layers, and

$$\Delta_{\text{thermo}} = -\frac{\sigma_4 \sigma_{3\ell}}{m_4 c_{30}^2} \frac{\partial}{\partial \sigma_4} \left(\frac{\partial \mu_4}{\partial \sigma_3}\right)_0 . \tag{22}$$

We can avoid a full mixture calculation with variable population of ^{3}He for calculating the derivative in Eq. (22) by taking advantage of the Maxwell relation

$$\left(\frac{\partial \mu_4}{\partial \sigma_3}\right)_{\mathcal{A},T,N_4} = \left(\frac{\partial \mu_3}{\partial \sigma_4}\right)_{\mathcal{A},T,N_3} . \tag{23}$$

Microscopic equations for the density derivatives are spelled out in Eqs. (11) and (15).

The starting points for the microscopic analysis are the ^{4}He and ^{3}He chemical potentials. Using these chemical potentials in Eq. (22) we can examine the experimental third sound results using microscopic theory and compare with the film–averaged hydrodynamic models and experiments. Using the pure ^{4}He third sound speed as a function of film thickness and Eq. (21), we can examine the effect of ^{3}He on the third sound speed in the low ^{3}He coverage limit.

One expects that film-averaged, hydrodynamic models should give the right answer in the limit of thick ^{4}He films where any surface-normal structure can be averaged out. In that limit, Eq. (21) simplifies to

$$m_4 c_3^2 = m_4 c_{30}^2 \left(1 - 4\frac{\sigma_{3\ell}}{\sigma_4}\theta_3\right) . \tag{24}$$

However, it is a matter of quantitative analysis to determine when the "thick film" behavior sets in. In Ref. 4, we showed that for ^{4}He films on strongly binding substrates this asymptotic regime is greater than six atomic layers.

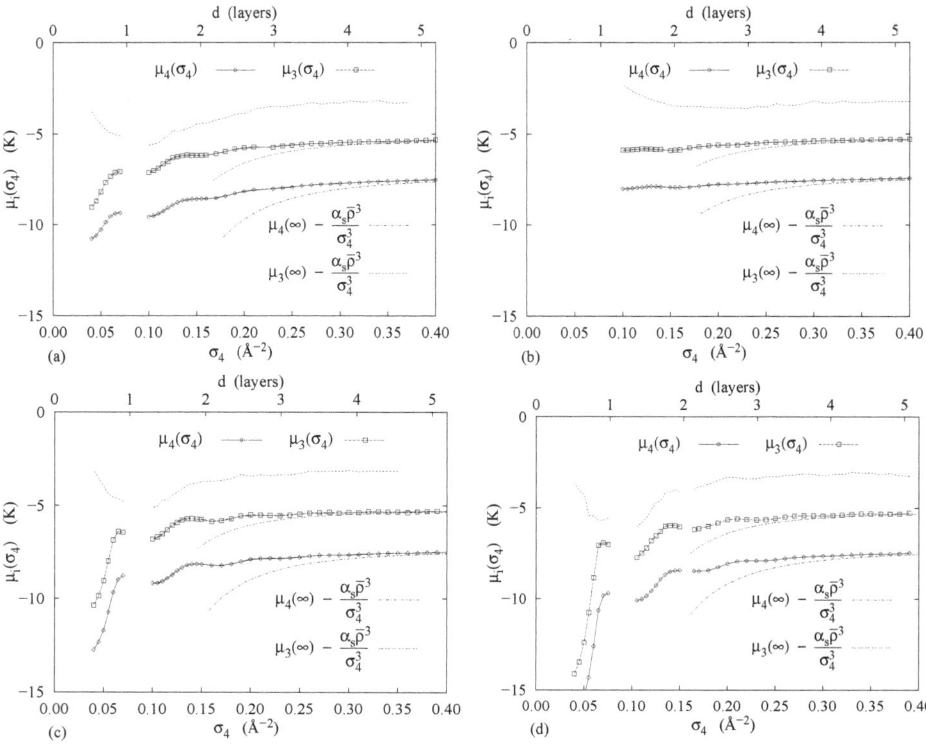

Fig. 1. The figure shows, for the lithium substrate (a), the sodium substrate (b), as well as the nuclepore (c) and the glass (d) substrates, the calculated ^{4}He chemical potential (diamonds), the ^{3}He chemical potential (boxes), and the ^{3}He first excited state energies (short–dashed line). Also shown are the full fits (25) to the ^{4}He (solid line) and the ^{3}He (dashed line) chemical potentials, as well as the asymptotic forms (26). The upper scale gives the thickness of the films in *liquid* layers, one ^{4}He layer is defined as representing an areal density of 0.077 Å^{-2}. A layering transition is present for the lithium system between $0.07 \lesssim \sigma_4 \lesssim 0.10$ Å^{-2}. There is no layering transition for the sodium substrate. These figures are from Ref. 6.

We first examine the key energetic quantities: the ^{3}He and ^{4}He chemical potentials and the ^{3}He first-excited state as functions of ^{4}He film coverage. These are shown, for the four substrates under consideration here, in Figs. 1(a)-(d), respectively. In these figures, we also show a fit to the numerical data of the form

$$\mu_i(\sigma_4) = \mu_i(\infty) - \frac{\alpha_s \bar{\rho}^3}{\sigma_4^3} + \frac{c}{\sigma_4^4} + \frac{d}{\sigma_4^5} + \frac{e\cos(k\sigma_4 - \gamma)}{\sigma_4^4}, \tag{25}$$

where $i = 3, 4$. The monolayer results are not well represented by this form, numerical fits for (25) were therefore carried out only in the coverage regime of two and more layers, $\sigma_4 > 0.10$ Å^{-2}. The term proportional to σ_4^{-3} contains the van der Waals constant α_s, and an average density $\bar{\rho}$. The last three terms are empirical. The values of the parameters obtained by fitting μ_4 and μ_3 for each of the substrates are givenin Table 1.

Table 1. Parameters for Eq. (25) determined by fitting calculated chemical potentials. The numbers under the substrate type give the Van der Waals coefficient of the respective potential in units of KÅ^{-3}.

Substrate		$\mu_i(\infty)$ (K)	$\bar{\rho}$ (Å^{-3})	k (Å^2)	$c \times 10^3$ (K Å^{-8})	$d \times 10^4$ (K Å^{-10})	$e \times 10^4$ (K Å^{-8})	γ rad.
Nuclepore	μ_4	-7.366	0.0189	96.85	2.13	-0.99	0.74	1.26
1900	μ_3	-5.229	0.0150	100.86	0.90	-0.35	0.76	1.28
Glass	μ_4	-7.305	0.0228	94.43	2.17	-0.89	1.20	1.27
1260	μ_3	-5.169	0.0197	96.92	1.19	-0.38	0.94	0.89
Li	μ_4	-7.343	0.0241	95.60	3.10	-1.39	0.88	1.00
1422	μ_3	-5.230	0.0200	100.27	1.56	-0.61	0.77	1.22
Na	μ_4	-7.348	0.0225	103.05	2.26	-1.09	0.19	-1.51
1070	μ_3	-5.195	0.0201	104.25	1.28	-0.45	0.75	1.33

The coefficients representing the van der Waals term and the oscillatory part of the density are directly related to the properties of the ^{4}He film, and are reasonably consistent. For all substrates, we find that for *thin* films the van der Waals form,

$$\mu_i(\sigma_4) \sim \mu_i(\infty) - \frac{\alpha_s \bar{\rho}^3}{\sigma_4^3} \qquad \text{for} \qquad \sigma_4 \to \infty \tag{26}$$

for $i = 3, 4$, is *not* in good agreement with the calculated, microscopic chemical potentials. In the regime $d \gtrsim 3$ layers, the agreement between the microscopic and van der Waals form is simply a statement that the bulk quantities $\mu_i(\infty)$ are being fit correctly. We conclude therefore that the films under consideration here are still far from the thickness where the chemical potential is dominated by the van der Waals term.

In order to gauge the accuracy of our calculations, we need to compare our results with experiment. In a series of papers, Hallock and co-workers[18–20] reported measurements of the ^{3}He ground-state and first excited state energies in the low ^{3}He concentration limit as a function of ^{4}He coverage on Nuclepore. In Ref. 4, we have shown that our calculated excited state energies are in excellent agreement with experiment, except in the region close to the layering transition where no agreement is expected. The ^{4}He and ^{3}He chemical potentials for the Li and Na substrates are basically constant as a function of ^{4}He coverage, with the exception of Li at the smallest ^{4}He coverages.

Figures 2 show that for all substrates the third sound speed oscillates in parallel with the layered structure of the ^{4}He film. In the regions of the layering transitions, the physical third sound speeds are determined by Maxwell-type constructions that are not shown in the figures. But note that the oscillations are clear even beyond the regions of the layering transitions as seen quite dramatically in Fig. 2(a) for lithium. It is even more evident than in the previous comparison of the chemical potentials that our highest coverage film barely approaches the thickness where asymptotic power laws are valid.

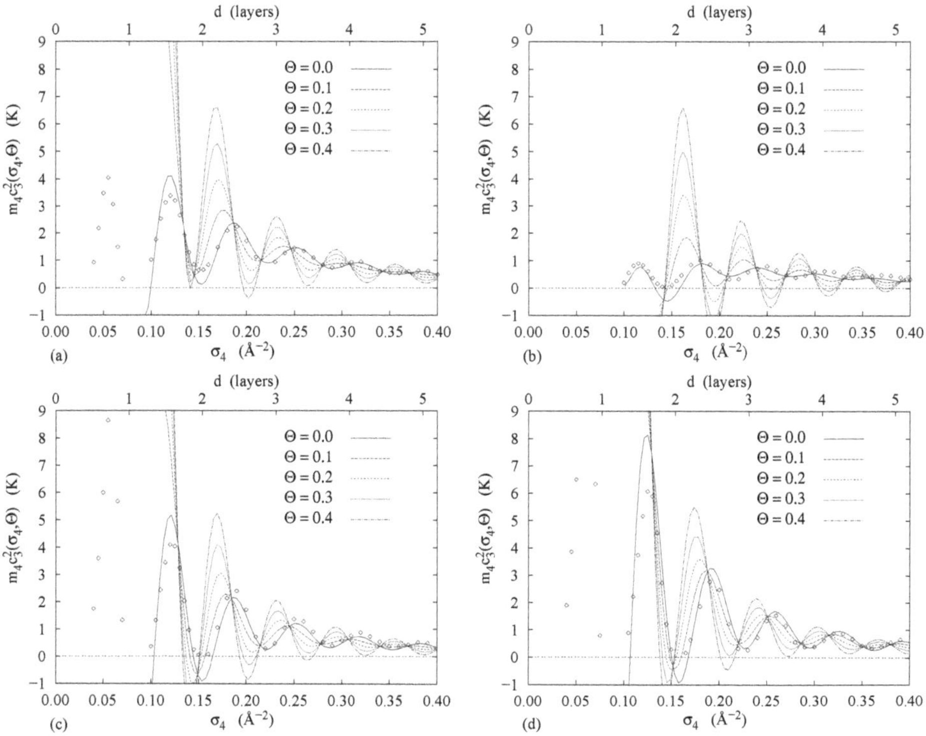

Fig. 2. The figure shows the total incompressibility of the mixture film on lithium (a), the sodium (b), the nuclepore (c) and the glass substrate (d) as a function of ^{3}He coverage in the low concentration regime, as given by Eq. (21) as calculated from the long–wavelength limit of the excitations (diamonds). Also shown are the results obtained by differentiating the fit (25) (solid line) and the asymptotic forms (26) (short–long dashed line). The figure also shows the corresponding quantity $\sigma_4\,(d\mu_3/d\sigma_4)$ obtained from the long–wavelength limit of the particle–hole interaction (Eq. (15), boxes) and from differentiating the fit (25) (dashed line). The upper scale gives the thickness of the films in *liquid* layers, one ^{4}He layer is defined as representing an areal density of 0.077 Å^{-2}. These figures are from Ref. 6.

Even more interesting is the effect of a small amount of ^{3}He on the third sound speed as predicted by Eq. (21) or (22). With detailed inspection of the calculated chemical potentials, we see that the *slope* of $d\mu_3/d\sigma_4$ *changes sign* in the region of the oscillations.[4,6] Thus, a further prediction of these calculations is that, for thin ^{4}He films on any of these substrates, regions may be found in which the third sound speed *increases* as a function of added ^{3}He and not just slows down as already seen on glass[7] and Nuclepore.[8]

The total, concentration dependent incompressibility is composed of the sum of the ^{4}He incompressibilities at zero ^{3}He coverage ($\theta_3 = 0$) plus an added contribution due to the ^{3}He in the low-concentration, linear limit. The result is shown in Figs. 2(a)-(d) for the four substrates under condsideration here. To the extent that the linearization of the ^{3}He dependence of the incompressibility is legitimate, the results are quite interesting. They quantify the above statement that, depending on the ^{4}He coverage, the incompressibility may *increase or decrease* with ^{3}He concentration. In the vicinity of layering transitions, the effect can be strong enough to make the film unstable. The effect is particularly pronounced in the triple–layer films with coverage between 0.165 Å^{-2} and 0.22 Å^{-2}. In the lower half of that coverage regime, the addition of ^{3}He would have the tendency to *stabilize* the films, whereas it *destabilizes* the translationally invariant configuration at coverages above 0.19 Å^{-2}. This is consistent with the drop in the speed of sound seen in the Nuclepore experiment of Sheldon and Hallock.[8] Of course, inspection of the Nuclepore data, Fig. 2(c), shows clearly that the drop in third sound speed depends quite sensitively on the coverage, and a quantitative comparison is not feasible.

4. Conclusion

Our calculations suggest a rather interesting scenario: *a ^{4}He film can undergo a layering transition which is driven by the presence of ^{3}He*. For Nuclepore, for example, our results indicate that this happens at modest ^{3}He concentrations of 0.25 monolayers, where the linearization (21) should still be permitted, and a ^{4}He coverage of about 0.21 Å^{-2}. The configuration of the film in this regime would be a mixture of regions that would consist of three layers of ^{4}He without ^{3}He in Andreev states, and double layers of ^{4}He with ^{3}He in surface states. It is interesting to note that this is the same regime in which phase separation has been reported by Bhattacharyya and Gasparini[21] for mixture films on Nuclepore.

Our results can be summarized as follows:

- Despite the lack of strong coverage dependence in the ^{4}He and ^{3}He chemical potentials for the lithium and sodium substrates, *cf.* Figs. 1(a) and 1(b), our microscopic analysis of the mixture–film third sound speeds, Figs. 2(a) and 2(b), show dramatic film–coverage dependence. In regions where the pure ^{4}He third sound speed is small, the contributions from the ^{3}He component can have critically important effects. We have shown above that this effect also ought to be visible on weak substrates.
- The layered structure of the ^{4}He film induces coverage dependent ^{3}He and ^{4}He chemical potentials that in turn yield non-monotonic third sound speeds as a function of ^{4}He film coverage. This response, predicted to be seen on Nuclepore and glass, should also be observable on much weaker substrates.
- For thin films, the film-averaged hydrodynamic formulas are in *qualitative* disagreement with the microscopic results. The formulas can be made to fit particular results with sufficient effort at modeling.
- The microscopic calculations also predict regions in which the third sound speed will increase with the addition of small amounts of ^{3}He . This response, already predicted to be observable in thin films on strong substrates, should likewise be observable on weaker substrates.

Acknowledgments

MDM would like to express his appreciation to the Institute for Theoretical Physics at Johannes-Kepler University for support during the period of a professional leave.

References

1. K. R. Atkins and I. Rudnick, *Progress in Low Temperature Physics* (North-Holland, Amsterdam, 1970), Amsterdam, ch. 2.
2. F. M. Ellis, R. B. Hallock, M. D. Miller and R. A. Guyer, *Phys. Rev. Lett.* **46**, p. 1461 (1981).
3. R. A. Guyer and M. D. Miller, *Phys. Rev. B* **25**, p. 5749 (1982).
4. E. Krotscheck and M. D. Miller, *Phys. Rev. B* **73**, p. 134514 (2006).
5. B. E. Clements, E. Krotscheck and H. J. Lauter, *Phys. Rev. Lett.* **70**, 1287 (1993).
6. E. Krotscheck and M. D. Miller, *J. Low Temp. Phys.* **141**, 1 (2005).
7. J. M. Valles, R. M. Heinrichs and R. B. Hallock, *Phys. Rev. Lett.* **56**, p. 1704 (1986).
8. P. A. Sheldon and R. B. Hallock, *Phys. Rev. B* **50**, p. 16082 (1994).
9. J. C. Noiray, D. Sornette, J. P. Romagnan and J. P. Laheurte, *Phys. Rev. Lett.* **53**, p. 2421 (1984).
10. B. E. Clements, J. L. Epstein, E. Krotscheck and M. Saarela, *Phys. Rev. B* **48**, 7450 (1993).
11. B. E. Clements, H. Forbert, E. Krotscheck and M. Saarela, *J. Low Temp. Phys.* **95**, 849 (1994).
12. V. Apaja and E. Krotscheck, A microscopic view of confined quantum liquids, in *Microscopic Approaches to Quantum Liquids in Confined Geometries*, eds. E. Krotscheck and J. Navarro (World Scientific, Singapore, 2002) pp. 205–268.
13. E. Krotscheck, Inhomogeneous quantum liquids: Statics, dynamics, and thermodynamics, in *Microscopic Quantum Many-Body Theories and their Applications*, eds. J. Navarro and A. Polls, Lecture Notes in Physics, Vol. 510 (Springer, Heidelberg, 1998) pp. 187–250.
14. B. E. Clements, E. Krotscheck and M. Saarela, *Phys. Rev. B* **55**, 5959 (1997).
15. E. Feenberg, *Theory of Quantum Fluids* (Academic, New York, 1969).
16. C. E. Campbell, The structure of quantum fluids, in *Progress in Liquid Physics*, ed. C. A. Croxton (Wiley, London, 1977) pp. 213–308.
17. J. W. Clark, Variational theory of nuclear matter, in *Progress in Particle and Nuclear Physics*, ed. D. H. Wilkinson (Pergamon Press Ltd., Oxford, 1979) pp. 89–199.
18. N. Alikacem, D. T. Sprague and R. B. Hallock, *Phys. Rev. Lett.* **67**, p. 2501 (1991).
19. D. T. Sprague, N. Alikacem, P. A. Sheldon and R. B. Hallock, *Phys. Rev. Lett.* **72**, 384 (1994).
20. P. A. Sheldon and R. B. Hallock, *Phys. Rev. Lett.* **77**, p. 2973 (1996).
21. B. Bhattacharyya and F. M. Gasparini, *Phys. Rev. Lett.* **49**, p. 919 (1982).

GENERIC ELECTRON MOBILITY IN SURFACE STATES ON HELIUM FILMS

E. KROTSCHECK[†] and M. D. MILLER[†,*]

[†]*Institut für Theoretische Physik, Johannes-Kepler-Universität, A-4040 Linz, Austria*
E-mail: eckhard.krotscheck@jku.at

[*]*Department of Physics and Astronomy, Washington State University, Pullman, WA 99164-2814, USA*
E-mail: mdm@wsu.edu

We study the mobility of electrons adsorbed to thin ^{4}He films. Utilizing the time-dependent version of the Euler-Lagrange, hypernetted chain variational theory, we compute the inelastic scattering rate of an electron due to collisions with film excitations (third sound). We obtain an analytic result valid in the long wavelength limit. In agreement with experiment, the mobility shows oscillations due to the underlying transverse film structure. The oscillations are explicitly due to the appearance of the third sound speed in the scattering rate since the third sound speed itself oscillates in conjunction with the ^{4}He film structure. The mobilities tend to be higher than reported mobilities on thin films, which we attribute to substrate structure. We interpret our results as *generic* mobilities that are valid in the limit of perfectly smooth, structureless substrates.

Keywords: Electron mobility; helium films.

1. Introduction

In 1964, Sommer[1] showed that there exists an appreciable barrier (on the order of 1 eV) for electron injection through a helium surface. Cole and Cohen[2] and Cole[3] pointed out that the external, attractive electrostatic image-potential coupled with the large repulsion localized at the helium surface would create a potential well that could trap the electron on the helium surface.

In 1987, Cieslikowski, Dahm, and Leiderer[4] reported mobility measurements for ^{3}He and ^{4}He films on a H_2 substrate. This experiment extended the work of Ref. 5 to thicker helium films. The authors found oscillations in the mobility that they attributed to layering in the mobile film. The mobilities were in the range of 10^3 cm^2/V·s. The remarkable aspect of this data is that they show evidence for at least nine oscillations in the mobility implying physical layering in the helium film extending quite far from the solid substrate. This in contrast to other features like adsorption isotherms and the superfluid fraction which allow one to resiolve perhaps three or four layers. The authors also noted that a ^{3}He sample displayed qualitatively similar modulations. However, as pointed out by Shikin, Klier, Doicescu, Würl, and Leiderer,[6] there is a many orders of magnitude difference between free surface (ripplon-limited) mobilities and thin film mobilities. Indeed, Shirahama, Ito, Suto, and Kono[7] report ^{4}He free-surface, ripplon-limited mobilities on the order of $10^7 - 10^8$ cm^2/V·s. Building on a previous model introduced by Monarkha, Albrecht, Kono, and Leiderer,[8] the authors of Ref. 6 argue that the mobilities being measured in the thin film experiments may be dominated by surface impurities or inhomogeneities.

We present in this manuscript a microscopic calculation of the electron mobility on thin ^{4}He films. The calculation is based on the time-dependent, variational theory developed for inhomogeneous, strongly-interacting boson systems.[9] Our model for the solid substrate is perfectly flat with no impurities and no periodic structure. We refer to our calculated mobilities as *generic* since the calculated electron scattering only includes coupling to third sound quanta and not any contributions from substrate defects. We derive an explicit result for the electron mobility that, in agreement with experiment, oscillates with the ^{4}He film thickness through the explicit appearance of the third sound speed. Our numerical results for the generic mobilities are the same order of magnitude as the ripplon-limited mobilities reported for the ^{4}He free surface.

In Sec. 2, we discuss the aspects of the theory necessary to calculate scattering times for impurities adsorbed in a ^{4}He film. Technically, we have to deal with two tasks: One is the calculation of coupling matrix elements between the electrons and the ^{4}He atoms, the other one is the calculation of inelastic scattering processes *inside* the medium. The case at hand is simplified by the fact that the physics is dominated by the long-wavelength properties of the electron-helium coupling. This will allow us to express the essential physics solely in terms of physical observables like the ^{4}He speed of sound and the coverage dependence of the electron binding energy. The necessary derivations are presented in Secs. 3 and 4. In Sec. 5, we show our numerical results and compare them with experiment. Sec. 6 is the conclusion.

2. Ground State and Excitations

2.1. *Energetics and structure*

The microscopic description of static and dynamic properties of helium films, including the dynamics of impurities, has been described in various places,[9] the reader is referred to these papers for general concepts and definitions. The background ^{4}He film is described by a Hamiltonian

$$H_N = \sum_i \left[-\frac{\hbar^2}{2m_4}\nabla_i^2 + U_{\rm sub}(\mathbf{r}_i) \right] + \sum_{i<j} V(|\mathbf{r}_i - \mathbf{r}_j|) \,, \tag{1}$$

where $V(|\mathbf{r}_i - \mathbf{r}_j|)$ is the ^{4}He-^{4}He interaction, and $U_{\rm sub}(\mathbf{r})$ is the external "substrate" potential. The ground-state wave function is written in the Jastrow-Feenberg form

$$\Psi_N(\mathbf{r}_1, \ldots, \mathbf{r}_N) = \exp\frac{1}{2}\left[\sum_i u_1(\mathbf{r}_i) + \sum_{i<j} u_2(\mathbf{r}_i, \mathbf{r}_j) + \sum_{i<j<k} u_3(\mathbf{r}_i, \mathbf{r}_j, \mathbf{r}_k)\right]. \quad (2)$$

The many-body correlations are optimized by solving the Euler equations

$$\frac{\delta E_N}{\delta u_n(\mathbf{r}_1, \ldots, \mathbf{r}_n)} = 0 \qquad (n = 1,\, 2,\, 3)\,, \quad (3)$$

where E_N is the energy expectation value of the N-particle Hamiltonian (1) with respect to the wave function (2). Executing the variations and manipulating the Euler equations to bring them in a numerically tractable form leads to an effective one-body Schrödinger equation

$$\left[-\frac{\hbar^2}{2m_4}\nabla^2 + U_{\rm ext}(\mathbf{r}) + V_H(\mathbf{r}) - \mu_4\right]\sqrt{\rho_1(\mathbf{r})} \equiv H_1(\mathbf{r})\sqrt{\rho_1(\mathbf{r})} = 0 \quad (4)$$

for the one body density $\rho_1(\mathbf{r})$ and the ^{4}He chemical potential μ_4, where $V_H(\mathbf{r})$ is a self-consistently determined "Hartree" potential. From Eq. (4) we can also write the one body Hamiltonian $H_1(\mathbf{r})$ as

$$H_1(\mathbf{r}) = -\frac{\hbar^2}{2m_4}\frac{1}{\sqrt{\rho_1(\mathbf{r})}}\nabla\rho_1(\mathbf{r})\cdot\nabla\frac{1}{\sqrt{\rho_1(\mathbf{r})}}\,. \quad (5)$$

Optimization of the two-body correlations leads to a normal mode equation

$$\int d^3r' \left[H_1(\mathbf{r})\delta(\mathbf{r}-\mathbf{r}') + 2\tilde{V}_{p-h}(\mathbf{r},\mathbf{r}')\right] H_1(\mathbf{r}')\phi^{(m)}(\mathbf{r}') = (\hbar\omega_m)^2\phi^{(m)}(\mathbf{r})\,, \quad (6)$$

where

$$\tilde{V}_{\rm p-h}(\mathbf{r},\mathbf{r}') = \sqrt{\rho_1(\mathbf{r})}V_{\rm p-h}(\mathbf{r},\mathbf{r}')\sqrt{\rho_1(\mathbf{r}')}\,, \quad (7)$$

is the "particle-hole interaction" that is determined self-consistently within the variational theory. In an exact theory, the particle-hole interaction can also be defined through $V_{p-h}(\mathbf{r},\mathbf{r}') = \delta V_H(\mathbf{r})/\delta\rho_1(\mathbf{r}')$. The eigenfunctions $\phi^{(m)}(\mathbf{r}')$ can be identified with the physical density fluctuations within a Feynman theory of excitations of the liquid.

The Hamiltonian of the $N+1$ particle system consisting of N ^{4}He atoms and one impurity (an electron in our case) located at $\mathbf{r}_0$ is

$$H_{N+1}^{(I)} = -\frac{\hbar^2}{2m_I}\nabla_0^2 + U_{\rm sub}^{(I)}(\mathbf{r}_0) + \sum_{i=1}^{N} V^{(I)}(|\mathbf{r}_0 - \mathbf{r}_i|) + H_N\,. \quad (8)$$

The variational wave function (2) for an inhomogeneous N-particle Bose system with a single impurity atom is

$$\Psi_{N+1}^{(I)}(\mathbf{r}_0, \mathbf{r}_1, \ldots, \mathbf{r}_N) = \exp\frac{1}{2}\left[u_1^{(I)}(\mathbf{r}_0) + \sum_{1\le i\le N} u_2^{(I)}(\mathbf{r}_0, \mathbf{r}_i) + \ldots\right]\Psi_N(\mathbf{r}_1, \ldots, \mathbf{r}_N)\,. \quad (9)$$

Analogous to Eq. (4), the impurity one-body density is determined by a one-body Schrödinger equation in terms of an effective "Hartree" potential,

$$\left[-\frac{\hbar^2}{2m_I}\nabla_0^2 + U_{\rm sub}^{(I)}(\mathbf{r}_0) + V_H^{(I)}(\mathbf{r}_0) - \mu_I\right]\sqrt{\rho_1^{(I)}(\mathbf{r}_0)} \equiv H_1^{(I)}(\mathbf{r}_0)\sqrt{\rho_1^{(I)}(\mathbf{r}_0)} = 0\,, \tag{10}$$

for the impurity density $\rho_1^{(I)}(\mathbf{r}_0)$ and binding energy μ_I. We can now introduce the impurity one-body operator, $H_1^{(I)}(\mathbf{r}_0)$, defined analogously to Eq. (5), and use it to generate a basis of impurity wave functions

$$H_1^{(I)}(\mathbf{r}_0)\eta^{(\alpha)}(\mathbf{r}_0) = \varepsilon_\alpha \eta^{(\alpha)}(\mathbf{r}_0)\,. \tag{11}$$

The spectrum is ε_α and the set of states are the $\{\eta^{(\alpha)}(\mathbf{r}_0)\}$. In the following, we will label impurity states with *Greek* letters and the background phonons with *Latin* letters.

We also need the two-body equation because it contains the basic ingredients needed for the dynamics of the system. The pair equation is a relationship between a "particle-hole" interaction $\tilde{V}_{\rm p-h}^{(I)}(\mathbf{r}_0, \mathbf{r}')$ and the "direct correlation function" $\tilde{X}^{(I)}(\mathbf{r}_0, \mathbf{r}')$ that is best written in a matrix representation in terms of the states $\phi^{(m)}$ and $\eta^{(\alpha)}$.

$$(\eta^{(\alpha)}|\tilde{X}^{(I)}|\phi^{(m)}) = -2\frac{(\eta^{(\alpha)}|\tilde{V}_{\rm p-h}^{(I)}|\phi^{(m)})}{\varepsilon_\alpha + \hbar\omega_m}\,, \tag{12}$$

where the tilde notation for the impurity particle-hole interaction is, in an exact theory, given by $\tilde{V}_{\rm p-h}^{(I)}(\mathbf{r}_0, \mathbf{r}_1) \equiv \sqrt{\rho_1^{(I)}(\mathbf{r}_0)}V_{\rm p-h}^{(I)}(\mathbf{r}_0, \mathbf{r}_1)\sqrt{\rho_1(\mathbf{r}')} = \sqrt{\rho_1^{(I)}(\mathbf{r}_0)}\delta V_H^{(I)}(\mathbf{r}_0)/\delta\rho(\mathbf{r}_1)\sqrt{\rho_1(\mathbf{r}_1)}$.

2.2. *Excitations and linear response*

Excitations and the response to perturbations are treated analogously within the theory. The system is exposed to a weak, time-dependent, external perturbation $\sum_i U_{\rm ext}(\mathbf{r}_i; t)$. As a response, the wave function becomes time-dependent. The natural generalization of the variational approach to excited states is to allow for *time-dependent* correlations $u_n(\mathbf{r}_0, \ldots, \mathbf{r}_n; t)$ of the form

$$\Phi(t) = e^{-iE_{N+1}^{(I)}t/\hbar}\psi^{(I)}(\mathbf{r}_0, \mathbf{r}_1, \ldots \mathbf{r}_N; t)/\sqrt{\langle\psi^I \mid \psi^{(I)}\rangle}\,, \tag{13}$$

where $\psi^{(I)}(\mathbf{r}_0, \mathbf{r}_1, \ldots\mathbf{r}_N; t)$ contains the dynamic, time-dependent correlations. These are again written in a Jastrow-Feenberg form

$$\psi^{(I)}(\mathbf{r}_0, \mathbf{r}_1, \ldots, \mathbf{r}_N; t) = \exp\left[\frac{1}{2}\delta U(\mathbf{r}_0, \mathbf{r}_1, \ldots\mathbf{r}_N; t)\right]\psi_{N+1}^{(I)}(\mathbf{r}_0, \mathbf{r}_1, \ldots, \mathbf{r}_N)\,, \tag{14}$$

where $\delta U(\mathbf{r}_0, \mathbf{r}_1, \ldots, \mathbf{r}_N; t) = \delta u_1(\mathbf{r}_0; t) + \sum_i \delta u_2(\mathbf{r}_0, \mathbf{r}_i; t)$. The time-dependent parts of the correlations, $\delta U(\mathbf{r}_0, \mathbf{r}_1, \ldots, \mathbf{r}_N; t)$, are determined by the stationary principle

$$\delta S = \delta\int_{t_0}^{t_1} \mathcal{L}(t)dt = \delta\int_{t_0}^{t_1}\left\langle\Phi(t)\left|H_{N+1}^{(I)} - i\hbar\frac{\partial}{\partial t}\right|\Phi(t)\right\rangle dt = 0\,. \tag{15}$$

The variations of the action integral with respect to the excitation amplitudes $\delta u_1^*(\mathbf{r}_0;t)$ and $\delta u_2^*(\mathbf{r}_0,\mathbf{r}_1;t)$ lead to coupled one- and two-particle continuity equations for the *first-order* density fluctuations. In the general impurity theory,[10] it was shown that the the equation of motion can be written in the form of a non-linear eigenvalue problem

$$\hbar\omega r_{\alpha} = \sum_{\beta}\left[\delta_{\alpha\beta}\varepsilon_{\alpha} + \Sigma_{\alpha\beta}(\omega)\right] r_{\beta}\,, \tag{16}$$

where the r_α are the expansion coefficients of the time-(or frequency-)dependent part of the impurity wave function in terms of the states $\eta^{(\alpha)}(\mathbf{r}_0)$, and

$$\Sigma_{\alpha\beta}(\omega) = -\sum_{\gamma m}\frac{W^{*(\alpha)}_{\gamma m}W^{(\beta)}_{\gamma m}}{\hbar\omega_m + \varepsilon_\gamma - \hbar\omega} \tag{17}$$

is the impurity's self-energy. The impurity-impurity-phonon vertex function occurring in the self-energy (17) is[10]

$$W^{(\alpha)}_{\gamma m} = -\frac{\hbar^2}{2m_I}\int d^3r_0\; \eta^{*(\gamma)}(\mathbf{r}_0)\sqrt{\rho_1^{(I)}(\mathbf{r}_0)}\nabla X_m(\mathbf{r}_0)\cdot\nabla\frac{\eta^{(\alpha)}(\mathbf{r}_0)}{\sqrt{\rho_1^{(I)}(\mathbf{r}_0)}}\,; \tag{18}$$

$$X_m(\mathbf{r}_0) \equiv \int d^3r_1 X^{(I)}(\mathbf{r}_0,\mathbf{r}_1)\sqrt{\rho_1(\mathbf{r}_1)}\,\phi^{(m)}(\mathbf{r}_1)\,, \tag{19}$$

where $X^{(I)}(\mathbf{r}_0,\mathbf{r}_1)$ is the "direct correlation function" and the three-body vertex function $W^{(\alpha)}_{\gamma m}$ describes an impurity atom scattering off of a background excitation and is given in terms of quantities obtained in the ground state calculation.

2.3. *Transport currents*

For the problem at hand we must introduce yet another quantity, namely the *second order current*. The flux of impurities (here, specifically, electrons) is given by the expectation value of the current operator

$$\hat{\mathbf{j}}^{(I)}(\mathbf{r}) = \frac{1}{2m_I}\left(\frac{\hbar}{\mathrm{i}}\nabla_0\delta(\mathbf{r}_0-\mathbf{r}) + \delta(\mathbf{r}_0-\mathbf{r})\frac{\hbar}{\mathrm{i}}\nabla_0\right)\,, \tag{20}$$

with respect to the wave function (14). This expectation value must be calculated to *second order* in the fluctuating correlations $\delta u_n(\mathbf{r}_1,\ldots,\mathbf{r}_n;t)$ because the first-order term is periodic in time and does not lead to mass transport, thus

$$\mathbf{j}_2^{(I)}(\mathbf{r}) \equiv \frac{1}{4}\frac{\langle\Psi_0|\,\delta U^*\,\hat{\mathbf{j}}^{(I)}(\mathbf{r})\,\delta U\;\;|\Psi_0\rangle}{\langle\Psi_0\mid\Psi_0\rangle}\,. \tag{21}$$

The excitation operator $\delta U(t)$ is understood to be obtained from the *first order* equations of motion (15). One finds[11]

$$\begin{aligned}\mathbf{j}_2^{(I)}(\mathbf{r}_0) &= \Re e\frac{\hbar}{4m_I\mathrm{i}}\left[\varphi_\omega^*(\mathbf{r}_0)\nabla\varphi_\omega(\mathbf{r}_0) + \int d^3r d^3r'\, S(\mathbf{r},\mathbf{r}')\,\delta\tilde{u}_2^*(\mathbf{r}_0,\mathbf{r})\nabla_0\delta\tilde{u}_2(\mathbf{r}_0,\mathbf{r}')\right]\\ &\equiv \mathbf{j}_{2,\mathrm{el}}^{(I)}(\mathbf{r}_0) + \mathbf{j}_{2,\mathrm{inel}}^{(I)}(\mathbf{r}_0)\,.\end{aligned} \tag{22}$$

Here, $S(\mathbf{r}_1, \mathbf{r}_2)$ is background static structure function. $\varphi_\omega(\mathbf{r_0})$ is the electron wave function that was obtained from the effective Schrödinger equation (16) and can, evidently, be identified with the elastic channel wave function. The form of (22) gives the decomposition of the current into a one-body term, which describes the *elastic* channels, and a many-body term contributing to *inelastic* scattering. The latter term contains the fluctuating pair correlations $\delta u_2(\mathbf{r}, \mathbf{r}_2)$, which demonstrates the importance of time-dependent pair correlations for describing inelastic processes. Using the two-body equations of motion, we can express the inelastic current as

$$\mathbf{j}^{(I)}_{2,\mathrm{inel}}(\mathbf{r}_0) = \frac{\hbar}{4m_I}\Im m\left[\sum_\gamma \eta^*_\gamma(\mathbf{r}_0)\nabla\eta_\gamma(\mathbf{r}_0)\sum_m \frac{W^{*(\boldsymbol{\alpha})}_{\boldsymbol{\gamma m}} W^{(\boldsymbol{\beta})}_{\boldsymbol{\gamma m}} r^*_{\boldsymbol{\alpha}} r_{\boldsymbol{\beta}}}{|\varepsilon_\gamma + \hbar\omega_m - \hbar\omega + \mathrm{i}\epsilon|^2}\right], \tag{23}$$

where the $r_{\boldsymbol{\beta}}$ are the amplitudes of the electron in the individual single particle states, *cf.* Eq. (16).

Eq. (23) also demonstrates why we identify the second term in Eq. (22) with an inelastic current. If the self-energy (17) is complex, the current (23) has a double pole and the expression must be re-interpreted as a rate-equation.

3. Electron: Self-Energy

The problem at hand differs from the previously studied problem of atomic impurities in the important aspect that the most visible effect, namely the oscillatory behavior of the electron mobility as a function of the film thickness, can be explained by looking at long-wavelength expansions for all quantities involved. This does not necessarily imply that other contributions are negligible, all we say is that these can at most contribute a smooth background.

The system fills a box of area L^2 with the z-axis oriented transverse to the substrate. The ^{4}He *coverage*, n, is defined by $n = N/L^2 = (1/L^2)\int d^3r\rho_1(\mathbf{r})$. In the plane surface geometry, the electron wave functions $\eta^{(\boldsymbol{\alpha})}(\mathbf{r}_0)$ and energies $\varepsilon_{\boldsymbol{\alpha}}$ are

$$\eta^{(\boldsymbol{\alpha})}(\mathbf{r}_0) = \eta^{(\alpha)}(z)e^{i\mathbf{q}_\parallel\cdot\mathbf{r}}, \qquad \varepsilon_{\boldsymbol{\alpha}} = \epsilon_\alpha + \frac{\hbar^2\mathbf{q}_\parallel^2}{2m_I}, \tag{24}$$

where $\eta^{(\alpha)}(z)$ and ϵ_α are the eigenfunction/eigenvalue pairs of Eq. (11) for $q_\parallel = 0$.

The electron remains in its transverse ground state but has a parallel momentum, and transitions to intermediate phonon states are negligible, *i.e.* we keep only

$$\eta^{(o,\mathbf{q}_\parallel)}(\mathbf{r}_0) = \frac{1}{L}\sqrt{\rho^{(I)}_1(z)}e^{i\mathbf{q}_\parallel\cdot\mathbf{r}}, \tag{25}$$

$\varepsilon_0 = \hbar^2 q_\parallel^2/2m_I$ and the third sound mode $\phi_{\mathbf{m}}(\mathbf{r})$, which will be labeled by $(\mathbf{m} \rightarrow$

$(R, \mathbf{q}_R))$. The coupling matrix elements then becomes

$$\begin{aligned} W^{(\boldsymbol{o},\boldsymbol{q}_\parallel)}_{(\boldsymbol{o},\boldsymbol{q}_\gamma)\,(R,\mathbf{q}_R)} &= -\frac{\hbar^2}{2m_I L^2}\int d^3r_0 \rho_1^{(I)}(z) e^{-i\boldsymbol{q}_\gamma\cdot\mathbf{r}_\parallel}\nabla X_R(\mathbf{r}_0)\cdot\nabla e^{i\mathbf{q}_\parallel\cdot\mathbf{r}_\parallel} \\ &= \frac{\hbar^2}{2m_I L}\delta_{\boldsymbol{q}_R+\mathbf{q}_\parallel-\boldsymbol{q}_\gamma}\mathbf{q}_R\cdot\mathbf{q}_\parallel\int dz_0\rho_1^{(I)}(z_0)X_R(z_0,q_R) \\ &\equiv \frac{\hbar^2}{2m_I L}\delta_{\boldsymbol{q}_R+\mathbf{q}_\parallel-\boldsymbol{q}_\gamma}\mathbf{q}_R\cdot\mathbf{q}_\parallel X_{\boldsymbol{o},R}(q_R)\,. \end{aligned} \tag{26}$$

We can also restrict the summation over intermediate states to the lowest (third sound) excitation and write the self-energy as

$$\begin{aligned} \Sigma(q_\parallel,\omega) \equiv \Sigma_{(o,\boldsymbol{q}_\parallel),(o,\boldsymbol{q}_\parallel)}(\omega) &= -\sum_{\boldsymbol{q}_\gamma\,\boldsymbol{q}_R}\frac{\left|W^{(o,\boldsymbol{q}_\parallel)}_{(o,\boldsymbol{q}_\gamma),(R,\boldsymbol{q}_R)}\right|^2}{\hbar\omega_R(q_R)+\frac{\hbar^2}{2m_I}q_\gamma^2-\hbar\omega} \\ &= -\left(\frac{\hbar^2}{2m_I}\right)^2\int\frac{d^2q_R}{(2\pi)^2}\frac{\left|(\boldsymbol{q}_R\cdot\boldsymbol{q}_\parallel)X_{\boldsymbol{o},R}(q_R)\right|^2}{\hbar\omega_R(q_R)+\frac{\hbar^2}{2m_I}(\boldsymbol{q}_R+\boldsymbol{q}_\parallel)^2-\hbar\omega}\,, \end{aligned} \tag{27}$$

where $\omega_R(q_R) \to c_3 q_R$ for $q_R \to 0$ is the ripplon/third sound dispersion relation. The long-wavelength expansion of the ripplon/third sound state $\phi_{R,\mathbf{q}_R}(z)$ has been worked out in Ref. 12. In the limit $\mathbf{q}_R \to 0$, the excitation function factorizes as

$$\phi_{R,\mathbf{q}_R}(z) = -\sqrt{\frac{n\hbar q_R^2}{2m_4\omega_R(q_R)}}\,\Phi_R(z) = \sqrt{\frac{2n\hbar q_R^2}{m_4\omega_R(q_R)}}\frac{d\sqrt{\rho_1(z)}}{dn}\,. \tag{28}$$

$\Phi_R(z)$ is the shape function of the long-wavelength density fluctuation; we have normalized it such that $\int dz\Phi_R(z)\sqrt{\rho_1(z)} = 1$. From the fact that $\tilde{V}^{(I)}_{\rm p-h}(\mathbf{r}_0,\mathbf{r}')$ is rigorously defined from the variation of the self-consistent one-body potential $V_H^{(I)}(\mathbf{r}_0)$, *cf.* Eq. (10), we have

$$\int d^3r' V^{(I)}_{\rm p-h}(\mathbf{r}_0,\mathbf{r}')\frac{d\rho_1(\mathbf{r}')}{dn} = \frac{dV_H^I(\mathbf{r}_0)}{dn} \quad\text{and}\quad \int dz\rho_1^{(I)}(z)\frac{dV_H^I(z)}{dn} = \frac{d\mu_I}{dn}\,. \tag{29}$$

We therefore obtain for the long wavelength limit of the direct correlation function

$$X_{\boldsymbol{o},R}(q_R) = \sqrt{\frac{2n\hbar q_R^2}{m_4\omega_R(q_R)}}\frac{d\mu_I}{dn}\frac{1}{\hbar\omega_R(q_R)+\frac{\hbar^2}{2m_I}q_R^2}\,, \tag{30}$$

which relates it directly to physical observables. Collecting everything, the electron dispersion relation is given by the implicit equation

$$\hbar\omega(q_\parallel) = \frac{\hbar^2 q_\parallel^2}{2m_I} - \left(\frac{\hbar^2}{2m_I}\right)^2\int\frac{d^2q_R}{(2\pi)^2}\frac{\left|(\boldsymbol{q}_R\cdot\boldsymbol{q}_\parallel)X_{\boldsymbol{o},R}(q_R)\right|^2}{\hbar\omega_R(q_R)+\frac{\hbar^2}{2m_I}(\boldsymbol{q}_R+\boldsymbol{q}_\parallel)^2-\hbar\omega(q_\parallel)}\,. \tag{31}$$

The self-energy is real in the long-wavelength limit $q_\parallel \to 0$ which determines the effective mass ratio $\frac{m_I}{m_I^*}$. To get a qualitative idea of the dependence of the effective mass on the various physical quantities, we insert the long-wavelength expansion (30)

$$\frac{m_I}{m_I^*} - 1 = \int_0^\infty \frac{dq_R q_R}{4\pi} \frac{\hbar^2 q_R^2}{2m_I} \frac{|X_{\mathbf{o},R}(q_R)|^2}{\hbar\omega_R(q_R) + \frac{\hbar^2 q_R^2}{2m_I^*}} \approx -\frac{1}{2\pi}\frac{m_I}{m_4}\left(n\frac{d\mu_I}{dn}\right)^2 \frac{1}{n\hbar^2 c_3^2} \,. \tag{32}$$

The expression is a bit more complicated if an effective mass is also allowed in the energy denominator of Eq. (31). Both the numerator and the denominator go to zero in the limit of large film thicknesses. Compared to the decay rate (see Eq. (36) below), which contains only one c_3 factor in the denominator, the dependence of the effective mass on the "softness" of the surface is obviously much stronger and can occasionally overcome the mass ratio of m_I/m_4 in front of the effective mass correction.

The calculation of the electron effective mass has two purposes: one, is that the decay rate will be seen to be directly proportional to the square of the effective mass ratio, a strongly fluctuating mass will therefore also enhance fluctuations of the decay rate; second, to assess the validity of our theoretical scheme, especially the self-energy expression (17). An effective mass ratio that is of the order of 1 is necessary for the validity of the approximation (17) of the electron self energy.

4. Electrons: Energy Loss

We introduce the same approximations on the inelastic currents as before: (1) we keep only the lowest state $\hbar\omega_R(q_R)$, and (2) assume that the electron moves with a wave vector $\mathbf{q}_\parallel$ in the lowest state, (*i.e.*, we keep only the wave function $\eta_0(\mathbf{r})$ or $r_{\boldsymbol{\alpha}} = \delta_{\boldsymbol{\alpha},0}$). Finally, if the electron wave number is such that the energy denominator has a double pole, the inelastic current must be reinterpreted as giving a *decay rate*: Then the inelastic current becomes

$$\begin{aligned}
\lim_{\epsilon\to 0+} \frac{d}{dt}\mathbf{j}^{(I)}_{2,\mathrm{inel}}(\mathbf{r};t) &= \lim_{\epsilon\to 0+} \frac{d}{dt}\frac{1}{2\hbar L^2}\left(\frac{\hbar^2}{2m_I}\right)^3 \rho_1^{(I)}(z)\int \frac{d^2 q_R}{(2\pi)^2}\left(\mathbf{q}_R + \mathbf{q}_\parallel\right) \\
&\quad \times \left|\frac{\mathbf{q}_R\cdot\mathbf{q}_\parallel X_{\mathbf{o},R}(q_R) e^{-\epsilon t/\hbar}}{\hbar\omega_R(q_R) + \frac{\hbar^2}{2m_I}(\mathbf{q}_R+\mathbf{q}_\parallel)^2 - \hbar\omega + i\epsilon}\right|^2 \\
&= \frac{1}{\hbar^2 L^2}\left(\frac{\hbar^2}{2m_I}\right)^3 \rho_1^{(I)}(z)\Im m\int \frac{d^2 q_R}{(2\pi)^2}\frac{\left(\mathbf{q}_R+\mathbf{q}_\parallel\right)\left|\mathbf{q}_R\cdot\mathbf{q}_\parallel X_{\mathbf{o},R}(q_R)\right|^2}{\hbar\omega_R(q_R) + \frac{\hbar^2}{2m_I}(\mathbf{q}_R+\mathbf{q}_\parallel)^2 - \hbar\omega - i\epsilon} \,.
\end{aligned} \tag{33}$$

Due to symmetry, the total decay rate can only be in the direction $\mathbf{q}_\parallel$ of the current. The decay rate relative to the elastic current

$$\mathbf{j}^{(I)}_{2,\mathrm{el}}(\mathbf{r};t) = \frac{\hbar\mathbf{q}_\parallel}{m_I L^2}\rho_1^{(I)}(z) \tag{34}$$

is then

$$\frac{\hbar}{\tau} = \frac{\hbar}{\left|\mathbf{j}^{(I)}_{2,\mathrm{el}}(\mathbf{r};t)\right|} \lim_{\epsilon\to 0+} \frac{d}{dt}\left|\mathbf{j}^{(I)}_{2,\mathrm{inel}}(\mathbf{r};t)\right|$$

$$= \frac{1}{2q_{\|}}\left(\frac{\hbar^2}{2m_I}\right)^2 \Im m \int \frac{d^2q_R}{(2\pi)^2} \frac{\left(q_{\|} + \mathbf{q}_R\cdot\hat{\mathbf{q}}_{\|}\right)\left|\mathbf{q}_R\cdot\mathbf{q}_{\|} X_{\mathbf{o},R}(q_R)\right|^2}{\hbar\omega_R(q_R) + \frac{\hbar^2}{2m_I^*}(\mathbf{q}_R + \mathbf{q}_{\|})^2 - \hbar\omega - \mathrm{i}\epsilon}\,. \tag{35}$$

Defining the ratio between the electron thermal velocity and the speed of third sound, $\lambda = m_I^* c_3/\hbar q_{\|} = c_3/v_e$, where $v_e = \hbar q_{\|}/m_I^*$ is the electron velocity, the leading term in λ^{-1} is

$$\frac{\hbar}{\tau} = \frac{1}{6\pi}\left(n\frac{d\mu_I}{dn}\right)^2 \left(\frac{m_I^*}{m_4}\right)^2 \frac{m_4}{\hbar^2 n}\frac{v_e}{c_3} + \mathcal{O}(\lambda^0)\,. \tag{36}$$

This expression gives the dependence of the decay rate on the essential physical parameters. Our result is the ratio of two terms that go to zero in the limit of thick films. The third sound mode with velocity c_3 will eventually turn into a ripplon mode with the familiar $q_{\|}^{3/2}$ dispersion relation for which our theory does not apply. Also, the electron chemical potential should approach a constant in the thick film limit. One expects, however, that the dependence of the electron chemical potential as a function of film thickness is relatively smooth, whereas the third sound speed is known to oscillate strongly.[13]

5. Results

5.1. *Substrate model*

We treat both the substrate and the helium films as homogeneous systems with individual dielectric constants ϵ_S for the substrate and ϵ_H for ^{4}He. We assume a ^{4}He film of thickness $h = n/\rho_\infty$ whose surface is at $z = 0$ and a flat substrate half-space between $-\infty < z < -h$. The potential seen by the electron is then

$$U_{\mathrm{sub}}(z) = \begin{cases} \infty & \text{for } z \le 0, \\ -\left[\frac{\epsilon_S - 1}{\epsilon_S + 1} - \frac{\epsilon_H - 1}{\epsilon_H + 1}\right]\frac{e^2}{2(z+h)} - \left[\frac{\epsilon_H - 1}{\epsilon_H + 1}\right]\frac{e^2}{2z} & \text{for } z > 0\,. \end{cases} \tag{37}$$

The hard core for helium potential barrier is a reasonable model because the measured potential barrier seen by the electron at the ^{4}He surface is about 1 eV.[1]

As the film thickness increases, the binding energy of the electron varies between the binding energy of the electron on the bare substrate and the one on a helium half-space. Since the binding on the ^{4}He half space is much weaker than on the substrate, the electron will, with increasing film thickness, also move away from the surface.

For the computation, we have assumed a glass substrate dielectric constant of $\epsilon_S = 1.3$ and a ^{4}He dielectric constant of $\epsilon_H = 1.055$. The speed of third sound as a function of film thickness was taken from Ref. 13, see also the accompanying paper.[14] For the purpose of calculating the electron holding potential, the film was modeled as having constant density over a width h, which was converted to ^{4}He coverage n. In the limit of zero film thickness, the binding energy to the substrate is approximately -169 K and the electron is approximately 27 Å away from the surface. In the limit of infinite film thickness the binding energy approaches -7.12 K and the distance from the ^{4}He surface is over 80 Å. Even a film of nominal thickness 100 Å is far the bulk half-space limit. Thus, the influence of the substrate reaches far into the regime where the ^{4}He film would otherwise be considered as "bulk". This observation is relevant because it demonstrates that thick-film approximations should not be used to estimate the term $d\mu_I/dn$ in Eq. (32) or (36).

5.2. *Electron mobility*

From the data on the energetics of electron bound states and the ^{4}He third sound speed we can calculate the effective mass ratio (32) and the contribution of the ^{4}He film to the loss rate (36) of the thermal electrons and, from that, the electron mobility

$$\mu = \frac{e}{m_I}\tau\,. \tag{38}$$

A quantitative calculation of the ^{4}He contribution to the electron mobility is evidently delicate because the electron lifetime contains the ratio of two quantities, $n(d\mu_I/dn)$ and c_3, that both go to zero in the thick film limit. One expects that the electron binding energy is relatively insensitive against the layer structure, because it is dominated, for thin films, by the substrate and layering effects in the density profile are negligible for film thicknesses of more than three or four layers. On the other hand, the speed of third sound shows, for practically all substrates, the pronounced oscillation seen in Fig. 1.

The generic quantities to calculate are the damping rate $1/\tau$ (36) and the effective mass ratio (32). Note that, in our units, the damping rate has units of an energy. For our calculations, we have assumed a thermal speed of the electrons of $v_e = 5000\,\mathrm{m/sec}$. Fig. 1 shows the decay-rate $1/\tau$ and the effective mass ratio m_I^*/m_I as a function of coverage. Of course, one must be cautious with the extrapolation of the fit of the calculated $n(d\mu_I/dn)$ to films that are much thicker than the regime from which the results were obtained; we therefore show in Fig. 1 only a conservative extrapolation up to a coverage of $n = 0.6$ Å^{-2}, corresponding to about nine layers. It appears the two terms, $(nd\mu_I/dn)^2$ and c_3, go to zero at about the same rate, leaving the decay rate oscillating with roughly the same amplitude. In the effective mass ratio, the third sound speed appears squared in the denominator, hence the oscillations increase with increasing coverage. This statement depends, of course, largely on the dielectric constant of the substrate; the effect is small in the model substrate model used here, but a slightly more polarizable substrate can cause much larger effective mass ration than those shown in Fig. 1.

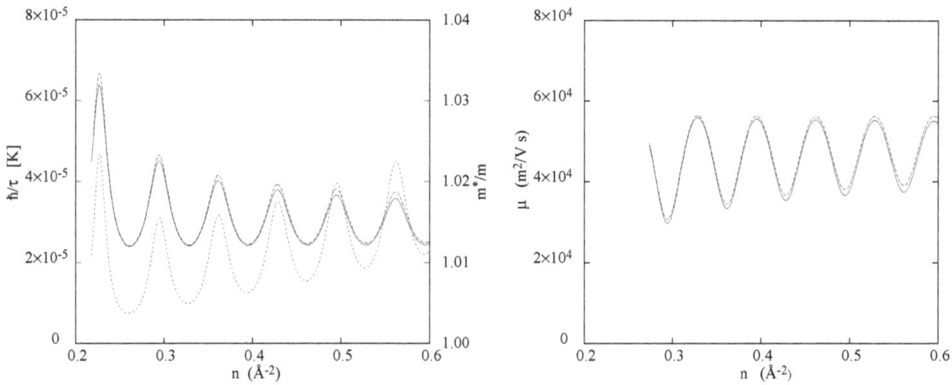

Fig. 1. The left figure shows the contribution of the ^{4}He film to the electron energy loss rate calculated from Eq. (36) as a function of ^{4}He coverage n (solid line, left scale). Also shown is the effective mass calculated from Eq. (32) (short dashed line, right scale) and the energy loss rate obtained by using that effective mass in Eq. 36 (long-dashed line, left scale). The right figure shows the contribution of the ^{4}He film to the electron mobility, calculated from Eq. (38). as a function of ^{4}He coverage n (solid line, left scale). Also shown is the mobility contribution obtained by using the effective mass in Eq. 36). Note that the mobility μ should not be confused with the ^{4}He chemical potential μ_4 or the electron binding energy μ_I.

6. Conclusion

In this paper, we have calculated the inelastic scattering rate due to ripplon/third sound coupling for an electron on ^{4}He films up to nine atomic layers. Our major result, Eq. (36), is a microscopic derivation of the scattering rate. We have identified the term that is responsible for the oscillatory behavior of the flux decay rate and, consequently, the electron mobility. The cause for this oscillatory behavior is the dependence of the third sound speed on the layered structure of the film, and the appearance of that third sound speed in the denominator of the decay rate. Transitions to higher-lying excitations are perhaps not negligible, but they can at most contribute a smooth background of the same order of magnitude.

The dominant contribution to the mobility is shown to depend explicitly on the ratio of the third sound speed to the average electron speed and also the derivative of the electron chemical potential with respect to the film coverage. The third sound speed oscillates synchronously with ^{4}He layer formation, but the electron chemical potential is a smooth function. Thus, we have shown explicitly that, in agreement with experiment,[4,5] the electron mobility on thin ^{4}He films should exhibit oscillations in concert with layer formation.

When the scattering rate shown in Fig. (1) is turned into a mobility via Eq. (38), we find results on the order of 10^4 m^2/V·s. These results are in rough agreement with measurements of ripplon-limited, electron mobilities on the free helium surface.[7] They are considerably larger however than reported electron mobilities on thin helium films.[4,5] It is our understanding that the mechanism sugested by Monarkha, Albrecht, Kono, and Leiderer[8] is at work and the existing two thin film measurements of elctron mobilities have been dominated by scattering due to substrate surface impurities and inhomogeneities. In Ref. 8 the authors report an orders of magnitude increase in the electron mobility on a hydrogen substrate by simply annealing the substrate. We specifically point out the fact that the experiments of Refs. 4 and 5 show small mobility oscillations on top of a smooth background whereas our results of Figs. 1 indicate that the oscillations of the mobility *due to coupling to the* ^{4}*He film* are in size comparable to the total scattering rate off the helium film. Experiments indicate that (a) the oscillatory terms are a small correction on a smooth background, and (b) that the background contribution can be changed by orders of magnitude by annealing the substrate. We conclude therefore that we have identified the essential cause of the mobility oscillations as being due to the form (36) of the helium-induced flux loss rate.

To conclude, we should mention the findings by Paalanen *et al.*[15] and by Cieslikowski *et al.*[16] (Fig. 9 of Ref. 16) who observe an decrease of the electron mobility with increasing film thickness. This effect is somewhat counter-intuitive, it has been interpreted as "...due to the stiffening of the film surface caused by the van der Waals term in the ripplon spectrum[15]". However, the ripplon stiffening is one of the outcomes of our description. A possible explanation is the softening of the ripplon mode with increasing film thickness; the limit of our equations (36) is delicate since both the numerator and the denominator go to zero and it might well be that the behavior of the mobility as the film thickens depends sensitively on the experimental setup.

References

1. W. T. Sommer, *Phys. Rev. Lett.* **12**, 271 (1964).
2. M. W. Cole and M. H. Cohen, *Phys. Rev. Lett.* **23**, 1238 (1969).
3. M. W. Cole, *Phys. Rev. B* **2**, 4239 (1970).
4. D. Cieslikowski, A. J. Dahm and P. Leiderer, *Phys. Rev. Lett.* **58**, 1751(Apr 1987).
5. M. A. Paalanen and Y. Iye, *Phys. Rev. Lett.* **55**, 1761(Oct 1985).
6. V. Shikin, J. Klier, I. Doicescu, A. Würl and P. Leiderer, *Phys. Rev. B* **64**, p. 073401(Jul 2001).
7. K. Shirahama, S. Ito, H. Suto and K. Kono, *J. Low Temp. Phys.* **101**, 439 (1995).
8. Y. P. Monarkha, U. Albrecht, K. Kono and P. Leiderer, *Phys. Rev. B* **47**, 13812 (May 1993).
9. V. Apaja and E. Krotscheck, A microscopic view of confined quantum liquids, in *Microscopic Approaches to Quantum Liquids in Confined Geometries*, eds. E. Krotscheck and J. Navarro (World Scientific, Singapore, 2002) pp. 205–268.
10. B. E. Clements, E. Krotscheck and M. Saarela, *Phys. Rev. B* **55**, 5959 (1997).
11. E. Krotscheck and R. Zillich, *J. Chem. Phys.* **115**, 10161 (2001).
12. B. E. Clements, E. Krotscheck and M. Saarela, *Z. Physik B* **94**, 115 (1994).
13. E. Krotscheck and M. D. Miller, *J. Low Temp. Phys.* **141**, 1 (2005).
14. M. D. Miller and E. Krotscheck, these proceedings.
15. M. A. Paalanen and Y. Iye, *Phys. Rev. Lett.* **55**, 1761 (1985).
16. D. Cieslikowski, P. Leiderer and A. J. Dahm, *Can. J. Phys.* **65**, p. 1525 (1987).

PAIRING IN ASYMMETRICAL FERMI SYSTEMS

KHANDKER F. QUADER and RENYUAN LIAO

Department of Physics, Kent State University,
Kent, OH 44242, USA

We consider inter- and intra-species pairing interactions in an asymmetrical Fermi system. Using equation of motion method, we obtain coupled mean-field equations for superfluid gap functions and population densities. We construct a phase diagram across BCS-BEC regimes. Due to intra-species correlations, the BCS singlet superfluid state can sustain finite polarizations, P. For larger P, we find phase separations in BCS and BEC regimes. A superfluid phase exists for all P deep in BEC regime. Our results may apply to pairing in ultracold fermions, nuclear and quark matter physics.

Keywords: Intra-species correlations; BEC-BCS phase diagram.

1. Introduction

Prompted by several unresolved issues regarding pairing in two-species Fermi systems with unequal population, there has been intense theoretical and experimental interest and activities in the area. Asymmetrical Fermi systems can occur in a variety of systems: superconductor in an external magnetic field[1] or a strong spin-exchange field;[2–4] superconductor with overlapping bands;[5,6] mixture of fermionic cold atoms with different densities;[7,8] isospin asymmetric nuclear matter with unequal isospin chemical potentials and possible proton-proton, neutron-neutron and proton-neutron pairing;[9] color superconductivity in dense quark matter[10] where quarks with different masses and charges have different chemical potentials.

One of the *outstanding* questions in asymmetrical Fermi systems is whether the BCS superfluid state can sustain any finite imbalance between the species. It is convenient to characterize two-species systems as two-spin systems and the species imbalance as spin polarization. It is believed that the BCS ground state in a finite magnetic field h, is robust against spin polarizations for $h \sim \Delta$, (Δ being the superconducting gap); beyond this it becomes unstable to a normal state. Another *intriguing issue* is the nature of fermion pairing and phase diagram as a function of pairing interaction strength and polarization. These issues have been brought to the fore by experiments on ultra-cold fermion atoms, wherein two hyperfine states with varying unequal populations have been achieved.[7,8] These systems can in effect be subjected to arbitrary repulsive and attractive interactions by sweeping across with a s-wave Feshbach resonance; thereby allowing the exploration of the fermion ground states in both BCS and BEC regimes. In the case of *equal population*, there is theoretical agreement with experiments that find superfluid states in both BCS and BEC regimes with a "smooth crossover" around the "unitarity limit" (diverging scattering length a_s).

In systems with *population imbalance*, several theoretical scenarios have been proposed.[11] There appears to be agreement of experiments with mean-field theoretical calculations that find that the superfluid state is unstable to phase separation into superfluid and normal states or a mixed phase;[12,13] a superfluid state stabilizes sufficiently deep in the BEC regime. Experimental studies however are not yet conclusive; e.g. one experiment[8] at a low polarization 1% near unitarity on the BCS side has reported signature of a superfluid state.

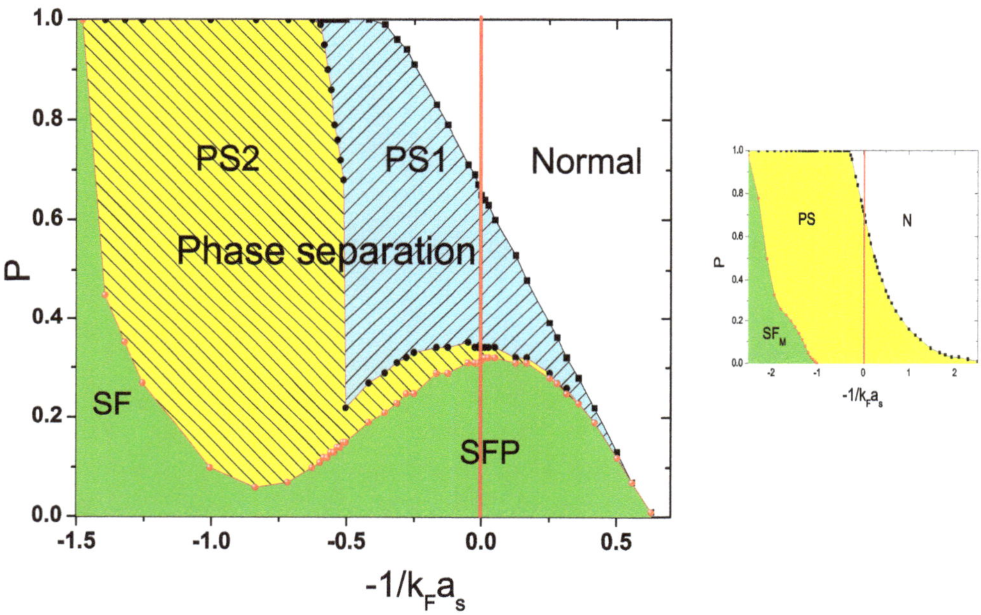

Fig. 1. Polarization P vs singlet coupling, $-1/(k_F a_s)$ phase diagram for asymmetrical fermions: (a) With a representative triplet correlation strength $g_1 = 50$; (b) Without triplet correlations, $g_1 = 0$. Vertical line refers to the unitarity limit; PS1, PS2, PS to phase separated regions; N to normal state.

To date, theoretical calculations have mostly considered inter-species singlet interaction, and have ignored the intra-species "spin-triplet" correlations. Ho et al[14] attempted to incorporate triplet correlations in a somewhat phenomenological manner, though it is not clear if the effects are considered in a self-consistent manner; Huang et al[15] recently explored the implications of such correlations on a FFLO state; Monte Carlo calculations[16] hint at a polarized superfluid phase near unitarity.

In this paper, we carry out a *detailed mean-field* study, including in a self-consistent manner both inter-species "spin-singlet" and intra-species "spin-triplet" correlations. Following Leggett[17] and Eagles,[18] our discussion is in terms of BCS-type pairing in both the BCS and BEC regimes. We ensure that the chemical potentials for the two species are determined self-consistently with the pairing gaps. We do a careful *stability analysis* of the multitude of states obtained from our equations.

Our findings are *dramatically different* from those without triplet correlations, Our phase diagram, shown in Fig. 1a, shows that in the BCS regime, at smaller polarizations, a singlet superfluid state (hereafter referred to as SFP) becomes *stable* for sufficiently strong triplet correlations; the largest polarizations being sustained on both BCS and BEC sides of the unitarity limit. Depending on the singlet interaction strength, at some polarization, SFP becomes unstable via a 1st-order transition to *phase separation*, denoted by PS1. PS1 is characterized by a negative "susceptibility", $\delta P/\delta h$; P being the spin-polarization, and h the difference of the chemical potentials, thereby playing the role of a "magnetic field". For a given triplet interaction, and for sufficiently weak singlet interaction, SFP and PS1 undergo transitions to the normal state on the BCS side. Beyond the unitarity limit, the gapped SFP persists into the BEC regime, sustaining progressively smaller polarizations. Deeper in the BEC regime, this goes smoothly over to a BEC singlet superfluid (SF) at all polarizations, except for full polarization, where it is a triplet superfluid. In the BEC regime, in addition to PS1, we find the existence of a somewhat different phase separated state, PS2, characterized by positive "susceptibility", but not satisfying requisite superfluid ground state stability criteria.

2. Mean-Field Model

For our detailed study, we consider a Fermi system containing two species with unequal "pseudo-spin" populations. To allow for both inter-species (spin-singlet) and intra-species (spin-triplet) correlations, we start with a pairing Hamiltonian given by

$$\begin{aligned} H = & \sum_{k\sigma} \xi_{k\sigma} c^+_{k\sigma} c_{k\sigma} \\ & + \sum_{kk'q\sigma} g^{\sigma\sigma}_{kk'} c^+_{k+q/2\sigma} c^+_{-k+q/2\sigma} c_{-k'+q/2\sigma} c_{k'+q/2\sigma} \\ & + \sum_{kk'q} g^{\uparrow\downarrow}_{kk'} c^+_{k+q/2\uparrow} c^+_{-k+q/2\downarrow} c_{-k'+q/2\downarrow} c_{k'+q/2\uparrow} \end{aligned} \tag{1}$$

where the pseudospin $\sigma = \uparrow, \downarrow$ denote for example the two hyperfine states of ultracold Fermi atoms. $c^+_{k\sigma}$ is the fermion creation operator with kinetic energy $\xi_{k\sigma} = \epsilon_{k\sigma} - \mu_\sigma$; μ_σ is the chemical potential of each of the species. $g^{\uparrow\uparrow}_{kk'}$, and $g^{\downarrow\downarrow}_{kk'}$ are the triplet interactions between the up and down spins res pectively. We take the singlet interaction, $g^{\uparrow\downarrow}_{kk'}$ to be a constant, g_o. This is usually expressed in terms of s-wave scattering length a_s using $(4\pi\hbar^2 a_s/m)^{-1} = g_o^{-1} + \sum_k (2\epsilon_k)^{-1}$. We do a mean-field decoupling of the interaction terms by introducing three order parameters or gap functions ($\sigma, \sigma' = \uparrow, \downarrow$) given by,

$$\Delta_{\sigma\sigma'}(k,q) = -\sum_{k'} g_{kk'}^{\sigma\sigma'} < c_{-k'+q/2\sigma} c_{k'+q/2\sigma'} > \tag{2}$$

This results in a mean-field Hamiltonian given by:

$$\begin{aligned} H^{MF} &= \sum_{k\sigma} \xi_{k\sigma} c_{k\sigma}^{+} c_{k\sigma} - \sum_{k,\sigma} \Delta_{\sigma\sigma}(k,q) c_{k+q/2\sigma}^{+} c_{-k+q/2\sigma}^{+} \\ &- \sum_{k,\sigma} \Delta_{\sigma\sigma}^{*}(k,q) c_{-k+q/2\sigma} c_{k+q/2\sigma} - \sum_{k.\sigma} |\Delta_{\sigma\sigma}(k,q)|^2 / g_{kk}^{\sigma\sigma} \\ &- \sum_{k} \Delta_{\downarrow\uparrow}(k,q) c_{k+q/2\uparrow}^{+} c_{-k+q/2\downarrow}^{+} - \sum_{k} \Delta_{\downarrow\uparrow}^{*}(k,q) c_{-k+q/2\downarrow} c_{k+q/2\uparrow} \\ &- \sum_{k} |\Delta_{\downarrow\uparrow}(k,q)|^2 / g_{kk}^{\downarrow\uparrow} \end{aligned} \tag{3}$$

3. Self-consistent Method and Solutions

We employ the *equation of motion* method using imaginary time normal and anomalous Matsubara Green's functions, $G_{\sigma\sigma'}(k,\tau)$, $F_{\sigma\sigma'}(k,\tau)$, respectively, and our mean-field Hamiltonian, H^{MF}. The coupled equations in terms of $\Delta_{\sigma\sigma'}$ are given by $G_{\sigma\sigma'}(k,\tau)$ and $F_{\sigma\sigma'}(k,\tau)$:

$$\begin{aligned} \partial_\tau G_{\sigma\sigma'}(k,\tau) &= -\delta(\tau)\delta_{\sigma\sigma'} - \xi_{k+q/2\sigma} G_{\sigma\sigma'}(k,\tau) \\ &\quad + \sum_{\sigma''} \Delta_{\sigma''\sigma}(k,q) F_{\sigma''\sigma'}(k,\tau) \end{aligned} \tag{4}$$

$$\begin{aligned} \partial_\tau F_{\sigma\sigma'}(k,\tau) &= \xi_{-k+q/2\sigma} F_{\sigma\sigma}(k,\tau) \\ &\quad + \sum_{\sigma''} \Delta_{\sigma\sigma''}^{*}(k,q) G_{\sigma''\sigma'}(k,\tau) \end{aligned} \tag{5}$$

where τ is the imaginary time variable. These equations may be Fourier transformed in the usual way with $\tau \to iw_n$, $\partial_\tau \to -iw_n$, where $iw_n = (2n+1)\pi/\beta$ are the Matsubara frequencies, n being an integer and $\beta = 1/k_B T$.

In this paper we focus on a superfluid condensate of pairs with *zero center-of-mass momentum*, q. Thus, we do not consider the $q \neq 0$ Fulde-Ferrel-Larkin-Ovchinnikov(FFLO) state,[2,3] but which may also be studied within this scheme. Solving the Fourier transformed equations at $q = 0$, we obtain the *2-point correlation functions*:

$$\begin{aligned} G_{\sigma\sigma'}(k,iw_n) &= \frac{\delta_{\sigma\sigma'} f_\sigma + \delta_{\sigma-\sigma'}(\Delta\Delta_\uparrow b_\downarrow - \Delta\Delta_\downarrow\Delta_\uparrow)}{D} \\ F_{\sigma\sigma'}(k,iw_n) &= \frac{\delta_{\sigma\sigma'} f_{\sigma\sigma} + \delta_{\sigma-\sigma'} f_{\sigma-\sigma}}{D} \end{aligned} \tag{6}$$

where $f_\sigma = \Delta^2 b_{-\sigma} + \Delta_{-\sigma}^2 b_\sigma - a_{-\sigma} b_\sigma b_{-\sigma}$;

$f_{\sigma\sigma} = \Delta^2\Delta_{-\sigma} + \Delta_\sigma\Delta^2_{-\sigma} - \Delta_\sigma a_{-\sigma}b_{-\sigma}$; $f_{\sigma-\sigma} = \Delta(\Delta^2 + \Delta_\sigma\Delta_{-\sigma} - a_\sigma b_{-\sigma})(\delta_{\sigma\downarrow} - \delta_{\sigma\uparrow})$; $D = (\Delta^2 + \Delta_\uparrow\Delta_\downarrow)^2 + a_\uparrow a_\downarrow b_\uparrow b_\downarrow - \sum_\sigma(\Delta^2 a_\sigma + \Delta^2_\sigma a_{-\sigma})b_{-\sigma}$;with $a_\sigma = \xi_{k+q/2\sigma} - iw_n$, $b_\sigma = -\xi_{-k+q/2\sigma} - iw_n$. We have set $\Delta_{\uparrow\uparrow} \equiv \Delta_\uparrow$; $\Delta_{\downarrow\downarrow} \equiv \Delta_\downarrow$; $\Delta_{\uparrow\downarrow} \equiv \Delta$. The *excitation spectrum* can be found by examining the poles of the Green's functions, yielding the quasiparticle energies,

$$E^2_{k\pm} = (iw_n)^2 = (\alpha \pm \sqrt{\beta})/2 \tag{7}$$

where $\alpha = \xi^2_{k\uparrow} + \xi^2_{k\downarrow} + 2\Delta^2 + \Delta^2_1 + \Delta^2_2$, and $\beta = \left[(\xi^2_{k\uparrow} - \xi^2_{k\downarrow}) + (\Delta^2_1 - \Delta^2_2)\right]^2 + 4\Delta^2\left[(\xi_{k\uparrow} - \xi_{k\downarrow})^2 + (\Delta_1 - \Delta_2)^2\right]$. Various quantities can now be obtained from our 2-point correlation functions. Thus, particle concentrations, n_σ for the two species ($\sigma =\uparrow,\downarrow$) are given by

$$\begin{aligned} n_\sigma &= \sum_k \langle c^+_{k\sigma} c_{k\sigma}\rangle = \sum_k \sum_{iw_n} \frac{1}{\beta} G_{\sigma\sigma}(k, iw_n)e^{iw_n 0^+} \\ &= \sum_k \sum_{l=\pm} (-1)^\lambda \left[\frac{n_F(E_{kl})f_\sigma(k, E_{kl}) - n_F(-E_{kl})f_\sigma(k, -E_{kl})}{2E_{kl}(E^2_{k+} - E^2_{k-})}\right] \end{aligned} \tag{8}$$

where λ is even for $l = +$, and odd for $l = -$; $n_F(E_{kl})$ are the Fermi functions. Likewise the three gaps equations are given by ($\sigma, \sigma' =\uparrow,\downarrow$):

$$\begin{aligned} \Delta_{\sigma\sigma'} &= -\sum_k g^{\sigma\sigma'}_{kk} \sum_{iw_n} \frac{1}{\beta} F^*_{\sigma\sigma'}(k, iw_n)e^{iw_n 0^+} \\ &= -\sum_k \sum_{l=\pm} (-1)^\lambda g^{\sigma\sigma'}_{kk} \left[\frac{n_F(E_{kl})f_{\sigma\sigma'}(k, E_{kl}) - n_F(-E_{kl})f_{\sigma\sigma'}(k, -E_{kl})}{2E_{kl}(E^2_{k+} - E^2_{k-})}\right] \end{aligned} \tag{9}$$

The above five equations are *coupled*, and can be solved self-consistently for the three gap functions, $\Delta, \Delta_\uparrow, \Delta_\downarrow$ for either fixed particle concentrations, $n_\uparrow$, $n_\downarrow$, or fixed chemical potentials, $\mu_\uparrow$, $\mu_\downarrow$.

We assume *equal masses* for the two species, and take the particle spectrum to be $\epsilon_k = \hbar^2k^2/2m$. We adopt standard definitions: polarization, $P = (n_1 - n_2)/(n_1 + n_2)$; mean chemical potential $\mu = (\mu_1 + \mu_2)/2$; chemical potential difference $h = (\mu_1 - \mu_2)/2$; Fermi momentum $k_{F\sigma} = (6\pi^2 n_\sigma)^{1/3}$. Since $\sum_k f(k) \to \int_0^\infty f(k)\frac{d^3k}{(2\pi)^3} \to \int_0^\infty k^3_F f(k/k_F)\frac{d^3(k/k_F)}{(2\pi)^3}$, we can scale quantities having dimension of energy to ϵ_F. The *inter-species* interaction g_o is represented by the coupling constant $\eta = -1/(k_F a_s)$. For the *intra-species* triplet interaction, we take the separable form $g^{\sigma\sigma}_{kk'} = g_\sigma\omega(k)\omega(k')Y_{10}(\hat{k})Y_{10}(\hat{k'})$, consistent with the lowest odd angular momentum, $\ell = 1$, m=0 order parameter symmetry. We explore three kinds of interaction, and find them to give qualitatively similar behavior: (i) $\omega(k) \propto$ *const*, corresponding to an interaction that is independent of the magnitude of momentum; (ii) $\omega(k) \propto k_0k/(k^2_0 + k^2)$, a generalization of the separable potential used by

Nozieres and Schmitt-Rink;[19] (3) $\omega(k) \propto \exp\left[-(k/k_0)^2\right]$, a Gaussian interaction; k_o being a cut-off momentum. The first two forms of interaction require regularization due to ultraviolet divergence, while the third does not. It is reasonable to set the triplet coupling coefficient $g_\uparrow = g_\downarrow \equiv g_1$.

4. Results

The key points of our paper can be made on considering our *T=0 results*, though we have also obtained interesting properties at finite-T. For arbitrary values of the singlet and triplet pairing interaction strengths, and population imbalance, we obtain self-consistent solutions of the $T = 0$ gap functions, $\Delta, \Delta_\uparrow, \Delta\downarrow$, and chemical potentials, μ_σ. On the BCS side, we find that for a given singlet interaction, g_o, the singlet pairing gap Δ *decreases* with increasing triplet strength g_1, while at the same time the triplet gaps $\Delta_\uparrow \neq \Delta_\downarrow$ increase, crossing at some value of g_1. The suppression of Δ is more pronounced at larger polarizations. Similar to the case without g_1, there exists a maximum polarization, P_{max} on the BCS side, beyond which we find no solution for Δ; this determines the SFP/PS1 $\leftrightarrow$ N boundary (Fig. 1a). P_{max} decreases with increasing $\eta = -1/k_F a_s$ (SPS/PS1 $\leftrightarrow$ N in Fig. 1a). Triplet correlations decrease P_{max}; e.g. for $g_1 = 50$ in Fig. 1a, at unitarity P_{max} decreases from 0.71 to 0.68 (Fig. 1).

A proper construction of the asymmetrical Fermi system *phase diagram* requires a determination of stable ground states out of the manifold of paired condensates given by our equations.[13,20] Accordingly, we carefully study several stability criteria. First we investigate the mean-field *ground state energies* as a function of the gap functions for different polarizations, P. This is given by:

$$\begin{aligned} E_G\left(\Delta, \Delta_\uparrow . \Delta_\downarrow\right) &= \langle \Psi | H^{MF} | \Psi \rangle \\ &= E_o + \sum_{k\sigma} \left[\xi_{k\sigma} G_{\sigma\sigma}(k, \tau = 0^-) - 2\Delta_\sigma F_{\sigma\sigma}(k, \tau = 0^-)\right] \\ &- \sum_{k} 2\Delta F_{\downarrow\uparrow}(k, \tau = 0^-) \end{aligned} \tag{10}$$

where $E_o = -|\Delta_\uparrow|^2/g_{kk}^{\uparrow\uparrow} - |\Delta_\downarrow|^2/g_{kk}^{\downarrow\downarrow} - |\Delta|^2/g$. To find the stability of the polarized singlet superfluid state SFP, we examine E_G versus the singlet gap function Δ ($\Delta_\uparrow, \Delta_\downarrow$ fixed) by evaluating this numerically for a large number of points, in particular around the SFP/SF $\leftrightarrow$ PS1/PS2 boundaries. This provides a robust determination of stability. We *supplement* this with analysis of $\frac{\partial^2 E_G}{\partial \Delta^2}$ and the "susceptibility" $\partial P/\partial h$. For a given triplet strength g_1, the stable singlet superfluid state SFP, in both BCS and BEC regimes, are characterized by E_G with a *global minimum* at a non-zero Δ, and self-consistently determined values of μ_σ's. Here both $\frac{\partial^2 E_G}{\partial \Delta^2} > 0$, and $\frac{\partial P}{\partial h} > 0$. SFP sustains larger polarizations for progressively larger g_1.

The region PS1 in Fig. 1a is characterized by a *negative* "susceptibility", $\frac{\partial P}{\partial h} < 0$ and $\frac{\partial^2 E_G}{\partial \Delta^2} < 0$. For a given P and h, $E_G(\Delta)$ is *maximum* at the non-zero gap

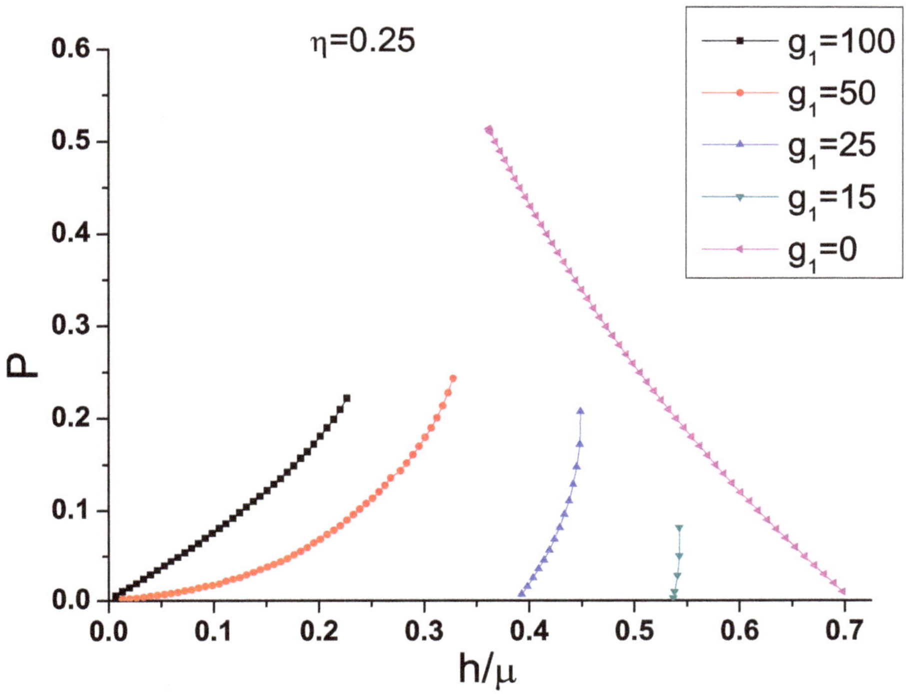

Fig. 2. Polarization $P = \frac{n_\uparrow - n_\downarrow}{n_\uparrow + n_\downarrow}$ vs $h/\mu = \frac{\mu_\uparrow - \mu_\downarrow}{\mu_\uparrow + \mu_\downarrow}$ (h acts like a "magnetic field") for different triplet correlation strengths g_1 at a fixed singlet coupling $\eta = 0.25$.

solution, separating two local minima – a feature of phase separation into a normal ($\Delta = 0$) and a superfluid ($\Delta \neq 0$) component by 1st-order phase transition. In this context, it is instructive to study P as a function of h/μ for different values of g_1, for a fixed singlet interaction strength, η (shown in Fig. 2 for $\eta = 0.25$ (BCS side)). For $g_1 \approx g_1^c$, the slope is vertical; g_1^c is the value of g_1 at which maximum polarization, P_{max} occurs for a given η, as shown in Fig. 3. For $g_1 \leq g_1^c$, the slope is *negative*, corresponding to the BCS superfluid state being unstable to the normal state for $h > \Delta$, but robust against polarization for $h < \Delta$. For $g_1 > g_1^c$, the singlet superfluid state can sustain a *finite polarization*, which exhibits a behavior over the range given by: $P \propto ah + bh^3 + c$; a,b,c being constants. The linear behavior is achieved for larger values of g_1, and at low polarizations. In examining P vs h/μ behavior *beyond* P_{max}, we find, for a given g_1, two solutions of P corresponding to one value of h/μ. To make a connection to Fig. 1a, obtained for $g_1 = 50$, we note that the allowed range of polarizations for SFP in P vs h/μ considerations correspond to the $\eta = 0.25$ vertical line, terminating at the SFP-PS1 boundary (Fig 1a). The same line extended from SFP-PS1 boundary to PS1-N boundary correspond to the polarization range bounded by the two solutions of P at a h/μ in Fig. 2. The region PS2 in BEC regime, though characterized by $\frac{\partial P}{\partial h} > 0$, does not form a stable superfluid phase; here $\frac{\partial^2 E_G}{\partial \Delta^2} < 0$. The line separating PS1 and

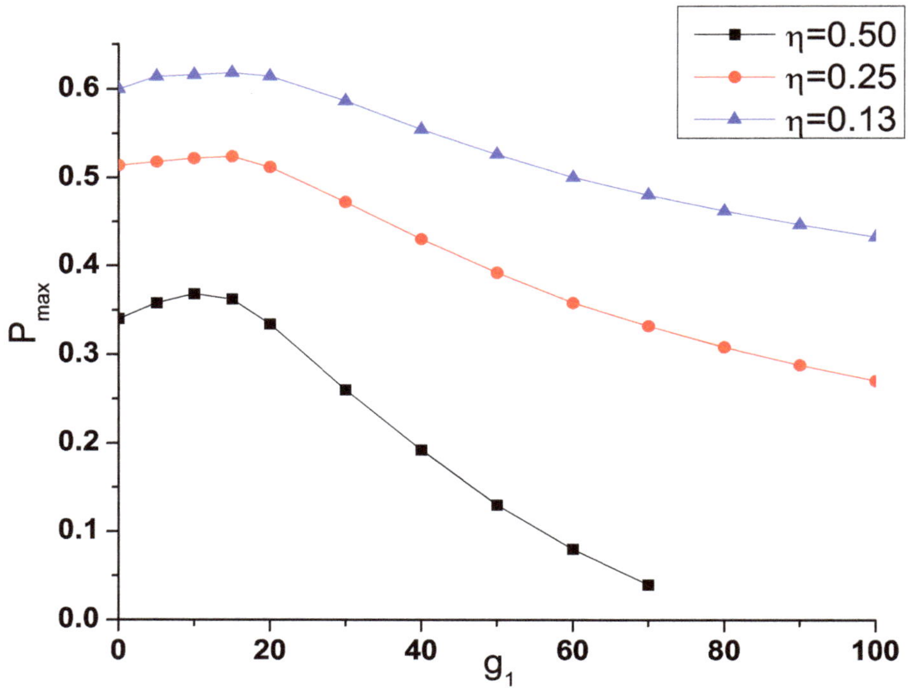

Fig. 3. Plot of maximum polarization, P_{max} vs. triplet correlation strength g_1 for different singlet couplings η. P_{max} decreases with increasing g_1, starting at the maximum value g_1^c. Solutions in the figure with $g_1 < g_1^c$ are un physical.

PS2 is probably a metastable line, the position of which depends on the triplet interaction strength, but would generally be controlled by dynamics not present in our calculations.

5. Discussions

We end with a few comments. The uniform "spin polarization" (non-zero "spin susceptibility") that our homogenous singlet superfluid state (SFP) exhibits is presumably facilitated by triplet "spin fluctuations" owing to triplet correlations. The quasiparticles forming SFP, being determined self-consistently, contain renormalizations due to these triplet spin fluctuations. The intra-species triplet interactions also result in the formation of triplet gaps, that can give rise to triplet pairing at appropriate temperatures. This results in interesting effects arising from the competition between singlet and triplet gaps, and singlet/triplet fluctuations. These and finite-T results that we obtain merit a separate discussion in a forthcoming paper. For a given singlet interaction strength ($1/(k_F a_s)$), the maximum polarization sustainable by SFP obviously depends on the strength of the triplet interaction, g_1. For example, in the ultracold ^6Li Fermi gas,[8] the corresponding g_1 would be expected to be less than $g_1 = 50$ in Fig. 1a.

Acknowledgments

We would like to thank D. Allender, K. Bedell, J. Engelbrecht, S. Gaudio, and M. Widom for fruitful discussions. We also acknowledge support from the U.S. Department of Army and ICAM.

References

1. *J. Phys. Chem. Solids* **24**, 1029, 1963.
2. A. I. Larkin and Y. Ovchinnikov, *Sov. Phys. JETP* **20** (1965).
3. P. Fulde and R. A. Ferrell, *Phys. Rev.* **A 135**, 550 (1964).
4. S. Takada and T. Izuyama, *Prog. Theo. Phys.* **41**, 635 (1969).
5. H. H. Suhl, B. T. Matthias, and L. R. Walker, *Phys. Rev. Lett.* **3**, 552 (1959).
6. J. Kondo, *Prog. Theo. Phys.* **29**, 1 (1963).
7. M. W. Zwierlein, A. Schirotzek, C. H. Schunck, and W. Ketterle, *Science* **311**, 492 (2006).
8. G. B. Partridge, W. Li, R. I. Kamar, Y. A. Liao, and R. G. Hulet, *Science* **311**, 503 (2006).
9. A. Sedrakian and U. Lombardo, *Phys. Rev. Lett.* **84**, 602 (2000); U. Lombardo, P. Nozieres, P. Schuck, H.- J. Schulze, and A. Sedrakian, *Phys. Rev.* **C 64**, 064314 (2001).
10. M. Alford, J. A. Bowers, and K. Rajagopal, *Phys. Rev.* **D 63** 074016 (2001); J. A. Bowers and K. Rajagopal, *Phys.Rev.* **D 66**, 065002 (2002); T. Schafer and F. Wilczek, *Phys. Rev.* **D 60**, 074014 (1999).
11. P. F. Bedaque, H. Caldas, and G. Rupak, *Phys. Rev.Lett.* **91**, 247002 (2003); H. Caldas, *Phys. Rev.* **A 69**, 063602 (2004); M. M. Forbes, E. Gubankova, M. V. Liu, and F. Wilczek, *Phys. Rev. Lett.* **94**, 017001 (2005); W. V. Liu and F. Wilczek, *Phys. Rev. Lett.* **90**, 047002 (2003); K. B. Gubbels, M. W. J. Romans, and H. T. C. Stoof (2006).
12. C.-H. Pao, S.-T. Wu, and S.-K. Yip, *Phys. Rev.* **B 73**, 132506 (2006); Q. Chen, Y. He, C.-C. Chien, and K. Levin, cond-mat/0608454 (2006); Z.-C. Gu, G. Warner, and F. Zhou, cond-mat/0603091 (2006); M. M. Parish, F. M. Marchetti, A. Lamacraft, and B. D. Simons, cond-mat/0605744 (2006); D. E. Sheehy and L. Radzihovsky, *Phys. Rev. Lett.* **96**, 060401 (2006).
13. D. E. Sheehy and L. Radzihovsky, cond-mat/0608172 (2006).
14. T.-L. Ho and H. Zhai, cond-mat/0602568 (2006).
15. X. Huang, X. Hao, and P. Zhuang, cond-mat/0610610 (2006).
16. J. Carlson and S. Reddy, *Phys. Rev. Lett.* **95**, 060401 (2005).
17. A. J. Leggett, *J. Phys.(Paris) Colloq.* **41** (1980).
18. D. M. Eagles, *Phys. Rev.* **186**, 456 (1969).
19. P. Noziers and S. Schmitt-Rink, *J. of Low Temp. Phys.* **59**, 195 (1985).
20. L. Viverit, C. J. Pethick, and H. Smith, *Phys. Rev.* **A 61**, 053605 (2000).

GROUND-STATE PROPERTIES OF SMALL ^{3}He DROPS FROM QUANTUM MONTE CARLO SIMULATIONS

ESTER SOLA, JOAQUIM CASULLERAS and JORDI BORONAT*

Departament de Física i Enginyeria Nuclear,
Campus Nord B4-B5, Universitat Politècnica de Catalunya,
E-08034 Barcelona, Spain
**E-mail: jordi.boronat@upc.edu*

We present recent results from a diffusion Monte Carlo study of small ^{3}He drops in the limit of zero temperature. Using the same methodology than in previous calculations carried out for bulk ^{3}He, we have obtained accurate results for their energy and structure properties. A relevant concern of this analysis has been the determination of the stability threshold for self binding. Our results show that the smallest self-bound drop is formed by $N = 30$ atoms, with preferred orbitals for open shells corresponding to maximum value of the spin. The quality of the upper bound energies provided by the fixed node approximation is analyzed by performing released-node estimations for short released times. Possible improvements of the nodal surface by including BCS-like correlation functions are also discussed.

Keywords: Helium drops; Quantum Monte Carlo.

1. Introduction

Helium drops are fascinating objects that have deserved continued interest of many-body theorists and effort of many experimental groups for several decades.[1,2] Composed by ^{4}He, by ^{3}He or by mixtures of both, helium drops are the most clean example of inhomogeneous quantum liquids with boson or fermion quantum statistics. The combination between their inhomogeneity and their unavoidable quantum nature makes them the best benchmark for quantum many-body techniques. The interest on helium drops has been even reinforced in the last years due to the increasing interest in the study of the properties of different molecules adsorbed in the drops. Helium drops offer ideal matrices for carrying out accurate spectroscopy of embedded molecules, and moreover one can gain access to microscopic properties of the medium such as its superfluidity. If the molecule is surrounded by ^{4}He, the rotational spectrum presents a sharp structure which has been attributed to the superfluid nature of ^{4}He, whereas for a ^{3}He drop a broad peak is observed.[3]

The smaller mass of the ^{3}He atom, and more fundamentally its Fermi statistics, introduce sizeable differences in the stability of the drop with respect to ^{4}He. Nowadays, it is well established that two ^{4}He atoms form a bound state but a minimum number of ^{3}He atoms is necessary to form a self-bound system.[4] This difference is observed in the experimental setup since very small ^{4}He drops are detected in the jet, starting from the dimer, whereas it has been proven to be difficult to observe ^{3}He drops with less than 1000 atoms.[5] In fact, reducing this number and generating drops with a number of atoms closer to the threshold limit is one of the most important challenges for the next future. Small ^{3}He drops are expected to present "magic numbers"[6] whose experimental determination would be a clear signature of their Fermi statistics.

The Fermi statistics of ^{3}He drops and their inherent inhomogeneity make their microscopic study a challenging program. The complexity of this task explains why the number of microscopic works on ^{3}He drops is significantly smaller than those devoted to ^{4}He drops. The first systematic study was carried out by Pandharipande *et al.*[4] in the eighties using the variational Monte Carlo (VMC) method. This calculation used a trial wave function incorporating backflow correlations to correct the nodal surface of the noninteracting system and predicted that a drop with 40 atoms is self bound. More recently, Guardiola and Navarro[7] carried out a detailed VMC calculation of small ^{3}He drops including in the trial wave function configuration-interaction-like correlations. These new correlations improved the energy in a significant way and the smallest bound drop was estimated to be the one formed by 35 atoms. Recently, the same authors[8] have obtained a lower value (32) for this upper-bound threshold in a diffusion Monte Carlo calculation (DMC) restricted to the range $N = 31 - 34$. Therefore, the improvement of both the trial wave function and the theoretical approach has progressively reduced the minimum number of atoms required for a self-bound drop. It is worth mentioning that using a non-local density-functional approach[9] this critical number was estimated to be slightly smaller (29) than these microscopic calculations.

In the present work, we report recent results[10] of the main properties of small ^{3}He drops obtained by means of the DMC method. Our aim has been to carry out a systematic study around the expected threshold limit for self binding using the methodology developed in our group for simulating bulk and two-dimensional ^{3}He. We use the fixed-node (FN) approximation,[11] which provides an upper bound to the exact eigenvalue, to get reliable results on the energy and density profiles and the release-node (RN) method[12] to estimate the quality of the FN upper-bound. In this way, we can not only obtain good upper bounds to the exact energies but analyze the quality of the nodal surface provided by the trial wave function used for importance sampling. The results obtained suppose the lowest energies obtained so far using quantum Monte Carlo and the RN analysis shows that there is still some bias due to the model wave function, which seems larger than in bulk.[13] Possible improvements coming from BCS-like wave functions are also discussed, and in fact a specific model for them has been developed. However, our preliminary results do not show sizeable improvements of the energies of the ^{3}He drops.

2. Quantum Monte Carlo Methods

The DMC method is nowadays a well-known tool devised to study quantum fluids and solids at zero temperature. Its starting point is the Schrödinger equation written in imaginary time,

$$-\hbar\frac{\partial\Psi(\boldsymbol{R},t)}{\partial t} = (H - E_{\rm r})\Psi(\boldsymbol{R},t) \ , \tag{1}$$

with an N-particle Hamiltonian

$$H = -\frac{\hbar^2}{2m}\sum_{i=1}^{N}\boldsymbol{\nabla}_i^2 + \sum_{i<j}^{N} V(r_{ij}) \ . \tag{2}$$

In Eq. (1), $E_{\rm r}$ is a constant acting as a reference energy and $\boldsymbol{R} \equiv (\boldsymbol{r}_1, \ldots, \boldsymbol{r}_N)$ is a *walker* in Monte Carlo therminology. DMC solves stochastically the Schrödinger equation (1) replacing $\Psi(\boldsymbol{R},t)$ by $\Phi(\boldsymbol{R},t) = \Psi(\boldsymbol{R},t)\psi(\boldsymbol{R})$, with $\psi(\boldsymbol{R})$ a trial wave function used for importance sampling. When $t \to \infty$ only the lowest energy eigenfunction, not orthogonal to $\psi(\boldsymbol{R})$, survives and then the sampling of the ground state is effectively achieved.

The application of DMC to fermions is more involved due to the well known sign problem caused by the non-positivity of their wave function. The standard approach to deal with the sign problem in DMC is the fixed-node (FN) approximation.[11] According to FN, the wave function $\Phi(\boldsymbol{R},t) = \psi(\boldsymbol{R})\Psi(\boldsymbol{R},t)$ evolves following the imaginary-time Schrödinger equation, with ψ acting as importance sampling function and nodal constraint. For long enough time $t \to \infty$, $\Psi(\boldsymbol{R})$ is the lowest energy state compatible with the nodal surface imposed by $\psi(\boldsymbol{R})$. It has been proved that the nodal constraint forced by the FN method leads to an upper bound to the eigenvalue.

FN is numerically stable and allows for a good estimation of the energy but the method itself does not provide information on the quality of the upper bound, i.e., the difference between the energy obtained and the exact eigenvalue. In order to estimate the bias due to the particular model nodal surface we have used the released node (RN) technique.[12] In the RN method, the walkers are allowed to cross the nodal surface determined by $\psi(\boldsymbol{R})$ for a finite lifetime $t_{\rm r}$ and a sign + or − is attached to each one. To this end, the importance sampling wave function is chosen positive

$$\psi(\boldsymbol{R}) = \left(\psi(\boldsymbol{R})^2 + a^2\right)^{1/2} \ , \tag{3}$$

with a a constant, and the fermionic energy is obtained by projecting on the antisymmetric component $\psi(\boldsymbol{R})$. The method would arrive to the exact ground-state energy for $t_{\rm r} \to \infty$, but this limit is not accesible in liquid ^{3}He due to the rapid emergence of bosonic noise. Nevertheless, the initial slope can be well determined and its value can be used for comparing different nodal surfaces and for an estimate of the magnitude of the bias introduced by the FN approximation.[13]

The trial wave function we have used is of Jastrow-Slater form,

$$\psi(\boldsymbol{R}) = \prod_{i<j}^{N} f(r_{ij})\, \Phi_\uparrow \Phi_\downarrow \ , \tag{4}$$

with a two-body correlation factor,

$$f(r) = \exp\left[-\frac{1}{2}\left(\frac{(\alpha r)^\gamma}{N} + \left(\frac{b}{r}\right)^\nu\right)\right] . \tag{5}$$

In Eq. (4), $\Phi_\uparrow$ ($\Phi_\downarrow$) is a Slater determinant for the spin-up (spin-down) particles. In the simplest model both Slater determinants are filled up with single-particle orbitals corresponding to the polynomial part of the harmonic oscillator basis. Using this basis is advantageous since the resulting Slater determinant is of the Vandermonde type and therefore translationally invariant. With this model, and considering the Jastrow part as a function of relative distances only, spurious energy contributions due to the movement of the center of mass are not present. The orbitals are chosen in its Cartesian coordinate representation, and for incomplete shells we have followed the prescription used in Ref. 7, which warrants invariance under 90^0 rotations with respect to the Cartesian axis.

The trial wave function (4) corresponds to a non-interacting system and therefore it is too simple for describing a correlated liquid like ^{3}He. Using the imaginary-time Schrödinger equation, one can prove that the first correction (corresponding to a short imaginary-time interval) to the non-interacting nodal surface corresponds to a displacement of the coordinates due to correlations with all the other particles.[20] These corrections are known as backflow correlations and are incorporated in ψ by replacing the coordinates $\boldsymbol{r}_i$ in the Slater determinants $\Phi_\uparrow$ and $\Phi_\downarrow$ by

$$\tilde{\boldsymbol{r}}_i = \boldsymbol{r}_i + \lambda_{\rm B} \sum_{j\neq i}^{N} \eta(r_{ij}) \boldsymbol{r}_{ij} \ . \tag{6}$$

Similarly to previous studies of the homogeneous liquid,[13,14] we have used for the function $\eta(r)$ a Gaussian,

$$\eta(r) = \exp\left[-\left(\frac{r - r_{\rm B}}{\omega_{\rm B}}\right)^2\right] , \tag{7}$$

with $\lambda_{\rm B}$, $r_{\rm B}$, and $\omega_{\rm B}$ variational parameters.

The Jastrow-Slater wave function corresponds to a description of normal liquid ^{3}He whereas at zero temperature ^{3}He is superfluid. Bouchaud and Luhillier[15] suggested a new proposal for the trial wave function to incorporate pairing through a BCS-like model. For an unpolarized bulk liquid ($N_\uparrow = N_\downarrow$) the antisymmetrized product of the pair functions can be written as a single determinant,

$$\psi_{\rm P}(\boldsymbol{R}) = \prod_{i<j}^{N} f(r_{ij})\, \Phi(\varphi(r_{ij'})) \tag{8}$$

with $i = 1, \ldots, N_\uparrow$ and $j' = 1, \ldots, N_\downarrow$. Previous applications[15,16] of this model to bulk ^{3}He did not show any improvements with respect to the Jastrow-Slater model. On the other hand, Bajdich *et al.*[17] have recently achieved excellent results in electronic-structure simulations with the BCS wave function. Also, its use has allowed an accurate description of the BEC-BCS crossover in dilute fermionic gases.[18]

The trial wave function (8) has not been used in the description of inhomogeneous Fermi liquids such as ^{3}He drops. We have carried out a first exploration for a drop with $N_\uparrow = N_\downarrow = 20$ using the following polynomial expansion for the pair function

$$\varphi(r_{ij'}) = \sum_{\alpha=0}^{M} c_\alpha H_\alpha(\beta r_{ij'}) \tag{9}$$

with $H_\alpha(\beta r_{ij'})$ Hermite polynomials and β a scaling parameter.

3. Results

We have carried out DMC simulations for ^{3}He drops near the expected limit for self-binding to study the dependence of the energy with the number of constituents N and establish as accurately as possible the smallest self-bound drop. In all the simulations, we have used the HFD-B(HE) Aziz potential,[19] which has proved high accuracy in the microscopic description of the bulk phases of liquid ^{4}He and ^{3}He.[20] The parameters entering into $\psi(\boldsymbol{R})$, Eqs. (5) and (6), are adjusted variationally. The dependence on the number of atoms in the drop is only significant for the

Table 1. DMC total (E) and kinetic (K) energies per particle for ^{3}He drops as a function of atoms N and with different spin configurations for open shells.

N	S_z	E/N (K)	K/N (K)
29	9/2	0.0194(10)	3.395(65)
30	5	-0.0006(11)	3.630(13)
	4	0.0067(12)	3.630(15)
	3	0.0184(12)	3.595(13)
31	9/2	-0.0078(12)	3.682(18)
	3/2	0.0056(12)	3.666(13)
32	4	-0.0258(18)	3.808(15)
	3	-0.0180(12)	3.729(11)
	0	0.0003(10)	3.751(12)
33	7/2	-0.0377(13)	3.822(13)
	5/2	-0.0340(11)	3.836(14)
	1/2	-0.0190(13)	3.857(15)
40	0	-0.1564(17)	4.442(17)

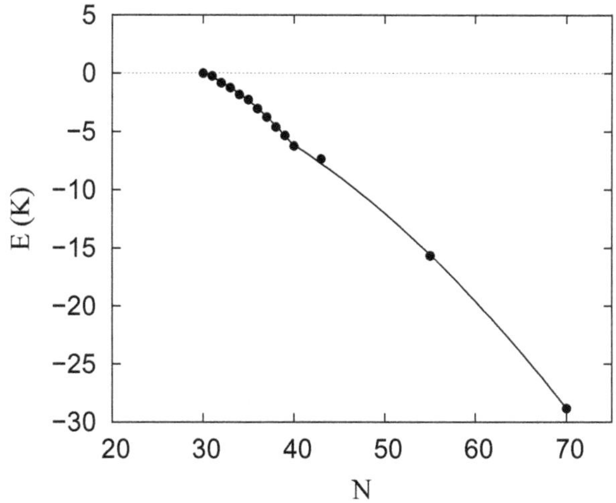

Fig. 1. DMC energies as a function of the number of atoms N. The line is to guide the eye.

parameter α (5): for $N = 30$, $\alpha = 4.0\ \sigma^{-1}$ and for $N = 40$, $\alpha = 4.4\ \sigma^{-1}$, increasing linearly with N ($\sigma = 2.556$ Å). The optimal values of the rest of parameters are: $\gamma = 1$, $\nu = 5$, $b = 1.14\ \sigma$, $\lambda_{\rm B} = 0.34$, $r_{\rm B} = 0.75\ \sigma$, and $\omega_{\rm B} = 0.54\ \sigma$.

FN-DMC results for the total (E/N) and kinetic energy (K/N) per particle as a function of the number of atoms N in the drop are reported in Table 1. According to our results, the threshold limit for a self-bound drop is $N = 30$ and all cases studied with equal N but different spin S_z show a preferred state corresponding to the maximum value of the spin. The latter result can be also interpreted taking into account the magic numbers which close a shell (of the spin-up or spin-down atoms), which in the range studied correspond to values 10, 20, and 35. The results contained in Table 1 show that the lowest energies follow the rule of having at least one closed shell, with preference for the smallest one: 10 for $N = 29$ and 20 for $N > 31$.

The complete set of our results, containing the optimal energies for each N, is shown in Fig. 1. The line on top of the DMC data correspond to polynomial fits and are only intended as to guide the eye. As one can see in the figure, the behavior of the energy with N is not monotonous in the regime studied showing a kink for $N = 40$, a doubly magic number $N_\uparrow = N_\downarrow = 20$. This shell effect is even more stressed by calculating the chemical potential as a function of N,

$$\mu(N) = E(N+1) - E(N) \ . \tag{10}$$

Results for $\mu(N)$ in the vicinity of $N = 40$ are shown in Fig. 2. The kink observed in the energy, which produces a peak in the chemical potential, is a remnant of the shell model chosen to describe the antisymmetry of the system in absence of correlations. Dynamical correlations induced by the interatomic potential smooth significantly this effect but, for these small drops, it is still clearly observable. On

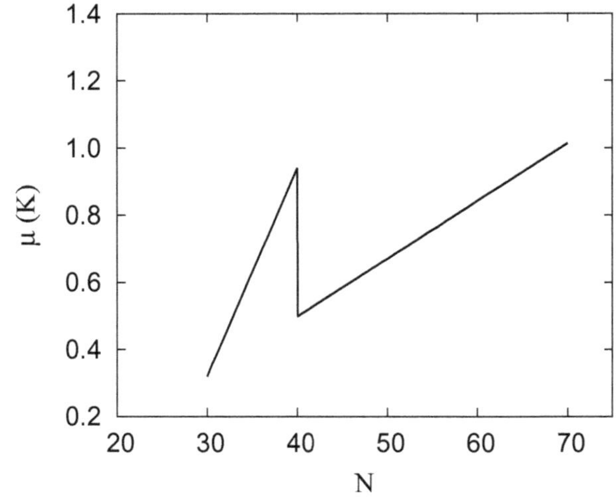

Fig. 2. Chemical potential μ as a function of the number of atoms N.

the other hand, Fig. 1 shows that in the regime $N = 30 - 40$ the DMC data display a regular behavior, which is well reproduced by a second-degree polynomial.

As commented in Sec. 2, we have used the RN technique to get some insight on the quality of the upper bound provided by the FN method. Although the RN method is unstable because the subjacent bosonic component asymptotically overwhelms the Fermi signal, the initial slope of the energy with the released time t_{r} can be determined with enough precision to make a comparison between different model nodal surfaces.[13] This kind of analysis is shown in Fig. 3, where the RN energies

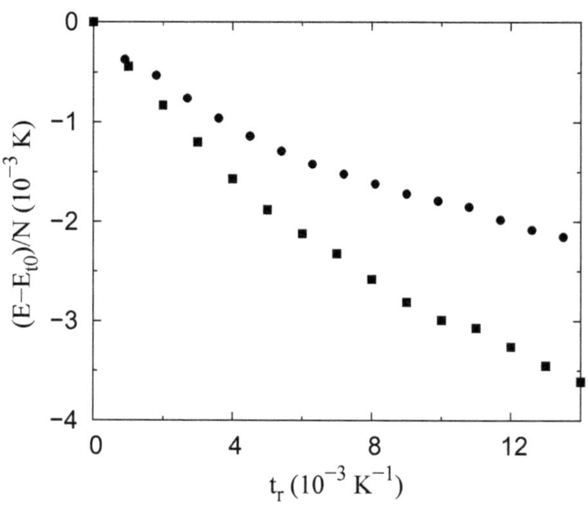

Fig. 3. RN energies as a function of the released time t_r for the $N = 40$ drop. Squares, without backflow; circles, with backflow.

Table 2. Optimal parameters of $\psi_{\rm P}$ (8,9) for a $N = 40$ ^{3}He drop ($\sigma = 2.556$ Å).

c_0	c_1	c_2	c_3	c_4	c_5	β
1.000	−1.567	0.670	−0.897	0.478	−0.076	$0.2279\sigma^{-1}$

calculated with a trial wave function ψ containing or not backflow correlations are plotted as a function of $t_{\rm r}$. Figure 3 shows a reduction of the slope by a factor of two when backflow correlations are present in ψ pointing to a significant improvement of the upper bound (an exact wave function would show zero slope). However, this improvement is worse than the one we observed in the past in a similar calculation of bulk ^{3}He.[13] In fact, also the variance observed in the present simulation of the drops is larger than the one obtained in bulk.

In order to further study the influence of the nodal constraint on the FN upper bounds for the energies we have carried out a preliminary study using a BCS-like wave function (8). The calculations correspond to $N = 40$ and do not incorporate backflow correlations. We have fixed the number of Hermite polynomials (9) to five, which supposes the search for a minimum in a space of six parameters, c_i ($i = 1, \ldots, 5$) and the scaling parameter β ($c_0 = 1$). To carry out this complex optimization, the simplex method has been used. After each iteration of the code a new set of parameters was obtained, and it was subsequently used to equilibrate the system and to accumulate some statistics before a new iteration of simplex was applied. This process is difficult due to the low level of statistical noise required at each iteration. The final parameters introduced in the function for $N = 40$ are reported in Table 2.

We compare the $N = 40$ results obtained with the BCS and JS models in Table 3. As can be observed, the JS energy is better than the BCS one using both the VMC and FN-DMC methods. However, VMC results differ in a 28% while DMC is able to give quite more similar results for both trial wave functions, differing only in a 4%. The difference between both DMC energies is small but we have not been able to go below the JS energy. It is worth noticing that contrarily to the harmonic oscillator basis of the JS model, which has been extensively used, the specific functional form

Table 3. Comparison between VMC and DMC energies for a $N = 40$ ^{3}He drop using the JS and BCS models. The first two rows correspond to VMC results and the second two to DMC.

Aprox.	E/N	V/N	K/N
VMC-J	0.108(3)	−3.964(8)	4.072(8)
VMC-BCS	0.1308(6)	−4.157(2)	4.288(4)
DMC-J	−0.124(4)	−4.18(3)	4.05(4)
DMC-BCS	−0.118(2)	−4.285(16)	4.167(20)

for the pairing functions $\varphi(r)$ (8) is much less known. Therefore it is not excluded that the energy corresponding to the BCS trial wave function can be improved by exploring a bigger functional space.

4. Discussion

To summarize, we have carried out a microscopic calculation of small ^{3}He drops using FN- and RN-DMC that has allowed for a reduction in the threshold limit for self-binding; our results show that this minimum number is $N = 30$. Similarly to previous density-functional[9] and microscopic calculations,[7] the present results confirm that the ground-state energy is achieved for maximum spin or, in other terms, for shell configurations where the number of atoms of one of the two species (up or down) closes a shell. In the N range studied, the energy shows kinks for $N = 30$ and $N = 40$ which correspond to magic numbers of the underlying shell model; its signal is however quite depressed by the relevance of ^{3}He-^{3}He dynamical correlations.

The RN analysis shows that there is still a bias in the calculation, which is rather small after the introduction of backflow correlations, but larger than the one observed in bulk ^{3}He. This conclusion and also the larger variance of the energy with respect to similar calculations for the bulk point to a somehow incomplete treatment of the inhomogeneity inherent to drops. Possible improvements on this line could be obtained by changing the functional dependence of correlations, from $(|\mathbf{r_i}-\mathbf{r_j}|)$ to $(\mathbf{r_i},\mathbf{r_j})$.[21] Notwithstanding, this modification would introduce additional complexity in the calculation and require from new correlation functions which at present are not very well known.

Concerning the BCS-like model used for the first time in the study of ^{3}He drops the results are not very exciting but additional work is required. We have realized that the optimization of the variational parameter set in this problem is rather involved and it is difficult to know if additional work could lead to better energies. Moreover, the search for an improved pairing function should be in a bigger functional space since we have manifestly not covered all possibilities.

Acknowledgments

We acknowledge partial support from DGI (Spain) Grant No. FIS2005-04181 and Generalitat de Catalunya Grant No. 2005SGR-00779.

References

1. See special number of *J. Chem. Phys.* **115**, 10065 (2001).
2. M. Barranco, R. Guardiola, S. Hernández, R. Mayol, J. Navarro, and M. Pi, *J. Low Temp. Phys.* **142**, 1 (2006).
3. S. Grebenev, J. P. Toennies, and A. F. Vilesov, *Science* **279**, 2083 (1998).
4. V. R. Pandharipande, S. C. Pieper, and R. B. Wiringa, *Phys. Rev. B* **34**, 4571 (1986).
5. J. Harms, J. P. Toennies, M. Barranco, and M. Pi, *Phys. Rev. B* **63**, 184513 (2001).

6. S. Stringari and J. Treiner, *J. Chem. Phys.* **87**, 5021 (1987).
7. R. Guardiola and J. Navarro, *Phys. Rev. Lett.* **84**, 1144 (2000).
8. R. Guardiola and J. Navarro, *Phys. Rev. A* **71**, 035201 (2005).
9. M. Barranco, J. Navarro, and A. Poves, *Phys. Rev. Lett.* **78**, 4729 (1997).
10. E. Sola, J. Boronat, and J. Casulleras, *Phys. Rev. B* **73**, 092515 (2006).
11. J. B. Anderson, *J. Chem. Phys.* **63**, 1499 (1975); **65**, 4121 (1976).
12. D. M. Ceperley and B. J. Alder, *Phys. Rev. Lett.* **45**, 566 (1980).
13. J. Casulleras and J. Boronat, *Phys. Rev. Lett.* **84**, 3121 (2000).
14. V. Grau, J. Boronat, and J. Casulleras, *Phys. Rev. Lett.* **89**, 045301 (2002).
15. J. P. Bouchaud and C. Luhillier, *Europhys. Lett.* **3**, 1273 (1987); *Z. Phys. B* **75**, 283 (1989).
16. S. Moroni, S. Fantoni, and A. Fabrocini, *Phys. Rev. B* **58**, 11607 (1998).
17. M. Badjich, L. Mitas, G. Drobný, L. K. Wagner, and K. E. Schmidt, *Phys. Rev. Lett.* **96**, 130201 (2006).
18. G. E. Astrakharchik, J. Boronat, J. Casulleras, and S. Giorgini, *Phys. Rev. Lett.* **93**, 200404 (2004).
19. R. A. Aziz, F. R. W. McCourt, and C. C. K. Wong, *Mol. Phys.* **61**, 1487 (1987).
20. J. Boronat in *Microscopic Approaches to Quantum Liquids in Confined Geometries*, edited by E. Krotscheck and J. Navarro (World Scientific, Singapore, 2002).
21. E. Krotscheck in *Microscopic Approaches to Quantum Liquids in Confined Geometries*, edited by E. Krotscheck and J. Navarro (World Scientific, Singapore, 2002).

GROUND-STATE ENERGY AND COMPRESSIBILITY OF A DISORDERED TWO-DIMENSIONAL ELECTRON GAS

B. TANATAR* and A. L. SUBAŞI

Department of Physics, Bilkent University,
Bilkent, Ankara, 06800, Turkey
**E-mail: tanatar@fen.bilkent.edu.tr*

K. ESFARJANI

Department of Physics, Sharif University of Technology,
Tehran, Iran
Department of Physics, University of California, Santa Cruz, CA 95064

S. M. FAZELI

Department of Physics, Sharif University of Technology,
Tehran, Iran

Two-dimensional (2D) electron systems in the presence of disorder are of interest in connection with the observed metal-insulator transition in such systems. We use density functional theory in its local-spin density approximation (LSDA) to calculate the ground-state energy of a 2D electron system in the presence of remote charged impurities which up on averaging provides disorder. The inverse compressibility calculated from the ground-state energy exhibits a minimum at a critical density controlled by the disorder strength. Our findings are in agreement with experimental results.

Keywords: 2D electron gas; disorder effects; compressibility.

1. Introduction

There has been an immense amount of experimental and theoretical activity in recent years to understand the ground state properties of two-dimensional (2D) electron systems.[1,2] Advances in fabrication techniques have made it possible to probe various quantities of interest in high quality and very low density samples. Most notably, the observation of a metal-insulator transition[1,2] in these systems provides a major motivation to study the various physical properties. Most experiments perform transport measurements obtaining resistivity or conductivity as a function of temperature at varying electron density to deduce the metallic or insulating phases.[3–6] In contrast, Ilani *et al.*[7] used the capacitance technique, a thermodynamic measurement, to measure the compressibility finding that it becomes spatially inhomogeneous as the density is lowered. Similar observations were also made by Dultz and Jiang[8] on a 2D hole system who noted that the inverse compressibility is minimum at the same density where the metal-insulator transition occurs. Recent compressibility and resistivity experiments by Allison *et al.*[9] on the same samples indicate that inverse compressibility minimum is not related to metal-insulator transition.

Most of our understanding of metal-insulator transition in 2D electron systems is based on various transport measurements. For instance, experiments at in-plane magnetic field concentrated on the spin susceptibility, Lande g-factor, and effective mass of the 2D electron systems present in Si-MOSFETS and GaAs quantum-well structures.[3–6,10,11] In particular, Shashkin *et al.*[12,13] reported a sharp increase of the effective mass near the critical density at which the system starts to show deviations from the metallic behavior. On the other hand, Pudalov *et al.*[3] have found only moderate enhancement of the spin susceptibility and effective mass in their samples. Thermodynamic measurements of magnetization of a dilute 2D electron system were reported by Prus *et al.*[10] and Shashkin *et al.*[11] Both experiments found large enhancement of the spin susceptibility χ_s over its Pauli value. Whereas the measurements of Prus *et al.*[10] found no indication toward a ferromagnetic instability, Shashkin *et al.*[11] observed diverging behavior in χ_s at a critical density coinciding with the metal-insulator transition density obtained from transport measurements.

As indicated above, there are very few thermodynamic measurements at densities around the metal-insulator transition. As recent theoretical work by Punnoose and Finkelstein[14] shows the observed behavior in 2D systems suggests an interaction driven mechanism with disorder playing an important role. On the theoretical side, calculations of compressibility for a 2D system of electrons in the presence of disorder predict the observed behavior of upturn and divergence.[15,16] Shi and Xie[17] performed density functional calculations based on the (unpolarized) local density approximation developed by Tanatar and Ceperley[18] within the Thomas-Fermi (TF) theory, and found similar results for the compressibility. They also identified the MIT point with the percolation transition point in this system. This was further developed by Das Sarma *et al.*[19] who measured the critical exponent for the conductivity and found it in agreement with that proposed by percolation theory. On the other hand, Fogler[20] argued against percolation transition by calculating the critical density to be much smaller.

In this work, we investigate the spatial distribution of carrier density and the compressibility of a 2D electron system using the local spin-density approximation (LSDA) both at the TF and Kohn-Sham (KS) levels. The correlation energy functional we use is the one constructed by Attaccalite *et al.*[21] from the very recent quantum Monte Carlo (QMC) calculations for correlation energy appropriate for uniform systems. An important feature of these simulations is that a transition to a ferromagnetic phase at low densities is built in the functional. A disorder potential due to remote impurities is included to make the calculation realistic. The density distribution of the system shows high and low density regions as reported earlier by Shi and Xie.[17] Our density functional theory calculations clearly show that the observed minimum behavior in inverse compressibility is the result of electron-impurity interaction. The critical density at which the minimum occurs is controlled by the disorder strength.

In the rest of this paper, we first outline the density functional theory as applied to 2D electron systems with impurity interaction, and introduce our computational method. We then present our results for the ground-state energy and compressibility for 2D electrons in the presence of disorder and discuss our findings in the light of recent experiments.

2. Theory

We consider a 2D electron system interacting via the long range Coulomb interaction whose Fourier transform is $V_q = 2\pi e^2/(\epsilon q)$ in which ϵ is the background dielectric constant. The system is characterized by the dimensionless interaction strength $r_s = 1/(\pi n a_B^{*2})^{1/2}$, where n is the 2D electron density and $a_B^* = \hbar^2\epsilon/(m^* e^2)$ is the effective Bohr radius defined in terms of the band mass m^* of electrons in the semiconductor structure and dielectric constant ϵ of the material.

Within the spin-density functional theory the total energy of an N-electron interacting system in a local external potential $V_{\rm ext}(\mathbf{r})$ is a unique functional of spin densities $n_\uparrow(\mathbf{r})$ and $n_\downarrow(\mathbf{r})$. The total energy functional can be expressed as

$$\begin{aligned} E_{\rm Total}[n_\uparrow, n_\downarrow] &= E_T[n_\uparrow, n_\downarrow] + E_H[n_\uparrow, n_\downarrow] \\ &+ E_x[n_\uparrow, n_\downarrow] + E_c[n_\uparrow, n_\downarrow] + E_{\rm ext}[n_\uparrow, n_\downarrow] \end{aligned} \tag{1}$$

We approximate the kinetic energy functional by the Thomas-Fermi-Weizäcker (TFW) form given by

$$E_T[n_\uparrow, n_\downarrow] = \sum_\sigma \int d\mathbf{r} \left[\pi n_\sigma^2(\mathbf{r}) + A \frac{|\nabla n_\sigma(\mathbf{r})|^2}{n_\sigma(\mathbf{r})} \right] \tag{2}$$

with $A = 1/8$.[22] The direct Coulomb energy is given by

$$E_H[n_\uparrow, n_\downarrow] = \frac{e^2}{2\epsilon} \int d^2\mathbf{r}\, d^2\mathbf{r}'\, \frac{n(\mathbf{r})\, n(\mathbf{r}')}{|\mathbf{r} - \mathbf{r}'|} \tag{3}$$

where $n(\mathbf{r}) = n_\uparrow(\mathbf{r}) + n_\downarrow(\mathbf{r})$ is the total density.

The exchange and correlation energy functionals within the local spin-density approximation are written as

$$E_{x,c}[n_\uparrow, n_\downarrow] = \int d^2\mathbf{r}\, \epsilon_{x,c}(n_\uparrow(\mathbf{r}), n_\downarrow(\mathbf{r}))\, [n_\uparrow(\mathbf{r}) + n_\downarrow(\mathbf{r})]\,. \tag{4}$$

Here ϵ_x is the exchange energy of the 2D homogeneous electron electron gas. For the correlation energy ϵ_c we use the recent QMC result of Attaccalite *et al.*[21] again for the 2D homogeneous electron gas which has been parametrized as a function of density and spin polarization.

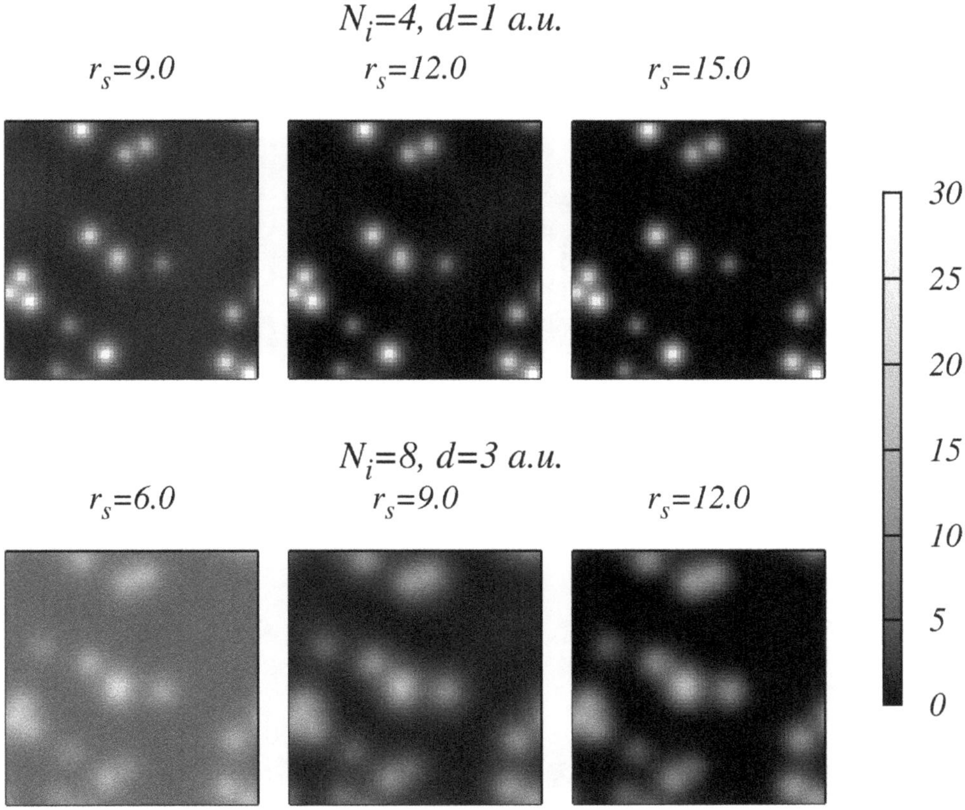

Fig. 1. Density distributions of a 2D electron system in a model disorder potential for several values of average electron density characterized by r_s. The numbers on the right scale are in units of 10^{-3} a.u. We choose the impurity parameters as upper panel: $N_i = 4$ and $d = 1$ a.u., lower panel: $N_i = 8$ and $d = 3$ a.u. Ten realizations of the impurity potential are averaged. (colour online)

The disorder studied in this work comes from a random distribution of charged impurities (with charge Ze) at a setback distance d from the electron layer. The energy functional due to the external potential is

$$E_{\mathrm{ext}}[n] = \int d^2\mathrm{r}\, V_{\mathrm{ext}}(\mathbf{r})\, [n(\mathbf{r})] \tag{5}$$

where the external potential, due to remote impurities, located at $\mathbf{r_i}$ at a distance d from the plane of 2D electrons is given by

$$V_{\mathrm{ext}}(\mathbf{r}) = -\sum_i \frac{Ze^2/\epsilon}{[(\mathbf{r}-\mathbf{r_i})^2 + d^2]^{1/2}}\,. \tag{6}$$

3. Method

The spin-densities $n_\uparrow(\mathbf{r})$ and $n_\downarrow(\mathbf{r})$ that extremize the total energy functional can be obtained by annealing from a Monte Carlo (MC) simulation. A sufficiently high

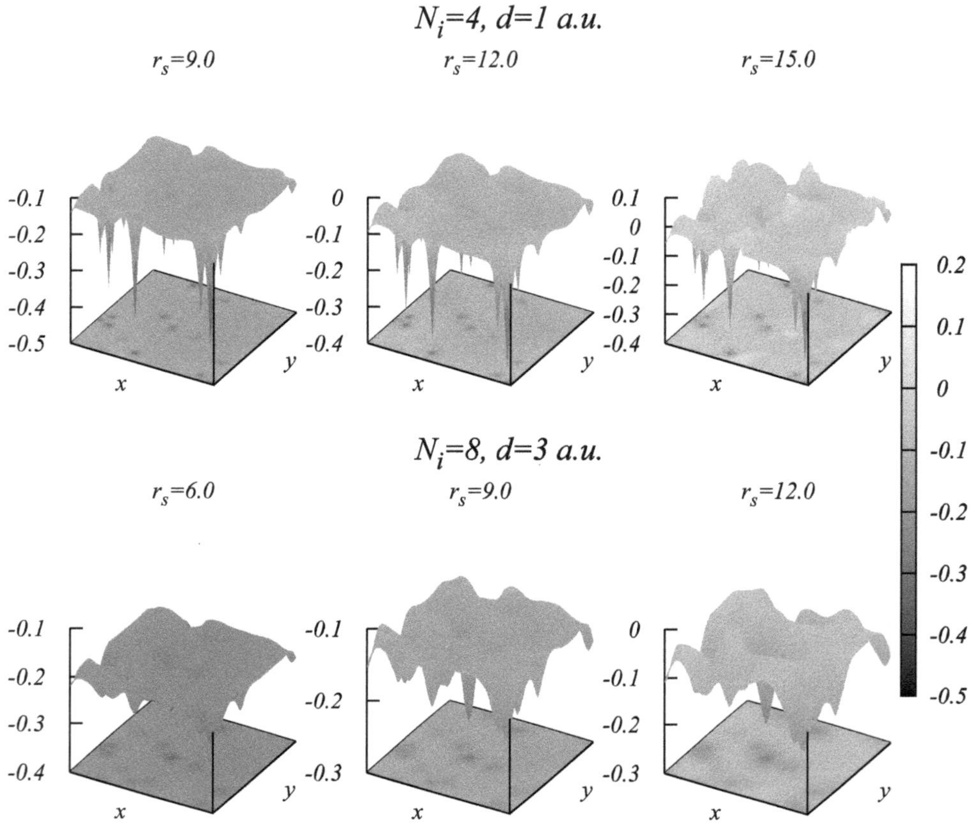

Fig. 2. The effective potential at different disorder strengths and average electron densities characterized b y r_s. We choose the impurity parameters as upper panel: $N_i = 4$ and $d = 1$ a.u., lower panel: $N_i = 8$ and $d = 3$ a.u. Ten realizations of the impurity potential are averaged.

temperature is first chosen and a Metropolis Monte Carlo run is performed long enough to reach thermodynamic equilibrium. Then the temperature is reduced and the run is repeated. This is continued until the final temperature is sufficiently low so that very little energy fluctuations occur during the last run. This simulation is done in order to reach the global minimum of the energy landscape. Once one is near the bottom of the valley, eventually a steepest-descent algorithm is applied to reach the minimum energy structure faster. The areal integral of the energy functional is approximated by a discrete sum: basically the density and potentials are discretized on a 64 by 64 mesh whose size is one effective Bohr radius being equal to 100 Å in GaAs samples. The long-range Coulomb potential is calculated using the Ewald sum method. In minimizing the total energy functional, if needed, we have considered several different realizations for the impurity atoms and have taken the statistical average.

To make sure of the validity of our analysis, the results from the TFW kinetic energy functional were also compared to the solution of the Kohn-Sham equations

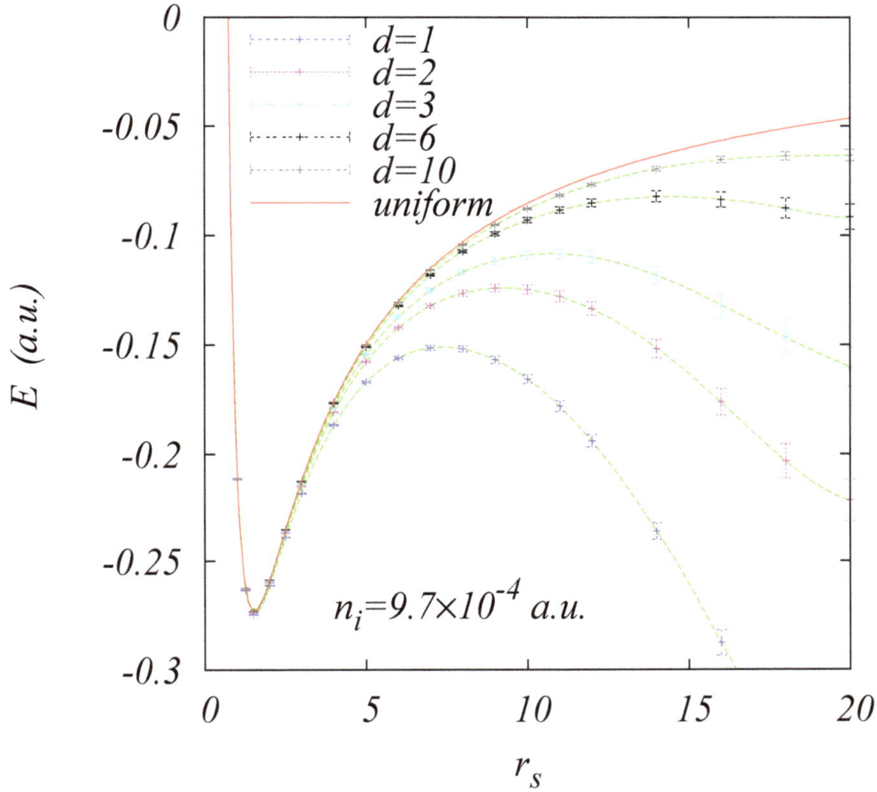

Fig. 3. Ground state energy (per particle) of a 2D electron system in the presence of charged impurities of concentration $n_i = 0.97 \times 10^{-3}$ a.u. Distance d of the impurities are given in the legend. Homogeneous system result (solid line) is the QMC calculation of Attaccalite *et al.* (Ref. 21)

where the kinetic energy is calculated exactly. Densities in the high effective potential regions were found to agree to within a few percent. Low potential regions showed good agreement between the densities as well as the effective exchange-correlation potential. The Kohn-Sham equations follow from the minimization of the total energy functional under the constraint of fixed total number of electrons. The effective single-particle Schrodinger equation is

$$[-\nabla^2 + V_{\text{eff}}(n, \mathbf{r})]\psi_{is}(\text{r}) = \epsilon_{is}\psi_{is}(\mathbf{r}) \tag{7}$$

where

$$V_{\text{eff}}(n, \mathbf{r}) = V_d(\mathbf{r}) + \int d\mathbf{r}' \, \frac{e^2}{|\mathbf{r} - \mathbf{r}'|} + \frac{\delta}{\delta n}[\epsilon_x(n)n + \epsilon_c(n)n] \tag{8}$$

is the effective potential self-consistently determined by the density distribution $n(\mathbf{r}) = \sum_{is} |\psi_{is}(\mathbf{r})|^2$ by summing over the occupied energy levels and the spin index s.

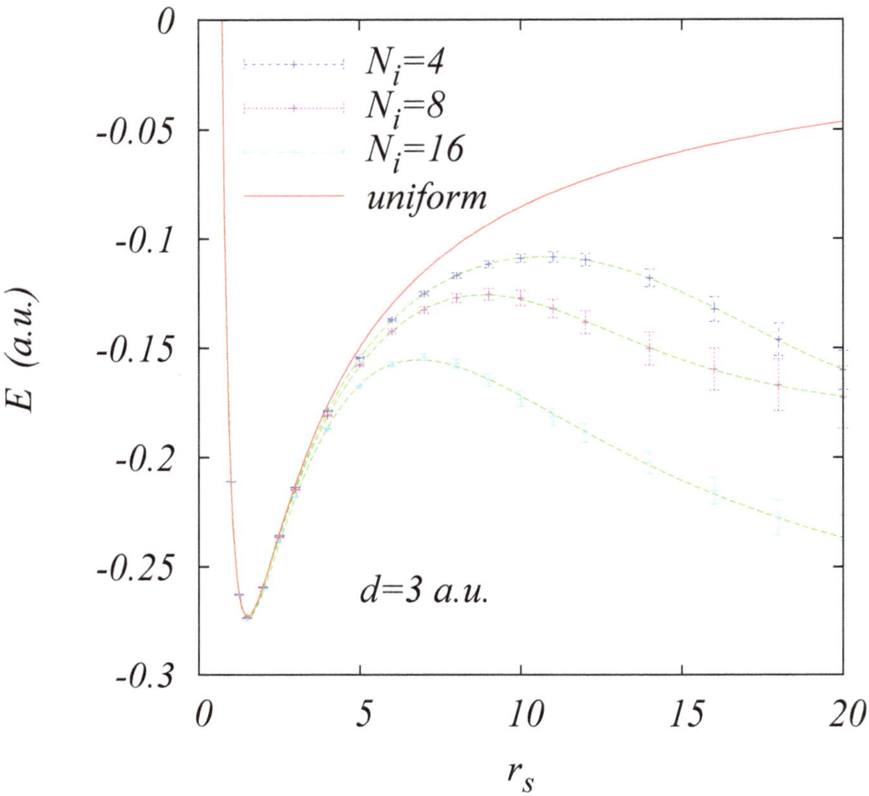

Fig. 4. Ground state energy (per particle) of a 2D electron system in the presence of charged impurities at a setback distance $d = 3\,\text{a.u.}$ Number of impurities are given in the legend. Homogeneous system result (solid line) is the QMC calculation of Attaccalite *et al.* (Ref. 21)

We also remark that a similar Hartree-Fock calculation was recently performed on a Mott insulator in the half-filled limit[23] where the ground state is antiferromagnetic. We believe that the exchange-correlation potential we are using is more realistic, and contains the correct physics, namely that at very low densities, before the Wigner crystallization, the system becomes ferromagnetic.

4. Results and Discussion

We have calculated the ground-state energy and corresponding density distribution of a 2D electron system for various impurity potentials characterized by n_i and d. We use typically 10 different realizations for the impurity potential to calculate the average value of the physical quantities. Our calculations are carried out in a 64×64 discrete space. The size of the system is set as $L = 64\,a_B^*$. The average electron density is adjusted by changing the electron number N. We use atomic units (a. u.) to measure energy and length, i.e. $a_B^* = 1$.

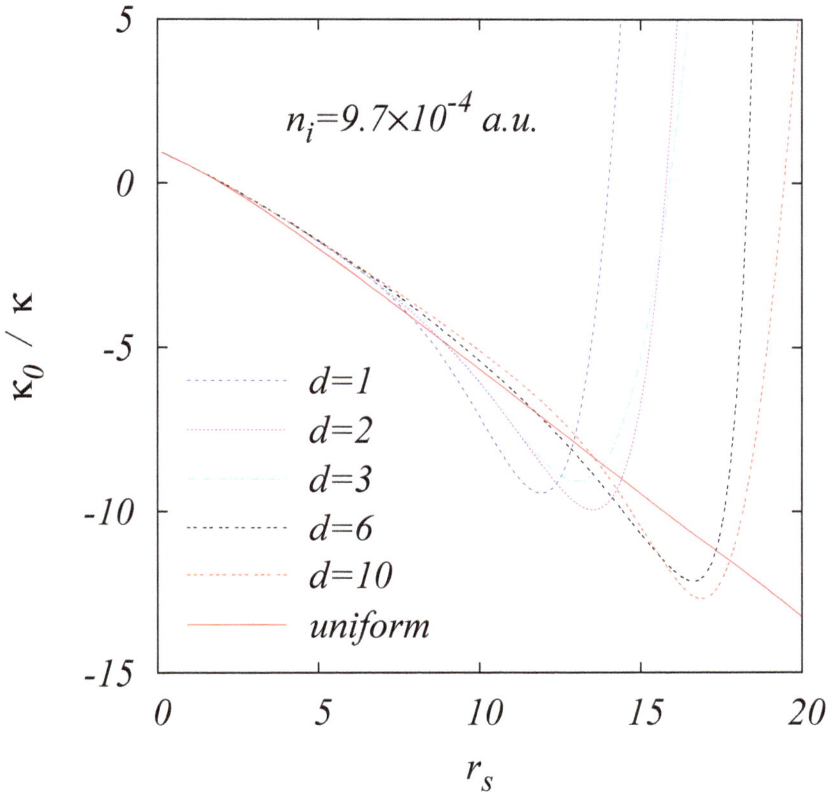

Fig. 5. Inverse compressibility scaled by its noninteracting value as a function of r_s for a 2D electron system in the presence of charged impurities of concentration $n_i = 0.97 \times 10^{-3}$ a.u. Distance d of the impurities are given in the legend. Homogeneous system result (solid line) is the QMC calculation of Attaccalite *et al.* (Ref. 21)

Illustrative examples of density distributions resulting from the minimization of total energy are shown in Fig. 1 for two sets of (n_i, d) values at various average electron densities described by r_s. We observe that electrons form high density and low density regions, an inhomogeneous distribution. Depending on the average density of the system, the high density regions may be connected to each other or form isolated patches. Approaching from the low average density side (large r_s), at some critical density the high density regions start to percolate through the system and form a conducting channel. This may be thought of a demonstration of metal-insulator transition in a 2D electron system as a percolation transition in electron density. Experiments on conductivity[19] and resistance and $1/f$ resistance noise[24] have been analyzed in terms of a percolation scenario. The calculations of Fogler[20] and analysis of Allison *et al.*[9] data, on the other hand, suggest otherwise. We have not performed any systematical analysis of our calculated density distributions in view of percolation transition, thus we are not able to infer any conclusions.

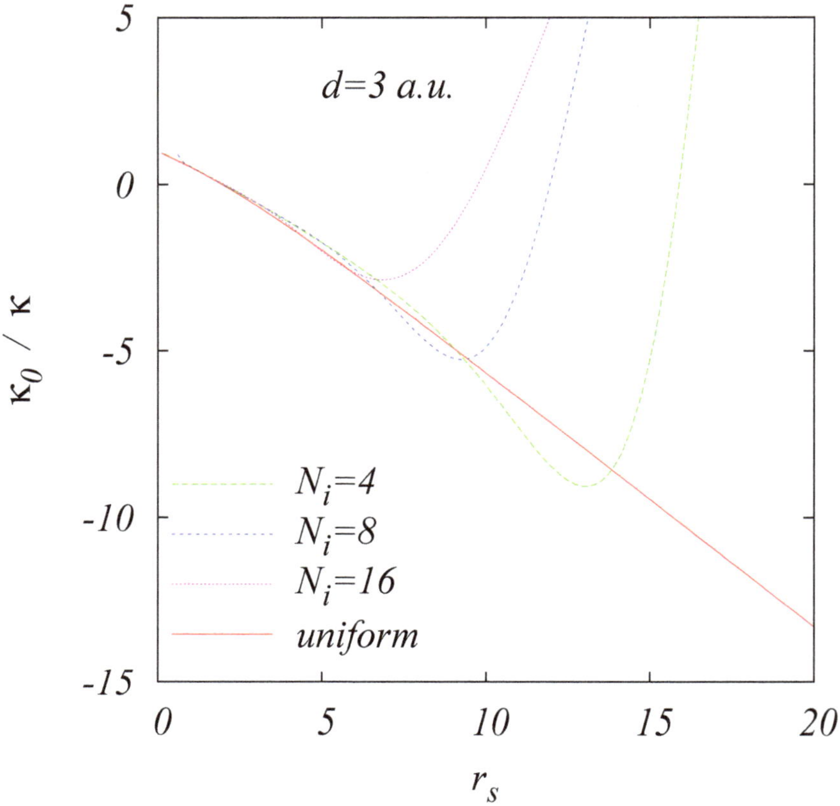

Fig. 6. Inverse compressibility scaled by its noninteracting value as a function of r_rs for a 2D electron system in the presence of charged impurities at a setback distance $d = 3$ a.u. Number of impurities are given in the legend. Homogeneous system result (solid line) is the QMC calculation of Attaccalite *et al.* (Ref. 21)

In Fig. 2 we display the effective potential [c.f. Eq. (8)] including the Hartree, exchange-correlation, and disorder contributions. As is the density distributions we average over 10 realizations of the impurity potential for each r_s value. It is seen that V_{eff} shows the largest fluctuations when the density distribution is highest.

In Fig. 3 we show the ground state energy (per particle) as a function of r_s for different disorder potentials. Here the impurity concentration is kept fixed at $n_i = 0.97 \times 10^{-3}/a_B^*$ and the setback distance d from the electron layer is varied. For large values of d the total energy approaches to that of the homogeneous system.[21] This demonstrates the reliability of our numerical approach. We also note that deviations from the clean system result manifest themselves at larger r_s values indicating the importance of interplay between interaction and disorder effects. That is, for the same strength of impurity potential the ground-state energy is more affected at a larger r_s value. In Fig. 4 the setback distance is fixed at $d = 3$ a.u.

and the number of charged impurities is varied. Similar qualitative effects are also observed on the ground-state energy as the disorder strength is increased.

Having calculated the ground-state energy for a 2D electron system in the presence of disorder, we now turn our attention to the isothermal compressibility $1/\kappa = (N^2/A)\partial\mu/\partial N$ where $\mu = \partial E/\partial N$ is the chemical potential. In terms of the noninteracting system compressibility κ_0 we obtain

$$\frac{\kappa_0}{\kappa} = -\frac{r_s^3}{4}\left[\frac{\partial E}{\partial r_s} - r_s\frac{\partial^2 E}{\partial r_s^2}\right] \tag{9}$$

where the energy E is in a.u. Using the ground-state energy curves for a 2D disordered electron system displayed in Figs. 3 and 4, we calculate the inverse compressibility. Figures 5 and 6 show the inverse compressibility as a function of r_s for different disorder strengths. We observe that when impurity scattering is present in the system, κ_0/κ exhibits a minimum at some critical r_s value, makes an upturn, becomes positive, and finally diverges at a larger r_s. This qualitative behavior seems to be quite general, in the sense that even a very weak disorder leads to it. In a fully interacting system without disorder, on the other hand, κ_0/κ is a monotone decreasing function of r_s. The minimum of κ_0/κ is controlled by the strength of electron-impurity interactions and through averaging by the disorder strength.

Our results for the inverse compressibility of a 2D electron system in the presence of disorder are in agreement with experimental findings.[7–9] In particular, the recent experiments of Allison *et al.*[9] performed on a variety of electron and hole systems have shown the existence of a minimum in κ_0/κ occurring quite generally. These experiments have also measured resistivity on the same samples and showed that the metal-insulator transition (determined from $d\rho/dT = 0$) is not related to the minimum behavior of κ_0/κ. Within our ground-state energy based calculations we cannot locate the metal-insulator transition density (or corresponding r_s). We expect the analysis of the solution of Kohn-Sham equations, for instance the statistical properties of single-particle energy levels may shed some light on this problem which we hope to study in future.

Acknowledgments

This work is partially supported by TUBITAK and TUBA. K. E. would like to thank TUBA for support and the hospitality of Bilkent University where part of this work was performed. We thank Dr. M. M. Fogler for discussions. We also dedicate this paper to Professor M. Tomak on the occasion of his sixtieth birthday.

References

1. E. Abrahams, S.V. Kravchenko, and M.P. Sarachik, *Rev. Mod. Phys.* **73**, 251 (2001).
2. S.V. Kravchenko and M.P. Sarachik, *Rep. Prog. Phys.* **67**, 1 (2004).
3. V. M. Pudalov, M. E. Gershenson, H. Kojima, N. Butch, E. M. Dizhur, G. Brunthaler, A. Prinz, and G. Bauer, *Phys. Rev. Lett.* **88**, 196404 (2002).
4. E. Tutuc, S. Melinte, and M. Shayegan, *Phys. Rev. Lett.* **88**, 036805 (2002).

5. J. Zhu, H. L. Stormer, L. N. Pfeiffer, K. W. Baldwin, and K. W. West, *Phys. Rev. Lett.* **90**, 056805 (2003).
6. W. Pan, D. C. Tsui, and B. L. Draper, *Phys. Rev. B* **59**, 10 208 (1999).
7. S. Ilani, A. Yacoby, D. Mahalu, and H. Shtrikman, *Phys. Rev. Lett.* **84**, 3133 (2000).
8. C. Dultz and H. W. Jiang, *Phys. Rev. Lett.* **84**, 4689 (2000).
9. G. Allison, E.A. Galaktionov, A.K. Savchenko, S.S. Safonov, M.M. Fogler, M.Y. Simmons, and D.A. Ritchie, *Phys. Rev. Lett.* **96**, 216407 (2006).
10. O. Prus, Y. Yaish, M. Reznikov, U. Sivan, and V. Pudalov, *Phys. Rev. B* **67**, 205407 (2003).
11. A. A. Shashkin, S. Anissimova, M. R. Sakr, S. V. Kravchenko, V. T. Dolgopolov, and T. M. Klapwijk, *Phys. Rev. Lett.* **96**, 036403 (2006).
12. A.A. Shashkin, S. V. Kravchenko, V. T. Dolgopolov, and T. M. Klapwijk, *Phys. Rev. B* **66**, 073303 (2002).
13. A. A. Shashkin, M. Rahimi, S. Anissimova, S. V. Kravchenko, V.T. Dolgopolov, and T. M. Klapwijk, *Phys. Rev. Lett.* **91**, 046403 (2003); A. A. Shashkin, S. V. Kravchenko, V. T. Dolgopolov, and T. M. Klapwijk, *J. Phys. A: Math. Gen.* **36**, 9237 (2003).
14. A. Punnoose and A.M. Finkelstein, *Science* **310**, 289 (2005).
15. Q. M. Si and C. M. Varma, *Phys. Rev. Lett.* **81**, 4951 (1998).
16. R. Asgari and B. Tanatar, *Phys. Rev B* **65**, 085311 (2002).
17. J. Shi and X.C. Xie, *Phys. Rev. Lett.* **88**, 086401 (2002).
18. B. Tanatar and D. M. Ceperley, *Phys. Rev. B* **39**, 5005 (1989).
19. S. Das Sarma, M.P. Lilly, E.H. Hwang, L.N. Pfeiffer, K.W. West, and J.L. Reno, *Phys. Rev. Lett.* **94**, 136401 (2005).
20. M.M. Fogler, *Phys. Rev. B* **69**, 121409 (2004).
21. C. Attaccalite, S. Moroni, P. Gori-Giorgi, and G. B. Bachelet, *Phys. Rev. Lett.* **88**, 256601 (2002).
22. Y. Wang, J. Wang, H. Guo, and E. Zaremba, *Phys. Rev. B* **52**, 2738 (1995).
23. D. Heidarian and N. Trivedi, *Phys. Rev. Lett.* **93**, 126401 (2004).
24. G. Deville, B. Leturcq, D. L'Hote, R. Tourbot, C. J. Mellor, and M. Henini, *Physica E* **34**, 252 (2006).

QUASIEXCITONS IN PHOTOLUMINESCENCE OF INCOMPRESSIBLE QUANTUM LIQUIDS

A. WÓJS* and A. GŁADYSIEWICZ

Institute of Physics, Wroclaw University of Technology,
Wroclaw 50-370, Poland
**E-mail: arkadiusz.wojs@pwr.wroc.pl*

J. J. QUINN

Department of Physics, University of Tennessee,
Knoxville, Tennessee 37996, USA

The problem of a trion (charged exciton) immersed in an incompressible quantum liquid formed by two-dimensional electrons in a high magnetic field is studied theoretically by combining the composite fermion model and exact numerical diagonalization. Laughlin correlation of the trion with the surrounding electrons is shown to lead to the formation of a family of novel many-body excitations called "quasiexcitons" (QXs). The residual charge of the most stable QX can be either zero or the fractional quantum $\pm e/3$, depending on the presence of negative or positive Laughlin quasiparticles in the electron liquid (controlled by the Landau level filling factor ν), but also on the width od the quantum well, Zeeman spin splitting, etc. The prediction of the magnetic-field driven transition from negative to positive QXs is argued to explain the recently observed minute discontinuity in the photoluminescence spectra of sufficiently wide ($w \geq 20$ nm) asymmetric GaAs quantum wells at $\nu = 1/3$. Conversely, the observed discontinuity is hence indirectly linked to the emergence of a fractional charge quantum in the electron liquid.

Keywords: Trion; Laughlin liquid; Photoluminescence; Fractional quantum Hall effect.

1. Introduction

High magnetic field B applied perpendicularly to a two-dimensional electron gas (2DEG) leads to the quantization of single-particle energy spectrum into the discrete and macroscopically degenerate Landau levels (LLs). At sufficiently large B, corresponding to the cyclotron energy $\hbar\omega_c \propto B$ being much higher than the Coulomb energy scale $e^2/\lambda \propto B^{1/2}$ (where $\lambda = \sqrt{\hbar c/eB}$ is the magnetic length), the low-energy many-body dynamics is completely determined by the electron–electron interaction within an isolated, partially filled LL. Thus, the 2DEG at high B is a unique setting for studying the nonperturbative many-body interaction phenomena, such as the fractional quantum Hall effect,[1] the emergence of incompressible quantum liquids,[2] or the formation of topological "skyrmion" excitations.[3]

Photoluminescence (PL) spectroscopy has been used extensively to study the interacting 2D carriers for at least two decades.[4–8] In a PL experiment, additional electron–hole (e–h) pairs are introduced into the system through the absorption of photons, and the response of the surrounding carriers makes the recombination spectrum sensitive to the many-electron dynamics. When electrons fill only a small fraction ν of the lowest LL, the PL spectrum is determined by the recombination of bound three-body complexes called trions or charged excitons.[9,10] A (negative) trion forms from an exciton ($\mathrm{X} = e + h$) by capturing another electron ($\mathrm{X}^- = \mathrm{X} + e$). It is analogous to a Hydrogen ion H^-, although with quite different energy and length scales, making the X^- much more susceptible to the external perturbations. The trion energy spectrum depends on B and also on the composition, width w, and doping of the quantum well.[11] In general, several bound trion states occur, distinguished by the pair-electron spin (singlet vs. triplet), total angular momentum, and optical lifetime (dark vs. bright). The binding energies are typically 1–2 meV.

The situation is more complicated at higher LL fillings $\nu = 2\pi n\lambda^2$ (where n is the 2D electron concentration). In this situation, the trions interact with the surrounding electrons and can no longer be regarded as isolated, well-defined three-body quasiparticles. In fact, it is precisely the coupling of trions with the 2DEG which allows (in principle) the many-electron dynamics to be indirectly probed by the trion recombination in a PL experiment.

The history of experiments showing features in the PL spectrum coincident with electron incompressibility goes back a long time.[4,5,7,8,10] However, no earlier realistic theory has (in our opinion) explained what happens to a trion when it is immersed into a 2DEG. Several ideas were proposed, including decoupling of neutral excitons[12,13] or formation of "anyon excitons".[14,15] Yet the simple models are not applicable to the experimental situation in which the anomalies in PL are observed, and not even qualitative agreement has been found.

In this note, we use a composite fermion (CF) model and the idea of magnetic flux attachment to explain how a trion correlates with the 2DEG to become a new, fractionally charged excitation called a quasiexciton (QX). In short, the QX is another kind of a CF, only made of a trion instead of an electron. The emergence of QXs under reasonable experimental conditions is predicted by the realistic numerical calculations carried out for doped GaAs 10–20 nm wells, with electron concentration $n = 2 \cdot 10^{11}$ cm^{-2} for which the Laughlin $\nu = 1/3$ state[2] occurs at a high yet accessible $B = 25$ T. The family of neutral, positive, and negative QX states is identified in the $Ne + h$ energy spectra ($N \leq 10$) in Haldane's spherical geometry. The QX binding energies and recombination rates depend on the involved trion state. As a result, the PL spectrum at $\nu \approx 1/3$ depends sensitively on the well width w. We predict no anomaly for $w = 10$ nm, a small (0.25–0.5 meV) splitting or discontinuity for $w = 20$ nm, and a different, larger splitting for $w \geq 40$ nm.

2. Numerics on a Haldane Sphere

In the numerical calculations we use spherical geometry,[16] with the magnetic field normal to the surface (i.e., radial) produced by a Dirac monopole of strength $2Q$

defined in the units of flux quantum, $2Q\,(hc/e) = 4\pi R^2 B$. The radius of the sphere R and the magnetic length λ are connected by $Q\lambda^2 = R^2$, which also relates $2Q$ with the surface curvature (in the units of λ^{-1}). The LLs on a sphere have the form of degenerate angular momentum shells with $l \geq Q$. To make calculations realistic, we take advantage of a 2D rotational symmetry of spheres with finite LL degeneracy, but use cyclotron energies and interaction matrix elements of an actual experimental system. The finite-size and surface-curvature errors are minimized by extrapolation of the results to $\lambda/R \to 0$. More details were given elsewhere.[11,17]

3. Isolated Trion

The X^- binding energies Δ in GaAs wells were obtained[17] from diagonalization of the $e + h$ and $2e + h$ hamiltonians at $2Q \leq 25$. For $e + h$, the lowest-energy band is the exciton energy dispersion, energy E_X as a function of wave vector k. In closed geometry, k must be quantized. On a sphere, it is represented by the total angular momentum related to k through $L = kR$. In particular, the $k = 0$ excitonic ground state appears at $L = 0$. For $2e + h$, the spectrum splits into the bound and unbound states. The bound states are those with the positive binding energy, $\Delta = E_X - E_{X^-}$ (where the electron energy at the lowest LL is taken as zero). Due to nonzero electric charge, bound trion states in a magnetic field are labeled by the relative angular momentum M rather than by a wave vector. On a sphere, different values of M correspond to different total angular momenta L, according to the following relation $M = Q - L$. Based on L (i.e., on M), the trions are either "bright" or "dark", meaning that they either can or cannot recombine radiatively ($X^- \to e+\gamma$ where γ is a photon) in the absence of an additional symmetry breaking collision. The criterion for a "bright" trion state is $L = Q$ (i.e., $M = 0$). Another good trion quantum number is the pair electron spin $S = 0$ or 1.

In the realistic calculation, five LLs and two quantum well subbands for both e and h were included, and the quasi-2D interaction matrix elements were integrated taking the actual e and h subband wave functions $\chi(z)$ in the normal direction. The dependence of $\chi_e(z)$ and $\chi_h(z)$ on well width w and electron concentration n was calculated self-consistently,[18] assuming doping on one side of the well.

Some results are shown in Fig. 1. Despite significant separation of e and h layers, the X^- occurs even in a fairly wide ($w = 20$ nm) and fairly strongly (asymmetrically) doped ($n = 2 \cdot 10^{11}$ cm^{-2}) quantum well. Remarkably, neither the lowest-subband, lowest-LL, or symmetric-well approximations are satisfactory for this well. For the purpose of the following discussion, let us observe that, assuming $n = 2 \cdot 10^{11}$ cm^{-2} and $B = 25$ T, the ground state in a narrow, $w = 10$ nm well is the "bright singlet" X_s^-, while for a wider, $w = 20$ nm well it is the "dark triplet" X_t^- (note that the plotted Δ excludes the Zeeman term that weakens binding of the singlet state). The other trion in Fig. 1 is the marginally bound "bright triplet" X_{tb}^-.

Remarkably, the trion spectrum for $w = 20$ nm resembles that of an "ideal" system ($w = 0$, $B = \infty$) more than the spectrum for $w = 10$ nm. The ideal system

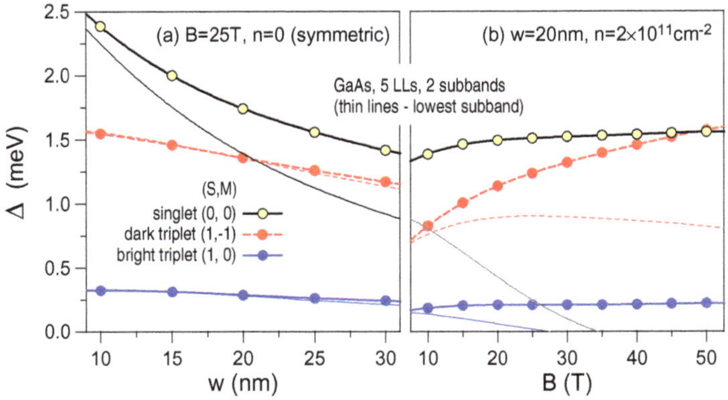

Fig. 1. (a) The X^- Coulomb binding energy Δ as a function of width w of a symmetric (or undoped) GaAs quantum well, at the magnetic field $B = 25$ T, calculated including five Landau levels and one (thin lines) or two (thick lines with symbols) well subbands for both electrons and the hole. Different trion states are labeled by spin S and relative angular momentum M. (b) Δ as a function of B, in a $w = 20$ nm well doped on one side to electron concentration $n = 2 \cdot 10^{11}$ cm^{-2}.

obeys the "hidden" (particle–hole) symmetry,[12] which implies the occurrence of "multiplicative" states,[13] only one bound trion state (X_t^-),[19] and simple selection rules for e–h recombination. In narrow wells, the hidden symmetry is broken by a stronger confinement of the hole (in all three dimensions) causing imbalance between the e–h and e–e interactions and, for example, binding of the X_s^-. In wider wells, the e–h separation reduces the interaction imbalance and favors the X_t^-.

4. Trion in 2DEG

When the X^- is immersed in a 2DEG, its effective, residual electric charge $\mathcal{Q}$ (counted from the uniform electron density n) depends on the exact form of e–X^- correlations. This is shown schematically in Fig. 2(a). In the lowest LL, the similar (i.e., strongly repulsive at short range) e–e and e–X^- Haldane interaction pseudopotentials[20] cause similar (Laughlin) e–e and e–X^- correlations.[11] They are described in a generalized[21] composite fermion (CF) picture[22] involving attachment of $2p$ flux quanta to each e and X^-.

At a Laughlin–Jain (LJ) series of $\nu_{\rm LJ} = s/(2ps+1)$, electrons converted to CF$_e$'s fill the lowest s LLs in effective magnetic field $B^* = B - 2pn(hc/e) = B/(2ps+1)$. At ν larger or smaller than $\nu_{\rm LJ}$, the LJ quasiparticles (QPs) occur in form of quasielectrons (QEs) in the $(s+1)^{\rm st}$ or quasiholes (QHs) in the $s^{\rm th}$ CF$_e$-LL, each carrying an effective charge $\varepsilon = \pm e/(2ps+1)$. Similarly, an X^- in a LJ liquid is converted to a CF$_{X^-}$ with the reduced charge $\mathcal{Q} = -\varepsilon$. In contrast to the CF$_e$'s, there is only few CF$_{X^-}$'s so they all fall into the lowest, nearly empty CF$_{X^-}$-LL. Thus, the strongly correlated 2DEG with an immersed trion is mapped onto the system of a smaller number of weakly correlated QPs and CF$_{X^-}$'s. The number (effective filling factor) and type of the QPs (QEs, QHs, or both) depends in a

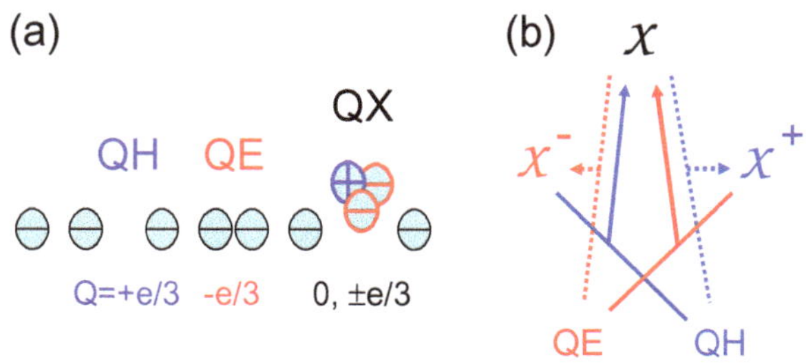

Fig. 2. (a) Partial screening of a trion immersed in a 2D electron liquid with Laughlin correlations, leading to the formation of a family of neutral and fractionally charged quasiexciton (QX) states. (b) Transitions between different QXs ($\mathcal{X}^- \leftrightarrow \mathcal{X} \leftrightarrow \mathcal{X}^+$) in the presence of free QEs or QHs.

nontrivial manner on ν and also to some extent on temperature. On the other hand, the number of $\mathrm{CF_{X^-}}$'s is controlled by the power density of the optical excitation. It can be typically assumed negligible, thus allowing for considering an isolated $\mathrm{CF_{X^-}}$ interacting with a varied type and concentration of QPs.

This result is independent of the particular X^- state or the filling factor, as long as the correlations are described by the CF model. The fact that $\mathcal{Q} \neq 0$ means that the electrons surround the X^- more closely than they surround each other. In other words, because of the severe restriction of the electron motion to the lowest LL (caused by the presence of a strong magnetic field), the LJ liquid cannot screen completely a foreign charge (e.g., an X^-). Instead, it leaves a residual fractional charge $\mathcal{Q} \neq 0$, equal to the charge of one CF quasiparticle. The nonzero value of $\mathcal{Q}$ can also be linked to the fact that the X^- is distinguishable from the macroscopic number of the surrounding electrons, reflected in the macroscopic degeneracy of the 2DEG-X^- ground state (i.e., of the $\mathrm{CF_{X^-}}$-LL). Moreover, the same value of $\mathcal{Q}$ results for any distinguishable charge $-e$ immersed in a LJ liquid, if it induces Laughlin correlations around itself.[23]

5. Neutral and Charged Quasiexcitons

A trion coupled to a LJ liquid and carrying a reduced charge $\mathcal{Q} = -\varepsilon$ is a many-body excitation. To distinguish it from the isolated three-body state $\mathrm{X}^- = 2e + h$, we call it a charged quasiexciton (QX) and denote by $\mathcal{X}^- \equiv \mathcal{X}^{-\varepsilon}$.

Being negatively charged, an $\mathcal{X}^-$ interacts with the QPs of the LJ liquid, attracting the QHs and repelling the QEs. Indeed, as shown schematically in Fig. 2(b), the problem of a trion coupled to a 2DEG can be reformulated as that of the QXs interacting with the QPs.

At $\nu < \nu_{\mathrm{LJ}}$, there are free QHs in the LJ liquid, making the negative $\mathcal{X}^-$ unstable. By means of direct Coulomb attraction, the $\mathcal{X}^-$ binds a QH to become a neutral $\mathcal{X}^-\mathrm{QH} = \mathcal{X}$, with a binding energy called Δ^0. The neutral $\mathcal{X}$ may or may not bind an additional QH to form a positively charged $\mathcal{X}^-\mathrm{QH}_2 = \mathcal{X}^+$, with binding energy Δ^+. Whether it does (i.e., whether $\Delta^+ > 0$) depends on the details of the $\mathcal{X}$ wave function, that is on the spin of the "parent" trion X^-, which in turn depends on various parameters of the quantum well.

At $\nu > \nu_{LJ}$, there are free QEs in the LJ liquid. This makes the positive $\mathcal{X}^+$ attract and annihilate a QE: $\mathcal{X}^+ + \text{QE} \rightarrow \mathcal{X}$. Such process releases the energy $\Delta_{LJ} - \Delta^+$ (where $\Delta_{LJ} = \mathcal{E}_{QE} + \mathcal{E}_{QH}$ is the LJ gap). The $\mathcal{X}$ may or may not (again, depending on the "parent" X^-, that is on the specific quantum well parameters) annihilate another QE: $\mathcal{X} + \text{QE} \rightarrow \mathcal{X}^-$, with the energy gain

$$\Delta^- = \Delta_{LJ} - \Delta^0 \tag{1}$$

that can be interpreted as $\mathcal{X}^-$ binding energy.

The neutral $\mathcal{X}$ and charged $\mathcal{X}^\pm$ are the three possible states in which a valence hole can exist in a LJ fluid. If $\Delta^\pm > 0$, then depending on ν, either $\mathcal{X}^-$ or $\mathcal{X}^+$ is the most strongly bound state. If $\Delta^- \neq \Delta^+$, they will recombine at a different energy and the PL spectrum will be discontinuous at $\nu = \nu_{LJ}$. For long-lived $\mathcal{X}^\pm$ (e.g., made of a dark X_t^-), recombination of the $\mathcal{X}$ is also possible, especially at $\nu \approx \nu_{LJ}$ (within a Hall plateau), when the QP localization impedes the $\mathcal{X}^\pm$ binding.

The QXs resemble normal excitons and trions in n- or p-type systems, with the QPs (QEs and QHs) playing role of conduction electrons and valence holes. Remarkably, the type and concentration of the QPs can be controlled in the same sample by tuning the magnetic field. In particular, the QP gas converts from effectively n- to p-type (from consisting of QEs to QHs) when the magnetic field is increased so that the filling factor passes through $\nu = 1/3$. Another important difference between the normal trions and charged QXs is the fractional charge of the constituent QPs, responsible for the adequately smaller binding energies. Also, the kinetics of QXs ($\mathcal{X} \leftrightarrow \mathcal{X}^\pm$) is more complicated because of the involved QE–QH annihilation.

6. QX Spectra for $w = 20$ nm

The above reasoning based on a CF model has been tested by realistic numerical calculations for the incompressible Laughlin $\nu = 1/3$ quantum liquid. We have calculated the energy and recombination spectra of $N \leq 10$ electrons and one hole assuming concentration $n = 2 \cdot 10^{11}$ cm^{-2} for which $\nu = 1/3$ occurs at $B = 25$ T.

For $w = 20$ nm, the most strongly bound trion is the X_t^-. This is apparent from Fig. 1, after the plotted Coulomb binding energy of the singlet trion is decreased compared to the triplet trion by the electron Zeeman term $E_Z = g^* \mu_B B = 0.75$ meV. From the diagonalization of a $2e + h$ hamiltonian we have established that the X_t^- has a high (94%) squared projection onto the subspace of the lowest LL and the lowest subband. This allows one to restrict the $Ne + h$ calculation to this smaller and thus numerically tractable (spin-polarized) subspace. The results for $N = 9$ are shown in Fig. 3, where the energy E is plotted as a function of the total angular momentum L, and the PL oscillator strengths τ^{-1} of radiative states are indicated by the diameter of the open circles.

These spectra can be understood using a CF picture generalized to a two-component fluid[21] and the addition rules for angular momentum. On a sphere, the

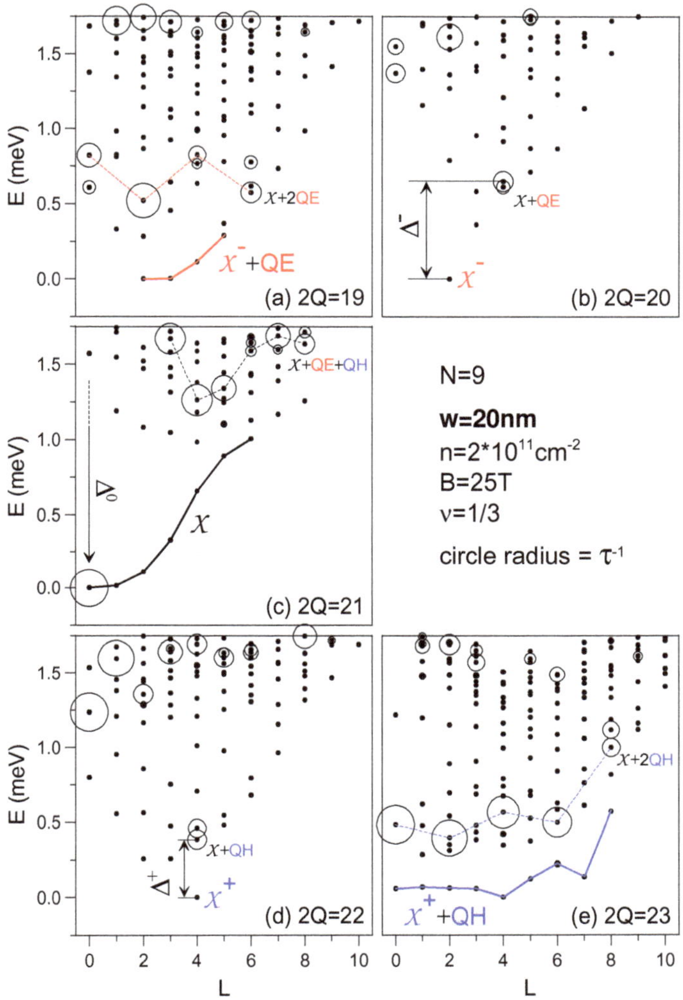

Fig. 3. Excitation spectra (energy E as a function of angular momentum L with the optical oscillator strength indicated by the diameter of the circles) of $9e+h$ systems on a Haldane sphere, at the values of $2Q$ corresponding to up to two QEs or QHs in the Laughlin $\nu = 1/3$ liquid.

CF transformation introduces an effective monopole strength $2Q^* = 2Q - 2(K-1)$, where $K = N - 1$ is the total number of free charges (electrons and trions). The angular momenta of Laughlin QPs are $l_{\rm QH} = Q^*$ and $l_{\rm QE} = Q^* + 1$. And since the X_t^- had $l_{X^-} = Q - 1$, the effective $\mathcal{X}^-$ angular momentum is $l_{\mathcal{X}^-} = Q^* - 1$ (in the following discussion for $w = 20$ nm we use $\mathcal{X} \equiv \mathcal{X}_t$).

The $\mathcal{X}^-$ is found at $2Q = 20$, as a dark ground state at $L = l_{\mathcal{X}^-} = 2$. At $2Q = 22$, there are two QHs in addition to the $\mathcal{X}^-$, and the $\mathcal{X}^+$ bound state is identified at $L = l_{\mathcal{X}^+} = |(2l_{\rm QH} - 1) - l_{\mathcal{X}^-}| = 4$.

The bands of interacting $\mathcal{X}^-$–QE and $\mathcal{X}^+$–QH pairs are marked at $2Q = 19$ and 23 (they terminate at $L < l_{\mathcal{X}^\pm} + l_{\rm QP}$ because of a hard-core reflecting the composite

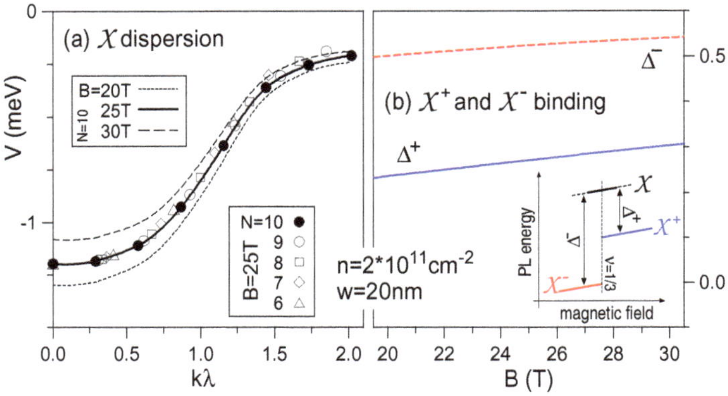

Fig. 4. Dispersion (interaction energy V as a function of wave vector k) of the neutral quasiexciton $\mathcal{X}$ (a) and the binding energies $\Delta^{\pm}$ of charged quasiexcitons $\mathcal{X}^{\pm}$ (b) in a Laughlin $\nu = 1/3$ liquid. Inset: predicted discontinuous field-dependence of the PL energy. λ is the magnetic length.

nature of the QX and the Pauli exclusion principle for its fermion constituents[21]). The increase of $E(L)$ in these bands confirms $\mathcal{X}^-$–QE and $\mathcal{X}^+$–QH repulsion (on a sphere, larger L corresponds to a smaller average squared separation), with the strength comparable to the known QP–QP interactions.

At $2Q = 21$, there is one $\mathcal{X}^-$ and one QH. The radiative ground state at $L = 0$ is (approximately) a multiplicative state. This means that it contains an X which has $k = 0$ and is decoupled from the Laughlin state of $K = 8$ electrons. It opens a band of $\mathcal{X}^-$–QH pair states (the sum of $l_{\rm QH}$ and $l_{\mathcal{X}^-}$ giving $L = 1$ to 6).

This can be understood as follows. An isolated, stationary X (with wave vector $k = 0$) has no dipole moment ($d = 0$). When it moves ($k > 0$), Lorentz force acts on its constituents and induces charge separation ($d \propto k$), and the X splits into e and h, each carrying charge $\pm e$. In a LJ liquid, charge quantum is reduced to ε. This has no consequence at $k = 0$, and the $\mathcal{X}$ is equivalent to an X decoupled from the remaining K electrons. But a moving $\mathcal{X}$ acquires dipole moment in a different way than X, by splitting into $\mathcal{X}^-$ and QH, each carrying only one small quantum $\pm\varepsilon$. The continuous dispersion of $\mathcal{X}$ (the $\mathcal{X}^-$–QH pseudopotential) at $\nu = 1/3$, obtained by the extrapolation of data for $N \leq 10$, is shown in Fig. 4(a). Its energy and length scales are comparable to the dispersion of X rescaled to account for the $e \to \varepsilon$ charge reduction.

By identifying the multiplicative states (from comparison with the spectra of ideal systems) and knowing that $\mathcal{X} \equiv$ X at $k = 0$, one can estimate the $\mathcal{X}^-$ binding energies $\Delta^{\pm}$ as marked in frames (b) and (d). More accurate values can be obtained from the comparison of energies obtained from different spectra in which there is always only one particular constituent ($\mathcal{X}^{\pm}$, $\mathcal{X}$, or QP) in the Laughlin liquid. These values were further extrapolated to $N \to \infty$, and the result is plotted in Fig. 4(b) as a function of B. The $\mathcal{X}$ binding energies Δ^0 were obtained in the same way, and the $V(0) \equiv \Delta^0$ equivalence was used to vertically shift dispersion curves in Fig. 4(a).

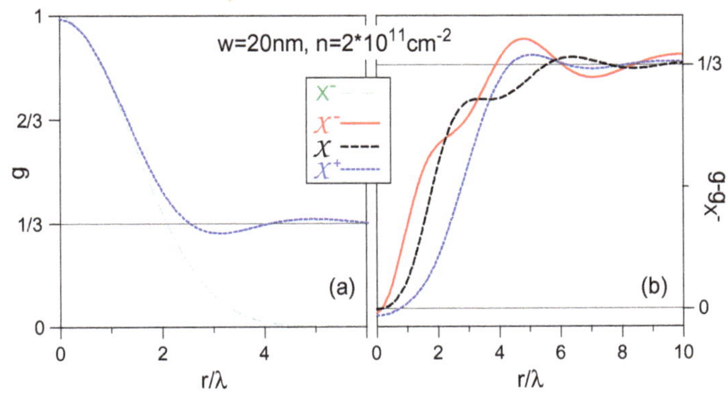

Fig. 5. (a) Comparison of the e–h pair-distribution functions (electron density g normalized so as to converge to the filling factor ν far away from the hole, as a function of the e–h distance r) of the positive quasiexciton $\mathcal{X}^+$ and the bare trion X_t^-. (b) The e–X^- pair-distribution functions calculated as the difference between the $\mathcal{X}/\mathcal{X}^\pm$ and X_t^- curves in (a). λ is the magnetic length.

Reasonable accuracy of our estimates for QX binding energies is confirmed by Eq. (1). Our values (GaAs, $w = 20$ nm, $n = 2 \cdot 10^{11}$ cm^{-2}, $\nu = 1/3$, $B = 25$ T) are: $\mathcal{E}_{\rm QH} = 0.73$ meV, $\mathcal{E}_{\rm QE} = 1.05$ meV, $\Delta^0 = 1.20$ meV, $\Delta^- = 0.52$ meV, and $\Delta^+ = 0.27$ meV. The $\Delta^+ \neq \Delta^-$ asymmetry is expected to make emission energy discontinuous at $\nu = 1/3$. PL spectra qualitatively similar to our prediction shown in the inset in Fig. 4(b) were already been observed[5,8] at $\nu \approx 1/3$.

7. QX Wave Functions for $w = 20$ nm

The QXs are defined through the following sequence of gedanken processes: (i) trion binding: $2e + h \to X^-$, (ii) charge reduction: $X^- \to \mathcal{X}^-$, (iii) QH capture: $\mathcal{X}^- \to \mathcal{X}$ or $\mathcal{X}^+$. Hence, the $\mathcal{X}^-$, $\mathcal{X}$, and $\mathcal{X}^+$ states in fact involve the same X^-, only more or less spatially separated from the surrounding electrons.

This is evident in Fig. 5(a) showing the e–h pair-distribution function $g(r)$ of the positive QX, calculated for $N = 10$ and normalized so as to measure electron concentration near the hole in the units of ν. At short range, it is quite similar to $g_{X^-}(r) = \exp(-r^2/4)$ accurately describing an X_t^-. This confirms that the $\mathcal{X}^+$ is an X^- well separated from the 2DEG. In Fig. 5(b) we plotted $\delta g = g - g_{X^-}$, measuring e–X^- correlations in the QX states. Clearly, $\delta g_{\mathcal{X}^+}$ resembles the e–e pair-distribution function of a Laughlin $\nu = 1/3$ liquid, with shoulders in $\delta g_{\mathcal{X}}$ and $\delta g_{\mathcal{X}^-}$ reflecting additional charge quanta pushed onto the hole in $\mathcal{X}$ and $\mathcal{X}^-$. Moreover, integration of $[g(r) - 1/3]$ reveals fractional electron charge of $-(4/3)e$, $-e$, and $-(2/3)e$ bound to the hole in the $\mathcal{X}^-$, $\mathcal{X}$, and $\mathcal{X}^+$ states.

8. QX Spectra for $w = 10$ nm

The QX concept can in principle be applied to any incompressible CF state. This might seem to imply complicated, nonmonotonic dependence of $\mathcal{Q}$ on B, and the entire LJ series of discontinuities in PL of a 2DEG, with the energy jumps $\propto \varepsilon^2/\lambda$. However, a discontinuity at each particular $\nu_{\rm LJ}$ depends on the (sample-dependent) relation between $\Delta_{\rm LJ}$ and Δ^0, and must be established separately.

The other quantum well we studied had the same n, but a smaller $w = 10$ nm. It is apparent from the trion spectra in Fig. 1 (we have also checked that in narrow wells, such as with $w = 10$ nm, the effect of electron concentration n or doping asymmetry on Δ is negligible) that the lowest (QX) states at $\nu \approx 1/3$ will contain an X_s^-. This singlet trion state involves a spin-flip, and its binding depends critically on the LL mixing (only the X_t^- remains bound in the lowest LL). This complicates calculations and led us to the following approximation.

In the computation of Coulomb matrix elements, the highest Haldane e–e pseudopotential,[20] V_0, was reduced by 10%. We have checked that this only affects interactions within the trion, and induces an (85% accurate) X_s^- ground state *within the lowest LL*. Critically for the purpose of our calculation, it has the correct pair-distribution functions g_{eh} and g_{ee}, which determine the investigated coupling od the trion to the surrounding 2DEG. Hence, the $Ne + h$ spectra calculated with one reversed-spin electron and using the reduced V_0 capture the essential difference between QX dynamics in 20 and 10 nm wells.

Such spectra for $N = 8$ are presented in Fig. 6. ¿From the comparison of PL oscillator strengths τ^{-1} (separate for recombination of a spin-$\uparrow$ or $\downarrow$ electron) we identified the (approximately) multiplicative states with a stationary $\mathcal{X} \equiv \mathcal{X}_{\rm s}$. An expected difference from Fig. 3 is that since low-energy states involve a (bright) X_s^-, they are all radiative. But surprisingly, charged QXs are now the excited states at $2Q = 17$ and 19 (when calculating L, note that $l_{\mathcal{X}^-} = Q^*$ for X_s^-). The multiplicative states with an $\mathcal{X}$ and two QPs (rather than an $\mathcal{X}^-$–QE or $\mathcal{X}^+$–QH pair) also form the lowest bands at $2Q = 16$ and 20.

The origin of such different behavior in narrow and wide quantum wells is explained as follows. The X_s^-, which at $B = 25$ T is the trion ground state in the absence of significant e–h layer separation (i.e., in narrow or symmetric wells), is smaller and has different charge distribution than the X_t^-. This difference has little effect on Laughlin correlation with electrons, which avoid small relative e–X^- pair angular momenta rather than approach closely to probe the X^-'s internal structure. However, it affects interactions with the QPs.

Indeed, the $\mathcal{X}_{\rm s}$ dispersion in Fig. 6(c) differs from $\mathcal{X}_{\rm t}$ of Fig. 3(c) and indicates stronger $\mathcal{X}^-$–QH attraction. We compare $\Delta^0 \sim 2$ meV with $\Delta_{\rm LJ} = 2.02$ meV ($\mathcal{E}_{\rm QE} = 1.38$ meV and $\mathcal{E}_{\rm QH} = 0.64$ meV) using Eq. (1) to find that Δ^- is very small (to prevent distinction of the $\mathcal{X}$ and $\mathcal{X}^-$ emission peaks) or even negative.

Negative Δ^-, as in Fig. 6(b), means that the $\mathcal{X}^-$ binds the first QH so strongly that even in the absence of free QHs it is unstable toward creation of a new QE–QH pair ($\mathcal{X}^- \to \mathcal{X} +$ QE). And negative Δ^+ in Fig. 6(d) means that a second QH is not bound at all. As a result, the (neutral) $\mathcal{X}$ is the most strongly bound state regardless of the presence of either QEs or QHs. This precludes discontinuity in PL at $\nu = 1/3$ in a 10 nm quantum well.

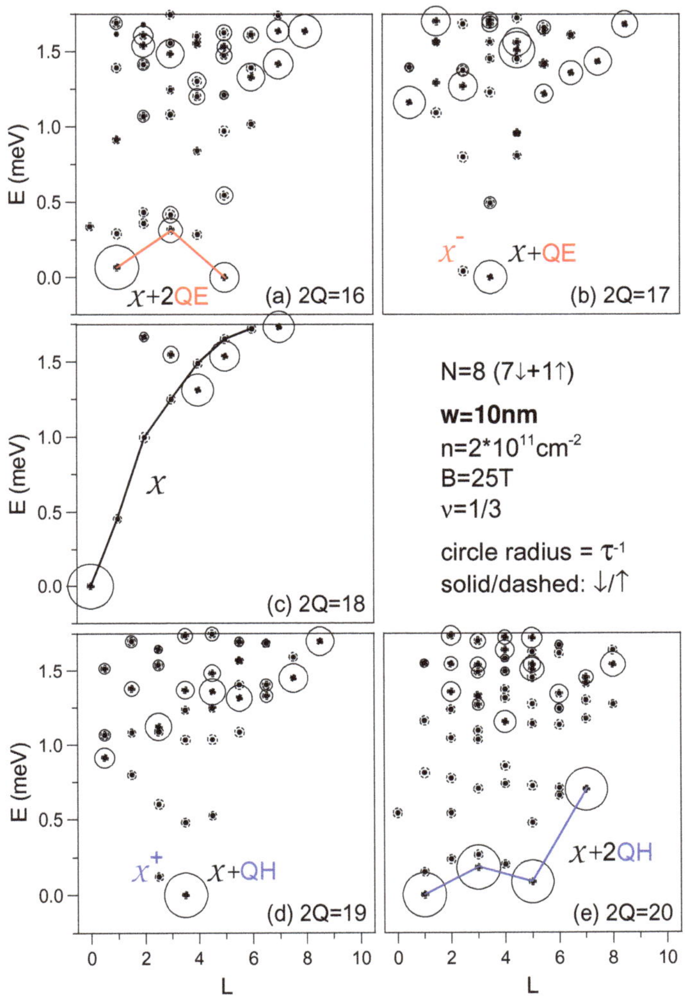

Fig. 6. Excitation spectra similar to Fig. 3, but calculated for $8e + h$ systems, with a modified e–e pseudopotential (see text), with one reversed-spin electron, and for the well width $w = 10$ nm.

9. Wider Wells

For $n = 2 \cdot 10^{11}$ cm^{-2} and $w \geq 40$ nm, the QXs are replaced by "anyon excitons" $h\text{QE}_\mu$ (proposed earlier in ideal bilayer systems[15]). Their recombination also causes a jump in PL energy at $\nu = 1/3$, but it can be easily distinguish from that due to QXs since it is much larger ($\sim \Delta_{\text{LJ}}$) and goes in the *opposite direction*.

10. Conclusion

We studied PL of incompressible electron liquids formed in a 2DEG in high magnetic fields. The emission spectrum is found due to the recombination of neutral and fractionally charged *quasiexcitons* (QXs). These novel bound state are formed from trions by Laughlin correlation with the surrounding electrons. The energy and recombination spectrum of the QXs critically depends on the well width, electron concentration, and the magnetic field, offering explanation for the discontinuities observed in PL in coincidence with the fractional quantum Hall effect.

Acknowledgment

The authors gratefully acknowledge discussions with M. Byszewski, M. Potemski, and P. Hawrylak, and financial support from grants DE-FG 02-97ER45657 of US DOE and N20210431/0771 of the Polish MNiSW.

References

1. D. C. Tsui, H. L. Störmer, and A. C. Gossard, *Phys. Rev. Lett.* **48**, 1559 (1982).
2. R. B. Laughlin, *Phys. Rev. Lett.* **50**, 1395 (1983).
3. S. L. Sondhi, A. Karlhede, S. A. Kivelson, and E. H. Rezayi, *Phys. Rev. B* **47**, 16419 (1993).
4. D. Heiman, B. B. Goldberg, A. Pinczuk, C. W. Tu, A. C. Gossard, and J. H. English, *Phys. Rev. Lett.* **61**, 605 (1988).
5. B. B. Goldberg, D. Heiman, A. Pinczuk, L. N. Pfeiffer, and K. West, *Phys. Rev. Lett.* **65**, 641 (1990).
6. I. V. Kukushkin and V. B. Timofeev, *Adv. Phys.* **45**, 147 (1996).
7. C. Schüller, K.-B. Broocks, P. Schröter, Ch. Heyn, D. Heitmann, M. Bichler, W. Wegscheider, T. Chakraborty, and V. M. Apalkov, *Phys. Rev. Lett.* **91**, 116403 (2003).
8. M. Byszewski, B. Chwalisz, D. K. Maude, M. L. Sadowski, M. Potemski, T. Saku, Y. Hirayama, S. Studenikin, D. G. Austing, A. S. Sachrajda, and P. Hawrylak, *Nature Physics* (London) **2**, 239 (2006).
9. K. Kheng, R. T. Cox, Y. Merle d'Aubigne, F. Bassani, K. Saminadayar, and S. Tatarenko, *Phys. Rev. Lett.* **71**, 1752 (1993).
10. G. Yusa, H. Shtrikman, and I. Bar-Joseph, *Phys. Rev. Lett.* **87**, 216402 (2001).
11. A. Wójs, J. J. Quinn, and P. Hawrylak, *Phys. Rev. B* **62**, 4630 (2000).
12. I. V. Lerner and Yu. E. Lozovik, *Zh. Eksp. Teor. Fiz.* **80**, 1488 (1981).
13. A. H. MacDonald and E. H. Rezayi, *Phys. Rev. B* **42**, 3224 (1990).
14. X. M. Chen and J. J. Quinn, *Phys. Rev. Lett.* **70**, 2130 (1993).
15. A. Wójs and J. J. Quinn, *Phys. Rev. B* **63**, 045303 (2001); *ibid.* **63**, 045304 (2001).
16. F. D. M. Haldane, *Phys. Rev. Lett.* **51**, 605 (1983).
17. A. Wójs and J. J. Quinn, cond-mat/0609103.
18. I. H. Tan, G. L. Snider, L. D. Chang, and E. L. Hu, *J. Appl. Phys.* **68**, 4071 (1990).
19. A. Wójs and P. Hawrylak, *Phys. Rev. B* **51**, 10880 (1995).
20. F. D. M. Haldane, *The Quantum Hall Effect*, eds. R. E. Prange and S. M. Girvin (Springer-Verlag, New York, 1987), chapter 8, pp. 303–352.
21. A. Wójs, I. Szlufarska, K. S. Yi, and J. J. Quinn, *Phys. Rev. B* **60**, 11273 (1999).
22. J. K. Jain, *Phys. Rev. Lett.* **63**, 199 (1989).
23. E. H. Rezayi and F. D. M. Haldane, *Phys. Rev. B* **32**, 6924 (1985).

Part B

Bose Liquids

QUANTUM BOLTZMANN LIQUIDS

K. A. GERNOTH

School of Physics and Astronomy, University of Manchester, United Kingdom

MANFRED L. RISTIG and THOMAS LINDENAU

Institut für Theoretische Physik, Universität zu Köln, Germany

We study homogeneous normal systems of bosons under the influence of interparticle forces with a strongly repulsive component at short relative particle-particle distances. The repulsion prevents short-ranged exchange between the bosonic constituents in the quantum fluid. Consequently, the bosons remain distinguishable at temperatures far below the classical high-temperature regime. At these low temperatures such fluids and liquids display nevertheless distinct quantum effects due to quantum-mechanical phase-phase correlations. Typical examples are liquid *para*-hydrogen and fluid ^{4}He under certain thermodynamic conditions. The study employs Correlated Density-Matrix theory and Path- Integral Monte-Carlo simulations.

Keywords: Exchange screening; radial distribution function; exchange correlation function; kinetic energy distribution; phase-phase correlations.

1. Introduction

Employing the binary-collision expansion[1] for the virial Larsen et al.[2] showed for a real gas that the presence of a strongly repulsive core in the particle-particle interaction entails a drastic suppression of exchange effects. We demonstrate in this contribution that Correlated Density-Matrix (CDM) theory and Path-Integral Monte-Carlo (PIMC) simulations corroborate this property also in the case of quantum fluids such as liquid *para*-hydrogen and Helium I at low temperatures but somewhat above the Bose-Einstein transition line.

The screening effect has the consequence that the constituents of such a liquid remain to be distinguishable at these temperatures and obey classical Boltzmann statistics, although the system exhibits significant dynamical quantum effects. These properties are seen in relevant spatial correlation functions, structure functions, and distributions in momentum space. Recent precision measurements of neutron-scattering cross sections in liquid *para*-hydrogen close to the triple point temperature confirmed the existence of such particle screening effects.[3,4]

CDM theory[5–7] and PIMC calculations[8,9] provide reliable mathematical tools for a numerically accurate analysis of strongly correlated quantum fluids, in particular of the more specialized quantum Boltzmann fluids or liquids.

Our present objective focusses on the spatial structure of the one-body and two-body reduced density-matrix elements associated with the system that characterize the exchange and dynamic properties. The corresponding Fourier transforms provide detailed information on the liquid structure function $S(k)$, the exchange structure function $S_{cc}(k)$, the momentum distribution $n(k)$ of a particle in the medium, and on related quantities such as the total kinetic energy and its distribution in momentum space, $\epsilon_0(k)n(k)$, with $\epsilon_0(k) = \hbar^2k^2/2m$. Our formal analysis is accompanied by a presentation and discussion of some numerical results on the quantum Boltzmann behavior of liquid hydrogen and fluid helium. A comprehensive study of the properties of quantum Boltzmann systems and specifically of liquid *para-* hydrogen is reported in a forthcoming publication.[10]

We mention, that the relaxation of the particle screening in a correlated boson fluid by a further lowering of the temperature toward the Bose-Einstein transition region can be theoretically and experimentally studied in normal liquid ^{4}He. Decreasing the temperature the Boltzmann behavior is weakened and the screening is only partial. Finally, the screening effect disappears at the phase boundary of the normal liquid to the Bose-Einstein condensed system. This is caused by the development of long-range cyclic exchange correlations which win over the short-ranged screening localized in the hard-core region of the particle-particle potential.

In Section 2 we collect some basic elements of CDM theory. The suppression of exchange correlations in a normal Bose fluid, where the particles encounter strong repulsion at short relative distances, is studied in Section 3. Here, we focus on the effects on the radial distribution function and the liquid structure function. The analysis of exchange properties is continued in Section 4 by exploring the cyclic exchange structure function and the correlation function. These quantities govern the screening impact and its effect on the statistical properties. The momentum distribution and kinetic energy distribution of a quantum Boltzmann fluid is analysed in Section 5. In Sections 3–5 we also present numerical results on some correlation functions of liquid hydrogen and fluid helium in the quantum Boltzmann thermodynamic regime. A short summary is given in Section 6.

2. Elements of CDM Theory

Our theoretical studies of quantum Boltzmann fluids employ CDM theory, which provides a formal and numerical analysis of strongly correlated quantum fluids at nonzero temperatures.[5–7,11] This formalism is based on an appropriate trial ansatz for the N-body density matrix of the N-body quantum system in the thermodynamic limit (particle number N and volume V to infinity with density $\rho = N/V$ held constant). The exact expression of the matrix contains complete information on the equilibrium properties of the many-body system considered, but is for a correlated fluid in general unknown. However, the Ansätze adopted in CDM theory provide adequate approximations in a first step of a systematic procedure. Experience gained in many theoretical studies, in particular at zero temperature where CDM theory specializes to the Correlated Basis Functions (CBF) theory,[12] tells that the first step already incorporates the most important correlation effects.

The CDM algorithm starts from the exact decomposition of the N-body density matrix of correlated bosons[5,7]

$$\rho_N(\mathbf{R}, \mathbf{R}') = \Phi(\mathbf{R}) \, Q(\mathbf{R}, \mathbf{R}') \, \Phi(\mathbf{R}') \tag{1}$$

with $\mathbf{R}$ spanning the $3N$-dimensional coordinate space. The approximations begin with the adoption of trial Ansätze for the incoherence factor $Q(\mathbf{R}, \mathbf{R}')$ and the correlated wave function $\phi(\mathbf{R})$. At the present level CDM theory employs the Jastrow ansatz

$$\Phi(\mathbf{R}) = \exp\left[(1/2)\sum_{i<j}^{N} u\left(|\mathbf{r}_i - \mathbf{r}_j|\right)\right] \tag{2}$$

with a trial pseudopotential $u(r)$. The ansatz for quantity $Q(\mathbf{R}, \mathbf{R}')$ reads

$$Q(\mathbf{R}, \mathbf{R}') = Q_{\text{coll}}(\mathbf{R}, \mathbf{R}') \, \text{Perm}_{i,j}\left[\Gamma_{\text{cc}}\left(|\mathbf{r}_i - \mathbf{r}'_j|\right)\right] . \tag{3}$$

The factor $Q_{\text{coll}}(\mathbf{R}, \mathbf{R}')$ deals with the contributions of collective excitations (phonons and rotons). The permanent of the statistical factor $\Gamma_{\text{cc}}(r)$ accounts for the exchange effects of quasiparticles at nonzero temperatures. The pseudopotential and the exchange factor may be optimally determined by a minimum principle for the Helmholtz free energy.[6] Numerical calculations have shown that the optimal statistical factor is well approximated by a Gaussian function.[13] In momentum space it reads

$$\Gamma_{\text{cc}}(k) \approx \Gamma_{\text{cc}}(0) \exp\left[-\beta\epsilon_0(k)\right] . \tag{4}$$

Finally, similar to ansatz (2), the collective factor is assumed to be of the Jastrow form

$$Q_{\text{coll}}(\mathbf{R}, \mathbf{R}') = \exp\left[\sum_{i,j}^{N} \gamma\left(|\mathbf{r}_i - \mathbf{r}'_j|\right)\right] . \tag{5}$$

The CDM formalism then provides explicit expressions and structural relations for the diagonal elements of the reduced density matrices and their Fourier transforms. The off-diagonal elements of the reduced quantities, in particular of the one-body reduced density matrix, may be analysed by an analogous algorithm.[11]

3. Screening of Particle Exchange

Under appropriate thermodynamic conditions the strongly correlated constituents of a Bose fluid can be distinguished at low temperatures where quantum effects are already significant. This behavior is caused by the strong repulsion acting at short relative distances between the bosons.

The most important information on this screening effect is stored in the off-diagonal elements of the one-body reduced density matrix $n(r)$ and in the diagonal elements of the two-body reduced density matrix represented by the radial distribution function $g(r)$. The structure of these elements has been analysed by CDM theory.[5]

For the radial distribution function the CDM formalism yields the product decomposition

$$g(r) = [1 + G_{\rm dd}(r)]\, F(r)\,. \tag{6}$$

The first factor incorporates the dynamic direct-direct (dd) contributions and the factor $F(r)$ embodies the explicit exchange effects. In the normal phase of a Bose fluid the latter quantity comprises the cyclic exchange (cc) contributions, the direct-exchange (de) portions, and the exchange-exchange (ee) parts. The factor $F(r)$ reads explicitly

$$F(r) = F_{\rm cc}^2(r) + [1 + N_{\rm de}(r) + E_{\rm de}(r)]^2 + N_{\rm ee}(r) + E_{\rm ee}(r) \tag{7}$$

with the cyclic component

$$F_{\rm cc}(r) = \Gamma_{\rm cc}(r) + N_{\rm cc}(r) + E_{\rm cc}(r)\,. \tag{8}$$

The nodal functions $N_{\rm de}(r)$, $N_{\rm ee}(r)$, and the elementary functions $E_{\rm de}(r)$, $E_{\rm ee}(r)$, and $E_{\rm cc}(r)$ are functions of short range. In a quantum Boltzmann fluid these components have no effect on the distribution $g(r)$, since they are completely suppressed in Eq. (6) by the factor $1 + G_{\rm dd}(r)$ that vanishes in the region of strong repulsion. Expression (7) therefore reduces to $F(r) = 1 + F_{\rm cc}^2(r)$. Moreover, if the system is not too close to the boundary of the normal phase to the Bose- Einstein condensed phase the statistical and the nodal terms $\Gamma_{\rm cc}(r)$ and $N_{\rm cc}(r)$, respectively, are also sufficiently short-ranged. We may therefore characterize a quantum Boltzmann fluid by the properties

$$[1 + G_{\rm dd}(r)]\, [F(r) - 1] \approx 0 \tag{9}$$

and

$$[1 + G_{\rm dd}(r)]\, F_{\rm cc}(r) \approx 0\,. \tag{10}$$

In this case the radial distribution function is completely determined by the dynamic effects and Eq. (6) reduces to $g(r) = 1 + G_{\rm dd}(r)$.

Within CDM theory the optimal expression for $g(r)$ of a quantum Boltzmann system can be calculated by solving an effective Schrödinger equation

$$\left\{ - \left(\hbar^2/m\right) \nabla^2 + v(r) + w(r) + v_{\rm coll}(r) \right\} \sqrt{g(r)} = 0 \tag{11}$$

with the interparticle potential $v(r)$ as input. The induced potential $w(r)$ is defined as the Fourier inverse of

$$w(k) = - (1/2)\, \epsilon_0(k)\, [1 + 2S(k)]\, \left[1 - S^{-1}(k)\right]^2\,. \tag{12}$$

The potential $v_{\rm coll}(r)$ describes the coupling of the medium to the collective excitations.[5]

An equivalent possibility to calculate the optimal radial distribution function $g(r)$ is to solve a renormalized Bogoliubov equation for its Fourier transform $S(k)$ in conjunction with a generalized Feynman equation,[14]

$$\omega_{\rm coll}^2(k) = \epsilon_0(k)\left[\epsilon_0(k) + 2v_{\rm ph}(k)\right] \tag{13}$$

and

$$S(k) = \epsilon_0(k)\,\omega_{\rm coll}^{-1}(k)\coth\left[(1/2)\,\beta\omega_{\rm coll}(k)\right] . \tag{14}$$

We have employed the CDM formalism for a numerical analysis of the various correlation functions and structural properties of fluid ^{4}He and liquid *para*-hydrogen in a temperature and density range where they can be described as perfect quantum Boltzmann systems. The input potential $v(r)$ for helium is the Aziz HFDB2 potential.[15] The particle-particle interaction given by Silvera and Goldman[16] is adopted for the hydrogen liquid. We have calculated the exchange contributions discussed above and found that they are indeed very small for the systems considered. They fulfill the conditions (9) and (10) for being quantum Boltzmann fluids almost perfectly.

Figure 1 displays numerical results on the optimal radial distribution function $g(r)$ for Helium I at a temperature $T = 12\,\mathrm{K}$ and particle number density $\rho = 0.02185\,\AA^{-3}$. The full line shows the results of a calculation based on the full expression (6) including the contribution of the exchange factor $F(r)$. On the scale of this diagram these results are not distinguishable from the results of a CDM calculation where F(r) is set to unity in Eqs. (6) and (7) at the outset. This finding demonstrates the complete suppression of particle exchange.

We have checked the accuracy of the CDM results by comparing them with data from PIMC simulations. These numerical results are plotted in Figure 1 by the dotted line. We find very good agreement between the CDM and the PIMC results and conclude that CDM theory provides an efficient and accurate tool for a quantitative evaluation of the correlation properties of quantum Boltzmann fluids.

Figure 2 exhibits the results of an analogous numerical study of liquid *para*-hydrogen at $T = 16.5\,\mathrm{K}$ and a pressure of 40 bar (see also Ref. 10).

The quantum Boltzmann behavior of the helium and hydrogen systems is also reflected in the numerical results on the corresponding liquid structure functions. The results on $S(k)$ are presented in Figures 3 and 4. As for $g(r)$ we find very good agreement between the CDM results with or without assuming conditions (9) and (10). At wavenumbers $k > 1.5\,\AA^{-1}$ these results also agree with the results of PIMC calculations. At smaller wavenumbers the PIMC data show oscillations due to finite box sizes, which we have not corrected by employing conventional tail correction procedures. A more detailed discussion is given in Ref. 10.

4. Distinguishable Bosons

In this section we concentrate on a study of various cyclic exchange components related to the cyclic correlation function $G_{\rm cc}(r)$ and its Fourier transform $S_{\rm cc}(k)$. A

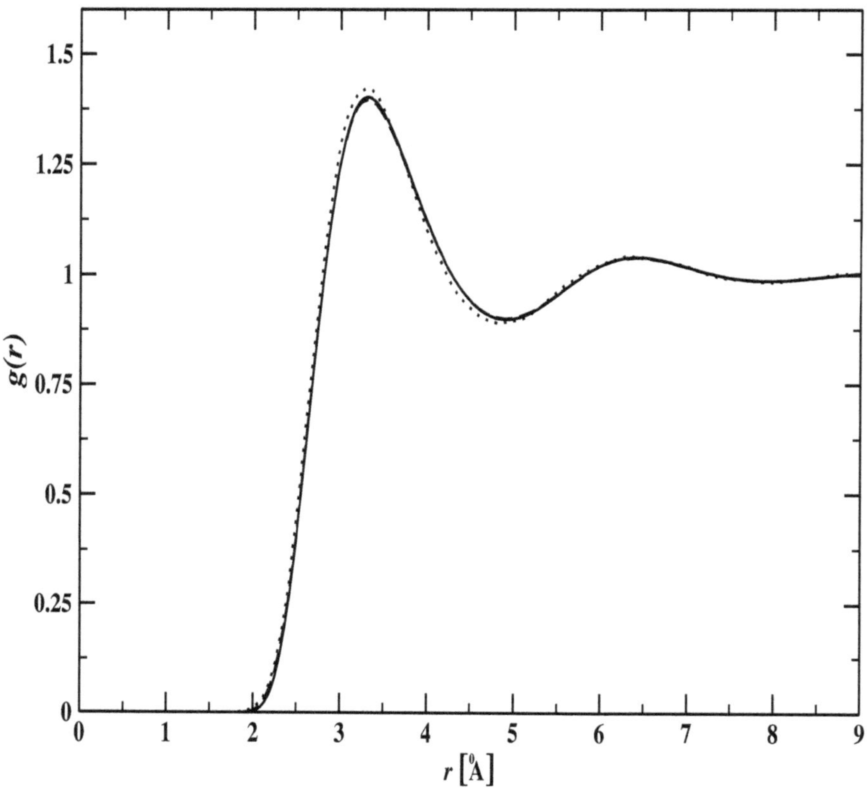

Fig. 1. Numerical results on the radial distribution function $g(r)$ of fluid ^{4}He at temperature $T = 12\,\mathrm{K}$ and particle-number density $\rho = 0.02185\,\AA^{-3}$ calculated within CDM theory and by PIMC simulations. The full line displays the results assuming the system to be a quantum Boltzmann fluid. The results indicated by the broken line have been calculated by accounting for the statistical function $\Gamma_{\mathrm{cc}}(r)$ and the nodal components appearing in Eqs. (7) and (8) but ignoring the elementary pieces (HNC/0 approximation). The close agreement between the two curves demonstrates the quantum Boltzmann behavior of the fluid. The results also agree closely with the corresponding results of PIMC simulations (dotted line).

hypernetted-chain analysis within CDM theory yields the decompositions[4,5]

$$S_{\mathrm{cc}}(k) = \Gamma_{\mathrm{cc}}(k) + X_{\mathrm{cc}}(k) + N_{\mathrm{cc}}(k) \tag{15}$$

and

$$S_{\mathrm{cc}}(k) = [\Gamma_{\mathrm{cc}}(k) + X_{\mathrm{cc}}(k)]\,[1 - \Gamma_{\mathrm{cc}}(k) - X_{\mathrm{cc}}(k)]^{-1}\,. \tag{16}$$

For non-interacting bosons the portion $X_{\mathrm{cc}}(k)$ vanishes identically and expression (16) specializes to the familiar Bose-function that represents the statistical momentum distribution $n_0(k)$ of free bosons with the Gaussian function $\Gamma_{\mathrm{cc}}(k) = \exp\left[-\beta\,(\epsilon_0(k) - \mu_0)\right]$.

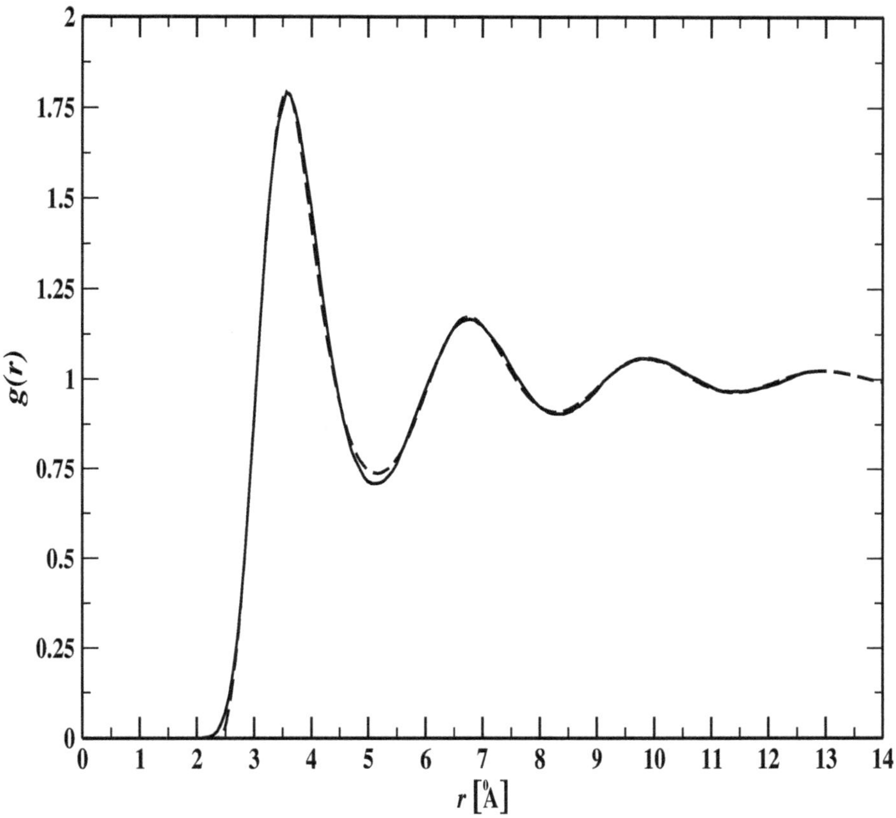

Fig. 2. The curves shown represent the results for the radial distribution function $g(r)$ of liquid *para*-hydrogen at $T = 16.5\,\mathrm{K}$ and a pressure of 40 bar. Results of CDM theory: broken line. PIMC data: full line. The system exhibits complete screening of particle exchange.

However, for interacting bosons CDM theory adequately generalizes function $n_0(k)$ by introducing the momentum distribution

$$n_{\mathrm{qp}}(k) = \Gamma_{\mathrm{cc}}(k)\,[1 + S_{\mathrm{cc}}(k)] \tag{17}$$

of quasiparticles having the energy $\epsilon_{\mathrm{cc}}(k)$. Thus, the thermodynamic statistics of these quasiparticles is governed by the exchange structure function $S_{\mathrm{cc}}(k)$ or its corresponding exchange correlation function $G_{\mathrm{cc}}(r)$. The latter function is of the product form

$$G_{\mathrm{cc}}(r) = [1 + G_{\mathrm{dd}}(r)]\,F_{\mathrm{cc}}(r)\,. \tag{18}$$

Consequently, a perfect quantum Boltzmann fluid has the property $G_{\mathrm{cc}}(r) \approx 0$ at each fixed distance r. Provided that this function is not only very small but also

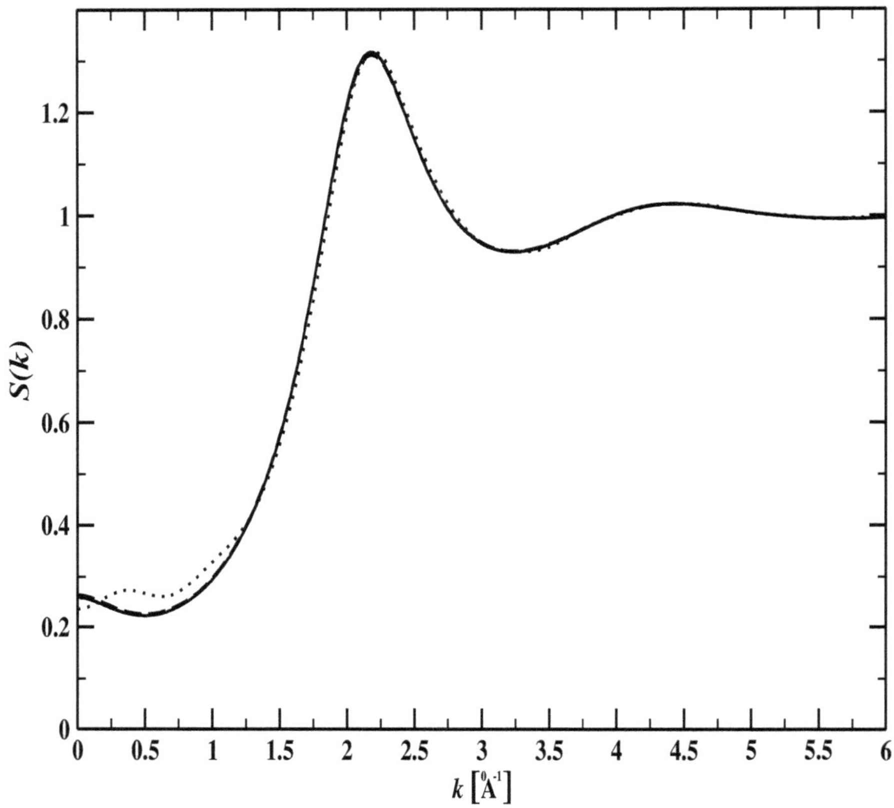

Fig. 3. Same as Figure 1 but for the liquid structure function S(k) of fluid helium. The oscillations of the PIMC data (dotted line) at small wavenumbers k are artifacts, since the simulations are performed in a box that is small compared to the box adopted in the CDM calculations. The plotted Monte Carlo data are not smoothed by conventional finite-size tail corrections.

of short range we may further conclude that the Fourier transform $S_{\rm cc}(k)$ is also practically zero. For a quantum Boltzmann fluid relation (17) reduces therefore to the simple form $n_{\rm qp}(k) \approx \Gamma_{\rm cc}(k)$.

The properties just discussed are supported by numerical calculations, for example, on Helium I at a density $\rho = 0.02185\, \AA^{-3}$ and at various temperatures. Some numerical results on the optimal correlation function $G_{\rm cc}(r)$ are plotted in Figure 5. At $T = 12\,$K the range and the magnitude of this function in coordinate space are small. Its maximum value is only about 0.02 compared with unity and its range is less than $4\,\AA$. By decreasing the temperature we see that the statistical correlations increase in magnitude and range and the screening of particle exchange becomes incomplete.

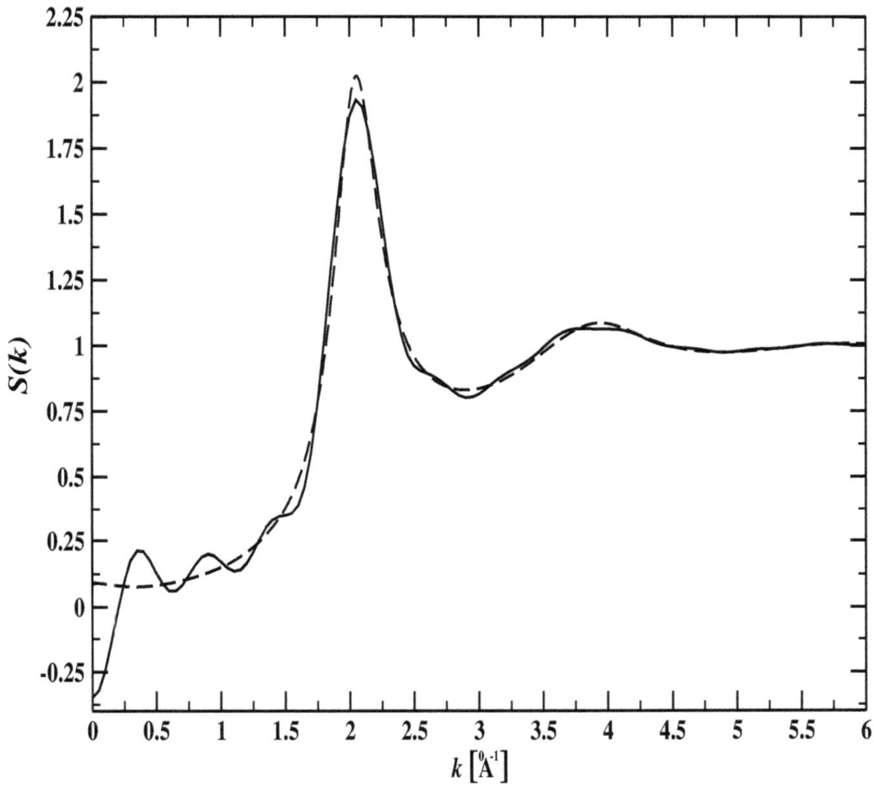

Fig. 4. Same as Figure 3 but for liquid *para*-hydrogen at $T = 16.5\,\mathrm{K}$ and pressure 1 bar. CDM theory: broken line. PIMC data: full line.

Turning to a study of the dependence of the factor $\Gamma_{\rm cc}(r)$ on the relative distance r we observe that we have in the classical thermodynamic limit (temperature T to infinity)

$$\Gamma_{\rm cc}(r) = \exp\left[-\pi\,(r/\lambda)^2\right] \tag{19}$$

with a thermal wavelength $\lambda = \sqrt{2\pi\beta\hbar^2/m}$. Numerical calculations within CDM theory reveal that the optimal solutions $\Gamma_{\rm cc}(r)$ are still proportional to the Gaussian dependence (19). We may therefore conclude that the quasiparticles in a quantum Boltzmann fluid are distinguishable, follow classical Boltzmann statistics, and move with momentum $\mathbf{k}$, and have an energy $\epsilon_0(k)$.

5. Momentum Distribution

The theoretical momentum distribution of one of the constituents in the fluid medium may be extracted from the off-diagonal elements of the one-body reduced

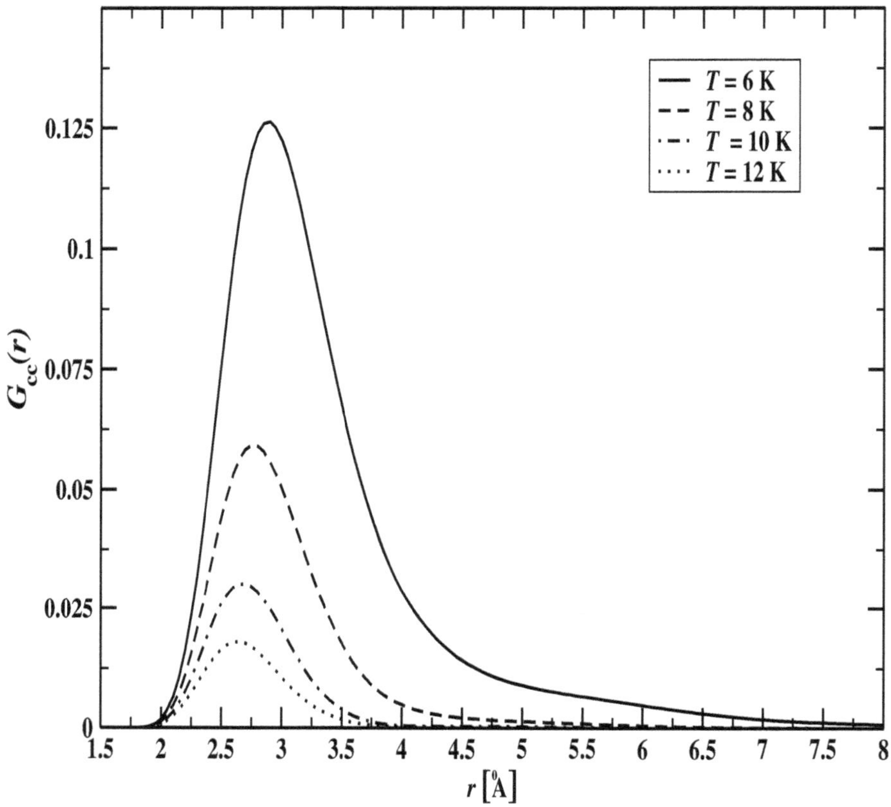

Fig. 5. CDM results on the optimal cyclic-exchange correlation function $G_{cc}(r)$ of normal helium at a particle-number density $\rho = 0.02185\ \AA^{-3}$ and at various low temperatures. It characterizes the exchange correlations. For a perfect quantum Boltzmann fluid $G_{cc}(r)$ vanishes. We see from the plot that function $G_{cc}(r)$ is of very short range and very small at $T = 12\,\mathrm{K}$, thus demonstrating that Helium I is practically a Boltzmann fluid at this temperature. The screening of particle exchange weakens if the temperature is decreased.

density matrix. For a normal Bose fluid CDM theory yields the factor decomposition[4,11]

$$n(r) = N_0(r) \exp \{p_0 [P(r) - 1]\} . \tag{20}$$

In diagrammatic language the factor $N_0(r)$ is characterized by cyclic exchange contributions and function $P(r)$ is composed of dynamical direct-direct diagrams. Thus, the first factor in Eq. (20) comprises the effects of particle exchange. The unit-normalized factor $P(r)$ represents the off-diagonal quantum-mechanical phase-phase correlations induced by the interparticle potential $v(r)$. The strength of these correlations is measured by the amplitude p_0.

Employing the structural decomposition (20) we may appropriately decompose the total kinetic energy[10]

$$E_{\text{kin}} = E_0 + p_0 E_{\text{p}} \tag{21}$$

with the exchange and dynamical components

$$E_0/N = -\left(\hbar^2/2m\right)\left[\nabla^2 N_0(r)\right]_{r=0} \tag{22}$$

and

$$E_{\text{p}}/N = -\left(\hbar^2/2m\right)\left[\nabla^2 P(r)\right]_{r=0}\,, \tag{23}$$

respectively.

This structural decomposition specializes for a quantum Boltzmann liquid due to the screening of particle exchange. The exchange function $N_0(r)$ simplifies to

$$N_0(r) \approx \Gamma_{\text{cc}}(r) \approx n_{\text{qp}}(r)\,, \tag{24}$$

where the right-hand side is explicitly given by Eq. (19). Consequently, the component E_0 is the classical kinetic energy $(3/2)Nk_{\text{B}}T$ with the Boltzmann constant k_{B} set to unity.

The quantum effects observed in a quantum Boltzmann fluid originate therefore from the phase-phase correlations. They influence the momentum dependence of the total kinetic energy E_{kin}, the momentum distribution $n(k)$, and the kinetic energy distribution $\epsilon_0(k)n(k)$. For this reason they differ from their classical counterparts.[3,10]

6. Summary

We have reported on some of our formal and numerical results on the properties of quantum Boltzmann fluids such as liquid *para*-hydrogen and Helium I. We have discussed characteristic properties and pointed out some significant features of such correlated Bose fluids: (i) Complete screening of bosonic exchange in the spatial distribution function $g(r)$; (ii) Absence of bosonic correlations in the cyclic components of the quasiparticle momentum distribution with the consequence that the particle excitations of the system follow classical Boltzmann statistics; (iii) Presence of quantum-mechanical effects in the total kinetic energy, in the momentum distribution of a particle in the medium, and in the kinetic energy distribution in momentum space caused by the existing phase-phase correlations in the medium.

References

1. T. D. Lee and C. N. Yang, *Phys. Rev.* **113**, 1165 (1959).
2. S. F. Larsen, J. E. Kilpatrick, E. H. Lieb, and H. F. Jordan, *Phys. Rev. A* **129**, 140 (1965).
3. J. Dawidowski, F. J. Bermejo, M. L. Ristig, C. Cabrillo, and S. M. Bennington, *Phys. Rev. B* **73**, 144203 (2006).

4. T. Lindenau, M. L. Ristig, K. A. Gernoth, J. Dawidowski, and F. J. Bermejo, in *Advances in Quantum Many-Body Theory*, edited by E. Susana Hernandez and Horacio M. Cataldo (World Scientific, Singapore), in press.
5. G. Senger, M. L. Ristig, C. E. Campbell, and J. W. Clark, *Ann. Phys. (N.Y.)* **218**, 160 (1992).
6. T. Lindenau, M. L. Ristig, J. W. Clark, and K. A. Gernoth, *J. Low Temp. Phys.* **129**, 143 (2002).
7. M. L. Ristig, G. Senger, M. Serhan, and J. W. Clark, *Ann. Phys. (N.Y.)* **243**, 247 (1995).
8. K. A. Gernoth, *Ann. Phys. (N.Y.)* **285**, 61 (2000).
9. K. A. Gernoth, *Ann. Phys. (N.Y.)* **291**, 202 (2001).
10. K. A. Gernoth, T. Lindenau, and M. L. Ristig, *Phys. Rev. B*, in press.
11. R. Pantfoerder, T. Lindenau, and M. L. Ristig, *J. Low Temp. Phys.* **108**, 245 (1997).
12. E. Feenberg, *Theory of Quantum Fluids* (Academic Press, New York, 1969).
13. M. L. Ristig and T. Lindenau, in *Condensed Matter Theories*, Vol. 20 (Nova Science Publisher, N.Y., 2006).
14. G. Senger, M. L. Ristig, K. E. Kuerten, and C. E. Campbell, *Phys. Rev. B* **33**, 7562 (1986).
15. R. A. Aziz, M. J. Slaman, A. Koide, A. R. Allnat, W. J. Meath, *Mol. Phys.* **77**, 321 (1992).
16. I. Silvera and V. Goldman, *J. Chem. Phys.* **69**, 4209 (1978).

CONDENSATE FRACTION IN THE DYNAMIC STRUCTURE FUNCTION OF BOSE FLUIDS

M. SAARELA*

Department of Physical Sciences ,University of Oulu,
P.O. Box 3000, FIN-90014 University of Oulu, Finland
**E-mail: Mikko.Saarela@oulu.fi*

F. MAZZANTI

Departament de Fısica i Enginyeria Nuclear, Universitat Politècnica de Catalunya
Comte Urgell 187, E-08036 Barcelona, Spain
E-mail: Ferran.Mazzanti@upc.edu

V. APAJA

Department of Physics, University of Jyväskylä
Jyväskylä, Finland
E-mail: Vesa.Apaja@gmail.com

We present results on the behavior of the dynamic structure function in the short wave length limit using the equation of motion method. The one-body continuity equation defines the self-energy, which becomes a functional of the fluctuating two-body correlation function. We evaluate the self-energy in this limit and show that sum rules up to the second moment, which requires the self-energy in the short wave length limit and zero frequency to be proportional to the kinetic energy per particle, are exactly satisfied. We compare our results with the impulse approximation and calculate the condensate fraction. An analytic expression for the momentum distribution is also derived.

Keywords: Superfluid helium; dynamic structure function; condensate fraction; equation of motion.

1. Introduction

The dynamic structure function $S(k, \omega)$ reveals information on the excitations of the system under study. Experimentally it can be determined by analysing scattering data, as this function is known to contain all the information about the dynamic properties of the system an inelastic scattering experiment can provide.[1,2] Neutron scattering cross section measurements in liquid helium have accumulated the main body of data for this analysis. At low energy and low momentum transfers the elementary excitation modes, phonon, roton and maxon dispersions are accurately determined. Their structure is also well understood theoretically using many-body techniques like Monte Carlo[3] or variational methods.[4–7]

The behavior of the dynamic structure function is also well determined in the high energy and high momentum transfers limit, where the impulse approximation holds.[8,9] In this limit $S(k,\omega)$ can be written in terms of the Compton Profile $J(y)$

$$S(k,\omega) \to \frac{1}{\hbar v} J(y) \quad \text{when} \quad k \to \infty\,, \tag{1}$$

which is a function of a single variable y known as the West scaling variable,[10]

$$y = \frac{1}{v}\left(\omega - \frac{\hbar k^2}{2m}\right)\,, \tag{2}$$

and $v = \hbar k/m$ is the velocity of the helium atom with momentum $\hbar k$. The leading contribution to the impulse approximation is determined by the one-body density matrix $\rho_1(z)$,

$$J(y) = \frac{1}{2\pi}\int_{-\infty}^{\infty} dz e^{-iyz}\rho_1(z)\,, \tag{3}$$

provided that the bare interaction between particles has a converging Fourier transform when $k \to \infty$.[8,11] The direction of the momentum transform is chosen here as the z-axis.

The momentum distribution, which is the Fourier transform of the one-body density matrix,

$$n(k) = \int d^3 r e^{-i\mathbf{k}\cdot\mathbf{r}}\rho_1(r)\,, \tag{4}$$

is not a directly measurable quantity in strongly correlated systems like liquid helium, although it can be accessed indirectly by inspecting the Compton profile at high momentum transfer as the previous expressions show. However, still at the highest momenta available at the laboratory the inclusion of corrections to the impulse approximation is required. These corrections are introduced in the form of *final state effects* and are harder to evaluate from the theoretical point of view as one requires precise knowledge of complex ground state quantities like the off-diagonal two-body density matrix. It has been a long standing goal in the experimental studies to reach such high momenta that these effects can be ignored or at least well parametrized. Theoretical analysis also indicate that final state effects are required to achieve a good description of $S(k,\omega)$ at almost any momentum when the system presents a Bose condensate as in pure He^4 or He^3-He^4 mixtures.[12–14]

In Bose systems the condensate fraction n_c is the value of the momentum distribution at $k = 0$ and tells how many particles are in the condensate state, $n_c = N_0/N$, where N is the total number of particles and N_0 the number of particles in the condensate. In scattering experiments at high momentum transfers the condensate fraction should show up as a delta-like peak on top of the non-condensate distribution broadened by final state effects and the instrumental resolution. In ^{4}He the highest momenta achieved while still keeping a good resolution is $k \approx 30\text{Å}^{-1}$. With experience a detailed parametrization of the final state function has been suggested and an estimate of the condensate fraction given.[15] Theoretical studies starting from a microscopic description of the quantities involved have also been carried out. One can conclude from all this that the condensate fraction in ^{4}He is slightly below ten percent at zero temperature.

Theoretical methods to calculate the one-body density matrix[16–18] and from that the scaling function J(y) are based on summation of infinite series of diagrams starting from the Jastrow-Feenberg wave function or the spectral function analysis where integration over all excited state energies is required.[19] Also the diffusion Monte Carlo method has been successfully implemented for the one-body density matrix calculation.[20,21] In this work we want to approach the problem from a different point of view and calculate directly the high energy and high momentum limit of $S(k,\omega)$ using equations of motions and the linear response theory.[22–25] The key quantity in this method is the self energy, $\Sigma(k,\omega)$, which is a complex quantity that scales in a similar way as the dynamic structure function

$$\Sigma(k,\omega) \to \hbar v(R(y) + iI(y)) \quad \text{when} \quad k \to \infty \,. \tag{5}$$

As an important check we use the second moment sum rule, which tells that at $k \to \infty$ and $\omega = 0$ the self energy should be proportional to the kinetic energy of the system.

2. Linear Response and Equation of Motion Method

Let us assume that a quantum fluid at zero temperature is driven out of the ground state by an infinitesimal external interaction $\tilde{U}_{\text{ext}}(k,\omega)$, with a given frequency ω and wave number k. It induces a change in the density, $\delta\tilde{\rho}_1(k,\omega)$, of an originally homogeneous system. The response of the system is assumed to be linear and the information on its dynamic properties, excitations and decay modes, is contained in the density–density linear–response function defined as

$$\chi(k,\omega) = \frac{\delta\tilde{\rho}_1(k,\omega)}{\rho_0 \tilde{U}_{\text{ext}}(k,\omega)} \,. \tag{6}$$

This is a complex quantity and its imaginary part is connected to the dynamic structure function by the fluctuation–dissipation theorem

$$S(k,\omega) = -\frac{1}{\pi}\Im m\left[\chi(k,\omega)\right]. \tag{7}$$

Hence, $S(k,\omega)$ is known once the relation between the one–body density fluctuations and the perturbation has been established, and one can then compare the theoretical results with experiments.

2.1. *Elementary excitation modes*

At low momenta the dynamic structure function typically consists of a low–frequency sharp peak and a high–frequency broad contribution.[2,26] It is therefore customary to write

$$S(k,\omega) = Z(k)\delta(\hbar\omega - \hbar\omega_0(k)) + S_{\text{mp}}(k,\omega) \,. \tag{8}$$

This immediately suggests that the density–density response function can be written in a form familiar from the random phase approximation,

$$\chi(k,\omega) = S(k)\left[\frac{1}{\hbar\omega - \varepsilon(k) - \Sigma(k,\omega)} - \frac{1}{\hbar\omega + \varepsilon(k) + \Sigma^*(k,-\omega)}\right], \tag{9}$$

where $\varepsilon(k)$ is the energy of a single collective mode and $\Sigma(k,\omega)$ is the complex self–energy,

$$\Sigma(k,\omega) = A(k,\omega) - iB(k,\omega)\,. \tag{10}$$

The elementary excitation modes are determined by the positive frequency poles $\omega = \omega_0(k)$ of Eq. (9), which lie on the real axis

$$\hbar\omega_0(k) = \varepsilon(k) + A(k,\omega_0(k)) \tag{11}$$

and $B(k,\omega_0(k)) = 0$. The strength $Z(k)$ of that pole can then be evaluated from the derivative of the self–energy,

$$Z(k) = S(k)\left[1 - \left.\frac{dA(k,\omega)}{d(\hbar\omega)}\right|_{\omega=\omega_0}\right]^{-1}. \tag{12}$$

The second term in Eq. (8), $S_{\mathrm{mp}}(k,\omega)$, is the multi–phonon background, *i.e.* the contribution in which a projectile like a neutron probing the system exchanges energy with two or more excitation modes. This contribution is determined by the imaginary part of the self–energy

$$S_{\mathrm{mp}}(k,\omega) = \frac{1}{\pi}S(k)\frac{B(k,\omega)}{[\hbar\omega - \varepsilon(k) - A(k,\omega)]^2 + [B(k,\omega)]^2}\,. \tag{13}$$

The dynamic structure function is a positive function for positive values of ω and that is why $B(k,\omega)$ must also be a non–negative function. The condition $B(k,\omega) > 0$ implies that the energy $\hbar\omega$ transferred into the system has created excitation modes that can decay into other lower lying excitations within finite lifetime.

In addition, the relative weight, $Z(k)/S(k)$, gives the efficiency of scattering processes from a single collective excitation, as seen from the (zeroth–moment) sum rule

$$\int_0^\infty S(k,\omega)d(\hbar\omega) = S(k)\,. \tag{14}$$

In other words, it gives the fraction of available scattering processes going through a single collective mode at a given wave vector. If the only excitation in the system were the collective mode $\varepsilon(k)$, like in the Feynman approximation, the ratio $Z(k)/S(k)$ would be one.

2.2. *Impulse approximation*

As mentioned in the introduction, the impulse approximation is the $k \to \infty$, $\hbar\omega \to \frac{\hbar^2 k^2}{2m}$ limit of $S(k,\omega)$ for those systems where the inter atomic potential is bounded from above and below, such that its Fourier transform exists. In that limit the dynamic structure function is completely determined by the one-body density matrix given in Eq. (3) and thus by the momentum distribution,[8]

$$S^{IA}(k,\omega) = \int \frac{d^3p}{(2\pi)^3 \rho_0} n(\mathbf{p}) \delta(t(|\mathbf{k}+\mathbf{p}|) - t(k) - \hbar\omega) \tag{15}$$

where $t(k) = \hbar^2 k^2/2m$.

It is customary to write the one-body density matrix in an exponential form[16,27]

$$\rho_1(|\mathbf{r}-\mathbf{r}'|) = \rho_0 n_c e^{-Q(|\mathbf{r}-\mathbf{r}'|)} \tag{16}$$

where ρ_0 is the density of the fluid and $Q(|\mathbf{r}-\mathbf{r}'|)$ contains all the correlations between the external points $\mathbf{r}$ and $\mathbf{r}'$, such that $\lim_{r\to\infty} Q(r) = 0$. The presence of the long range order in the one-body density matrix is due to the macroscopic occupation of the Bose condensate and thus the condensate fraction n_c can be recovered from it

$$\lim_{r\to\infty} \rho_1(r) = \rho_0 n_c = \frac{N_0}{V} \, . \tag{17}$$

This leads to a δ-function in the momentum distribution and one usually separates it from the non-condensate part

$$\begin{aligned} n(\mathbf{p}) &= (2\pi)^3 \rho_0 n_c \delta(\mathbf{p}) + n_c n_{nc}(\mathbf{p}) \\ n_{nc}(\mathbf{p}) &= \rho_0 \int d^3r \, e^{i\mathbf{p}\cdot\mathbf{r}} \left(e^{-Q(r)} - 1 \right) \, . \end{aligned} \tag{18}$$

With the normalization

$$\int \frac{d^3p}{(2\pi)^3 \rho_0} n(\mathbf{p}) = 1 \, , \tag{19}$$

one gets

$$\int \frac{d^3p}{(2\pi)^3 \rho_0} n_{nc}(\mathbf{p}) = \frac{1-n_c}{n_c} \tag{20}$$

and

$$e^{-Q(0)} = \frac{1}{n_c} \, . \tag{21}$$

With these definitions the impulse approximation $S^{IA}(k,\omega)$ of Eq. (15) splits into two parts

$$S^{IA}(k,\omega) = n_c \delta(\hbar\omega - t(k)) + n_c S^{NC}(k,\omega) \, , \tag{22}$$

where the first term comes from the fraction of particles in the condensate and the second term gives the contribution of the rest,

$$
\begin{aligned}
S^{NC}(k,\omega) &= \int \frac{d^3p}{(2\pi)^3\rho_0} n_{nc}(\mathbf{p})\delta(\hbar\omega - t(|\mathbf{k}+\mathbf{p}|) - t(p)) .\\
&\rightarrow \frac{m}{\hbar^2 k}\,\frac{1}{(2\pi)^2\rho_0}\int_{|y|}^{\infty} pdp\; n_{nc}(p)\ .
\end{aligned}
\tag{23}
$$

In this way, the dynamic structure function can be expressed as a function of the West scaling variable defined in Eq. (2)

$$
J(y) \equiv \lim_{k\to\infty} \frac{\hbar^2 k}{m} S(k,\omega) = n_c\delta(y) + n_c J^{NC}(y)\ . \tag{24}
$$

Now in the large k limit $S(k) \rightarrow 1$ and $\epsilon(k) \rightarrow \hbar^2k^2/2m$ which means that $\hbar\omega - \epsilon(k) \rightarrow y(\hbar^2 k/m)$. The requirement that $(\hbar^2k/m)S(k,\omega)$ scales on y implies therefore that the self energy must also scale in y

$$
\lim_{k\to\infty} \frac{m}{\hbar^2 k}\Sigma(k,\omega) = R(y) - iI(y)\ , \tag{25}
$$

and $J(y)$ can be written in the form

$$
J(y) = -\frac{1}{\pi}\Im m \frac{1}{y - R(y) + iI(y)}\ . \tag{26}
$$

If this has a pole at $y = 0$ then the strength of the pole gives the condensate fraction

$$
n_c = \left[1 - \frac{d}{dy} R(y) \bigg|_{y=0} \right]^{-1}\ , \tag{27}
$$

and the non-condensate contribution becomes

$$
n_c J^{NC}(y) = \frac{1}{\pi}\frac{I(y)}{(y - R(y))^2 + I^2(y)}\ . \tag{28}
$$

Differentiating Eq. (23) with respect to y one gets

$$
n_c \frac{dJ^{NC}(y)}{dy} = \frac{1}{(2\pi)^2\rho_0} y\, n_{nc}(y) \quad \text{when} \quad y > 0\ , \tag{29}
$$

which shows that the complete momentum distribution of the system can be recovered once the real and imaginary parts of the scaled self energy are known.

In the following we show explicitly using equations of motion that the self energy scales in y and that a pole exists in the dynamic structure function. The location of this pole is different from the location of the poles generated by elementary excitations at low k as this one emerges in the middle of the continuum. This means that both the imaginary and real parts must vanish at $y = 0$. Around the pole the real part must be linear and the imaginary part quadratic in y.

3. Self Energy from Equations of Motion

In the derivation of the equations of motion we assume that the time evolution of the wave function is governed by the least–action principle[28,29]

$$\delta \mathcal{S} = \delta \int_{t_0}^{t} dt \mathcal{L}(t) \equiv \delta \int_{t_0}^{t} dt \left\langle \Psi(t) \left| H(t) - i\hbar \frac{\partial}{\partial t} \right| \Psi(t) \right\rangle = 0 \tag{30}$$

where the Hamiltonian

$$H(t) = H_0 + \hat{U}_{\rm ext}(t) \tag{31}$$

contains the ground–state Hamiltonian H_0 and a time dependent one-body external potential $\hat{U}_{\rm ext}(t) = \sum_{i=1}^{N} U_{\rm ext}(\mathbf{r}_i;t)$ that creates an infinitesimal disturbance into the system. This is exactly the term that provides the relation between the one–body density fluctuations and the perturbation, needed to calculate the dynamic linear response.

The presence of a time-dependent perturbation modifies the ground state wave function. The time dependence of the new wave function can be separated into three parts: a phase that comes from the time evolution of the unperturbed ground state $\Psi_0(\mathbf{r}_1, \mathbf{r}_2, \ldots, \mathbf{r}_N)$, a normalization factor that is modified due to the perturbation and a set of correlation functions that become time dependent. Putting all this together we get

$$\Psi(\mathbf{r}_1, \ldots, \mathbf{r}_N; t) = e^{-iE_0 t/\hbar}\Phi(t) = e^{-iE_0 t/\hbar} \frac{1}{\sqrt{\mathcal{N}(t)}} \, e^{\frac{1}{2}\delta U(\mathbf{r}_1, \ldots, \mathbf{r}_N; t)} \Psi_0(\mathbf{r}_1, \ldots, \mathbf{r}_N) \,. \tag{32}$$

One can think about $\delta U(\mathbf{r}_1, ..., \mathbf{r}_N; t)$ as a complex–valued excitation operator

$$\delta U(\mathbf{r}_1, ..., \mathbf{r}_N; t) = \sum_i \delta u_1(\mathbf{r}_i; t) + \sum_{i<j} \delta u_2(\mathbf{r}_i, \mathbf{r}_j; t) \tag{33}$$

expanded in terms of one– and two–body correlation functions. The time–dependent one–body function $\delta u_1(\mathbf{r}_i; t)$ must be included in the description because dynamics will normally break the translational invariance of the system, but restricting the time dependence to the one–body component only would lead directly to the Feynman theory of one–phonon excitations. The time–dependent two–body component is significant in situations where the external field excites fluctuations of wavelengths comparable to the inter–particle distance.

Using the ground state Schrödinger equation we can write the expectation value entering in the action integral in the form

$$\begin{aligned}
\mathcal{L}(t) &= \left\langle \Phi(t) \left| H - E_0 - \frac{\hbar}{2}\left(i\frac{\partial}{\partial t} + h.c.\right) \right| \Phi(t) \right\rangle \\
&= \frac{1}{\mathcal{N}(t)} \left\langle \Psi_0 \left| e^{\frac{1}{2}\delta U^*(t)} [H_0, e^{\frac{1}{2}\delta U(t)}] \right| \Psi_0 \right\rangle \\
&+ \left\langle \Phi(t) \left| -\frac{\hbar}{2}\left(i\frac{\partial}{\partial t} + h.c.\right) + \hat{U}_{ext}(t) \right| \Phi(t) \right\rangle .
\end{aligned}$$

where the notation h.c. indicates the hermitian conjugation of the time–derivative operator.

The potential energy term commutes with $\delta U(t)$ and thus only the kinetic energy gives contribution to the commutator. The algebra to evaluate both matrix elements is straight forward giving

$$\mathcal{L}(t) = \left\langle \Phi(t) \left| \frac{\hbar^2}{8m} \sum_{j=1}^{N} |\nabla_j \, \delta U(t)|^2 + \frac{1}{2}\hbar \, \Im m[\delta \dot{U}(t)] + \hat{U}_{ext}(t) \right| \Phi(t) \right\rangle . \tag{34}$$

In the least–action principle we search for the correlation functions that minimize the action integral (30). The variation with respect to $\delta u_1(\mathbf{r}_1;t)$ and $\delta u_2(\mathbf{r}_1,\mathbf{r}_2;t)$ gives two Euler equations, which can be written as one- and two-particle continuity equations — the equations of motion of the system,[22–24]

$$\nabla_1 \cdot \mathbf{j}_1(\mathbf{r}_1;t) + \delta\dot{\rho}_1(\mathbf{r}_1;t) = D_1(\mathbf{r}_1;t) \tag{35}$$

$$\left[\ \nabla_1 \cdot \mathbf{j}_2^{(1)}(\mathbf{r}_1,\mathbf{r}_2;t) + \text{same for } \ (1 \leftrightarrow 2)\right] + \delta\dot{\rho}_2(\mathbf{r}_1,\mathbf{r}_2;t) = D_2(\mathbf{r}_1,\mathbf{r}_2;t)\,, \tag{36}$$

where $\mathbf{j}_1(\mathbf{r}_1;t)$ and $\mathbf{j}_2^{(1)}(\mathbf{r}_1,\mathbf{r}_2;t)$ are the one- and two-body currents, while $\delta\rho_1(\mathbf{r}_1;t)$ and $\delta\rho_2(\mathbf{r}_1,\mathbf{r}_2;t)$ are the one- and two-body density fluctuations, which are related to the time dependent n–particle densities

$$\bar{\rho}_n(\mathbf{r}_1,\ldots\mathbf{r}_n;t) = \frac{N!}{(N-n)!}\int d^3r_{n+1}\ldots d^3r_N |\Psi(\mathbf{r}_1,\ldots\mathbf{r}_N;t)|^2\,. \tag{37}$$

In linear response theory one assumes that the time dependent perturbation is infinitesimal and hence it is natural to separate the time dependent and independent parts of the density,

$$\bar{\rho}_n(\mathbf{r}_1,\ldots\mathbf{r}_n;t) = \rho_n(\mathbf{r}_1,\ldots\mathbf{r}_n) + \delta\rho_n(\mathbf{r}_1,\ldots\mathbf{r}_n;t)\,. \tag{38}$$

Without loss of generality we can write the density fluctuations in terms of relative changes

$$\begin{aligned}
\delta\rho_1(\mathbf{r}_1;t) &\equiv \rho_0\xi(\mathbf{r}_1;t)\,,\\
\delta\rho_2(\mathbf{r}_1,\mathbf{r}_2;t) &\equiv \rho_0^2 g_2(\mathbf{r}_1,\mathbf{r}_2)\Big[\xi(\mathbf{r}_1;t) + \xi(\mathbf{r}_2;t) + \Delta(\mathbf{r}_1,\mathbf{r}_2;t)\Big]\,,\\
\delta\rho_3(\mathbf{r}_1,\mathbf{r}_2,\mathbf{r}_3;t) &\equiv \rho_0^3 g_3(\mathbf{r}_1,\mathbf{r}_2,\mathbf{r}_3)\Big[\xi(\mathbf{r}_1;t) + \xi(\mathbf{r}_2;t) + \xi(\mathbf{r}_3;t)\\
&\quad + \Delta(\mathbf{r}_1,\mathbf{r}_2;t) + \Delta(\mathbf{r}_1,\mathbf{r}_3;t) + \Delta(\mathbf{r}_2,\mathbf{r}_3;t)\\
&\quad + \Delta_3(\mathbf{r}_1,\mathbf{r}_2,\mathbf{r}_3;t)\Big]\,.
\end{aligned} \tag{39}$$

which define the functions $\xi(\mathbf{r}_1;t)$, $\Delta(\mathbf{r}_1,\mathbf{r}_2;t)$ and $\Delta_3(\mathbf{r}_1,\mathbf{r}_2,\mathbf{r}_3;t)$ in terms of one-, two- and three-body density fluctuations, and two- and three-particle ground state distribution functions $g_2(\mathbf{r}_1,\mathbf{r}_2)$ and $g_2(\mathbf{r}_1,\mathbf{r}_2,\mathbf{r}_3)$.

With these substitutions and using the BGY equations one gets explicit expressions for the currents[30]

$$\mathbf{j}_1(\mathbf{r}_1;t) = \frac{\hbar\rho_0}{2mi}\Big\{\nabla_1\xi(\mathbf{r}_1;t) \qquad (40)$$

$$-\rho_0\int d^3r_2 g_2(\mathbf{r}_1,\mathbf{r}_2)\nabla_1 u_2(\mathbf{r}_1,\mathbf{r}_2)\left[\xi(\mathbf{r}_2;t)+\Delta(\mathbf{r}_1,\mathbf{r}_2;t)\right]\Big\}$$

$$\mathbf{j}_2^{(1)}(\mathbf{r}_1,\mathbf{r}_2;t) = \frac{\hbar\rho_0^2}{2mi}\Big\{g_2(\mathbf{r}_1,\mathbf{r}_2)\nabla_1\left[\xi(\mathbf{r}_1;t)+\Delta(\mathbf{r}_1,\mathbf{r}_2;t)\right]$$

$$-\rho_0\int d^3r_3\, g_3(\mathbf{r}_1,\mathbf{r}_2,\mathbf{r}_3)\nabla_1 u_2(\mathbf{r}_1,\mathbf{r}_3)\Big[\xi(\mathbf{r}_3;t)+\Delta(\mathbf{r}_1,\mathbf{r}_3;t)$$

$$+\Delta(\mathbf{r}_2,\mathbf{r}_3;t)+\Delta_3(\mathbf{r}_1,\mathbf{r}_2,\mathbf{r}_3;t)\Big]\Big\}\,. \qquad (41)$$

These results are exact within the assumption that the wave function contains only one– and two–particle correlations, but the currents contain now three unknown quantities. As we have ignored fluctuations of the triplet correlation functions it is natural to set $\Delta_3(\mathbf{r}_1,\mathbf{r}_2,\mathbf{r}_3;t) = 0$. In the superposition approximation limit that would be an exact result. Then we are left with two continuity equations which solve the remaining two unknown quantities.

The terms which depend on the external potential are collected into the functions $D_1(\mathbf{r}_1;t)$ and $D_2(\mathbf{r}_1,\mathbf{r}_2;t)$ and they are the terms which drive excitations into the system,

$$D_1(\mathbf{r}_1;t) = \frac{2\rho_0}{i\hbar}\Big\{U_{\rm ext}(\mathbf{r}_1;t)+\rho_0\int d^3r_2\Big[g_2(\mathbf{r}_1,\mathbf{r}_2)-1\Big]U_{\rm ext}(\mathbf{r}_2;t)\Big\} \qquad (42)$$

$$D_2(\mathbf{r}_1,\mathbf{r}_2;t) = \frac{2\rho_0^2}{i\hbar}\Big\{g_2(\mathbf{r}_1,\mathbf{r}_2)\left[U_{\rm ext}(\mathbf{r}_1;t)+U_{\rm ext}(\mathbf{r}_2;t)\right]$$

$$+\rho_0\int d^3r_3\Big[g_3(\mathbf{r}_1,\mathbf{r}_2,\mathbf{r}_3)-g_2(\mathbf{r}_1,\mathbf{r}_2)\Big]U_{\rm ext}(\mathbf{r}_3;t)\Big\}\,. \qquad (43)$$

The continuity equations for a homogeneous system are more easily solved in momentum space. The first continuity equation (35) can then be written in a more familiar form

$$\hbar\omega - \varepsilon(k) - \Sigma(k,\omega) = \frac{2S(k)\tilde{U}_{\rm ext}(k,\omega)}{\tilde{\xi}(k,\omega)} \qquad (44)$$

in terms of the collective reference mode $\varepsilon(k)$ and the self energy $\Sigma(k,\omega)$,

$$\Sigma(\mathbf{k},\omega) = -\frac{\hbar^2}{2m}\int\frac{d^3p}{(2\pi)^3\rho_0}\,\mathbf{k}\cdot\left(\tfrac{\mathbf{k}}{2}-\mathbf{p}\right)Q\left(\left|\tfrac{\mathbf{k}}{2}-\mathbf{p}\right|\right)\frac{\tilde{\Delta}(\mathbf{k},\mathbf{p};\omega)}{\xi(\mathbf{k};\omega)}\,. \qquad (45)$$

Here $Q(q)$ is a shorthand notation for the Fourier transform

$$\mathbf{q}Q(q) = F[g_2(r)\nabla u_2(r)]\,. \qquad (46)$$

The Fourier transform of the two-particle density fluctuation, $\tilde{\Delta}(\mathbf{k},\mathbf{p};\omega)$, is determined by the solution of the second continuity equation. On the right hand side of

Eq. (44) we have the inverse of the linear response function $\chi(k,\omega)$ defined in Eq. (6).

In this work we are interested in the $k \to \infty$ limit where the second continuity equation can be solved analytically

$$\frac{\tilde{\Delta}(\mathbf{k},\mathbf{p};\omega)}{\xi(\mathbf{k};\omega)} = -\frac{\hbar^2}{2m}\frac{\left[\mathbf{k}\cdot(\frac{\mathbf{k}}{2}+\mathbf{p})\tilde{L}\left(|\frac{\mathbf{k}}{2}+\mathbf{p}|\right)+\mathbf{k}\cdot(\frac{\mathbf{k}}{2}-\mathbf{p})\tilde{L}\left(|\frac{\mathbf{k}}{2}-\mathbf{p}|\right)\right]}{\hbar\omega - t\left(|\frac{\mathbf{k}}{2}+\mathbf{p}|\right) - t\left(|\frac{\mathbf{k}}{2}-\mathbf{p}|\right)} \tag{47}$$

with $L(q)$ the Fourier transform of $\log g_2(r)$.

4. Results for the Kinetic Energy and Condensate Fraction

Now we can obtain results for two interesting limits when $k \to \infty$. In the $\omega \to 0$ limit the self energy must be proportional to the kinetic energy per particle $\langle T_{\rm kin}\rangle$. ¿From Eqs. (45) and (47) we get

$$\lim_{k\to\infty}\Sigma(k,\omega=0) = -\frac{\hbar^2}{6m}\int\frac{d^3q}{(2\pi)^3\rho}q^2Q(q)\tilde{L}(q) = -\frac{4}{3}\langle T_{\rm kin}\rangle\ . \tag{48}$$

This result is exact if we assume that the ground state wave function contains only the Jastrow two-particle correlation function $u_2(r)$. In that case

$$\langle T_{\rm kin}\rangle = \frac{\hbar^2}{8m}\rho_0\int d^3r\nabla g_2(r)\cdot\nabla u_2(r) = \frac{\hbar^2}{8m}\rho_0\int d^3r\nabla\log g_2(r)\cdot g_2(r)\nabla u_2(r) \tag{49}$$

which has the Fourier transform of Eq. (48).

The second limit is the impulse approximation where we replace the energy $\hbar\omega$ by $(\hbar^2k/m)y - \hbar^2k^2/2m$ and let k go to infinity. After the substitution we can write the self energy of Eq. (45) as

$$\begin{aligned}\Sigma(\mathbf{k},y) &= \frac{\hbar^2}{4m}\int\frac{d^3q}{(2\pi)^3\rho_0}\frac{[\mathbf{k}\cdot\mathbf{q}]^2Q(q)\tilde{L}(q)}{ky+\mathbf{k}\cdot\mathbf{q}}\\ &= \frac{\hbar^2k^2}{4m}\frac{1}{(2\pi)^2\rho_0}\int_0^\infty dqq^4\ Q(q)\tilde{L}(q)\int_{-1}^1 dt\frac{t^2}{ky+kqt}\\ &= A(k,y) - iB(k,y)\end{aligned} \tag{50}$$

The imaginary part comes from the pole in the integrand.

$$\begin{aligned}B(k,y) &= \frac{\hbar^2k}{4m}\frac{1}{4\pi\rho}\int_0^\infty q^3dq\ Q(q)\tilde{L}(q)\int_{-1}^1 dt\ t^2\delta(t+\frac{y}{q})\\ &= \frac{\hbar^2}{4m}\frac{1}{4\pi\rho}ky^2\int_y^\infty qdq\ Q(q)\tilde{L}(q) \equiv \frac{\hbar^2k}{m}I(y)\,.\end{aligned} \tag{51}$$

The real part is then the principle value

$$A(\mathbf{k},y) = -\frac{\hbar^2}{2m}\frac{ky}{(2\pi)^2\rho_0}\int_0^\infty dqq^2Q(q)\tilde{L}(q)\Big[1-\frac{y}{2q}\log\left|\frac{y/q+1}{y/q-1}\right|\Big] \equiv \frac{\hbar^2k}{m}R(y)\,. \tag{52}$$

With these results for the self energy we can calculate the impulse approximation of the dynamic structure function $S(k,\omega)$

$$J(y) = \frac{\hbar^2 k}{m} S(k,\omega) = -\frac{1}{\pi} \Im m \, \frac{1}{y - R(y) + iI(y)} \,. \tag{53}$$

The denominator is zero at $y = 0$ and thus $J(y)$ has the structure of Eq.(22) with the pole at $y = 0$, while the strength of the pole is proportional to the condensate fraction

$$\frac{1}{n_c} = 1 + \frac{1}{4} \int \frac{d^3q}{(2\pi)^3 \rho} Q(q) \tilde{L}(q) \,. \tag{54}$$

Inserting the ground state results for the radial distribution function $g_2(r)$ and the correlation function $u_2(r)$ we can evaluate the condensate fraction, resulting in a value $n_c = 10.1\%$ in reasonable agreement with other estimates.

¿From the above real and imaginary parts we can also calculate the non-condensate contribution to the scaled structure function,

$$J^{NC}(y) = \frac{1}{n_c \mathcal{D}(y) \pi} \frac{1}{(2\pi)^2 \rho_0} \int_{|y|}^{\infty} dqq \frac{1}{4} Q(q) \tilde{L}(q) \tag{55}$$

with

$$\mathcal{D}(y) = \left[1 - \frac{1}{y} R(y)\right]^2 + \left[\frac{1}{y} I(y)\right]^2 \,. \tag{56}$$

Differentiating $J^{NC}(y)$ with respect to y and using Eq.(29) one gets the non-condensate part of the momentum distribution, while the one-dimensional Fourier transform gives the exponential function $e^{Q(r)} - 1$ appearing in the one-body density matrix as defined in Eq. (16).

5. Summary

We have shown in this work how to connect two different methods to calculate the dynamic structure function. The variational method based on equations of motion built from the correlated wave function has been very successful in describing the low momentum excitations. In the high momentum limit the impulse approximation is strictly valid and assumes knowledge of the one-body density matrix. Its systematic diagrammatic expansion has been carefully studied[16,17] and the one-dimensional Fourier transform gives the scaled dynamic structure function, $J(y)$. We have shown that $J(y)$ can also be evaluated if the $k \to \infty$ limit of the self energy as a function of y is known. The accuracy of the method has been tested at two frequencies. At $\omega = 0$ we get the kinetic energy, which is equal to the one calculated directly from the Jastrow-type correlated wave function. At $\omega = \hbar k^2/(2m)$ we get the condensate fraction, which is in reasonable agreement with experimental estimates and Monte Carlo simulations. The effect of triplet correlations into these quantities will be studied in a separate work.

Acknowledgments

We thank Eckhard Krotscheck for many lively discussions we had in Linz where all of us have enjoyed his hospitality during this work.

References

1. P.C.Hohenberg and P.M.Platzman, *Phys. rev.* **152**, 198 (1966).
2. H. Glyde, *Excitations in liquid and solid helium* (Oxford University Press, Oxford, 1994).
3. J. Boronat and J. Casulleras, *Europhys. Lett.* **38**, p. 291 (1997).
4. C. C. Chang and C. E. Campbell, *Phys. Rev. B* **13**, 3779 (1976).
5. V. Apaja and M. Saarela, *Phys. Rev. B* **57**, 5358 (1998).
6. W. Wu, S. A. Vitiello, L. Reatto and M. H. Kalos, *Phys. Rev. Lett.* **67**, p. 1446 (1991).
7. F. Pistolesi, *Phys. Rev. Lett.* **81**, 397 (1998).
8. H. A. Gersch, L. J. Rodriguez and P. N. Smith, *Phys. Rev. A* **5**, p. 1547 (1972).
9. S.W.Lovessey, *Theory of neutron scattering from condensed matter* (Clarendon Press, Oxford, 1984).
10. G. B. West, *Phys. Rep. C* **18**, p. 263 (1975).
11. J. J. Weinstein and J. W. Negele, *Phys. Rev. Lett.* **49**, p. 1016 (1982).
12. C. Carraro and S. E. Koonin, *Phys. Rev. Lett.* **65**, 2792 (1990).
13. F.Mazzanti, J.Boronat and A.Polls, *Phys. Rev. B* **53**, 5661 (1996).
14. F.Mazzanti, A.Polls and J.Boronat, *Phys. Rev. B* **63**, 054521/1 (2001).
15. H. R. Glyde, R. T. Azuah and W. G. Stirling, *Phys. Rev. B* **62**, p. 14337 (2000).
16. M. L. Ristig and J. W. Clark, *Phys. Rev. B* **14**, 2875 (1976).
17. S. Fantoni, *Nuovo Cimento* **44A**, 191 (1978).
18. E. Krotscheck, *Phys. Rev. B* **32**, 5713 (1985).
19. O. Benhar, A. Fabrocini and S. Fantoni, *Nucl.Phys. A* **505**, p. 267 (1989).
20. P. M. Whitlock and R. M. Panoff, *Can. J. Phys.* **65**, 1409 (1987).
21. E. L. Pollock and D. M. Ceperley, *Phys. Rev. B* **36**, p. 8343 (1987).
22. M. Saarela and J. Suominen, Dynamic structure function of the quantum boson fluid in the jastrow theory, in *Condensed Matter Theories*, eds. J. S. Arponen, R. F. Bishop and M. Manninen (Plenum, New York, 1988).
23. B. E. Clements, H. Forbert, E. Krotscheck, H. J. Lauter, M. Saarela and C. J. Tymczak, *Phys. Rev. B* **50**, 6958 (1994).
24. B. E. Clements, E. Krotscheck and C. J. Tymczak, *Phys. Rev. B* **53**, 12253 (1996).
25. M. Saarela, *Elementary excitations and dynamic structure of quantum fluids*, in *Introduction to modern methods of quantum many-body theory and their applications*, eds. A. Fabrocini, S. Fantoni and E. Krotscheck (World-Scientific, London, 2002), pp. 205–264.
26. R. A. Cowley and A. D. B. Woods, *Can. J. Phys.* **49**, 177 (1971).
27. M. L. Ristig, G. Senger, M. Serhan and J. W. Clark, *Ann. Phys. (NY)* **243**, 247 (1995).
28. A. K. Kerman and S. E. Koonin, *Ann. Phys. (NY)* **100**, 332 (1976).
29. P. Kramer and M. Saraceno, *Geometry of the time-dependent variational principle in quantum mechanics*, Lecture Notes in Physics, Vol. 140 (Springer, Berlin, Heidelberg, and New York, 1981).
30. J. Suominen and M. Saarela, Elementary excitations and the phase transition in liquid ^{4}he, in *Condensed Matter Theories*, ed. J. Keller (Plenum, New York, 1989).

Part C

Strongly-Correlated Electronic Systems

ELECTRON GAS IN HIGH-FIELD NANOSCOPIC TRANSPORT: METALLIC CARBON NANOTUBES

F. GREEN[1] and D. NEILSON[1,2,3,*]

[1] *School of Physics, University of New South Wales, Sydney, Australia*
[2] *Dipartimento di Fisica, Università di Camerino, Camerino, Italy*
[3] *NEST CNR-INFM, 56126 Pisa, Italy*
[*] *E-mail: david.neilson@unicam.it*

Nanotube structures are unprecedented in their stability and current-carrying capacity at intense driving fields. A comprehensive understanding of electron conduction from equilibrium through to the high-driving-field regime is needed. We present a microscopically conserving quantum-kinetic description of transport for ohmically contacted carbon nanotubes. The approach is computationally straightforward and can describe nonequilibrium response over a wide range of parameters. We have analyzed the interplay of degeneracy and scattering dynamics on gate-controlled conduction in the one-dimensional channel, and have determined transconductances.

Keywords: Carbon nanotubes; transconductance; transport theory.

PACS numbers: 73.63.Fg, 72.10.-d, 73.63.-b

1. Introduction

For transport physicists and microelectronics designers alike, carbon nanotubes have three attractive properties: (a) they are remarkably versatile conductors, (b) they are surprisingly tough, and (c) they have a rich surface chemistry both inside and outside their tubular scaffold. It is the combination of these, in one extremely compact and controllable package, that makes the carbon nanotube unique.[1]

Here we propose a kinetically based approach to the electrical response of carbon nanotubes. We aim to predictively explore the distinct aspects of carrier behavior in these structures not only severally, but in the rich patterns of their interplay. Our main tool will be the kinetic equation for the carrier distribution function along the device. A calculable and microscopically conserving model of this kind can be straightforwardly adapted from standard quantum kinetics. With it, we address the wide spectrum of conductive behaviors that are of current concern to the meso- and nanoscopic design communities.

Prototype devices designed around carbon nanotubes as the active channels have already been successfully demonstrated. More are rapidly emerging. Hence the evolution of nanotube physics may best be seen in the context of the constant drive to ever smaller devices at the highest possible packing densities, and exhibiting ever broader functionalities. The attributes (a)-(c) above, embodied in carbon-nanotube architectures, are clearly ideal to such demands. This is so, at least in principle; in practice, the underlying physics is likely to stretch every capacity for experimental and theoretical analysis at the near-to-molecular scale of these structures.

Nanotube electronics has some formidable challenges of fabrication. The very qualities that make it so desirable entail renewed levels of ingenuity from both the physical-chemistry and condensed-matter communities. The day is still early in harvesting the benefits of nanotubes, and the fruits need to be ripened by every reasonable means. Theoretical understanding of the physics is crucial to that process.

To reach an adequate understanding, the task of describing and controlling current flow in nanotubes has been addressed with almost the whole repertoire of transport analysis. Applications have been made of single-particle quantum transmission, kinetic transport equations, Monte Carlo simulations, density-matrix and Wigner-function formalisms, density functional theory, and Green-function analysis. Regardless of its origins and its strengths, however, each methodology faces the same set of fundamental issues: (i) the boundary conditions at the nanotube-reservoir interfaces and their maximal openness to the macroscopic environment; (ii) the inevitability of operation far from equilibrium; and (iii) the enhancement of carrier correlations by quasi-one-dimensional confinement.

While any approach to electrical conduction in nanotubes must deal with each of these, the larger challenge lies in the concurrent and strongly interactive fashion in which all the issues come into play. For example, a key requirement is to describe the device response from equilibrium through the highest sustainable fields in the nanotube (which are very high indeed). In the process, collision mechanisms that dominate in distinct regimes come to overshadow, and then overtake, one another to various degrees and in qualitatively different ways as the system crosses over from one stage to the next. One of our aims is to set up a microscopically tight model with enough flexibility to cover the important physics, in its most basic aspects, over a very wide operating range.

Let us briefly review some relevant developments. Although it remains essential to cover and understand the weak-field domain, much of the attention must fall on nonequilibrium transport. A pioneering study of conduction in a zero-gap (metallic) single-walled nanotube was made by Yao, Kane, and Dekker[2] who applied a form of the Boltzmann equation to the current response of their nanotube at high source-drain voltages. Their work identified a significant role for inelastic phonon emission in high-field current saturation, and compared their results with experimental measurements whose source-drain fields were up to the order of 50 kV cm^{-1}. However, the Yao *et al.* calculation did not account for elastic scattering, which must dominate at lower fields, where the electron-phonon mean free path readily exceeds the device length.

A detailed examination of strong inelastic electron-phonon collisions, also based on a Boltzmann transport equation, was made by Perebeinos *et al.*[3] for semiconducting tubes. This model too does not include low-field elastic scattering.

Thus, a major simulation goal that remains largely unexplored is how to capture the physics in the wide transitional regime where the system passes from its low-field quasi-ballistic limit to its high-field dissipative limit, in which energy loss by phonon scattering hastens the onset of current saturation.

The demand for calculable models that are sufficiently versatile to describe transport all the way from linear response to strong departures from equilibrium, means that the formalism must be (i) well controlled everywhere, (ii) computationally tractable, and (iii) as faithful as possible to the leading physical effects in each subregion of interest. For our analytical tool we choose the quantum version of the Boltzmann kinetic equation, whose capacity to meet criteria (i)-(iii) has been amply demonstrated in other systems exhibiting one or more of the physical attributes of nanotubes (though very rarely all of them at once).

2. Control of Carrier Density by Gating

High-current-gain transistors use the properties of the confined metallic electron gas trapped at III-V semiconductor heterojunctions. These devices dominate millimeter-wave signal processing technology thanks to (i) the high carrier densities attainable in a quantum-well confined metallic channel; (ii) the ability to sensitively tune these densities by simple gate control; and (iii) reliable, low-resistance contacts at the channel's source and drain.

Metallic Carbon nanotubes have a great deal in common with conventional heterojunctions and they promise a novel class of amplifier exploiting the unique behavior of 1D confined carriers. They differ markedly in having (i) much stronger confinement; (ii) much larger energy scale; (iii) smaller dimensions; and (iv) greater stability.

We recall nanotubes can have a zero or non-zero gap between their valence and conduction bands, depending on the alignment of the tube relative to the atomic planes of the graphite sheets from which they are fabricated. Those with zero gap are termed metallic nanotubes while with non-zero gap they are called semiconductor nanotubes.

Semiconductor nanotube devices are known to perform below their theoretical potential because of difficulties with fabricating source and drain contacts that are reproducible and that have low access resistance. This issue degrades the maximum achievable current-voltage response and the signal gain. However in metallic nanotube devices it has been shown that these issues can be resolved.[4] In the fields of analog signal detection and amplification it has already been demonstrated that metallic nanotubes outperform their semiconductor counterparts.[5]

In semiconductor nanotube devices a gate voltage can control the current flow along the tube in analogy with III-V Field Effect Transistors (FET)s (see Fig. 1). Given the advantages of metallic nanotube devices, it is interesting to establish whether control of carrier density by gating is also possible for metallic nanotubes.

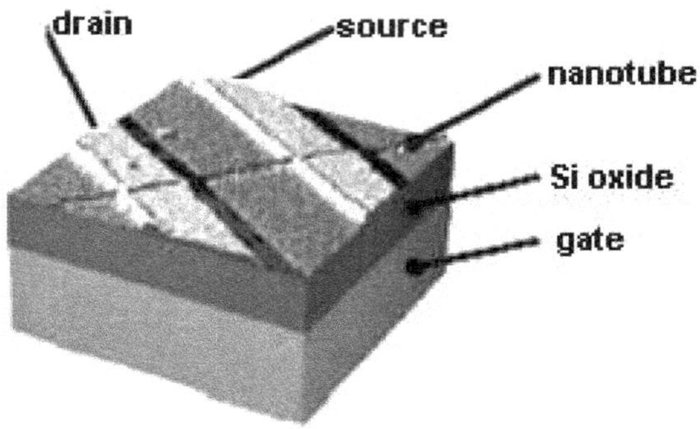

Fig. 1. In semiconductor nanotube devices a gate voltage can control the current flow along the tube in analogy with III-V MOSFETs.

3. Transport Theory

Our aim is to represent conduction across the full operating range of the nanotube. The system goes from the near-ballistic limit at low driving fields to the high-field dissipative limit where energy loss by phonon scattering causes the current to saturate or even to drop. We calculate the current gain of metallic nanotube devices using a quantum Boltzmann kinetic equation approach which (i) is easily integrated with current-voltage measurements, and (ii) allows straightforward extraction of basic scattering parameters.

We consider electrons in 1D conduction subband of uniform metallic single-wall nanotube with a steady-state momentum distribution $f_k(\mu)$. Recent I-V data from Ref. 4 for supported and suspended metallic tubes (with no gate control) is shown in Fig. 2.

We start with the quantum transport equation at a driving voltage V over a length L. The time-dependent Boltzmann equation for distribution function $f_k(\mu)(x,t)$ in a linearly dispersive band is

$$\frac{\partial f}{\partial t} + v_{\rm k}\frac{\partial f}{\partial x} + \frac{eE}{\hbar}\frac{\partial f}{\partial k} = \\ -\frac{1}{\tau_{\rm in}(E_k)}\left(f_k(x,t) - \langle f(x,t)\rangle \frac{f_k^{\rm eq}}{n}\right) - \frac{1}{\tau_{\rm el}(E_k)}\frac{f_k(x,t) - f_{-k}(x,t)}{2}, \quad (1)$$

where $v_{\rm k} = v_{\rm b}[\Theta(k) - \Theta(-k)]$, with $v_{\rm b}$ the group velocity. We take the magnitude of $v_{\rm b}$ as a constant, $v_{\rm b} \approx 5 \times 10^7$ cm.s^{-1}. The scattering dynamics is contained in the inelastic and elastic relaxation rates $\tau_{\rm in}^{-1}$ and $\tau_{\rm el}^{-1}$, respectively, which can depend on E_k. The brackets $\langle ... \rangle$ represent the average over wavevector, spin and valley degeneracy. $f_k^{\rm eq}(\mu)$ is Fermi-Dirac equilibrium function.

The form of the collision terms renders Eq. (1) gauge invariant and ensures that it satisfies detailed balance. It also implies that functional derivatives of f_k such as $\delta f_k/\delta\mu$, which underlie the transconductance, are also microscopically conserving. For collision times independent of E_k, Eq. (1) can be solved analytically at high and low fields

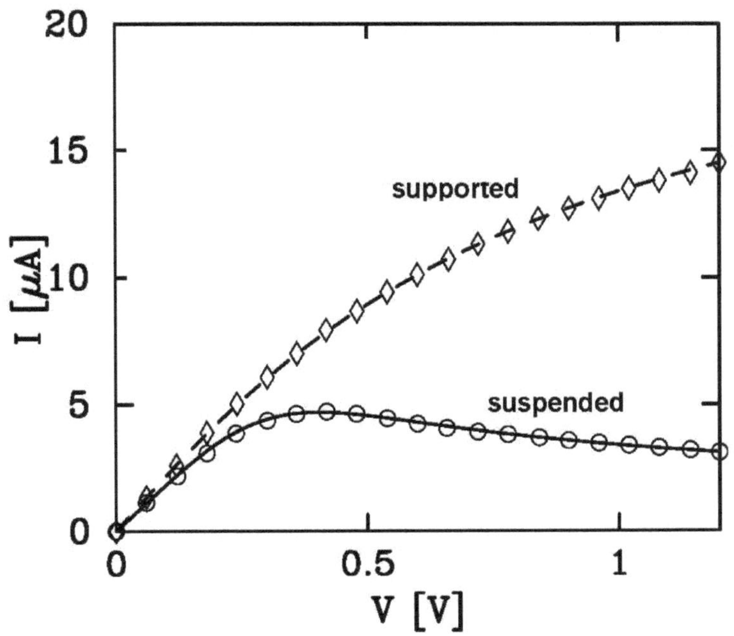

Fig. 2. Current I as a function of source-drain voltage V in two types of metallic carbon nanotubes. Channels are identical; only their heat-sinking differs. Data are from Fig. 2(a) of Ref. 4: *diamonds* for heat-sunk tube, *circles* for suspended tube.

We take into account only single-band scattering. This is consistent if the external driving potentials are less than the subband spacing, which is ~ 0.5 eV. We assume collision times are independent of carrier energy $E_k = \hbar v_b|k|$, and that the channel is free of defects over its operational length L.

The characteristic velocity of the system at density $n = \langle f^{eq}(\mu, k_BT)\rangle$ is the group velocity at the characteristic wave vector $k_{av} = \langle 2|k|f^{eq}(\mu, k_BT)\rangle$. This gives $v_{av}(\mu) = \hbar^{-1}[dE/dk]_{k=k_{av}}$. At large carrier densities $k_{av} \to k_F$ and $v_{av} \to v_F$. The elastic collision time is then $\tau_{el} = \lambda_{el}/v_{av}(\mu) = L/v_{av}(\mu)$. We conclude that elastic scattering is insensitive to field effects.

However, even if inelastic collisions act only at the leads, at high driving fields the accelerated electrons will emit enormous numbers of longitudinal optical phonons. This has the effect of reducing the inelastic mean free path $\lambda_{in}(\mu, eV)$ to be less than L, and it makes λ_{in} sensitive to the driving voltage. To take into account the dependence of the inelastic collision time τ_{in} on the driving field V and on the carrier degeneracy, we introduce a dependence of τ_{in} on V with the criterion that the working length L, comprising the clean channel plus its interfaces with the macroscopic leads, is the upper limit for the elastic and inelastic mean free paths λ_{el} and λ_{in}.

To determine the inelastic mean free path $\lambda_{in}(\mu, V)$, we start with the expecta-

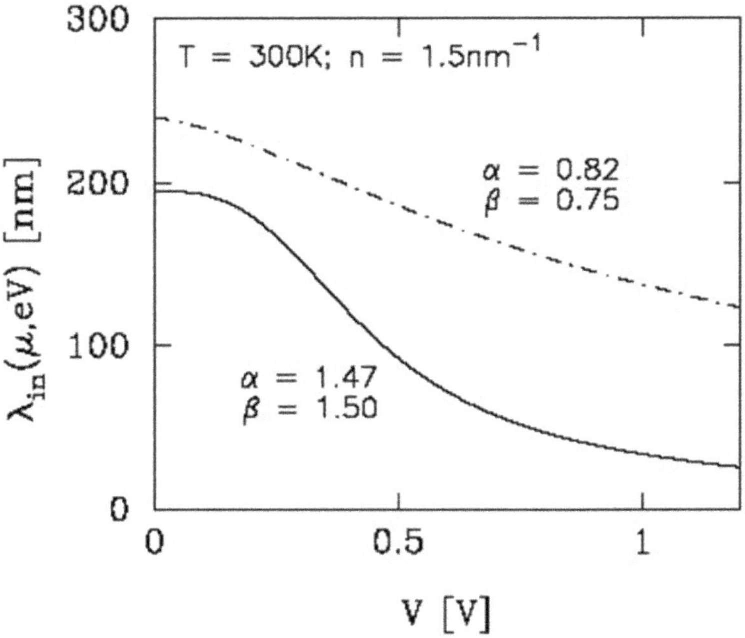

Fig. 3. Inelastic mean free path $\lambda_{in}^{(0)}(\mu, V)$ is set by slope of I-V curve at low-field; α and β are set by I-V curve at large V. Characteristic energy $E_{\rm av} = E_{k_{\rm av}}$.

tion value of the solution[6] to Eq. (1),

$$I = \langle -ev_{\rm k} f_k \rangle = 4G_0 V \frac{\tau k_{\rm F}}{\tau_{\rm el} k_{\rm av}} \left[1 - \exp\left(-\frac{\hbar L k_{\rm av}}{eV\sqrt{\tau\tau_{\rm in}}} \right) \right] \tag{2}$$

where $G_0 = e^2/\pi\hbar$ is the Landauer conductance, $\tau_{\rm in} = \lambda_{\rm in}(\mu, V)/v_{\rm av}(\mu)$, and τ is the total relaxation time given by $\tau^{-1} = \tau_{\rm el}^{-1} + \tau_{\rm in}^{-1}$.

Let us consider the behavior of currents I at high fields from Eq. (2). Supposing $\tau_{\rm in} \sim V^{-\beta} \to \tau$, then for A a constant

$$\begin{aligned} &\lim_{v\to\infty} I = nev_b \sqrt{\frac{\tau}{\tau_{\rm in}}} \frac{1-\exp(-AV^{\beta-1})}{AV^{\beta-1}} \to nev_b \sqrt{\frac{\tau}{\tau_{\rm in}}} && \beta \le 1 \\ &\lim_{v\to\infty} \quad I \sim V^{1-\beta} \to 0. && \beta > 1 \end{aligned} \tag{3}$$

For $\beta \le 1$ the current saturates, while for $\beta > 1$ the current dies off at large V. Since I must vanish linearly as $V \to 0$, for $\beta > 1$ there is thus a critical field at which I peaks.

It straightforward to evaluate $\lambda_{\rm in}(\mu, V)$ by inverting Eq. (2) to solve for the inelastic MFP and fitting the result to the data in Fig. 2. Figure 3 plots the resulting V-dependence of $\lambda_{\rm in}$ so obtained, given the I–V data (see Fig. 2). In the degenerate regime the outcome is not sensitive to n. We take $n = 1.5{\rm nm}^{-1}$, equivalent to $E_{\rm F} = 400{\rm meV}$.

For the heat-sunk device of Fig. 2, the high-field inelastic mean free path obeys a power-law falloff with exponent $\beta = 0.75$. For all the suspended samples studied

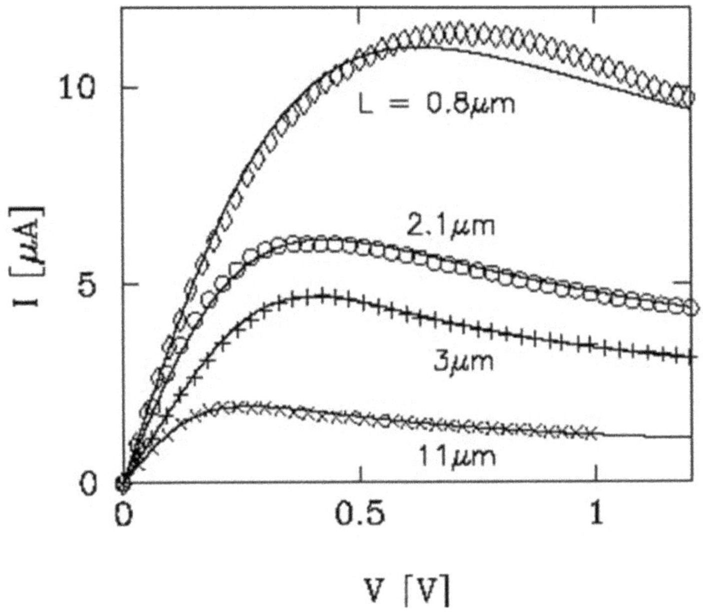

Fig. 4. Data points from Ref. 4 for suspended tubes of different L, and our fits to the data.

in Ref. 4, there is a clear falloff with $\beta = 1.5$. In the low-field regime, $\lambda_{\rm in}$ for each sample goes to a constant.

With the extracted results we can replicate the strong field-dependent inelastic effects by the analytic fit

$$\lambda_{\rm in}(\mu, eV) = \lambda_{\rm in}^{(0)} \tanh(u^\beta)/u^\beta; \quad u = \alpha eV/E_{\rm av}. \tag{4}$$

Figure 4 compares our results from Eq. (2) with experimental data for different length channels from Ref. 4. The quality of fit to the measured I–V curves from is good. The asymptotic tails are faithfully rendered. In the shortest channel shortest channel $L = 0.8$ μm we see some departure from measured data. We ascribe this to (a) the proximity of the bulk leads, which provide efficient heat sinking and thus qualitatively change the hot-phonon behavior, and (b) nonuniform densities in very short carbon nanotubes.

4. Gated Metallic Nanotube Device Characteristics

In signal detection the transconductance G_m is central. We recall the transconductance $G_{\rm m} = \delta I/\delta V_{\rm g}|_V$ is the signal gain from modulation of the carrier density by a gate-bias voltage $V_{\rm g}$ at the fixed operating point $V, I(eV, \mu)$. The differential conductance $G_{\rm diff} = \delta I/\delta V|_{V_{\rm g}}$ is used when analyzing small-signal response about a fixed operating point.

We take the bias gate to be a metallic plate[5,7] separated by a distance $R = 10$ nm from the nanotube axis (see Fig. 5). The nanotube has a radius $r = 1.2$ nm.

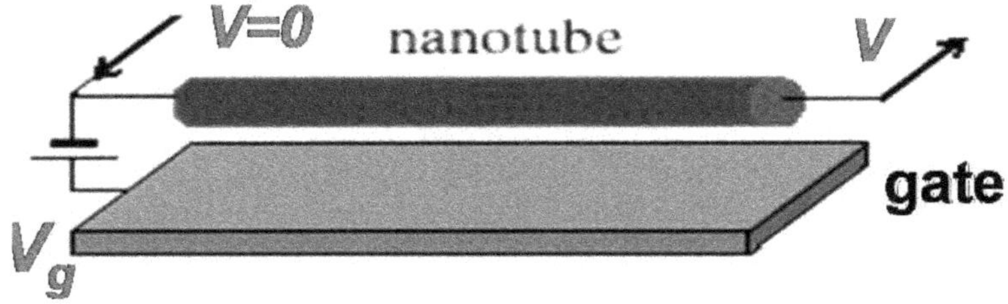

Fig. 5. Bias gate.

The capacitive loop then comprises the bias voltage source $V_{\rm g}$, the tube, and the gap to the gate. Thus $e\delta V_{\rm g} = -\delta\mu + 2e^2\delta n \log(r/2R)/\epsilon$, where δn is the induced change in carrier density

The bias loop's inverse differential capacitance is

$$C_{\rm g}^{-1} = -\delta V_{\rm g}(eL\delta n) = (2/\epsilon L)\log(2R/r) + C_{\rm q}^{-1} \tag{5}$$

where $C_{\rm q}^{-1} = e^2 L\partial n/\partial\mu$ is the quantum capacitance. Then

$$G_{\rm m} = e\frac{\partial I}{\partial\mu}\frac{C_{\rm g}}{C_{\rm q}} \tag{6}$$

In our case, with a large group velocity $v_{\rm b} = 5 \times 10^{14}$ nm s^{-1}, and for a dielectric constant $\epsilon \sim 175$ times the vacuum value, means that $C_{\rm g} \approx C_{\rm q}$ within 5%. Then,

$$G_{\rm m} = e\frac{\partial I}{\partial\mu} \tag{7}$$

Since the current I is a strong function of $\tau_{\rm in}(\mu, eV)$, the derivative $\partial I/\partial\mu$ has a rich structure with μ and V entering more than once:

$$G_{\rm m}(\mu, eV) = e\frac{\partial I}{\partial\mu} = eI\frac{\partial}{\partial\mu}\left[\ln\left(\frac{\tau(\mu, eV)}{\tau_{\rm el}(\mu)}\right) + \ln\left(\frac{k_{\rm F}(\mu)}{k_{\rm av}(\mu)}\right)\right.$$
$$\left. + \ln\left(1 - \exp\left(-\frac{\hbar L k_{\rm av}(\mu)}{eV\sqrt{\tau(\mu, eV)\tau_{\rm in}(\mu, eV)}}\right)\right)\right] \tag{8}$$

The time scale for electron-hole recombination is very much greater than the transit times (typically, 6 ps for a channel of length 3 μm), so that the electron and hole transconductances will add.

Figure 6 shows the transconductance G_m for the supported metallic nanotube. Note there is no overshoot. For good signal gain we need the operating region of maximum G_m. We find that this region lies well away from threshold where for a semiconducting carbon nanotube there would be pinch off. The maximum $G_{\rm m} \approx 17$ S.mm^{-1} should be compared with the maximum $G_{\rm m}$ in state of the art InGaAs heterostructure transitors, $G_{\rm m} \approx 1.6$ S.mm^{-1}. The cut-off frequency of a gated nanotube is another major performance measure in field-effect applications. The cut-off frequency for the supported metallic nanotube, $f_{\rm T} = G_{\rm m}L/2\pi C_{\rm g} \approx 140$ GHz, indicated by the dot, is high.

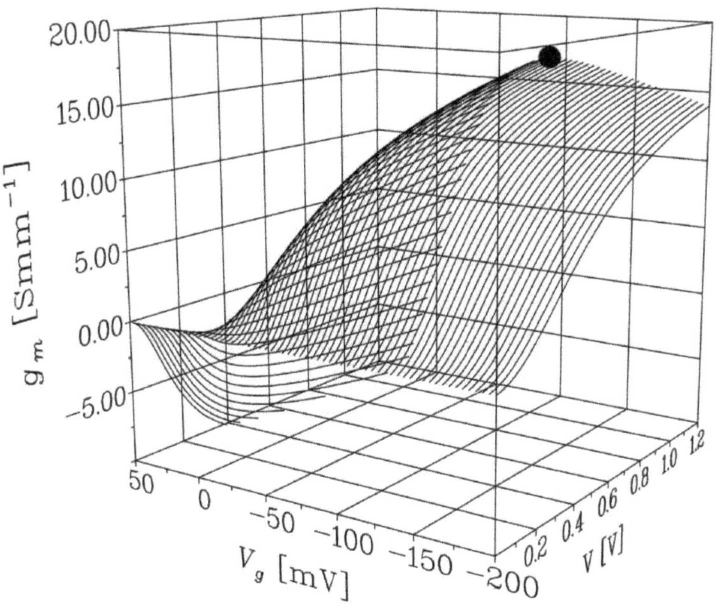

Fig. 6. Transconductance G_m for supported metallic nanotube.

Figure 7 shows the transconductance $G_{\rm m}$ for the suspended metallic nanotube. There is now significant overshoot. The maximum $G_{\rm m} \approx 6$ S.mm^{-1} is a factor of three smaller than for the supported metallic nanotube. The value of the cut-off frequency, indicated by the dot, is $f_T = 50$ GHz.

5. Conclusion

We have analyzed the interplay of degeneracy and scattering dynamics on gate controlled conduction in the one-dimensional channel of ohmically contacted metallic nanotubes, and have determined the transconductance using a microscopically conserving quantum-kinetic equation. We find that features of degenerate carrier kinetics enable a novel class of quantum-confined devices. These devices can have current gains exceeding the best attained in existing transistor technology based on III-V heterojunction quantum wells. Additionally, suspended metallic nanotubes are unusual in that both their high transconductance $G_{\rm m}$ and their negative differential conductance coexist in the same operating region.

The carbon nanotube's promise as a wide-ranging useful tool both in electronics and sensing, depends on being able to accurately and efficiently model its dynamic response. This is an intriguing and potentially rich new area for many-body physics since the reduced dimensionality of the nanotubes makes many-body concepts indispensable in any nanotube modeling project. Our approach demonstrates that a relatively straightforward analysis can be extremely fruitful, provided only that some fundamental conserving rules are respected.

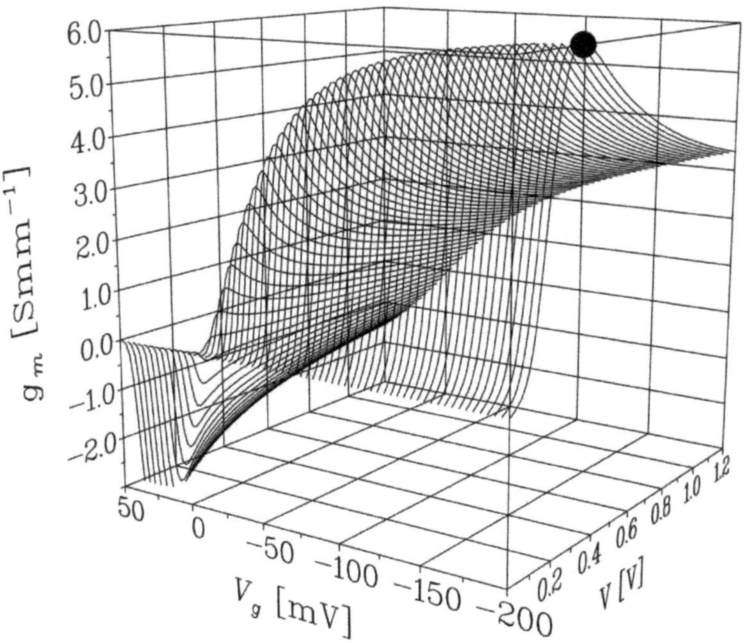

Fig. 7. Transconductance G_m for suspended metallic nanotube.

Acknowledgments

Thanks go to Fabio Beltram and Stefano Roddaro (NEST, Pisa) and to Mukunda Das (Australian National University) for helpful discussions on 1D conduction. The form of kinetic equations that we use were first analyzed by Fred Green and Mukunda Das in a different context.

The research described herein was supported by an Australian Research Council Grant, and a PRIN (Programmi di Ricerca Scientifica di Rilevante Interesse Nazionale) Grant, Italy.

References

1. M.S. Dresselhaus, G. Dresselhaus, and A. Jorio, Annual Review of Materials Research **34** 247 (2004).
2. Z. Yao, C.L. Kane, and C. Dekker, Phys. Rev. Lett. **84**, 2941 (2000).
3. V. Perebeinos, J. Tersoff, and P. Avouris, Phys. Rev. Lett. **94**, 086802 (2005).
4. E. Pop, D. Mann, J. Cao, Q. Wang, K. Goodson, and H. Dai, Phys. Rev. Lett. **95**, 155505 (2005).
5. B. M. Kim, T. Brintlinger, E. Cobas, M.S. Fuhrer, Haimei Zheng, Z. Yu, R. Droopad, J. Ramdani, K. Eisenbeiser, App. Phys. Lett. **84**, 1946 (2004).
6. F. Green, J.S. Thakur, and M.P. Das, Phys. Rev. Lett. **92**, 156804 (2004)
7. Michael Fuhrer, Hongkun Park, and Paul L. McEuen, IEEE Trans. on Nanotech. **1**, 78 (2002).

EVOLUTION AND DESTRUCTION OF THE KONDO EFFECT IN A CAPACITIVELY COUPLED DOUBLE DOT SYSTEM

DAVID E. LOGAN* and MARTIN R. GALPIN

Physical and Theoretical Chemistry Laboratory,
University of Oxford, Oxford, OX1 3QZ, UK
**E-mail: dlogan@physchem.ox.ac.uk*

We consider the simplest example of a double quantum dot system: two equivalent, capacitively coupled single-level dots, each coupled to its own lead. Focussing on a regime of two electrons on the double dot, a rich range of behaviour is found on increasing the interdot coupling: first a crossover from spin- to charge-Kondo physics, via an intermediate $SU(4)$ state with entangled spin and charge degrees of freedom; followed by a quantum phase transition of Kosterlitz-Thouless type to a non-Fermi liquid 'charge-ordered' phase with finite residual entropy and anomalous transport properties. The numerical renormalization group method, coupled with simple physical arguments, is employed to obtain a detailed understanding of the problem.

Keywords: Kondo effect; quantum dots; quantum phase transition; renormalization group.

1. Introduction

The emergence of semiconducting quantum dot materials has led to a strong revival of interest in the Kondo effect,[1] one of the most venerable paradigms of condensed matter theory.[2] The canonical example here is the spin-Kondo effect,[3–5] in which the spin in an odd-electron dot is quenched by coupling to the conduction electrons of a metallic lead, as manifest e.g. in recovery of the unitarity limit for the zero-bias conductance at temperatures below the characteristic Kondo scale.[5]

There is naturally great current interest in *coupled* quantum dot devices, where both spin *and* orbital degrees of freedom are now relevant and the prospect of creating novel strongly correlated electron states arises. Double quantum dot (DQD) systems in particular are under active investigation both theoretically, see e.g. [6-22], and experimentally.[23–31] Here we consider probably the simplest such case: two equivalent semiconducting dots, each consisting in effect of a single level and separately connected to their own leads, are mutually coupled by a capacitive interaction embodied in an *inter*dot interaction U', in addition to the usual intradot interaction U. The isolated DQD can contain up to $n = 4$ electrons, controlled by varying the dot levels relative to the Fermi level, e.g. by application of suitable gate voltages. This problem has been studied quite extensively in the $n = 1$-electron sector,[6–8,22] where the degenerate DQD states $(n_L, n_R) = (1, 0)$ and $(0, 1)$ dominate and lead to a striking example of the $n = 1$ $SU(4)$ Kondo effect involving spin and orbital degrees of freedom.[7,8]

Here by contrast we review recent work on the n = 2-electron regime[10,19,20] where a rich range of physical behaviour, qualitatively distinct from that of the $n = 1$ sector, arises on increasing the ratio U'/U of interdot to intradot interaction strengths. In the uncoupled limit $U' = 0$, the DQD states $(n_L, n_R) = (1,1)$ naturally dominate and the normal spin-Kondo effect prevails. That essential behaviour persists on increasing U'/U, until close to $U'/U = 1$ where we find a rapid but continuous crossover to the $n = 2$ $SU(4)$ Kondo state occurring at $U'/U = 1$. Here the six DQD states $(1,1), (2,0)$ and $(0,2)$ are degenerate, spin and charge degrees of freedom are entangled, and the characteristic low-energy Kondo scale is strongly enhanced. On increasing U'/U just beyond the $SU(4)$ point the system then enters a charge-Kondo state, likewise a Fermi liquid, where a charge pseudospin arising from the DQD states $(2,0)/(0,2)$ is Kondo quenched on coupling to the leads.

That Kondo quenching is however quite fragile: for only a small increase in U'/U, charge-pseudospin tunneling is suppressed, the associated Kondo scale collapses rapidly but continuously, and the system undergoes a Kosterlitz-Thouless quantum phase transition[10,13,32,33] at a critical U'_c from the Fermi liquid phase to a charge ordered phase, a broken-symmetry non-Fermi liquid state with $\ln 2$ residual entropy reflecting the degenerate, unquenched $(2,0)$ and $(0,2)$ DQD configurations. Here we study the problem using Wilson's numerical renormalization group (NRG) approach[34] as the natural method of choice, combined with simple physical arguments to elucidate the essential underlying physics.

2. Model

The capacitively coupled double dot system is shown schematically in Fig. 1. In standard notation, the Hamiltonian can be written as $H = H_0 + H_{\text{int}}$ where

$$H_0 = \sum_{i,\mathbf{k},\sigma} \epsilon_{\mathbf{k}} a^\dagger_{\mathbf{k}i\sigma} a_{\mathbf{k}i\sigma} + \sum_i \epsilon \hat{n}_i + \sum_{i,\mathbf{k},\sigma} V(a^\dagger_{\mathbf{k}i\sigma} c_{i\sigma} + \text{h.c.}), \tag{1}$$

$$H_{\text{int}} = \sum_i U \hat{n}_{i\uparrow}\hat{n}_{i\downarrow} + U'\hat{n}_L \hat{n}_R \tag{2}$$

and $\hat{n}_i = \sum_\sigma \hat{n}_{i\sigma} = \sum_\sigma c^\dagger_{i\sigma} c_{i\sigma}$ is the number operator for dot $i \in \{L, R\}$ (referring to the 'left' and 'right' leads). The first two terms of H_0 describe the isolated lead and dot levels respectively, while the third term contains the one-electron hopping processes between each dot and its associated lead. H_{int} contains the interaction terms that render the problem non-trivial; here we include both an intradot Coulomb repulsion U between the $\sigma =\uparrow$- and $\downarrow$-spin electrons on each dot, and the interdot U' that embodies the capacitive coupling of the L and R channels. Despite its relative simplicity, the model described above displays a very wide range of physics, some of which is rather subtle. We consider the $n = \sum_i n_i = 2$ regime, and focus in practice on the particle-hole (p-h) symmetric limit (the mid-point of the $n = 2$

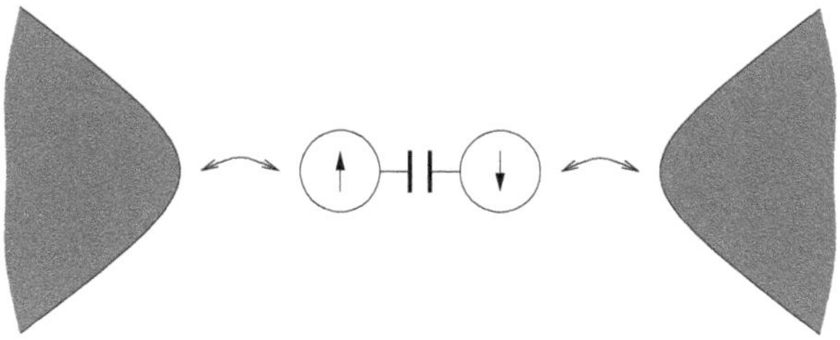

Fig. 1. Schematic picture of the model.

regime); emphasising from the outset that the essential physics is wholly robust to departure from p-h symmetry and that the qualitative conclusions drawn from our results remain valid throughout the $n \simeq 2$ domain. Particle-hole symmetry is imposed in practice by requiring $\epsilon = -U/2 - U'$, which is implicit in what follows.

Consider as a starting point the 'atomic' limit, where one sets $V = 0$ and thereby decouples the dots from their leads. There are three distinct possibilities for the ground state of the dots: $(n_L, n_R) = (1,1)$ if $U' < U$, $(0,2)$ or $(2,0)$ if $U' > U$, and all three degenerate when $U' = U$. When a finite V is then switched on, one might expect (by analogy with the Anderson impurity model itself) that for $U' < U$ the $(1,1)$ states will be quenched by the Kondo effect—and indeed they are. But what happens to the degenerate pair of $(0,2)$ or $(2,0)$ ground states arising when $U' > U$ is rather less obvious.

We thus give a simple physical argument to explain the physics as U' is increased from zero, which will be confirmed by our results below. The basic idea is to consider the effective Hamiltonian of the system as a function of U' (for some $U \gg \Gamma = \pi V^2 \rho$, with ρ the leads' density of states), to order V^2 in the dot-lead coupling.

For $U' = 0$, the result is well-known:[2] it is the Schrieffer-Wolff transformation of the two Anderson impurities of $\hat{H}$ into a pair of spin-$\frac{1}{2}$ Kondo models [denoted $SU(2) \times SU(2)$ Kondo in this paper]. The idea is that one retains in the effective model only the manifold of states in which $(n_L, n_R) = (1,1)$; the others (of $\mathcal{O}(U)$ higher in energy) are taken into account via virtual excitations of the $(1,1)$ states to second order in V. These virtual processes generate the antiferromagnetic exchange coupling of the Kondo model, and hence lead ultimately to the Kondo effect and its associated Fermi liquid (or 'strong coupling', SC) behaviour.

When $U' \gg U$, the lowest manifold of states instead involves the $(0,2)/(2,0)$ configurations, with $(1,1)$ now lying $\mathcal{O}(U' - U)$ higher in energy. The effective Hamiltonian obtained to $\mathcal{O}(V^2)$ is found to contain only simple potential scattering from the atomic limit states: *no terms that mix* $(0,2)$ *and* $(2,0)$ *can be generated to* $\mathcal{O}(V^2)$. The ground state of the effective Hamiltonian is therefore doubly-degenerate and corresponds to a non-Fermi-liquid, 'charge ordered' (CO) phase. Since the $U' = 0$ physics is known to be Fermi-liquid like, the model must therefore exhibit a quantum phase transition from SC to CO phases at some critical U'_c.

One might wonder whether U'_c is equal to U, since this is the point at which

$(1,1)$ and $(0,2)/(2,0)$ are degenerate in the atomic limit. However, when $U' = U$ the effective model is the $SU(4)$ Coqblin-Schrieffer model[2] in which the $(1,1)$ states connect back to themselves *and* to $(0,2)$ or $(2,0)$. This, like the $SU(2) \times SU(2)$ Kondo model at $\tilde{U}' = 0$, has a Fermi liquid ground state and—by continuity—will be connected to it via a simple crossover; the quantum phase transition between SC and CO phases must therefore lie at some $U'_{\rm c} > U$.

To pin down the location of $\tilde{U}'_{\rm c}$ further still, we argue that the relevant energy scale to consider near the transition is the $SU(4)$ Kondo scale $T_{\rm K}^{SU(4)}$: the stabilisation energy gained by involving all three configurations, $(1,1)$, $(0,2)$ and $(2,0)$ in the formation of a Kondo singlet state. Once U' is increased above U, the $(1,1)$ states will be raised above $(0,2)/(2,0)$ by an amount $U'-U$. But for $U'-U < \mathcal{O}(T_{\rm K}^{SU(4)})$, forming a singlet ground state by interconverting $(0,2)$ and $(2,0)$ *via* $(1,1)$ will remain favourable. The FL phase will thus persist above $U' = U$—in the form of a *charge Kondo* state where the $(0,2)/(2,0)$ 'charge pseudospin' is quenched—until $U' - U \sim \mathcal{O}(T_{\rm K}^{SU(4)})$ where the mixing of $(0,2)$ and $(2,0)$ via $(1,1)$ becomes unfavourable and the quantum phase transition to the CO phase takes place.

3. NRG Approach and Fixed Points

The physical picture described above is confirmed by our numerical calculations. The technique of choice here is the numerical renormalization group (NRG),[34] which provides accurate results for many quantum impurity models over all energy scales of interest. While the full details of the calculation itself are rather lengthy (see [19] for a more complete discussion), its essential structure can be understood quite simply without the need for detailed mathematics.

One begins by recasting the conduction band part of the Hamiltonian Eq. (1) as a pair of infinite linear chains, each coupled at one end to the respective dot. Using a logarithmic discretization of the bands (parametrised as usual[34] by $\Lambda > 1$), this can be done in such a way that the chain hopping matrix elements decrease exponentially as one moves away from the dot along the chain. Starting then with the Hamiltonian for just the dots and the *first* sites of the chains ('H_0'), the problem is solved iteratively: the remaining conduction band sites of each chain are coupled in one at a time and the Hamiltonian rediagonalised at every step.

Since the energy splittings at each iteration are exponentially smaller than those of the last, the effect of successive iterations is, in effect, to 'refine' systematically the eigenvalue spectrum of H_0 into that of the full Hamiltonian. The key point is that the 'coarse-grained' energy levels obtained from a given iteration N are entirely sufficient for the calculation of thermodynamics at some sufficiently-high temperature T_N (or dynamics at a frequency ω_N), and thus the full temperature/frequency dependence of the problem can be built up by combining results from multiple iterations. While the number of states in the calculation scales as 16^N in principle, it can be reduced to a manageable level by discarding all but the lowest N_s states of each iteration: the inherent energy-scale separation ensures that this does not significantly affect the final result, since the neglected high-energy states of one iteration would effectively be exponentially-high at the next, in which case they would be thermally unpopulated anyway.

By formulating this iterative procedure as a renormalization group transformation, the stepwise evolution of the Hamiltonian can be interpreted as flows between various fixed points. Each such fixed point corresponds to a limit of $\hat{H}$ in which its parameters Γ, U and U' are set to values of 0 or ∞: the complete set pertinent to this work are listed in table 1. The table shows also the stability of each fixed

Table 1. Fixed points of the NRG. The quantity α is an 'anomalous' exponent defined in Eq. (5).

Label	Γ	U	U'	Stability
FO	0	0	0	Unstable
$\mathrm{LM}^{SU(2)}$	0	∞	0	Unstable
$\mathrm{LM}^{SU(4)}$	0	∞	∞	Unstable
SC	∞	0	0	Stable
CO	0	0	∞	Stable for $\alpha > 0$

point, which can be determined by analysing flows in its vicinity to leading order. Since there are two stable fixed points, one can immediately see that there will be a quantum phase transition in the system, in agreement with the prediction of the previous section.

The three fixed points with $U' = 0$ in table 1 can be viewed quite simply as 'doubled' versions of the Anderson model free orbital (FO), local moment (LM) and strong coupling (SC) fixed points,[34] with the SC fixed point in particular corresponding to the quenching of the dot degrees of freedom by the Kondo effect. The additional $\mathrm{LM}^{SU(4)}$ fixed point arises in this model for $U' = U \to \infty$, and represents the atomic limit physics of $(1, 1)$, $(2, 0)$ and $(0, 2)$ states decoupled from the leads.

The charge ordered (CO) fixed point in table 1 deserves more detailed comment. It corresponds to the situation where $U' \to \infty$, where the dots adopt only the $(0, 2)$ and $(2, 0)$ configurations. What makes this fixed point so interesting is the existence of the marginal correction

$$\tilde{K}\Lambda^{(N-1)/2} \sum_{i,\sigma} (\hat{n}_i - 1) f_{0i\sigma}^{\dagger} f_{0i\sigma}, \tag{3}$$

(where $f_{0i\sigma} = 2^{-1/2} \sum_{\mathbf{k},\sigma} a_{\mathbf{k}i\sigma}$), which describes equal and opposite potential scattering (PS) of magnitude $\tilde{K} \in [0, \infty)$ on the leads. (Note that this effectively turns the CO fixed 'point' into a *line* of fixed points, one for each $\tilde{K}$.) The key point to notice is that the sign of the PS on a particular lead is *correlated* with the charge state of the respective dot, such that interchanging the $(0, 2)$ and $(2, 0)$ dot configurations instantaneously reverses the signs of the potential scattering felt by all the conduction electrons in the system. This physics, reminiscent of that seen in the well-known X-ray edge problem,[35] renders analysis of the fixed point stability rather more complicated than usual—and is ultimately the mechanism that drives the quantum phase transition.

To determine the stability of the CO fixed point, it is in fact necessary to consider the effect of the following four-electron cotunneling term in its vicinity:

$$\delta\hat{H} \sim \Lambda^{(N-1)/2}\left(c^\dagger_{L\uparrow}c^\dagger_{L\downarrow}c_{R\uparrow}c_{R\downarrow}f^\dagger_{0R\uparrow}f^\dagger_{0R\downarrow}f_{0L\uparrow}f_{0L\downarrow} + \text{h.c.}\right) \tag{4}$$

Simple power counting of the type normally used in NRG would suggest that this correction is always irrelevant (i.e. diminishes with increasing iteration number N). This is incorrect however: a bosonization approach[13,19,36] shows that Eq. (4) scales as $\Lambda^{-\alpha N}$, with α an anomalous exponent given by

$$\alpha = -\frac{1}{2} + \left(\frac{2\delta}{\pi} - 1\right)^2 \tag{5}$$

with δ the magnitude of the conduction band phase shifts produced by Eq. (3),

$$\delta = \tan^{-1}\left(\frac{\pi(1-\Lambda^{-1})\tilde{K}}{2\ln\Lambda}\right). \tag{6}$$

For sufficiently large $\tilde{K}$, it can be seen from Eqs. (5) and (6) that α will be negative and $\delta\hat{H}$ will be *relevant*. Moreover, as the other corrections to the CO fixed point all turn out to be irrelevant, it follows that $\delta\hat{H}$ alone controls the stability of the CO fixed point. Hence the line of CO fixed points divides into unstable ($\alpha < 0$) and stable ($\alpha > 0$) regions. RG flows near the former will ultimately cross over to the stable SC fixed point (and thus lead to the SC phase), whereas those near the latter will of course give rise to the CO phase. We remark here that the behaviour described above is quite different from that found for many other quantum phase transitions, where a single, unstable fixed point controls the physics near criticality. Indeed, the existence of a line of fixed points divided into stable and unstable regions, as seen also in the anistropic Kondo model[2] for example, suggests a quantum phase transition of Kosterlitz-Thouless[37] type. This is confirmed in the following section, where the results of our numerical calculations are examined.

4. Results

4.1. *Phase diagram*

We outlined above how the quantum phase transition between SC and CO phases is readily interpreted in terms of flows to the corresponding stable fixed point. Since these two fixed points have different eigenvalue spectra, it is relatively straightforward to locate the phase boundary from the numerics. In Fig. 2(a) we show the position of the SC/CO phase boundary in the $(\tilde{U},\tilde{U}')$ plane, where the (dimensionless) $\tilde{U} = U/\pi\Gamma$ and $\tilde{U}' = U'/\pi\Gamma$ are thus defined and used in the following.

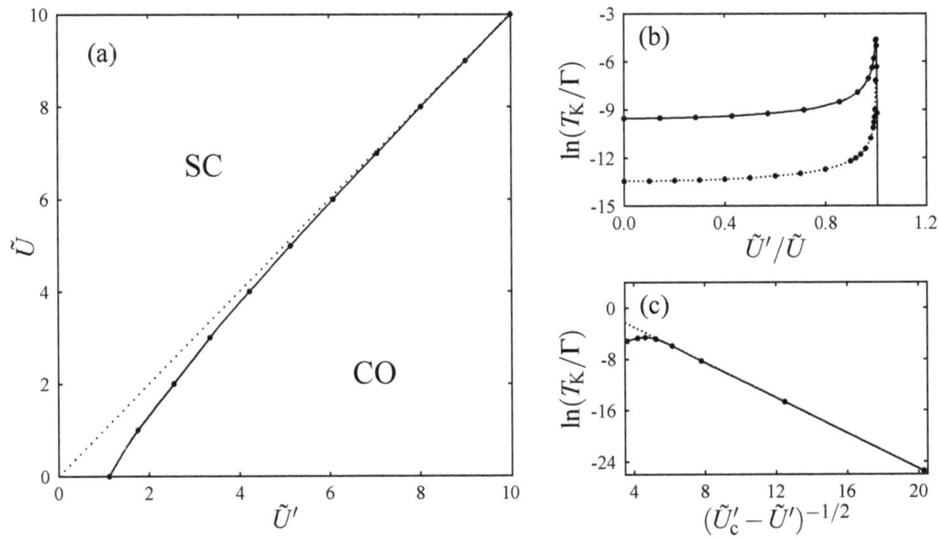

Fig. 2. (a) Phase diagram as a function of $\tilde{U}$ and $\tilde{U}'$. (b) Kondo scale $\ln(T_K/\Gamma)$ vs. $\tilde{U}'/\tilde{U}$ for $\tilde{U} = 7$ (solid line) and $\tilde{U} = 10$ (dotted line). (c) $\ln(T_K/\Gamma)$ vs. $(\tilde{U}_c' - \tilde{U}')^{-1/2}$ near the phase boundary for $\tilde{U} = 7$ (solid line), with $\tilde{U}_c' \simeq 7.046$. The dotted line shows the fit to the Kosterlitz-Thouless form discussed in the text.

We find, in agreement with the arguments of Sec. 2, that the phase boundary satisfies $\tilde{U}_c'(\tilde{U}) > \tilde{U}$ for all $\tilde{U} \geq 0$, with $\tilde{U}_c' - \tilde{U}$ becoming exponentially-small for $\tilde{U} \gg 1$. To interpret this latter result more quantitatively, it is helpful to bring into the discussion the variation of the Kondo scale T_K within the SC phase.

We define T_K throughout the SC phase as $1/\gamma$, with γ the linear coefficient of the dots' contribution to the specific heat $C(T)$. As a function of $\tilde{U}'/\tilde{U}$ for fixed $\tilde{U} \gg 1$, it shows the characteristic behaviour illustrated in Fig. 2(b) for $\tilde{U} = 7$ and 10. The scale rises initially very slowly from its value at $\tilde{U} = 0$ (denoted $T_K^{SU(2)}$) until very close to the $SU(4)$ point $\tilde{U}' = \tilde{U}$, where it then rapidly crosses over to the maximum of $T_K^{SU(4)}$ arising at $\tilde{U}' = U$. Thereafter, T_K decreases extremely quickly to zero as $\tilde{U}' \to \tilde{U}_c'^{-}$, and of course remains at zero throughout the whole CO phase.

As a function of $\tilde{U}$ for fixed $\tilde{U}'/\tilde{U}$, we find that the well-known result[34] $T_K^{SU(2)}/\Gamma \sim \tilde{U}^{1/2} \exp(-\pi^2 \tilde{U}/8)$ for $\tilde{U}' = 0$ is complemented by a similar result for $T_K^{SU(4)}/\Gamma \sim \tilde{U}^{3/4} \exp(-\pi^2 \tilde{U}/16)$ at the $\tilde{U}' = \tilde{U}$ point. In terms of this $SU(4)$ scale, the phase boundary tends to the asymptotic result $\tilde{U}_c'/\tilde{U} - 1 \sim T_K^{SU(4)}/(\Gamma \tilde{U}^{1/2})$, and hence its location is consistent with the physical arguments of Sec. 2.

Finally, we find that the critical behaviour of T_K as $\tilde{U}' \to \tilde{U}_c'^{-}$ can be used to get a handle on the nature of the transition. As seen in Fig. 2(c), it closely follows

$$\frac{T_K}{\Gamma} \sim \exp\left(-\frac{a}{\sqrt{U_c' - \tilde{U}'}}\right) \tag{7}$$

(with $a > 0$ a constant for the chosen $\tilde{U}$) over most of the 'charge Kondo' region above $\tilde{U}' = \tilde{U}$. This behaviour is observed for all $\tilde{U} \geq 0$, not just in strong coupling ($\tilde{U} \gg 1$); hence the physics near the phase boundary is generic. Eq. (7) is moreover of the form one would expect for a Kosterlitz-Thouless transition (c.f. the divergence of the correlation length ξ of the two-dimensional xy model[37]), and thus provides further evidence that this transition is of the same fundamental type.

4.2. *Thermodynamics*

The non-monotonic variation of $T_{\rm K}$ with increasing $\tilde{U}'$ in Fig. 2(b) reflects changes in the way the dots' degrees of freedom are quenched as $T \to 0$. These can be examined directly by calculating 'impurity' susceptibilities that probe the different spin and charge excitations of the system. We consider the impurity spin susceptibility

$$\chi_{\rm s} = (g\mu_{\rm B})^2 \left\langle [\hat{S}_z - \langle \hat{S}_z \rangle]^2 \right\rangle_{\rm imp}, \tag{8}$$

and the uniform and staggered impurity charge susceptibilities

$$\chi_{\rm c}^{\pm} = \frac{1}{4} \left\langle [\hat{N}_L \pm N_R - \langle \hat{N}_L \pm N_R \rangle]^2 \right\rangle_{\rm imp}. \tag{9}$$

Here, $\hat{S}_z$ is the total spin operator, N_i is the total charge of the ith half of the system ($i = L, R$), and $\langle \ldots \rangle_{\rm imp}$ denotes a thermal average in which the contribution from the leads alone has been subtracted.[34]

The behaviour of $T\chi_{\rm s}(T)$ with increasing $\tilde{U}'$ is indicated in Fig. 3(a): the curves shown are for $\tilde{U}' = 6, 6.9, 6.97, 7, 7.03, 7.1$ and 8 (from top to bottom), with $\tilde{U} = 7$ in each case. The figure illustrates three key points:

(1) The $SU(2) \times SU(2)$ spin-Kondo physics known for $\tilde{U}' = 0$ (i.e. the quenching of two free spin-$\frac{1}{2}$ states on the scale $T_{\rm K}$) persists into most of the SC phase up to $\tilde{U}' \lesssim \tilde{U}$. The $\tilde{U}' = 6$ curve shown in the figure is in fact virtually indistinguishable from what we find for $\tilde{U}' = 0$ (on an absolute scale), and the low-temperature behaviour of the $\tilde{U}' = 6.9$ and 6.97 curves also collapses onto the $\tilde{U}' = 0$ result if one scales temperature as $T/T_{\rm K}$.

(2) At the $SU(4)$ point (dashed line in Fig. 3(a)) where $\tilde{U}' = \tilde{U}$, the spin susceptibility is quite distinct from the universal $SU(2) \times SU(2)$ behaviour described above. Here, the entangled spin and charge degrees of freedom of the LM$^{SU(4)}$ fixed point [i.e. the $(1, 1)$, $(2, 0)$ and $(0, 2)$ atomic limit states] are together quenched by the $SU(4)$ Kondo effect, such that $T\chi_{\rm s}$ drops from $\simeq \frac{1}{3}$ to zero on the scale $T_{\rm K}^{SU(4)}$ with a low-temperature form entirely different from that described above.

(3) For $\tilde{U}' > \tilde{U}$ (dotted lines), the spin susceptibility drops to zero on the scale $\tilde{U}' - U$ and, moreover, *shows no sign of the quantum phase transition* at $\tilde{U}'_{\rm c} \simeq 7.046$.

Point (3) above is quite natural if one remembers the physical arguments of Sec. 2. The spin susceptibility $\chi_{\rm s}$ probes the *magnetic* degrees of freedom associated with the $(1, 1)$ states, which act only as the virtual intermediaries between $(0, 2)$

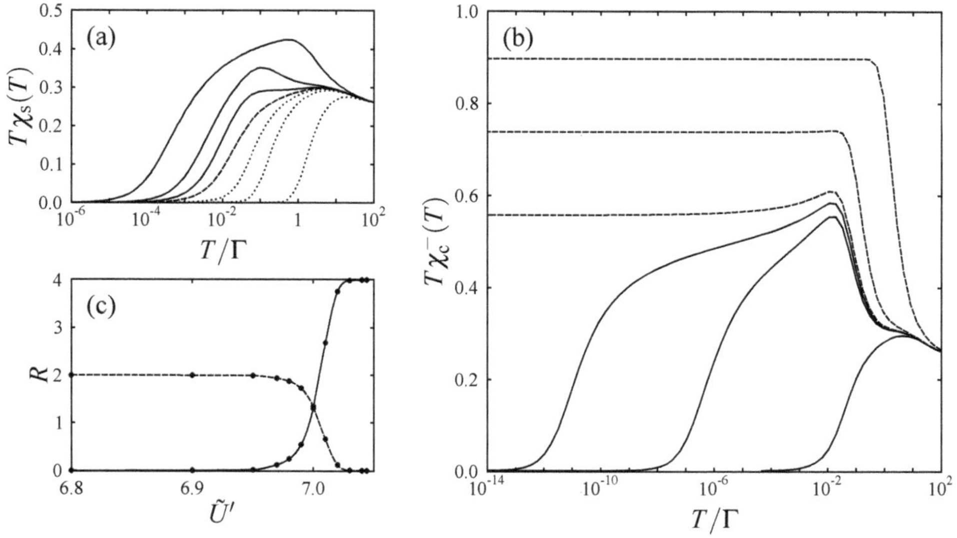

Fig. 3. (a) Spin susceptibility $T\chi_s(T)$ vs. T/Γ for $\tilde{U} = 7$ with $\tilde{U}' = 6$, 6.9, 6.97, 7, 7.03, 7.1 and 8 (top to bottom). (b) Staggered charge susceptibility $T\chi_c^-$ vs. T/Γ for $\tilde{U} = 7$ with $\tilde{U}' = 7$, 7.04, 7.044, 7.048, 7.1 and 8 (bottom to top). (c) Wilson ratios R_c^- (solid) and R_s (dashed) vs. $\tilde{U}'$.

and $(2, 0)$ when $U' > U$. Thermal excitations to these states are therefore frozen out on the scale $U' - U$, regardless of whether or not the low-energy charge degrees of freedom of $(0, 2)$ and $(2, 0)$ are quenched by a charge Kondo effect. To see signs of the quantum phase transition in the finite-T thermodynamics, one must instead look at the staggered charge susceptibility, shown in Fig. 3(b) for $\tilde{U}' = 7$, 7.04, 7.044, 7.048, 7.1 and 8 (bottom to top), again all with $\tilde{U} = 7$.

As seen clearly in the figure, the charge susceptibility is qualitatively distinct in the two phases. In the CO phase (dashed lines), $T\chi_c^-(T)$ tends to a constant as $T \to 0$, reflecting the Curie-like behaviour of a free charge-pseudospin [corresponding essentially to the $(0, 2)$ and $(2, 0)$ atomic limit states]. On lowering $\tilde{U}'$ into the SC phase (solid lines), this charge-pseudospin is quenched by the charge Kondo effect on the scale T_K, and hence $T\chi_c^-(T)$ vanishes as $T \to 0$. In other words, $\chi_c^-(0)$ is the analogue here of the zero-temperature *magnetic* susceptibility of the 2-dimensional *xy* model, diverging as one goes from the SC ("vortex unbound") phase into the CO ("vortex bound") phase, and thereafter remaining infinite as $U' \to \infty$. The $T \to 0$ thermodynamics of the SC phase are neatly summed up by three 'Wilson' ratios, $R_s = \chi_s/\gamma$ and $R_c^\pm = \chi_c^\pm/\gamma$. In the strong coupling regime of most interest (i.e. $\tilde{U} \gg 1$), the uniform charge Wilson ratio R_c^+ is exponentially-small as total charge fluctuations away from $n = 2$ are negligible. The other two Wilson ratios are non-zero however, and shown in Fig. 3(c) as a function of $\tilde{U}'$ for $\tilde{U} = 7$ (solid line: R_c^-, dashed line: R_s). Each of the $SU(2) \times SU(2)$ spin-Kondo, $SU(4)$ spin/charge-Kondo and charge-Kondo regimes leads to a characteristic set of Wilson ratios, and the figure emphasises how rapidly the crossovers from one regime to another occurs in practice. By considering the leading corrections to the SC fixed point, one can in fact derive the following results of Fermi liquid theory[19]

$$R_s = 2 \text{ and } R_c^- = 0 \text{ for } U' = 0 \tag{10}$$

$$R_s = R_c^- = \tfrac{4}{3} \text{ for } U' = U \tag{11}$$

in the strongly correlated regime $\tilde{U} \gg 1$ and, more generally, that

$$R_s = 0 \text{ and } R_c^- = 4 \text{ as } U' \to U_c'^- \tag{12}$$

$$2R_s + R_c^+ + R_c^- = 4 \tag{13}$$

in the SC phase. The agreement of the numerical results in Fig. 3(c) with Eqs. (10)–(12) is clearly excellent.

4.3. *Dynamics*

The physics of the different Kondo effects arising in the SC phase, and of the quantum phase transition to the CO phase, are equally apparent in dynamics.[20] We focus on the low-frequency behaviour of the $T = 0$ single-particle spectrum of (either) dot i, $D_i(\omega)$, related to the retarded Green function $G_i(\omega)$ via $D_i(\omega) = -\pi^{-1}\text{Im}G_i(\omega)$. NRG spectra are calculated using the standard approach.[38]

The SC phase spectrum is naturally characterized by a Kondo resonance at the Fermi level, with HWHM proportional to the Kondo scale T_K: this is shown in Fig. 4(a) for fixed $\tilde{U} = 7$ and $\tilde{U}' = 0$, 6.5, 6.9 and 7 (bottom to top). The inherent Fermi liquid nature of the SC phase, together with particle-hole symmetry, leads to the conclusion that $\pi\Gamma D(\omega = 0) = 1$ is 'pinned' throughout the phase. As with the

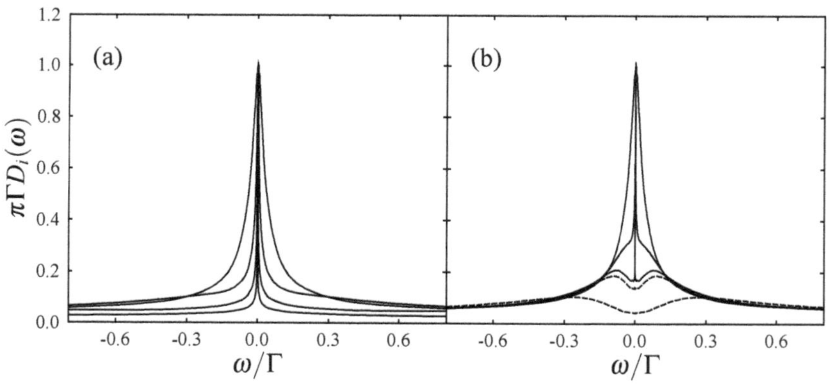

Fig. 4. Single particle spectra, $\pi\Gamma D_i(\omega)$ vs. ω/Γ for $\tilde{U} = 7$ with (a) $\tilde{U}' = 0$, 6.5, 6.9 and 7 (top to bottom); and (b) $\tilde{U}' = 7$, 7.03 and 7.044 (solid lines, top to bottom), 7.048 and 7.1 (dashed lines from top to bottom). The critical $\tilde{U}_c' = 7.046$.

thermodynamics discussed in the previous section, the spectrum is seen to change very little until $\tilde{U}$ is close to $\tilde{U}'$, at which point T_K then rises rapidly and the Kondo resonance is significantly broadened.

However on moving past the $SU(4)$ line ($\tilde{U}' = \tilde{U}$), the Kondo scale $T_{\rm K}$ decays rapidly according to Eq. (7), vanishing as $\tilde{U}' \to \tilde{U}_{\rm c}'^{-}$. The Kondo resonance thus narrows in fashion (its $\omega = 0$ value remaining pinned), as seen in Fig.4(b) for $\tilde{U}' = 7$, 7.03 and 7.044 (solid lines, top to bottom). But on entering the CO phase the resonance vanishes 'on the spot', leaving behind a incoherent continuum around the Fermi level (see Fig. 4(b) for $\tilde{U}' = 7.048$ and 7.1, dashed lines from top to bottom), which itself slowly vanishes as $\tilde{U}' \to \infty$ and the dots become locked into the $(0,2)$ and $(2,0)$ 'atomic limit' states.

The single-particle spectrum at the Fermi level, $D_i(0)$, is particularly important, as it is related to the zero-bias differential conductance (G) of dot i by $G = \left(\frac{2e^2}{h}\right)\pi\Gamma D_i(0)$ with e the electronic charge and h Planck's constant. (This result is obtained formally by splitting each of the leads in Fig. 1 as a pair of identical leads of coupling strength $\Gamma/2$, so that an infinitesimal voltage can be applied across one of the dots.) As seen in the spectra in Fig. 4, the destruction of the Kondo resonance on entering the CO phase produces a discontinuous drop in G; this is shown more clearly in Fig. 5 where we plot $\pi\Gamma D_i(0)$ as a function of $\tilde{U}'$ for $\tilde{U} = 7$ (thick solid line).

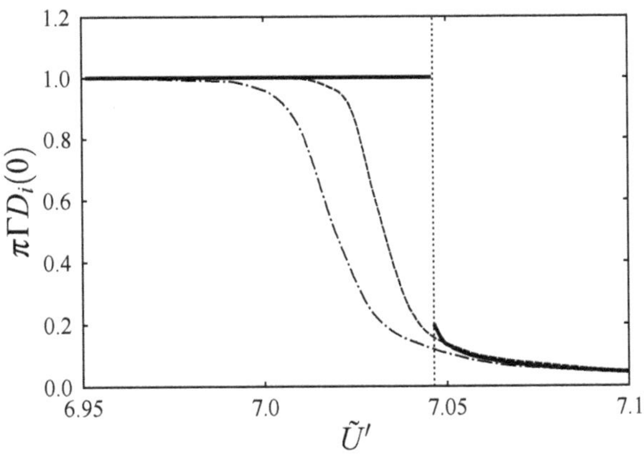

Fig. 5. Zero frequency spectrum $\pi\Gamma D_i(0)$ as a function of $\tilde{U}'$ for $\tilde{U} = 7$ (thick solid line). The dashed and dot-dashed lines show the effect of detuning the dot levels, as discussed in the text.

The discontinuity predicted in G across the transition is perhaps the most experimentally accessible feature of the model. One might, however, wonder whether the physics of the transition is robust to deviations from the symmetries of the Hamiltonian. Insofar as particle-hole symmetry is concerned the transition is wholly robust to departure from it, and the discontinuity persists away from the particle-hole symmetric limit as long as one remains relatively 'deep' within the $n = 2$ regime.[19,20] Breaking left-right symmetry on the other hand, will smooth the discontinuity into a crossover between a predominantly '$(1,1)$' Fermi liquid regime and a predominantly '$(0,2)$' or '$(2,0)$' Fermi liquid regime (depending on which way the symmetry is broken). But as long as any new energy scales introduced by such deviations are less than or comparable to $T_K^{SU(4)}$, the transition will still be very much observable.

To illustrate this latter point briefly, we show in Fig. 5 the effect[20] of detuning the dot levels by applying a splitting $\epsilon_{L,R} \rightarrow \epsilon \pm \delta\epsilon$, where $\delta\epsilon / T_K^{SU(4)} = 0.3$ (dashed line) and 2 (dot-dashed line): the underlying transition, occurring strictly for $\delta\epsilon = 0$, indeed remains apparent as a crossover even for $\delta\epsilon \sim T_K^{SU(4)}$.

References

1. L. P. Kouwenhoven and L. Glazman, *Physics World* **14**, 33 (2001).
2. A. C. Hewson, *The Kondo Problem to Heavy Fermions* (Cambridge: Cambridge University Press) (1993).
3. D. Goldhaber-Gordon *et al.*, *Nature* **391**, 156 (1998).
4. S. M. Cronenwett , T. H. Oosterkamp and L. P. Kouwenhoven, *Science* **281**, 540 (1998).
5. W. G. van der Wiel *et al.*, *Science* **289**, 2105 (2000).
6. T. Pohjola, H. Schoeller and G. Schön, *Europhys. Lett.* **54**, 241 (2001).
7. D. Boese, W. Hofstetter and H. Schoeller, *Phys. Rev. B* **66**, 125315 (2002).
8. L. Borda *et al.*, *Phys. Rev. Lett.* **90**, 026602 (2003).
9. R. López *et al.*, *Phys. Rev. B* **71**, 115312 (2005).
10. M. R. Galpin, D. E. Logan and H. R. Krishnamurthy, *Phys. Rev. Lett.* **94**, 186406 (2005).
11. J. M. Golden and B. I. Halperin, *Phys. Rev. B* **53**, 3893 (1996).
12. N. Andrei, G. T. Zimanyi and G. Schön, *Phys. Rev. B* **60**, R5125 (1999).
13. M. Garst *et al.*, *Phys. Rev. B* **69**, 214413 (2004).
14. M. R. Li and K. Le Hur, *Phys. Rev. Lett.* **93**, 176802 (2004).
15. W. Izumida and O. Sakai, *Phys. Rev. B* **62**, 10260 (2000).
16. M. Vojta, R. Bulla and W. Hofstetter, *Phys. Rev. B* **65**, R140415 (2002).
17. J. Mravlje, A. Ramsak and T. Rejec, *Phys. Rev. B* **73**, 241305 (2006).
18. K. Kikoin and Y. Avisai, *Phys. Rev. B* **65**, 115329 (2002).
19. M. R. Galpin, D. E. Logan and H. R. Krishnamurthy, *J. Phys.: Condens. Matter* **18**, 6546 (2006).
20. M. R. Galpin, D. E. Logan and H. R. Krishnamurthy, *J. Phys.: Condens. Matter* **18**, 6571 (2006).
21. R. Zitko and J. Bonca, *Phys. Rev. B* **74**, 045312 (2006).
22. A. K. Mitchell, M. R. Galpin and D. E. Logan, *Europhys. Lett.* **76**, 95 (2006).
23. F. R. Waugh *et al.*, *Phys. Rev. Lett.* **75**, 705 (1995).
24. L. W. Molenkamp, K. Flensburg and M. Kemerink, *Phys. Rev. Lett.* **75**, 4282 (1995).
25. R. H. Blick *et al.*, *Phys. Rev. B* **53**, 7899 (1996).
26. A. W. Holleitner *et al.*, *Phys. Rev. Lett.* **87**, 256802 (2001).
27. U. Wilhelm, J. Schmid, J. Weis and K. von Klitzing, *Physica* **14E**, 385 (2002).
28. I. H. Chan *et al.*, *Appl. Phys. Lett.* **80**, 1818 (2002).
29. A. W. Holleitner *et al.*, *Phys. Rev. B* **70,** 075204 (2004).
30. H. Jeong, A. M. Chang and M. R. Melloch, *Science* **293**, 2221 (2001).
31. S. Sasaki *et al.*, *Phys. Rev. Lett.* **93**, 017205 (2004).
32. W. Hofstetter and H. Schoeller, *Phys. Rev. Lett.* **88**, 016803 (2002).
33. R. Bulla, N.-H. Tong and M. Vojta, *Phys. Rev. Lett.* **91**, 170601 (2003).
34. H. R. Krishnamurthy *et al,*, *Phys. Rev. B* **21**, 1003 (1980); *ibid* **21**, 1044 (1980).

35. P. Nozières and C. T. de Dominicis, *Phys. Rev.* **178**, 1097 (1969).
36. K. D. Schotte and U. Schotte, *Phys. Rev.* **182**, 479 (1969).
37. J. M. Kosterlitz, *J. Phys. C: Solid State Phys.* **7**, 1046 (1974).
38. T. A. Costi, A. C. Hewson and V. Zlatić, *J. Phys.: Condens. Matter* **6**, 2519 (1994).

THE METHOD OF INCREMENTS — A WAVEFUNCTION-BASED AB-INITIO CORRELATION METHOD FOR SOLIDS

BEATE PAULUS

Max-Planck-Institut für Physik komplexer Systeme
Nöthnitzer Straße 38, D-01187 Dresden, Germany
**E-mail: beate@mpipks-dresden.mpg.de*

The method of increments is a wavefunction-based ab initio correlation method for solids, which explicitly calculates the many-body wavefunction of the system. After a Hartree-Fock treatment of the infinite system the correlation energy of the solid is expanded in terms of localised orbitals or of a group of localised orbitals. The method of increments has been applied to a great variety of materials with a band gap, but in this paper the extension to metals is described. The application to solid mercury is presented, where we achieve very good agreement of the calculated ground-state properties with the experimental data.

Keywords: Ab-initio calculation; electron correlations; method of increments; mercury.

1. Introduction

Ab-initio electronic structure methods aim to solve the full electronic Hamiltonian (in atomic units):

$$H = \sum_i \left(-\frac{1}{2}\Delta_i - \sum_\alpha^K \frac{Z_\alpha}{|\vec{r}_i - \vec{R}_\alpha|} \right) + \frac{1}{2} \sum_{i \neq j} \frac{1}{|\vec{r}_i - \vec{r}_j|}. \tag{1}$$

The first two terms are the kinetic energy and the electron-nucleus interaction, both one-particle terms and properly described within a mean-field approach. The last term, the electron-electron interaction is a real many-body term, the two-body interaction can be solved only approximatively within a mean-field description.

Besides the very successful method to solve this Hamiltonian within the density-functional approach[1,2] (DFT), which relies on the ground-state density of the system and avoids the calculation of the many-body wavefunction of the system, it is desirable for many systems to calculate the many-body wave-function of the system explicitly yielding a deeper insight to the correlation effects of the system. The advantage of these methods, sometimes called quantum chemical methods, is that they can be improved systematically. As the starting point for such a calculation a single-particle wave-function is necessary. Nearly all methods rely on the Hartree-Fock wavefunction[3–5] of the systems. This independent electron approximation yields the lowest ground-state energy which is possible for independent electrons. "Independent electrons" means that the electron experiences an interaction field which is calculated as the spatial average over the positions of all other electrons. This field depends on the Hartree-Fock wavefunction, therefore the problem is solved in a self-consistent manner, until both the wavefunction and the field is converged. For a closed shell system the solution is a single Slater determinant, which is for weakly correlated systems a good starting point for successive correlation calculations. Moreover, the amount of electronic correlations, i.e. the instantaneous reaction of an electron due to the movement of any other electron, is defined as the difference between the Hartree-Fock solution and the true ground-state solution.

Quantum-chemical correlation methods rely on the Hartree-Fock solution of the system. Starting from the Hartree-Fock orbitals, which are divided into occupied and unoccupied ones with respect to the ground-state Slater determinant, in the configuration interaction (CI) ansatz[6,7] additional determinants are considered. In these determinants one, two or more occupied orbitals are replaced by unoccupied ones. The weights of these determinants in the ansatz for the many-body wavefunction are optimised by minimising the total energy of the system.

In order to apply the quantum-chemical correlation methods to an extended system the property that the correlation hole (i.e. the region, where the electrons interact instantaneously) is fairly local can be used. Therefore for a solid it is useful to switch from the description with delocalised Bloch orbitals to localised Wannier orbitals.[8] But even if the correlation hole extends only e.g. to the fourth-nearest neighbours, that would correspond to 64 atoms in a dense-packed three-dimensional solid and e.g. for carbon with its four valence electrons this would result in the need to correlate 256 electrons. It is hard to handle so many electrons with explicitly correlated methods, because the computational demand increases with N^5 to N^8, where N is the number of electrons, depending on the specific method used.

Therefore further approximations are necessary. For this purpose the method of increments was developed by Stoll.[9–11] In this approach the correlation energy is calculated in the first step as the sum of independent contributions of translationally invariant localised orbitals or groups of localised orbitals. As corrections the non-additive parts of two, three or more localised orbitals are added. If the group of localised orbitals is chosen reasonably, this partitioning of the correlation energy of the solid yields a fast converging series. Due to the partitioning it is necessary that the correlation method used for the individual increments is size-consistent and size-extensive.

The method of increments has been applied to ground-state properties of various material classes: From insulators[12–16] over semiconductors[9,10,17–21] to metals,[11,22,23,25] from strongly bound ionic or covalent systems to weakly bound van der Waals solids,[26–28] from large molecules[29–31] over polymers[32–39] to three-dimensional solids, from weakly correlated systems to strongly correlated ones such as transition-metal oxides[40,41] and rare-earth nitrides and oxides.[42–44] This large variety of applications is possible, because the method of increments allows for an individual selection of the localised orbital group for each system and for the choice of the suitable correlation method used for the ansatz of the many-body wavefunctions. The results allow us to discuss the influence of electronic correlation on the cohesive energy, on the lattice constants, the bulk moduli and other ground-state properties. In addition we can determine the origin of the electronic correlations and the range of the correlation hole.

The generalisation to metals is discussed in this paper for the example of solid mercury.[23–25] Mercury condenses at 233 K into the rhombohedral structure with a bond length $a = 3.005$ Å and an angle of 70.53^{o}[45] (see Fig. 1). This is in contrast to

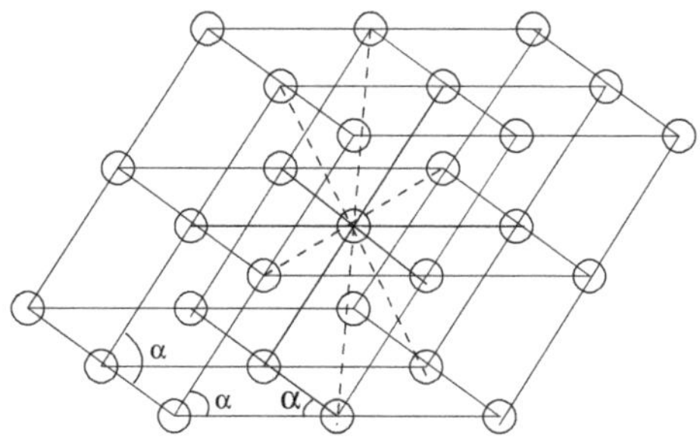

Fig. 1. The rhombohedral structure occurring in solid mercury. The solid lines indicate the nearest neighbour distance a and the dashed lines show the second nearest neighbours, which form a regular hexagon.

zinc and cadmium, which adopt the hexagonal close-packed (hcp) structure, with an anomalous c/a ratio which is far from ideal hcp. Density functional methods fail to describe either of these structures accurately. Even in mercury some functionals yield a closed-packed fcc structure, and others a simple cubic structure. An application of the method of increments to mercury, including correlation via coupled cluster calculations on finite fragments of the solid, allows the systematic inclusion and comparison of the competing effects that leads to the observed structure.

The paper is organised as follows: In the next section the method of increments is presented and the extension to metals is described. In Sec. 3 the results for the ground-state properties of mercury are discussed. Conclusion follows in Sec. 4.

2. The Method of Increments

The method of increments expands the correlation energy of the systems in terms of local increments (for a sketch see Fig.2). It is formally similar to treating the hierarchy of nth order atomic Bethe-Goldstone equations.[46] Here we want to sketch the basic ideas and some important formulae (for more details see Ref. [47,48]). The method relies on localised orbitals or a group of localised orbitals generated in a Hartree-Fock reference calculation. One-body correlation-energy increments ϵ_i are obtained by correlating each of the localised orbital groups separately while keep-

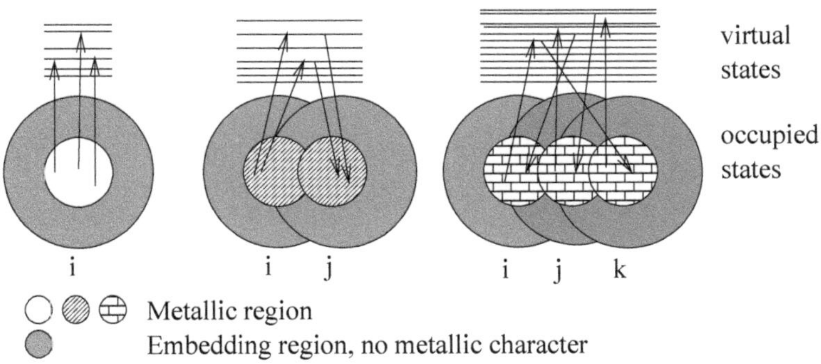

Fig. 2. Schematic overview of the procedure of the method of increments for metals.

ing the other ones inactive. In the present work we are using the coupled-cluster approach with single and double substitutions with perturbative triples (CCSD(T)) as implemented in Molpro,[49–51] but every size-consistent correlation method is possible. This yields a first approximation of the correlation energy

$$E_{\text{corr}}^{(1)} = \sum_i \epsilon_i, \tag{2}$$

which corresponds to the correlation energy of independent orbital groups.

In the next step we include the correlations of pairs of orbital groups. Only the non-additive part $\Delta\epsilon_{ij}$ of the two-body correlation energy ϵ_{ij} is needed.

$$\Delta\epsilon_{ij} = \epsilon_{ij} - (\epsilon_i + \epsilon_j). \tag{3}$$

Higher order increments are defined analogously, e.g. the three-body increment

$$\Delta\epsilon_{ijk} = \epsilon_{ijk} - (\epsilon_i + \epsilon_j + \epsilon_k) - (\Delta\epsilon_{ij} + \Delta\epsilon_{jk} + \Delta\epsilon_{ik}). \tag{4}$$

The correlation energy of the system is finally obtained by adding up all the increments with appropriate weight factors:

$$E_{\text{corr}} = \sum_i \epsilon_i + \frac{1}{2}\sum_{\substack{ij\\ i\neq j}} \Delta\epsilon_{ij} + \frac{1}{6}\sum_{\substack{ijk\\ i\neq j\neq k}} \Delta\epsilon_{ijk} + \cdots. \tag{5}$$

It is obvious that by calculating higher and higher increments the exact correlation energy is determined within the correlation method applied.

This partitioning of the correlation energy is only sensible if the series converges quickly enough, both with the order of increments (i.e., 3-body increments should be significantly smaller than two-body increments, etc.) and with the distance of the orbital groups involved in the increments. Increments involving distant orbital groups must decay faster than the number of increments increases in a 3-dimensional system. For a semiconductor or an insulator, increments with distant orbital groups interact only via van der Waals and thus decay like r^{-6}, whereas the number of pairs grows with r^2, so an overall decay of r^{-4} guarantees the convergence with respect to the distance of orbital groups.

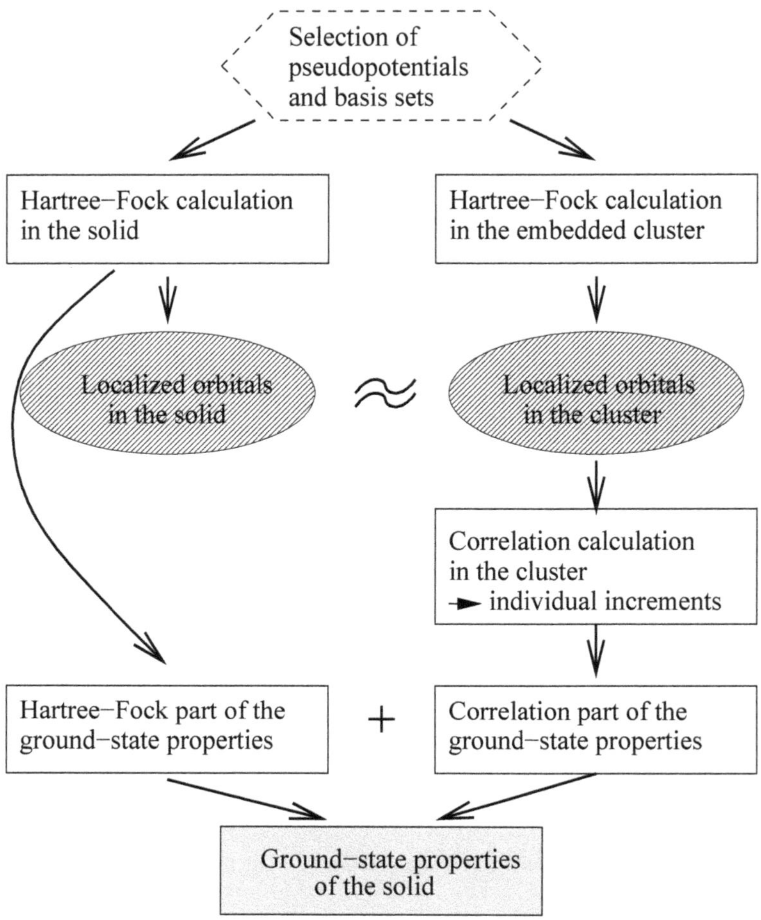

Fig. 3. Schematic overview of the procedure of the ground-state calculations with the method of increments.

A direct transfer of this approach to metallic systems is not possible, however, since localised orbitals become very long-range entities here and a many-body expansion in terms of such orbitals cannot be expected to have useful convergence characteristics. In order to make the expansion still computationally feasible, we have suggested[23,30] to start from suitable model systems where long-range orbital tails are absent, and to allow for delocalisation only successively in the course of the expansion; more specifically, when calculating pair contributions for a given orbital group combination (i, j), we allow for delocalisation $i \rightarrow j$ and $j \rightarrow i$, and similarly with the 3-body terms we allow for delocalisation over the triples of atoms, etc. It is clear that the final result is not affected, only the convergence properties of the many-body expansion are changed. As an additional advantage, we can calculate individual terms of the expansion from (suitably modelled/embedded) finite clusters of reasonable size. In the case of mercury, for example, we can force localisation of the solid by using a *s*-type atomic basis set for describing the valence-electron system. This way, delocalisation due to *sp*-mixing is avoided, but still each atom has its correct crystal surroundings concerning the electrostatic interaction. Due to the metallic embedding scheme even the conduction bands are localised in the central region. The incremental expansion has to take care subsequently of the non-additive effects of correlation and delocalisation.

Here we want only briefly sketch the technical details necessary to apply the method of increments to real systems. The first question is the selection of the pseudopotentials, if necessary, and the basis set which is usually a set of contracted Gaussian functions centred at the atoms involved in the system. Because we want to use standard correlation methods as implemented in quantum chemical programs we have to calculate the correlation energy increments in finite fragments of the solid. Therefore we generate a proper embedding, so that the localised orbitals in the cluster are very simular to the one in the solid. The localisation procedure is mostly according to Foster and Boys.[52] An schematic overview of the procedure is given in Fig.3.

3. Application to Solid Mercury

The Hartree-Fock ground-state properties are calculated with CRYSTAL[53] supplying a 20-valence-electron pseudopotential[54] for mercury and the corresponding crystal-optimised basis set.[30] The value for the cohesive energy is corrected for the basis set superposition error[55] by a counterpoise correction with the atomic energy calculated in the presence of basis sets placed at the positions of neighbouring atoms in the solid (convergence required 12 nearest neighbours for Hg). As shown in Fig.4,5 the HF treatment yields no binding for mercury, the cohesive energy at the experimental structure is repulsive by about the same amount as the experimental cohesive energy is binding.

For the correlation calculation we use an embedding scheme where the part of the cluster which is not correlated is described with 2-valence-electron scalar relativistic pseudopotentials[56] which simulate the Hg $5s^2 5p^6 5d^{10}$ shells within the atomic core.

Thus, only the $6s$ shell is explicitly treated in the embedding region. The critical part in the application of the method of increments to metals is the possibility of describing the metallic orbitals calculated in the embedding region in a local way. This localization is done in our approach by a unitary transformation of the occupied canonical orbitals according to the criterion of Foster and Boys.[52] Good localisation characteristics are achieved by using an *s*-type atomic basis set ($2s$ basis, with contraction coefficients optimised for the free atom) on the embedding and thereby avoiding delocalisation due to *sp*-mixing. Within this localised environment we can use the full basis for the 1, 2, or 3 atoms that we are calculating the correlation energy of, for the 1-, 2-, or 3-body increments, respectively. This approach allows us to successively include metallic delocalisation in an incremental way.

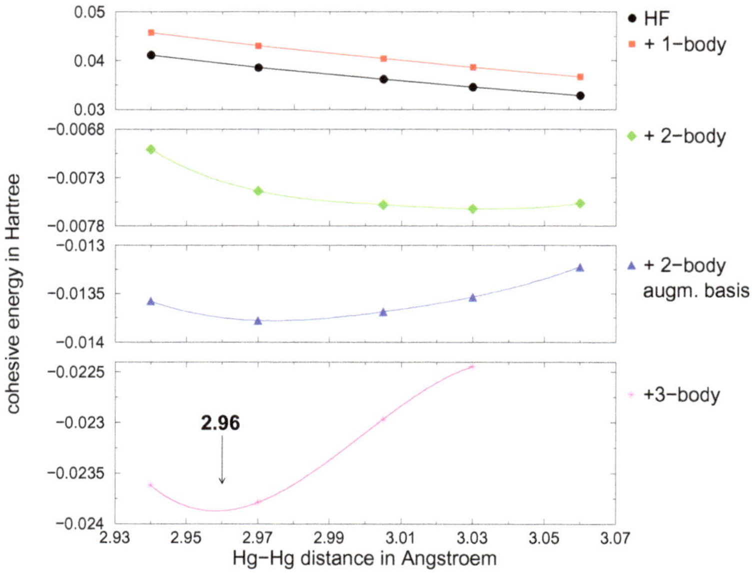

Fig. 4. The cohesive energy of Hg with respect to the nearest neighbour-distance a.

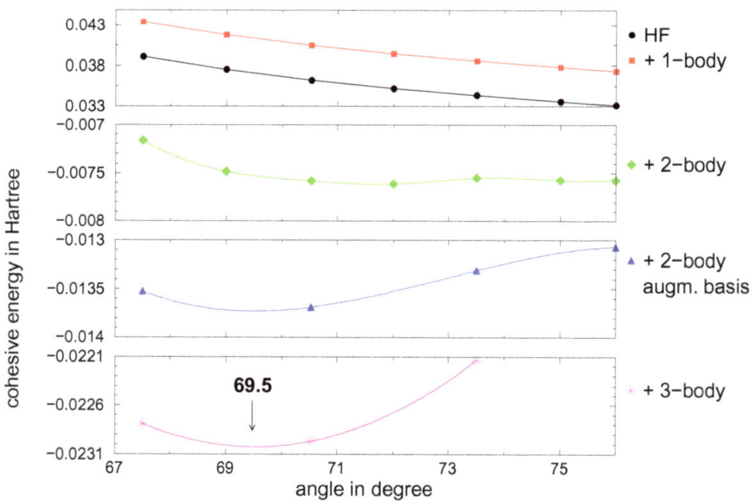

Fig. 5. The cohesive energy of Hg with respect to the angle in the rhombohedral structure.

The increments are calculated for selected cluster models which reflect the geometry of the Hg crystal. The rhombohedral structure of the infinite crystal can be viewed as a central atom surrounded by atom shells of various size. We select for the embedding the first shell containing 12 atoms, 6 of them at distance a(nn) (=3.005 Å) and 6 at 1.155 a(nn); For calculating a few-body term, we include all atoms in the embedding which are in the first shell of one of the atoms to be correlated.

The description of the atoms to be correlated is much more important for the final correlation energy, as we will keep frozen the localised orbitals of the atoms of the embedding region when calculating correlation energies.

The basis sets of the correlated atoms were described in Ref. 30. A $(10s9p7d2f1g)/[8s7p6d2f1g]$ set is used for the 20 valence electrons of mercury, and an even tempered augmented version of the same basis to gain an impression of the basis set completeness for these calculations. Using these basis sets, we recalculate the integrals and reoptimise the orbitals of the atoms to be correlated, in a HF calculation, within the frozen environment of the embedding. Thus we have reasonably delocalised orbitals within the interior of our cluster over the atoms $i, j, ..$ to be correlated, but localised orbitals with respect to the embedding. Then we calculate the correlation energy of these atoms in a coupled-cluster calculation with single and double excitations and perturbative treatment of the triples (CCSD(T)).[50,51] We can do this with different definitions of the core in order to see the different contributions of the valence $6s$, $5d$, and core $5sp$ electrons. All these calculations are performed using the MOLPRO suite of ab-initio programs.[49]

The results for the lattice parameters, the cohesive energy and the bulk modulus are summarised in Table 1. In order to discuss the dependence of the energy

Table 1. Results for the structure and bulk modulus of solid mercury. a is in Å, α in degrees, E_{coh} in eV and B in Mbar. For the LDA value in the hcp structure, $a = a(\text{nn})$ is given.

Method	a	α	E_{coh}	B
Hartree-Fock	—	—	+0.985	—
Incremental correlation (2b:aug. basis, s-only)	—	—	+0.245	—
Incremental correlation (2b:aug. basis)	2.97	70.0	-0.375	0.132
Incremental correlation (3b:s-only)	2.97	69.2	-0.561	0.383
Incremental correlation (3b)	2.96	69.5	-0.649	0.360
Expt.[45,57]	3.005	70.53	-0.670	0.322

of the lattice on its structure, we vary the lattice distance and angle of the structure around the experimental lattice parameters. The HF energy decreases with increasing distance (see Fig. 4) as is normal for a purely repulsive potential. It also decreases with increasing angle (see Fig. 5) which corresponds to an opening of the structure and lowering of the density. The one-body terms of the correlation-energy expansion are repulsive for mercury and have nearly no dependence on the lattice parameter. The first term in the expansion to examine closely with respect to the effect of the lattice parameter is therefore the two-body increment. Here the potential is actually very flat, and thus the minimum can be shifted noticeably by the use of an augmented basis set, if we are only considering the two-body increments. In Fig. 5 the angular dependence of the two-body part is shown to be even weaker.

Only with the augmented basis set does a shallow minimum appear. With the augmented basis, we find a minimum with $a = 2.97$ Å, and $\alpha = 70.0°$. If we include only the correlation of the valence s-electrons in the two-body increments we still have no minimum. Only with the inclusion of d-correlation in the two-body part do we find a bound solid. We have chosen to truncate the expansion of the three-body correlation after 8 increments. We concentrate on these eight geometries because they contribute about 80% to the 3-body part of the cohesive energy. These 8 clusters all have 2 nearest or second-nearest-neighbour distances connecting the atoms. Including the contributions of the three-body increments the rhombohedral lattice structure is stabilised with only a very small shift to a smaller interatomic distance and somewhat smaller angle. Overall the lattice parameter agree within 1.4% with the experimental values.

The results for the cohesive energy are summarised in the third column of Table 1. The HF part of the cohesive energy is repulsive, the 1-body correlation contribution to the cohesive energy is repulsive, too. The main contribution to the binding comes from the 2-body increments, with about half of this originating from the core-valence correlation of the d shell. Without correlating the d shell solid mercury would not be bound. The 3-body contributions are attractive, both for the valence as for the d shell, whereas the latter one contribute by about one third to the total 3-body contributions. The most important four-body contributions have been calculated, they are small (≤ 0.08 eV) and repulsive. The spin-orbit contribution to the binding and the zero-point energy can be neglected. In summary, the method of increments allows us to determine the cohesive energy of solid mercury within the same accuracy compared to the experimental value (error less than 15% within the given finite basis set) as was achieved for semiconductors and insulators.

Our calculated values for the bulk modulus are given in Tab. 1. The bulk modulus calculated with two-body increments only is 0.132 Mbar, considerably lower than the experimental value. When only the s-correlation of the three-body increments is included, the bulk modulus increases to a value of 0.383 Mbar. The final result of the method of increments, with the inclusion of d-correlation for the three-body increments, gives 0.360 Mbar in good agreement with experiment.

4. Conclusion

We have presented here the generalisation of the method of increments to metals. Only the correlation part of the energy is treated by means of the incremental scheme. To obtain localised orbitals, we start from embedded clusters without any metallic character. Within the incremental scheme we allow for a delocalisation of the orbitals and therefore account for the metallic character of the systems. With this approach the convergence rate, especially with the order of increments, is not as good as for insulators, but still three- and higher-body terms account for only about 15% of the correlation contribution to the binding. For the first time we could solve the puzzling problem of the binding in solid mercury, where the binding is entirely due to electronic correlations. The lattice structure of solid mercury has been calculated using the method of increments. With this wavefunction-based method, we can reproduce the lattice constant and rhombohedral angle to within 99% of the experimental values, and obtain a cohesive energy of -0.65 eV, 97% of the experimental value, and a bulk modulus well within the experimental error bars.

Acknowledgment

The author would like to thank Dr. Nicola Gaston and Prof. Dr. Krzysztof Rościszewski taking part in this project and Prof. Dr. Hermann Stoll for useful discussions.

References

1. R. M. Dreizler and E. K. U. Gross, Density Functional Theory, Springer, Berlin, 1990.
2. H. Eschrig, The Fundamentals of Density Functional Theory, B. C. Teubner, Stuttgart, 1996.
3. D. R. Hartree, Proc. Cambridge Philos. Soc., 24 (1928) 89.
4. V. Fock, Z. Phys., 61 (1930) 126.
5. J. C. Slater, Phys. Rev., 35 (1930) 210.
6. P. O. Löwdin, J. Phys. Chem., 61 (1957) 55.
7. R. K. Nesbet, Rev. Mod. Phys., 33 (1961) 28.
8. G. H. Wannier, Phys. Rev., 52 (1937) 191.
9. H. Stoll, Phys. Rev. B, 46 (1992) 6700.
10. H. Stoll, Chem. Phys. Lett., 191 (1992) 548.
11. H. Stoll, J. Chem. Phys., 97 (1992) 8449.
12. K. Doll, M. Dolg, P. Fulde, and H. Stoll, Phys. Rev. B, 52 (1995) 4842.
13. K. Doll, M. Dolg, and H. Stoll, Phys. Rev. B, 54 (1996) 13529.
14. K. Doll and H. Stoll, Phys. Rev. B, 56 (1997) 10121.
15. K. Doll and H. Stoll, Phys. Rev. B, 57 (1998) 4327.
16. K. Doll, P. Pyykkö, and H. Stoll, J. Chem. Phys., 109 (1998) 2339.
17. B. Paulus, P. Fulde, and H. Stoll, Phys. Rev. B, 51 (1995) 10572.
18. B. Paulus, P. Fulde, and H. Stoll, Phys. Rev. B, 54 (1996) 2556.
19. S. Kalvoda, B. Paulus, P. Fulde, and H. Stoll, Phys. Rev. B, 55 (1997) 4027.
20. B. Paulus, F.-J. Shi, and H. Stoll, J. Phys.: Condens. Matter, 9 (1997) 2745.
21. M. Albrecht, B. Paulus, and H. Stoll, Phys. Rev. B, 56 (1997) 7339.
22. B. Paulus, Chem. Phys. Lett., 371 (2003) 7.
23. B. Paulus and K. Rościszewski, Chem. Phys. Lett., 394 (2004) 96.
24. B. Paulus, K. Rościscewski, N. Gaston, P. Schwerdtfeger, and H. Stoll, Phys. Rev. B, 70 (2004) 165106.
25. N. Gaston, B. Paulus, K. Rościscewski, P. Schwerdtfeger, and H. Stoll, Phys. Rev. B, 74 (2006) 094102.
26. K. Rościszewski, B. Paulus, P. Fulde, and H. Stoll, Phys. Rev. B, 60 (1999) 7905.
27. K. Rościszewski, B. Paulus, P. Fulde, and H. Stoll, Phys. Rev. B, 62 (2000) 5482.
28. K. Rościszewski and B. Paulus, Phys. Rev. B, 66 (2002) 092102.
29. S. Kalvoda, B. Paulus, M. Dolg, H. Stoll, and H.-J. Werner, Phys. Chem. Chem. Phys., 3 (2001) 514.
30. B. Paulus, Int. J. of Quant. Chem., 100 (2004) 1026.
31. H. Stoll, B. Paulus, and P. Fulde, J. Chem. Phys., 123 (2005) 144108.

32. M. Yu, S. Kalvoda, and M. Dolg, Chem. Phys., 224 (1997) 121.
33. A. Abdurahman, M. Albrecht, A. Shukla, and M. Dolg, J. Chem. Phys., 110 (1999) 8819.
34. A. Abdurahman, A. Shukla, and M. Dolg, J. Chem. Phys., 112 (2000) 4801.
35. A. Abdurahman, A. Shukla, and M. Dolg, Chem. Phys., 257 (2000) 301.
36. A. Abdurahman, A. Shukla, and M. Dolg, Phys. Rev. B, 65 (2002) 115106.
37. C. Willnauer and U. Birkenheuer, J. Chem. Phys., 120 (2004) 11910.
38. C. Buth and B. Paulus, Phys. Rev. B, accepted (2006).
39. C. Buth and B. Paulus, Chem. Phys. Lett., 389 (2004) 44.
40. K. Doll, M. Dolg, P. Fulde, and H. Stoll, Phys. Rev. B, 55 (1997) 10282.
41. K. Rościszewski, K. Doll, B. Paulus, P. Fulde, and H. Stoll, Phys. Rev. B, 57 (1998) 14667.
42. S. Kalvoda, M. Dolg, H.-J. Flad, P. Fulde, and H. Stoll, Phys. Rev. B, 57 (1998) 2127.
43. E. Voloshina and B. Paulus, Theor. Chem. Acc., 114 (2005) 259.
44. E. Voloshina and B. Paulus, J. Chem. Phys., 124 (2006) 234711.
45. CRC Handbook of Chemistry and Physics, CRC Press, New York (1997).
46. R. K. Nesbet, Adv. Chem. Phys., 14 (1969) 1.
47. P. Fulde, Adv. in Physics, 51 (2002) 909.
48. B. Paulus, Phys. Rep., 421 (2006) 1.
49. R. D. Amos, A. Bernhardsson, A. Berning, P. Celani, D. L. Cooper, M. J. O. Deegan, A. J. Dobbyn, F. Eckert, C. Hampel, G. Hetzer, P. J. Knowles, T. Korona, R. Lindh, A. W. Lloyd, S. J. McNicholas, F. R. Manby, W. Meyer, M. E. Mura, A. Nicklass, P. Palmieri, R. Pitzer, G. Rauhut, M. Schütz, U. Schumann, H. Stoll, A. J. Stone, R. Tarroni, T. Thorsteinsson, and H.-J. Werner, MOLPRO, a package of ab initio programs designed by H.-J. Werner and P. J. Knowles, Birmingham, 2002.
50. C. Hampel, K. Peterson, and H.-J. Werner, Chem. Phys. Lett. 190 (1992) 1.
51. M. J. O. Deegan and P. J. Knowles, Chem. Phys. Lett 227 (1994) 321.
52. J. M. Foster and S. F. Boys, Rev. Mod. Phys., 32 (1960) 296.
53. V. R. Saunders, R. Dovesi, C. Roetti, M. Causa, N. M. Harrison, R. Orlando, and C. M. Zicovich-Wilson, CRYSTAL, Version 98 is designed by the Theoretical Chemistry Group, University of Torino, Torino, Italy, 1998.
54. D. Andrae, U. Häussermann, M. Dolg, H. Stoll, and H. Preuss, Theor. Chim. Acta 77 (1990) 123. (The f and g projectors are omitted in the Crystal code.)
55. S. F. Boys and F. Bernardi, Mol. Phys., 19 (1970) 553.
56. W. Küchle, M. Dolg, H. Stoll, and H. Preuss, Mol. Phys., 74 (1991) 1245.
57. H. B. Huntingdon, Solid State Phys., 7 (1958) 282.

FRACTIONALLY CHARGED EXCITATIONS ON FRUSTRATED LATTICES

E. RUNGE

Technische Universität Ilmemau, Fakultät für Mathematik und Naturwissenschaften, FG Theoretische Physik I, 98693 Ilmenau, Germany

F. POLLMANN and P. FULDE

Max-Planck-Institut für Physik komplexer Systeme, Nöthnitzer Straße 38, 01187 Dresden, Germany

Systems of strongly correlated fermions on certain geometrically frustrated lattices at particular filling factors support excitations with fractional charges $\pm e/2$. We calculate quantum mechanical ground states, low–lying excitations and spectral functions of finite lattices by means of numerical diagonalization. The ground state of the most thorough-fully studied case, the criss-crossed checkerboard lattice, is degenerate and shows long–range order. Static fractional charges are confined by a weak linear force, most probably leading to bound states of large spatial extent. Consequently, the quasi-particle weight is reduced, which reflects the internal dynamics of the fractionally charged excitations. By using an additional parameter, we fine–tune the system to a special point at which fractional charges are manifestly deconfined—the so–called Rokhsar–Kivelson point. For a deeper understanding of the low–energy physics of these models and for numerical advantages, several conserved quantum numbers are identified.

Keywords: Fractional charge; strongly correlated particles; confinement.

1. Introduction

Quantization of charge is a very basic feature in the description of the physical world. Therefore, fractionally charged excitations came as a surprise to physicists. Already back in the year 1979, Su, Schrieffer, and Heeger[1] showed that a model describing the one-dimensional (1D) chain molecule trans–polyacetylene $(CH)_n$ supports excitations with spin–charge separation. This is not yet charge fractionalization, but when the model is considered at different electron densities (corresponding to extremely high doping), it turns out that at certain filling factors elementary excitations with fractional charge exist [in the simplest case $q = \pm e/3$ and $q = \pm 2e/3$]. A few years later, Laughlin citelaughlin1983 interpreted the much celebrated fractional quantum Hall effect (FQHE) in terms of fractionally charged (quasi-) particles (fcp) and fractionally charged (quasi-) holes (fch). Thereby, he firmly established the idea of fractional charges in our understanding of solid–state physics. Direct electron–electron interactions are crucial in the FQHE: Correlations become strong in an applied magnetic field because the kinetic energy of the electrons is quenched. Of course, no one will ever extract a fraction of an electron from an FQHE sample. But thinking of an extra electron or an extra hole near fill factor $\nu = 1/3$ as decaying into three separate entities of charge $q = \pm e/3$ each proved enormously helpful for a qualitative and quantitative understanding of the FQHE—and was rewarded by the Nobel Prize in Physics 1998.

The question was left open whether or not fractionally charged excitations exist in 2D or 3D systems without a magnetic field. In 2002, it was suggested by one of us[3] that excitations with charge $\pm e/2$ do exist in certain 2D and 3D lattices, e.g., the pyrochlore lattice, which is a prototype 3D structure with geometrical frustration. The original work was motivated by the transition metal compound LiV_2O_4, which surprisingly shows heavy–fermion behavior with, e.g., a large γ coefficient in the low-temperature specific heat $C = \gamma T$.[4] However, we will not address the issue whether the models discussed here apply to any particular material or artifical systems such as optical lattices, see e.g. Ref. 5 and citations therein. Instead, we try to contribute to the very general question whether or not fractional charges can exist at all in truly 2D or 3D systems in the absence of magnetic fields.

We would like to argue that general prerequisites for fractional charges are—as in the FQHE case—strong short–range correlations and certain band fillings. Furthermore, the short–range correlations should be somehow incompatible with the lattice structure in order to prevent the development of long–range order. Following the general usage, we will simply call the lattice "frustrated" even though calling the interaction "frustrated by the lattice" would be the more accurate terminology.

In this contribution, we study the charge degrees of freedom on such lattices systematically and consider a class of models of strongly correlated spinless fermions. Most of our numerical calculations were done for the 2D checkerboard lattice, which is easier to deal with than the even more interesting 3D cases. For future work, one can hope to learn from comparison with spin systems where numerous studies exist for the pyrochlore lattice and its 2D relatives, the crisscrossed checkerboard lattice and the kagome lattice.[6–10]

2. Fractional Charges on Frustrated Lattices

In order to illustrate the concept of fractional charges on frustrated lattices, we focus here on the (crisscrossed) 2D checkerboard lattice. Figure 1(a) illustrates that the checkerboard lattice can be thought of as a projection of the pyrochlore lattice onto a plane. In the following, we adopt the ideas of Ref. 3 and consider a model Hamiltonian of spinless fermions

$$H = -t \sum_{\langle i,j \rangle} \left(c_i^\dagger c_j + \text{H.c.} \right) + V \sum_{\langle i,j \rangle} n_i n_j \qquad (1)$$

on a checkerboard lattice. The operators $c_i (c_i^\dagger)$ annihilate (create) fermions on sites i. The density operators are $n_i = c_i^\dagger c_i$. We assume half filling, $\sum_i n_i = \frac{N}{2}$ for a

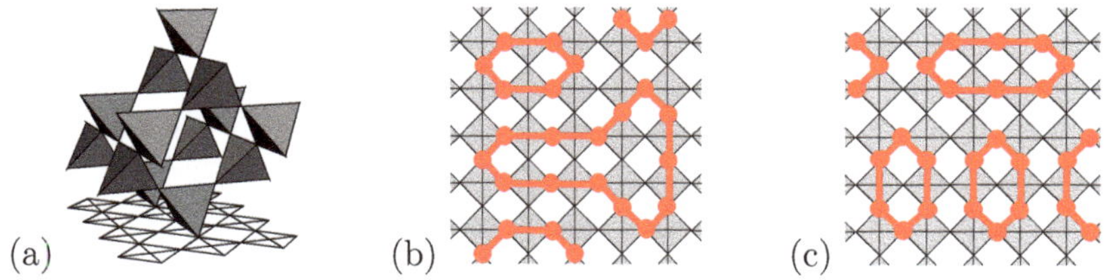

Fig. 1. (a) Checkerboard lattice as 2D projection of the 3D pyrochlore lattice.[11] (b,c) Two examples of allowed configurations on a checkerboard lattice at half filling. Occupied sites are connected by thick lines as guides to the eye.

system with N sites. Our main interest is the regime where the nearest–neighbor hopping t is much smaller than the nearest–neighbor repulsion V, i.e., $|t|/V \ll 1$. Henceforth, we assume $t > 0$. In analogy to the tetrahedra in a pyrochlore lattice, all bonds in a crossed square are equivalent.

Classical correlations and ground–state degeneracy. For a moment, let us set the hopping–matrix element t to zero. The ground–state manifold is then *macroscopically degenerate*: Every configuration that satisfies the so–called tetrahedron rule of having *exactly* two particles on *each* tetrahedron (crisscrossed square) is a (classical) ground state[12] and will henceforth be referred to as an "allowed configuration." Examples are shown in Fig. 1(b) and (c). Note that our coordinate system is rotated by 45° relative to that of, e.g., Ref. 13. The resulting difference in boundary conditions can lead to noticeable numerical differences in particular for small cluster sizes.

One can visualize the origin of the macroscopic degeneracy as follows: Take the set of the six allowed crisscrossed squares and construct row by row a larger allowed configuration. Whenever a crisscrossed square is added, we can choose between one, two, or three different possibilities, depending on the neighboring crisscrossed squares. Since there often *is* a choice to make, an exponential number of different allowed configurations can be constructed. Thus, the system has a finite entropy at $T = 0$. The exact value of the ground-state degeneracy can be obtained from a mapping to the so–called six–vertex model.[14] Its solution is highly non–trivial due to the existence of long–range correlations, which seem to be a generic feature seen in many frustrated lattice models.[15]

All classical ($t = 0$) ground states have the important property of being incompressible in the sense that no fermion can hop to another empty site without creating defects and thereby increasing the repulsion energy. In other words, we have to violate the tetrahedron rule in intermediate states if we want to connect via hopping processes one allowed configuration with another.

Fractional charges. Placing one additional particle with charge e onto an empty site of an allowed configuration leads to a violation of the tetrahedron rule on two

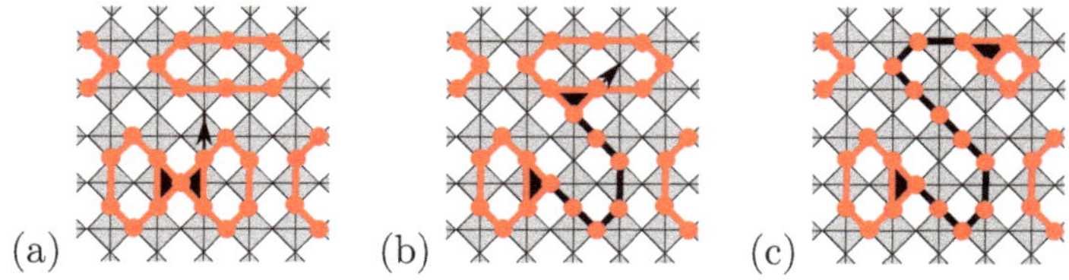

Fig. 2. (a) Adding one fermion to the half–filled checkerboard lattice leads to two defects (marked by black triangles) on adjacent crisscrossed squares. (b), (c) Two defects with charge $e/2$ can separate without creating additional defects. They are connected by a string consisting of an odd number of fermions.

adjacent crisscrossed squares, see Fig. 2(a). The energy is increased by $4V$ since the added particle has four nearest neighbors (charge gap). There is no way to remove the violations of the tetrahedron rule by moving electrons. However, fermions on a crisscrossed square with three particles can hop to another neighboring crisscrossed square without creating *additional* violations of the tetrahedron rule, i.e., without increase of the repulsion energy [see Fig. 2(b,c)]. By these hopping processes, two local defects (violations of the tetrahedron rule) can separate and the added fermion with charge e breaks into two pieces. They carry a fractional charge of $e/2$ each. In the quantum mechanical case ($t \neq 0$), the separation leads to a lowering of the kinetic energy of order $|t|$.

Energy and momentum must be conserved by the decay processes $1e \to 2\text{fcp}'\text{s}$. If we associate momentum $\mathbf{k}$ and energy $E(\mathbf{k})$ with the inserted fermion, they must now be shared between the resulting two fcp's into which it has decayed

$$E(\mathbf{k}) = 4V + \epsilon(\mathbf{k}_1) + \epsilon(\mathbf{k}_2). \tag{2}$$

Here $\mathbf{k} = \mathbf{k}_1 + \mathbf{k}_2$ and $\epsilon(\mathbf{k})$ is the energy dispersion of a fcp. Even though $\epsilon(\mathbf{k})$ is at present completely unknown, Eq. (2) allows to predict that for deconfined fcp's the electronic spectral function should not contain a Fermi–liquid peak, but should show a broad continuum instead.

Figure 2 demonstrates that the two defects can be thought of as always being connected by a string of occupied sites consisting of an *odd* number of sites. The fractional charges can thus be alternatively interpreted as the ends of a string–like excitation. In this picture, the connection is not static, as two pairs of fcp's can exchanges partners when the connecting strings come close.

Quantum Fluctuations. If we relax the constraint of having two fermions on each crisscrossed square and consider a small but finite ratio t/V, quantum fluctuations come into play. The quantum fluctuations lead also to fractional charges, but do not change the net charge of the systems. Starting from an allowed configuration, the hopping of a fermion to a neighboring site increases the energy by V. One crisscrossed square contains three fermions while the other has only one fermion, see Fig. 3(a)–(c). These so-called vacuum fluctuations (virtual fcp–fch pairs) lead to two mobile fractional charges with opposite charges $+e/2$ and $-e/2$, which are

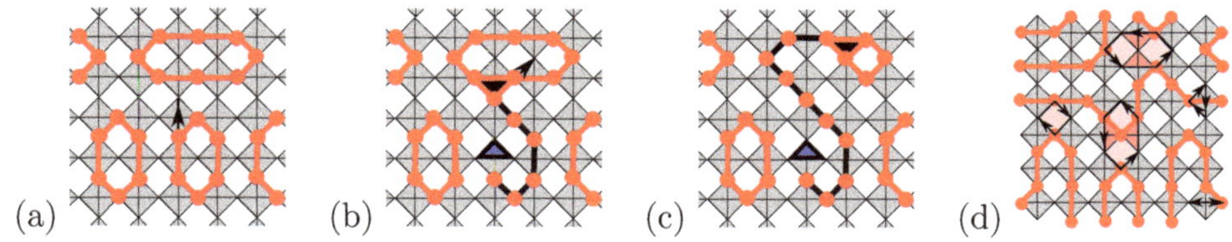

Fig. 3. (a)–(c) Hopping of a fermion to a neighboring site in an allowed configuration generates a fluctuation: (a) A fractional charged particle (fcp) and fractional charged hole (fch) are generated. (b)–(c) The two defect (marked by triangles) with charge $\pm e/2$ can separate without creating additional defects and are connected by a string consisting of an even number of fermions. (d) Example of an allowed configuration on a checkerboard lattice at half filling with possible low–order hopping processes.

connected by a string of an *even* number of fermions. The energy associated with a free fcp and a free fch is $\Delta E_{\text{vac}} = V + \epsilon\,(\mathbf{k}) + \bar{\epsilon}\,(-\mathbf{k})$, where $\bar{\epsilon}\,(-\mathbf{k})$ denotes the kinetic energy of a fch. Such virtual processes connect different allowed configurations, lower the total energy, and reduce the macroscopic degeneracy, as we will see more explicitely in the next section.

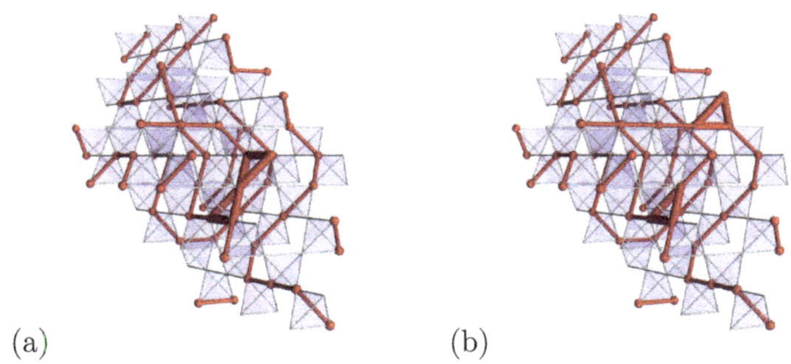

Fig. 4. Adding one fermion to the half–filled three-dimensional pyrochlore lattice leads to two defects on adjacent tetrahedra. The two defects with charge $e/2$ can separate without creating additional defects and are connected by a string consisting of an odd number of fermions.

Fractional charges in 3D. All arguments mentioned above for the existence of fcp's on a 2D checkerboard lattice can be directly transferred to a 3D pyrochlore lattice at half filling, see Fig. 4. Fcp's correspond to tetrahedra with three fermions and fch's to tetrahedra with only one fermion.

3. Effective Hamiltonians

Ring exchange processes in the undoped case. The numerical and analytical work is greatly simplified, if a simpler Hamiltonian $\mathcal{H}$ can be derived that shows in the limit $|t| \ll V$ the same low–energy physics as the model Hamiltonian (1). A down–folding procedure can be used to define $\mathcal{H}$: Per definition, it acts only on the subspace of allowed configurations and includes all virtual processes up to the lowest non–trivial order, which is t^3/V^2. The relevant processes are shown in see Fig. 3(d) and can be classified as self–energy contributions H_Σ and ring–exchange processes H_{eff}, $\mathcal{H} = H_\Sigma + H_{\text{eff}}$. The former comprises the terms which are diagonal in the real space basis. The latter includes those which connect different allowed configurations. H_Σ contains fermion hops to an empty neighboring site and back again as well as hops around an adjacent triangle. These contribute only a constant, configuration–independent energy shift. It does not lift the macroscopic degeneracy and will hence be ignored furtheron. The total amplitude of ring–exchange processes around empty squares is proportional to t^2/V. It vanishes for spinless fermions because the amplitudes for clockwise and counter–clockwise ring–exchange cancel each other due to fermionic anti–commutation relations. Thus, the macroscopic ground–state degeneracy is first lifted by ring exchanges $\sim t^3/V^2$ around hexagons, and the looked–for effective Hamiltonian reads

$H_{\text{eff}} = -g\sum_{\{⬭,⬯\}}\big(\,|$ 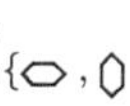 $\rangle\langle$  $| - |$  $\rangle\langle$  $| + \text{H.c.}\big),$ (3)

with the effective hopping–matrix element $g = 12\,t^3/V^2 > 0$ and the sum taken over all vertical and horizontal oriented hexagons. The pictographic operators represent the hopping around hexagons which have either an empty or an occupied central site. The signs of the matrix elements depend on the representation and the sequence in which the fermions are ordered. When the sites are enumerated along diagonal rows, an exchange process commutes an odd number of fermionic operators if the site in the center of the hexagon is empty and an even number if the center is occupied. In Ref. 13, it has been shown that the effective Hamiltonian (3) gives a good approximation of the low–energy excitations of the full Hamiltonian (1) in the limit considered.

Propagation of defects and extra charges. We have derived H_{eff} as effective Hamiltonian for local rearrangement processes as resulting from the (virtual/intermediate) generation of a defect pair which subsequently recombines, leaving behind a different allowed generation. For many questions of physical interest, it is necessary to consider the propagation of defects over large distances. Analogously to conventional semiconductors, this refers to long–lived thermally generated defects (e–h pairs in the semiconductor analogy) as well as to the consequences of slight doping, i.e., addition of two fcp's by addition of one extra fermion. The natural generalization of the effective Hamiltonian (3) to these cases includes besides the lowest order ring–exchange processes H_{eff} a projected hopping term that moves the defects

$$H_{\text{tg}} = H_{\text{eff}} - t\sum_{\langle i,j\rangle} P\left(c_i^\dagger c_j + \text{H.c.}\right) P. \tag{4}$$

The projector P ensures that H_{tg} acts only on the subspace of configurations with the smallest possible number of violations of the tetrahedron rule which is compatible with the number of particles, e.g., two in the case of one added fermion. We refer to H_{tg} as the t–g model and consider the parameters t and g as independent (i.e., not restricted to the regime $g \ll t$ as enforced by $t \ll V$). This allows to study the effect of ring exchange onto the dynamics of fractionally charged excitations. In particular, the question of confinement can be studied even on rather small clusters by increasing the ring exchange strength g relative to t.

4. Height Representation, Conserved Quantities, and Gauge Symmetries

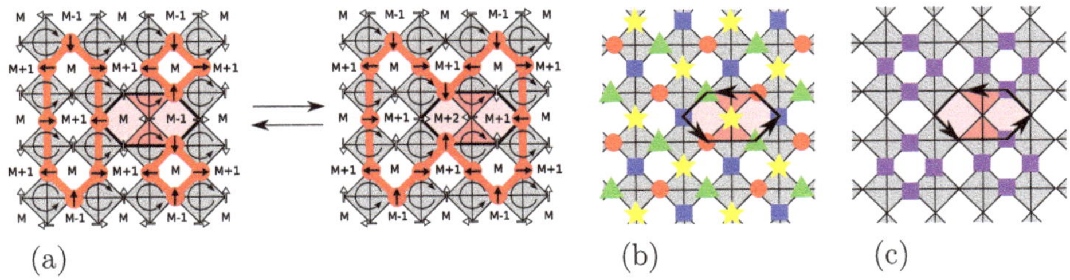

Fig. 5. (a) Height representation for examples of allowed configurations of a $\sqrt{32} \times \sqrt{32}$ checkerboard lattice with periodic boundary conditions at half filling. The height field h (numbers in the non–crossed squares) is uniquely defined for a given configuration up to an additive constant M. The field $\mathbf{f} = \nabla h$ is indicated by small arrows on the lattice sites. Details of the mapping can be found in the text. The effect on the height fields of a ring–exchange process around a hexagon is shown explicitely. (b) The effective Hamiltonian conserves the number of fermions on each of the four sublattices of the checkerboard lattice which are labeled by "blue squares", "red circles", "yellow stars" and "green triangles." (c) Ring–exchange processes change the number of fermions on sites marked by dark "purple squares" by two.[16]

Conserved quantities of a Hamiltonian allow for a reduction of the numerical effort by exploiting the resulting block–diagonal form of the matrix representation. The real space configurations spanning the subspaces corresponding to these blocks will be referred to as "subensembles." Eigenstates can conveniently be classified by the eigenvalues of the conserved quantities as quantum numbers for our model. We will now identify several such quantum numbers in order to characterize different subensembles. A topological quantity, which is conserved by all local processes, i.e., ring–exchange processes, is the average tilt of a scalar height field which will be introduced in the next paragraph. Another useful set of quantum numbers are the number $(N_{\mathcal{B}}, N_{\mathcal{Y}}, N_{\mathcal{G}}, N_{\mathcal{R}})$ of particles on the four sublattices referred to as blue, yellow, green and red as shown in Fig. 5(b). They are conserved by H_{eff}.

Any allowed configurations of the half–filled checkerboard can uniquely be represented by a vector field $\mathbf{f}$ for which the discretized lattice version of the (discrete) curl vanishes, i.e., a pure (discrete) gradient of a scalar field (height field) $\mathbf{f} = \nabla h$, see Fig. 5. The height field h, is derived from the local constraint expressed by the tetrahedron rule as follows: A clockwise or counter–clockwise orientation is assigned alternatingly to the crisscrossed squares. Arrows of unit length are placed on the lattice sites. The arrows point along (against) the orientation of the adjacent crisscrossed squares if the site is occupied (empty). It is easily checked that allowed configurations are those for which the discretized line integral of $\mathbf{f}$ around every closed loop vanishes. Thus, $\mathbf{f} = \nabla h$ defines a height field h up to an arbitrary constant M.

The height at the upper (right) and at the lower (left) boundary of a finite lattice of $N_x \times N_y$ squares with periodic boundary conditions can differ only by an integer $-N_{y(x)} \leq \kappa_{y(x)} \leq N_{y(x)}$, which is the same for all columns (rows). This defines topological quantum numbers (κ_x, κ_y). They remain unchanged by all *local* processes that transform one allowed configuration into another, i.e., by ring–exchange processes along contractible loops. In particular, H_{eff} merely lowers or raises the local height of two adjacent plain squares by ± 2, as illustrated in Fig. 5(a). We refer to $(\kappa_x/N_x, \kappa_y/N_y)$ as global slope.

The lattice symmetry is broken for states with $(\kappa_x, \kappa_y) \neq (0,0)$:[13] E.g., a finite positive value of κ_x/N_x implies a charge modulation along diagonal stripes. Similarly, a charge density modulation is present if the condition $N_\mathcal{B} = N_\mathcal{Y} = N_\mathcal{G} = N_\mathcal{R}$ is violated.

For the identification of exactly solvable points in parameter space and for future quantum Monte Carlo simulations, it would be very advantageous if all matrix elements had the same sign. As first steps in this direction, we describe a gauge transformation that changes the global sign of g in the effective Hamiltonian (3) and then show conditions under which it is possible to remove minus signs for the half-filled checkerboard lattice.[17]

Consider a "purple" sublattice $\mathcal{P}$ that contains the sites around every second plaquette as shown in Fig. 5(c). Define $\sigma_\mathcal{P}$ as the number of fermions on purple sites for a given configuration: $\sigma_\mathcal{P} = \sum_{i \in \mathcal{P}} n_i$. One observes that ring–exchange processes change $\sigma_\mathcal{P}$ by two. Thus, if all configurations are multiplied by a factor of $i^{\sigma_\mathcal{P}}$, the sign of all ring–exchange matrix–elements are changed. The invariance with respect to this gauge transformation proves a global $g \leftrightarrow (-g)$ symmetry.

However, a fermionic sign problem remains: Ring–exchange processes around empty and occupied hexagons carry opposite signs in Eq. (3). We argue that it can be avoided in certain (but not all) cases.[17] In order to do so, we represent the ground–state manifold by ensembles of fully–packed loops as exemplified in Fig. 1. We notice that a ring exchange around a hexagon with an occupied center site does not change the loop topology, whereas a ring exchange around an empty hexagon always does cause changes in one of the three topological different ways shown in Fig. 6(a).

Let us consider allowed configurations with "fixed" boundary conditions with an even number of fermions on the four boundaries. They are represented by closed loops in the interior and loops terminating at a boundary. We orient the closed loops as follows: (i) Color the areas separated by the loops alternatively white and grey, with white being the outmost color; (ii) orient all loops so that the white regions

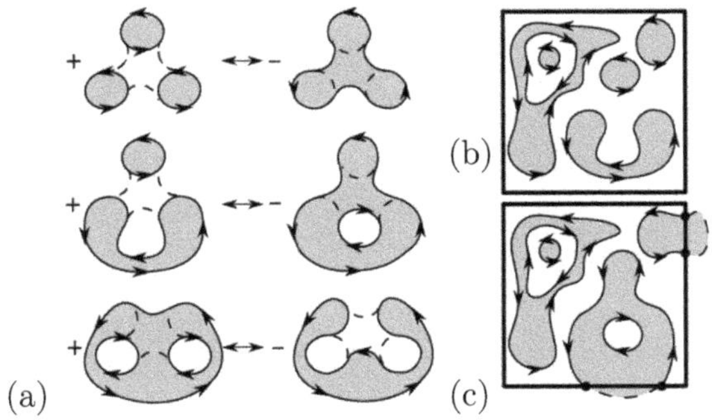

Fig. 6. The three different actions of the effective Hamiltonian on the topology of loop configurations are shown in panel (a). Panels (b) and (c) show representations of two configurations by fully–packed directed loops.[17]

are always to the right, see Fig. 6(b). If open loops are present [Fig. 6(c)], these are closed arbitrarily but intersection-free outside the sample and colored as described above. Let us assume without loss of generality that the color at infinity is white. We now notice by inspection of Fig. (6) that the relative signs resulting from the exchange processes around empty hexagons are consistent with multiplying each loop configuration by $i^r(-i)^l$, where r and l are the total number of the clockwise and counter–clockwise winding loops, respectively. Hence, by simultaneously changing the sign of the exchange–processes around empty hexagons and transforming the loop states

$$|\mathcal{L}\rangle \to i^{l(\mathcal{L})}(-i)^{r(\mathcal{L})}|\mathcal{L}\rangle, \tag{5}$$

we cure the sign problem, thus making the system effectively bosonic.

This construction need not work for periodic boundary conditions: Firstly, only even–winding subensembles (sectors) on a torus allow for such a two–color coloring. Secondly, even then it might be possible to dynamically reverse the coloring while returning to the same loop configuration. However, the exact diagonalization results presented below (see Fig. 8) suggest that for periodic boundary conditions on *even* tori (preserving the bipartiteness of the lattice), the lowest–energy states belong to a sector where such a transformation works. We remark that the presented non–local loop–orienting construction is restricted to the effective Hamiltonian (3), i.e., to the ring–exchange processes of length six.

5. Ground States and Lowest Excitations in the Undoped Case

In order to discuss the possible confinement of fcp's, we have to investigate the nature of the quantum–mechanical ground state of the undoped system. We do so in the approximation of an effective ring-exchange Hamiltonian (3) acting only on the subset of allowed configurations. Following Rokhsar and Kivelson,[18] we add to H_{eff} an extra term that counts the number of flippable hexagons. The extended

Hamiltonian reads

$$H_{g\mu} = H_{\text{eff}} + \mu \textstyle\sum_{\{\hexagon,\hexagon\}} \left(|\,\text{[pictograph]}\,\rangle\langle\,\text{[pictograph]}\,| + |\,\text{[pictograph]}\,\rangle\langle\,\text{[pictograph]}\,| \right), \tag{6}$$

where the pictographic operators with grey–colored dots in the center are summed over all flippable hexagons, independent of the occupancy of the site in the center. Next, we discuss some limiting cases of the Hamiltonian (6):

(i) $\mu \to +\infty$: All configurations which contain no flippable hexagons (frozen con-

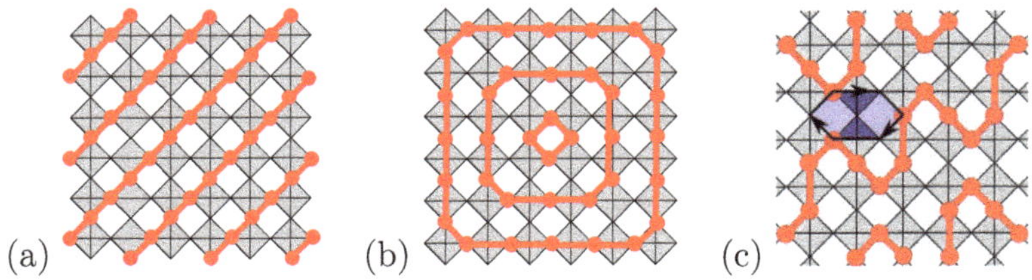

Fig. 7. (a)–(b) Fragments of possible "frozen" ground states (no flippable hexagons). (c) One of several configurations of the half–filled checkerboard lattice that maximize the number of flippable hexagons. The unit cell contains $\sqrt{20} \times \sqrt{20}$ sites.

figurations) become ground states with $E = 0$. Some are shown as Fig. 7 (a,b).

(ii) $\mu \to -\infty$: Ground states are configurations with maximal number of flippable hexagons N_{fl}. Using a simple Metropolis–like Monte Carlo algorithm for lattices with up to 1000 sites, we always find that configurations of the type shown in Fig. 7(c) or slight variations thereof maximize N_{fl}. Such configurations will be referred to as "squiggle" configurations.[19] Our numerical results suggest that in the thermodynamic limit the degenerate ground–state configurations all lie in the $(\kappa_x, \kappa_y) = (0, 0)$ subspace.

(iii) $\mu = g > 0$: This is the exactly solvable Rokhsar–Kivelson (RK) point.[18] Following the original RK construction, we rewrite the Hamiltonian (6) for $\mu = g$ in a way which explictly shows that some liquid–like ground states have energy $E = 0$ and, thus, become degenerate with the frozen states, i.e., the $\mu \to +\infty$ solutions. We assume that we can use the gauge transformation (5) to change the sign in the second term and to rewrite the Hamiltonian as

$$H_{g=\mu} = g \sum_{\{\hexagon,\hexagon\}} \left[\left(|\,\text{[pictograph]}\,\rangle - |\,\text{[pictograph]}\,\rangle \right) \times \left(\langle\,\text{[pictograph]}\,| - \langle\,\text{[pictograph]}\,| \right) \right]. \tag{7}$$

Since this is a sum over projectors, all eigenvalues are non–negative. Furthermore, after re–gauging, *all* off–diagonal elements are non–positive and for each subensemble ℓ an exact ground–state wavefunction is given by the equally weighted superposition of its configurations $|C_i^{(l)}\rangle$, i.e., $|\psi_{0,RK}^{(l)}\rangle \sim \sum_i |C_i^{(l)}\rangle$. These coherent superpositions are the analogs of the Resonating Valence Bond (RVB) state, originally discussed by L. Pauling[20] and P. W. Anderson.[21] Their energy is easily computed: Each flippable hexagon contributes $-g$ and $+\mu$, thus for $g = \mu$: $\langle\psi_{0,RK}^{(l)}|H_{g\mu}|\psi_{0,RK}^{(l)}\rangle = 0$. Note that for fermionic systems, we find a well defined RK point only if a gauge transformation exists such that all off–diagonal matrix elements are non–positive. Otherwise the energy is most likely larger than zero and the subensembles consequently do not form a ground state.

Next, we explore by means of numerically exact diagonalization the eigenstates of small clusters general μ values. For the actual calculations, we first generate all configurations that fulfill the tetrahedron rule, then group them according to quantum numbers and generate a sparse block–diagonal matrix representation of the Hamiltonian (6). For a 72–site checkerboard cluster with periodic boundary conditions, the 16 448 400 dimensional low–energy Hilbert space of allowed configurations can be decomposed into a few hundred subspaces, where the largest one has 1 211 016 dimensions. The low–energy states of the sparse block–diagonal Hamiltonian matrix and physical properties such as charge density distribution and density–density correlations can easily be obtained on a 64–bit workstation. Figure 8(a) shows energies of the ground-state and the lowest excited states of all subensembles.

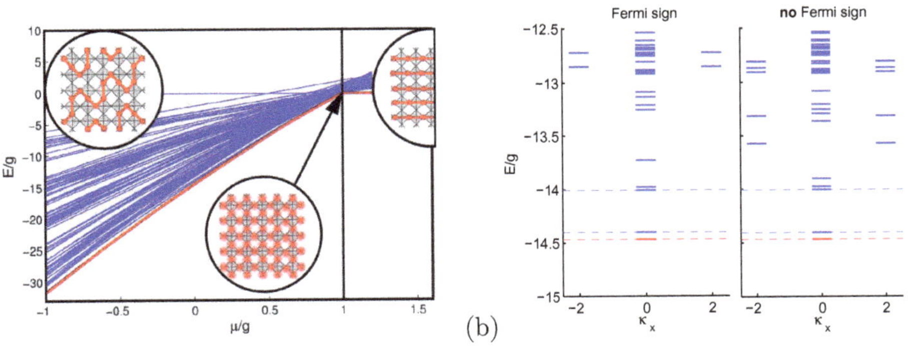

Fig. 8. (a) Energies of the ground state and lowest excited states in each subensemble of a 72–site half–filled checkerboard cluster for different values of μ of the g–μ Hamiltonian. Level crossing of ground states occurs only at $\mu = g$. The insets indicate different phases: Maximal flippable plus fluctuations for $\mu < g$, a critical point $\mu = g$ where the ground state is an equally weighted superposition of all configurations, and frozen configurations as ground states for $\mu > g$. (b) Left side: Ground–state energy and energies of the lowest excited states of the effective Hamiltonian H_{eff} in subspaces with different global slopes $(\kappa_x, \kappa_y) = (\kappa_x, 0)$ for 72–site cluster. Right side: Same system, but assuming same signs for all matrix elements ("bosonic calculation").[17]

(iv) For $\mu < g$, two ground states are found in subensembles with $(\kappa_x, \kappa_y) = (0, 0)$ and $(N_{\mathcal{B}}, N_{\mathcal{Y}}, N_{\mathcal{G}}, N_{\mathcal{R}}) = (6, 6, 12, 12)$ and $(12, 12, 6, 6)$, respectively. These are superpositions of configurations with the maximal number of flippable hexagons. At the physical point $\mu = 0$, the 72–site system is in a crystalline and confining phase. In the thermodynamic limit, we expect to recover the 10–fold degeneracy of the squiggle phase instead of the two–fold degeneracy. For small values $\mu > 0$, the average weight of configurations in the ground state with the maximum number N_{fl} is large (not shown), but decreases with increasing μ until at the RK point, i.e., μ=g, the ground state is formed by an equally weighted superposition of all configurations of a certain subensemble.

(v) For $\mu > g$, we find essentially the same properties as described above in the limit $\mu \to \infty$.

In summary, we find a confining phase and a phase in which the ground states are given by static isolated configurations. The two phases are separated by a point with deconfined excitations, i.e., the Rokhsar–Kivelson (RK) point.[18] The main finding is that the original effective Hamiltonian is in a confining phase with a long–range ordered ground state (squiggle phase). This phase maximizes the gain in kinetic energy and is stabilized by quantum fluctuations (order from disorder). The results of the exact diagonalization on small samples indicate that the symmetry remains broken all the way along the μ–axis up to the RK point. This observation is also strongly disfavoring a deconfining phase to the left of the RK point. Hence, the fermionic RK point is likely to be an isolated quantum critical point just as it is for the bosonic model.[10]

We come back to the fermionic sign problem. Figure 8(b) compares energies of a system in which the Fermi sign is taken into account to those from calculations which exclude the Fermi sign. The ground–state energy, the first excited states in the $(\kappa_x, \kappa_y) = (0,0)$ sector, and the weights of the different configurations in the corresponding eigenstates are the same in both case. In some subensembles with $(\kappa_x, \kappa_y) \neq (0,0)$ the energies of the ground states including the Fermi sign are higher than the ones for bosons (no Fermi sign).

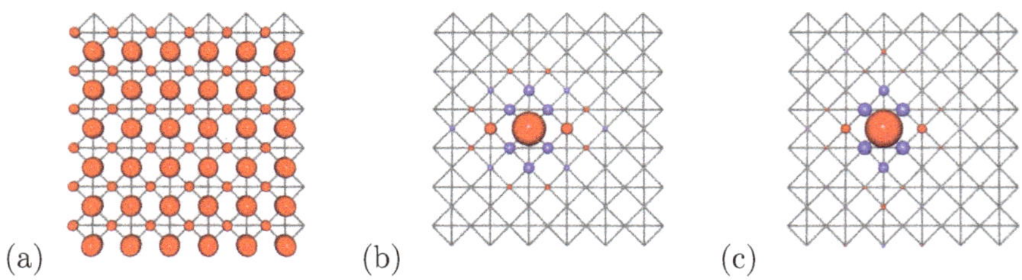

Fig. 9. Half filling: (a) Charge density distribution for one of the two ground sates. (b) Corresponding density–density correlation function $C^{(l)}_{i_0 i}$. The site i_0 with average density 2/3 shows up as the largest dot in the panels (b)–(c). The radius of the dots is proportional to the absolute value. A red or blue color represents a positive or negative value, respectively. (c) Classical density–density correlation function.

At the physical point $\mu = 0$, the 72–site ground state is two-fold degenerate with quantum numbers $(N_\mathcal{B}, N_\mathcal{Y}, N_\mathcal{G}, N_\mathcal{R}) = (6,6,12,12)$ or $(12,12,6,6)$. The resulting charge order with alternating stripes of average occupation 1/3 and 2/3 is shown in Fig. 9(a). The density–density correlation function in the quantum–mechanical ground states $|\psi_0^{(l)}\rangle$ [Fig. 9(b)]

$$C^{(l)}_{i_0 i} = \langle\psi_0^{(l)}|n_i n_{i_0}|\psi_0^{(l)}\rangle - \langle\psi_0^{(l)}|n_i|\psi_0^{(l)}\rangle\langle\psi_0^{(l)}|n_{i0}|\psi_0^{(l)}\rangle\,, \tag{8}$$

is best understood as reflecting the algebraic correlations present in the average over all classically degenerate configurations, see Fig. 9(c) and Ref. 15. Specific quantum–mechanical features resulting from ring hopping become visible in the difference of the actual correlations and have been discussed in Ref. 13.

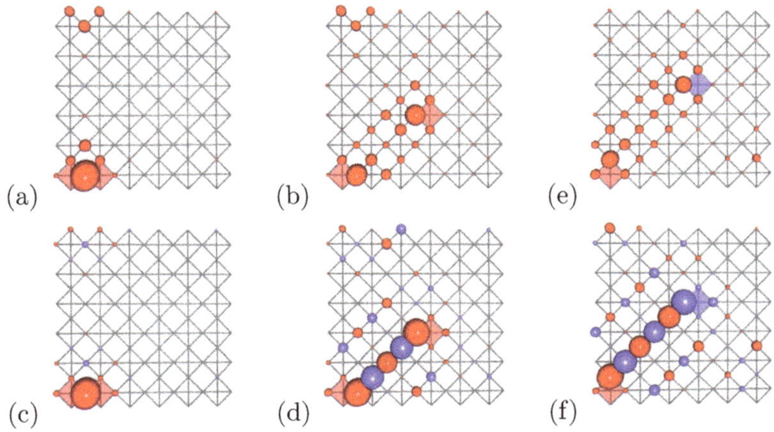

Fig. 10. (a), (b), and (e): Local loss of kinetic energy due to the separation of two fractionally (static) charged defects (fcp's marked by light red squares or fch's marked by dark blue squares). The radii of the circles are proportional to the local energy loss. (c), (d), and (f): Red (blue) circles show an increase (decrease) of the local density (vacuum polarization due to the two fcp's or the fcp–fch pair).

6. Static and Dynamical Properties of the Doped System

Next, we turn to the important issue of confinement/deconfinement of defects generated either particle–hole excitations or by weak doping. Two different kinds of calculations can be done: In the first case—referred to as *static* charges—the defects are fixed to some lattice positions and ground–state calculations involving H_{eff} are performed in the restricted subspace spanned by the configurations with exactly those defects at given positions. The total (free) energy is recorded as function of the defect positions. Its spatial derivative is interpreted as attractive or repulsive force. In the second case—referred to as *dynamic* charges—the defects propagate according to the Hamiltonian (4). The existence or non–existence of bound states is interpreted as confinement or deconfinement.

Static charges. If an additional particle is added/doped to a half-filled checkerboard lattice, at least two crisscrossed squares are occupied by three particles. It is easy to see that, e.g., for the t–μ Hamiltonian in the limit of large positive μ (frozen configurations) the ground–state energy of two *static* fractional charges is independent of the distance between. Thus static charges are not attracted by a force, which suggests that dynamic fcp's are deconfined at zero temperature in that limit.

Also, it is obvious that at the RK point the total energy of a system with two static charges is independent of the distance between them. Thus dynamic fcp's are expected to be deconfined.

We will not discuss systems with a general values of μ further, but focus on the physical case $\mu = 0$ and discuss the changes of the kinetic energy density in the presence of two static charges $e/2$.[22] The energy change can be decomposed into local contributions ϵ_i from all ring–exchange processes involving a given site i. An increase in kinetic energy in the region between the two fractional charges, i.e., along the connecting string, is found and illustrated in Fig. 10(a,b). The total energy increase is approximately proportional to the length of the generated string and implies at large distances a constant confining force. The changes of the local energy density goes along with density changes, as illustrated in Fig. 10(c,d). Similar results are found when a particle is removed from the ground state or when a particle–hole excitation is generated out of the ground state, see Fig. 10(c,d,f).
The calculations have been performed in a reduced Hilbert space in which only two static defects are present. In a calculation within the full Hilbert space, or a subspace allowing at least an additional fcp–fch pair, the energy would *not* increase linearly to infinity, but the connecting string is expected to break by creating additional pairs of defects when this energetically favorable. For the relevant parameters, e.g., $g = 0.01t$ and $V = 10t$, this occurs when two fcp's have separated over 1000 lattice sites. This effect is well known for the case of confined quarks where pair production (quark anti–quark pairs) occurs before the quarks have been separated to an observable distance.

The attractive constant force acting between two fractional charges in the confining phase results from a reduction of vacuum fluctuations and a polarization of the vacuum in the vicinity of the connecting strings. These findings suggest that a number of features known from QCD are also expected to occur in a modified form in solid–state physics. Conversely, one would hope that by studying frustrated lattices or dimer models one might be able to obtain better insight into certain aspects of QCD.

Dynamic charges. We turn now to the dynamical properties of fcp's, in particular spectral functions and optical conductivity for the half–filled checkerboard lattice. Numerical studies of finite clusters are again the method of choice,[17] because conventional approximation schemes such as mean–field theories or Green's function decoupling schemes are unable to describe the strong local correlations expressed by the tetrahedron rule. Diagonalizations with up to 50 sites were performed within the minimal Hilbert space spanned by the configurations with the smallest possible number of violations of the tetrahedron rule [half–filled system: all allowed configurations plus those with one fcp–fch pair; doped case: 2 fcp's or 2 fch's].

The spectral function $A(\mathbf{k}, \omega) = A^-(\mathbf{k}, \omega) + A^+(\mathbf{k}, \omega)$ of an interacting many–particle system is the sum of the probability amplitudes for adding (+) to the N–particle ground–state system $|\psi_0^N\rangle$ or removing (–) from it a particle with momentum $\mathbf{k}$ and energy ω ($\hbar = 1$)

$$A^+(\mathbf{k},\omega) = \lim_{\eta\to 0^+} -\frac{1}{\pi}\mathrm{Im}\langle\psi_0^N|c_\mathbf{k}\ \frac{1}{\omega + i\eta + E_0 - H}\ c_\mathbf{k}^\dagger|\psi_0^N\rangle \tag{9}$$

$$A^-(\mathbf{k},\omega) = \lim_{\eta\to 0^+} -\frac{1}{\pi}\mathrm{Im}\langle\psi_0^N|\ c_\mathbf{k}^\dagger\ \frac{1}{\omega + i\eta - E_0 + H}\ c_\mathbf{k}|\psi_0^N\rangle. \tag{10}$$

We use operators $c_\mathbf{k}^\dagger = \frac{1}{\sqrt{N}}\sum_j e^{i\mathbf{r}_j\mathbf{k}}c_j^\dagger$ in the extended Brillouin zone.

The spectral functions yield direct insight into the dynamics of a many–body system, as seen, e.g., in angular–resolved photoemission spectroscopy (ARPES). Expectation values of the form $G(z) = \langle\psi_0|A\ (z-H)^{-1}\ A^\dagger|\psi_0\rangle$ can conveniently be calculated numerically by the Lanczos continued fraction method[23] or kernel polynomial expansion.[24] We found essentially identical results for both algorithms.[16] However, the implementation of the Lanczos method turned out to be slightly faster. Well converged results were obtained already after several hundred iterations.

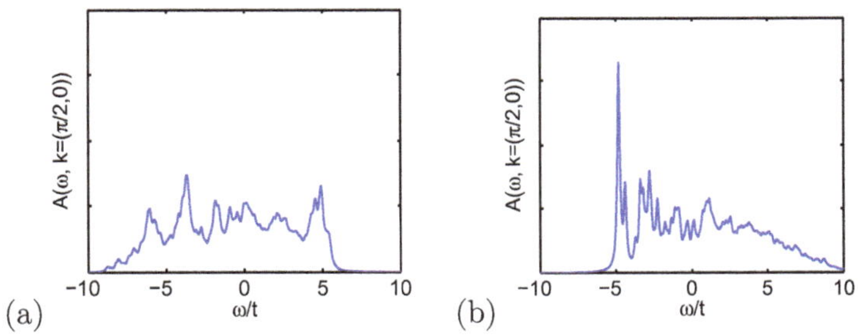

Fig. 11. (a) Spectral function $A(\mathbf{k} = \pi/2, \omega)$ for $V = 25t$ calculated for a $\sqrt{32}\times\sqrt{32}$ cluster for the effective t–g Hamiltonian with (a) $g = 0.01$ and (b) $g = 1$. A Lorentzian broadening $\eta = 0.1t$ is used.

We checked numerically that in the limit of large V, it is sufficient to calculate the spectral functions within the minimal Hilbert space. This enables us to study rather large systems and to address the question whether or not the two defects created by injecting one particle are closely bound to one another or not.

The concept of a spatial separation of fcp's can be confirmed by comparing results within the minimal Hilbert space with those from an artificially restricted calculation in which a particle is prevented from decaying into two defects with charge $e/2$ each. In our finite–cluster calculations, a broad low–energy continuum is seen in the spectral functions for the unrestricted case, which is missing when the restriction is imposed.[22] The respective bandwidths are about $13t$ and $8t$. This suggests a simple interpretation: The dynamics of two separated fcp's having a bandwidth of $\approx 6t$ each due to six nearest neighbors would explain the calculated $13t$, while a confined added particle has a much smaller bandwidth $8t$, which is close to $6t$.

Further insight can be obtained if the amplitude of the ring exchange $g{=}12t^3/V^2$

in the effective Hamiltonian is considered an independent parameter which is no longer restricted to the regime $g \ll t$ as enforced by $t \ll V$. Figure 11 compares results for the "physical" regime corresponding to the previously considered parameters $g \sim t^3/V^2 = 0.01t$ with those for $g = t$. In the latter case, the broad continuum at the bottom of the spectral density vanishes and a sharp δ–peak evolves instead. The latter should be viewed as a Landau quasi-particle peak. This suggests the following interpretation: The ring exchange term leads for the undoped case to charge order, which in finite–cluster calculations is destroyed when a particle is added by the separation of the two fcp's. These are weakly bound to each other. If $g \ll t$, the distance of the two fcp's is larger than the system size considered and thus the excitations seem to be deconfined. An artificially increased g leads to much stronger confinement and the diameter of the bounded pair becomes smaller or comparable to the system size. The pair acts at low energies as one entity. The electron as a (however strongly modified) "particle" is recovered and a finite weight of the quasi-particle peak results. These findings suggest that for the physical regime $V/t \approx 10$ quasi-particles with a spatial extent over more than hundred lattice sites are formed. The huge spatial extent is expected to lead to interesting effect. E.g., one may find with increasing doping concentration a transition from a "confined" phase to an "fcp plasma" phase with yet unknown properties when the average distance of bound fcp–fcp pairs falls below the diameter of a single bound fcp pair. Other open questions for future research include a thorough investigation of the 3D pyrochlore lattice. Even though the checkerboard lattice and pyrochlore lattice show many similarities, there exist certain differences between them, e.g., due to the higher spatial dimensionality.[8,25] Here, differences of the two lattices arise in the U(1) gauge theory which describes the low energy excitations of the considered systems. The related compact electrodynamic is always confining in 2+1 dimensions while it allows for the existence of a deconfined phase in 3+1 dimension.[26] Systematic exact diagonalization studies as well as the application of Monte Carlo techniques will provide a deeper insight and we hope to confirm the expected existence of a deconfined phase in a 3D pyrochlore lattice. In addition, a natural extension of the model considered in this paper is the inclusion of spin. This leads to a more realistic model and could provide a better link to experiments.

References

1. W. P. Su, J. R. Schrieffer and A. J. Heeger, *Phys. Rev. Lett.* **42**, p. 1698 (1979).
2. R. B. Laughlin, *Phys. Rev. Lett.* **50**, p. 1395 (1983).
3. P. Fulde, K. Penc and N. Shannon, *Ann. Phys. (Leipzig)* **11**, p. 893 (2002).
4. S. Kondo, D. C. Johnston, C. A. Swenson, F. Borsa, A. V. Mahajan, L. L. Miller, T. Gu, A. I. Goldman, M. B. Maple, D. A. Gajewski, E. J. Freeman, N. R. Dilley, R. P. Dickey, J. Merrin, K. Kojima, G. M. Luke, Y. J. Uemura, O. Chmaissem and J. D. Jorgensen, *Phys. Rev. Lett.* **78**, p. 3729 (1997).
5. W. Hofstetter, J. I. Chirac, P. Zoller, E. Demler and M. D. Lukin, *Phys. Rev. Lett.* **89**, p. 220407 (2002).
6. G. Misguich and C. Lhuillier, *cond-mat/0310405* (October 2003), in Ref. 7, p. 229ff.

7. H. T. Diep (ed.), *Frustrated Spin Systems* (World Scientific, Singapore, 2005).
8. M. Hermele, M. P. A. Fisher and L. Balents, *Phys. Rev. B* **69**, p. 064404 (2004).
9. A. Läuchli and D. Poilblanc, *Phys. Rev. Lett.* **92**, p. 236404 (2004).
10. N. Shannon, G. Misguich and K. Penc, *Phys. Rev. B* **69**, p. 220403(R) (2004).
11. R. Moessner, O. Tchernyshyov and S. L. Sondhi, *J. Stat. Phys.* **116**, p. 755 (2004).
12. P. W. Anderson, *Phys. Rev.* **102**, p. 1008 (1956).
13. E. Runge and P. Fulde, *Phys. Rev. B* **70**, p. 245113 (2004).
14. R. Baxter, *Exactly Solved Models in Statistical Mechanics* (Academic Press, San Diego, 1982).
15. F. Pollmann, J. J. Betouras and E. Runge, *Phys. Rev. B* **73**, p. 174417 (2006).
16. F. Pollmann, *Charge Degrees of Freedom on Frustrated Lattices*, Doctoral Dissertation, Technische Universität Ilmenau, (Ilmenau, Germany, 2006), `http://nbn-resolving.de/urn/resolver.pl?urn=nbn:de:gbv:ilm1-2006000140`.
17. F. Pollmann, J. Betouras, K. Shtengel and P. Fulde, *Phys. Rev. Lett.* **97**, p. 170407 (2006).
18. D. S. Rokhsar and A. A. Kivelson, *Phys. Rev. Lett.* **61**, p. 2376 (1988).
19. K. Penc and N. Shannon, private communication and to be published (2006).
20. L. Pauling, *Proc. Natl. Acad. Sci.* **39**, p. 551 (1953).
21. P. W. Anderson, *Mater. Res. Bull.* **8**, p. 173 (1973).
22. F. Pollmann, E. Runge and P. Fulde, *Phys. Rev. B* **73**, p. 125121 (2006).
23. E. R. Gagliano and C. A. Balseiro, *Phys. Rev. Lett.* **59**, p. 2999 (1987).
24. R. N. Silver, H. Roeder, A. F. Voter and J. D. Kress, *J. Comp. Phys.* **124**, p. 115 (1996).
25. F. Pollmann, J. J. Betouras, E. Runge and P. Fulde, Charge degrees in the quarter-filled checkerboard lattice, in *Proceedings of ICM2006* (Kyoto, Aug. 20-25, 2006), (in print).
26. A. M. Polyakov, *Nucl. Phys. B* **120**, p. 429 (1977).

5f ELECTRONS IN ACTINIDES: DUAL NATURE AND PHOTOEMISSION SPECTRA

GERTRUD ZWICKNAGL*

Institut für Mathematische Physik, Techn. Universität Braunschweig, 38106 Braunschweig, Germany
**E-mail: g.zwicknagl@tu-bs.de*

Increasing experimental evidence points towards a two-fluid model for the low-temperature behavior of 5f systems. Assuming some 5f electrons to be localized atomic-like and others to be band-like itinerant can explain various unusual low-energy properties of U compounds including the formation of heavy fermion and their unconventional superconductivity. The dual model, however, provides an effective Hamiltonian which is valid only for the low-energy regime. It has been shown that intra-atomic correlations among the 5f electrons are a possible mechanism for orbital-selective (partial) localization. In this paper we theoretically derive criteria and specify properties which experimentally characterize correlation-induced partial localization in actinide compounds. Of particular interest are the 5f spectral function which can be related to photoemission data. Results for the U-based heavy fermion superconductor UPd_2Al_3 are presented emphasizing the consequences for the interpretation of experimental data.

Keywords: 5f systems; strongly correlated electrons; spectral functions; dual nature; photoemission.

1. Introduction

Actinide-based intermetallic compounds exhibit highly complex properties at low temperatures.[1] The anomalies observed in numerous properties originate from the partially filled 5f shells of the actinide ions. For these states we face the fundamental question which picture provides the better starting point for theoretical models, a delocalized description in terms of energy bands or a localized representation which accounts for the local atomic properties. The general problem is well known from the transition metals whose d-electrons exhibit both aspects. This fact was the origin of the historical controversy between Slater emphasizing a band picture and VanVleck who advocated a description of the d-electrons in terms of local moments. It is now generally accepted that the d-electrons in transition metals are delocalized despite the fact that some atomic features persist. The correlations simply restrict the energy or temperature range where the Fermi liquid picture is valid. Similar behavior is found for the 4f-electrons whose wave functions are much more localized than the highest atomic d-levels. The local Coulomb repulsion is much larger than the estimated band widths which seems to favor a localized picture. It was shown that the 4f-electrons in Ce-based heavy fermion compounds can also be described within a band picture as delocalized states as far as the low-energy excitations are concerned.[2] As the extension of the 5f electron wave functions is intermediate between that of the 3d- and 4f wave functions it is not surprising that we find 5f-derived bands at low temperatures in actinide compounds.

For the low-energy regime of actinide compounds, increasing experimental evidence points towards a two-fluid model implying the co-existence of both localized atomic-like and itinerant band-like 5f states.[3,4] It has been shown recently that the dual model allows for a quantitative description of the renormalized quasi-particles - the heavy fermions - in UPd_2Al_3 and UPt_3. The measured dH-vA frequencies for the heavy quasiparticle portions as well as the large anisotropic effective masses can be explained very well by treating two of the 5f electrons as localized.[5–7] Experimentally, the co-existence of 5f-derived quasi-particles and local magnetic excitations in this compounds has been confirmed by recent neutron scattering and tunnelling experiments.[8,9]

Considering the success of the dual model in describing the low-energy behavior of actinide heavy fermion compounds[10] we have to explain how the splitting of the 5f shells arises and propose means of identifying the underlying microscopic mechanism. At this point we would like to emphasize that the dual model for 5f electrons as sketched above is an effective low-energy Hamiltonian designed for excitation energies $\hbar\omega \leq 5$–10 eV . It arises — to borrow the language from Wilson's renormalization group — from a more microscopic model by integrating out effects at higher energies. The dual model conjectures that the complicated excitations of the strongly correlated 5f electron system can be approximately decomposed in the low-energy regime into coherent dispersive quasiparticle bands and incoherent local excitations . The central task is to identify the microscopic mechanism which causes the effective band widths to renormalize to zero for certain orbital symmetries while staying finite for others. The focus of the present paper is the influence of intra-atomic correlations on the 5f excitations. The study has been triggered by recent model calculations for the ground state properties of small clusters. The latter show that intra-atomic correlations as described by Hund's rules may strongly enhance anisotropies in the kinetic energy and thus lead to an orbital-selective Mott transition in 5f-systems.[6,11] The result, i.e., the splitting of the correlated 5f electrons into itinerant and localized ones parallels the picture postulated recently to explain the behavior of transition metal oxides.[12–15] Nevertheless the physics of partial localization in transition metal oxides and 5f systems is quite different. In compounds with d electrons the large crystalline electric field (CEF) set up by the surroundings of a transition metal ion plays a major role. It is often larger than the band width. In case of a cubic lattice it splits the five d orbitals into a t_{2g} triplet and an e_g doublet and the corresponding subbands are well separated. When the Hund's rule energy is larger than this splitting and when the orbital energy of the t_{2g} is lower than that of the e_g states the t_{2g} states will be occupied by the first three d electrons. Those three d electrons remain localized in a high-spin state with $S = 3/2$. Additional d electrons occupy e_g states and remain delocalized. The situation differs when the CEF splitting is larger than Hund's rule coupling. In that case the t_{2g} subband can accommodate six electrons. When the d electron count per ion n_d is larger than six, only (n_d-6) d electrons will be itinerant and contribute to metallic behavior.

In 5f compounds we are facing a different situation. Since the 5f atomic wave functions are closer to the nuclei than d electron wave functions are, CEF splittings are smaller and less important. But Hund's rule energies are larger. Therefore we deal with a situation where the 5f count per actinide ion n_f exceeds two only those 5f electrons will delocalize which enable the remaining ones to form a Hund's rule state. Otherwise the Coulomb interation is increased so much that delocalization is disadvantageous as far as energy is concerned. Therefore Hund's rule correlations may strongly enhance anisotropies in the kinetic energy and eventually lead to the co-existence of band-like itinerant 3d or 5f states with localized atomic-like ones.

Having identified intra-atomic correlations as possible mechanism for the dual nature of 5f electrons we next turn to the question of how the validity of the underlying microscopic model can be assessed experimentally. The central idea is that strong Hund's rule type correlations and the resulting formation of atomic multiplets should be reflected in the single-particle spectral functions which can be measured by angle-resolved photoemission (ARPES). Characteristic features are (a) the co-existence of dispersive renormalized quasiparticle peaks[2] with incoherent low-energy excitations in the low-energy regime and (b) structures associated with the formation of atomic multiplets in the high-energy regime. First theoretical results for UPd_2Al_3 are discussed here. Comparison with recent experiments[16] qualitatively confirm the characteristic features which highlight the presence of atomic multiplets.

The paper is organized as follows: We begin the discussion in Sect. 2 by presenting the results of for the heavy quasiparticles in UPd_2Al_3 as obtained from the dula model. In the subsequent Sect. 3 we introduce the microscopic model Hamiltonian: The correlated 5f spectra and the validity of cluster-based methods are discussed in Sect. 4. Adopting a Hubbard-I-type approximation the multiplet effects in the spectra of UPd_2Al_3 are calculated in Sect. 5. We conclude by summarizing the results and giving an outlook in Sect. 6.

2. Heavy Quasiparticles in UPd_2Al_3

We adopt the dual model to calculate the strongly renormalized heavy quasiparticles in UPd_2Al_3. The central assumption is that we deal with localized as well as delocalized 5f electrons and that the coupling between the two subsystems results in the heavy fermions. The mass enhancement of the quasiparticles is shown to follow from the local exchange interaction of the delocalized 5f electrons with localized $5f^2$ configurations. The situation resembles that in Pr metal where a mass enhancement of the conduction electrons by a factor 5 results from virtual crystal-field (CEF) excitations of the localized $4f^2$ configurations. We refer to[1,7] for a detailed description of the procedure which is schematically summarized in Fig. 1.

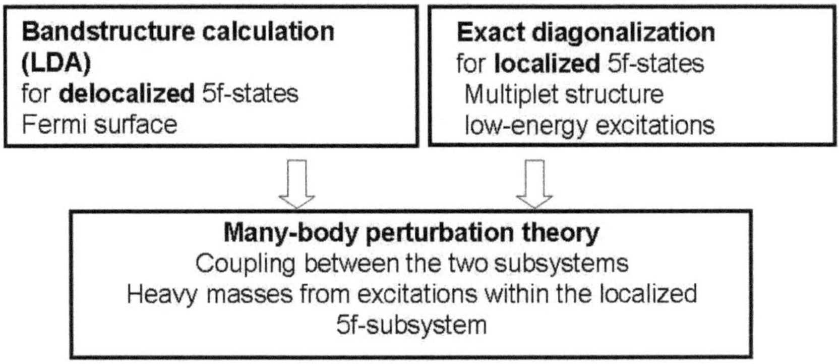

Fig. 1. Schematic summary of the computational procedure for calculating the quasiparticle dispersion within the dual model.

We should like to emphasize that we treat all 5f electrons as quantum mechanical particles obeying Fermi anticommutation relations.

Here we consider two of the 5f electrons as localized in agreement with the apparent absence of any Kramers doublets causing large magnetic moments. We put the localized electrons into the 5f, j=5/2 orbitals with $j_z = \pm 5/2$ and $\pm 1/2$ while the $j_z = \pm 3/2$ states are treated as itinerant band electrons. As it turned out only this particular choice yields the observed deHaas-vanAlphen (dH-vA) frequencies when subsequently a band structure calculation is done. It is worth pointing out that the $j_z = \pm 3/2$ states hybridize rather strongly with the conduction states in a conventional LDA calculation. The scheme has been successfully applied to UPd_2Al_3 and UPt_3 as can be seen from the comparison between the calculated and measured dH-vA frequencies and the effective masses in Fig. 2 and Table 1, respectively. It is important to note that the data are derived from a parameter-free calculation.

3. Microscopic Model Hamiltonian

A characteristic feature of actinide-based heavy fermion compounds are large distances between actinide ions. This structural property implies that the direct overlap between 5f wave functions at neighboring sites is rather small. The 5f electrons can become itinerant by (weak) hybridization with weakly correlated conduction electrons. For a microscopic description we therefore adopt the Anderson-like model

$$H = H_f + H_c + H_{cf} \tag{1}$$

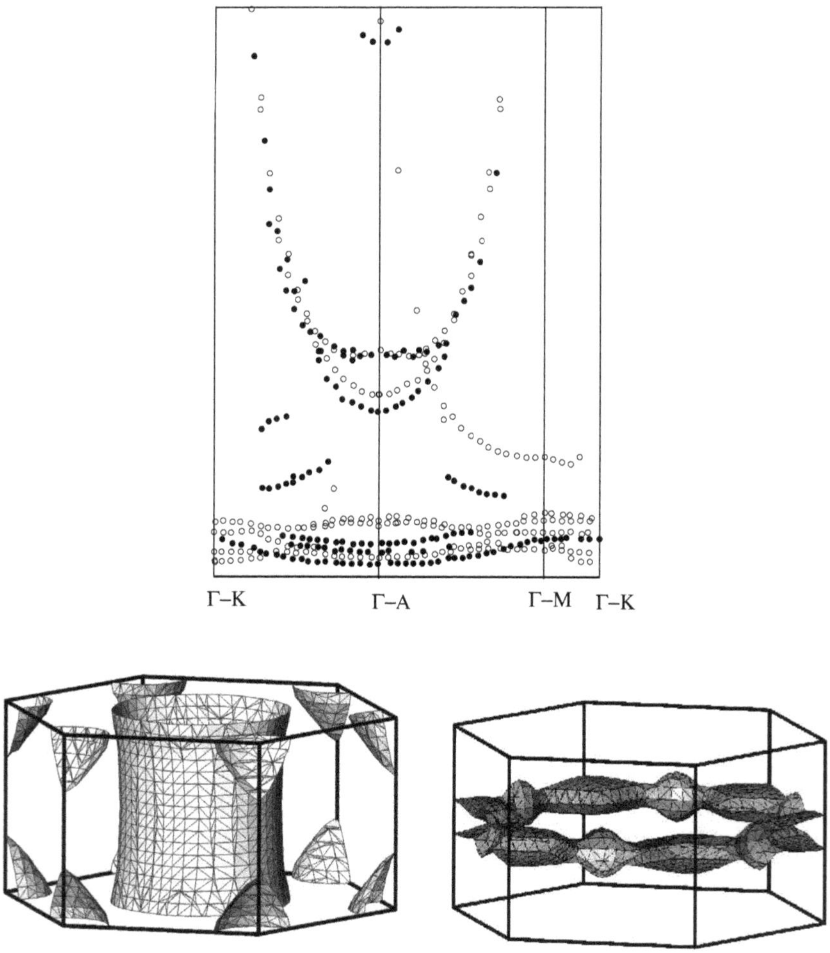

Fig. 2. Upper panel: Comparison of experimental dHvA frequencies (black symbols) from (Inada et al. 1999) and calculated frequencies (open symbols) from the dual approach (Zwicknagl et al 2003). Large parabola corresponds to the main FS cylinder.Lower panel: Fermi surface of UPd_2Al_3 in the antiferromagnetic state calculated within the dual approach (Zwicknagl et al 2003). The main cylinder part has effective masses with $m^*= 19 - 33$ m, the highest masses are found on the torus.

where the three components H_f, H_c, and H_{cf} describe the 5f states, the conduction electrons and the mixing or hybridization between the two. Let us first discuss H_f which describes the dynamics of the 5f electrons. It is given by

$$H_f = \sum_a \epsilon_f f^\dagger_{j_z}(a) f_{j_z}(a) + H_{f\text{Coul}} \quad (2)$$

where the orbital energy ϵ_f determines the f-valence of the ground state of the isolated ion. Throughout the calculations we concentrate on the intermediate-valent regime and assume a 5f valence of $n_f \sim 2.5$ per actinide site which is appropriate for

Table 1. Effective masses in UPd_2Al_3 for **H** $\|$c. Notation for FS sheets and experimental values from (Inada et al. 1999). Theoretical values from (Zwicknagl et al. 2003)

FS sheet	m*/m (exp.)	m*/m (theory)
ζ	65	59.6
γ	33	31.9
β	19	25.1
ϵ_2	18	17.4
ϵ_3	12	13.4
β	5.7	9.6

U heavy fermion compounds. The usual fermionic operators $f^\dagger_{j_z}(a)$ ($f_{j_z}(a)$) create (annihilate) an electron at site a in the 5f-state with total angular momentum j=5/2 and z-projection j_z. Considering the fact that the spin-orbit splitting is large we neglect contributions from the excited spin-orbit multiplet j=7/2 and adopt the j-j-coupling scheme. The local Coulomb repulsion is then given by

$$H_{f\text{Coul}} = \frac{1}{2}\sum_a \sum_{j_{z1},\ldots,j_{z4}} U_{j_{z1}j_{z2}j_{z3}j_{z4}} f^\dagger_{j_{z1}}(a) f^\dagger_{j_{z2}}(a) f_{j_{z3}}(a) f_{j_{z4}}(a)\,, \tag{3}$$

where the Coulomb matrix elements $U_{j_{z1}j_{z2}j_{z3}j_{z4}}$ for $j_{zi} = -5/2,\ldots,5/2$,are expressed in terms of the usual Clebsch-Gordan coefficients $C^{\cdots}_{\cdots}$ and the Coulomb parameters U_J.

We do not attempt at an ab initio calculation of the isotropically averaged Coulomb repulsion in a metal which is reduced as compared to its value in an atom. We leave it as a parameter, $U_{J=0}$, whose value can be estimated from the positions of the valence peaks in photoemission and inverse photoemission. The intra-atomic correlations which are the focus of interest in the present paper involve the anisotropic parts of the Coulomb interaction. The latter give rise to the multiplet structure. It is important to note that the corresponding interactions are (usually) not screened and hence retain their atomic values. As we expect the anisotropic Coulomb parameters to be rather robust with respect to changes in the chemical composition or to pressure we use fixed values for the differences $U_{J=4}$-$U_{J=0}$=-3.79 eV and $U_{J=2}$-$U_{J=0}$=-2.72 eV which were determined from an ab initio calculation for UPt_3.[5]

The hybridization model Eq. (1) which explicitly accounts for the dynamics of the conduction electrons will be used to study high-energy effects involving excitation energies of the order of ~ 1 eV. For the low-energy regime, however, we replace it by a simpler effective Hubbard-type Hamiltonian for the 5f electrons

$$H_f + H_c + H_{cf} \rightarrow H_f + H^{eff}_{f,kin} \tag{4}$$

where the hybridization with the conduction states is accounted for by effective hopping between all pairs of neighboring sites $\langle ab \rangle$

$$H^{eff}_{f,kin} = -\sum_{\langle ab\rangle, j_z, j'_z} t_{j_z,j'_z}(a,b)\left(f^\dagger_{j_z}(a) f_{j'_z}(b) + \text{h.c.}\right). \tag{5}$$

The replacement Eq. (5) is justified for those systems where LDA band structure calculations yield (almost) pure 5f states. This is the case for the well-known heavy fermion superconductor UPt_3[17] and the U dipnictides[18] where indirect hopping is mediated by the p states of Pt and the pnictogen ions, respectively. We are mainly interested in the strong-coupling regime where the interaction effects (largely) exceed the bare effective band widths as measured by the transfer integrals t_{j_z,j'_z}. The effective transfer integrals will sensitively depend upon the chemical composition and the structure of the system under consideration. In particular, they can be varied by applying external pressure. For this reason we treat them as parameters throughout the studies.

Correlation-induced partial localization is predicted for a rather wide range of hopping parameters by the mean-field approximation. This theory describes the low-energy excitations of the interacting system in terms of an effective single-particle Hamiltonian where the interactions enter through the orbital-dependent renormalization of the kinetic energy

$$\tilde{t}_{jz}(a,b) = \tilde{z}_{j_z}(a) t_{j_z} \tilde{z}_{j_z}(b) \tag{6}$$

and orbital-dependent shifts of the band centers

$$\tilde{\epsilon}_{fj_z}(a) = \epsilon_f + \lambda_{j_z}(a) - \nu(a) \quad . \tag{7}$$

The latter can be interpreted as effective dynamically generated Crystalline Electric Field terms. The effective CEF terms a well as the renormalization of the transfer integrals are determined self-consistently from the requirement that the mean-field free energy be minimal in the space spanned by the local 2- and 3-electron states $|a; j_{z1} j_{z2}\rangle$ and $|a; j_{z1} j_{z2} j_{z3}\rangle$ under the constraint that the number of electrons is fixed. For a discussion we refer to[19] and references therein.

4. Correlated 5f Spectra

The central quantity measured in ARPES is related to the **k**-resolved density of states

$$N(\mathbf{k}, \omega) = -\frac{1}{\pi} \lim_{\eta \to 0^+} \text{ Im Tr}\, \mathbf{G}(\mathbf{k}, \omega + i\eta) \tag{8}$$

which yields the probability for adding ($\omega > 0$) or removing ($\omega < 0$) an electron with energy ω in a state characterized by momentum $\mathbf{k}$ to the ground state. It is calculated from the Green' s function $\mathbf{G}(\mathbf{k}, \omega + i\eta)$. For the effective Hubbard-type Hamiltonian Eq. (5) we focus on the **k**-resolved f density of states

$$N_f(\mathbf{k}, \omega) = -\frac{1}{\pi} \lim_{\eta \to 0^+} \text{ Im Tr}\, \mathbf{G}_f(\mathbf{k}, \omega + i\eta) \tag{9}$$

with the f-Green's function $\mathbf{G}_f$. The latter is a 6×6 matrix function in the general case where thr angular momentum projection is not a good quantum number.

The Green's functions $\mathbf{G}$ and $\mathbf{G}_f$ are calculated within Cluster Perturbation Theory (CPT).[20,21] The central idea is to model the infinite periodic lattice as a superlattice of finite clusters whose complex dynamics is fully accounted for. For a general review of cluster methods applied to correlated electron systems see Maier et al.[22] an references therein. Due to the large degeneracy of the 5f orbitals the dimension of the Hilbert space increases dramatically with the cluster size. As a result, one is restricted to rather small clusters in the numerical calculations. The dynamics, however, is dominated by the intra-atomic correlations which are preserved in a rather wide parameter range. This can be seen from Fig. 3 which analyses the contributions from the f^2 and f^3 configurations with respect to the total angular momentum J^2. The rather small deviations from the Hund's rule values show that the corresponding correlations are rather insensitive to hopping, i.e., they are not broken by kinetic energy.

5. Multiplet Effects in 5f Spectral Functions

To get a first impression of the behavior we adopt a Hubbard-I-type approximation for the intra-atomic correlations within the 5f shells. Technically we determine the full matrix Green's function

$$\mathbf{G}(\mathbf{k},z) = \begin{pmatrix} G_{cc}(\mathbf{k},z) & G_{cf}(\mathbf{k},z) \\ G_{fc}(\mathbf{k},z) & G_{ff}(\mathbf{k},z) \end{pmatrix} = \left\{ \left(\mathbf{G}^{(0)}(\mathbf{k},z) \right)^{-1} - \mathbf{V}(\mathbf{k}) \right\}^{-1} \tag{10}$$

which results from hybridization

$$\mathbf{V}(\mathbf{k}) = \begin{pmatrix} 0 & V_{cf}(\mathbf{k}) \\ V_{fc}(\mathbf{k}) & 0 \end{pmatrix} \tag{11}$$

between the ones of the conduction states and the local 5f states in the atomic limit.

$$\mathbf{G}^{(0)}(\mathbf{k},z) = \begin{pmatrix} G_{cc}^{(0)}(\mathbf{k},z) & 0 \\ 0 & G_{ff}^{(at)}(z) \end{pmatrix} \tag{12}$$

The indices $c = (n_1, n_{2,}, \ldots)$ and $f = (-5/2, \ldots, 5/2)$ denote the band indices of the unhybridized bands and the j_z quantum numbers of the 5f states, respectively.

To calculated the spectra in the atomic limit we model the system as a statistical ensemble of isolated 5f atoms which have either 2 or 3 electrons, respectively. We assume the energies of these two configurations to be (almost) degenerate, i. e., $E(5f^2)=E(5f^3)$. The lowest-energy states of an intermediate-valent system in the zero-band width limit are formed by atomic f^2- and f^3-configurations. Due to the rotational invariance of the Coulomb interaction Eq. (3) the Green's function $\mathbf{G}_f^{at}(z)$ is diagonal in j_z. As a consequence, the corresponding spectral functions $A_f^{at}(\omega)$ will not depend on the magnetic quantum number j_z. The spectral function is a sum of δ-functions centered at the discrete energies which correspond to the atomic excitations. The spectral weights reflect the f-valence of the ground state. In the actual calculations we assume the latter to be $n_f = 2.5$. The low-energy behavior is determined by the transitions $f^2 \leftrightarrow f^3$ within the f^2- and f^3-configurations. This low-energy peak is a direct consequence of the intermediate-valent ground state. It is further split by intra-atomic correlations which are usually described by Hund's

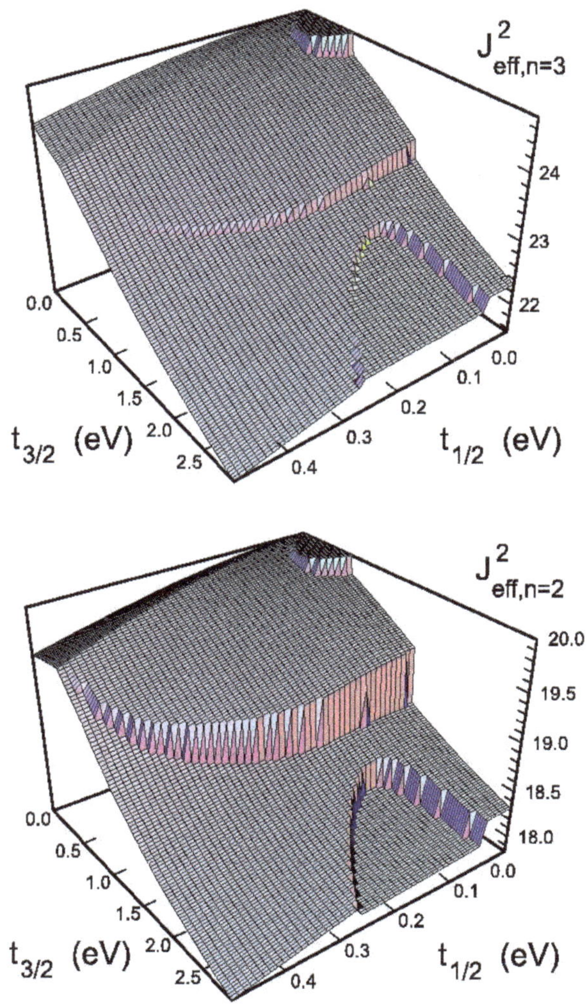

Fig. 3. Total angular momentum for the contributions from f^3-configurations (upper panel) and f^2-configurations (lower panel). Note the small deviations from the Hund's rule values J(J+1)=24.75 and J(J+1)=20, respectively.

rules. In addition to the central peak, one finds transitions to the excited multiplets $|f^3, J = 9/2\rangle \rightarrow |f^2, J = 2\rangle$ and $|f^2, J = 4\rangle \rightarrow |f^3, J = 3/2\rangle$, $|f^3, J = 5/2\rangle$ occurring at the corresponding multiplet excitations. The general structure of the spectral functions is displayed in Fig. 4. In a previous paper we analyzed (a) the evolution of the central peak at low energies in a periodic lattice and (b) the side band at $\sim$ 1eV binding energy. The investigation of the low-energy regime provides a microscopic basis for the dual model while the presence of 5f spectral weight at $\sim$ 1 eV binding energy is a hallmark of the conjectured microscopic mechanism, i.e., the correlation-driven partial localization. Concerning the low-energy regime we

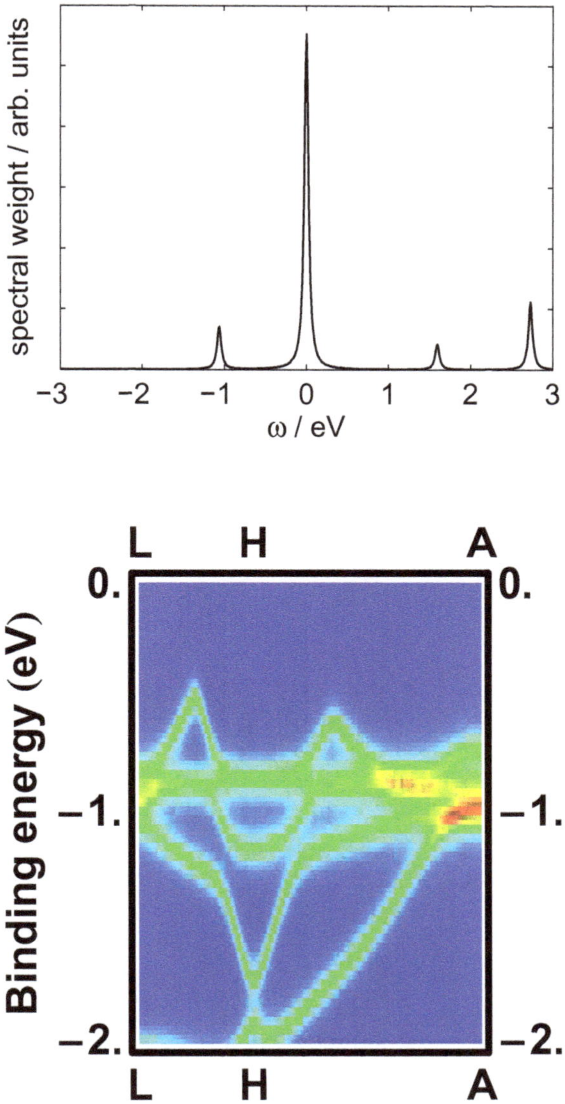

Fig. 4. 5f spectral function for intermediate-valent model systems with $n_f \sim 2.5$ Upper panel: Atomic limit. The low-energy peak corresponds to transitions between the Hund's rule ground states of the f^2- and f^3- configurations while the high-energy peaks involve transitions into excited er give rise to coherent quasiparticle bands and the incoherent low-energy background. The peak at $\sim$ 1eV binding energy leads to the high-energy feature. Lower panel:Multiplet side bands in UPd_2Al_3. The spectra are calculated as described in the text using a Lorentzian broadening of $\sim$ 0.05eV. The spectral weight of the weakly dispersive 5f side band is much smaller than that associated with the strongly dispersive non-f (conduction) bands.

find narrow, renormalized quasiparticle bands which co-exist with local (magnetic) excitations. The momentum distribution function exhibits a discontinuity.[21]

A hallmark of systems with strong intra-atomic correlations are side bands in the spectral function originating from transitions into excited multiplets. Fig. 4 displays these high-energy features which can be observed in photoemission. The weak hybridization with the conduction bands introduces dispersion. As a result, one finds narrow "f bands" at binding energies of $\sim$ 1eV. The band widths reflect a reduced effective hybridization. The renormalization is given by the spectral weight of the atomic transition which, in turn, depends upon the 5f valence.[2] Changing the 5f valence will affect the dispersion of the high-energy satellite bands.

As a first example, we calculate the high-energy satellite bands for the heavy fermion compound UPd_2Al_3 . Inserting $\mathbf{G}_{cc}(\mathbf{k}, z)$ and $\mathbf{V}(\mathbf{k})$ from an LDA calculation we find the UPd_2Al_3 spectral functions displayed in Figure 4. The peaks are broadened with an imaginary part of $\sim$ 0.05eV to simulate instrumental resolution in a soft X-ray photoemission experiment. The important features of the theoretical result are a strongly dispersive peak with high intensity and a rather flat band with low-intensity. While we assign the former to a non-f band we interpret the latter as multiplet side band. The theoretical predictions are qualitatively confirmed by experiment.[16]

6. Summary and Outlook

We have discussed properties of strongly correlated 5f electrons. Of particular interest are the strong intra-atomic correlations which are usually summarized in terms of Hund's rules. The latter are responsible for the "dual character". This model allows for a quantitative calculation of the low-energy properties of the heavy-fermion compound UPd_2Al_3.

A fingerprint of the underlying microscopic model is the appearance of f-spectral weight at 0.5 eV - 1eV binding energies which reflects the existence of atomic multiplets. Due to hybridization with broad conduction bands the corresponding peaks acquire dispersion. The band width, however, is renormalized reflecting the spectral weight of the atomic transition. The spectra obtained for the U-based heavy fermion superconductor UPd_2Al_3 within a local Hubbard-I-type approximation agree qualitatively well with recent ARPES data. The experimentally observed strong variation with wave vector of the spectral weight as well as the variation with temperature point towards the importance of coherence effects. In forthcoming studies, the latter will be deduced from cluster calculations. This important generalization is currently in progress.

Acknowledgment

Financial support in terms of COST P16 is gratefully acknowledged.

References

1. P. Thalmeier and G. Zwicknagl. *Handbook on the Physics and Chemistry of Rare Earths*, volume **34**, page 135. Elsevier B. V., 2005.
2. G. Zwicknagl. *Adv. Phys.* **41**, 203, 1992.
3. A. Grauel, A. Böhm, H. Fischer, C. Geibel, R. Köhler, R. Modler, C. Schank, F. Steglich, and G. Weber. *Phys. Rev. B* **46**, 5818, 1992.
4. Andrei Galatanu, Yoshinori Haga, Tatsuma D. Matsuda, Shugo Ikeda, Etsuji Yamamoto, Dai Aoki, Tetsuya Takeuchi, and Yoshichika Onuki. *J. Phys. Soc. Jpn.* **74**, 1582, 2005.
5. G. Zwicknagl, A. N. Yaresko, and P. Fulde. *Phys. Rev. B* **65**, 081103(R), 2002.
6. G. Zwicknagl and P. Fulde. *J. Phys., Condens. Matter* **15**, S1911, 2003.
7. G. Zwicknagl, A. Yaresko, and P. Fulde. *Phys. Rev. B* **68**, 052508, 2003.
8. A. Hiess, N. Bernhoeft, N. Metoki, G. H. Lander, B. Roessli, N. K. Sato, N. Aso, Y. Haga, Y. Koike, T. Komatsubara, and Y. Onuki. *J. Phys. Condens. Matter* **18**, R437, 2006.
9. N. K. Sato, N. Aso, K. Miyake, R. Shiina, P. Thalmeier, G. Varelogiannis, C. Geibel, F. Steglich, P. Fulde, and T. Komatsubara. *Nature* **410**, 340, 2001.
10. Peter Fulde, Peter Thalmeier, and Gertrud Zwicknagl. In F. Seitz, D. Turnbull, and H. Ehrenreich, editors, *Solid State Physics*, volume **60**, page 1. Academic Press, New York, 2006.
11. D. V. Efremov, N. Hasselmann, E. Runge, P. Fulde and G. Zwicknagl. *Phys. Rev. B* **69**, 115114, 2004.
12. V. I. Anisimov, I. A. Nekrasov, D. E. Kondakov, T. M. Rice, and M. Sigrist. *Eur. Phys. J. B*, 25, 191, 2002.
13. Akihisa Koga, Norio Kawakami, T. M. Rice, and Manfred Sigrist. *Phys. Rev. Lett.* **92**, 216402, 2004.
14. A. Liebsch. *cond-mat/0505393*, 2005.
15. M. S. Laad, L. Craco, and E. Müller-Hartmann. *Europhys. Lett.* **69**, 984, 2005.
16. Shin ichi Fujimori, Kota Terai, Yukiharu Takeda, Tesuo Okane, Yuji Saitoh, Yasuji Muramatsu, Atsushi Fujimori, Hiroshi Yamagami, Yoshifumi Tokiwa, Shugo Ikeda, Tatsuma D. Matsuda, Yoshinori Haga, Etsuji Yamamoto, and Yoshichika Onuki. *Phys. Rev.* **73**, 125109, 2006.
17. R.C. Albers, A. M. Boring, and N. E. Christensen. *Phys. Rev. B* **33**, 8116, 1986.
18. S. Lebègue, P. M. Oppeneer, and O. Eriksson. *Phys. Rev. B* **73**, 045119, 2006.
19. Gertrud Zwicknagl. *J. Phys. Soc. Jpn.* **75**, Supplement, 226, 2006.
20. D. Sénéchal, D. Perez, and Dany Plouffe. *Phys. Rev. B* **66**, 075129, 2002.
21. Frank Pollmann and Gertrud Zwicknagl. *Phys. Rev. B* **73**, 035121, 2006.
22. Thomas Maier, Mark Jarrell, Thomas Pruschke, and Matthias H. Hettler. *Rev. Mod. Phys.* **77**, 1027, 2005.

Part D

Magnetism

MAGNETISM IN THE DISORDERED TWO-DIMENSIONAL KONDO-NECKLACE

WOLFRAM BRENIG

Technische Universität Braunschweig,
Mendelssohnstr. 3,
D-38106 Braunschweig, Germany,
E-mail: w.brenig@tu-bs.de

We analyze the effects of site-dilution disorder on the thermodynamic properties of the two-dimensional Kondo necklace using finite-temperature stochastic series expansion. Results will be discussed for the dependence on dilution concentration, temperature, and Kondo exchange-coupling strength of the uniform susceptibility, the staggered structure factor, and the Chakravarty-Halperin-Nelson ratio. Dilution is shown to induce effective free-spin clusters in the gapped phase of the clean system with a low-temperature Curie constant renormalized below 1/4. Furthermore, dilution is demonstrated to suppress the quantum critical point of the clean system and to generate antiferromagnetic order in the quantum disordered phase of the clean system, i.e. order-from-disorder.

Keywords: Quantum spin models; Quantum critical phenomena; Disordered systems.

1. Introduction

Quantum critical points (QCPs), i.e. zero temperature phase transitions as a function of some control parameter are likely to be at the core of unconventional finite temperature behavior of many novel materials.[1,2] Quantum antiferromagnets (AFMs) with an intrinsic spin dimerization and weak inter-dimer exchange like $K(Tl)CuCl_3$[3,4] or $BaCuSi_2O_6$[5] are of particular interest here, since they allow for switching between quantum disordered spin-gapped phases and states with magnetic long-range order (LRO). QCPs have been induced in these materials both, by tuning the inter-dimer exchange via pressure,[6–8] as well as by applying external magnetic fields.[9–11]

Combining the physics of quantum critical spin systems with that of disorder is an open issue. Site dilution with non-magnetic impurities has been observed to induce LRO in the spin gapped phases of several dimerized quantum AFMs.[12–15] In spin ladders and dimerized spin chains a picture of weakly interacting 'defect moments' has emerged with enhanced AFM correlations in the vicinity of the non-magnetic sites.[16–19] In dimensions $D \geq 2$ and at finite impurity concentration this may trigger LRO in spin-gapped systems on bipartite lattices and remove a QCP.[20] In the context of the cuprate parent compounds,[21] site dilution of the 2D Heisenberg AFM has been shown to lead to a QCP at percolation where LRO is

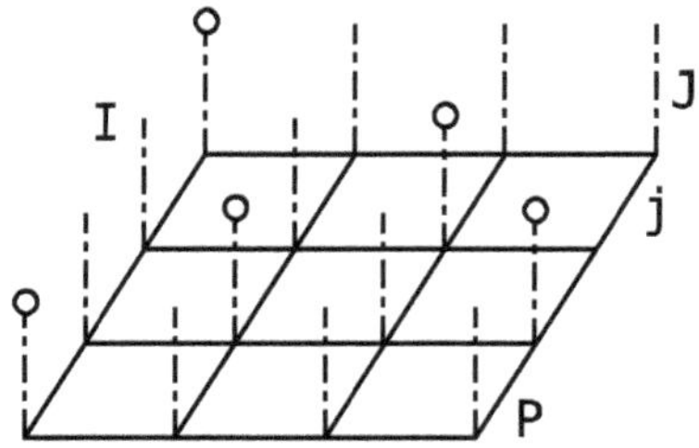

Fig. 1. 2D Kondo-necklace. Spin-1/2 moments coupled by j and J and located at vertices in 'conduction-plane' P and on 'Kondo-sites' in plane I at end of dash-dotted lines. 'o' labels randomly diluted 'Kondo-sites' with no magnetic moment.

suppressed.[22,23] In addition to site dilution several other forms of disorder are of interest. Early on random exchange has been investigated in a variety of 1D and 2D Heisenberg and Ising AFMs where it can lead to random singlet fix-points[24–26] and Griffiths phases.[27] Very recently 'dimer dilution' has been studied in bilayer Heisenberg models yielding multicritical points at percolation.[28–30]

In this work we focus on random site dilution in the $SU(2)$ symmetric 2D spin-1/2 Kondo-necklace (SKN) which is shown in Fig. 1

$$H_{SKN} = j \sum_{lm} \mathbf{S}_{Pl} \cdot \mathbf{S}_{Pm} + J \sum_{l} \mathbf{S}_{Pl} \cdot \mathbf{S}_{Il} \tag{1}$$

with $j \equiv 1$ hereafter. This model can be viewed as the strong coupling limit at large Coulomb correlations in the conduction electron band of the Kondo-lattice model at half filling.[31] In turn, the two layers P and I in Fig. 1, will be loosely referred to as 'conduction-band' and 'Kondo-site' layer hereafter. The clean limit of the 2D SKN has been studied at finite temperatures recently.[31,32] The model exhibits a QCP between AFM-LRO and a dimer phase at $J_c \approx 1.4$. It shows temperature-scaling and a Chakravarty-Halperin-Nelson ratio consistent with the $O(3)$ non-linear sigma model. The aim of this work is to shed light onto the effect of randomly substituting non-magnetic ions into the Kondo-layer, as shown in Fig. 1. In that case the site index l in in the second sum of Eqn. (1) runs only over those sites which are *occupied by a spin* in layer I.

2. Quantum Monte-Carlo Approach

Our analysis is based on the stochastic series expansion (SSE) with loop-updates introduced by Sandvik and Syljuasen.[33,34] Averaging over disorder configurations is performed by rapid thermalization using the temperature-halving scheme proposed by Sandvik.[35] We refer the reader to the latter three references for details. Here we only summarize the main ingredients of this method. Starting from the S^z representation $\{|\alpha\rangle\} = \{|S_1^z, S_2^z, \ldots, S_N^z\rangle\}$ of the Hilbert space, the partition function $Z_r = Tr[\exp(-\beta H_{SKN})]_\alpha$ of a particular disorder realization r of the SKN is expressed as

$$Z_r = \sum_\alpha \sum_{S_M} \frac{\beta^n (M-n)!}{M!} \left\langle \alpha \left| \prod_{i=0}^{M} H_{a_i,b_i} \right| \alpha \right\rangle \ . \tag{2}$$

Here S_M denotes the so-called operator string, i.e. all sequences of pairs

$$S_M = [a_1, b_1] \cdots [a_M, b_M] \tag{3}$$

where $b_i \in \left\{1, \ldots, \tilde{N}\right\}$ labels bonds in the lattice of Fig. 1 which are linked by exchange interactions and a_i labels three types of operators H_{a_i,b_i} on these bonds, namely the contributions from H_{SKN} which are diagonal and non-diagonal w.r.t. to S^z for $a_i = 1, 2$ and the unity operator for $a_i = 0$. The number of unity operators in S_M is given by $M - n$. This form of the partition function is directly suitable for Markovian chain Monte-Carlo sampling by a Metropolis scheme as described in refs.[33,34] To account for the disorder, and in addition to the Quantum-Monte-Carlo sampling of Z_r a classical averaging has to be performed over a sufficiently large number R of dilution realizations[35]

$$Z = \frac{1}{R} \sum_{r=1}^{R} Z_r \tag{4}$$

In the following, all of the SSE results reported for a finite dilution concentration comprise of an average over 1000 disorder configurations on a system of $N = 24 \times 24 \times 2$ sites. For the clean SKN, finite size effects beyond this point have been observed to be rather weak[31] and are also not expected to modify the essential conclusions of the present work.

3. Uniform Susceptibility

To begin we discuss the uniform susceptibility

$$\chi_u = \langle m^2 \rangle / T \ , \tag{5}$$

where T is the temperature and $m = \sum_{n=P+I,l} S^z_{nl}/N^o$ is the total spin z-component with N^o being the total number of sites *occupied* by a spin. The defect concentration is $c = 2N^e_I/N$, where N^e_I is the number of empty site in layer I, i.e. $N^o = (2-c)\,N/2$. In Fig. 2 we summarize several aspects of χ_u versus temperature at $J > J_c$ which refers to the gapped phase of the clean system. The figure contrasts the clean system for $0.05 \le T \le 1$ against three impurity concentrations $c = 0.03$, 0.1, and 0.2 for $1/1024 \le T \le 1$. First, Fig 2(a) clearly demonstrates that the non-magnetic defects induce low-energy magnetic density of states in the spin gap. Second, the log-log plot of Fig. 2(b) shows, that the susceptibility of these states is similar to that of 'free' spins, i.e. it follows almost a Curie law in the temperature range below the spin gap.

To further elucidate the free spin behavior, we start from the strong coupling limit, i.e. $J \to \infty$, of the Kondo-necklace. In that case and in the clean system the

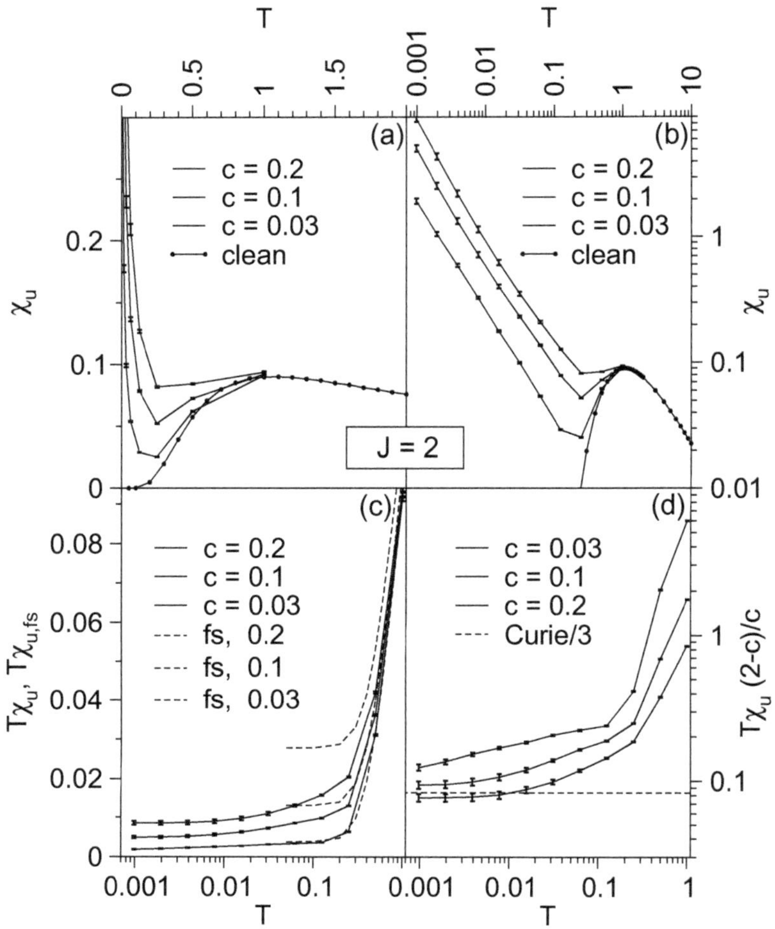

Fig. 2. (a), (b): Uniform susceptibility χ_u at $J = 2 > J_c$ vs. temperature for c=0.03, 0.1 and 0.2 and $1/1024 \leq T \leq 1$ and for $c = 0$ and $0.05 \leq T \leq 1$ on a lin-lin (a) and a log-log scale (b). (c): $T\chi_u$ product at $c \neq 0$ as compared with 'free spin' approximation $T\chi_{u,fs}$ from Eqn. (6), for identical parameters as in (a). (d): $T\chi_u$ product at $c \neq 0$ as compared to Curie behavior renormalized by 1/3. Statistical errors in all panels either less than the solid-circle marker size or depicted by error bars. Legends label plots from top to bottom.

ground state is a product of singlets on the dash-dotted bonds of Fig. 1. Removing a 'Kondo' spin leaves a free spin in the 'conduction' plane. Denoting the uniform susceptibility per occupied site by $\chi_{u,fs}$, with the index *fs* referring to the free spins, one has

$$\chi_{u,fs} = \frac{2\,(1-c)}{(2-c)} \chi_u\,(c=0) + \frac{c}{(2-c)} \frac{1}{4T} \quad . \tag{6}$$

The first term is the contribution from the singlets and the second term stems from the free spins. To assess the relevance of this free spin picture - other than in the strong coupling limit - we may use Eqn. (6) with $\chi_u\,(c=0)$ taken from the SSE at $c = 0$ but for J other than ∞. In Fig. 2(c) $T\chi_{u,fs}$ from this approximation

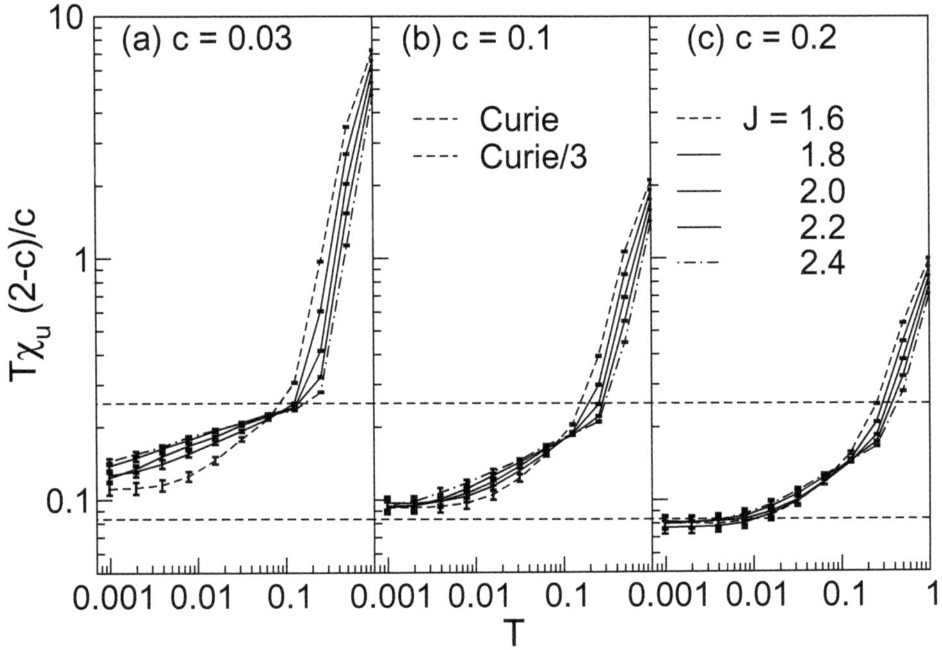

Fig. 3. $T\chi_u$ vs. T at $c = 0.03$ (a), 0.1 (b), and 0.2 (c) for various values of $J > J_c$. Dashed straight lines correspond to (renormalized) Curie constant. Legends label plots according to order listed between dashed and dashed-dotted. Statistical errors are depicted by vertical bars.

is compared with $T\chi_u$ from Eqn.(5) at $J = 2$. First, at low temperatures $T\chi_u$ is smaller than $T\chi_{u,fs}$. Second, while $T\chi_{u,fs}$ saturates on an energy scale set by the spin gap, $T\chi_u$ continues to decrease as a function of T and seem to saturate at far lower temperatures. This is a clear indication of low-energy correlations which develop between the impurity induced degrees of freedom. This also clarifies the need for SSE results at far lower values of T at finite disorder as compared to the clean case.

For site diluted spin-ladders Sigrist and Furusaki have proposed[17] that below a crossover temperature, and due to long-range, higher order exchange the free spins will form clusters of a size set by the inverse temperature. These clusters lead to a Curie-type of susceptibility per defect, however with a Curie constant reduced from $1/4$ to $1/(3 \cdot 4)$. The argument of ref.[17] is not restricted to ladders. To consider this, Fig. 2(d) shows the Curie contribution normalized to the defect concentration on a log-log scale. For $c \geq 0.1$ saturation of $T\chi_u$ can be anticipated for $T < 1/500$. Consistently with ref.[17] $\lim_{T\to 0} [T\chi_u (2-c)/c] < 1/4$. Moreover, while a renormalization by a factor of exactly $1/3$ can not be read off from this figure, the reduction is very close to this value. The kink in $T\chi_u$ observable at elevated temperatures and small concentrations in this log-log plot is related to the opening of the spin gap.

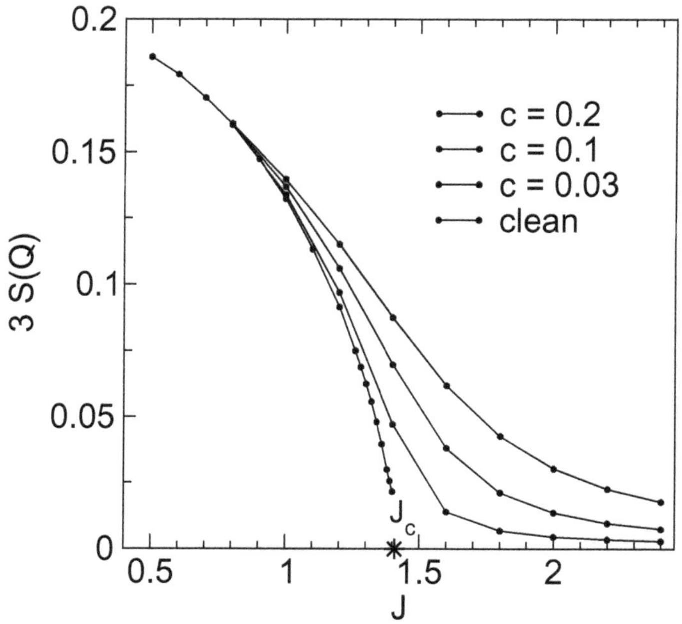

Fig. 4. Low-temperature staggered magnetization $3S(\mathbf{Q})$ versus J for various $c = 0$, 0.03, 0.1, and 0.2. Statistical errors are either less than the solid-circle marker size or are depicted by vertical bars. Legends label plots from top to bottom.

In Fig. 3 we summarize $T\chi_u$ versus T for $c = 0.03$, 0.1, and 0.2 over a finite range of coupling constants J corresponding to the gapped regime of the clean system. First, while low-temperature convergence to a constant $T\chi_u$ is observed for $c = 0.1$ and 0.2, Fig. 3 allows for an approximate extrapolation to $T = 0$ at $c = 0.03$ only. Second, it is tempting to claim that at fixed c the low temperature Curie constant depends only weakly on J. Third, the low temperature Curie constant is less than $1/4$ and close to $1/(3 \cdot 4)$ for all cases depicted. However, the figure suggests that the Curie constant is a *decreasing* function of c and cannot be described by a single renormalization factor at all c. This is consistent with the fact, that at $c = 1$ the SKN is identical to the 2D AFM Heisenberg model, for which $\lim_{T\to 0} T\chi_u = 0$.

4. Order from Disorder

Now we turn to the longitudinal staggered structure factor, i.e. the order parameter for AFM-LRO

$$S(\mathbf{Q}) = \left\langle \left(m^z_{\mathbf{Q}}\right)^2 \right\rangle , \tag{7}$$

where $m^z_{\mathbf{Q}} = \sum_{n=P+I,l} S^z_{nl} \exp(i\mathbf{Q} \cdot \mathbf{r}_l)/N^o$ is the staggered magnetization with $\mathbf{Q} = (\pi, \pi, \pi)$. All results discussed correspond to values of T such that low temperature saturation of $S(\mathbf{Q})$ has been reached. Fig. 4 shows the squared staggered moment $M^2_{\mathbf{Q}} = 3S(\mathbf{Q})$ versus J for the clean case and for c=0.03, 0.1 and 0.2.

In the clean case $M_{\mathbf{Q}}^2$ is finite below the critical value of $J = J_c$ and, apart from finite size effects, drops to zero for $J > J_c$. This corresponds to a QCP at $J_c \approx 1.4$ separating AFM-LRO for $J < J_c$ from a spin gapped state for $J > J_c$. This QCP has been investigated in detail in refs.[31,32] Generating SSE data of smaller value of $M_{\mathbf{Q}}^2$ for $J \sim J_c$ at $c = 0$ requires 'fine-tuning' of J which we refrain from here.

The main result of Fig. 4 is contained in the change of $M_{\mathbf{Q}}^2$ versus J upon doping. First, for all concentrations investigated the QCP disappears. Second, a finite staggered moment can be observed at all values of J and in particular also in the formerly gapped phase of the clean system. This implies that doping by non magnetic impurities induces AFM-LRO in the quantum disorder phase of the SKN, i.e. *order-from-disorder*. It is tempting to speculate, that this behavior is true for all c.

We may also discuss these results from a different point of view, i.e. by considering $M_{\mathbf{Q}}^2$ versus the impurity concentration for various exchange coupling constants. As is obvious from Fig. 4. $M_{\mathbf{Q}}^2$ increases at small c, both below and above J_c. At J_c the induced AFM order is most sensitive to the impurity concentration leading to the largest absolute changes of $M_{\mathbf{Q}}^2$ vs. c for $c \ll 1$. For $c \to 1$ all curves are expected to level off at a single line corresponding to the staggered magnetization of the planar Heisenberg AFM. The asymptotic behavior of $M_{\mathbf{Q}}^2$ for $c \to 0$ and $J \to \infty$ is an open issue which needs further analysis.

5. Renormalized Classical Behavior

Finally we discuss the Chakravarty-Halperin-Nelson ratio

$$R = \frac{S(\mathbf{Q})}{T\chi(\mathbf{Q})}, \tag{8}$$

which, in addition to $S(\mathbf{Q})$ includes information on the longitudinal staggered susceptibility

$$\chi(\mathbf{Q}) = \int_0^{1/T} d\tau \left\langle m_{\mathbf{Q}}^z(\tau)\, m_{\mathbf{Q}}^z \right\rangle \tag{9}$$

This ratio relates our analysis to that of the AFM non-linear σ-model (NLσM).[1,2,36] Regarding clean systems with a QCP between AFM-LRO and dimerization, this ratio has been studied in detail for the bilayer Heisenberg AFM[37] and for the SKN.[31] In agreement with the NLσM it was found that $R = 1$ in the (classical) high-T regime, as well as in AMF-LRO (renormalized classical) regime for $T \to 0$. At the QCP the NLσM requires $R(T \to 0) \simeq 1.09$, which is also consistent with SSE at $J = J_c$. While - to our knowledge - rigorous results are absent for a disorder induced AFM-LRO phase, as for $J > J_c$, it is yet tempting to speculate that renormalized classical behavior will re-emerge. Since in the gaped phase of the clean system R diverges as T^{-1} this will lead to drastic variations in R versus c for $J > J_c$. Indeed this is observed in Fig. 5, where R is shown versus T as c varies

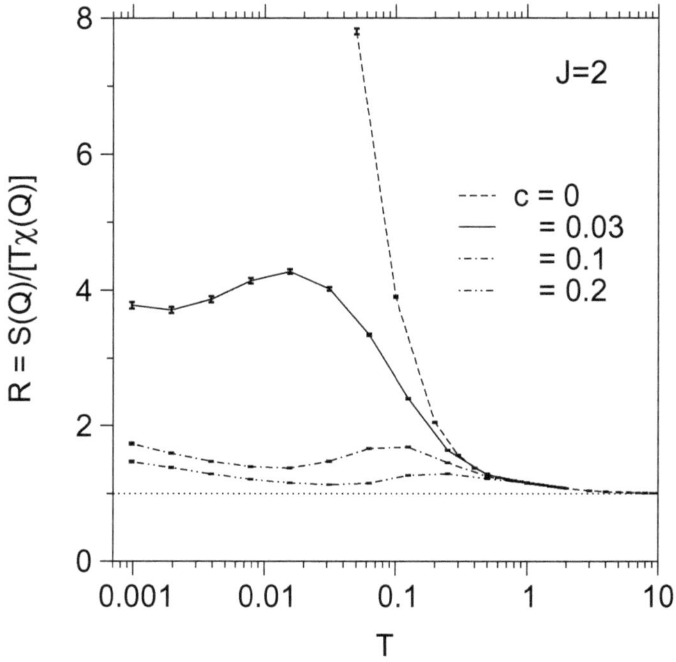

Fig. 5. Chakravarty-Halperin-Nelson ratio R of the total staggered structure factor and susceptibility vs. temperature at $J = 2 > J_c$ for $c = 0$, 0.03, 0.1, and 0.2, corresponding to dashed, solid, dashed-dotted and dashed-double-dotted. Result for $c = 0$ from ref.[31] Temperatures are $1/1024 \leq T \leq 1$ ($0.05 \leq T \leq 10$) for $c \neq 0$ ($c = 0$). Dotted line: unity. Size of statistical errors is given by vertical bars in all panels.

from 0 to 0.2. Obviously R decreases markedly with increasing c and almost reaches 1 for $c \geq 0.1$ and $T \sim 0.01$. As for $c = 0$ the zero temperature limit of R will be very sensitive to finite size gaps.[31,37] This may cause the low-T increase seen in this figure, i.e. we speculate that for larger systems $R = 1$ as $T \to 0$ for all $c \neq 0$.

Acknowledgments

Part of this work has been supported by DFG grant No. BR 1084/4-1.

References

1. S. Chakravarty, B. I. Halperin and D. R. Nelson, *Phys. Rev. B* **39**, p. 2344 (1989).
2. A. V. Chubukov, S. Sachdev and J. Ye, *Phys. Rev. B* **49**, p. 11919 (1994).
3. H. Tanaka, K. Takatsu, W. Shiramura and T. Ono, *J. Phys. Soc. Jpn.* **65**, p. 1945 (1996).
4. K. Takatsu, W. Shiramura and H. Tanaka, *J. Phys. Soc. Jpn.* **66**, p. 1611 (1997).
5. Y. Sasago, K. U. A. Zheludev and G. Shirane, *Phys. Rev. B* **55**, p. 8357 (1997).
6. K. Goto, M. Fujisawa, Ono, Tanaka and Y. Uwatoko, *J. Phys. Soc. Jpn.* **73**, p. 3254 (2004).

7. A. Oosawa, K. Kakurai, T. Osakabe, M. Nakamura, M. Takeda and H. Tanaka, *J. Phys. Soc. Jpn.* **73**, p. 1446 (2004).
8. C. Rüegg, A. Furrer, D. Sheptyakov, T. Strässle, K. W. Krämer, H.-U. Güdel and L. Mélési, *Phys. Rev. Lett.* **93**, p. 257201 (2004).
9. A. Oosawa, M. Ishii and H. Tanaka, *J. Phys.: Condens. Matter* **11**, p. 265 (1999).
10. T. Nikuni, M. Oshikawa, A. Oosawa and H. Tanaka, *Phys. Rev. Lett.* **84**, p. 5868 (2000).
11. M. Jaime, V. F. Correa, N. Harrison, C. D. Batista, N. Kawashima, Y. Kazuma, G. A. Jorge, R. Stern, I. Heinmaa, S. A. Zvyagin, Y. Sasago, and K. Uchinokura, *Phys. Rev. Lett.* **93**, p. 087203 (2004).
12. M. . Hase, I. Terasaki, Y. Sasago, K. Uchinokura and H. Obara, *Phys. Rev. Lett.* **71**, p. 4059 (1993).
13. M. Azuma, Y. Fujishiro, M. Takano, M. Nohara and H. Takagi, *Phys. Rev. B* **55**, p. R8658 (1997).
14. A. Oosawa, M. Fujisawa, K. Kakurai and H. Tanaka, *Phys. Rev. B* **67**, p. 184424 (2003).
15. H. Fujiwara, Y. Shindo and H. Tanaka, *Prog. Theor. Phys. Suppl.* **159**, p. 392 (2005).
16. G. B. Martins, E. Dagotto and J. A. Riera, *Phys. Rev. B* **54**, p. 16032 (1996).
17. M. Sigrist and A. Furusaki, *J. Phys. Soc. Jpn.* **65**, p. 2385 (1996).
18. T. Miyazaki, M. Troyer, M. Ogata, K. Ueda and D. Yoshioka, *J. Phys. Soc. Jpn.* **66**, p. 2580 (1997).
19. M. Laukamp, G. B. Martins, C. Gazza, A. L. Malvezzi, E. Dagotto, P. M. Hansen, A. C. López and J. Riera, *Phys. Rev. B* **57**, p. 10755 (1998).
20. S. Wessel, B. Normand, M. Sigrist and S. Haas, *Phys. Rev. Lett.* **86**, p. 1086 (2001).
21. S. W. Cheong, A. S. Cooper, L. W. Rupp, B. Batlogg, J. D. Thompson and Z. Fisk, *Phys. Rev. Lett.* **44**, p. 9739 (1991).
22. K. Kato, S. Todo, K. Harada, N. Kawashima, S. Miyashita and H. Takayama, *Phys. Rev. Lett.* **84**, p. 4204 (2000).
23. A. W. Sandvik, *Phys. Rev. Lett.* **86**, p. 3209 (2001).
24. S. K. Ma, C. Dasgupta and C. Hu, *Phys. Rev. Lett.* **43**, p. 1434 (1979).
25. D. S. Fisher, *Phys. Rev. B* **50**, p. 3799 (1994).
26. D. S. Fisher, *Phys. Rev. B* **51**, p. 6411 (1995).
27. R. B. Griffiths, *Phys. Rev. Lett.* **23**, p. 17 (1969).
28. A. W. Sandvik, *Phys. Rev. Lett.* **89**, p. 177201 (2002).
29. O. P. Vajk and M. Greven, *Phys. Rev. Lett.* **89**, p. 177202 (2002).
30. Y. C. Lin, H. Rieger, N. Laflorencie and F. Igloi, preprint: cond-mat/0604126.
31. W. Brenig, *Phys. Rev. B* **73**, p. 104450 (2006).
32. W. Ling, K. S. D. Beach and A. W. Sandvik, *Phys. Rev. B* **73**, p. 014431 (2006).
33. A. W. Sandvik, *Phys. Rev. B* **59**, p. R14157 (1999).
34. O. F. Syluåsen and A. Sandvik, *Phys. Rev. E* **66**, p. 046701 (2002).
35. A. W. Sandvik, *Phys. Rev. B* **66**, p. 024418 (2002).
36. A. Sokol, R. Glenister and R. Singh, *Phys. Rev. Lett.* **72**, p. 1549 (1994).
37. P. V. Shevchenko, A. W. Sandvik and O. P. Sushkov, *Phys. Rev. B* **61**, p. 3475 (2000).

ON THE DE HAAS-VAN ALPHEN OSCILLATION IN 2D

S. FUJITA*

Department of Physics, University at Buffalo, State University of New York, Buffalo, New York 14260, USA
**E-mail: fujita@buffalo.edu*

D. L. MORABITO

Department of Physics, Rochester Institute of Technology, Rochester, New York 14623, USA

The statistical weight W as a function of energy E for quasielectrons with mass m^* subject to a fixed magnetic field $\mathbf{B}$ is $W/A = (m^*/\pi\hbar^2)E + (eB/\pi^2\hbar)\sum_{\nu=1}^{\infty}(-1)^{\nu}\nu^{-1}\sin(2\pi\nu E/\hbar\omega_c)$, where A is the sample area, and $\omega_c \equiv eB/m^*$ the cyclotron frequency. Significantly, there is no Landau term proportional to B^2 in 2D. This leads to the conclusion that the 2D electron system is always paramagnetic, but shows a magnetic oscillation.

Keywords: Magnetic oscillations; Susceptibility; 2D system; absence of Landau magnetism.

1. The Statistical Weight

Landau,[1] Sondheimer and Wilson[2] discussed the de Hass-van Alphen (dHvA) oscillation[3] of a three-dimensional (3D) system of quasi-free electrons. Nakamura[4] calculated the statistical weight W associated with the Landau states and treated the dHvA oscillation. We extend Nakumura's theory to a 2D system.

Let us take a dilute system of electrons moving in a plane. Applying a magnetic field $\mathbf{B}$ perpendicular to the plane, each electron will be in the Landau state with the energy given by

$$E = \left(N_L + \frac{1}{2}\right)\hbar\omega_c, \quad N_L = 0, 1, \cdots, \quad \omega_c \equiv eB/m^*. \tag{1}$$

The degeneracy of the Landau level (LL) is

$$\frac{eBA}{2\pi\hbar}, \quad A = \text{sample area.} \tag{2}$$

We introduce kinetic momenta

$$\Pi_x = p_x + eA_x, \quad \Pi_y = p_y + eA_y, \tag{3}$$

in terms of which the Hamiltonian $\mathcal{H}$ for the electron is

$$\mathcal{H} = \frac{1}{2m^*}(\Pi_x^2 + \Pi_y^2) \equiv \frac{1}{2m^*}\Pi^2. \tag{4}$$

After simple calculations, we obtain

$$dxd\Pi_x dyd\Pi_y = dxdp_x dydp_y. \tag{5}$$

We can represent quantum states by small quasi-phase space elements $dxd\Pi_x \cdot dyd\Pi_y$. The Hamiltonian $\mathcal{H}$ in Eq. (4) does not depend on the position (x, y). Assuming large normalization lengths (L_1, L_2), we can represent the Landau states by the concentric shells of the phase space having the statistical weight $2\pi\Pi\Delta\Pi \cdot L_1L_2(2\pi\hbar)^{-2} = eBA/2\pi\hbar$, with $A = L_1L_2$ and $\hbar\omega_c = \Delta(\Pi^2/2m^*) = \Pi\Delta\Pi/m^*$. Hence, LL degeneracy is given by Eq. (2). Figure 1 represents a typical Landau state in the Π_x-Π_y space.

As the field B is raised, the separation $\hbar\omega_c$ increases, and the quantum states are bunched together. As a result of the bunching, the density of states $\mathcal{N}(\varepsilon)$ should change periodically.

The electrons obey the Fermi-Dirac statistics. Considering a system of quasi-free electrons, we define the free energy $\mathcal{F}$ by

$$\mathcal{F} = N\mu - 2k_BT \cdot \sum_i \ln\left[1 + e^{(\mu - E_i)/k_BT}\right], \tag{6}$$

where μ is the chemical potential and the factor 2 arises from the spin degeneracy. The chemical potential μ is determined from the condition

$$\frac{\partial \mathcal{F}}{\partial \mu} = 0. \tag{7}$$

The magnetization $\mathcal{M}$ for the system can be found from

$$\mathcal{M} = -\frac{1}{A}\frac{\partial \mathcal{F}}{\partial B}. \tag{8}$$

Equation (7) is equivalent to the usual condition that the number of the electrons, N, can be obtained in terms of the Fermi distribution function F:

$$N = 2 \cdot \sum_i F(E_i), \quad F(E) \equiv \left[e^{\beta(E-\mu)} + 1\right]^{-1}. \tag{9}$$

The LL E_i is characterized by the Landau oscillator quantum number N_L.

Let us introduce the density of state $dW/dE \equiv \mathcal{N}(E)$ such that $\mathcal{N}(E)dE =$ the number of states having an energy between E and $E + dE$. We write Eq. (6) in the form

$$\begin{aligned} \mathcal{F} &= N\mu - 2k_BT \int_0^\infty dE \frac{dW}{dE} \ln\left[1 + e^{(\mu - E)/k_BT}\right] \\ &= N\mu - 2\int_0^\infty dE W(E)F(E). \end{aligned} \tag{10}$$

The statistical weigth W is the total number of states having energies less than $E = (N_L + 1/2)\hbar\omega_c$. This W is from Fig. 1

$$W = \frac{L_1L_2}{(2\pi\hbar)^2} 2\pi\Pi\Delta\Pi \cdot 2 \sum_{N_L=0}^{\infty} \Theta\left[E - \left(N_L + \frac{1}{2}\right)\hbar\omega_c\right], \tag{11}$$

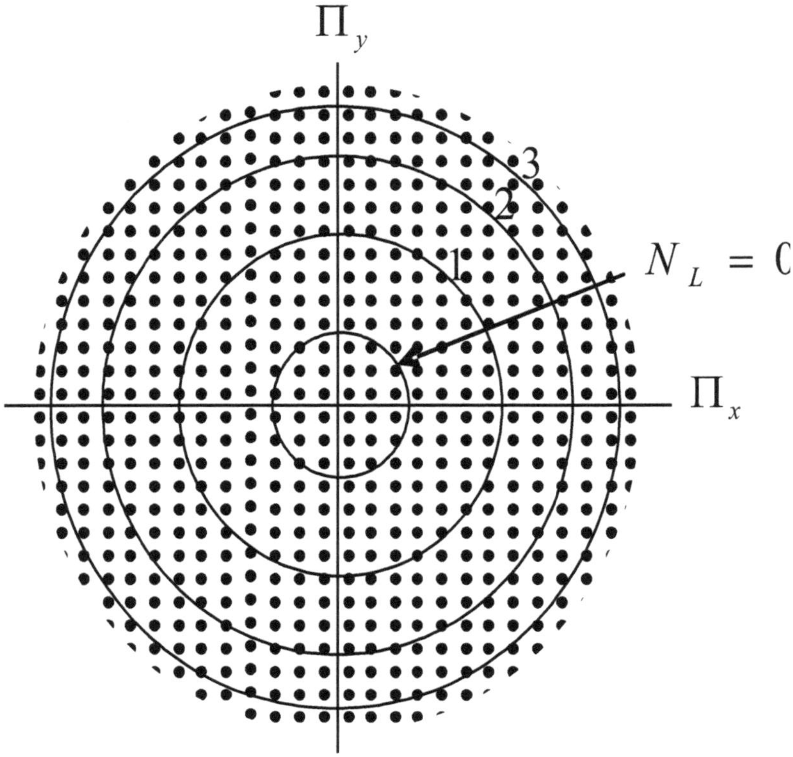

Fig. 1. Quantization scheme for free electrons: without magnetic field (dots) and in a magnetic field (circles).

where $\Theta(x)$ is the Heaviside step function:

$$\Theta(x) = \begin{cases} 0 \text{ if } x > 0 \\ 1 \text{ if } x < 0 \end{cases} . \tag{12}$$

We introduce the dimensionless variable $\varepsilon \equiv 2\pi E/\hbar\omega_c$, and write W as

$$W(E) = C(\hbar\omega_c)2 \cdot \sum_{N_L=0}^{\infty} \Theta\left[\varepsilon - (2N_L + 1)\pi\right], \tag{13}$$

with $C = 2\pi m^* A(2\pi\hbar)^{-2}$. We assume a high Fermi-degeneracy such that $\mu \simeq \varepsilon_F \gg \hbar\omega_c$. The sum in Eq. (13) will be computed by using Poisson's summation formula:[5]

$$\sum_{n=-\infty}^{\infty} f(2\pi n) = \frac{1}{2\pi} \sum_{m=-\infty}^{\infty} \int_{-\infty}^{\infty} d\tau f(\tau) e^{-im\tau}, \tag{14}$$

where $\sum_{n=-\infty}^{\infty} f(2\pi n + t)$, $0 \le t < 2\pi$ is by assumption uniformly convergent and is periodic with the period 2π. We write the sum in Eq. (13) as

$$2 \cdot \sum_{n=0}^{\infty} \Theta\left[\varepsilon - (2n+1)\pi\right] = \Theta(\varepsilon - \pi) + \phi(\varepsilon; 0),$$

$$\phi(\varepsilon; x) \equiv \sum_{n=-\infty}^{\infty} \Theta\left(\varepsilon - \pi - 2\pi\left|n + x\right|\right). \tag{15}$$

Note that $\phi(\varepsilon; x)$ is periodic in x. After the Fourier expansion, we set $x = 0$ and obtain Eq. (15). By taking the real part (Re) of Eq. (15) and using Eq. (14), we obtain

$$\text{Re}\{\text{Eq. (15)}\} = \frac{1}{\pi}\int_0^{\infty} d\tau \Theta(\varepsilon - \tau) + \frac{2}{\pi}\sum_{\nu=1}^{\infty}(-1)^{\nu}\int_0^{\infty} d\tau \Theta(\varepsilon - \tau)\cos\nu\tau, \tag{16}$$

where we assumed $\varepsilon \equiv 2\pi E/\hbar\omega_c \gg 1$ and neglected π against ε. The integral in the first term in Eq. (16) yields ε. The integral in the second term is

$$\int_0^{\varepsilon} d\tau \Theta(\varepsilon - \tau)\cos\nu\tau = \frac{1}{\nu}\sin\nu\varepsilon.$$

We obtain

$$\text{Re}\{\text{Eq. (15)}\} = \frac{1}{\pi}\varepsilon + \frac{2}{\pi}\sum_{\nu=1}^{\infty}\frac{(-1)^{\nu}}{\nu}\sin\nu\varepsilon. \tag{17}$$

Using Eqs. (13) and (17), we obtain

$$W(E) = W_0 + W_{\text{OSC}} \tag{18}$$

$$W_0 = C\hbar\omega_c(\varepsilon/\pi) = A\left(m^*/\pi\hbar^2\right)E \tag{19}$$

$$W_{\text{OSC}} = C\hbar\omega_c\frac{2}{\pi}\sum_{\nu=1}^{\infty}\frac{(-1)^{\nu}}{\nu}\sin\left(\frac{2\pi\nu E}{\hbar\omega_c}\right). \tag{20}$$

The oscillatory term W_{OSC} contains an infinite sum with repsect to ν but only the first term $\nu = 1$ is important in practive as we see later. This term W_{OSC} can generate magnetic oscillations. There is no term proportional to B^2 generating the Landau diamagnetism. This is unexpected. We shall briefly discuss the difference between 2D and 3D.

In 3D, the LL E is given by

$$E = \left(N_L + \frac{1}{2}\right)\hbar\omega_c + \frac{p_z^2}{2m^*}. \tag{21}$$

The energy E is continuous in the bulk limit. The statistical weight W' is the total number of states having energies less than E. The allowed value of p_z are distributed over the range in which $|p_z|$ does not exceed $\{2m^*\left[E - (N_L + 1/2)\hbar\omega_c\right]\}^{1/2}$. We obtain

$$W'(E) = C'\frac{(\hbar\omega_c)^{3/2}}{\sqrt{2\pi}}2\sum_{N_L=0}^{\infty}\sqrt{\varepsilon - (2N_L + 1)\pi}, \tag{22}$$

where $C' \equiv V(2\pi m^*)^{3/2}/(2\pi\hbar)^3$, $\varepsilon \equiv 2\pi E/\hbar\omega_c$, and $V \equiv L_1L_2L_3$ is the sample volume. We procees similarly and obtain

$$W'(E) = W'_0 + W'_L + W'_{\rm OSC} \tag{23}$$

$$W'_0 = C'\frac{4}{3\sqrt{\pi}}E^{3/2} \tag{24}$$

$$W'_L = -C'\frac{1}{24\sqrt{\pi}}\frac{(\hbar\omega_c)^2}{E^{1/2}} \tag{25}$$

$$W'_{\rm OSC} = C'\frac{(\hbar\omega_c)^{3/2}}{\sqrt{2}\pi^{3/2}}\sum_{\nu=1}^{\infty}\frac{(-1)^\nu}{\nu^{3/2}}\sin\left(\frac{2\pi\nu E}{\hbar\omega_c} - \frac{\pi}{4}\right). \tag{26}$$

In detail, we write the sum in Eq. (24) as

$$2\cdot\sum_{N_L=0}^{\infty}\sqrt{\varepsilon-(2N_L+1)\pi} = (\varepsilon-\pi)^{1/2} + \psi(\varepsilon;0) \tag{27}$$

$$\psi(\varepsilon;x) \equiv \sum_{n=-\infty}^{\infty}(\varepsilon-\pi-2\pi\,|n+x|)^{1/2}. \tag{28}$$

By taking the real part of Eq. (24) and using Eq. (14), we obtain

$$\text{Re}\{\text{Eq. (27)}\} = \frac{1}{\pi}\int_0^\varepsilon d\tau(\varepsilon-\tau)^{1/2} + \frac{2}{\pi}\sum_{\nu=1}^{\infty}(-1)^\nu\int_0^\varepsilon d\tau(\varepsilon-\tau)^{1/2}\cos\nu\tau, \tag{29}$$

where we neglected π against ε. The integral in the first term in Eq. (29) yields $(2/3)\varepsilon^{3/2}$, leading to W'_0 in Eq. (24). The integral in the second term can be written after integrating by part, changing the variable ($\nu\varepsilon - \nu\tau = t$) and using $\sin(A-B) = \sin A\cos B - \cos A\sin B$ as

$$\frac{1}{2\nu^{3/2}}\left[\sin\nu\varepsilon\int_0^{\nu\varepsilon}dt\frac{\cos t}{\sqrt{t}} - \cos\nu\varepsilon\int_0^{\nu\varepsilon}dt\frac{\sin t}{\sqrt{t}}\right]. \tag{30}$$

We now use asymptotic expansion for $\nu\varepsilon = x \gg 1$:

$$\int_0^x dt\frac{\sin t}{\sqrt{t}} \sim \sqrt{\frac{\pi}{2}} - \frac{\cos x}{\sqrt{x}} - \cdots$$

$$\int_0^x dt\frac{\cos t}{\sqrt{t}} \sim \sqrt{\frac{\pi}{2}} + \frac{\sin x}{\sqrt{x}} - \cdots. \tag{31}$$

The second terms in the expansion lead to W'_L in Eq. (25), where we used $\sin^2 A + \cos^2 A = 1$ and

$$\sum_{\nu=1}^{\infty}\frac{(-1)^{\nu-1}}{\nu^2} = \frac{\pi^2}{12}. \tag{32}$$

The first terms lead to the oscillatory term $W'_{\rm OSC}$ in Eq. (26). The term W'_0, which is independent of B, gives the weight equal to that for a free-electron system with no field. The term W'_L proportional to B^2 is negative (diamagnetic), and can generate a Landau diamagnetic moment.[1] The main difference is that the energy E is continuous in 3D while it is discrete in 2D. Hence, the sum in Eq. (23) is more singular than the sum in Eq. (22), and cannot generate the Landau-like term.

2. The Thermal Average

The statistical weight $W_{\rm OSC}$ in Eq. (20) has a sine term. Hence, the density of states, $\mathcal{N} = dW/dE$, has an oscillatory part of the form

$$\sin\left(\frac{2\pi E}{\hbar\omega_c}\right), \quad E \equiv \frac{\Pi^2}{2m^*}. \tag{33}$$

If the density of states oscillates violently in the drop of Fermi distribution function: $F(E) \equiv \left[e^{\beta(E-\mu)} + 1\right]^{-1}$, one cannot use the delta-function replacement formula

$$-\frac{dF}{dE} = \delta(E - \mu). \tag{34}$$

The width of $-dF/dE$ is of the order k_BT. The critical temperature T_c below which the oscillations can be observed is $k_BT_c \sim \hbar\omega_c$. For $T < T_c$, we may proceed as follows. Let us consider the integral

$$I = \int_0^\infty dEF(E)\sin(2\pi E/\hbar\omega_c). \tag{35}$$

We introduce the new variable $\zeta \equiv \beta(E-\mu)$ and extend the lower limit to $-\infty$ (low temperature limit):

$$\int_0^\infty dE \cdots \frac{1}{e^{\beta(E-\mu)}+1} = \frac{1}{\beta}\int_{-\mu\beta}^\infty d\zeta \cdots \frac{1}{e^\zeta + 1} \to \frac{1}{\beta}\int_{-\infty}^\infty d\zeta \cdots \frac{1}{e^\zeta+1}. \tag{36}$$

Using $\sin(A+B) = \sin A\cos B + \cos A \sin B$ and

$$\int_{-\infty}^\infty d\zeta \frac{e^{ia\zeta}}{e^\zeta + 1} = \frac{\pi}{i\sinh \pi a}, \tag{37}$$

whose derivation is given in Appendix, we obtain

$$I = \frac{\pi k_BT\cos(2\pi\varepsilon_F/\hbar\omega_c)}{\sinh(2\pi^2 k_BTm^*/\hbar eB)}. \tag{38}$$

For very low fields the oscillation number in the range k_BT becomes great, and hence the sinusoidal contribution must cancel out. This effect is represented by the factor $\left[\sinh(2\pi^2k_BTm^*/\hbar eB)\right]^{-1}$.

We calculate the free energy $\mathcal{F}$ in Eq. (10), using the statistical weight W in Eqs. (18)–(20), and obtain

$$\mathcal{F} = N\mu + A2\frac{m^*}{\pi\hbar^2}\varepsilon_F + A\frac{e}{\pi\hbar}Bk_BT \cdot \sum_{\nu=1}\frac{(-1)^\nu}{\nu}\frac{\cos(2\pi\nu\varepsilon_F/\hbar\omega_c)}{\sinh(2\pi^2\nu k_BTm^*/\hbar e)}, \tag{39}$$

where we used the integration formula in Eq. (38). We took the low-temperature limit except for the oscillatory terms. The magnetization $\mathcal{M}$ can be obtained using Eq. (8). So far, we have not considered the Pauli spin magnetization:[6]

$$\mathcal{M}_{\text{Pauli}} = \frac{2\mu_B^2 B}{A}\mathcal{N}_0(\varepsilon_F) = 2n\mu_B^2 B/\varepsilon_F, \tag{40}$$

where $\mu_B = e\hbar/(2m)$ is a Bohr magneton and n the electron density. Using Eqs. (8), (9) and (32), we obtain the total magnetization $\mathcal{M}_{\text{tot}} = \mathcal{M}_{\text{Pauli}} + \mathcal{M}_{\text{OSC}}$:

$$\mathcal{M}_{\text{tot}} = 2n\mu_B \frac{\mu_B B}{\varepsilon_F}\left[1 - \frac{k_B T}{\mu_B B}\frac{m}{m^*}\frac{\cos(2\pi\varepsilon_F/\hbar\omega_c)}{\sinh(2\pi^2 m^* k_B T/\hbar e B)}\right]. \tag{41}$$

In the calculation, we neglected the contribution of the B-derivatives of the quantities inside the cosines $(2\pi\nu\varepsilon_F/\hbar\omega_c)$. The variation of the statistical weight W is periodic in B^{-1} but it is far from sinusoidal. Only the first oscillatory term, $\nu = 1$, is important and kept in Eq. (34) since $\sinh(2\pi^2 m^* k_B T/\hbar e B) \gg 1$. The width of dF/dE is finite for finite T. In this E-range, many oscillations occur if the field B is made low. We assumed this condition to obtain Eq. (41). The magnetic susceptibility χ is defined by the ratio $\chi = \mathcal{M}/B$.

In conclusion, the 2D system is intrinsically paramagnetic since Landau diamagnetic term is absent. But it can exhibit a dHvA oscillation.

The authors are pleased to acknowledge a partial travel support from the U.S. Army Research Office. The views, opinions, and/or findings contained in this article, are those of the authors and should not be construed as an official department of the army position, policy or decision, unless so designated by other documentation.

Appendix. Derivation of Eq. (37)

Let us consider an integral

$$I(\alpha, R) = \int_{-R}^{R} dz \frac{e^{i\alpha(x+iy)}}{e^z + 1}, \qquad \alpha, R > 0. \tag{42}$$

We add an integral over a semicircle of the radius R in the upper z-plane, and take the limit $= R \to \infty$. The integral over the semicircle vanishes in this limit if $\alpha > 0$. The integral on the real axis, $I(\alpha, \infty)$, becomes the desired integral in Eq. (37). The integral over the closed contour can be evaluated by using the residue theorem. Note that $(e^z + 1)^{-1}$ has simple poles at $z = i\pi, i3\pi, \cdots, i(2n-1)\pi, \cdots$. We then obtain

$$\begin{aligned} I(\alpha, \infty) &= 2\pi i \sum_{n=1}^{\infty} \text{Residue}\left\{\frac{e^{i\alpha z}}{e^z + 1}, z_n = i(2n-1)\pi\right\} \\ &= 2\pi i \sum_{n=1}^{\infty} \frac{e^{i\alpha[i(2n-1)\pi]}}{e^{i(2n-1)\pi}} \\ &= -2\pi i \frac{e^{-\alpha\pi}}{1 - e^{-2\alpha\pi}} = \frac{\pi}{i}\frac{1}{\sinh \alpha\pi}. \end{aligned}$$

References

1. L. D. Landau, *Z. Physik* **64**, 629 (1930).
2. A. H. Wilson, *The Theory of Metals* (Cambridge University Press, London, 1954); E. H. Sondheimer and A. H. Wilson, *Proc. Roy. Soc. London, Ser. A* **210**, 173 (1951).
3. W. J. de Haas and P. M. van Alphen, *Leiden Comm.*, pp. 208d, 212a (1930); *ibid*, p. 220d (1932).
4. T. Nakamura, *Magnetism*, (Makishoten, Tokyo, Japan, 1965), pp. 75–82.
5. R. Courant and D. Hilbert, *Methods of Mathematical Physics*, Vol. I (Interscience-Wiley, New York, 1953), pp. 76–77.
6. W. Pauli, *Z. Physik* **41**, 81 (1927).

DYNAMICS IN ONE-DIMENSIONAL SPIN SYSTEMS – DENSITY MATRIX RENORMALIZATION GROUP STUDY

S. NISHIMOTO

Max-Planck-Institut für Physik Komplexer Systeme, Dresden, D-01187, Germany
E-mail: nishim@mpipks-dresden.mpg.de
www.mpipks-dresden.mpg.de

M. ARIKAWA

Yukawa Institute for Theoretical Physics, Kyoto University, Kyoto, 606-8502, Japan
E-mail: arikawa@yukawa.kyoto-u.ac.jp
www.yukawa.kyoto-u.ac.jp

We study the one-dimensional $S = 1/2$ Heisenberg model with an uniform and a staggered magnetic fields, using the dynamical density-matrix renormalization group (DDMRG) technique. The DDMRG enables us to investigate the dynamical properties of chain with lengths up to a few hundreds, and the results are numerically exact in the same sense as 'exact diagonalization'results are. Thus, we can analyze the low-energy spectrum almost in the thermodynamic limit. In this work, we calculate the dynamical spin structure factor and demonstrate the performance of the DDMRG method applying the open-end boundary conditions as well as the periodic boundary conditions.

Keywords: Dynamical spin structure factor; Heisenberg model; DDMRG; Bethe Ansatz; magnetic field.

1. Introduction

In one-dimensional electron systems, the conformal field theory has succeeded in the description of the low-energy physics of the Tomonaga-Luttinger liquid. Recent experimental progress, such as angle resolved photoemission and neutron scattering, demands the understanding of the elementary excitation pictures of the dynamics not only in the low-energy and long-wave length limit, but over the wide range of frequency and momentum.

In no magnetization case, Müller *et al*[1] proposed the approximate conjecture of the spin structure factor $S(q,\omega)$ in the one-dimensional (1D) $S = 1/2$ antiferromagnetic Heisenberg model with nearest neighbor interaction. By usage of the mathematical works,[2] the exact expression for two-spinon contributions to $S(q,\omega)$ in the thermodynamic limit was found.[3] The exact result and Müller ansatz for $S(q,\omega)$ have the same singularity at the lower spectral boundary, which is called the des Cloizeaux-Pearson mode. Remarkable progress has been made on the Bethe Ansatz (BA),[4] which enables us to numerically calculate the $S(q,\omega)$ for the Heisenberg model in a magnetic field for relatively large system size.[5]

Recently, a dynamical density-matrix renormalization group (DDMRG) method has been developed for calculating dynamical correlation functions at zero temperature in quantum lattice models. It is based on a variational formulation of the correction vector technique.[6,7] This method is an extension of the standard DMRG method[8] which is a powerful numerical technique for a variety of 1D systems, whereby we can obtain very accurate ground state and low-lying excited states. So far, It has been shown that the momentum-dependent dynamical quantities can be calculated with good resolution in finite open Hubbard chains by the DDMRG method.[9] We could also expect comparable performance for the same model in magnetic fields. In this work, we apply the DDMRG method to the 1D $S = 1/2$ Heisenberg model with an uniform and a staggered magnetic fields and calculate the spin structure factor $S(q,\omega)$. The accuracy of the DDMRG method is checked by comparing to some analytic solutions. In addition, the difference of performance by the boundary conditions is demonstrated. We note that the Heisenberg model can be dealt with the DMRG technique much easier than the Hubbard model.

This paper is organized as follows. In the next section, we introduce the model and the spin structure factor. Section 3 is devoted to show the brief outline of the DDMRG method. In Section 4, we demonstrate the DDMRG calculations of the spin structure factor for various magnetic fields. The performance with open boundary conditions (OBCs) as well as with periodic boundary conditions (PBCs) is examined in comparison with the exact solutions. Some remarks are given in the final section.

2. Model

We consider the 1D $S = 1/2$ Heisenberg model in magnetic fields:

$$\hat{H} = J\sum_i \vec{S}_i \cdot \vec{S}_{i+1} - H\sum_i S_i^z + h\sum_i (-1)^i S_i^z, \tag{1}$$

where J is an antiferromagnetic exchange interaction between neighboring sites. Henceforth we take J as the unit of energy. H is a uniform magnetic field and h is a staggered magnetic field, along the z-direction.

The dynamical spin structure factor is defined as

$$S^{\alpha\beta}(q,\omega) = \sum_{n>0} \langle\psi_0|\hat{S}^\alpha_{-q}|\psi_n\rangle\langle\psi_n|\hat{S}^\beta_q|\psi_0\rangle\delta(\omega - E_n + E_0), \tag{2}$$

where $\hat{S}^\beta_q$ $(= \hat{S}^{\alpha}_{-q}{}^{\dagger})$ is the Fourier transform of the spin operator $\hat{S}^\beta_i$ at site i, and E_n and $|\psi_n\rangle$ are, respectively, the n-th eigenenergy and eigenstate of the system (with the ground state denoted by $n = 0$). Note that the definition of the momentum-dependent spin operator $\hat{S}^\beta_q$ depends on choosing the boundary conditions. The definitions are given in Section 4.

3. DDMRG Method

In this section, the DDMRG method is briefly summarized for calculating the dynamical spin structure factor (2). Thus, we are interested in the momentum-dependent spectral functions,

$$A(q,\omega) = \lim_{\eta\to 0} \frac{1}{\pi} \mathrm{Im}\langle\psi_0|\hat{S}^{\alpha}_{-q} \frac{1}{\hat{H}+\omega-E_0-i\eta} \hat{S}^{\beta}_{q}|\psi_0\rangle. \tag{3}$$

The dynamical spin structure factor is then obtained as

$$\begin{aligned} S(q,\omega) &= -\frac{1}{\pi}\mathrm{Im}A(q,\omega) \\ &= \lim_{\eta\to 0} \mathrm{Im}\langle\psi_0|\hat{S}^{\alpha}_{-q} \frac{\eta}{(\hat{H}+\omega-E_0)^2+\eta^2} \hat{S}^{\beta}_{q}|\psi_0\rangle. \end{aligned} \tag{4}$$

To compute the dynamic quantity such as Eq. (3), we use the DDMRG method. This approach is based on a variational principle. One can easily show that for $\eta > 0$ and fixed frequency ω the minimum of the functional

$$W(\psi) = \langle\psi|(E_0+\omega-\hat{H})^2+\eta^2|\psi\rangle + \eta\langle\psi_0|\hat{S}^{\alpha}_{-q}|\psi\rangle + \eta\langle\psi|\hat{S}^{\beta}_{q}|\psi_0\rangle \tag{5}$$

with respect to all quantum states $|\psi\rangle$ is

$$W(\psi_{\min}) = \langle\psi_0|\hat{S}^{\alpha}_{-q} \frac{-\eta^2}{(E_0+\omega-\hat{H})^2+\eta^2} \hat{S}^{\beta}_{q}|\psi_0\rangle. \tag{6}$$

The functional minimum is related to the convolution of the dynamical spin structure factor (4) with a Lorentz distribution of width η by

$$W(\psi_{\min}) = -\pi\eta S(q,\omega). \tag{7}$$

The DDMRG method consists essentially of minimizing the functional (5) numerically using the standard DMRG algorithm. Thus the DDMRG provides the dynamical spin structure factor $S^{\eta}(q,\omega)$ for a finite broadening η. The full spectrum (4) convolved with the Lorentz distribution is given as

$$S^{\eta}(q,\omega) = \int_{-\infty}^{\infty} d\omega' S(q,\omega') \frac{\eta}{\pi[(\omega-\omega')^2+\eta^2]} \tag{8}$$

The real part of Eq. (3) can be calculated with no additional computational cost but is generally less accurate. The necessary broadening of spectral functions in DDMRG calculations is actually very useful for studying continuous spectra or for doing a finite-size scaling analysis.[7]

If we would like to obtain the spectrum in the limits of $L \to \infty$ and $\eta = 0$, this can be done by carrying out a deconvolution of the DDMRG data.[10] In theory, a deconvolution amounts to solving (8) for $S_(q,\omega)$ using the DDMRG data on the left-hand side.[10] We also know that the broadened spectrum of the system on an infinite lattice ($L \to \infty$) is usually almost identical to the spectrum on a finite lattice ($N < \infty$) if $\eta \geq c/L$ (the constant c is comparable to the width of the spectrum). Therefore, one can make the approximation that the DDMRG data

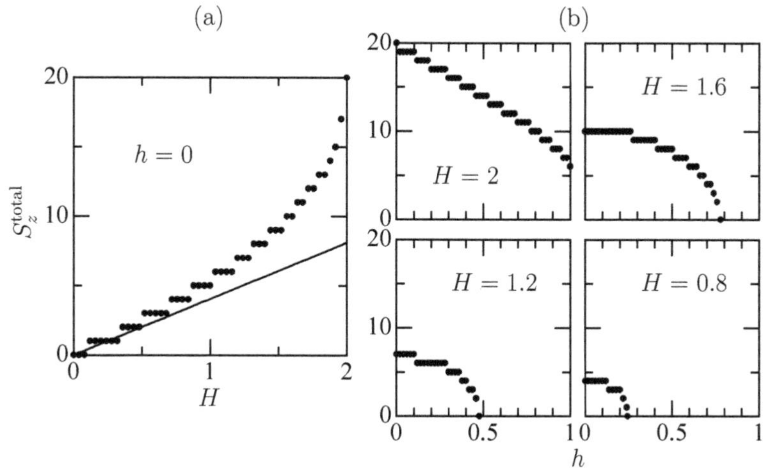

Fig. 1. z-component of the total-spin quantum number as a function of H (a), and as a function of h for various fixed H (b). The solid line denotes the exact behavior for $H \to 0$.

for $S^\eta(q,\omega)$ describes the broadened spectrum for $N \to \infty$ and can then solve (8) approximately under the condition that $S(q,\omega)$ is the exact convolved spin structure factor. For instance, one can require that $S(q,\omega)$ is a continuous and relatively smooth function. To obtain quantitatively accurate spectra after deconvolution, we need to take η smaller than the width of the spectra in $\eta = 0$.

4. Results

4.1. *Field-induced magnetization*

The z-component of the total-spin, S_z^{tot}, is now a good quantum number. It is quite efficient for the DMRG calculation to classify the Hamiltonian by S_z^{tot}. Therefore, let us first check S_z^{tot} of the ground state in the system with magnetic fields before starting the dynamical calculations. In Fig. 1(a), we show S_z^{tot} of the ground state as a function of the uniform magnetic field H. The calculations are carried out with the DMRG method in a chain with $L = 40$ sites under the PBCs. Note that S_z^{tot} is equivalent to the magnetization of the system via the relation $M = S_z^{\text{tot}}/L$.

For small H, the magnetization is induced proportionately with the uniform field, $M = (2/\pi^2)H$.[11] The slope becomes sharper with increasing H since the antiferromagnetic spin fluctuations are weakened. Then, the system is fully polarized at a critical uniform field $H_c = 2$ where the spin susceptibility diverges. We also show the dependence of S_z^{tot} on the staggered magnetic field h for various H values in Fig. 1(b). The magnetization is rapidly suppressed by the staggered field and it becomes zero at a critical value h_c. The critical staggered field is roughly scaled as $h_c \sim H/2$.

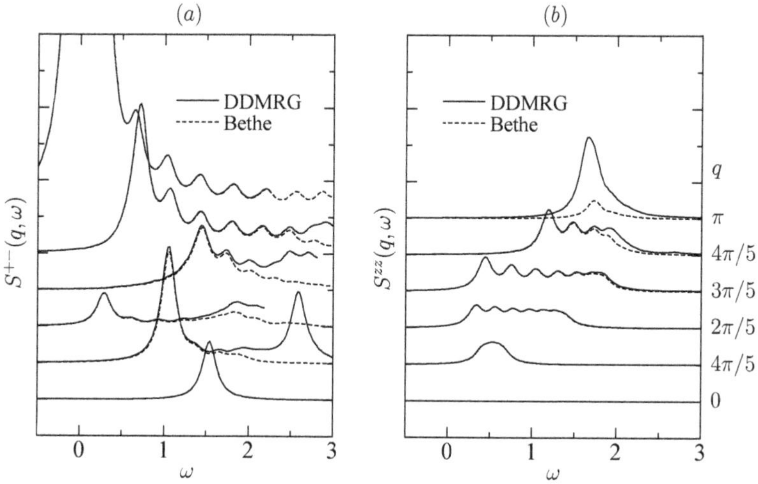

Fig. 2. Spin structure factor $S^{+-}(q,\omega)$ (a) and $S^{zz}(q,\omega)$ (b) for $L = 40$, $H = 1.537$, and $\eta = 0.1$ applying the PBCs. Dotted curves show the BA solution with $\eta = 0.1$ for comparison: up to two-spinon and two-antispinon contributions (a) and up to two-spinon and one-antispinon contributions (b).

4.2. *Dynamical spin structure factor*

4.2.1. *Periodic boundary conditions*

Normally, the OBCs are applied in the (D)DMRG method to achieve an accurate treatment. We can also study systems with the OBCs on much larger lattices than systems with the PBCs. Especially in interacting fermion systems such as the Hubbard and the $t-J$ models, the OBCs seems to be rather essential to truncate their large Hilbert space with keeping enough accuracy. However, a balk system is generally better described using the PBCs, and moreover the exact definition of the momentum-dependent operators is quite difficult with the OBCs.

Here, an idea will naturally arise as follows: it may be that the PBCs can be applicable in a spin system where the Hilbert space is much more reduced than that in the fermion systems. Motivated by this idea, we check the DDMRG performance with the PBCs in the 1D $S = 1/2$ Heisenberg model. We now start to calculate the dynamical spin structure factor (2). With the PBCs, the spin operators S_q^α are defined by

$$S_q^\alpha = \frac{1}{\sqrt{L}} \sum_l e^{iql} S_l^\alpha, \tag{9}$$

with momentum $q = 2\pi z/L$ for integers $-L/2 < z \leq L/2$.

Using BA solutions on a finite lattice, the spectral weight are available in the case with the uniform magnetic field.[4,5] Comparison with the BA solution will provide an opportunity to test the performance of the DDMRG method. In Fig. 2, we compare the DDMRG spectra with the line shapes based on the BA solutions in $H = 1.537$,

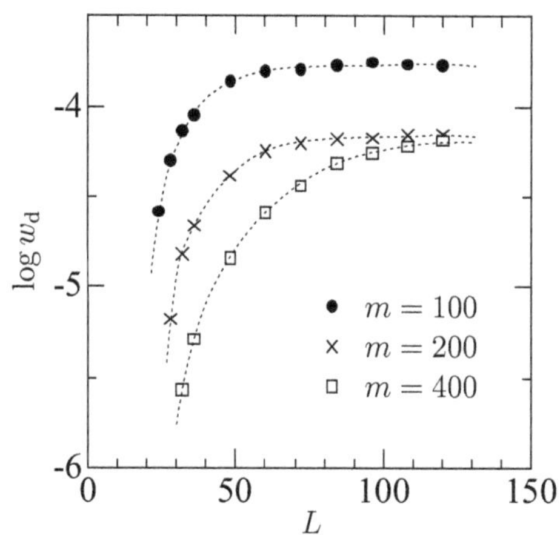

Fig. 3. Average discarded weight w_{d} as a function of system size L for the truncation number $m = 100$, 200, and 400 in the calculation of $S^{+-}(0, \omega)$. Dotted lines are guides for eyes.

where the system size is $L = 40$ sites and the broadening is taken as $\eta = 0.1$ for both of the methods. We can see excellent agreements in the low-energy excitations, whereas the BA spectra deviate from the DDMRG data in the high-frequency range. It means that the DDMRG method can take into account higher-order terms than the BA treatment. If the discarded weight w_{d} is sufficiently small in the DMRG truncation procedure, e.g., $w_{\mathrm{d}} < 10^{-5}$, the DDMRG results for finite systems are numerically exact in the same sense as 'exact diagonalization' results are. In the present calculations, we keep up to $m = 300$ density-matrix eigenstates and the typical discarded weight is less than 5×10^{-6}. Thus, we are confident that the DDMRG method using the PBCs works well in the $S = 1/2$ Heisenberg chain with at least up to several tens of sites.

Let us now mention how large system can be dealt with using the DDMRG method. The performance depends mainly on the number of the density-matrix eigenstates truncated m. As the truncation number m increases, the accuracy of calculation is improved but more CPU time and disk space are required. Practically, we could keep a several hundreds density-matrix eigenstates with the present typical cluster machine, e.g., Pentium 4 3.2GHz or Opteron 252 2.6GHz. In Fig. 3, we show the logarithm of average discarded weight w_{d} as a function of system size L for $m = 100$, 200, and 400 in the calculation of $S^{+-}(0, \omega)$. Naturally, w_{d} decreases as m increases at a fixed system size. On the other hand, for a fixed m, w_d increases rapidly with increasing L for smaller systems and is almost constant for larger systems. Empirically, we can no longer expect a good DMRG performance in this 'constant' region. For example, we estimate $L = 40$, 60, and 80 as the optimum system sizes for $m = 100$, 200, and 400, respectively. Therefore, the DDMRG calculation with about a hundred lattice sites would be adequately possible in the $S = 1/2$ Heisenberg chain using the PBCs if we take $m \approx 600$.

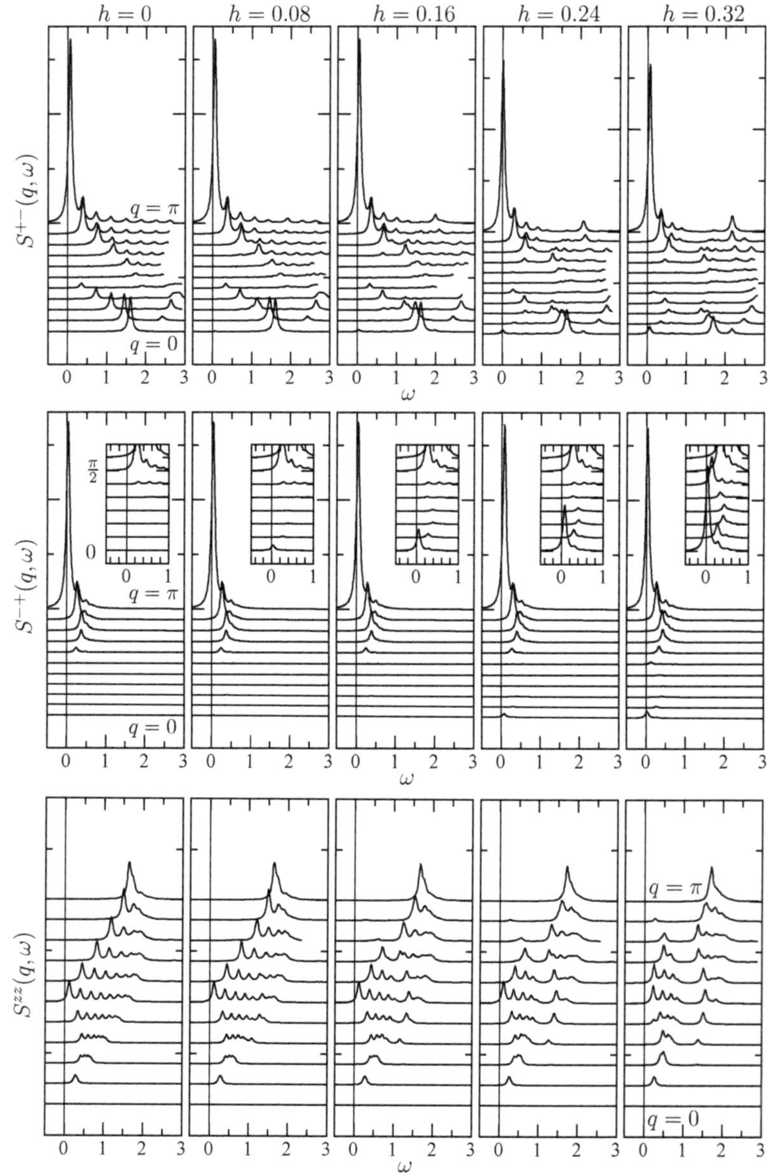

Fig. 4. Spin structure factor $S^{+-}(q,\omega)$ (upper), $S^{-+}(q,\omega)$ (middle), and $S^{zz}(q,\omega)$ (bottom) calculated with the DDMRG method for $L = 40$ and $\eta = 0.05$ applying the PBCs. The staggered magnetic field h is varied with fixing the uniform magnetic field at $H = 1.6$.

We then add the staggered magnetic field to the system under the fixed uniform magnetic field. In Fig. 4, we show the spin structure factors $S^{\alpha\beta}(q,\omega)$ for $\alpha\beta = +-, -+, zz$ at $h = 0$, 0.08, 0.16, 0.24, and 0.31 with fixing $H = 1.6$, calculated with the DDMRG method for $L = 40$ and $\eta = 0.05$ applying the PBCs. The high-energy

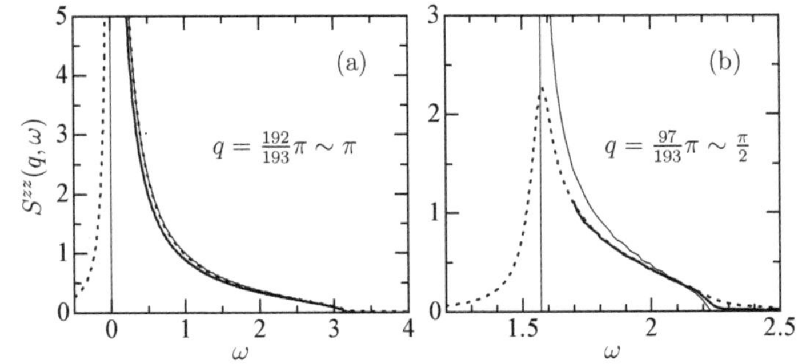

Fig. 5. Spin structure factor $S^{zz}(q, \omega)$ in a chain with $L = 192$ sites for $q = (192/193)\pi \sim \pi$ (a) and $q = (97/193)\pi \sim \pi/2$ (b), calculated with the DDMRG method for $\eta = 0.05$ (dotted lines) applying the OBCs and then deconvoluted (bold lines). The thin lines show the analytic solutions in the thermodynamic limit.[3]

structures, which cannot be obtained by the conformal field theory, are clearly seen.

For $h = 0$ [see also Fig. 2], the feature of $S(q, \omega)$ is very similar to that of Haldane-Shastry model ($1/r^2$ interaction).[12–15] In the lower energy edge of the small q ($0 \leq q < 2\pi S_z^{\text{tot}}/L$) in the $S^{+-}(q, \omega)$, the one antispinon (magnon) contributes as a δ-function peak. At $q = 0$, the exact form is known, $S^{+-}(0, \omega) = (4\pi S_z^{\text{tot}}/L)\delta(\omega - H)$, which presents a resonant mode of the magnetization induced by the uniform magnetic field.[a] The spectra are changed gradually with increasing h and we can see some predominant features for $h = 0.32$ as the following:

- The low-energy structures of $S^{+-(-+)}(q, \omega)$ around $q = 0$ arise. They are associated with increase of the weight of unfluctuating antiferromagnetic configuration in the ground state.
- The one-magnon peaks of $S^{+-}(q, \omega)$ in the small momentum region are smeared and they will disappear if $h > h_c$.
- In the large momentum ($q \sim \pi$) region of $S^{+-}(q, \omega)$, the continuum structure when $h = 0$ turn to a few discrete peaks due to decrease of the spin fluctuations.
- The band structure of $S^{zz}(q, \omega)$ seems to be split into two small-dispersive bands around $\omega \lesssim 0.5$ and $\omega \sim H$.

4.2.2. *Open-end boundary conditions*

We now turn to calculation with the OBCs. A DDMRG result with the OBCs is less exact than that with the PBCs for the same system size. However, the OBCs enable us to carry out a calculation for much larger systems, and then we may obtain

[a]The one-antispinon and the two-spinon plus two-antispinon contribution to $S^{+-}(q, \omega)$ in Haldane-Shastry model can be expressed in terms of the analytic expression of the advanced Green function for the spinless Sutherland model with coupling parameter $\lambda = 2$.[16] Other components, $S^{-+}(q, \omega)$ and the two-spinon plus one-antispinon contribution to $S^{zz}(q, \omega)$ for Haldane-Shastry model also can be expressed by the correlation function of the spinless Sutherland model.[17,18]

the dynamical quantities in the limits of $L \to \infty$ and $\eta \to 0$ through additional techniques such as the finite-size scaling or the deconvolution techniques. Hence, it would be also useful to check a performance with the OBCs.

When the OBCs are applied, we usually use the eigenstates of the particle-in-a-box problem to define the operators

$$S_q^\alpha = \sqrt{\frac{2}{L+1}} \sum_l \sin(ql) S_l^\alpha, \tag{10}$$

with quasi-momentum $q = \pi z/(L+1)$ for integers $1 < z \leq L$. This definition of S_q^α should be equivalent to that for the PBCs in the thermodynamic limit, but the agreement may not necessarily good in finite systems. In Fig. 5, we show the DDMRG results of $S^{zz}(q,\omega)$ without magnetic field for $L = 192$ and $\eta = 0.05$ with OBCs, as well as the exact line shape of the two-spinon excitation contribution for $L \to \infty$.[3] In order to compare the finite-size DDMRG spectrum with the exact solution without broadening, we need to eliminate the broadening η from the 'convolved' DDMRG spectrum.

An approach for obtaining the spectrum in the $\eta = 0$ limit is the deconvolution of the DDMRG spectrum. Theoretically, a deconvolution means solving (8) for $S(q,\omega)$ using the DDMRG data in the left-hand side. If it was possible to do this calculation exactly, one would obtain the discrete spectrum on a finite lattice of L sites. Nevertheless, the broadened spectrum on a infinite lattice ($L \to \infty$) is usually almost identical to the discretized DDMRG spectrum ($L < \infty$) under the condition $\eta \geq c/L$. Since the width of $S^{zz}(q,\omega)$ is always less than $c = 4$ according to the exact solution, our choice $\eta = 0.05$ indeed satisfies the condition.

At $q = (192/193)\pi$, the deconvoluted DDMRG data seems to agree well with the exact solution in the presented scale [see Fig. 5 (a)]. Actually, the deconvoluted DDMRG data is a shade thinner than the exact solution. It might come from the difference of momentum taken in the calculations, i.e., $q = (192/193)\pi$ in the DDMRG and π in the exact solutions. Nevertheless, it can be rather hard to find the existence of $1/\omega$-divergence for the lower edge $\omega = 0$ expected from the exact solution. If the width of a peak for $\eta = 0$ is smaller than η, the we cannot obtain it accurately by the deconvolution technique. Hence, we could not see the very sharp structure of $S^{zz}(\pi,\omega)$ at $\omega \leq 0.05$ correctly. On the other hand, the agreement is not very good at $q = (97/193)\pi \sim \pi$ [see Fig. 5 (b)]. Occasionally, quantitative estimation would be somewhat problematic. We find that the accuracy of spectrum is usually the worst around $q = \pi/2$ using the OBCs. However, we suggest that the definition of operators with the quasi-momentum gives a good approximation at least qualitatively.

The other approach is the finite-size scaling analysis of the DDMRG data. In Fig. 6 (a), we show the spin structure factor $S^{zz}(q,\omega)$ at $q = (192/193)\pi$ for several system sizes L. We keep a relation $\eta L = 9.6$ for systematic extrapolation. We clearly see the convergence of the finite-size spectra toward the exact spectrum as η decreases. We also can study the lower-edge behavior more quantitatively by

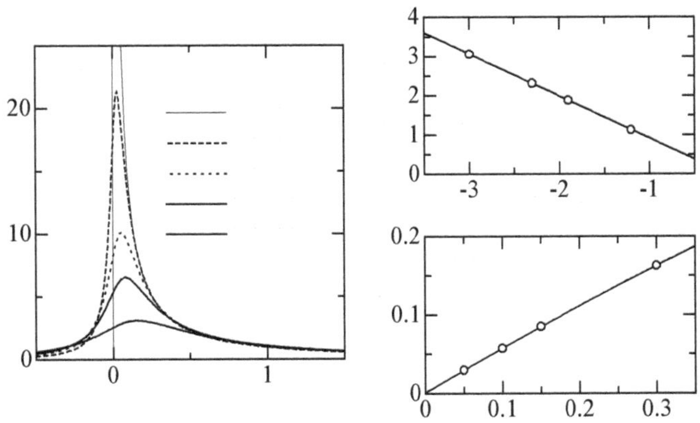

Fig. 6. (a) Spin structure factor $S^{zz}(q,\omega)$ at $q = (L - 1/L)\pi \sim \pi$ for various system lengths L and $\eta = 9.6/L$. (b) Extrapolation of $S_{\max}$ to $\eta \to 0$. (c) Extrapolation of $E_{\max}$ to $\eta \to 0$.

a scaling analysis of the maximal value in the DDMRG spectrum $S^{zz}(q,\omega)$. For instance, the hight of the low-energy maximum $\log_{10} S_{\max}$ can be scaled linearly as a function of $\log_{10}\eta$ [see Fig. 6 (b)] and the slope is -1. It means that the spectrum diverges as η^{-1} for $\eta \to 0$. Moreover, the position of the low-energy maximum, $E_{\max}$, approaches to 0 as η goes to 0. We then confirm that $S^{zz}(\pi,\omega)$ has a singularity with exponent -1 at $\omega = 0$ in the thermodynamic limit $L \to \infty$. This is consistent with the exact result which shows the $1/\omega$-divergence of $S^{zz}(\pi,\omega)$ at the lower edge $\omega = 0$.

5. Remarks and Outlook

In this work, we demonstrate successful application of both the OBCs and the PBCs to the DDMRG calculation in the 1D $S = 1/2$ Heisenberg model. Each of the boundary conditions has advantages and disadvantages, and then we need to make proper use of the boundary conditions depending on the situation.

(i) With the PBCs, the system size must be restricted practically up to about a hundred but the result is numerically exact. Therefore, the PBCs are suited to study

- quantitative estimation of the whole spectrum, and
- accurate picture of complex dispersive structure.

(ii) With the OBCs, we can set the system size to be two hundreds or more, and thus we can easily obtain the spectrum in the thermodynamic limit via the finite-size scaling or the deconvolution techniques. However, quantitative accuracy is occasionally missing on a finite lattice. Therefore, the OBCs are suited to study

- the behavior and position of a singularity in the spectrum, and
- rough picture of simple dispersive structure.

Possibly, the finite-size scaling or the deconvolution techniques will be applicable to the DDMRG data with the PBCs if the spectrum had relatively simple form.

Finally, we add that the method can easily be extended to other systems: for example,

- $S = 1/2$ Heisenberg ladder systems,
- more realistic models which include the transverse staggered magnetic field or the Dzyaloshinsky-Moriya interaction, and
- spin systems where the total spin is more than $S = 1/2$, etc.

References

1. G. M. Müller, H. Thomas, H. Beck, and J.C. Bonner, Phys. Rev. B **24**, 1429 (1981).
2. M. Jimbo and T. Miwa, *Algebraic Analysis of Solvable Lattice Models*, (American Mathematical Society, 1994).
3. A.H. Bougourzi, M. Couture and M. Kacir: Phys. Rev. B **54** (1996) R12669; M. Karbach, G. Müller, A.H. Bougourzi, A. Fledderjohann, and K-H. Mütter, Phys. Rev. B **55** 12510 (1997).
4. N. Kitanine, J.M. Maillet, V. Terras, Nucl. Phys. B **554** 647 (1999).
5. D. Biegel, M. Karbach, and G. Müller, Europhys Lett. **59** 882 (2002).
6. T. Kühner and S.R. White, Phys. Rev. B **60** 335 (1999).
7. E. Jeckelmann, Phys. Rev. B **66** 045114 (2002).
8. S.R. White, Phys. Rev. Lett. **69** 2863 (1992).
9. H. Benthien, F. Gebhard, and E. Jeckelmann, Phys. Rev. Lett. **92** 256401 (2004).
10. S. Nishimoto and E. Jeckelmann, J. Phys.: Condens. Matter **16** 613 (2004).
11. R.B. Griffiths, Phys. Rev. **133** A768 (1964).
12. F. D. M. Haldane, Phys. Rev. Lett. **60**, 635 (1988).
13. B. S. Shastry, Phys. Rev. Lett. **60**, 639 (1988).
14. F. D. M. Haldane and M. R. Zirnbauer, Phys. Rev. Lett. **71**, 4055 (1993).
15. J. C. Talstra and F. D. M. Haldane, Phys. Rev. B **50**, 6889 (1994): (Erratum) Phys. Rev. B **54**, 12594 (1996).
16. D.Serban, F.Lesage and V.Pasquier, Nucl. Phys. B **466**, 499 (1996).
17. Y. Saiga and Y. Kuramoto, J. Phys. Soc. Jpn **69** 3917 (2000).
18. M. Arikawa and Y. Saiga, J. Phys. A Gen **39** 10603 (2006).

FRUSTRATED QUANTUM ANTIFERROMAGNETS: APPLICATION OF HIGH-ORDER COUPLED CLUSTER METHOD

J. RICHTER, R. DARRADI and R. ZINKE

Institut für Theoretische Physik, Otto-von-Guericke Universität Magdeburg, P.O.B. 4120, 39016 Magdeburg, Germany
www.uni-magdeburg.de/itp

R. F. BISHOP

School of Physics and Astronomy, The University of Manchester, Sackville Street Building, P.O. Box 88, Manchester, M60 1QD, United Kingdom
www.physics.man.ac.uk

We report on recent results for strongly frustrated quantum J_1–J_2 antiferromagnets in dimensionality $d = 1, 2, 3$ obtained by the coupled cluster method (CCM). We demonstrate that the CCM in high orders of approximation allows us to investigate quantum phase transitions driven by frustration and to discuss novel quantum ground states. In detail we consider the ground-state properties of (i) the Heisenberg spin-1/2 antiferromagnet on the cubic lattice in $d = 1, 2, 3$, and use the results for the energy, the sublattice magnetization and the spin stiffness as a benchmark test for the precision of the method; (ii) coupled frustrated spin chains (the quasi-one-dimensional J_1–J_2 model) and discuss the influence of the quantum fluctuations and the interchain coupling on the incommensurate spiral state present in the classical model; (iii) the Shastry-Sutherland antiferromagnet on the square lattice; and (iv) a stacked frustrated square-lattice Heisenberg antiferromagnet (the quasi-two-dimensional J_1–J_2 model), and discuss the influence of the interlayer coupling on the quantum paramagnetic ground-state phase that is present for the strictly two-dimensional model.

Keywords: Frustration; quantum antiferromagnets; quantum phase transitions; coupled cluster method.

1. Introduction

Strongly frustrated quantum magnets have attracted much attention in recent years, both theoretically and experimentally.[1–3] In particular, quantum phase transitions have very much become the focus of interest (and see, e.g., Refs. [4–6]). At zero temperature, $T = 0$, there are no thermal fluctuations present and the transitions between ground-state (GS) phases are driven purely by the interplay of quantum-mechanical fluctuations and competition between interactions (e.g., frustration). In particular, novel quantum GS phases, such as valence-bond or spin-liquid phases may appear[6,7] which do not possess semiclassical magnetic long-range order (LRO).

A basic model which shows strong quantum fluctuations is the spin-1/2 pure Heisenberg antiferromagnet (HAFM). The GS ordering of this quantum HAFM strongly depends on the dimensionality.[3] For example, unlike its three-dimensional (3d) and two-dimensional (2d) model counterparts on the cubic or square lattice, respectively, the one-dimensional (1d) pure HAFM (i.e., with nearest-neighbor antiferromagnetic bonds only, and all of equal strength) does not have a Néel-ordered GS.[6,8] Although the tendency to order is more pronounced in 3d quantum spin systems than in lower-dimensional ones, a magnetically disordered phase can also be observed for 2d and 3d HAFM's when strong frustration is present, e.g., for the J_1–J_2 model on the square lattice (see, e.g., Refs. [9–21] and references contained therein); or the 2d Shastry-Sutherland antiferromagnet;[6,15,22–33] or, in three dimensions, for the HAFM on the pyrochlore lattice[34,35] or on the stacked kagomé lattice.[36]

Besides the general interest in frustrated quantum antiferromagnets as rich examples of quantum many-body systems, there is a strong motivation for their theoretical study which is driven by the many recent experiments on quasi-low-dimensional materials that are well described by a frustrated spin-1/2 Heisenberg model. Among many others we mention here (i) the quasi-1d edge-sharing copper oxides like $LiCuVO_4$, $LiCu_2VO_2$ and $NaCu_2O_2$,[37–40] which were identified as frustrated (with ferromagnetic nearest-neighbor (NN) and antiferromagnetic next-nearest-neighbor (NNN) interactions) quantum helimagnets with an incommensurate spiral GS; (ii) the quasi-2d $SrCu_2(BO_3)_2$,[41,42] which is well described by the Shastry-Sutherland antiferromagnet, and which exhibits a gapped quantum paramagnetic GS; and (iii) the quasi-2d Li_2VOSiO_4[43,44] which can be described by the J_1–J_2 model on the square lattice.

The theoretical treatment of the frustrated quantum antiferromagnets is far from being trivial. Though, surprisingly, one can find exact GS's of a simple product nature in some exceptional cases,[22,45,46] many of the standard many-body methods, such as quantum Monte Carlo techniques, may fail or become computationally infeasible to implement if frustration is present. Other methods, such as density-matrix renormalization group (DMRG) or exact diagonalization techniques, are essentially restricted to low-dimensional systems, at least for the present. Hence, there is considerable interest in any method that can deal with frustrated spin systems in any number of dimensions, including magnetic systems with incommensurate spiral GS's. A method fulfilling this requirement is the coupled cluster method (CCM). This approach, introduced many years ago by Coester and Kümmel,[47] is one of the most universal and most powerful methods of quantum many-body theory (and for a review of which see, e.g., Ref. [48]). The CCM has previously been applied to various quantum spin systems with much success.[14,33,49–62] The application to frustrated spin systems was started in the 1990's,[14,52] and has been developed in more recent years to the point where it has become a powerful tool in this field by including higher orders of approximations in a well-defined truncation scheme.[33,54,56,57,60,62]

In this paper, we review some recent applications of the CCM to the generic frustrated quantum spin-1/2 Heisenberg model

$$H = \sum_{i,j} J_{ij}\, \mathbf{s}_i \cdot \mathbf{s}_j \;, \tag{1}$$

where the $\mathbf{s}_i$ are the spin operators fulfilling $\mathbf{s}_i^2 = 3/4$, and the J_{ij} are the (competing) exchange coupling parameters.

The paper is organized as follows. In Sec. 2 we illustrate the main features of the CCM, paying special attention to the methodological aspects relevant for the application of the CCM to frustrated magnets. In Sec. 3 we report (as a benchmark test) high-order CCM results for standard unfrustrated lattices in $d = 1, 2, 3$ and compare them with other accurate methods. In Sec. 4 we discuss the influence of quantum fluctuations on GS spiral ordering for quasi-1d J_1–J_2 spin systems, and in Sec. 5 we report on a CCM treatment of the 2d Shastry-Sutherland model. Finally, Sec. 6 contains a discussion of the GS ordering of a stacked frustrated square-lattice HAFM (the quasi-2d J_1–J_2 model).

2. The Coupled Cluster Method

The CCM formalism is now briefly outlined. For further details the interested reader is referred to Refs. [50,53–55,59–61]. The starting point for the CCM calculation is the choice of a normalized reference or model state $|\Phi\rangle$, together with a set of (mutually commuting) multi-configurational creation operators $\{C_L^+\}$ and the corresponding set of their Hermitian adjoints $\{C_L\}$,

$$\langle\Phi|C_L^+ = 0 = C_L|\Phi\rangle \quad \forall L \neq 0\,, \quad C_0^+ \equiv 1\,; \quad [C_L^+, C_J^+] = 0 = [C_L, C_J]\,. \tag{2}$$

The operators C_L^+ (C_L) are defined over a complete set of many-body configurations denoted by the set of set-indices $\{L\}$. With the set $\{|\Phi\rangle, C_L^+\}$ the CCM parametrizations of the exact ket and bra GS eigenvectors $|\Psi \rightarrow$ and $\langle\tilde{\Psi}|$ of our many-body system are then given by

$$|\Psi \rightarrow = e^S|\Phi \rightarrow, \; S = \sum_{L\neq 0} a_L C_L^+\;; \quad \langle\tilde{\Psi}| = \langle\Phi|\tilde{S}e^{-S}\;, \; \tilde{S} = 1 + \sum_{L\neq 0} \tilde{a}_L C_L\;. \tag{3}$$

The CCM correlation operators, S and $\tilde{S}$, contain the correlation coefficients, a_L and $\tilde{a}_L$, which have to be calculated. Once known, all GS properties of the many-body system can clearly be found in terms of them. To find the GS correlation coefficients a_L and $\tilde{a}_L$, we simply require that the GS energy expectation value $\bar{H} = \langle\tilde{\Psi}|H|\Psi \rightarrow$ is a minimum with respect to the entire set $\{a_L, \tilde{a}_L\}$, which leads to the GS CCM ket-state and bra-state equations

$$\langle\Phi|C_L^- e^{-S} H e^S|\Phi\rangle = 0\,, \; \langle\Phi|\tilde{S}e^{-S}[H, C_L^+]e^S|\Phi\rangle = 0 \quad \forall L \neq 0\,. \tag{4}$$

For frustrated spin systems an appropriate choice for the CCM model state $|\Phi\rangle$ is often a classical spiral spin state, (i.e., pictorially, $|\Phi\rangle = |\uparrow\nearrow\rightarrow\searrow\downarrow\swarrow\cdots\rangle$), which is characterized by a pitch angle α. Such states include the Néel state, for which $\alpha = \pi$. In the quantum model the pitch angle may be modified by quantum fluctuations. Hence, we do not choose the classical result for the pitch angle but, rather, we consider it as a free parameter in the CCM calculation, which is to be determined by minimization of the CCM GS energy.

In order to find an appropriate set of creation operators it is convenient to perform a rotation of the local axes of each of the spins, such that all spins in the reference state align in the negative z-direction. This rotation by an appropriate angle δ_i of the spin on lattice site i is equivalent to the spin-operator transformation

$$s_i^x = \cos\delta_i \hat{s}_i^x + \sin\delta_i \hat{s}_i^z \; ; \quad s_i^y = \hat{s}_i^y \; ; \quad s_i^z = -\sin\delta_i \hat{s}_i^x + \cos\delta_i \hat{s}_i^z \; . \tag{5}$$

In this new set of local spin coordinates the reference state and the corresponding creation operators C_L^+ are given by

$$|\hat{\Phi}\rangle = |\downarrow\downarrow\downarrow\downarrow\cdots\rangle \; ; \quad C_L^+ = \hat{s}_i^+ \, , \; \hat{s}_i^+\hat{s}_j^+ \, , \; \hat{s}_i^+\hat{s}_j^+\hat{s}_k^+ \, , \; \ldots \, , \tag{6}$$

where the indices $i, j, k, \ldots$ denote arbitrary lattice sites. In the new coordinates the initial Heisenberg Hamiltonian of Eq. (1) becomes

$$\begin{aligned}\hat{H} = \; & \frac{1}{2}\sum_{i,j} J_{ij} \left\{ [\cos(\alpha_{ij}) + 1](\hat{s}_i^+\hat{s}_j^- + \hat{s}_i^-\hat{s}_j^+) + [\cos(\alpha_{ij}) - 1](\hat{s}_i^+\hat{s}_j^+ + \hat{s}_i^-\hat{s}_j^-) \right. \\ & \left. + \, 2\sin(\alpha_{ij})[\hat{s}_i^+\hat{s}_j^z - \hat{s}_i^z\hat{s}_j^+ + \hat{s}_i^-\hat{s}_j^z - \hat{s}_i^z\hat{s}_j^-] + 4\cos(\alpha_{ij})\hat{s}_i^z\hat{s}_j^z \right\} ,\end{aligned} \tag{7}$$

where $\alpha_{ij} \equiv \delta_j - \delta_i$ is the angle between the two interacting spins, and $\hat{s}_i^\pm \equiv \hat{s}_i^x \pm i\hat{s}_i^y$. In the case of the Néel model state $\hat{H}$ becomes much simpler, so that, for example, for the case of NN bonds only, we have $\hat{H} = -J\sum_{\langle i,j\rangle} \left(\hat{s}_i^+\hat{s}_j^+ + \hat{s}_i^-\hat{s}_j^- + 2\hat{s}_i^z\hat{s}_j^z\right)$.

For the ensuing discussion of the GS properties we concentrate on the GS energy E, the order parameter m and the spin stiffness ρ_s. Within the CCM scheme we have $E = \langle\Phi|e^{-S}He^{S}|\Phi\rangle$ and $m = -\frac{1}{N}\langle\tilde{\Psi}|\sum_{i=1}^{N}\hat{s}_i^z|\Psi\rangle$. The spin stiffness ρ_s can be calculated by imposing a twist on the order parameter of a magnetically long-range ordered system along a given direction, i.e.,

$$\frac{E(\theta)}{N} = \frac{E(\theta=0)}{N} + \frac{1}{2}\rho_s\theta^2 + \mathcal{O}(\theta^4) \, , \tag{8}$$

where $E(\theta)$ is the GS energy as a function of the twist angle θ, and N is the number of sites (and where the interested reader is referred to Ref. [61] for further details).

The CCM formalism is exact if we take into account all possible multispin configurations in the correlation operators S and $\tilde{S}$. However, in general, this is impossible to do in practice for a quantum many-body system. Hence, it is necessary to use approximation schemes in order to truncate the expansions of S and $\tilde{S}$ in Eq. (3) in any practical calculation. A quite general approximation scheme is the so-called SUBn-m approximation. In this approximation all correlations in the correlation operators S and $\tilde{S}$ are taken into account which span a range of no more than m contiguous sites and contain only n or fewer spins. In most cases the SUBn-n scheme is used (i.e., with $n = m$), and it is then (for spin-1/2 systems) called the LSUBn scheme. To find all different fundamental configurations entering S and $\tilde{S}$ at a given level of LSUBn approximation we use lattice symmetries and, where possible, any exact conservation laws.

Since the LSUBn approximation becomes exact in the limit $n \to \infty$, it is useful to extrapolate the 'raw' LSUBn results to the limit $n \to \infty$. Although an exact scaling theory for the LSUBn results is not known, there is some empirical experience[33,53–55,62] indicating how the physical quantities for spin models might scale with n. For the GS energy we employ[54,55]

$$E(n) = a_0 + a_1 \frac{1}{n^2} + a_2 \left(\frac{1}{n^2}\right)^2 . \tag{9}$$

Furthermore, we note that it may be useful to discard the LSUB2 results for the extrapolation, because generally they fit poorly to the asymptotic behavior.[55] For the order parameter and the stiffness one utilizes[55,61] an extrapolation law with leading power $1/n$, i.e.,

$$A(n) = b_0 + b_1 \frac{1}{n} + b_2 \left(\frac{1}{n}\right)^2 . \tag{10}$$

However, there is some experience that when applied to systems showing an order-disorder quantum phase transition this kind of extrapolation tends to overestimate the parameter region where magnetic LRO exists, i.e. to yield too large critical values for the exchange parameter driving the transition.[33,54,58,61] The reason for such behavior might derive from the change of the scaling near a critical point. Hence, in addition to the extrapolation rule of Eq. (10) for the order parameter m, we also use a leading 'power-law' extrapolation[33,55,61] given by

$$m(n) = c_0 + c_1 \left(\frac{1}{n}\right)^{c_2} , \tag{11}$$

where the leading exponent c_2 is determined directly from the LSUBn data.

3. The CCM for the Pure HAFM on Cubic Lattices for $d = 1, 2, 3$

During the last few years the running time and memory requirements of the original CCM code have been considerably improved.[60,63] Consequently, it is now possible to run higher levels of approximation by using an improved parallelization procedure. In this Section we present a collection of CCM results for the (unfrustrated) pure HAFM (i.e., with the Hamiltonian of Eq. (1) with $J_{ij} = 1$ for NN bonds, and $J_{ij} = 0$ otherwise) on some basic lattices, and compare them with the most accurate results obtained by other methods. While some of the CCM results have already been published elsewhere,[55,61] we also present here the new unpublished results from higher levels of approximation. The results are shown in Tables 1, 2 and 3, in which the entries shown in boldface are the new ones.

4. The Quasi-One-Dimensional J_1–J_2 Model

One-dimensional quantum spin systems like the frustrated J_1–J_2-model have attracted much attention over many years.[8] The physics of such quantum spin systems

Table 1. Data for the spin-1/2 linear-chain pure HAFM. N_f is the number of fundamental configurations for the Néel reference state, E/N is the GS energy per spin and m is the sublattice magnetization. The LSUBn results are extrapolated using Eq. (9) for E/N and Eq. (10) (in parentheses, Eq. (11))for m.

linear chain	N_f	E/N	m
LSUB2	1	-0.41667	0.33333
LSUB4	3	-0.43627	0.24839
LSUB6	9	-0.44002	0.20789
LSUB8	26	-0.44137	0.18297
LSUB10	81	-0.44200	0.16562
LSUB12	267	-0.44234	0.15263
LSUB14	**931**	**-0.44255**	**0.14240**
LSUB16	**3362**	**-0.44269**	**0.13408**
Extrapolated CCM	-	**-0.44315**	**0.07737 (-0.01086)**
Bethe ansatz[66]	-	-0.44315	0.0

Table 2. Data for the spin-1/2 square-lattice pure HAFM. N_f is the number of fundamental configurations for the Néel (in parenthesess, for the spiral) reference state, E/N is the GS energy per spin, m is the sublattice magnetization and ρ_s is the spin stiffness. The LSUBn results are extrapolated using Eq. (9) for E/N, Eq. (10), (in parentheses, Eq. (11)) for m and Eq. (10) for ρ_s.

square lattice	N_f	E/N	m	ρ_s
LSUB2	1 (3)	-0.64833	0.42071	0.2574
LSUB4	7 (40)	-0.66366	0.38240	0.2310
LSUB6	75 (828)	-0.66700	0.36364	0.2176
LSUB8	1287 (21124)	-0.66817	0.35242	0.2097
LSUB10	**29605 (586787)**	**-0.66870**	**0.34483**	-
Extrapolated CCM	-	**-0.66936**	**0.31024 (0.28073)**	0.1812
3rd order SWT[64]	-	-0.66931	0.3069	0.1747
QMC[67]	-	-0.66944	0.3070	0.1852

Table 3. Data for the spin-1/2 simple-cubic lattice pure HAFM. N_f is the number of fundamental configurations for the Néel (in parenthesess, for the spiral) reference state, E/N is the GS energy per spin, m is the sublattice magnetization and ρ_s is the spin stiffness. The LSUBn results are extrapolated using Eq. (9) for E/N, Eq. (10), (in parentheses, Eq. (11)) for m and Eq. (10) for ρ_s.

cubic lattice	N_f	E/N	m	ρ_s
LSUB2	1 (4)	-0.89076	0.45024	0.2527
LSUB4	9 (106)	-0.90043	0.43392	0.2416
LSUB6	181 (5706)	-0.90180	0.42860	0.2380
LSUB8	**8809 (444095)**	**-0.90214**	**0.42626**	-
Extrapolated CCM	-	**-0.90247**	**0.42054 (0.42141)**	0.2312
3rd order SWT[68]	-	-0.9025	0.4227	0.2343

is often remarkably different from that of their corresponding classical counterparts, with a rich variety of different quantum GS's. As already mentioned in Sec. 1, novel edge-sharing copper oxides like $LiCuVO_4$ and $NaCu_2O_2$,[37–40] that were identified as quantum helimagnets with ferromagnetic NN and frustrating antiferromagnetic

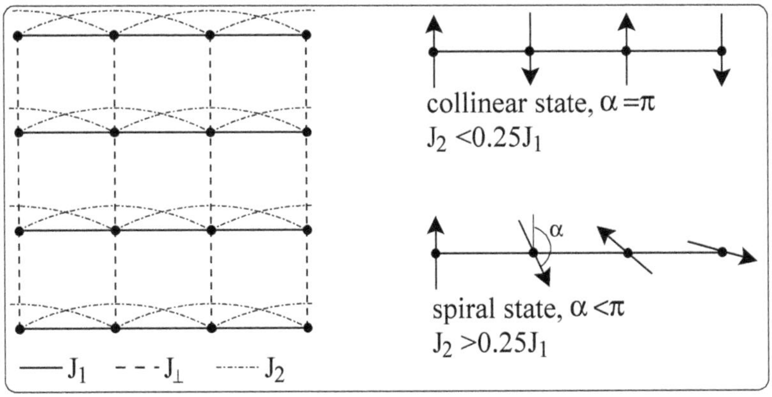

Fig. 1. Illustration of the quasi-1d J_1–J_2 model and the reference states $|\Phi\rangle$ used for our CCM calculations.

NNN bonds, have stimulated a great deal of renewed interest in frustrated Heisenberg chains. In these materials a finite interchain coupling is also present, which in turn may lead to a low-temperature phase transition to a magnetically ordered phase.

Therefore, we consider in this section the following quasi-1d spin-1/2 J_1–J_2 Heisenberg model,

$$H = \sum_n \left(J_1 \sum_i \mathbf{s}_{i,n} \cdot \mathbf{s}_{i+1,n} + J_2 \sum_i \mathbf{s}_{i,n} \cdot \mathbf{s}_{i+2,n} \right) + J_\perp \sum_{i,n} \mathbf{s}_{i,n} \cdot \mathbf{s}_{i,n+1} \,, \tag{12}$$

where n labels the chains, $J_\perp$ is the interchain coupling, J_1 is the in-chain NN coupling and J_2 the in-chain NNN coupling, as shown in Fig. 1. We fix J_1 to $J_1 = 1$ (for the antiferromagnetic case) or to $J_1 = -1$ (for the ferromagnetic case) and consider $J_2 \geq 0$ (which is frustrating in both cases) and $J_\perp \geq 0$. The case $J_\perp = 0$ corresponds to the strictly 1d case.

The classical system is characterized by a second-order transition from a collinear phase to a non-collinear spiral phase at $|\frac{J_2}{J_1}| = 0.25$. For $J_2 \geq |J_1|$ the classical spiral (pitch) angle $\alpha_{\rm cl}$ is given by $\alpha_{\rm cl} = \arccos(-0.25J_1/J_2)$. Note that in the classical model neither the pitch angle $\alpha_{\rm cl}$ nor the transition point $|\frac{J_2}{J_1}| = 0.25$ depends on the interchain coupling $J_\perp$.

From the experimental point of view it is of interest to discuss the influence of quantum fluctuations on the pitch angle. Furthermore the question arises whether in the quantum model the interlayer coupling $J_\perp$ does or does not influence either or both of the pitch angle and the transition point. While the influence of quantum fluctuations on the pitch angle for $J_\perp = 0$ has been discussed previously in the literature,[52,69–71] this question has not been considered for finite $J_\perp > 0$ so far. Furthermore, any effect which a finite $J_\perp$ may have on the transition point has also not been considered in the literature. To discuss this point we use the CCM for the model of Eq. (12) at the same level of approximation (viz., SUB2-3) as in Ref. [52].

In that paper it was demonstrated that the SUB2-3 approximation for strictly 1d systems leads to results of comparable accuracy to those obtained using the DMRG method. For the SUB2-3 approximation the relevant CCM equations can be found in closed analytical form, even for nonzero values of $J_\perp$.

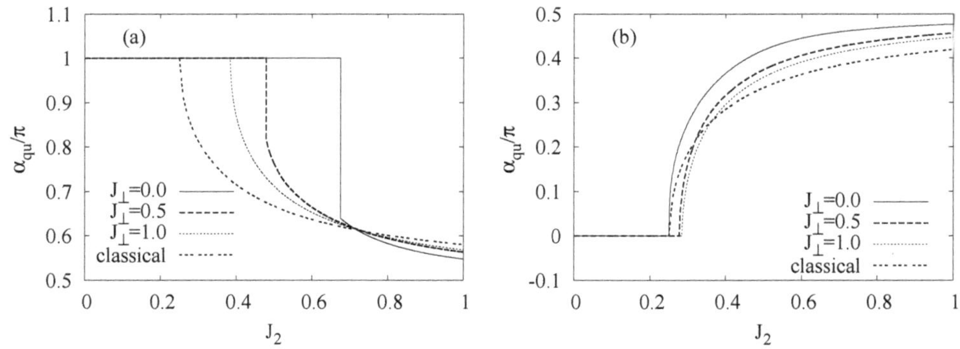

Fig. 2. The quantum pitch angle $\alpha_{\rm qu}$ versus J_2 for the two cases (a) $J_1 = 1$ and (b) $J_1 = -1$ in the CCM SUB2-3 approximation, compared with the classical result.

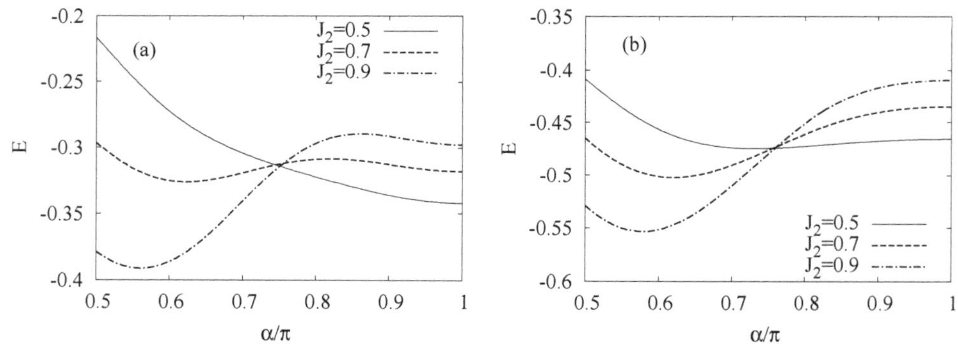

Fig. 3. Ground-state energy per spin in the CCM SUB2-3 approximation as a function of the pitch angle α for different (fixed) values of the NNN frustration strength J_2 and antiferromagnetic NN bonds with $J_1 = 1$, for the two cases (a) $J_\perp = 0$ and (b) $J_\perp = 0.7J_1$. The quantum pitch angle $\alpha_{\rm qu}$ shown in Fig. 2 corresponds to the absolute minimum of the energy E.

As quantum fluctuations may lead to a 'quantum' pitch angle $\alpha_{\rm qu}$ that is different from the classical value $\alpha_{\rm cl}$, we consider the pitch angle in the reference state as a free parameter. We then determine $\alpha_{\rm qu}$ by minimizing the CCM energy $E(\alpha)$ with respect to α. Some results for the pitch angle $\alpha_{\rm qu}$ of the quantum model are presented in Fig. 2. We have checked explicitly, that for $J_\perp = 0$ our results coincide with those of Ref. [52]. (We note that due to the different definitions of the Hamiltonian parameters that agreement is not obvious). For antiferromagnetic J_1

and weak interchain coupling $J_\perp$ we find, contrary to the behavior of the classical system, a first-order transition from a collinear phase to a spiral phase, i.e., the quantum pitch angle jumps from $\alpha_{\text{qu}} = \pi$ (in the collinear phase) to $\alpha_{\text{qu}} < \pi$ (in the spiral phase). This is illustrated in Fig. 3 by the energy curves $E(\alpha)$, which show a typical first-order scenario for $J_\perp = 0$ compared with a typical second-order scenario for $J_\perp = 0.7J_1$. Furthermore, the transition point is both shifted to larger values of $J_2 > 0$ than in the classical case (and cf. Refs. [52,69,70]), and does now depend on $J_\perp$. For increasing $J_\perp$ the transition point is shifted towards the classical value (but even for $J_\perp = J_1$ it remains significantly above the classical value), and for large enough $J_\perp$ the transition between the collinear and the spiral phase becomes continuous.

For ferromagnetic $J_1 = -1$ one has, in complete analogy to the classical system, a second-order transition from the collinear to a spiral phase at $J_2 = 0.25$ when $J_\perp = 0$, which is in agreement with earlier considerations.[56,71] However, for $J_2 > 0.25|J_1|$ the quantum pitch angle deviates from the classical value and depends on the interlayer coupling $J_\perp$. There is also a shift of the transition point for increasing values of $J_\perp$ towards larger values of J_2.

A common feature for both cases (i.e., $J_1 = -1$ and $J_1 = +1$) is, that for increasing values of J_2 the quantum pitch angle approaches its limiting value $\pi/2$ much faster than for the classical model.

The general behavior discussed above can be qualitatively related to the strength of quantum fluctuations (and see, e.g., the discussion in Refs. [33,54,56)], which themselves depend on the values of the exchange parameters of the model. In general, quantum fluctuations tend to stabilize collinear phases. Consequently, for the model with the strongest fluctuations, namely the strictly 1d antiferromagnetic model ($J_\perp = 0, J_1 = 1$), the transition to the non-collinear GS takes place for largest J_2 (cf. Fig. 2a). Increasing the interchain coupling then reduces the strength of the quantum fluctuations, which leads to a shift of the transition point to smaller values of J_2. In the case of the strictly 1d chain with ferromagnetic $J_1 = -1$ and $J_2 < 0.25|J_1|$, the GS is the fully polarized ferromagnetic state, which does not exhibit any quantum fluctuations. As a result the transition point is the same as in the classical model, but the quantum pitch angle starts to deviate from the classical value immediately if $J_2 > 0.25|J_1|$.

Finally, we emphasize that we did not speculate on possible spiral LRO. For strictly 1d systems it is clear that there is no magnetic LRO and the quantum pitch angle discussed above corresponds to incommensurate short-range correlations.

5. A Frustrated Heisenberg Antiferromagnet on the Square Lattice: The Shastry-Sutherland Model

As already mentioned in Sec. 1, the Shastry-Sutherland model, first introduced some 25 years ago,[22] has attracted much attention in connection with experiments on $SrCu_2(BO_3)_2$. The model is characterized by a special arrangement of frustrating

NNN J_2 bonds on the square lattice, as shown in Fig. 4. Its Hamiltonian reads

$$H = J_1 \sum_{\langle i,j \rangle} \mathbf{s}_i \cdot \mathbf{s}_j + J_2 \sum_{\{i,k\}} \mathbf{s}_i \cdot \mathbf{s}_k \,, \tag{13}$$

where the sum on $\langle i,j \rangle$ runs over all NN bonds and the sum on $\{i,k\}$ runs only over the selected NNN bonds shown in Fig. 4. In what follows we set $J_1 = 1$ and consider

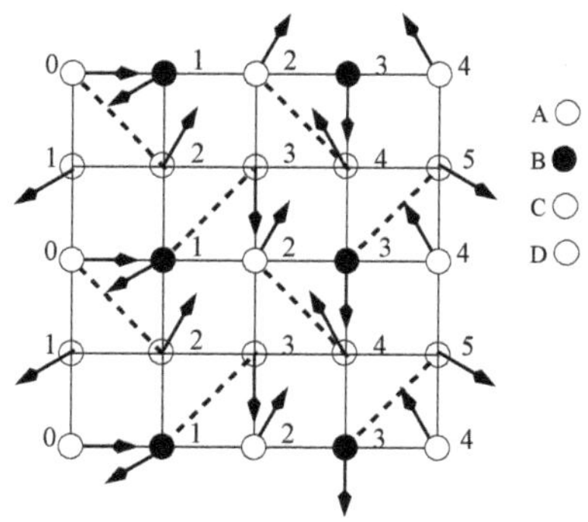

Fig. 4. Illustration of the Shastry-Sutherland model with antiferromagnetic NN bonds $J_1 = 1$ (solid lines) and NNN bonds J_2 (dashed lines), together with its classical spiral state. The spin orientations at lattice sites n are given by the angles $\theta = n\alpha_{\rm cl}$, where $n = 0, 1, 2, \ldots,$ and $\alpha_{\rm cl}$ is the characteristic pitch angle of the classical spiral state. The state is shown for $\alpha_{\rm cl} = 5\pi/6$ and $n = 0, 1, \ldots, 5$.

the case of (frustrating) $J_2 > 0$. Although the GS of this model is well understood in both the limits of small J_2 and large J_2, the GS phase at intermediate values of J_2 is still a matter of discussion.[6,23–33] The CCM treatment of this model, briefly reviewed in this Section, is explained in more detail in Ref. [33].

We start with the classical GS of the Shastry-Sutherland model. It is the collinear Néel state for $J_2/J_1 \leq 1$, but a non-collinear spiral state for $J_2/J_1 > 1$ (and see Fig. 4 and Refs. [23,25]), with a characteristic pitch angle $\alpha_{\rm cl}$ given by $\alpha_{\rm cl} = \pi$ for $J_2 \leq J_1$ and $\alpha_{\rm cl} = \pi - \arccos(J_1/J_2)$ for $J_2 > J_1$. The transition from the collinear Néel state to the non-collinear spiral state at $J_2/J_1 = 1$ is of second order for the classical model. We note further that there are only two different angles between interacting spins, namely $\alpha_{\rm cl}$ for the J_1 couplings and $-2\alpha_{\rm cl}$ for the J_2 couplings.

Similarly as we did for the quasi-1d J_1–J_2 model we now calculate the GS energy as a function of J_2 using as reference state a spiral state as shown in Fig. 4. Again, we find that due to quantum fluctuations the onset of the spiral phase in the quantum model is shifted to higher values of J_2, and the transition between the collinear and the spiral states becomes discontinuous, as seen from Fig. 3a. We conclude, that the collinear Néel state is the favored CCM reference state up to values $J_2 \approx 1.5J_1$.

It is known from the early paper of Shastry and Sutherland[22] that for large J_2 the quantum GS of the model of Eq. (13) is a rotationally-invariant orthogonal-dimer state given by $|\Psi\rangle_{\rm dimer} = \prod_{\{i,j\}_{J_2}} [| \uparrow_i\rangle | \downarrow_j\rangle - | \downarrow_i\rangle | \uparrow_j\rangle]/\sqrt{2}$, where i and

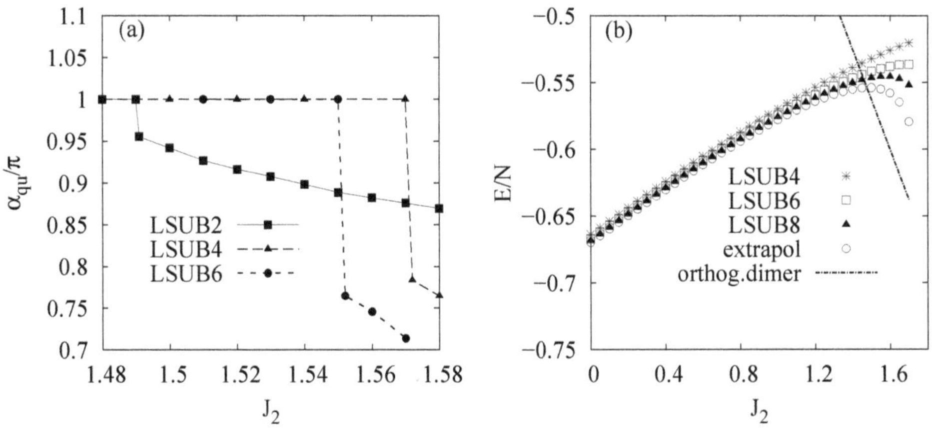

Fig. 5. (a) The quantum pitch angle α_{qu} for the Shastry-Sutherland model with $J_1 = 1$ as a function of J_2, calculated within the CCM-LSUBn approximation with $n = 2, 4, 6$; and (b) the GS energy per spin as a function of J_2 of (i) the collinear Néel phase obtained within the CCM LSUBn approximation scheme with $n = 4, 6, 8$ and its extrapolated $n \to \infty$ value using Eq. (9), and (ii) the orthogonal-dimer state.

j correspond to those sites which cover the J_2 bonds. The energy per site of this state is $E_{\text{dimer}}/N = -3J_2/8$. We compare its energy in Fig. 5b with that of the CCM GS obtained with the collinear Néel state as reference state. Our results demonstrate that the orthogonal-dimer state has lower energy than the Néel phase for $J_2 \gtrsim 1.477J_1$, i.e., significantly before the point where the spiral CCM phase has lower energy than the collinear CCM phase. We note that we also checked that $|\Psi\rangle_{\text{dimer}}$ similarly remains the state of lowest energy in the region where the non-collinear spiral phase has lower energy than the Néel phase. We conclude that there is no spiral GS phase in the quantum model. The critical value $J_2^d = 1.477J_1$ where the transition to the orthogonal-dimer phase takes place obtained by the CCM is in good agreement with results obtained by other methods (and see, e.g., Table 2 in Ref. [31]).

So far we have mainly discussed the energy of competing GS phases. The next question we would like to discuss is the question of the stability of the Néel LRO in the frustrated regime. For that purpose we calculate the order parameter (viz., the sublattice magnetization) m within the LSUBn approximation scheme up to $n = 8$ and extrapolate to $n \to \infty$ using the extrapolation scheme of Eq. (11). The results are shown in Fig. 6a. The extrapolated data clearly demonstrate that the LRO vanishes before the orthogonal-dimer state becomes the GS. The transition from Néel LRO to magnetic disorder is of second order. Hence, in agreement with previous investigations (and see, e.g., Ref. [31]), we come to the second important conclusion that there exists an intermediate magnetically disordered phase. The critical value $J_2^c \approx 1.14J_1$ where the Néel LRO breaks down agrees well with the corresponding value calculated by series expansion techniques as given, for example, in Table 2 of Ref. [31].

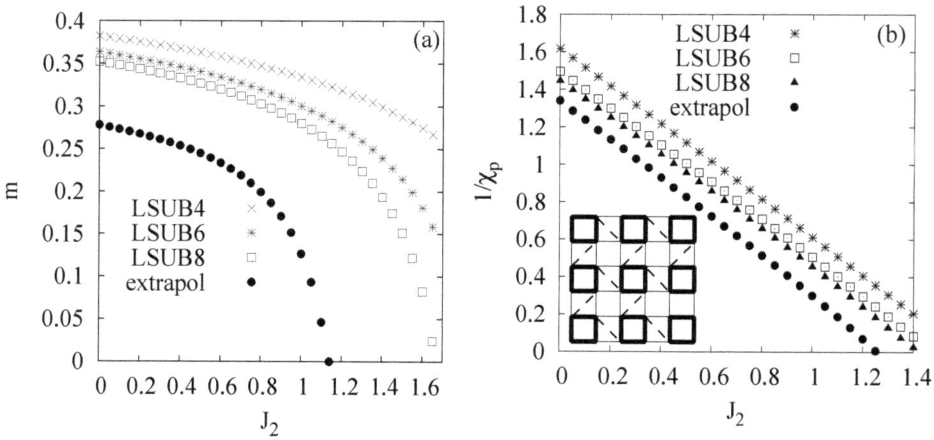

Fig. 6. (a) The sublattice magnetization m versus J_2 obtained within the CCM LSUBn approximation with $n = 4, 6, 8$ and extrapolated to $n \to \infty$ using Eq. (11); and (b) the inverse susceptibility $1/\chi_p$ versus J_2 obtained within the CCM LSUBn approximation with $n = 4, 6, 8$ and extrapolated to $n \to \infty$ using Eq. (10). Inset: Pattern of plaquette valence bond order.

The nature of the nonmagnetic GS in the region $1.14J_2 \lesssim J_2 \lesssim 1.477J_2$ is a matter of some controversy in the literature. A favored candidate is a plaquette singlet phase.[25,26] To address this question we follow similar reasoning to that used in Ref. [21] and consider the response of the spin system to a field F_p given by

$$F_p = \delta \sum_{x,y} \left[(-1)^x \mathbf{s}_{x,y} \cdot \mathbf{s}_{x+1,y} + (-1)^y \mathbf{s}_{x,y} \cdot \mathbf{s}_{x,y+1} \right] , \tag{14}$$

where x, y are components (integer numbers) of the lattice vectors of the square lattice. This field corresponds to a plaquette valence bond order (as illustrated in the inset of Fig. 4b), which breaks the lattice symmetry. Thus, we use the CCM with the Néel state as reference state to calculate the energy per site $e(J_1, J_2, \delta)$ for $H + F_p$, namely the Hamiltonian of Eq. (13) perturbed by the additional term of Eq. (14). The susceptibility χ_p is then defined as[21]

$$\chi_p = - \left. \frac{\partial^2 e}{\partial \delta^2} \right|_{\delta=0} . \tag{15}$$

In Fig. 6b we present the results for the inverse susceptibility, $1/\chi_p$, as a function of J_2. The extrapolation to $n \to \infty$ is performed using an extrapolation scheme with leading power $1/n$ as in Eq. (10). Clearly, in the magnetically ordered Néel phase χ_p is finite as it should be. However, close to the transition to the magnetically disordered phase at $J_2^c \approx 1.14J_1$ the susceptibility becomes very large and diverges at $J_2 \approx 1.26J_1$, which is close to J_2^c. Hence, we conclude from our CCM data that there exists a valence-bond phase between the Néel-ordered phase and the orthogonal-dimer phase.

6. The Quantum J_1–J_2 Antiferromagnet on the Stacked Square Lattice

As already mentioned in Sec. 1 the J_1–J_2 model on the square lattice is a canonical model to study quantum phase transitions in $d = 2$.[9–21] However, in experimental realizations of the J_1–J_2 model the magnetic couplings are expected to be not strictly 2d, since a nonzero interlayer coupling $J_\perp$ is always present. For example, recently Rosner *et al.*[44] have found $J_\perp \approx 0.07J_1$ for Li_2VOSiO_4, a material which can be described by a square lattice J_1–J_2 model with large J_2.[43,44]

Therefore, in this Section we consider the influence of an interlayer coupling on the GS phases of the the J_1–J_2 spin-1/2 HAFM, i.e., we consider the HAFM on the stacked square lattice described by

$$H = \sum_n \left(J_1 \sum_{\langle ij \rangle} \mathbf{s}_{i,n} \cdot \mathbf{s}_{j,n} + J_2 \sum_{[ij]} \mathbf{s}_{i,n} \cdot \mathbf{s}_{j,n} \right) + J_\perp \sum_{i,n} \mathbf{s}_{i,n} \cdot \mathbf{s}_{i,n+1} \,, \tag{16}$$

where n labels the layers and $J_\perp \geq 0$ is the interlayer coupling. The expression in parentheses represents the J_1–J_2 model of the layer n with intralayer NN bonds $J_1 = 1$ and NNN bonds $J_2 \geq 0$.

The classical GS's of the model are the Néel state for $J_2 < 0.5J_1$ and another particular collinear state for $J_2 > 0.5J_1$. The latter state (which we henceforth refer to as the collinear-columnar or, simply the collinear state) is a columnar $(\pi, 0)$ state characterized by a parallel spin orientation of nearest neighbors along the direction of one axis (say, the vertical or columnar direction) in each layer, and an antiparallel spin orientation of nearest neighbors along the perpendicular (say, horizontal or row) direction. It is well known[9–21] that for $J_\perp = 0$ the quantum model has two corresponding GS phases with semi-classical magnetic LRO, one (Néel-like) for small $J_2 \lesssim 0.4J_1$ and one (collinear-columnar-like) for large $J_2 \gtrsim 0.6J_1$, which are separated by a magnetically disordered (quantum paramagnetic) GS phase.

For the treatment of the model of Eq. (16) with arbitrary $J_\perp$ we apply the CCM and use both classical GS's (Néel and collinear-columnar) as reference states. Here we illustrate the CCM approach to this model only very briefly, and refer the interested reader to Ref. [62] for more details. In order to determine the GS phase transition points we calculate the order parameters for various values of $J_\perp$ and determine those values $J_2 = \gamma_{\rm Neel}(J_\perp)$ and $J_2 = \gamma_{\rm col}(J_\perp)$ where the order parameters vanish. In Fig. 7a we present some typical curves showing the extrapolated order parameters (according to Eq. (11)) versus J_2 for some values of $J_\perp$. The magnetic order parameters of both magnetically long-range ordered phases vanish continuously as is typical for second-order transitions. We note, however, that there are arguments[11] that the transition from the collinear-columnar phase to the quantum paramagnetic phase should be of first order. The order parameters for both phases are monotonically increasing functions of $J_\perp$, and the transition points $\gamma_{\rm Neel}$ and $\gamma_{\rm col}$ also move together as $J_\perp$ increases. In Fig. 7b we present the dependence on $J_\perp$ of these transition points. Close to the strictly 2d case (i.e., for small $J_\perp \ll J_1$) the influence of the interlayer coupling is largest. For a characteristic value of $J_\perp^* \approx 0.19 J_1$ the two transition points $\gamma_{\rm Neel}$ and $\gamma_{\rm col}$ meet each other. Hence, we conclude that already for quite weak interlayer coupling the magnetically disordered quantum phase that is present in the strictly 2d model disappears, and a direct transition between the two semi-classically ordered magnetic phases occurs. This conclusion is also in agreement with the statement that there is no magnetically disordered phase in the 3d J_1–J_2 model on the bcc lattice.[72,73]

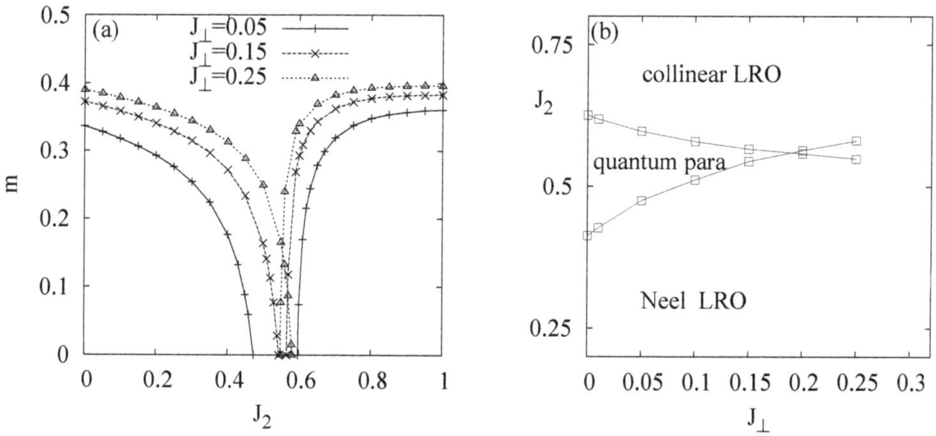

Fig. 7. (a) The magnetic order parameter m versus J_2 for various strengths of the interlayer coupling $J_\perp$ (with $J_1 = 1$); and (b) the ground-state phase diagram (where the solid lines show those values of J_2 for which the order parameters vanish). Note that $J_1 = 1$.

References

1. S.T. Bramwell and M.J.P. Gingras, *Science* **294**, 14 (2001); T. Senthil, A. Vishwanath, L. Balents, S. Sachdev, and M.P.A. Fisher, *Science* **303**, 1490 (2004); G.C. Lau, R.S. Freitas, B.G. Ueland, B.D. Muegge, E.L. Duncan, P. Schiffer, and R.J. Cava, *Nature Physics* **2**, 249 (2006); R. Moessner and A.P. Ramirez, *Phys. Today* **59**, 24 (February 2006).
2. H.T. Diep (ed.), *Frustrated Spin Systems* (World Scientific, Singapore, 2004).
3. U. Schollwöck, J. Richter, D.J.J. Farnell and R.F. Bishop (eds.), *Quantum Magnetism*, Lecture Notes in Physics **645** (Springer-Verlag, Berlin, 2004).
4. S. Sachdev, *Quantum Phase Transitions* (Cambridge University Press, Cambridge, 1999).
5. S. Sachdev, in *Quantum Magnetism*, eds. U. Schollwöck, J. Richter, D.J.J. Farnell, and R.F. Bishop, Lecture Notes in Physics **645** (Springer-Verlag, Berlin, 2004), p. 381.
6. J. Richter, J. Schulenburg, and A. Honecker, in *Quantum Magnetism*, eds. U. Schollwöck, J. Richter, D.J.J. Farnell, and R.F. Bishop, Lecture Notes in Physics **645** (Springer-Verlag, Berlin, 2004), p. 85.
7. G. Misguich and C. Lhuillier, in *Frustrated Spin Systems*, ed. H.T. Diep, (World Scientific, Singapore, 2004), p. 229.

8. H.-J. Mikeska and A.K. Kolezhuk, in *Quantum Magnetism*, eds. U. Schollwöck, J. Richter, D.J.J. Farnell, and R.F. Bishop, Lecture Notes in Physics **645** (Springer-Verlag, Berlin, 2004), p. 1.
9. P. Chandra and B. Doucot, *Phys. Rev. B* **38**, 9335 (1988).
10. E. Dagotto and A. Moreo, *Phys. Rev. Lett.* **63**, 2148 (1989).
11. H.J. Schulz and T.A.L. Ziman, *Europhys. Lett.* **18**, 355 (1992); H.J. Schulz, T.A.L. Ziman, and D. Poilblanc, *J. Phys. I* **6**, 675 (1996).
12. J. Richter, *Phys. Rev. B* **47**, 5794 (1993); J. Richter, N.B. Ivanov, and K. Retzlaff, *Europhys. Lett.* **25**, 545 (1994).
13. L. Siurakshina, D. Ihle, and R. Hayn, *Phys. Rev. B* **64**, 104406 (2001).
14. R.F. Bishop, D.J.J. Farnell, and J.B. Parkinson, *Phys. Rev. B* **58**, 6394 (1998).
15. R.R.P. Singh, Zheng Weihong, C.J. Hamer, and J. Oitmaa, *Phys. Rev. B* **60**, 7278 (1999).
16. L. Capriotti and S. Sorella, *Phys. Rev. Lett.* **84**, 3173 (2000).
17. O.P. Sushkov, J. Oitmaa, and Zheng Weihong, *Phys. Rev. B* **63**, 104420 (2001).
18. L. Capriotti, F. Becca, A. Parola, and S. Sorella, *Phys. Rev. Lett.* **87**, 097201 (2001).
19. L. Capriotti, *Int. J. Mod. Phys. B* **15**, 1799 (2001).
20. R.R.P. Singh, Zheng Weihong, J. Oitmaa, O.P. Sushkov, and C.J. Hamer, *Phys. Rev. Lett.* **91**, 017201 (2003).
21. J. Sirker, Zheng Weihong, O.P. Sushkov, and J. Oitmaa, *Phys. Rev. B* **73**, 184420 (2006).
22. B.S. Shastry and B. Sutherland, *Physica B* **108**, 1069 (1981).
23. M. Albrecht and F. Mila, *Europhys. Lett.* **34**, 145 (1996).
24. S. Miyahara and K. Ueda, *Phys. Rev. Lett.* **82**, 3701 (1999).
25. Zheng Weihong, J. Oitmaa, and C.J. Hamer, *Phys. Rev. B* **65**, 014408 (2001).
26. A. Läuchli, S. Wessel, and M. Sigrist, *Phys. Rev. B* **66**, 014401 (2002).
27. Zheng Weihong, C.J. Hamer, and J. Oitmaa, *Phys. Rev. B* **60**, 6608 (1999).
28. E. Müller-Hartmann, R.R.P. Singh, C. Knetter, and G.S. Uhrig, *Phys. Rev. Lett.* **84**, 1808 (2000).
29. A. Koga and N. Kawakami, *Phys. Rev. Lett.* **84**, 4461 (2000).
30. C.H. Chung, J.B. Marston, and S. Sachdev, *Phys. Rev. B* **64**, 134407 (2001).
31. S. Miyahara and K. Ueda, *J. Phys.: Condens. Matter* **15**, R327 (2003).
32. M. Al Hajj, N. Guihéry, J.-P. Malrieu, and B. Bocquillon, *Eur. Phys. J. B* **41**, 11 (2004).
33. R. Darradi, J. Richter, and D.J.J. Farnell, *Phys. Rev. B* **72**, 104425 (2005).
34. B. Canals and C. Lacroix, *Phys. Rev. Lett.* **80**, 2933 (1998).
35. R. Moessner, *Can. J. Phys.* **79**, 1283 (2001).
36. D. Schmalfuß, J. Richter, and D. Ihle, *Phys. Rev. B* **70**, 184412 (2004).
37. M. Enderle, C. Mukherjee, B. Fak, R.K. Kremer, J.-M. Broto, H. Rosner, S.-L. Drechsler, J. Richter, J. Malek, A. Prokofiev, W. Assmus, S. Pujol, J.-L. Raggazoni, H. Rakato, M. Rheinstädter, and H.M. Ronnow, Europhys. Lett. **70**, 237 (2005).
38. T. Masuda, A. Zheludev, A. Bush, M. Markina, and A. Vasiliev, *Phys. Rev. Lett.* **92**, 177201 (2004); S.-L. Drechsler, J. Málek, J. Richter, A.S. Moskvin, A.A. Gippius, and H. Rosner, *Phys. Rev. Lett.* **94**, 039705 (2005).
39. S.-L. Drechsler, J. Richter, J. Málek, A.S. Moskvin, A. Klingeler, and H. Rosner, *J. Magn. Magn. Mater.* **290-291**, 345 (2005).
40. S.-L. Drechsler, J. Richter, A.A. Gippius, A. Vasiliev, A.S. Moskvin, J. Málek, Y. Prots, W. Schnelle, and H. Rosner, Europhys. Lett. **73**, 83 (2006).
41. S. Taniguchi, T. Nishikawa, Y. Yasui, Y. Kobayashi, M. Sato, T. Nishioka, M. Kontani, and K. Sano, J. Phys. Soc. Jpn. **64**, 2758 (1995).

42. H. Kageyama *et al.*, *Phys. Rev. Lett.* **82**, 3168 (1999).
43. R. Melzi, P. Carretta, A. Lascialfari, M. Mambrini, M. Troyer, P. Millet, and F. Mila, *Phys. Rev. Lett.* **85**, 1318 (2000).
44. H. Rosner, R.R.P. Singh, Zheng Weihong, J. Oitmaa, S.-L. Drechsler, and W.E. Pickett, *Phys. Rev. Lett.* **88**, 186405 (2002).
45. C.K. Majumdar and D.K. Gosh, *J. Math. Phys.* **10**, 1388 (1969).
46. J. Schulenburg, A. Honecker, J. Schnack, J. Richter, and H.-J. Schmidt, *Phys. Rev. Lett.* **88**, 167207 (2002); J. Richter, J. Schulenburg, A. Honecker, J. Schnack, and H.-J. Schmidt, *J. Phys.: Condens. Matter* **16**, 779 (2004).
47. F. Coester, *Nucl. Phys.* **7**, 421 (1958); F. Coester and H. Kümmel, *ibid.* **17**, 477 (1960).
48. R.F. Bishop, in *Microscopic Quantum Many-Body Theories and Their Applications*, eds. J. Navarro and A. Polls, Lecture Notes in Physics **510** (Springer-Verlag, Berlin, 1998), p. 1.
49. M. Roger and J.H. Hetherington, *Phys. Rev. B* **41**, 200 (1990).
50. R.F. Bishop, J.B. Parkinson, and Y. Xian, *Phys. Rev. B* **44**, 9425 (1991).
51. R.F. Bishop, R.G. Hale, and Y. Xian, *Phys. Rev. Lett.* **73**, 3157 (1994).
52. R. Bursill, G.A. Gehring, D.J.J. Farnell, J.B. Parkinson, T. Xiang, and C. Zeng, *J. Phys.: Condens. Matter* **7**, 8605 (1995).
53. C. Zeng, D.J.J. Farnell, and R.F. Bishop, *J. Stat. Phys.* **90**, 327 (1998).
54. S.E. Krüger, J. Richter, J. Schulenburg, D.J.J. Farnell, and R.F. Bishop, *Phys. Rev. B* **61**, 14607 (2000).
55. R.F. Bishop, D.J.J. Farnell, S.E. Krüger, J.B. Parkinson, and J. Richter, *J. Phys.: Condens. Matter* **12**, 6887 (2000).
56. S.E. Krüger and J. Richter, *Phys. Rev. B* **64**, 024433 (2001).
57. N.B. Ivanov, J. Richter, and D.J.J. Farnell, *Phys. Rev B* **66**, 014421 (2002).
58. R. Darradi, J. Richter, and S.E. Krüger, *J. Phys.: Condens. Matter* **16**, 2681 (2004).
59. D.J.J. Farnell and R.F. Bishop, in *Quantum Magnetism*, eds. U. Schollwöck, J. Richter, D.J.J. Farnell, and R.F. Bishop, Lecture Notes in Physics **645** (Springer-Verlag, Berlin, 2004), p. 307.
60. D.J.J. Farnell, J. Schulenburg, J. Richter, and K.A. Gernoth, *Phys. Rev. B* **72**, 172408 (2005).
61. S.E. Krüger, R. Darradi, J. Richter, and D.J.J. Farnell, *Phys. Rev. B* **73**, 094404 (2006).
62. D. Schmalfuß, R. Darradi, J. Richter, J. Schulenburg, and D. Ihle, *Phys. Rev. Lett.* **97**, 157201 (2006).
63. For the numerical calculations we use the program package *CCCM* (by D.J.J. Farnell and J. Schulenburg).
64. C.J. Hamer, Zheng Weihong, and P. Arndt, *Phys. Rev. B* **46**, 6276 (1992); C. J. Hamer, Zheng Weihong and J. Oitmaa *Phys. Rev. B* **50**, 6877 (1994).
65. B.S. Shastry and B. Sutherland, *Phys. Rev. Lett.* **47**, 964 (1981).
66. H.A. Bethe, *Z. Phys.* **71**, 205 (1931).
67. M.S. Makivic and H.-Q. Ding, *Phys. Rev. B* **43**, 3562 (1991); J.-K. Kim and M. Troyer, *Phys. Rev. Lett.* **80**, 2705 (1998).
68. J. Oitmaa, C.J. Hamer, and Zheng Weihong, *Phys. Rev. B* **50**, 3877 (1994).
69. S.R. White and I. Affleck, *Phys. Rev. B* **54**, 9862 (1996).
70. A.A. Aligia, C.D. Batista, and F.H.L. Eßler, *Phys. Rev. B* **62**, 3259 (2000).
71. D.V. Dmitriev and V.Ya. Krivnov, *Phys. Rev. B* **73**, 024402 (2006).
72. R. Schmidt, J. Schulenburg, J. Richter, and D.D. Betts, *Phys. Rev. B* **66**, 224406 (2002).
73. J. Oitmaa and Zheng Weihong, *Phys. Rev. B* **69**, 064416 (2004).

VORTICITY AND ANTIVORTICITY IN SUBMICRON FERROMAGNETIC FILMS

HAO WANG, M. YAN* and C. E. CAMPBELL

School of Physics and Astronomy, University of Minnesota,
116 Church St. SE Minneapolis, Mn 55455, USA

The existence of stable or metastable vortex and antivortex states in submicron ferromagnetic films is reviewed. Using standard simulation methods, the dynamics of an antivortex in a permalloy asteroid geometry is explored. As in the case of the vortex in similarly sized permalloy disks and squares, there is a sub-GHz frequency gyrotropic mode of the core of the antivortices, and a rich spin wave spectrum in the multi-GHz range. The spin waves are generated in the simulations by using magnetic pulses similar to those available experimentally, which have been applied to circular and square particles of similar sizes in previous research.

Keywords: Hyperchaotic LC oscillator system; active tracking control; different-structure synchronization.

1. Introduction

Vorticity in ferromagnets is a natural consequence of the role of flux closure in establishing stable and metastable states as energy minima. In certain geometries or with certain boundary conditions, one can also produce antivortex states, as well as multi-vortex arrays and arrays which contain both vortices and antivortices. We report here on recent results on antivortex states and their spin and gyrotropic dynamics.[1–3] By way of background, we review related results for vortex states.[4–16]

A vortex structure in a two-dimensional disk is shown in Fig. 1, where it is seen that it has a winding number of +1. This figure shows only the projection of the magnetic moments into the plane of the disk; but as one nears the core, the magnetization begins to have a component perpendicular to the plane, and is fully perpendicular to the plane at the center of the vortex (which we will refer to below as a polarized core). An antivortex structure in the square crossing section of a cross structure is shown in Fig. 2, where it is seen that it has a winding number of −1. Its core is also a singular structure with magnetization perpendicular to the plane at the center, i.e. it also has a polarized core.[3]

*Present address: Institut für Festkörperforschung IFF-9 "Elektronische Eigenschaften", Forschungszentrum Jölich GmbH, D-52425 Jülich, Germany.

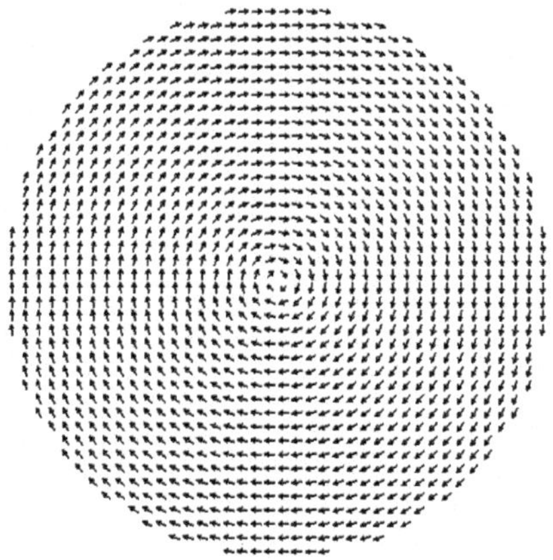

Fig. 1. A magnetic vortex structure in a disk, showing only the components of the magnetic moments that lie in the plane of the disk.

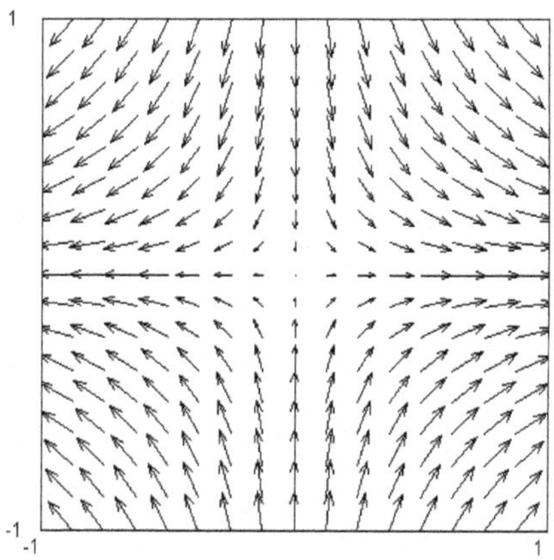

Fig. 2. Square crossing section of a cross shaped quasi-two dimensional ferromagnet in an antivortex structure. Only the in-plane components of the magnetization vectors are shown. Note that, at the center of the sample, the magnetization is perpendicular to the plane. Only a fraction of the magnetic moments is included in this figure and Figs. 3–5. The magnetization in the horizontal arms of the cross points outward but inward in the vertical arms.[3]

Structures such as the planar vortex[17] and the planar antivortex[18] are observed in ferromagnetic films of lateral dimensions order from order 10^2 nm to microns. In thick disks it is possible to find more complex structures, including the appearance of vortex ring structures (similar to smoke rings) in, e.g., nickel cylinders with high perpendicular anisotropy.[19] However in this report we focus on thin films with high isotropy, such as permalloy, which support the planar topological singularity states shown above if the lateral dimensions are sufficiently large.

2. Micromagnetics

The length scale of samples that we are presenting here is the micromagnetic scale, which is intermediate between the microscopic lattice size (angstroms) and the typical size of magnetic domains; this is the scale range of current interest for magnetic information storage. For an accurate description of real magnetic materials, the underlying theory must include both the dipolar interaction (demagnetization energy) and the ferromagnetic exchange interaction. However, for the systems of interest here, the magnetic curvature is on a long enough length scale (greater than the exchange length) so that the Heisenberg exchange contribution to the energy can be replaced by the leading non-trivial term in the cosine of the angle between adjacent moments, leading to a continuum theory that includes a second order gradient term as the representation of the exchange energy.20 Thus the torque equation — or Landau–Lifshitz–Gilbert equation (LLG) — for the local magnetization $\mathbf{M}(r,t)$ is written conveniently in terms of an effective field $\mathbf{H}_{\text{eff}}(r,t)$ as[20]

$$\frac{dM(r,t)}{dt} = -\gamma \mathbf{M}(r,t) \times \mathbf{H}_{\text{eff}}(r,t)\,, \tag{1}$$

where this effective field is the negative functional derivative of the energy of the system with respect to the magnetization. The vector field $H_{\text{eff}}(r,t)$ contains a dipolar field $\mathbf{H}_d$ due to the magnetization of the system at other points, an effective exchange field which is proportional to the Laplacian of the magnetization, an applied field $\mathbf{H}_{\text{ext}}$, and an anisotropy field due to the lattice structure and sample growth conditions:

$$\begin{aligned} H_{\text{eff}} &= H_{\text{ext}} + H_{\text{exch}} + H_d + H_{\text{an}} \\ &= H_{\text{ext}} + \frac{2A}{M_s^2}\nabla^2 M + H_d + H_{\text{an}}\,. \end{aligned} \tag{2}$$

M_S is the saturation magnetization of the material and A is given in terms of the Heisenberg coupling constant J by $A = 2JS^2/a$, where a is the microscopic lattice constant. In the present case where we are considering only permalloy, the anisotropy field can be ignored.

Analytic or semi-analytic solutions of the LLG equation are possible only in the simplest cases. As is more common, we must resort to a discretization and numerical simulation for the systems and situations reported here. The discretization scale must be smaller than the exchange length $l_{\text{ex}} = \sqrt{A/2\pi M_s^2}$ in order to obtain physically meaningful simulations. Including a phenomenological damping parameter α, the discretized LLG equation, Eq. (1), becomes

$$\frac{\partial \mathbf{M}_i}{\partial t} = -\frac{\gamma}{(1+\alpha^2)}\mathbf{M}_i \times \mathbf{H}_{\text{eff},i} - \frac{\alpha\gamma}{M_2(1+\alpha^2)}\mathbf{M}_i(\mathbf{M}_\times \mathbf{H}_{\text{eff},i}) \tag{3}$$

3. Stable and Metastable States

Stable and metastable solutions of these equations may be obtained by choosing an initial magnetic configuration — which may or may not be accomplished with the aid of a formative external magnetic field — and permitting the system to evolve according to Eq. (3) until it reaches a stationary state. This resulting state is largely due to a competition between the exchange energy and the dipolar energy when there is no external field. Smaller systems tend to be essentially single domain states of nearly uniform magnetization, distorted somewhat due to the system shape in order to minimize edge "charges." Interesting examples of winding number +1 states are shown in Figs. 3–5, each of which is the square crossing section of a cross structure with no external field (remanent state).[3] Initializing the system in a vortex configuration, as in Fig. 3, with magnetization pointing outward in both the horizontal and vertical arms of the cross, it evolves into a stable state that

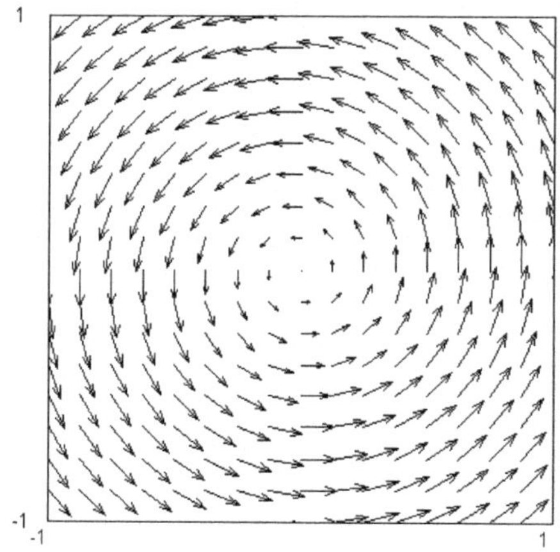

Fig. 3. Square crossing section of a cross shaped quasi-two dimensional ferromagnet in a conventional vortex structure.[3]

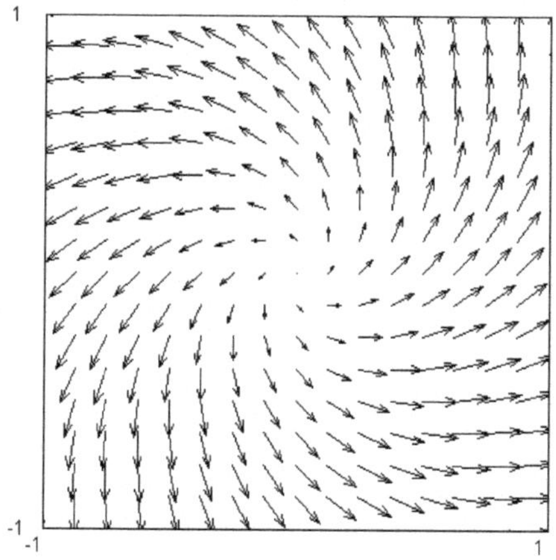

Fig. 4. Square crossing section of a cross shaped quasi-two dimensional ferromagnet in an unconventional vortex structure.[3]

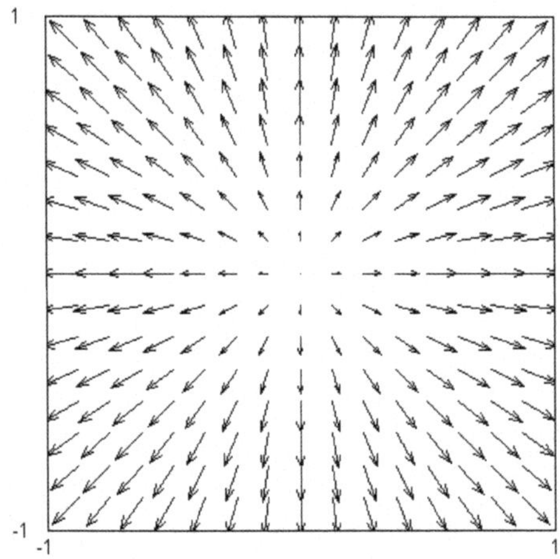

Fig. 5. Square crossing section of a cross shaped quasi-two dimensional ferromagnet in an unconventional vortex structure which has all magnetic moments radially oriented.[3]

retains this winding number, but becomes an "unconventional" vortex in the sense that all of the magnetic moments are rotated relative to the conventional vortex configuration, as shown in Fig. 4 where the magnetization vectors are rotated by an angle of approximately $\pi/4$ with respect to the conventional vortex of Fig. 3, and Fig. 5 where the magnetization vectors are rotated by an angle of approximately $\pi/2$ with respect to the conventional vortex.[3] Figure 5 bears an obvious resemblance to the electric field of an electrical charge, (though it should be kept in mind that this is mostly a two-dimensional structure, whereas the electric field of an isolated electric charge is a three-dimensional, spherically symmetric field).

The degree of rotation of the magnetic moments depends upon the competition between the demagnetization energy and the exchange energy. Indeed, without the demagnetization energy, all of the structures shown in these three figures would have the same (exchange) energy in the absence of an applied magnetic field. The balance is also affected by the choice of material, which in the present case is permalloy, and the size of the system. The conventional vortex structure shown schematically in Fig. 3 was obtained from a simulation for a cross of thickness 10 nm with the square crossing section of side length 200 nm, while the "charge" structure of Fig. 5 was obtained for a much smaller cross, namely one of thickness 1 nm and crossing section of side length 20 nm.[3] As one moves progressively between these extreme sizes, the typical magnetic moment rotates by an increasing angle, with an intermediate size example shown in Fig. (4).[3]

The antivortex structure at the crossing section of a cross shaped particle, shown in Fig. 2, is much more robust than the vortex counter part of Fig. 4 in the sense that it remains very close to a conventional antivortex structure throughout the size range explored; i.e., we found no evidence from simulations for a stable unconventional antivortex state in the cross geometry.[3] This is understood qualitatively by the fact that, if one rotates all of the magnetic moments of an antivortex structure by a common angle, the effect is to essentially rotate the major axes of the antivortex. Thus rotating the in-plane components of all of the moments of the antivortex of Fig. 2 by an angle of $\pi/2$ would result in a "conventional" antivortex with its major axes rotated from the x and y axes to the diagonal axes, while a rotation by π would return the major axes to the x and y directions, though rotating the entire Fig. 2 by $\pi/2$ (or, equivalently, $-\pi/2$). Thus the robustness of the antivortex with the orientation shown in Fig. 2 in the cross structure is due to the fact that the magnetization in the arms of the cross (not shown in the figure) is parallel to the arms, while there would be significant edge charging in the close-in part of the arms if the antivortex axes were diagonal.

In the next section we focus on the dynamics of an antivortex state in an asteroid shaped particle, which is of particular interest because it is, to our knowledge, the only experimentally known single antivortex state;[18] this particle is illustrated schematically in Fig. 6(d). To conclude this section, we consider the stable and metastable states of this particle.[1,2] As noted previously, under some circumstances stable and metastable remanent states are found which are single domain states, and thus have winding number zero. In particular, in our micromagnetic simulations we found four stable or metastable states of a permalloy particle of this shape in the size range studied, with lateral size L (the size of the encompassing square) ranging from 200 nm to 1 micron and thicknesses ranging from 10 to 30 nm. In addition to the antivortex state, we found a vortex state [shown schematically in Fig. 6(c)] and

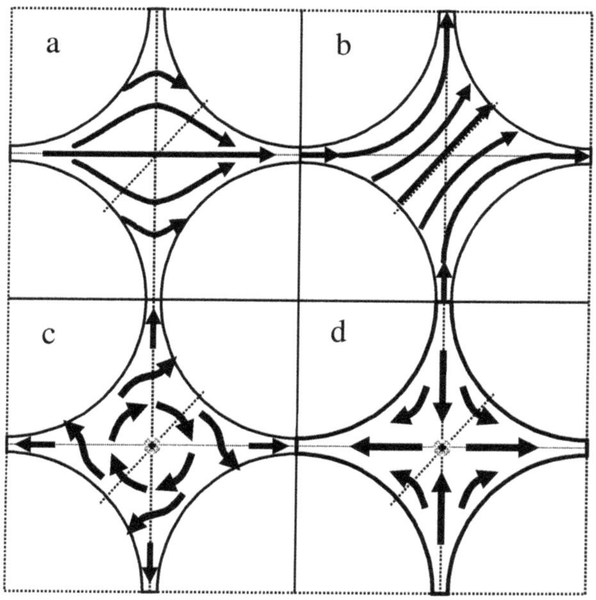

Fig. 6. Metastable states of a permalloy asteroid. Each square (a)–(d) contains a single asteroid. The arrows represent the general direction of the planar components of the magnetic moments.[1,2]

two single domain states, shown in Figs. 6(a) and 6(b). In all cases, the diagonal single domain state similar to Fig. 6(b) was most stable, i.e., had lowest energy. All three of the other states had size ranges where they were the second lowest energy state and the third lowest energy state. The antivortex states were the second most stable at the largest sizes simulated. However, all were metastable throughout this size range.[1,2]

It is interesting to note that the vortex state for a submicron asteroid, as shown in Fig. 6(c) has conventional vortex structure similar to Fig. 3 near the core, while it evolves to the unconventional, charge-like structure of Fig. 5 in the asteroid arms as one moves away from the core.

4. Spin and Gyrotropic Dynamical Modes

Even an infinite sized ferromagnet has a somewhat complicated spin-wave spectrum due to the relative orientation of the spin wave vector to the magnetization or internal magnetic field. E.g., when these two vectors are parallel, the spin wave dispersion has negative group velocity at long wavelengths, and thus is called the backward volume mode (BVM).[21] In the Damon-Eshbach (DE) mode (DEM), where the vectors are perpendicular, the mode has positive group velocity.[22] Intermediate between these two extremes, the long wavelength character of the mode changes continuously from BVM to DEM.[21] Although spin waves in finite systems such as those discussed above have discrete frequencies, it has sometimes been illuminating to interpret the modes in terms of an effective wave vector and its orientation to the local magnetization, which in the simpler cases may permit the mode to be interpreted as "predominantly BVM" or "predominantly DEM." E.g., this has been particularly useful in the analysis of low-frequency spin wave modes in wires,[23–35] disks,[4–14] and squares.[15–16]

The presence of vorticity in the equilibrium state profoundly affects the spin-wave spectrum.[1,3,4–16] Moreover, a new dynamical mode — the gyrotropic mode — appears at the lowest frequencies (less than 1 GHz in the sizes we have discussed), in which the vortex core orbits its equilibrium position, i.e. the center of the particle in the disk vortex state as well as the asteroid vortex and antivortex states discussed above. This mode is generated by first applying an in-plane magnetic field which displaces the core of the vortex from the center of the system, and then returning the applied field to zero. In the presence of damping, as represented in Eq. (3), the vortex core spirals back to its equilibrium position.

The gyrotropic motion couples to the azimuthal spin-wave modes of the vortex state in a circular disk, lifting a chiral degeneracy of the azimuthal modes.[10,11]

Not surprisingly, qualitatively similar behavior was obtained in our simulations of the dynamics of the antivortex state in the asteroid.[1,2] We summarize the results of those simulations in the remainder of this report.

The asteroids in which we simulated the dynamics are the same that we discussed in the last section. To complete their description, the radius of the circle

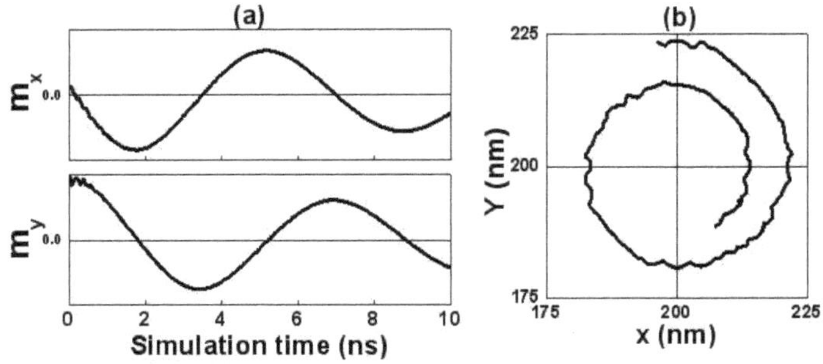

Fig. 7. (a) The time dependence of the x and y magnetization components for the gyrotropic motion of an initially displaced core in remanence in a 400 nm × 20 nm permalloy asteroid. (b) The trajectory of the antivortex core for the same system. In this case, the core polarity is $P = +1$.

sectors excised from the corners of the squares to make the asteroid shape (see Figure 6) was (96/200)L. To simulate permalloy particles, we used material parameters of a saturation magnetization M_s of 800 emu/cc, exchange stiffness A of 1.05 micro ergs per centimeter, and the gyromagnetic ratio γ of 17.6 MHz/Oe. The damping constant α for the simulations reported here was taken to be 0.04 for the gyrotropic modes. Spin-wave dynamics were obtained with a damping constant of 0.005. The discretization cubes were 4 nm or 5 nm, and thus smaller than the exchange length. Considerable previous experience with permalloy simulations has shown this discretization to be sufficient for present purposes.

Figure 7 shows the time dependence of the average of the in-plane components of the dimensionless magnetization $m(r,t) = m(r,t)/M_S$. The frequency of this gyrotropic mode is 0.14 GHz. The gyrotropic motion may be modeled by Thiele's equation[36]

$$G \times \frac{d\mathbf{X}}{dt} - \frac{\partial W(\mathbf{X})}{\partial \mathbf{X}} = 0 \tag{4}$$

where $W(\mathbf{X})$ is a phenomenological potential energy well for the core, X is the displacement of the core from its equilibrium position, and the gyrovector is given by $G = -Gv P\hat{z}$, where the magnitude is $G = 2\pi T M_S/\gamma$, T is the sample thickness, v is negative one for an antivortex and positive one for a vortex, and P is the polarization of the antivortex or vortex core. The clockwise spiral in Fig. 7(b) may be reversed to a counter-clockwise spiral by reversing the core polarization, which changes the sign of P. Note also that the difference in sign of v between vortex and antivortex means that the antivortex spirals in the opposite direction of a vortex with the same core polarization. This is unimportant for present purposes, but may have significant consequences for vortex/antivortex arrays.[2]

Portions of the spin-wave power spectrum are shown in Figs. 8 and 9 for a 400 nm × 20 nm permalloy asteroid. Figure 8 is the response to an in-plane, spatially uniform magnetic field pulse in the $\hat{x}$ direction, which is Gaussian in time with a width of 30 ps and an amplitude of 5 Oe. The gyrotropic mode also appears in

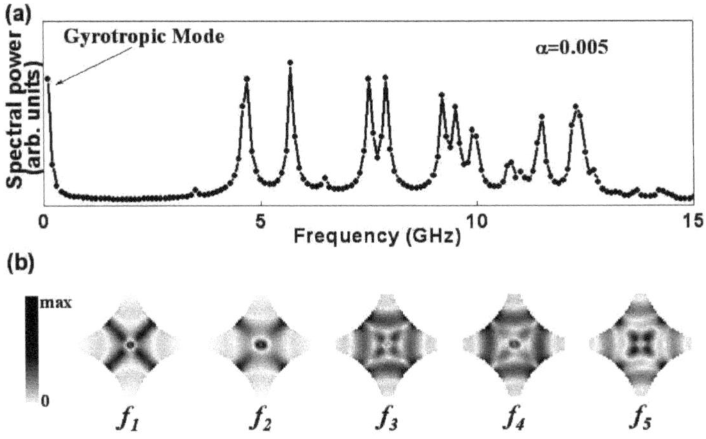

Fig. 8. Spectrum of azimuthal spin wave modes of a 400 × 400 × 20 nm asteroid with core polarization $P = +1$ generated by an in-plane magnetic field pulse described in the text: (a) Power spectrum at resonance for the azimuthal spin wave modes; (b) Spectral amplitudes at resonance for the first five azimuthal spin wave modes at the resonant frequencies of Fig. (a) given in the text.

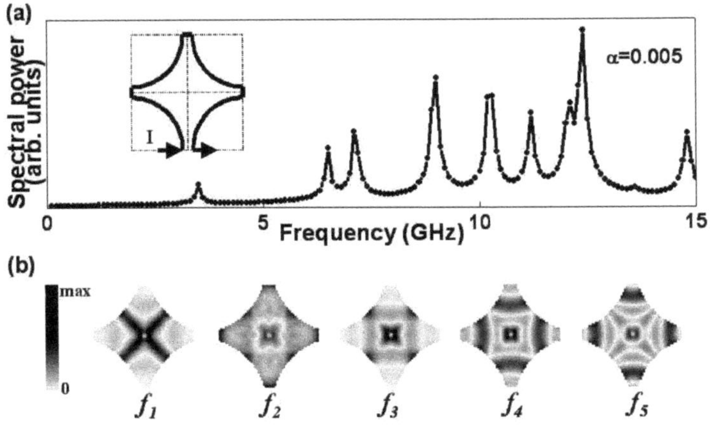

Fig. 9. Spectrum of radial spin wave modes of a 400 × 400 × 20 nm asteroid: (a) Power spectrum at resonance for the radial spin wave modes after excitation by an out-of-plane pulse field from a surrounding current, shown in the inset: (b) Spectral amplitude at resonance for first five radial spin wave modes, with frequencies 3.5, 6.5, 7.1, 9.0, 10.3 GHz, respectively.

this power spectrum as the very low frequency peak, whose peak position cannot be resolved with the limited frequency resolution due to our finite simulation time. The lowest spin-wave resonance generated by this field pulse is at a frequency of $f_1 = 4.7$ GHz. The spectral amplitudes of the lowest five modes are shown in Figure 8(b). Phase plots would show clearly that these are the analogues of the azimuthal modes with effective wave vectors curling around the center, as would be expected from the torque exerted by the in-plane pulse. The amplitudes of the two lowest modes are very similar, although the frequency of the second mode is $f_2 = 5.7$ GHz. This was also a feature of the azimuthal modes in the system of a vortex in a disk geometry where it was shown that these modes would be degenerate were it not coupled to the gyrotropic mode.[10,11] Clearly there is a similar lifting of degeneracy of the third and fourth modes, which have frequencies of 7.5 GHz and 7.9 GHz, respectively, but are otherwise nearly identical. The dynamics of these modes show that, in each case, the phase of the lower frequency mode (not shown) of the nearly degenerate pair sweeps in the clockwise direction and the phase of the higher frequency mode of the pair sweeps in the counter-clockwise direction. This orientation is due to fact that the core polarization is $P = +1$; an identical situation obtains in the case of opposite core polarization except of course the helicity of the modes is reversed. As in the case of the disk-vortex system, inserting a hole at the center of the system, i.e., the equilibrium position of the core, suppresses the gyrotropic mode and the coupling of the azimuthal spin-wave modes to the gyrotropic motion, which restores the degeneracy of each of the pairs of modes.

Low frequency radial spin-wave modes may be generated by a magnetic field pulse perpendicular to the plane of the film. Following the experimental setup in Ref. 14, we simulated a pulse due to an electric current loop around the asteroid sample as shown in the inset in Fig. 9(a). This out-of-plane pulse will not shift the antivortex core away from the center since there is no additional torque exerted on the core. Consequently, the lower frequency resonances in the power spectrum in Fig. 9(a) are predominantly radial modes involving no coupling to the core's gyrotropic motion. The spectral amplitude of each spin wave mode exhibits a symmetry about the origin as shown in the Fig. 9(b). The lowest radial mode is mainly distributed in the "X" shaped region as in the azimuthal mode case. The spectral amplitude of the lowest mode exhibits nodes along the four long arms of the asteroid as seen by the variation of the image brightness in Fig. 9(b), while no nodes are found along the four diagonal directions. The amplitudes of higher frequency radial modes are distributed over the entire asteroid and show more nodes along both the diagonal and the arm directions of the asteroid.

The difference between the spectrum of the long arm direction and the diagonal direction of the asteroid can be ascribed to the different angles between the wave vector and the equilibrium internal magnetic field. For radial modes, the wave vector is perpendicular to the internal field in the diagonal direction, thus having the characteristics of a DE mode.[22] However, along the long arm directions, the wave vector of radial modes is (anti)parallel to the internal field, which has magnetostatic BVM character.[21] Thus, as a result of the unique spin configuration of a magnetic antivortex, both the azimuthal and the radial spin wave modes in the asteroid sample are hybrids of the DEM and BVM modes.

In conclusion, a rich spin-wave dynamics was found in a single antivortex metastable state in a submicron, asteroid shaped, permalloy particle. The similarity to the results for similar sized vortex states suggests a rich possibility of modes in vortex-antivortex arrays. An example of such an array is the cross-tie domain wall that separates a pair of anti-parallel domains of a ferromagnet.[37] Even more interesting is the possibility of collective core modes at low frequencies due to possible correlations between the cores of these topological charges.[2]

Acknowledgments

We are pleased to acknowledge numerous helpful discussions and collaborations with C. Bayer, P. A. Crowell, E. D. Dahlberg, P. Eames, A. B. Kunz, J. Park and G. Skidmore on the research reported in this paper and related research. This work was supported in part by the Office of Naval Research Grant No. ONR N/N00014-02-1-0815. One of the authors (CEC) is pleased to acknowledge a partial travel support from the U.S. Army Research office. The views, opinions, and/or findings contained in this article, are those of the authors and should not be construed as an official Department of the Army position, policy, or decision, unless so designated by other documentation.

References

1. H. Wang and C. E. Campbell, cond-mat/0608245 (2006).
2. H. Wang, University of Minnesota PhD Dissertation (2006).
3. M. Yan, University of Minnesota PhD Dissertation (2004).
4. K. Yu. Guslienko, B. A. Ivanov, V. Novosad, Y. Otani, H. Shima and K. Fukamichi, *J. Appl. Phys.* **91**, 8037 (2002).
5. J. P. Park, P. Eames, D. M. Engebretson, J. Berezovsky and P. A. Crowell, *Phys. Rev. B* **67**, 020403(R) (2003).
6. M. Buess, R. Hllinger, T. Haug, K. Perzlmaier, U. Krey, D. Pescia, M. R. Scheinfein, D. Weiss and C. H. Back, *Phys. Rev. Lett.* **93**, 077207 (2004).
7. R. Zivieri and F. Nizzoli, *Phys. Rev. B* **71**, 014411 (2005).
8. B. A. Ivanov and C. E. Zaspel, *Phys. Rev. Lett.* 94, 027205 (2005).
9. M. Buess, T. Haug, M. R. Scheinfein and C. H. Back, *Phys. Rev. Lett.* **94**, 127205 (2005).
10. X. Zhu, Z. Liu, V. Metlushko, P. Grütter and M. R. Freeman, *Phys. Rev. B* **71**, 180408(R) (2005).
11. J. P. Park and P. A. Crowell, *Phys. Rev. Lett.* **95**, 167201 (2005).
12. B. E. Argyle, E. Terrenzio, and J. C. Slonczewski, *Phys. Rev. Lett.* **53**, 190 (1984).
13. S.-B. Choe, Y. Acremann, A. Scholl, A. Bauer, A. Doran, J. Stöhr and H. A. Padmore, *Science* **304**, 420 (2004).
14. K. Perzlmaier, M. Buess, C. H. Back, V. E. Demidov, B. Hillebrands and S. O. Demokritov, *Phys. Rev. Lett.* **94**, 057202 (2005).
15. M. Yan, G. Leaf, H. Kaper, R. Camley, and M. Grimsditch, *Phys. Rev. B* **73**, 014425 (2006).
16. M. Bolte, G. Meier, and C. Bayer, *Phys. Rev. B* **73**, 52406 (2006).
17. T. Shinjo, T. Okuno, R. Hassdorf, K. Shigeto and T. Ono, *Science* **289**, 930 (2000).
18. K. Shigeto, T. Okuno, K. Mibu, T. Shinjo and T. Ono, *Appl. Phys. Lett.* **80**, 4190 (2002).
19. G. D. Skidmore, A. Kunz, C. E. Campbell and E. D. Dahlberg, *Phys. Rev. B* **70**, 012410 (2004).
20. W. F. Brown, Jr., *Micromagnetics* (Interscience Publishers, New York, 1963); A. H. Morrish, *The Physical Principles of Magnetism* (IEEE, Inc., New York, 2001); R.

C. Handley, *Modern Magnetic Materials Principles and Applications* (John Wiley & Sons, Inc., New York, 2000).
21. B. A. Kalinikos and A. N. Slavin, *J. Phys. C* **19**, 7013 (1986).
22. R. W. Damon and J. R. Eshbach, *J. Phys. Chem. Solids* **19**, 308 (1961).
23. J. Jorzick, S. O. Demokritov, B. Hillebrands, D. Berkov, N. L. Gorn, K. Guslienko and A. N. Slavin, *Phys. Rev. Lett.* **88**, 047204 (2002).
24. J. P. Park, P. Eames, D. M. Engebretson, J. Berezovsky and P. A. Crowell, *Phys. Rev. Lett.* **89**, 277201 (2002).
25. C. Bayer, S. O. Demokritov, B. Hillebrands and A. N. Slavin, *Appl. Phys. Lett.* **82**, 607 (2003).
26. C. Bayer, J. P. Park, H. Wang, M. Yan, C. E. Campbell and P. A. Crowell, *Phys. Rev. B* **69**, 134401 (2004).
27. C. Bayer, J. Jorzick, B. Hillebrands, S. O. Demokritov, R. Kouba, R. Bozinoski, A. N. Slavin, K. Y. Guslienko, D. V. Berkov, N. L. Gorn and M. P. Kostylev, *Phys. Rev. B* **72**, 064427 (2005).
28. M. Grimsditch, G. K. Leaf, H. G. Kaper, D. A. Karpeev and R. E. Camley, *Phys. Rev. B* **69**, 174428 (2004).
29. M. Grimsditch, L. Giovannini, F. Montoncello, F. Nizzoli, G. K. Leaf and H. G. Kaper, *Phys. Rev. B* **70**, 054409 (2004).
30. K. Yu. Guslienko, S. O. Demokritov, B. Hillebrands and A. N. Slavin, *Phys. Rev. B* **66**, 132402 (2002).
31. S. O. Demokritov, *J. Phys. Cond. Matt.* **15**, s2575 (2003).
32. R. Skomski, M. Chipara and D. J. Sellmyer, *J. Appl. Phys.* **93**, 7604 (2003).
33. Y. Roussigne, S. M. Cherif and P. Moch, *J. Magn. Magn. Mater.* **263**, 289 (2003); and **268**, 89 (2004).
34. M. Bailleul, D. Olligs and C. Fermon, *Phys. Rev. Lett.* **91**, 137204 (2003).
35. M. Yan, H. Wang, P. A. Crowell, C. E. Campbell and C. Bayer, *Condensed Matter Theories* **20**, 251 (2006).
36. A. A. Thiele, *Phys. Rev. Lett.* **30**, 230 (1973); D. L. Huber, *Phys. Rev. B* **26**, 3758 (1982).
37. K. L. Metlov, arXiv:cond-mat/0012146, (2002).

Part E

Superconductivity

D-WAVE CHECKERBOARD BOSE CONDENSATE OF MOBILE BIPOLARONS

A. S. ALEXANDROV

Physics Department, Loughborough University,
Loughborough, LE11 3TU, United Kingdom
E-mail: a.s.alexandrov@lboro.ac.uk
www.lboro.ac.uk/departments/ph/staff/asa.html

Strong finite-range electron-phonon interaction (EPI) bounds carriers into small mobile bipolarons. The Bogoliubov-de Gennes (BdG) equations are formulated and solved in the strong-coupling limit, when small bipolarons bose-condense with finite center-of-mass momentum, $\mathbf{K} \neq 0$. There are two energy scales in this regime, a temperature independent incoherent gap Δ_p and a temperature dependent coherent gap $\Delta_c(T, \mathbf{R})$, modulated in the real space, $\mathbf{R}$. The order parameter has d-wave symmetry in $\mathbf{R}$-space and the single-particle density of states (DOS) reveals checkerboard modulations similar to the checkerboard tunnelling DOS observed in cuprates.

Keywords: Bipolarons; Bose condensate; d-wave; checkerboard; cuprates.

1. Mobile Bipolarons in High-Temperature Superconductors

While at present no consensus exists concerning the microscopic mechanism of high-temperature superconductivity there is substantial experimental evidence for (bi)polaronic carries in high-temperature superconductors from isotope effects on the carrier mass[1] (predicted for (bi)polaronic conductors in Ref. 2), high resolution angle-resolved photoemission spectroscopies,[3] and also from optical (see, for example[4–9] and references therein), neutron scattering,[10] and magneto-transport measurements.[11–13] These experiments show that EPI is very strong in high temperature superconductors, and acting together with the strong Coulomb repulsion (Hubbard U) it could rise to small but yet mobile bipolarons.[14–16]

A parameter-free estimate of the Fermi energy[17] clearly supports the real-space (i.e individual) pairing in cuprate superconductors. The band structure of cuprates is quasi-two-dimensional. Applying the parabolic approximation for the band dispersion one readily estimates the Fermi energy in 2D as $\epsilon_F = d/4e^2\lambda_H^2$, where d is the distance between copper-oxygen planes, $\lambda_H^{-2} = 4\pi e^2 n/m^*$ is the in-plane magnetic-field penetration depth at low temperatures, and n, m^* are the density of hole polarons and their effective mass, respectively (we take $\hbar = c = 1$). Here one assumes that the 'superfluid' density at zero temperature is about the same as the normal state density of holes, as it must be in any clean superfluid. The low-temperature penetration depth is unusually large, $\lambda_H \gtrsim 150$ nm, in cuprates, so that the renormalised Fermi energy turns out to be surprisingly low, $\epsilon_F \lesssim 100$ meV, rendering the Migdal-Eliashberg theory inadequate. In fact, ϵ_F is so small that the individual pairing is very likely. Such pairing will occur when the size of the pair, $r_b = 1/\sqrt{m^*\Delta}$ is small compared with the inter-pair separation, $r = \sqrt{\pi/m^*\epsilon_F}$, so that the condition for real-space pairing is $\epsilon_F \lesssim \pi\Delta$. There is strong experimental evidence for a gap in the normal-state electron density of states of cuprates,[11] which is known as the pseudogap. Experimentally measured pseudogaps of many cuprates are about 50 meV or larger. If one accepts that the pseudogap is about half of the pair binding energy[18] then the condition for real-space pairing is well satisfied in most cuprates (typically $r_b \approx 0.2 - 0.4$ nm).

Here I discuss the $d - wave$ symmetry of the superconducting order parameter and the checkerboard real-space DOS modulations in cuprates in the framework of the bipolaron theory (see also[13]).

2. D-Wave Bose-Einstein Condensation

Many independent observations show that the superconducting state of cuprates is as anomalous as the normal state. In particular, there is strong evidence for a d-like order parameter, which changes sign when the CuO_2 plane is rotated by $\pi/2$.[19] The phase-sensitive experiments[20] provide unambiguous evidence in this direction. The d-wave gap may appear in the two-dimensional Hubbard model near half filling, as suggested by Scalapino, Loh, and Hirsch[21] in the framework of the BCS theory concurrently with the discovery of novel superconductors. On the other hand, high-precision magnetic measurements,[22] photo-excited quasi-particle relaxation dynamics[23] and some other experiments[24] support more conventional anisotropic s-like gap.

In fact, as mentioned above, there are stronger deviations from the canonical Fermi/BCS-picture than the gap symmetry. In particular, tunnelling and photoemission gaps are almost temperature independent in the superconducting state[25,26] and survive in the normal state well above T_c.[25,27,28] Their maximum magnitudes are often several times larger than expected in the weak and intermediate-coupling[29] BCS theory. Kinetic[30] and thermodynamic[31] data suggest that the gap opens both in charge and spin channels in a wide range of doping. A plausible explanation is that the normal state(pseudo)gap, Δ_p, is half of the bipolaron binding energy,[18] although alternative models have been proposed.[32]

Further studies of the gap function revealed even more complicated physics. Andreev reflection experiments, in which an incoming electron from the normal side of a normal/superconducting contact is reflected as a hole along the same trajectory,[33] showed a much smaller gap edge than the bias at the tunnelling conductance maxima.[34] Two distinctly different gaps with different magnetic field and temperature dependence were observed in the c-axis I(V) characteristics.[35] They were also observed with the femtosecond time-resolved optical spectroscopy.[36] More recent STM experiments revealed checkerboard spatial modulations of the tunnelling DOS, with[37] and without[38,39] applied magnetic fields.

We have proposed a simple model[40] explaining two different gaps in cuprates. The main assumption, supported by the parameter-free estimate of the Fermi energy (see above) is that the attractive potential is large compared with the renormalised Fermi energy, so that the ground state is the Bose-Einstein condensate of real-space small bipolarons. The anomalous Bogoliubov-Gor'kov average

$$F_{ss'}(\mathbf{r}_1, \mathbf{r}_2) = \left\langle \hat{\Psi}_s(\mathbf{r}_1)\hat{\Psi}_{s'}(\mathbf{r}_2)\right\rangle,$$

is the superconducting order parameter in this strong-coupling regime as in the weak-coupling BCS theory. It depends on the relative coordinate $\rho = \mathbf{r}_1 - \mathbf{r}_2$ of two electrons (holes), described by field operators $\hat{\Psi}_s(\mathbf{r})$, and on the center-of-mass coordinate $\mathbf{R} = (\mathbf{r}_1 + \mathbf{r}_2)/2$. Its Fourier transform, $f(\mathbf{k}, \mathbf{K})$, depends on the relative momentum $\mathbf{k}$ and on the center-of-mass momentum $\mathbf{K}$. In the BCS theory $\mathbf{K} = 0$, and the Fourier transform of the order parameter is proportional to the gap in the quasi-particle excitation spectrum, $f(\mathbf{k}, \mathbf{K}) \propto \Delta_{\mathbf{k}}$. Hence the symmetry of the order parameter and the symmetry of the gap are identical in the weak-coupling regime. Under the rotation of the coordinate system $\Delta_{\mathbf{k}}$ changes its sign, if the Cooper pairing appears in the d-channel. The Cooper pairing might also take place with a finite center-of-mass momentum, if electrons are spin polarized.[41]

However, in the strong-coupling regime the symmetry of the order parameter can be different from the 'internal' symmetry of the pair wave-function, and from the symmetry of a single-particle excitation gap.[42] In principle real-space pairs could have the unconventional *internal* symmetry due to an unconventional symmetry of the pairing potential as in the case of the Cooper pairs.[21] The d-wave symmetry of the ground state could be also due to a topological degeneracy of inter-site pairs on a square lattice.[43] But in any case the ground state and DOS are homogeneous, if pairs are condensed with $\mathbf{K} = 0$. However, if a pair band dispersion has its minima at finite $\mathbf{K}$ in the center-of-mass Brillouin zone, the Bose condensate has the unconventional symmetry even when the internal wave-function of the pair is s-wave.

As an example, we consider the *apex* bipolaron tunnelling in cuprates.[14] High-T_c oxides are doped charged-transfer ionic insulators with narrow electron bands. Therefore, the interaction between holes can be analyzed minimizing the ground state energy of an ionic insulator with two holes, the lattice deformations and the Coulomb repulsion fully taken into account, but neglecting the kinetic energy terms. Using these techniques net inter-site interactions of the in-plane oxygen hole with the *apex* hole, Fig.1, and of two in-plane oxygen holes, were found to be attractive in La_2CuO_4[44] with the binding energies $\Delta = 119meV$ and $\Delta = 60meV$, respectively. All other interactions were found to be repulsive.

Both apex and in-plane inter-site bipolarons can tunnel from one unit cell to another via a *single*-polaron tunnelling from one oxygen ion to its neighbor in case of the apex bipolaron,[14] Fig.1, or via the next-neighbor hopping in case of the

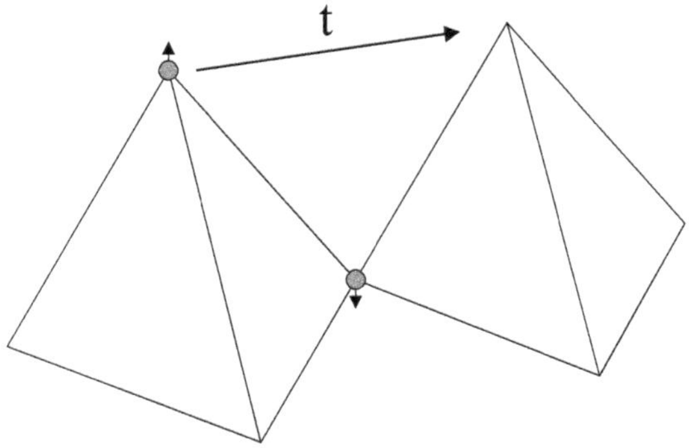

Fig. 1. Apex bipolaron tunnelling in perovskites (after[14]).

in-plane bipolaron.[15] Analytical[14,15] and numerical (QMC)[16] studies suggest that these intersite bipolarons are small but at the same time are quite mobile. The Bloch bands of these bipolarons are obtained using the polaronic canonical transformation, and projecting the transformed Hamiltonian onto a reduced Hilbert space of intersite pairs (see[12] for details). The wave function of the apex bipolaron localized, say in cell $\mathbf{m}$ is written as

$$|\mathbf{m}\rangle = \sum_{i=1}^{4} A_i c_i^\dagger c_{apex}^\dagger |0\rangle, \tag{1}$$

where i denotes $p_{x,y}$ orbitals and spins of the four plane oxygen ions in the cell, Fig.(1), and $c_{apex}^\dagger$ is the creation operator for the hole in one of the three apex oxygen orbitals with the spin, which is the same or opposite to the spin of the in-plane hole depending on the total spin of the bipolaron. The probability amplitudes A_i are normalized by the condition $|A_i| = 1/2$, if four plane orbitals p_{x1}, p_{y2}, p_{x3} and p_{y4} are involved, or by $|A_i| = 1/\sqrt{2}$ if only two of them are relevant. Then a matrix element of the transformed Hamiltonian describing the *bipolaron* tunnelling to the nearest neighbor cell $\mathbf{m}+\mathbf{a}$ is found as

$$t_b = \langle \mathbf{m}|\tilde{H}|\mathbf{m}+\mathbf{a}\rangle = |A_i|^2 t_{pp'}^{apex}, \tag{2}$$

where $t_{pp'}^{apex}$ is a *single polaron* hopping integral between two apex ions. The inter-site bipolaron tunnelling appears already in *the first order* with respect to the single-hole transfer $t_{pp'}^{apex}$, and the bipolaron energy spectrum consists of two subbands $E^{x,y}(\mathbf{K})$, formed by the overlap of p_x and p_y *apex* oxygen orbitals, respectively (here we take the lattice constant $a = 1$),

$$\begin{aligned} E^x(\mathbf{K}) &= t\cos(K_x) - t'\cos(K_y), \\ E^y(\mathbf{K}) &= -t'\cos(K_x) + t\cos(K_y). \end{aligned} \tag{3}$$

Fig. 2. D-wave Bose-Einstein condensate wave-function in the real (Wannier) space. The order parameter is zero in the blank cells and it has different signs in dark and grey cells.

They transform into one another under $\pi/2$ rotation. If $t, t' > 0$, "x" bipolaron band has its minima at $\mathbf{K} = (\pm\pi, 0)$ and y-band at $\mathbf{K} = (0, \pm\pi)$. In these equations t is the renormalized *single-particle* hopping integral between p orbitals of the same symmetry elongated in the direction of the hopping ($pp\sigma$) and t' is the renormalized hopping integral in the perpendicular direction ($pp\pi$). Their ratio $t/t' = 4$ as follows from the tables of hopping integrals in solids. Two different bands are not mixed because $t^{apex}_{p_x,p'_y} = 0$ for the nearest neighbors. A random potential does not mix them either, if it varies smoothly on the lattice scale. Hence, we can distinguish 'x' and 'y' bipolarons with a lighter effective mass in x or y direction, respectively. The apex z bipolaron, if formed, is *ca.* four times less mobile than x and y bipolarons. The bipolaron bandwidth is *of the same order* as the polaron one, which is a specific feature of the inter-site bipolaron. For a large part of the Brillouin zone near $(0, \pi)$ for 'x' and $(\pi, 0)$ for 'y' bipolarons, one can adopt the effective mass approximation

$$E^{x,y}(\mathbf{K}) = \frac{K_x^2}{2m^{**}_{x,y}} + \frac{K_y^2}{2m^{**}_{y,x}} \tag{4}$$

with $K_{x,y}$ taken relative to the band bottom positions and $m^{**}_x = 1/t$, $m^{**}_y = 4m^{**}_x$.

X and y bipolarons bose-condense at the boundaries of the center-of-mass Brillouin zone with $K = (\pm\pi, 0)$ and $K = (0, \pm\pi)$, respectively. These four states are degenerate, so that the condensate wave function $\psi(\mathbf{m})$ in the real (Wannier) space,

$\mathbf{m} = (m_x, m_y)$, is their superposition,

$$\psi(\mathbf{m}) = \sum_{\mathbf{K}=(\pm\pi,0),(0,\pm\pi)} b_{\mathbf{K}} e^{-i\mathbf{K}\cdot\mathbf{m}}, \tag{5}$$

where $b_{\mathbf{K}} = \pm\sqrt{n_c}/2$ are c-numbers, and $n_c(T)$ is the atomic density of the Bose-condensate. The superposition, Eq.(5), respects the time-reversal and parity symmetries, if

$$\psi(\mathbf{m}) = \sqrt{n_c}\left[\cos(\pi m_x) \pm \cos(\pi m_y)\right]. \tag{6}$$

Two order parameters, Eq.(6), are physically identical because they are related by the translation transformation. They have d-wave symmetry changing sign in the real space, when the lattice is rotated by $\pi/2$. This symmetry is entirely due to the center-of-mass-band energy dispersion with four minima at $\mathbf{K} \neq 0$, rather than due a specific pairing potential. It reveals itself as a *checkerboard* modulation of the hole density with two-dimensional patterns, oriented along the diagonals, Fig.(2). From this insight one can expect a fundamental connection between real-space checkerboard structures and the symmetry of the order parameter in cuprates.[42]

3. Checkerboard DOS

Now let us take into account that in the superconducting state ($T < T_c$) single-particle excitations interact with the pair condensate via the same short-range attractive potential, which forms the pairs.[40] The Hamiltonian describing the interaction of excitations with the pair Bose-condensate in the Wannier representation is

$$H = -\sum_{s,\mathbf{m},\mathbf{n}} [t(\mathbf{m}-\mathbf{n}) + \mu\delta_{\mathbf{m},\mathbf{n}}] c^{\dagger}_{s\mathbf{m}} c_{s\mathbf{n}} + \sum_{\mathbf{m}} [\Delta(\mathbf{m}) c^{\dagger}_{\uparrow\mathbf{m}} c^{\dagger}_{\downarrow\mathbf{m}} + H.c.], \tag{7}$$

where $s = \uparrow, \downarrow$ is the spin, $t(\mathbf{m})$ and μ are hopping integrals and the chemical potential, respectively, $c^{\dagger}_{s\mathbf{m}}$ and $c_{s\mathbf{m}}$ create (annihilate) an electron or hole at site $\mathbf{m}$, and $\Delta(\mathbf{m}) \propto \psi(\mathbf{m})$. Applying equations of motion for the Heisenberg operators $\tilde{c}^{\dagger}_{s\mathbf{m}}(t)$ and $\tilde{c}_{s\mathbf{m}}(t)$, and the Bogoliubov transformation[45]

$$\tilde{c}_{\uparrow\mathbf{m}}(t) = \sum_{\nu} [u_{\nu}(\mathbf{m})\alpha_{\nu} e^{-i\epsilon_{\nu}t} + v^*_{\nu}(\mathbf{m})\beta^{\dagger}_{\nu} e^{i\epsilon_{\nu}t}], \tag{8}$$

$$\tilde{c}_{\downarrow\mathbf{m}}(t) = \sum_{\nu} [u_{\nu}(\mathbf{m})\beta_{\nu} e^{-i\epsilon_{\nu}t} - v^*_{\nu}(\mathbf{m})\alpha^{\dagger}_{\nu} e^{i\epsilon_{\nu}t}], \tag{9}$$

one obtains BdG equations describing the single-particle excitation spectrum,

$$\epsilon u(\mathbf{m}) = -\sum_{\mathbf{n}} [t(\mathbf{m}-\mathbf{n}) + \mu\delta_{\mathbf{m},\mathbf{n}}] u(\mathbf{n}) + \Delta(\mathbf{m}) v(\mathbf{m}), \tag{10}$$

$$-\epsilon v(\mathbf{m}) = -\sum_{\mathbf{n}} [t(\mathbf{m}-\mathbf{n}) + \mu\delta_{\mathbf{m},\mathbf{n}}] v(\mathbf{n}) + \Delta(\mathbf{m}) u(\mathbf{m}), \tag{11}$$

where excitation quantum numbers ν are omitted for transparency. These equations are supplemented by the sum rule $\sum_\nu [u_\nu(\mathbf{m})u^*_\nu(\mathbf{n}) + v_\nu(\mathbf{m})v^*_\nu(\mathbf{n})] = \delta_{\mathbf{m},\mathbf{n}}$, which provides the Fermi statistics of single particle excitations α and β. Different from the conventional BdG equations in the weak-coupling limit, there is virtually no feedback of single particle excitations on the off-diagonal potential, $\Delta(\mathbf{m})$, in the strong-coupling regime. The number of these excitations is low at temperatures below Δ_p/k_B, so that the coherent potential $\Delta(\mathbf{m})$ is an external (rather than a self-consistent) field, solely determined by the pair Bose condensate.[40]

While the analytical solution is not possible for any arbitrary off-diagonal interaction $\Delta(\mathbf{m})$, one can readily solve the infinite system of discrete equations (10,11) for a periodic $\Delta(\mathbf{m})$ with a period commensurate with the lattice constant, for example

$$\Delta(\mathbf{m}) = \Delta_c[e^{i\pi m_x} - e^{i\pi m_y}], \tag{12}$$

which corresponds to the pair condensate at $\mathbf{K} = (\pm\pi, 0)$ and $(0, \pm\pi)$, Eq.(6), with a temperature dependent (coherent) $\Delta_c \propto \sqrt{n_c(T)}$. In this case the quasi-momentum $\mathbf{k}$ is the proper quantum number, $\nu = \mathbf{k}$, and the excitation wave-function is a superposition of plane waves,

$$u_\nu(\mathbf{m}) = u_{\mathbf{k}}e^{i\mathbf{k}\cdot\mathbf{m}} + \tilde{u}_{\mathbf{k}}e^{i(\mathbf{k}-\mathbf{g})\cdot\mathbf{m}}, \tag{13}$$

$$v_\nu(\mathbf{m}) = v_{\mathbf{k}}e^{i(\mathbf{k}-\mathbf{g}_x)\cdot\mathbf{m}} + \tilde{v}_{\mathbf{k}}e^{i(\mathbf{k}-\mathbf{g}_y)\cdot\mathbf{m}}. \tag{14}$$

Here $\mathbf{g}_x = (\pi, 0)$, $\mathbf{g}_y = (0, \pi)$, and $\mathbf{g} = (\pi, \pi)$ are reciprocal doubled lattice vectors. Substituting Eqs.(13) and (14) into Eqs.(10,11) one obtains four coupled algebraic equations,

$$\epsilon_{\mathbf{k}}u_{\mathbf{k}} = \xi_{\mathbf{k}}u_{\mathbf{k}} - \Delta_c(v_{\mathbf{k}} - \tilde{v}_{\mathbf{k}}), \tag{15}$$

$$\epsilon_{\mathbf{k}}\tilde{u}_{\mathbf{k}} = \xi_{\mathbf{k}-\mathbf{g}}\tilde{u}_{\mathbf{k}} + \Delta_c(v_{\mathbf{k}} - \tilde{v}_{\mathbf{k}}), \tag{16}$$

$$-\epsilon_{\mathbf{k}}v_{\mathbf{k}} = \xi_{\mathbf{k}-\mathbf{g}_x}v_{\mathbf{k}} + \Delta_c(u_{\mathbf{k}} - \tilde{u}_{\mathbf{k}}), \tag{17}$$

$$-\epsilon_{\mathbf{k}}\tilde{v}_{\mathbf{k}} = \xi_{\mathbf{k}-\mathbf{g}_y}\tilde{v}_{\mathbf{k}} - \Delta_c(u_{\mathbf{k}} - \tilde{u}_{\mathbf{k}}), \tag{18}$$

where $\xi_{\mathbf{k}} = -\sum_{\mathbf{n}} t(\mathbf{n})e^{i\mathbf{k}\cdot\mathbf{n}} - \mu$. The determinant of the system (15-18) yields the following equation for the energy spectrum ϵ:

$$\begin{aligned}(\epsilon - \xi_{\mathbf{k}})(\epsilon - \xi_{\mathbf{k}-\mathbf{g}})(\epsilon + \xi_{\mathbf{k}-\mathbf{g}_x})(\epsilon + \xi_{\mathbf{k}-\mathbf{g}_y})\\ = \Delta_c^2(2\epsilon + \xi_{\mathbf{k}-\mathbf{g}_x} + \xi_{\mathbf{k}-\mathbf{g}_y})(2\epsilon - \xi_{\mathbf{k}} - \xi_{\mathbf{k}-\mathbf{g}}).\end{aligned} \tag{19}$$

Two positive roots for ϵ describe the single-particle excitation spectrum. Their calculation is rather cumbersome, but not in the extreme strong-coupling limit, where the pair binding energy $2\Delta_p$ is large compared with Δ_c and with the single-particle bandwidth w. The chemical potential in this limit is pinned below a single-particle band edge, so that $\mu = -(\Delta_p + w/2)$ is negative, and its magnitude is large compared with Δ_c. Then the right hand side in Eq.(19) is a perturbation, and the spectrum (for dispersion curves see Ref. 13) is

$$\epsilon_{1\mathbf{k}} \approx \xi_{\mathbf{k}} - \frac{\Delta_c^2}{\mu}, \tag{20}$$

$$\epsilon_{2\mathbf{k}} \approx \xi_{\mathbf{k}-\mathbf{g}} - \frac{\Delta_c^2}{\mu}. \tag{21}$$

If a metallic tip is placed at the point $\mathbf{m}$ above the surface of a sample, the STM current $I(V, \mathbf{m})$ creates an electron (or hole) at this point. Applying the Fermi-Dirac golden rule and the Bogoliubov transformation, Eq.(8,9), and assuming that the temperature is much lower than Δ_p/k_B one readily obtains the tunnelling conductance

$$\sigma(V, \mathbf{m}) \equiv \frac{dI(V, \mathbf{m})}{dV} \propto \sum_{\nu} |u_\nu(\mathbf{m})|^2 \delta(eV - \epsilon_\nu), \tag{22}$$

which is a local excitation DOS. The solution Eq.(13) leads to a spatially modulated conductance,

$$\sigma(V, \mathbf{m}) = \sigma_{reg}(V) + \sigma_{mod}(V)\cos(\pi m_x + \pi m_y). \tag{23}$$

The smooth (regular) contribution is

$$\sigma_{reg}(V) = \sigma_0 \sum_{\mathbf{k}, r=1,2} (u_{r\mathbf{k}}^2 + \tilde{u}_{r\mathbf{k}}^2)\delta(eV - \epsilon_{r\mathbf{k}}), \tag{24}$$

and the amplitude of the modulated contribution is

$$\sigma_{mod}(V) = 2\sigma_0 \sum_{\mathbf{k}, r=1,2} u_{r\mathbf{k}}\tilde{u}_{r\mathbf{k}}\delta(eV - \epsilon_{r\mathbf{k}}), \tag{25}$$

where σ_0 is a constant. Conductance modulations reveal a checkerboard pattern, as the Bose condensate itself, Eq.(6),

$$\frac{\sigma - \sigma_{reg}}{\sigma_{reg}} = A\cos(\pi m_x + \pi m_y), \tag{26}$$

where

$$A = 2\sum_{\mathbf{k}} \left[u_{1\mathbf{k}}\tilde{u}_{1\mathbf{k}}\delta(eV - \epsilon_{1\mathbf{k}}) + u_{2\mathbf{k}}\tilde{u}_{2\mathbf{k}}\delta(eV - \epsilon_{2\mathbf{k}})\right] /$$
$$\sum_{\mathbf{k}} \left[(u_{1\mathbf{k}}^2 + \tilde{u}_{1\mathbf{k}}^2)\delta(eV - \epsilon_{1\mathbf{k}}) + (\tilde{u}_{2\mathbf{k}}^2 + u_{2\mathbf{k}}^2)\delta(eV - \epsilon_{2\mathbf{k}})\right]$$

is the amplitude of modulations depending on the voltage V and temperature.

An analytical result can be obtained in the strong-coupling limit with the excitation spectrum, Eqs. (20,21), for the voltage near the threshold, $eV \approx \Delta_p$. In this case only states near bottoms of each excitation band contribute to the integrals in Eq.(26), so that

$$\tilde{u}_{1\mathbf{k}} = \frac{\xi_{\mathbf{k}} - \epsilon_{1\mathbf{k}}}{\epsilon_{1\mathbf{k}} - \xi_{\mathbf{k}-\mathbf{g}}} u_{1\mathbf{k}} \approx -u_{1\mathbf{k}}\frac{\Delta_c^2}{\mu w} \ll u_{1\mathbf{k}}, \tag{27}$$

and

$$u_{2\mathbf{k}} = \frac{\xi_{\mathbf{k-g}} - \epsilon_{2\mathbf{k}}}{\epsilon_{2\mathbf{k}} - \xi_{\mathbf{k}}} \tilde{u}_{2\mathbf{k}} \approx -\tilde{u}_{2\mathbf{k}} \frac{\Delta_c^2}{\mu w} \ll \tilde{u}_{2\mathbf{k}}. \tag{28}$$

Substituting these expressions into A, Eq.(26), yields in the lowest order of Δ_c,

$$A \approx -\frac{2\Delta_c^2}{\mu w}. \tag{29}$$

Our result, Eq.(26) generally agrees with the STM experiments,[37–39] where the spatial checkerboard modulations of σ were observed in a few cuprates. The period of the modulations was found either commensurate or non-commensurate depending on a sample composition. In our model the period is determined by the center-of mass wave vectors $\mathbf{K}$ of the Bose-condensed preformed pairs. While the general case has to be solved numerically, the perturbation result, Eq.(26), is qualitatively applied for any $\mathbf{K}$ at least close to T_c, where the coherent gap is small, if one replaces $\cos(\pi m_x + \pi m_y)$ by $\cos(K_x m_x + K_y m_y)$. The period of DOS modulations does not depend on the voltage in the perturbation regime, as observed,[46] but it could be voltage dependent well below T_c, where higher powers of Δ_c are important. Different from any other scenario, proposed so far,[47] the hole density, which is about twice of the condensate density at low temperatures, is spatially modulated with the period determined by the inverse wave vectors corresponding to the center-of-mass pair band-minima. This 'kinetic' interpretation of charge modulations in cuprates, originally proposed[42] before STM results became available, is consistent with the inelastic neutron scattering, where incommensurate inelastic peaks were observed *only* in the *superconducting* state.[48] The vanishing at T_c of the incommensurate peaks is inconsistent with any other stripe picture, where a characteristic distance needs to be observed in the normal state as well. In our model the checkerboard charge modulations should disappear above T_c, where the Bose-condensate evaporates and the coherent gap $\Delta_c(T)$ vanishes, so that $A = 0$ in Eq.(26). While some STM studies[46] report incommensurate DOS modulations somewhat above T_c, they might be due to extrinsic inhomogeneities. In particular, preformed pairs in the surface layer could bose-condense at higher temperatures compared with the bulk T_c.

4. Conclusion

The bipolaron model has been microscopically derived as the strong-coupling extension of the BCS theory.[12] If EPI is sufficiently strong so that the BCS coupling constant $\lambda > 1$, the carrier bandwidth is exponentially reduced. This polaron band-narrowing explains the extremely low Fermi energy as estimated from the experimental London penetration depth in cuprates.[17] Carriers form small but mobile bipolarons well above T_c, which bose-condense below T_c in this strong-coupling regime.

Here, I have discussed the strong-coupling BdG equations with the periodic off-diagonal potential describing the interaction between unpaired single-particle excitations and the d-wave Bose-Einstein condensate of bipolarons with non-zero center-of-mass momenta. The checkerboard modulations of the single-particle DOS similar to those observed in cuprates have been found. While evidence for an exceptionally strong EPI in high-temperature superconductors is now overwhelming, yet, more generally, the model describes d-wave symmetry and checkerboard charge modulations of any preformed pairs bose-condensed with non-zero center-of-mass momenta.

I would like to thank Mukunda Das, Manuel deLlano, Victor Khodel, Mikko Saarela and other participants of XXX International Workshop on Condensed Matter Theories (June 2006, Dresden, Germany) for useful discussions. This work was supported by EPSRC (UK) (grant numbers EP/C518365/1 and EP/D07777X/1).

References

1. G. Zhao and D. E. Morris Phys. Rev. B **51**, 16487 (1995); G.-M. Zhao, M. B. Hunt, H. Keller, and K. A. Müller, Nature (London) **385**, 236 (1997); R. Khasanov, D. G. Eshchenko, H. Luetkens, E. Morenzoni, T. Prokscha, A. Suter, N. Garifianov, M. Mali, J. Roos, K. Conder, and H. Keller Phys. Rev. Lett. **92**, 057602 (2004).
2. A.S. Alexandrov, Phys. Rev. B **46**, 14932 (1992).
3. A. Lanzara, P. V. Bogdanov, X. J. Zhou, S. A. Kellar, D. L. Feng, E. D. Lu, T. Yoshida, H. Eisaki, A. Fujimori, K. Kishio, J. I. Shimoyana, T. Noda, S. Uchida, Z. Hussain and Z. X. Shen, Nature (London) **412**, 510 (2001); G-H. Gweon, T. Sasagawa, S. Y. Zhou, J. Craf, H. Takagi, D. H. Lee, and A. Lanzara, Nature (London) **430**, 187 (2004); X. J. Zhou, J. Shi, T. Yoshida, T. Cuk, W. L. Yang, V. Brouet, J. Nakamura, N. Mannella, S. Komiya, Y. Ando, F. Zhou, W. X. Ti, J. W. Xiong, Z. X. Zhao, T. Sasagawa, T. Kakeshita, H. Eisaki, S. Uchida, A. Fujimori, Z. Y. Zhang, E. W. Plummer, R. B. Laughlin, Z. Hussain, and Z. X. Shen, Phys. Rev. Lett. **95**, 117001 (2005).
4. D. Mihailovic, C.M. Foster, K. Voss, and A.J. Heeger, Phys. Rev. B **42**, 7989 (1990).
5. P. Calvani, M. Capizzi, S. Lupi, P. Maselli, A. Paolone, P. Roy, S.W. Cheong, W. Sadowski, and E. Walker, Solid State Commun. **91**, 113 (1994).
6. R. Zamboni, G. Ruani, A.J. Pal, and C. Taliani, Solid St. Commun. **70**, 813 (1989).
7. A. S. Alexandrov and A. M. Bratkovsky, Phys. Rev. Lett. **84**, 2043 (2000).
8. S. Lupi, S. P. Maselli, M. Capizzi, P. Calvani, P. Giura, and P. Roy, Phys. Rev. Lett. **83**, 4852 (1999).
9. J. Tempere and J. T. Devreese, Phys. Rev. B **64**, 104504 (2001).
10. T. R. Sendyka, W. Dmowski, and T. Egami, Phys. Rev. B **51**, 6747 (1995); T. Egami, J. Low Temp. Phys. **105**, 791 (1996).
11. A. S. Alexandrov and N. F. Mott, Rep. Prog. Phys. **57**, 1197 (1994).
12. A. S. Alexandrov, *Theory of Superconductivity: from Weak to Strong Coupling* (IoP Publishing, Bristol-Philadelphia, 2003).
13. A. S. Alexandrov, in *Studies in High Temperature Superconductors,* ed. A.V. Narlikar (Nova Science Pub., NY 2006), v.**50**, pp. 1-69.
14. A. S. Alexandrov, Phys. Rev. B **53**, 2863 (1996).
15. A. S. Alexandrov and P. E. Kornilovitch, J. Phys.: Condens. Matter **14**, 5337 (2002).

16. J. P. Hague, P. E. Kornilovitch, J. H. Samson, and A. S. Alexandrov, Phys. Rev. Lett. **98**, 037002 (2007).
17. A. S. Alexandrov, Physica C **363**, 231 (2001).
18. A. S. Alexandrov, Physica C **182**, 327 (1991); ibid. J. Low Temp. Phys. **87**, 721 (1992).
19. J. Annett, N. Goldenfeld and A. J. Legget, in *Physical Properties of High Temperature Superconductors* **5** (ed. Ginsberg D M, World Scientific, Singapore) 375 (1996).
20. C. C. Tsuei and J. R. Kirtley, Rev. Mod. Phys. **72**, 969 (2000).
21. D. J. Scalapino, J. E. Loh, and J. E. Hirsch, Phys. Rev. B **34**, 8190 (1986).
22. K. A. Müller, Phil. Mag. Lett. **82**, 279 (2002).
23. V. V. Kabanov, J. Demsar, B. Podobnik, and D. Mihailovic, Phys. Rev. B **59**, 1497 (1999).
24. G. M. Zhao, Phys. Rev. B **64**, 024503 (2001).
25. Ch. Renner, B. Revaz, J. Y. Genoud, K. Kadowaki, and O. Fischer, Phys. Rev. Lett. **80**, 149 (1998).
26. H. Hancotte, R. Deltour, D. N. Davydov, A. G. M. Jansen, and P. Wyder, Phys. Rev. B **55**, R3410 (1997).
27. H. Ding, T. Yokoya, J. C. Campuzano, T. Takahashi, M. Randeria, M. R. Norman, T. Mochiku, K. Kadowaki, and J. Giapintzakis, Nature (London) **382**, 51 (1996).
28. Z. X. Shen and J. R. Schrieffer, Phys. Rev. Lett. **78**, 1771 (1997).
29. G. M. Eliashberg, Zh. Eksp. Teor. Fiz. **38**, 966 (1960) [Sov. Phys. JETP **11** 696 (1960)].
30. B. Batlogg, H. Y. Hwang, H. Takagi, R. J. Cava, H. L. Kao, and J. Kwo, Physica C **135-140**, 130 (1994).
31. J. W. Loram, K. A. Mirza, J. M. Wade, J. R. Cooper, and W. Y. Liang, Physica C **235**, 134 (1994).
32. T. Timisk and B. Statt, Rep. Prog. Phys. **62**, 61 (1999).
33. A. F. Andreev, Zh. Eksp. Teor. Fiz. **46**, 1823 (1964) [Sov. Phys. JETP **19**, 1228 (1964)].
34. G. Deutscher, Nature **397**, 410 (1999).
35. V. M. Krasnov, A. Yurgens1, D. Winkler, P. Delsing, and T. Claeson, Phys. Rev. Lett. **84**, 5860 (2000).
36. J. Demsar, R. Hudej, J. Karpinski, V. V. Kabanov, and D. Mihailovic, Phys. Rev. B **63**, 054519 (2001).
37. J. E. Hoffman, E. W. Hudson, K. M. Lang, V. Madhavan, H. Eisaki, S. Uchida, and J. C. Davis, Science **295**, 466 (2002).
38. J. E. Hoffman, K. McElroy, D. H. Lee, K. M Lang, H. Eisaki, S. Uchida, and J. C. Davis, Science **297**, 1148 (2002).
39. C. Howald, H. Eisaki, N. Kaneko, M. Greven, and A. Kapitulnik, Phys. Rev. B **67**, 014533 (2003).
40. A. S. Alexandrov and A. F. Andreev, Europhys. Lett. **54**, 373 (2001).
41. A. Larkin and Yu. Ovchinnikov, Zh. Eksp. Teor. Fiz. **44**, 1136 (1964): P. Fulde and R. Ferrel, Phys. Rev. **153**, 550 (1964).
42. A. S. Alexandrov, Physica C **305**, 46 (1998).
43. A. F. Andreev, Pis'ma Zh. Eksp. Teor. Fiz. **79**, 100 (2004).
44. C. R. A. Catlow, M. S. Islam, and X. Zhang, J. Phys.: Condens. Matter **10**, L49 (1998).
45. N. Bogoliubov, J.Phys. USSR **11**, 23 (1947).
46. M. Vershinin, S. Misra, S. Ono, Y. Abe, Y. Ando, and A. Yazdani, Science **303**, 1995 (2004).
47. The Wigner crystal of doped holes or hole Cooper pairs is refutable because their

Coulomb repulsion is relatively weak. The relevant dimensionless parameter is $r_s = m^* e^2/\epsilon_0(4\pi n/3)^{1/3}$. The Wigner crystallization could appear at $r_s \gtrapprox 100$, which corresponds to an unrealistically small atomic density of holes $n \leq 10^{-6}$ with the static dielectric constant $\epsilon_0 = 30$ and the effective mass $m^* = 5m_e$.

48. P. Bourges, Y. Sidis, H. F. Fong, L. P. Regnault, J. Bossy, A. Ivanov, and B. Keimer, Science, **288** 1234 (2000); J. Supercond. **13**, 735 (2000).

FIVE POSSIBLE REASONS WHY HIGH-T_c SUPERCONDUCTIVITY IS STALLED

M. GRETHER

Facultad de Ciencias, UNAM, 04510 México, DF, Mexico
E-mail: mdgg@hp.fciencias.unam.mx

M. DE LLANO

H.H. Wills Physics Laboratory, University of Bristol, Bristol BS8 1TL, UK
and Instituto de Investigaciones en Materiales, UNAM
*Apdo. Postal 70-360, 04510 México, DF, Mexico**
E-mail: dellano@servidor.unam.mx

Five commonly held premises considered questionable assumptions in the microscopic theory of superconductivity are discussed as possible reasons why the search appears to be stalled for a theoretical framework, admittedly ambitious, capable of predicting materials with critical temperatures T_c higher than the 1993 record of $164K$ in $HgTlBaCaCuO$ (under pressure). We focus the dilemma as a whole in terms of a generalized Bose-Einstein condensation (GBEC) interpretation that includes and further extends BCS theory, as well as substantially enhancing its predicted T_cs *within*the electron-phonon mechanism producing pairing. The new GBEC model is an extension of the Friedberg-T.D. Lee 1989 boson-fermion BEC theory of high-$T_c$superconductors in that it includes hole pairs as well as electron pairs.

Keywords: Superconductivity; Bose-Einstein condensation; electron-phonon coupling.

1. Introduction

We discuss at least five possible reasons why a first-principles theory of superconductivity appears to be stalled, namely the following premises or questionable assumptions that:

♦ Cooper pairs[1] (CPs) and BCS[2] pairs are the same thing and are *not* bosons, so that a Bose-Einstein condensation (BEC) picture of the superconducting state is out of the question. Though no one to our knowledge has been able to construct CP operators that obey Bose commutation relations, as we discuss below, they in fact at least obey Bose statistics which is all that matters for a legitimate BEC scenario of superconductivity.

♦ CPs are negative-energy stable (i.e., stationary, ∞-lifetime) total (or center-of-mass) wavevector $\mathbf{K} \equiv \mathbf{k} + \mathbf{k}_2 = 0$ bound states whose energy for the

*Permanent address.

Cooper two-parameter (V and $\hbar\omega_D$) model interaction (later employed by BCS), with respect to two free electrons on the Fermi surface of energy E_F, is given by[1]

$$\mathcal{E}_0 = -2\hbar\omega_D/(e^{2/\lambda} - 1) \underset{\lambda \to 0}{\longrightarrow} -2\hbar\omega_D e^{-2/\lambda} \tag{1}$$

where the dimensionless coupling $\lambda \equiv N(0)V \geqslant 0$, with $N(0)$ being the density of electronic states for one spin at the Fermi surface. In fact, including both 2e- and 2h-CPs, as consistency would seem to demand, leads to a purely imaginary result as we discuss below.

♦ BEC is impossible in 2D. This is true for bosons propagating like $\hbar^2 K^2/4m$, with $\mathbf{K} \equiv \mathbf{k}_1 + \mathbf{k}_2$ and m is the effective electron mass. This is in fact the correct energy-momentum (dispersion) relation for a composite particle of mass $2m$ *in vacuo*, but it is linear in K when propagation takes place in the Fermi sea.

♦ Another questionable assumption is the assertion that "...the picture...by Schafroth (1955)...of electron pairs...which at low T undergo a BEC, is not valid."[3]

♦ Since BCS predicts a ceiling value of $T_c \simeq 1.13\Theta_D e^{-1/\lambda} \lesssim 45$ K with the electron- *phonon* mechanism, where $\Theta_D \equiv \hbar\omega_D/k_B$ is the Debye temperature $\simeq 300K$ and $\lambda \lesssim \frac{1}{2}$,[4] higher T_cs require: excitons or magnons or plasmons or otherwise purely-electronic mechanisms. We show below how one can surpass the BCS ceiling on T_c and in principle even reach room-temperature superconductivity without abandoning phonons.

2. Are Cooper and BCS Pairs Really Identical?

The BCS-pair annihilation and creation operators for any total (or center-of-mass momentum) wavevector $\mathbf{K} \equiv \mathbf{k}_1 + \mathbf{k}_2 \geqslant 0$ can be defined[5] as

$$b_{\mathbf{kK}} \equiv a_{\mathbf{k}_2\downarrow} a_{\mathbf{k}_1\uparrow} \text{ and } b^\dagger_{\mathbf{kK}} \equiv a^\dagger_{\mathbf{k}_1\uparrow} a^\dagger_{\mathbf{k}_2\downarrow} \tag{2}$$

where $\mathbf{k} \equiv \frac{1}{2}(\mathbf{k_1} - \mathbf{k_2})$ is the relative momentum wavevector, and the a's are the electron operators satisfying the usual Fermi anticommutation relations. Clearly, $\mathbf{k}_1 = \mathbf{K}/2 + \mathbf{k}$ and $\mathbf{k}_2 = \mathbf{K}/2 - \mathbf{k}$. Our only restriction is $\mathbf{K} \equiv \mathbf{k}_1 + \mathbf{k}_2 = \mathbf{k}'_1 + \mathbf{k}'_2$. Straightforward commutator algebra then gives

$$\left[b_{\mathbf{kK}}, b^\dagger_{\mathbf{k'K}}\right] = (1 - n_{\mathbf{K}/2-\mathbf{k}\downarrow} - n_{\mathbf{K}/2+\mathbf{k}\uparrow})\delta_{\mathbf{kk'}} \tag{3}$$

$$\left[b^\dagger_{\mathbf{kK}}, b^\dagger_{\mathbf{k'K}}\right] = [b_{\mathbf{kK}}, b_{\mathbf{k'K}}] = 0 \tag{4}$$

where $n_{\mathbf{K}/2\pm\mathbf{k}s} \equiv a^\dagger_{\mathbf{K}/2\pm\mathbf{k}s} a_{\mathbf{K}/2\pm\mathbf{k}s}$ are Fermi number operators. Since (3) are not precisely the Bose commutation rules it has been claimed,[2] correctly, that the b's of (2) do *not* stand, strictly speaking, for true bosons. Moreover, (2) also give the "pseudo-fermion" relations

$$\{b_{\mathbf{kK}}, b_{\mathbf{k'K}}\} = 2b_{\mathbf{kK}} b_{\mathbf{k'K}}(1 - \delta_{\mathbf{kk'}}) \tag{5}$$

so that BCS pairs are not fermionic either. If $\mathbf{K} = 0$ so that $\mathbf{k}_1 = -\mathbf{k}_2 = \mathbf{k}$ (the only case considered by BCS), and calling $b_{\mathbf{kK=0}} \equiv b_{\mathbf{k}}$, etc., (3) to (5) become Eqs.

(2.11)-(2.13) of Ref. 2. In addition, it has been shown[6] that the assumption[7] that nonzero-K CPs are negligible in number, compared with $K = 0$ CPs, is not entirely justified. Moreover, it is natural to conjecture that the so-called "preformed" pairs referred to in the literature[8,9] are mostly *precisely these* $K > 0$ *pairs* neglected in BCS theory.

On the other hand, while BCS pairs evidently depend on *both* $\mathbf{K}$ and $\mathbf{k}$, CPs depend *only*[1,10–12] on $\mathbf{K}$. Thus, in strict adherence to the Pauli Exclusion Principle the Fermi number operators $n_{\mathbf{k_1}}, n_{\mathbf{k_2}}$ have eigenvalues $0, 1$ which implies that the BCS-pair number operator $N_{\mathbf{kK}} \equiv n_{\mathbf{K}/2+\mathbf{k}} n_{\mathbf{K}/2-\mathbf{k}}$ has eigenvalues $0, 1$ but an appropriate CP number operator $N_{\mathbf{K}} \equiv \sum_{\mathbf{k}} N_{\mathbf{kK}}$ evidently has eigenvalues $0, 1, ...\infty$.[13] Hence, CPs obey Bose-Einstein statistics even though BCS pairs do not as no two of them can sit *precisely* on top of each other in real space. The BCS pairs represented by (2) are thus sometimes referred to as "hard-core bosons"[14,15]—but note that this is a "zero-measure" physical effect as the "hard-core" radii involved are of magnitude 0^+ . A rigorous finding by Girardeau[16] that "Bose condensation of fermion pairs...is impossible...in a Fermi system" and that it is rather a kind of "Fermi condensation" in fact refers to BCS, rather than, Cooper, pairs—as they have been distinguished in Ref. 5.

3. Are CPs Really Negative-Energy Bound States?

Originally, the equation for the eigenvalue $\mathcal{E}_K$ of "ordinary" or two-electron (2e) CPs was derived[1] based on a Schrödinger-like equation in the momentum-representation to be

$$\sum_{\mathbf{k}}{}' (\hbar^2 k^2/m + \hbar^2 K^2/4m - 2E_F - \mathcal{E}_K)^{-1} = 1/V \tag{6}$$

for the Cooper/BCS model interaction of strength $-V$ which is attractive. The prime on the summation sign means restriction to the two electrons lying within the energy shell E_F and $E_F + \hbar\omega_D$. This equation later sometimes became known as the Bethe-Goldstone equation.[17] Its solution for $K = 0$ is exact in 2D for all coupling; it is otherwise a good approximation provided $\hbar\omega_D \ll E_F$ and reduces to (1). Thus, a negative-energy follows from the original Cooper problem of two electrons above the Fermi sea. For $K \geq 0$ the solution of (6) is

$$\mathcal{E}_K = \mathcal{E}_0 + b(d)\hbar v_F K + O(K^2) \tag{7}$$

with $b(d) = 1$, $2/\pi$, $1/2$ in $d = 1, 2, 3$ dimensions.[10,18] Clearly, since $\mathcal{E}_0 < 0$ from (1) for large enough K (7) changes sign signaling pair breakup. The result in 3D was cited long ago by Schrieffer (Ref. 19, p. 33) without proof; it is also mentioned in Ref. 20 (p. 336) albeit with a misprint of 1 instead of the correct coefficient 1/2.

If both electron- and hole-pairs are considered simultaneously, as consistency would apparently demand, a Bethe-Salpeter[21] integral equation treatment, in the ladder approximation, leads *in any dimensionality* d, to the $K = 0$ eigenvalue

$\mathcal{E}_0$ equation $1 = \frac{1}{2}\lambda \ln[1 - (2\hbar\omega_D/\mathcal{E}_0)^2]$ which on solving for $\mathcal{E}_0$ gives the purely-imaginary result

$$\mathcal{E}_0 = \pm i2\hbar\omega_D/\sqrt{e^{2/\lambda} - 1} \underset{\lambda \to 0}{\longrightarrow} \pm i2\hbar\omega_D e^{-1/\lambda}. \tag{8}$$

This key result is already found in Refs. 22 and 23, as well as in Ref. 19 (pp. 167-8) where the same result is arrived at *without* explicitly mentioning hole pairs. However, its interpretation in all these references is at best *ambiguous.* In contrast to what might be termed a *benign* instability given by (1), (8) is a *true* or "malignant" instability. The CP concept defined thusly is meaningless but to discard it would be disastrous for the very notion of superconductivity.

To avoid scrapping the CP concept altogether—which would be disastrous in trying to construct a microscopic theory of superconductors, or even to interpret associated experimental data—we have proposed[24–26] replacing the ideal Fermi gas (IFG) "sea" wrt which CPs are referred by a BCS-correlated ground-state sea. The general result, in $d = 3, 2$ and 1, is a 4×4 (*not* merely a 3×3 one as in Ref. 27) which can be shown to reduce to a 3×3 *and* a 1×1 determinant which reduce to eigenvalue equations for $\mathcal{E}_K$. The former, besides giving the BCS $T = 0$ gap equation when $\mathcal{E}_0 = 0$ at $K = 0$, leads to what might be called a trivial Anderson-Bogoliubov-Higgs (ABH) sound-excitation solution $\mathcal{E}_K = \hbar v_F K/\sqrt{d} + o(\lambda)$, with the last term representing correction terms due to interactions between the fermions. The first term corresponds to the IFG; it can also be derived using elementary thermodynamics for a d-dimensional IFG ($\lambda = 0$) from $mc^2 = \partial P/\partial n$ where P is the pressure at $T = 0$ and n the number density of the N fermions of mass m. Since $P = n^2[\partial(E/N)/\partial n]$ and the total energy-per-fermion $E/N = dE_F/(d+2)$ where E_F is the Fermi energy, then $P = 2nE_F/(d+2)$ which in turn implies that the sound speed $c = v_F/\sqrt{d}$.

In addition to this (trivial) sound-excitation solution resulting from the 3×3 determinant, the 1×1 determinant leads to a nontrivial solution for the eigenvalue energy $\mathcal{E}_K$ of what we have called a "moving CP" and given, e.g., in 2D,[25] by

$$\frac{1}{2\pi}\lambda\hbar v_F \int_{k_F-k_D}^{k_F+k_D} dk \int_0^{2\pi} d\varphi u_{\mathbf{K}/2+\mathbf{k}} v_{\mathbf{K}/2-\mathbf{k}} \{u_{\mathbf{K}/2-\mathbf{k}} v_{\mathbf{K}/2+\mathbf{k}} - u_{\mathbf{K}/2+\mathbf{k}} v_{\mathbf{K}/2-\mathbf{k}}\}$$

$$\times \frac{E_{\mathbf{K}/2+\mathbf{k}} + E_{\mathbf{K}/2-\mathbf{k}}}{-\mathcal{E}_K^2 + (E_{\mathbf{K}/2+\mathbf{k}} + E_{\mathbf{K}/2-\mathbf{k}})^2} = 1 \tag{9}$$

where $v_F \equiv \sqrt{2E_F/m}$; φ is the angle between $\mathbf{K}$ and $\mathbf{k}$; $k_D/k_F \equiv \hbar\omega_D/2E_F \ll 1$; $E_{\mathbf{k}} \equiv \sqrt{\xi_k{}^2 + \Delta^2}$ with $\xi_k \equiv \hbar^2k^2/2m - E_F$ and Δ is electronic gap at $T = 0$. Also, $v_k^2 \equiv \frac{1}{2}(1 - \xi_k/E_{\mathbf{k}})$ and $u_k^2 \equiv 1 - v_k^2$ are the usual $T = 0$ BCS-Bogoliubov-Valatin functions. The moving CP solution is then

$$\pm\mathcal{E}_K = 2\Delta + c(d)\lambda\hbar v_F K + O(K^2) + i\Gamma(K) \tag{10}$$

with the upper and lower signs referring to 2e- and 2h-CPs, respectively, and $c(d) \equiv 1,\ 1/2\pi,\ 1/4$ for $d = 1,\ 2,\ 3$. The *positive*-energy 2e-CP resonance in 2D has an energy width $\Gamma(K)$ and thus a lifetime $\tau_K \equiv \hbar/2\Gamma(K) = \hbar/2\left[(\lambda/\pi)\hbar v_F K + (\hbar v_F/12k_D)e^{1/\lambda}K^2\right]$ that diverges at $K=0$, falling to zero as K increases. The gapped aspect of (10) agrees with an entirely different study in 2D in Ref. 28, while the gapped structure *and* linearly-dispersive leading term in (10) are consistent with two other studies in 2D by Traven[29,30] . The results sketched in this and the previous section are summarized in Table 1.

Even in the simplest, original conception of Cooper pairing as embodied in (6), in addition to the bound-state solution (1), there is another, *positive*-energy, solution given by (1) with the sign changed due to the elementary fact that $\int x^{-1}dx = \ln|x|$ and *not* $\ln x$ as commonly assumed. This solution may be interpreted as a "bound-state in the continuum" analogous to (10) with the upper sign.

Table 1. Leading term in the K-expansion of the CP dispersion relation from (7), first column, and from (10), second column, where Δ is the $T=0$ BCS single-electron gap.

d	Bethe-Goldstone ($2e$-CPs only)	Bethe-Salpeter ($2e$- & $2h$-CPs)
1D	$\hbar v_F K + ...$Ref.[18]	$2\Delta + \lambda\hbar v_F K + ...$Ref.[26]
2D	$(2/\pi)\hbar v_F K + ...$Ref.[10]	$2\Delta + \frac{1}{2\pi}\lambda\hbar v_F K + ...$Ref.[25]
3D	$\frac{1}{2}\hbar v_F K + ...$Ref.[10]	$2\Delta + \frac{1}{4}\lambda\hbar v_F K + ...$Ref.[24]

4. Is BEC Really Impossible in 2D?

With the same "regularized" interaction $-v_0\delta(r)$ used by Miyake[32] to obtain in 2D at $T=0$ closed expressions for the gap Δ and the fermion chemical potential μ as *explicit* functions of the two-body binding energy *in vacuo* $B_2 \geqslant 0$, namely $\Delta = \sqrt{2B_2E_F}$ and $\mu = E_F - \frac{1}{2}B_2$, we have deduced[11] the 2e-CP nonnegative excitation energy ε_K, again neglecting hole pairs, to be

$$\varepsilon_K \equiv (\mathcal{E}_K - \mathcal{E}_0) = \frac{2}{\pi}\hbar v_F K + \left[1 - \left\{2 - \left(\frac{4}{\pi}\right)^2\right\}\frac{E_F}{B_2}\right]\frac{\hbar^2K^2}{2(2m)} + O(K^3). \tag{11}$$

In spite of being, strictly speaking, incomplete as it neglects hole pairs, it serves as a vivid illustration of two points: a) The relation (11) becomes quadratic in K for any coupling if and only if $E_F \to 0$ (i.e., the vacuum limit of no Fermi sea present) for any fixed $B_2 > 0$, namely $\to \hbar^2K^2/2(2m)$. b) On the other hand, for any $E_F > 0$ (i.e., in the presence of a Fermi sea) (11) $\to (2/\pi)\hbar v_F K$. Thus, a nonzero BEC $T_c > 0$ is possible[33] for all $d > s$ where s is the exponent of the leading term K^s in the bosonic dispersion relation, where $s=1$ for linear and $s=2$ for quadratic dispersion. Diverse empirical data in support of linearly-dispersive CPs is collected in Ref. 34.

5. Is the Schafroth BEC Picture Invalid?

Not at all. A recent[35,36] generalization (GBEC) of BEC theory starts from a Hamiltonian $H = H_0 + H_{int}$ where H_0 represents ternary (2e-CPs, 2h-CPs and unpaired electrons) BF gas "normal" state that is precisely "an appropriate 'zeroth-order' starting point" for "pairs of electronic excitations with charge 2e that all have the same ground-state wavefunction."[37] Further, H_{int} stands for BF interactions causing 2e- and 2h-CP formations/disintegrations and consists of four terms. This differs markedly from the BCS H in that the normal state associated with H_0 is not a Fermi but a non-Fermi liquid. It also differs from BCS theory in that $K \neq 0$ pairs are neglected *only* in H_{int} but not in H_0. As a special case, it leads to the BCS gap equation for all temperatures T if the BF interaction parameter f is related to the two BCS coupling parameters V and $\hbar\omega_D$ through $f^2 \equiv 2\hbar\omega_D V$. Lifting the BCS assumption that the chemical potential μ equals the Fermi energy E_F,the GBEC theory also leads to the two coupled equations for $\Delta(T)$ and $\mu(T)$—as in the Miyake[32] 2D case just mentioned—that define the "BCS-Bose crossover picture" developed by Friedel *et al.* (1967); Eagles (1969), who first dealt with its BEC aspect explicitly; Leggett (1980); Miyake (1983); Nozières *et al.* (1985); Ranninger, Micnas and Robaszkiewicz (1988); T.D. Lee *et al.* (1989); Randeria *et al.* (1989); van der Marel (1990); Bar-Yam (1991); Drechsler and Zwerger (1992); Hausmann (1993); Pistolesi and Strinati (1996); and many others. For a review of the crossover work see Ref. 38. Note that the GBEC theory is also a generalization of the Friedberg-T.D. Lee BEC theory[39] of superconductors in that hole pairs are not ignored.

Besides the BCS gap equation for all T, the GBEC theory also contains[40] the $T = 0$ BCS condensation energy (Ref. 2, Eq. 2.42)

$$\frac{E_s - E_n}{L^d} = -N(0)(\hbar\omega_D)^2 \left[\sqrt{1 + (\Delta/\hbar\omega_D)^2} - 1\right] \underset{\Delta \to 0}{\longrightarrow} -\frac{1}{2}N(0)\Delta^2 + O(\Delta^4) \qquad (12)$$

where $\Delta(T = 0) \equiv \Delta$. In other words, it yields the *full* condensation energy for all coupling $\Delta \geqslant 0$ as given in the first expression on the rhs, instead of the only leading term as in the last expression on the rhs of (12) as was reported earlier.[36] It can also be seen[40] that neglecting hole-pairs gives only *half* the condensation energy (12). The GBEC model yields a nonzero BEC T_cs under diverse conditions, thus completely vindicating the Schafroth picture[41–44] in essence. The GBEC model connects the order parameters of the superconductor $\Delta(T)$ with that of the 2e-CP and 2h-CP BE condensate number densities $n_0(T)$ and $m_0(T)$, respectively, through the relation between BCS and BEC order parameters

$$\Delta(T) = f\sqrt{n_0(T)} = f\sqrt{m_0(T)}.$$

The first equality of this expression was first apparently suggested in Ref. 45. The relation yields the same T_c value as the solution of $0 = \Delta(T_c) = n_0(T_c) = m_0(T_c)$. Indeed, the GBEC theory suggests that for all intents and purposes a BCS condensate is *precisely* a BE condensate, implying that BCS theory is truly an "independent-pair model" as it has been referred to occasionally in the literature.

6. Are Phonons Sufficient for T_cs Higher than 45K?

Yes. Applying the GBEC theory in 3D for the Cooper/BCS model interaction between fermions but for zero boson-fermion interaction f (for which there is *no mixed phase* as opposed to when $f \neq 0$ which has both a mixed phase as well as two pure phases) one gets[46] two *analytical* working number equations critical temperature T_c. One is for the *pure 2e-CP BEC phase*

$$\begin{aligned}1/3 = &-\frac{\sqrt{\pi}\tilde{T}_c^{\ 3/2}}{4} PolyLog[3/2, -\exp\{(\tilde{n}^{-2/3} + \nu/2)/\tilde{T}_c\}] \\ &+ \frac{16\tilde{T}_c^{\ 3}}{\lambda^3}\{PolyLog[3, \exp(-2\tilde{\Delta}/\tilde{T}_c)] - PolyLog[3, \exp\{-(2\nu + 2\tilde{\Delta})/\tilde{T}_c\}]\} \quad (13)\end{aligned}$$

where $\tilde{n} \equiv n/n_f$, $\tilde{T}_c \equiv T_c/T_F$, $\nu \equiv \hbar\omega_D/E_F$, and we assumed the gapped and linearly-dispersive $\varepsilon_K \simeq 2\Delta + \lambda\hbar v_F K/4$ for $d = 3$, Ref. 24 Eq. (12), and where $PolyLog[\sigma, z]$ are the well-known polylogarithm functions. This can now be solved numerically for $\tilde{T}_c$ as a function of $\tilde{n}$ which is plotted as the thick dashed curve in Fig. 1 for $\lambda = 2$ and $\nu \equiv \hbar\omega_D/E_F = 0.005$. All curves are essentially BEC phase boundaries. The other equation is for the *pure 2h-CP BEC phase*

$$\begin{aligned}1/3 = &-\frac{\sqrt{\pi}\tilde{T}_c^{\ 3/2}}{4} PolyLog[3/2, -\exp\{(\tilde{n}^{-2/3} - \nu/2)/\tilde{T}_c\}] \\ &+ \frac{16\tilde{T}_c^{\ 3}}{\lambda^3}\{PolyLog[3, \exp\{-(2\nu + 2\tilde{\Delta})/\tilde{T}_c\}] - PolyLog[3, \exp(-2\tilde{\Delta}/\tilde{T}_c)]\}. \quad (14)\end{aligned}$$

Its solution $\tilde{T}_c$ as a function of $\tilde{n}$ is plotted as the thick full curves in Fig. 1 for $\lambda = 1/2$, 1, and 2. The BCS values are shown as large black dots; they correspond to a BE condensate which is a mixed 50/50 phase of 2e- and 2h-CPs. For all pure 2h-CP BEC phase curves there is a "catastrophe" in that T_c/T_F *diverges* as $n/n_f \to 0$. This is evident for $\lambda = 1/2$ and 1; for $\lambda = 2$ we found, e.g., that $T_c/T_F \simeq 113$ for $n/n_f = 0.1$, implying that this curve too eventually bends to the left and also gives a catastrophe. For lack of a more rigorous basis to do so, we shall discard these catastrophic solutions as unphysical. Empirical data[47] for "exotic" SCs fall within the range $T_c/T_F \simeq 0.01 - 0.05$, shown shaded. The curves follow on setting equal to zero the first derivative (wrt to n_0 or m_0, respectively, for the dashed and full curves) of the associated Helmholtz free energy per unit volume. But this is only a necessary, not sufficient, condition for state in thermodynamic equilibrium. The *second* derivative must be positive for that state to be a true minimum. For all the $f = 0$ cases graphed in the figure this second derivative turns out to be zero, i.e., it is not a minimum but rather an inflection point and thus corresponds to a thermodynamically unstable state.

Fortuitously, *all* the phase-boundary curves will stabilize by their Helmholtz free energy acquiring a positive second-derivative (again, wrt to n_0 or m_0, respectively,

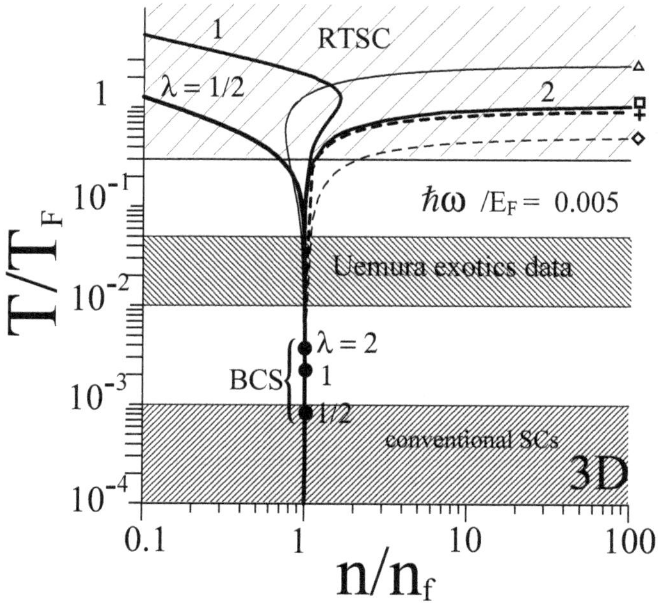

Fig. 1. BEC-phase boundaries for zero BF interaction ($f = 0$) in the GBEC theory for 2h-CPs (full curves) from (14) and 2e-CPs (dashed curves) from (13) using $\Theta_D/T_F = 0.005$, a typical value for 3D superconductors, and $\lambda = 1/2, 1$, and 2. Thick curves refer to gapped and linearly-dispersive $\varepsilon_K = 2\Delta + \lambda\hbar v_F K/4^{24}$ bosons moving in the Fermi sea, while thin curves refer to quadratically-dispersive $\varepsilon_K = \hbar^2K^2/4m$ bosons (as, e.g., in Refs. 39 and 36) and is obviously independent of coupling λ. Black dots are the BCS $T_c/T_F = 1.134(\Theta_D/T_F)\exp(-1/\lambda)$ for the λ values indicated. Four symbols on the extreme right give $n/n_f \to \infty$ limit of 2e- (dashed curves) and 2h- (full curves) CP BE condensate T_c/T_F values. RTSC stands for room-temperature superconductivity in a material with $T_F = 10^3K$. Curves for 2h-CP BE condensate phase boundaries that twist to the left are "catastrophes" in that T_c *diverges* as $n/n_f \to 0$. These unphysically divergent curves can be discarded entirely on physical grounds.

for the dashed and full curves) if f, the BF coupling, is nonzero and identified with the Cooper/BCS model interaction parameters V and $\hbar\omega_D$ via $f = \sqrt{2V\hbar\omega_D}$. This is required for the GBEC theory to give the BCS gap equation as a special limit. The corresponding T_c formulae in this case are, of course, more complicated than (13) and (14). However, the T_c results of Fig. 1 with $f = 0$ differ very little from those with $f \neq 0$: the difference in T_c/T_F between the zero- and nonzero-f cases is less than 2×10^{-4} for $n/n_f \gtrsim 2$. This insensitivity of T_c to f appears to be highly significant for it implies that, at least for an accurate determination of T_c, most of the physics is already contained in the *non*-interacting boson-fermion ternary mixture described by the GBEC theory. Thus, moderate departures from perfect 2e/2h-CP symmetry enable the GBEC model, unlike BCS theory, empirical T_c values for exotic SCs *without* abandoning electron-phonon dynamics. If we discard the pure 2h-CP phase-boundary (full) curves on the basis that they all lead to "catastrophes" since T_c/T_F diverges when $n/n_f \to 0$, one can then say that finite T_cs for thermodynamically stable states are achieved *only* for the pure 2e-CP BE condensate phase, which in turn seems to be consistent with at least one interpretation[48] of London magnetic-moment measurements[49] apparently establishing *electron* pairs as the charge carriers in the condensates of superconductors as diverse as Pb (p-type single charge carriers above T_c), $BaPb_{0.8}Bi_{0.2}O_3$ (n-type) and $YBa_2Cu_3O_{7-\delta}$ (p-type), i.e., independently of whether the normal state carriers are electrons or holes.

If we assume quadratically-dispersive CPs by taking $\varepsilon_K = \hbar^2K^2/4m$ instead, one obtains the two thin curves of the figure (clearly independent of λ), full for the 2h-CP BEC and dashed for the 2e-CP BEC. But quadratic dispersion is only appropriate *in vacuo* in view of the $k_F \to 0$ limit of (11). Their associated Helmholtz free energies were confirmed as minima. The first predicted T_c instability for quadratically-dispersive bosons is associated with a pure condensate of 2h-CPs, now no longer to be discarded as it lacks the divergence associated with the linearly-dispersive curves. This is *not* consistent with the interpretation[48] we assumed for the findings of Ref. 49 and may serve as further empirical proof for linearly-dispersive CPs in addition to the criteria referenced in 50 and presented in Ref. 34.

7. Conclusions

We have submitted five possible reasons why the search for a first-principles microscopic theory of superconductivity has been stalled. Such theories, to date, have largely failed to be predictive, as opposed to merely descriptive. There seems to be a consensus that the Cooper pair (CP) concept is central to any theory and/or experimental interpretation of the phenomena. We have sketched how two-hole CPs *must* be accounted for, along with the usual two-electron pairs. This lead to purely-imaginary pair energies and thus spoil at first, but then clarify, the CP concept as a finite-lifetime, positive-energy *resonance* when the ideal-Fermi-gas sea is replaced by a BCS-correlated sea. In the many-fermion problem two-hole CPs account for a full *half* of the BCS zero-temperature condensation energy. A Bose-Einstein condensation (BEC) picture of the superconducting condensate can be shown to be valid, at least in zeroth order, in terms of a *ternary* boson-fermion (BF) gas where the bosons are CPs (that obey Bose statistics) as opposed to BCS pairs (that do not obey Bose commutation rules). In addition to two normal phases above T_c (a bosonless phase followed at lower temperatures by a ternary BF but still incoherent phase), the ternary BF gas in turn gives rise to three condensed phases: two pure BEC phases with either kind of pairs in the BE condensate, and a mixed phase with any proportion of either kind of pairs in the BE condensate. Indeed, a BCS condensate is a special case of a generalized BE condensate (GBEC) when the number of hole- and electron-pairs in the BE condensate are equal. Critical temperatures are largely insensitive to whether or not BF interactions are taken into account, showing that at least insofar as calculating BEC T_c's most of the physics is contained in the zeroth-order Hamiltonian describing the ideal ternary BF gas. Finally, room-temperature T_c's *are* possible without abandoning the usual attractive electron-phonon dynamics to produce pairing at moderate coupling.

Acknowledgments

MdeLl thanks UNAM-DGAPA-PAPIIT (Mexico) IN106401 and CONACyT (Mexico) 41302F for partial support. MdeLl thanks the Max Planck Institute for the Physics of Complex Systems, Dresden, Germany, for its support and hospitality.

References

1. L.N. Cooper, Phys. Rev. **104**, 1189 (1956).
2. J. Bardeen, L.N. Cooper and J.R. Schrieffer, Phys. Rev. **108**, 1175 (1957).
3. J. Bardeen, Physics Today (Jan. 1963) p. 25.
4. A.B. Migdal, JETP **7**, 996 (1958).
5. M. de Llano, F.J. Sevilla, and S. Tapia, Int. J. Mod. Phys. B **20**, 2931 (2006).
6. M.G. López-Arrieta, M. de Llano and M.A. Solís, Rev. Mex. Fís. **47**, 54 (2001).
7. L.N. Cooper, Am. J. Phys. **28**, 91 (1960).
8. T. Timusk and B. Statt, Rep. Prog. Phys. **62**, 61 (1999).
9. M. Kugler, Ø. Fischer, Ch. Renner, S. Ono, and Y. Ando, Phys. Rev. Lett. **86**, 4911 (2001).
10. M. Casas, S. Fujita, M. de Llano, A. Puente, A. Rigo, M.A. Solís, Physica C **295**, 93 (1998).
11. S.K. Adhikari, M. Casas, A. Puente, A. Rigo, M. Fortes, M.A. Solís, M. de Llano, A.A. Valladares & O. Rojo, Phys. Rev. B **62**, 8671 (2000).
12. S.K. Adhikari, M. Casas, A. Puente, A. Rigo, M. Fortes, M. de Llano, M.A. Solís, A. A. Valladares & O. Rojo, Physica C **351**, 341 (2001).
13. S. Fujita and S. Godoy, *Theory of High Temperature Superconductivity* (Kluwer, NY, 2001) pp. 97-98.
14. R. Micnas, J. Ranninger, and S. Robaszkiewicz, Rev. Mod. Phys. **62**, 113 (1990).
15. R. Micnas, S. Robaszkiewicz, and A. Bussmann-Holder, Struct. Bond. **114,** 13 (2005).
16. M.D. Girardeau, J. Math. Phys. **11**, 684 (1970).
17. H.A. Bethe and J. Goldstone, Proc. Roy. Soc. (London) **A 238**, 551 (1957).
18. S. Fujita and S. Godoy, *Quantum Statistical Theory of Superconductivity* (Plenum, NY, 1996) p. 167.
19. J.R. Schrieffer, *Theory of Superconductivity* (Benjamin, Reading, MA, 1983).
20. A.L. Fetter and J.D. Walecka, *Quantum Theory of Many-Particle Systems* (McGraw-Hill, New York, 1971).
21. E.E. Salpeter and H.A. Bethe, Phys. Rev. **84**, 1232 (1951).
22. N.N. Bogoliubov, V.V. Tolmachev and D.V. Shirkov, Fortschr. Phys. **6**, 605 (1958); and also in *A New Method in the Theory of Superconductivity* (Consultants Bureau, NY, 1959).
23. A.A. Abrikosov, L.P. Gorkov, and I.E. Dzyaloshinskii, *Methods of Quantum Field in Statistical Physics* (Dover, NY, 1975) §33.
24. M. Fortes, M.A. Solís, M. de Llano, and V.V. Tolmachev, Physica C **364,** 95 (2001).
25. V.C. Aguilera-Navarro, M. Fortes, and M. de Llano, Sol. St. Comm. **129,** 577 (2004).
26. M. Fortes, M. de Llano, and M.A. Solís, *to be published.*
27. L. Belkhir and M. Randeria, Phys. Rev. B **49**, 6829 (1994).
28. É.V. Gorbar, V.P. Gusynin, and V.M. Loktev, Low Temp. Phys. **19**, 832 (1993).
29. S.V. Traven, Phys. Rev. Lett. **73**, 3451 (1994).

30. S.V. Traven, Phys. Rev. **51**, 3242 (1995).
31. T.A. Mamedov and M. de Llano, this Workshop.
32. K. Miyake, Prog. Theor. Phys. **69**, 1794 (1983).
33. M. Casas, A. Rigo, M. de Llano, O. Rojo, and M.A. Solís, Phys. Lett. A **245**, 5 (1998).
34. M. de Llano and J.J. Valencia, Mod. Phys. Lett. B **20**, 1067 (2006).
35. V.V. Tolmachev, Phys. Lett. A **266**, 400 (2000).
36. M. de Llano and V.V. Tolmachev, Physica A **317**, 546 (2003).
37. A.J. Leggett, Nature Phys. **2**, 134 (2006).
38. M. Randeria, in: *Bose-Einstein Condensation*, ed. A. Griffin *et al.* (Cambridge University, Cambridge, UK, 1995) p. 355.
39. R. Friedberg, T.D. Lee, and H.-C. Ren, Phys. Rev. B **45**, 10732 (1992) and refs. therein.
40. S.K. Adhikari, M. de Llano, F.J. Sevilla, M.A. Solís, and J.J. Valencia, *The BCS-Bose crossover theory* (to be published).
41. J.M. Blatt, *Theory of Superconductivity* (Academic, New York, 1964).
42. M.R. Schafroth, Phys. Rev. **96**, 1442 (1954).
43. M.R. Schafroth, S.T. Butler, and J.M. Blatt, Helv. Phys. Acta **30**, 93 (1957).
44. M.R. Schafroth, Sol. State Phys. **10**, 293 (1960).
45. J. Ranninger, R. Micnas, and S. Robaszkiewicz, Ann. Phys. Fr. **13**, 455 (1988).
46. M. de Llano and M. Grether, in *Lectures on the Physics of Highly-Correlated Electron Systems X,* AIP Conf. Proc. **846**, 221 (2006).
47. Y.J. Uemura, J. Phys.: Condens. Matter **16**, S4515 (2004).
48. J.E. Hirsch, Phys. Rev. B **68**, 012510 (2003).
49. A.A. Verheijen, J.M. van Ruitenbeek, R. de Bruyn Ouboter, and L.J. de Jongh, Nature **345**, 418 (1990).
50. J.A. Wilson, Phil. Mag. **84**, 2183 (2004), see esp. Fig. 2.

MULTISTABILITY AND MULTI 2π-KINKS IN THE FRENKEL–KONTOROVA MODEL: AN APPLICATION TO ARRAYS OF JOSEPHSON JUNCTIONS

KARL E. KÜRTEN[1] and CHRISTIAN KRATTENTHALER[2]

[1]*Institut für Experimentalphysik, Universität Wien, Boltzmanngasse 5, A-1090 WIEN, Austria*
E-mail: karl.kuerten@univie.ac.at

[2]*Fakultät für Mathematik, Universität Wien, Nordbergstraße 15, A-1090 Vienna, Austria*
WWW: http://www.mat.univie.ac.at/~kratt

A regular ring of Josephson junctions, connected in parallel, is studied analytically and numerically. We show that, depending on the strength of the r-well cosine potential the energy landscape of the Hamiltonian can have of the order of r^N/N locally stable minima separated by large barriers specified by unstable saddle points. The counting problem for the degeneracy of the total energy is equivalent to a wellknown necklace problem in combinatorial mathematics. We also demonstrate that the distribution of the phase differences as well as the energy spectrum is fractal provided that the strength of the cosine potential is sufficiently strong.

Keywords: Josephson junctions; Frenkel-Kontorova model.

1. Introduction

The Frenkel–Kontorova model,[1,2] also known as the discrete sine-Gordon model, describes a harmonic chain mimicking a layer of atoms in a spatially periodic potential playing the role of the substrate. The particles or atoms placed on a one domensional array interact via nearest neighbours and are subjected to a periodic external multi-well potential. The model was introduced more than 60 years ago with the intention to study crystal dislocation. Meanwhile it has been applied to numerous physical systems such as charge-density waves, magnetic spirals, disordered crystals, magnetic multilayers, and adsorbed monolayers. Due to competitions between two or more length scales or frequencies, locally stable spatially modulated structures arise, where the particles are non-trivially displaced from the sites of a reference lattice. To be more precise, the Frenkel–Kontorova model has become a favourite model for disordered systems and glassy materials because of its huge number of meta-stable states. More recently, the model has also found successful applications in the investigation of fluxon dynamics in systems consisting of superconducting wires interrupted by Josephson junctions, so-called Josephson arrays.[1,3,4] Here, a discrete chain of effective particles emerges when one considers the flux flow in discrete parallel arrays of weak links between superconductors. Note that these small Josephson junctions, in contrast to long Josephson junctions associated with the continuous sine-Gordon equation, can be modelled by Kirchhoff's circuit equations. In this case, they describe an array of N discrete Josephson elements connected by a parallel resistance/inductance combination. Moreover, Josephson devices are a candidate for the realization of models of superconducting interference devices (SQUIDS) used for experiments on quantum manipulation of qubits on which quantum computation can be realized. On the other hand, Josephson devices are also a suitable playground for fundamental investigations of soliton dynamics, chaos, and fractal behaviour.

2. The Model

We consider a regular ring of small Josephson tunnel junctions connected in parallel via superconducting leads. Specifying the gauge-invariant phase difference between the macroscopic wave functions across the i-th junction by the variable ϕ_i, and assuming periodic boundary conditions, the total energy of the system is given by the N-particle Hamiltonian

$$\mathbf{H}(\phi_1, \phi_2, \ldots, \phi_N) = \sum_{i=1}^{N} \left(\frac{1}{2}(\phi_{i-1} - \phi_i)^2 + a(1 - \cos\phi_i) \right). \tag{1}$$

The first term describes the inductance energy, specified by harmonic nearest-neighbour interactions, while the cosine potential in the second term stems from the supercurrent through the junction given by the Josephson relation. The control parameter a can be considered as a discreteness parameter, since for $a \to 0$ the model can be shown to be equivalent to the continuous sine-Gordon equation, where analytic solutions exist.

The control parameter $a > 0$ also specifies the height of the barrier which separates two wells of the cosine potential. The ferromagnetic ground state with zero energy is obtained when all units are placed in the same minimum of the Josephson potential.

In this report we mainly focus on analyzing the structure of the distribution of the phase differences as well as the structure of the energy distribution.

Another macroscopic quantity, the total phase M of the chain,

$$M = \sum_{i=1}^{N} \phi_i, \tag{2}$$

is also of interest in the context of experimental investigations of fluxon dynamics,[3] where the fluxon corresponds to a 2π-kink of a quantum phase difference ϕ_i between the two superconducting electrodes of the junction. Here the behaviour of travelling kinks and solitons, describing Josephson vortices, each of them carrying a magnetic flux quantum, is studied experimentally as well as theoretically.[3,4]

3. Few Particle Systems: Two Junctions

Analytic results obtained for few particle systems are often a rather useful guide for the understanding of the many-particle system, since characteristic features occurring in large systems often show up at the one- and two-particle level, respectively. In order to get some insight into the structure of the energy landscape of an N-particle system we first consider a small system consisting of only $N = 2$ interacting junctions.

The energy takes the form

$$E(\phi_1, \phi_2) = \frac{1}{2}(\phi_1 - \phi_2)^2 + a \ \left((1 - \cos\phi_1) + (1 - \cos\phi_2)\right) \tag{3}$$

and the necessary condition for equilibrium is given by the set of two coupled equilibrium equations

$$2(\phi_1 - \phi_2) + a \ \sin\phi_1 = 0, \qquad 2(\phi_2 - \phi_1) + a \ \sin\phi_2 = 0. \tag{4}$$

To give an illustrative example, we first restrict the values of ϕ_i, $i = 1, 2$, to the range $(-3\pi, 3\pi)$, where the cosine potential has only three single wells. In this case, Eqs. (4) can have up to nine distinct solutions representing locally stable minima. The homogeneous configurations $-(2\pi, 2\pi)$, $(0, 0)$, and $(2\pi, 2\pi)$, specify the three degenerate ground states with zero energy. They are always stable, independent of the discreteness parameter a. Fig. 1 shows the energy landscape $E(\phi_1, \phi_2)$ for the control parameter $a = 25$. We observe the nine valleys separated by saddle points and the corresponding maxima.

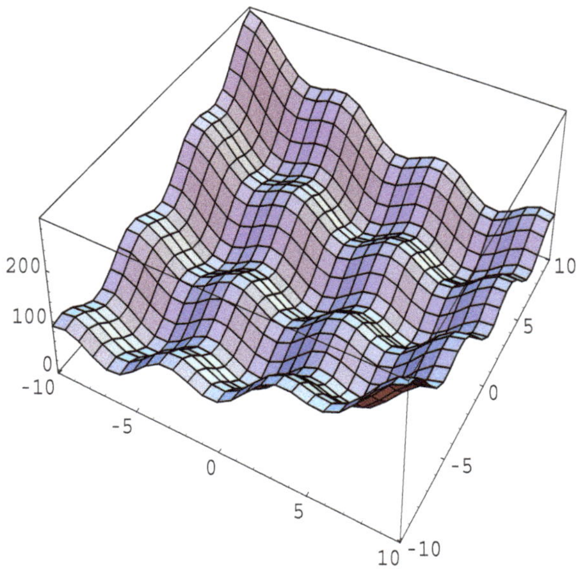

Fig. 1. Energy landscape $E(\phi_1, \phi_2)$ for $a = 25$, $\phi_i \in (-3\pi, 3\pi)$.

The six excited states which only exist for sufficiently large values of a cannot be calculated analytically. However, for $a \gg 1$, a Taylor expansion of first order in $\frac{1}{a}$ yields the configurations

$$\left(\phi_1^* - \frac{2}{a}(\phi_1^* - \phi_2^*), \phi_2^* - \frac{2}{a}(\phi_2^* - \phi_1^*)\right), \tag{5}$$

where the quantities ϕ_1^* and ϕ_2^* vary through all possible asymptotic values for $a \to \infty$, that is, $\phi_i^* \in \{-2\pi, 0, 2\pi\}$.

In total, these are nine configurations. The three configurations with zero energy have already been discussed above. The four configurations with $\phi_2 = 2\pi - \phi_1$ for $\phi_1 > 0$ as well as $\phi_2 = -2\pi - \phi_1$ for $\phi_1 < 0$, namely

$$2\pi\left(1 - \frac{2}{a}, \frac{2}{a}\right), \; 2\pi\left(-1 + \frac{2}{a}, -\frac{2}{a}\right), \; 2\pi\left(\frac{2}{a}, 1 - \frac{2}{a}\right), \; 2\pi\left(-\frac{2}{a}, -1 + \frac{2}{a}\right),$$

correspond to 2π-kinks. The remaining two antisymmetric configurations with $\phi_2 = -\phi_1$, namely

$$2\pi\left(1 - \frac{4}{a}, -1 + \frac{4}{a}\right), \; 2\pi\left(-1 + \frac{4}{a}, 1 - \frac{4}{a}\right),$$

represent 4π-kinks. They have the highest energy. While decreasing the control parameter a, the energy landscape is changed until the excited states eventually become unstable and we only have the three degenerate ground states.

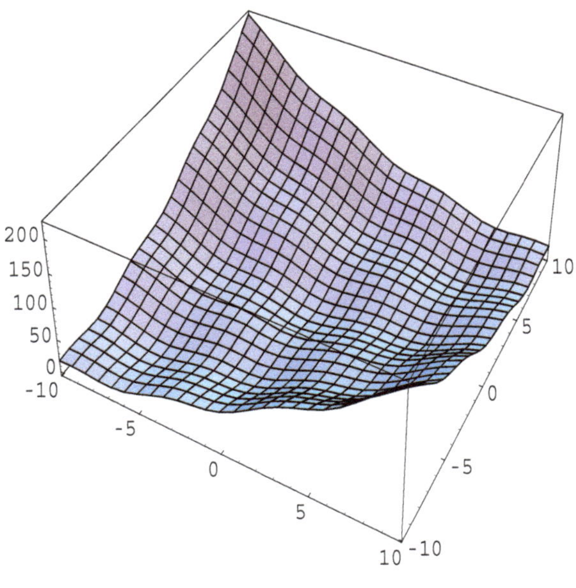

Fig. 2. Energy landscape $E(\phi_1, \phi_2)$ for $a = 3$, $\phi_i \in (-3\pi, 3\pi)$.

Figure 2 shows the energy landscape $E(\phi_1, \phi_2)$ for a small value of a. The saddle points act as energy barriers. Their height is proportional to the control parameter a, while the energy values of the excited states are all of the order of one. Accordingly, for large discreteness parameters a, the energy difference between these states is small compared to the energy required to overcome the barrier and switch to another state of local equilibrium. The same arguments hold for larger values of N, where we will show in the next sections that the energy landscape exhibits a highly complicated multi-valley structure with an exponentially increasing number of local minima and saddle points.

All possible phase differences ϕ_i as a function of the control parameter a are depicted in Fig. 3. Besides the lines with ordinates 2π, -2π, and 0, arising from the three-fold degenerate ground state, we see the curves constituing the 2π-kinks (second and fourth curve from above and from below) as well as the curves constituing the 4π-kinks (third curve from above and third curve from below). Inspection of the eigenvalues of the Hessian reveals that they are always stable for $a > 2\pi$ and for $a > 6\pi$, respectively.

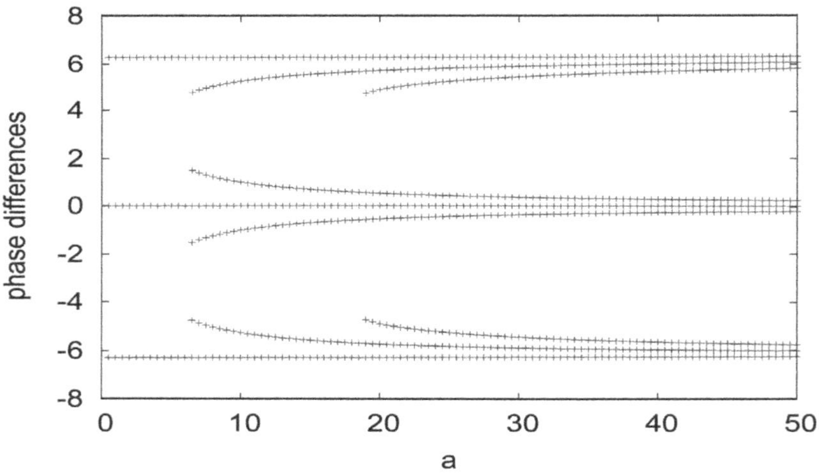

Fig. 3. Phase differences ϕ_i as a function of a for $N = 2$ and three potential wells.

Increasing the number of wells increases the number of bands. For a cosine potential with five wells, that is, if we restrict the values of ϕ_i, $i = 1, 2$, to the range $(-5\pi, 5\pi)$, all possible phase differences as a function of a are depicted in Fig. 4. In comparison with Fig. 3, here we see in addition the curves constituing the 6π-kinks, which only exist for $a > 10\pi$, and the two curves constituing the 8π-kinks, which only exist for $a > 14\pi$.

To be more general, if we restrict the values of ϕ_i, $i = 1, 2$, to the range $(-r\pi, r\pi)$, a stability analysis yields that a $k(2\pi)$-kink structure is only stable for $a \geq 2(2k-1)\pi$. These multi-2π-kinks were first discovered by Peyrard and Kruskal.[5]

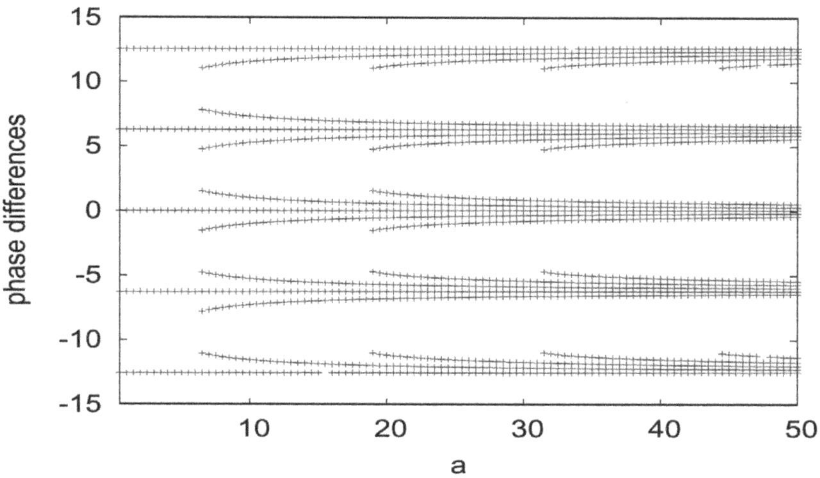

Fig. 4. Phase differences ϕ_i as a function of a for $N = 2$ and five potential wells.

Various bunched states of 2π-kinks as well as 4π-kinks were discovered later by Ustinov et al.[4] A general analytic theory about the origin of multikinks in dispersive nonlinear systems with a piecewise parabolic potential[6] has been presented by Champneys and Kivshar.[7]

4. Counting the Number of Locally Stable Energy Minima

There are only few results about the concrete number of excited equilibrium configurations, though it has often been mentioned that this number can be exponentially large, growing exponentially with the length N of the chain. According to the variational principle, stationary configurations obey the set of N coupled equations

$$-\phi_{i-1} + 2\phi_i - \phi_{i+1} + a\ \sin\phi_i = 0, \qquad i = 1, \ldots, N. \tag{6}$$

(As before, indices have to be taken mod N.) Inserting Eq. (6) into the energy function Eq. (1) the energy simplifies to

$$E = a \sum_{i=1}^{N} (-\phi_i \sin\phi_i + (1 - \cos\phi_i)) . \tag{7}$$

Let us now concentrate on large values of a, where the discreteness parameter is sufficiently strong. In the limit of large a, the stationary condition Eq. (6) reduces to

$$\sin\phi_i = 0, \qquad i = 1, \ldots, N. \tag{8}$$

If we restrict the ϕ_i's to the range $(-r\pi, r\pi)$, where r is odd, then it is obvious that the set of Eqs. (8) has exactly r^N distinct solutions, all describing locally stable minima in the limit $a \to \infty$. The solution vectors take the form

$\phi^* = (\phi_1^*, \phi_2^*, \ldots, \phi_N^*)$, where the components ϕ_i^* can only take one of the r discrete values $\phi_i^* \in \{-(r-1)\pi, \ldots, -2\pi, 0, 2\pi, \ldots, (r-1)\pi\}$, which correspond to the number of kinks trapped in the system. Provided that a is sufficiently large, solutions of Eq. (6) still exist near the ones in the limit $a \to \infty$ that we just described, and they are locally stable. They can be expanded in a Taylor series in $\frac{1}{a}$,

$$\phi_i = \phi_i^* - \frac{-\phi_{i-1}^* + 2\phi_i^* - \phi_{i+1}^*}{a} - \frac{-\phi_{i-2}^* + 4\phi_{i-1}^* - 6\phi_i^* + 4\phi_{i+1}^* - \phi_{i+2}^*}{a^2} + \cdots. \tag{9}$$

Inserting this expansion into Eq. (2), we immediately find

$$M = \sum_{i=1}^{N} \phi_i^*. \tag{10}$$

Hence, for sufficiently large values of a, the total phase can only take discrete values which are multiples of 2π ($M \in \{-(r-1)N\pi, \ldots, -2\pi, 0, 2\pi, \ldots, (r-1)N\pi\}$).

Since the junctions reside on a ring (of size N), we consider configurations that differ by a rotation $(\phi_i \to \phi_{s+i})_{1\le i\le N}$ (for some fixed s with $1 \le s \le N$) or by a mirror reflection (opposite orientations) $(\phi_i \to \phi_{s-i})_{1\le i\le N}$ (for some fixed s with $1 \le s \le N$) as equivalent. Note that, because of the periodic boundary conditions, the energy is invariant under these two types of transformations. Together, rotations and reflections form the dihedral group $\mathcal{D}_N$ which acts on the particles. The dihedral group is the symmetry group of an N-sided regular polygon.

Since, for large a, the stable configurations satisfying Eq. (6) can be identified with the r^N N-dimensional vectors whose components consist only of the value 0 and multiples of 2π in the range $\{-(r-1)\pi, \ldots, -2\pi, 0, 2\pi, \ldots, (r-1)\pi\}$, the number of inequivalent solutions will be equal to the number of orbits of these circular vectors under the action of the dihedral group $\mathcal{D}_N$. In more colloquial terms, the problem of counting the inequivalent stable configurations is equivalent to the problem of determining the number $D_r(N)$ of different necklaces composed of N beads, each of r possible colours. The reflection symmetry allows the necklace to be picked up out of the plane and turned over. It follows from Redfield–Pólya theory (cf. [8, Theorem 5.1] and [9,10,]) that the number of these necklaces is

$$D_r(N) = \begin{cases} \dfrac{1}{2} r^{(N+1)/2} + \dfrac{1}{2N} \sum_{d|N} \varphi(d)\, r^{N/d}, & \text{if } N \text{ is odd,} \\ \dfrac{1}{4} r^{N/2}(r+1) + \dfrac{1}{2N} \sum_{d|N} \varphi(d)\, r^{N/d}, & \text{if } N \text{ is even,} \end{cases} \tag{11}$$

where $\varphi(n)$ is the Euler totient function, which, by definition, is the number of positive integers $\le n$ that are relatively prime to n. To give an example, for three potential wells ($r = 3$), corresponding to necklaces made out of three colours, the first terms of the sequence $D_3(N)$ are given by

$$\{3, 6, 10, 21, 39, 92, 198, 498, 1219, 3210, \ldots\}, \tag{12}$$

$N = 1, 2, 3, \ldots$. For $N = p$ an odd prime number, Eq. (11) simplifies to

$$D_r(p) = \frac{r}{2}\left(\frac{r^{p-1} - 1}{p} + r^{\frac{p-1}{2}} + 1\right). \tag{13}$$

Moreover, since $\varphi(1) = 1$, the number of inequivalent locally stable configurations describing all local minima of the energy landscape $D_r(N)$ increases exponentially with N.

More precisely, as N tends to ∞, we have

$$D_r(N) \sim \frac{r^N}{2N}. \tag{14}$$

5. The Energy Spectrum

Let us now study the degeneracies of the energy. In the following we consider three potential wells, although the generalization to an arbitrary number r of wells is straightforward. Besides the rotations and mirror reflections described in the previous section, the Hamiltonian Eq. (1) is also invariant with respect to the transformation

$$\phi_i \to -\phi_i. \tag{15}$$

If we now consider configurations associated with the same energy as equivalent, for solving the corresponding counting problem we have to regard two necklaces also as equivalent if we keep one of the three colours fixed (ϕ_i close to 0; colour "0") and exchange the two remaining colours ($\phi_i \to -\phi_i$; that is, we identify colours "2π" and "-2π"). Thus, the situation that we encounter is that there is not only a symmetry group acting on the configurations (in our case: the dihedral group $\mathcal{D}_N$ acting on the beads), but there is also a symmetry group acting on the set of colours, namely the symmetric group S_2 which exchanges the two colours "2π" and "-2π".

It follows from Redfield–Pólya theory (cf. [8, Theorem 5.4, with $G = \mathcal{D}_N$ and $H = S_2$, the latter acting on *three* colours, that is, $P_H = \frac{1}{2}(x_1^3 + x_1 x_2)$]) that the number of inequivalent stable configurations equals

$$D_3^{S_2}(N) = \begin{cases} 3^{(N-1)/2} + \dfrac{1}{4} + \dfrac{1}{4N} \sum_{d|N} \varphi(d)\, 3^{N/d}, & \text{if } N \text{ is odd,} \\ 2 \cdot 3^{(N-2)/2} + \dfrac{1}{4 \cdot 2^{v_2(N)}} & \\ + \dfrac{1}{4N} \sum_{d|N} \varphi(d)\, (1 + \chi(d \text{ even }))3^{N/d}, & \text{if } N \text{ is even,} \end{cases} \tag{16}$$

for $N = 1, 2, 3, \ldots$, where $v_2(N)$ is the 2-adic valuation of N (that is, the largest v such that $2^v | d$), and where $\chi(\mathcal{A}) = 1$ if $\mathcal{A}$ is true and $\chi(\mathcal{A}) = 0$ otherwise.

For example, the first few values of this sequence are

$$\{2, 4, 6, 13, 22, 52, 106, 266, 630, 1652, \ldots\}. \tag{17}$$

This should be compared to Eq. (12).

Furthermore, for $\phi_i > 0$ ($0 < \phi_i < 2\pi$), the Hamiltonian is also invariant with respect to the transformation

$$\phi_i \to 2\pi - \phi_i, \tag{18}$$

as well as for $\phi_i < 0$ $(-2\pi < 0 < \phi_i)$ with respect to the complementary transformation

$$\phi_i \to -2\pi - \phi_i. \tag{19}$$

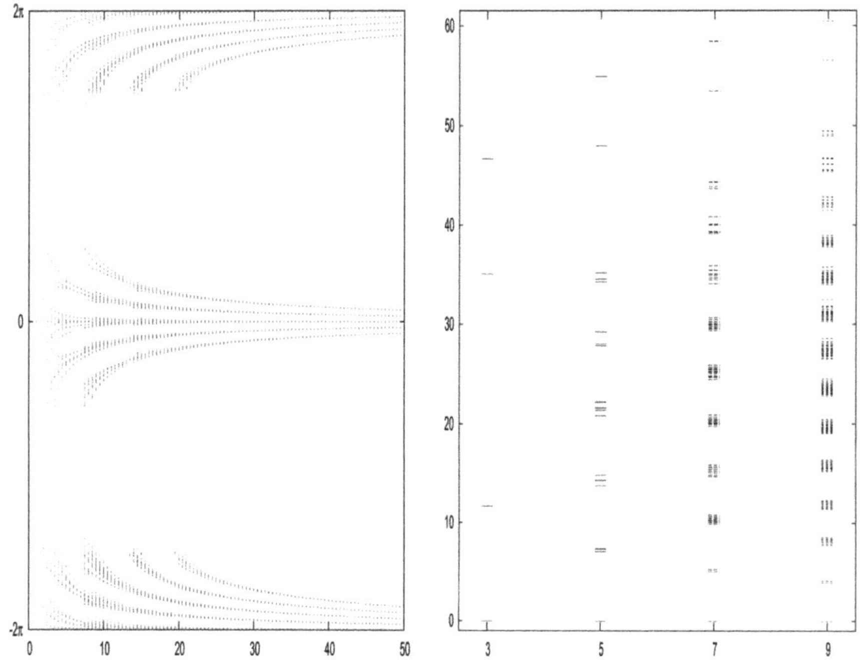

Fig. 5. Phase differences ϕ_i versus a for $N = 9$, and the energy spectrum versus N for $a = 25$.

In terms of our colouring problem, these two transformations translate into the transformation exchanging the two colours "0" and "2π" (but only for those necklaces featuring *only* these colours), respectively into the transformation exchanging the two colours "0" and "-2π" (but only for those necklaces featuring *only* these colours). Necklaces which are mapped to each other under these transformations are supposed to be equivalent.

It suffices to consider the transformation exchanging the two colours "0" and "2π". If we apply this transformation to a necklace featuring only these two colours then we may obtain the same necklace (modulo rotation and mirror reflection) or not. We must subtract the half of the number of those which are *not* mapped to itself from $D_3^{S_2}(N)$.

In order to determine this number, we consider equivalence classes of necklaces with beads which are coloured in only *two* colours, where we regard two necklaces as equivalent when they arise from each other by exchanging the two colours. Redfield–Pólya theory again (cf. [8, Theorem 5.4, with $G = \mathcal{D}_N$ and $H = S_2$, the latter acting on *two* colours, that is, $P_H = \frac{1}{2}(x_1^2+x_2)]$) yields that the number of these equivalence classes is

$$D_2^{S_2}(N) = \begin{cases} 2^{(N-3)/2} + \dfrac{1}{4N}\sum_{d|N}\varphi(d)\,2^{N/d}, & \text{if } N \text{ is odd,} \\ 2^{(N-2)/2} + \dfrac{1}{4N}\sum_{d|N}\varphi(d)\,(1+\chi(d \text{ even }))2^{N/d}, & \text{if } N \text{ is even.} \end{cases} \tag{20}$$

Then, one half of the number of necklaces with only these two colours which are not invariant under the exchange of "0" and "2π" equals $D_2(N) - D_2^{S_2}(N)$. Accordingly, if we write $D_3^E(N)$ for the number of inequivalent configurations associated with distinct energies, then we arrive at

$$D_3^E(N) = D_3^{S_2}(N) - (D_2(N) - D_2^{S_2}(N)), \tag{21}$$

$N = 1, 2, 3, \ldots$. The first few terms in this sequence are

$$\{1, 3, 4, 11, 18, 47, 97, 254, 607, 1618, \ldots\}. \tag{22}$$

The number of inequivalent locally stable minima $D_3^E(N)$ increases exponentially with N. More precisely, as N tends to ∞, we have

$$D_3^E(N) \sim \frac{3^N}{4N}. \tag{23}$$

Note that this is exactly half of the asymptotic value (14) with $r = 3$. In fact, the quotient of (23) by (14) (with $r = 3$), tends to $1/2$ exponentially fast (compare (12) and (22)).

The distribution of the phase differences as a function of the control parameter a is shown in Fig. 5 (left). We observe bands and "sub-bands," as well as gap structures. The bands cluster into groups, which themselves cluster again into sub-groups, and so on.

In other words, each band breaks up into sub-bands, which again break up into sub-sub-bands, and so on. Similarly, between the individual bands we find gap structures as well as sub-gap structures. These structures are self-similar in the sense that each portion is quasi a reduced-scale replica of the whole.[6,12] Moreover, exponentially many phase difference values come exponentially close to each other. Fig. 5 (right) depicts the energy spectrum for the control parameter $a = 25$ for $N = 3, 5, 7$, and 9 junctions, respectively. We clearly see how the band- and the fractal structure evolve with increasing number of junctions.

According to Eq. (17) the figure depicts the 4, 18, 97, and 607 distinct energies for $N = 3, 5, 7$ and $N = 9$, respectively. Also here, a thorough analysis reveals that the inner bands exhibit features of self-similarity.

6. Discussion

We have shown that, depending on the strength of the r-well cosine potential, the energy surface of the model can have of the order of $\frac{r^N}{N}$ locally stable minima separated by saddle points representing large barriers. The counting problem for the degeneracy of the total energy turned out to be equivalent to a well-known problem in combinatorial mathematics, in which colourings of necklaces are enumerated, with the dihedral group acting on the beads and another symmetry group acting on the colours, the solution of the problem being provided by Redfield–Pólya theory. Another important finding is that the distribution of the phase differences as well as the energy spectrum is fractal for a sufficiently large control parameter. We observed that the distribution of the phase differences form a hierarchical band structure such that exponentially many configurations become exponentially close to each other. Preliminary investigations also show a similar hierarchical band structure for the energy spectrum.[10–12]

The structure of the energy landscape is reminiscent of the appearance of glassy structures within the formation of domains, kinks, and solitons. However, in contrast to the familiar spin glass models, the model is described by a Hamiltonin without disorder.

Acknowledgments

The first author thanks P. Martin, S. Charpentier and M. Olivier of the Université du Littoral de Dunkerque for helpful computer assistance. We also thank F.V. Kusmartsev for useful discussions about experimental aspects. This work has been partially supported by the European Science Foundation (ESF) in the framework of the network program "Arrays of Quantum Dots and Josephson Junctions".

References

1. O. M. Braun and Y. S. Kivshar, *The Frenkel–Kontorova Model, Concepts, Methods, and Applications*, (Springer-Verlag, Berlin, Heidelberg, New York).
2. O. V. Zhirov, G. Casati, and D. L. Shepelyansky, Phys. Rev. E **65**, 026220 (2002).
3. J. Pfeiffer, M. Schuster, A. A. Abdumalikov, Jr., and A. V. Ustinov, Phys. Rev. Lett. **96**, 034103 (2006).
4. A. V. Ustinov, B. A. Malomed, and S. Sakai, Phys. Rev. B **57**, 11691 (1998).
5. M. Peyrard and M. D. Kruskal, Physica D **14**, 88 (1984).
6. K. E. Kürten, Fractal properties of small magnetic particles and magnetic multilayers: An exactly solvable model, *Condensed Matter Theories*, Vol. 20, edited by J. W. Clark, (Nova Science Publishers, New York, 2005).
7. A. Champneys and Y. S. Kivshar, Phys. Rev. E **61**(3), 2551 (2000).
8. N. G. de Bruijn, *Pólya's theory of counting*, in: Applied Combinatorial Mathematics, E. .F. Beckenbach, ed., Wiley, New York, 1964; reprinted by Krieger, Malabar, Florida, 1981.
9. E. N. Gilbert and J. Riordan, Illinois J. Math. **5**, 657 (1961).
10. K. E. Kürten and C. Krattenthaler, Multistability, Phase Transitions and Fractal Structures in Magnetic Multilayers, *Condensed Matter Theories*, Vol. 19, edited by E. Suraud, (Nova Science Publishers, New York, 2004).
11. K. E. Kürten and F. V. Kusmartsev, Phys. Rev. B **72**, 014433 (2005).
12. M. D. Forrester, K. E. Kürten and F. V. Kusmartsev, Phys. Rev. B **74**, 1 (2006).

LOWERING OF BOSON-FERMION SYSTEM ENERGY WITH A GAPPED COOPER RESONANT-PAIR DISPERSION RELATION

T. A. MAMEDOV

*Faculty of Engineering, Baskent University, 06530 Ankara, Turkey and Institute of Physics, Academy of Sciences of Azerbaijan Baku 370143, Azerbaijan**
E-mail: tmamedov@baskent.edu.tr

M. DE LLANO

*H.H. Wills Physics Laboratory, University of Bristol Bristol BS8 1TL, UK and Instituto de Investigaciones en Materiales, Universidad Nacional Autnoma de Mxico Apdo. Postal 70-360, 04510 Mxico, DF, Mexico**
E-mail: dellano@servidor.unam.mx

Applying two-time Green-function techniques to the Friedberg-T.D. Lee phenomenological Hamiltonian of a many-fermion system, it is shown that positive-energy resonant bosonic pairs associated with four-fermion excitations above the Fermi sea are energetically lower in a ground-state that is a mixture of two coexisting and dynamically interacting many-particle subsystems: a) unpaired fermions and b) composite bosons. It is argued that an interaction between free fermions and bosons excited *above* the Fermi sea in the mixture, namely, the continuous processes of pair-formation from, and disintegration into, two unpaired electrons, results in a substantially lowering the total system energy. The positive-energy composite bosons begin to appear incoherently below a depairing temperature T^* as their coupling- and temperature-dependent number density gradually increases from zero. This leads quite naturally to the pseudogap phenomenon observed in high-T_c cuprates

Keywords: High-T_c superconductivity; boson-fermion model; gapped Cooper-pair dispersion.

1. Introduction

We consider system of N-electrons with pairwise attractive interactions of strength $-V$ at finite temperature T and assume all particles in single-fermion states distributed around an appropriate Fermi energy $E_F \gg k_B T$. At sufficiently high T thermal effects are strong enough so that in spite of the attraction between electrons there are yet no *paired* electrons in the system. The interaction $-V$ acts only on electrons whose energies are within the interval $[E_F - \hbar\omega_D, E_F + \hbar\omega_D]$ with $\hbar\omega_D \ll E_F$ being the Debye cutoff energy. We then propose that below some temperature T^* some of fermions may pair up forming two mutually-confined electrons

*Permanent address.

(or/and holes) near the Fermi surface and considered here as composite bosons with a finite lifetime.

The Friedberg-T.D. Lee[1] phenomenological Hamiltonian describing binary boson-fermion (BF) mixture is written in the form

$$H = \sum_{\mathbf{k},\sigma} \epsilon_{\mathbf{k}} a^{+}_{\mathbf{k}\sigma} a_{\mathbf{k}\sigma} + \sum_{\mathbf{K}} E(\mathbf{K}) b^{+}_{\mathbf{K}} b_{\mathbf{K}} + H_{int} \tag{1}$$

$$H_{int} \equiv \frac{f}{L^{d/2}} \sum_{\mathbf{q},\mathbf{K}} \left(b^{+}_{\mathbf{K}} a_{\mathbf{q}+\mathbf{K}/2\uparrow} a_{-\mathbf{q}+\mathbf{K}/2\downarrow} + b_{\mathbf{K}} a^{+}_{-\mathbf{q}+\mathbf{K}/2\downarrow} a^{+}_{\mathbf{q}+\mathbf{K}/2\uparrow} \right) \tag{2}$$

where (2) represents the formation and disintegration processes of the two-electron (2e)-CPs. Here $a^{+}_{\mathbf{k}\sigma}$ and $a_{\mathbf{k}\sigma}$ are the usual fermion creation and annihilation operators for individual electrons of momenta $\mathbf{k}$ and spin $\sigma = \uparrow, \downarrow$ while $b^{+}_{\mathbf{K}}$ and $b_{\mathbf{K}}$ are *postulated* (as in Refs. 2,3) to be bosonic operators representing Cooper pairs (CPs) of total, or center-of-mass, momentum (CMM) wavevector $\mathbf{K}$. CP energies $E(\mathbf{K})$ in (2) are taken as

$$E(\mathbf{K}) = 2E_F + 2\Delta + \varepsilon_{\mathbf{K}} \tag{3}$$

with $\varepsilon_{\mathbf{K}}$ the boson dispersion relation depending on $\mathbf{K}$, and 2Δ is a coupling- and temperature-dependent energy necessary to create a single bosonic excitation. This is the energy for a composite boson of two confined electrons to appear above E_F. Finally, f in (2) is a phenomenological form factor responsible for *2*e-CP formation/disintegration processes related with the attractive inter-electron (*four-*fermion) electron-phonon strength parameter V of the Cooper/BCS model interaction through $f = \sqrt{2\hbar\omega_D V}$ (see in Refs. 2,3). To clarify the origin of the term 2Δ in (3) consider the familiar Cooper equation[4,5]

$$1 = \frac{VN(0)}{2} \int_{E_F}^{E_F+\hbar\omega_D} \frac{d\epsilon}{\epsilon - E_F - \varepsilon_{\mathbf{0}}/2}, \tag{4}$$

with its well-known negative-energy solution

$$\varepsilon^{-}_{\mathbf{0}} = -\frac{2\hbar\omega_D}{e^{2/\lambda} - 1} \rightarrow -2\hbar\omega_D e^{-2/\lambda} \tag{5}$$

for the binding energy of two electrons $\varepsilon_{\mathbf{0}}$ associated with the appearance of bound pairs. This solution $\varepsilon^{-}_{\mathbf{0}}$ played a key role in understanding the superconducting state and lead to the BCS model.[6] However, along with the bound-state solution $\varepsilon^{-}_{\mathbf{0}} < 0$ (4) possesses a second solution $\varepsilon^{+}_{\mathbf{0}} > 0$ given by

$$\varepsilon^{+}_{\mathbf{0}} = +\frac{2\hbar\omega_D}{e^{2/\lambda} + 1} \rightarrow +2\hbar\omega_D e^{-2/\lambda} \tag{6}$$

due to the elementary fact that $\int x^{-1}dx = \ln|x|$ and not $\ln x$ as commonly assumed. Such a solution corresponds to pairs with a positive energy gap analogous to 2Δ in (3), i.e., as the resonant modes with nonzero-CMM found in 3D and 2D in the more general treatment[7,8] of Cooper pairing via the Bethe-Salpeter (BS) integral equation that does not neglect hole-pairs. To understand what new physics emerges from the positive-energy boson modes suggested by (6), or/and by the BS equation results of Refs. 7,8, we have assumed that the energies $E(\mathbf{K})$ in (1) are V-dependent and gapped at zero K with a V-dependent gap 2Δ in (3). Excitations with positive 2Δ would seem to *increase* the contribution from the second term on the rhs of (1), i.e., from free bosons, and therefore their existence would seem to be energetically unfavorable. The main goal of this paper is to show that this assertion need not be true and that taking into account the corrections stemming from H_{int} in (1), which appear to be negative thus changing the situation substantially. We show that paired states excited above the Fermi sea (i.e., associated with the $\varepsilon_{\mathbf{0}}^{+}$ solution (6) or with the positive 2Δ as in the more complete treatment[7,8] of the fermionic subsystem) may in fact turn out to be energetically lower in a different ground-state, namely, one corresponding to a BF mixture given by (1) and which is distinct from the BCS ground-state. This is done by direct calculation of the energy corresponding to (1) and proving that the continuous processes of pair formation and disintegration into two unpaired electrons may result in lowering considerably the total system energy and therefore justifying a BF mixture with bosons excited *above* the Fermi sea. Another goal is to show that a conception based on the Hamiltonian (1) with positive-energy resonant pairs very naturally leads to the pseudogap[9,10] phenomena observed in high-T_c cuprates. We introduce a grand canonical Hamiltonian $\mathcal{H} \equiv H - \mu N$ where the electronic chemical potential μ is to be fixed by requiring a constant electron number

$$N \equiv \langle \sum_{\mathbf{k},\sigma} a_{\mathbf{k}\sigma}^{+} a_{\mathbf{k}\sigma} + 2\sum_{\mathbf{K}} b_{\mathbf{K}}^{+} b_{\mathbf{K}} \rangle$$

where the last term accounts for those electrons paired into composite bosons. Single angular brackets $\langle X \rangle$ of an operator X are T-dependent thermal averages over the Hamiltonian $\mathcal{H}$. The Hamiltonian $\mathcal{H}$ itself is

$$\mathcal{H} \equiv \mathcal{H}_0 + H_{int} = \sum_{\mathbf{k},\sigma} (\epsilon_{\mathbf{k}} - \mu) a_{\mathbf{k}\sigma}^{+} a_{\mathbf{k}\sigma} + \sum_{\mathbf{K}} [E(\mathbf{K}) - 2\mu] b_{\mathbf{K}}^{+} b_{\mathbf{K}} + H_{int}. \tag{7}$$

2. Total Boson-Fermion System Energy

The total BF system energy $\mathcal{E} \equiv \langle \mathcal{H} \rangle$ is

$$\mathcal{E} = \langle \mathcal{H}_F^o \rangle + \langle \mathcal{H}_B^o \rangle + \langle H_{int} \rangle . \tag{8}$$

It consists of three pieces corresponding to: a) unpaired fermions

$$\mathcal{H}_F^o = \sum_{\mathbf{k},\sigma} (\epsilon_{\mathbf{k}} - \mu) a_{\mathbf{k}\sigma}^{+} a_{\mathbf{k}\sigma}, \tag{9}$$

b) free composite bosons

$$\mathcal{H}_B^o = \sum_{\mathbf{K}} [E(\mathbf{K}) - 2\mu] b_{\mathbf{K}}^{+} b_{\mathbf{K}}, \tag{10}$$

and c) the BF interaction causing boson formations/disintegrations

$$H_{int} \equiv H_{int}^{(1)} + H_{int}^{(2)} = \frac{f}{L^{d/2}} \sum_{\mathbf{q},\mathbf{K}} \left(b_{\mathbf{K}}^{+} a_{\mathbf{q}+\mathbf{K}/2\uparrow} a_{-\mathbf{q}+\mathbf{K}/2\downarrow} + b_{\mathbf{K}} a_{-\mathbf{q}+\mathbf{K}/2\downarrow}^{+} a_{\mathbf{q}+\mathbf{K}/2\uparrow}^{+} \right). \tag{11}$$

The contributions $\langle \mathcal{H}_F^o \rangle$ and $\langle \mathcal{H}_B^o \rangle$ are easily calculated by first determining the (temperature- and coupling-dependent) average number of composite bosons with a fixed CMM wavevector $\mathbf{K}$ defined as $n_{B,\mathbf{K}}(T) \equiv \langle b_{\mathbf{K}}^{+} b_{\mathbf{K}} \rangle$ as well as the average number of electrons $n_{\mathbf{k}\sigma}(T) \equiv \langle a_{\mathbf{k}\sigma}^{+} a_{\mathbf{k}\sigma} \rangle$ in a state $\mathbf{k}$ and σ. These were determined in Ref. 11,12. As to the contribution of $\langle H_{int} \rangle$ to (8), we shall express it in terms of two-time *single-particle* Green functions.

3. Two-Time Green-Function Analysis

Retarded Green functions (GFs) at times t and t' for dynamical operators $A(t)$ and $B(t')$ (also known as two-time or double-time GFs) are defined[13] via double-angular brackets as

$$\left\langle \left\langle A(t) \mid B(t') \right\rangle \right\rangle \equiv \theta(t-t') \left\langle [A(t), B(t')]_\eta \right\rangle$$

where the square brackets $[A,B]_\eta \equiv AB + \eta BA$ denote the commutator ($\eta = -1$) or anticommutator ($\eta = +1$) of operators A and B, and $\theta(x)$ is the Heaviside unit step function. In this formalism any dynamical operator X is in the Heisenberg representation, i.e., is of the form $X(t) = \exp(i\mathcal{H}t) X \exp(-i\mathcal{H}t)$. It is not difficult to see that Fourier transforms defined as

$$\langle\langle A \mid B \rangle\rangle_\omega \equiv \frac{1}{2\pi} \int_{-\infty}^{\infty} \left\langle \left\langle A(t) \mid B(t') \right\rangle \right\rangle e^{i\omega(t-t')} d(t-t')$$

satisfy the chain of equations[13]

$$\omega \langle\langle A \mid B \rangle\rangle_\omega = \left\langle [A,B]_\eta \right\rangle + \langle\langle [A,\mathcal{H}]_- \mid B \rangle\rangle_\omega \tag{12}$$

where $\langle\langle [A,\mathcal{H}]_- \mid B \rangle\rangle_\omega$ is a higher-order GF in the variable ω. Knowing the Fourier transforms $\langle\langle A \mid B \rangle\rangle_\omega$ one can find expressions for the corresponding correlation functions $\left\langle A(t) B(t') \right\rangle$ from the relation[13]

$$\left\langle A(t) B(t') \right\rangle = \frac{1}{2\pi} \int_{-\infty}^{\infty} d\omega e^{i\omega(t-t')} J_{AB}(\omega), \tag{13}$$

where the so-called *spectral density* function $J_{AB}(\omega)$ is in turn determined from

$$\{ \langle\langle A \mid B \rangle\rangle_{\omega+i\varepsilon} - \langle\langle A \mid B \rangle\rangle_{\omega-i\varepsilon} \} = -i(e^{\hbar\omega/k_B T} + \eta) J_{BA}(\omega). \tag{14}$$

Note that the order of operators A and B in $J_{AB}(\omega)$ is important and that the relation

$$J_{AB}(\omega) = e^{-\hbar\omega/k_B T} J_{BA}(-\omega), \tag{15}$$

follows on transposing A and B.

Choosing $A \equiv b_{\mathbf{Q}}$ and $B \equiv b_{\mathbf{Q}}^{+}$ in (12) with $\eta = -1$ we write (12) as

$$\omega \left\langle \left\langle b_{\mathbf{Q}} \mid b_{\mathbf{Q}}^{+} \right\rangle \right\rangle_{\omega} = 1 + \left\langle \left\langle [b_{\mathbf{Q}}, \mathcal{H}]_{-} \mid b_{\mathbf{Q}}^{+} \right\rangle \right\rangle_{\omega} \tag{16}$$

since $[b_{\mathbf{Q}}, b_{\mathbf{K}}^{+}]_{-} = \delta_{\mathbf{Q},\mathbf{K}}$. As is customary in BF models, we assume that a and b operators in (1) commute with each other. Since $[b_{\mathbf{Q}}, b_{\mathbf{K}}]_{-} = 0$ while $[b_{\mathbf{Q}}, b_{\mathbf{K}}^{+} b_{\mathbf{K}}]_{-} = b_{\mathbf{K}} \delta_{\mathbf{Q},\mathbf{K}}$ we have that

$$[b_{\mathbf{Q}}, \mathcal{H}]_{-} = [E(\mathbf{Q}) - 2\mu] b_{\mathbf{Q}} + \frac{f}{L^{d/2}} \sum_{\mathbf{q}} \left\langle \left\langle a_{\mathbf{q}+\mathbf{Q}/2\uparrow} a_{-\mathbf{q}+\mathbf{Q}/2\downarrow} \mid b_{\mathbf{Q}}^{+} \right\rangle \right\rangle_{\omega}.$$

Inserting this into (16) yields

$$[\omega - E(\mathbf{Q}) + 2\mu] \left\langle \left\langle b_{\mathbf{Q}} \mid b_{\mathbf{Q}}^{+} \right\rangle \right\rangle_{\omega} = 1 + \frac{f}{L^{d/2}} \sum_{\mathbf{q}} \left\langle \left\langle a_{\mathbf{q}+\mathbf{Q}/2\uparrow} a_{-\mathbf{q}+\mathbf{Q}/2\downarrow} \mid b_{\mathbf{Q}}^{+} \right\rangle \right\rangle_{\omega}. \tag{17}$$

Let us now employ (14). Setting in (17) first $\omega = \omega + i\varepsilon$ and then $\omega \to \omega - i\varepsilon$, with a positive $\varepsilon \to 0$, consider the change in (17) that occurs on passing through the real axis ω from above ($\omega + i\varepsilon$) to below ($\omega - i\varepsilon$). It is easy to see that the lhs of (17) then changes by

$$\begin{aligned} &[\omega + i\varepsilon - E(\mathbf{Q}) + 2\mu] \left\langle \left\langle b_{\mathbf{Q}} \mid b_{\mathbf{Q}}^{+} \right\rangle \right\rangle_{\omega+i\varepsilon} - [\omega - i\varepsilon - E(\mathbf{Q}) + 2\mu] \left\langle \left\langle b_{\mathbf{Q}} \mid b_{\mathbf{Q}}^{+} \right\rangle \right\rangle_{\omega-i\varepsilon} \\ &\equiv [\omega - E(\mathbf{Q}) + 2\mu] \left\{ \left\langle \left\langle b_{\mathbf{Q}} \mid b_{\mathbf{Q}}^{+} \right\rangle \right\rangle_{\omega+i\varepsilon} - \left\langle \left\langle b_{\mathbf{Q}} \mid b_{\mathbf{Q}}^{+} \right\rangle \right\rangle_{\omega-i\varepsilon} \right\} \\ &= -i(e^{\hbar\omega/k_B T} - 1)[\omega - E(\mathbf{Q}) + 2\mu] J_{b_{\mathbf{Q}}^{+} b_{\mathbf{Q}}}(\omega) \end{aligned} \tag{18}$$

where according to (14) a spectral density $J_{b_{\mathbf{Q}}^{+} b_{\mathbf{Q}}}(\omega)$ for $\left\langle \left\langle b_{\mathbf{Q}} \mid b_{\mathbf{Q}}^{+} \right\rangle \right\rangle_{\omega}$ is introduced in the last line of (18). On the other hand, the change in the rhs of (17) on passing from $\omega + i\varepsilon$ to $\omega - i\varepsilon$ is just

$$-i(e^{\hbar\omega/k_B T} - 1) \frac{f}{L^{d/2}} \sum_{\mathbf{q}} J_{b_{\mathbf{Q}}^{+} a_{\mathbf{q}+\mathbf{Q}/2\uparrow} a_{-\mathbf{q}+\mathbf{Q}/2\downarrow}}(\omega). \tag{19}$$

Equating (18) and (19) we have

$$\frac{f}{L^{d/2}} \sum_{\mathbf{q}} J_{b_{\mathbf{Q}}^{+} a_{\mathbf{q}+\mathbf{Q}/2\uparrow} a_{-\mathbf{q}+\mathbf{Q}/2\downarrow}}(\omega) = [\omega - E(\mathbf{Q}) + 2\mu] J_{b_{\mathbf{Q}}^{+} b_{\mathbf{Q}}}(\omega). \tag{20}$$

Owing to (13) and summing (20) over $\mathbf{Q}$, followed by integration over ω and dividing by 2π, immediately gives

$$\frac{f}{L^{d/2}} \sum_{\mathbf{q},\mathbf{Q}} \left\langle b_{\mathbf{Q}}^{+} a_{\mathbf{q}+\mathbf{Q}/2\uparrow} a_{-\mathbf{q}+\mathbf{Q}/2\downarrow} \right\rangle = \frac{1}{2\pi} \sum_{\mathbf{Q}} \int_{-\infty}^{\infty} [\omega - E(\mathbf{Q}) + 2\mu] J_{b_{\mathbf{Q}}^{+} b_{\mathbf{Q}}}(\omega) d\omega \tag{21}$$

the rhs of which is $\left\langle H_{int}^{(1)}\right\rangle$ as in (11). The last equation then becomes

$$\left\langle H_{int}^{(1)}\right\rangle = -\sum_{\mathbf{K}}[E(\mathbf{K})-2\mu]\left\langle b_{\mathbf{K}}^{+}b_{\mathbf{K}}\right\rangle + \frac{1}{2\pi}\sum_{\mathbf{K}}\int_{-\infty}^{\infty}\omega J_{b_{\mathbf{K}}^{+}b_{\mathbf{K}}}(\omega)d\omega. \tag{22}$$

Considering $\left\langle\left\langle b_{\mathbf{Q}}^{+} \mid b_{\mathbf{Q}}\right\rangle\right\rangle_{\omega}$ and repeating calculations analogously as above (please see Appendix) leads to

$$\frac{f}{L^{d/2}}\sum_{\mathbf{q}} J_{b_{\mathbf{Q}}a^{+}_{-\mathbf{q}+\mathbf{Q}/2\downarrow}a^{+}_{\mathbf{q}+\mathbf{Q}/2\uparrow}}(\omega) = -[\omega+E(\mathbf{Q})-2\mu]J_{b_{\mathbf{Q}}b_{\mathbf{Q}}^{+}}(\omega) \tag{23}$$

which in turn gives

$$\left\langle H_{int}^{(2)}\right\rangle = -\sum_{\mathbf{K}}[E(\mathbf{K})-2\mu]\left\langle b_{\mathbf{K}}b_{\mathbf{K}}^{+}\right\rangle - \frac{1}{2\pi}\sum_{\mathbf{K}}\int_{-\infty}^{\infty}\omega J_{b_{\mathbf{K}}b_{\mathbf{K}}^{+}}(\omega)d\omega. \tag{24}$$

Summing (22) and (24) we have

$$\begin{aligned}\langle H_{int}\rangle \equiv \left\langle H_{int}^{(1)}\right\rangle + \left\langle H_{int}^{(2)}\right\rangle &= -\sum_{\mathbf{K}}[E(\mathbf{K})-2\mu]\left\langle b_{\mathbf{K}}^{+}b_{\mathbf{K}}+b_{\mathbf{K}}b_{\mathbf{K}}^{+}\right\rangle + \\ &\frac{1}{2\pi}\sum_{\mathbf{K}}\int_{-\infty}^{\infty}\omega[J_{b_{\mathbf{K}}^{+}b_{\mathbf{K}}}(\omega)-J_{b_{\mathbf{K}}b_{\mathbf{K}}^{+}}(\omega)]d\omega.\end{aligned} \tag{25}$$

In (25) we may use: a) $\left\langle b_{\mathbf{K}}^{+}b_{\mathbf{K}}+b_{\mathbf{K}}b_{\mathbf{K}}^{+}\right\rangle \equiv 1+2n_{B,\mathbf{K}}$ as follows from the commutation relation between b operators; and b) according to (15)

$$J_{b_{\mathbf{K}}b_{\mathbf{K}}^{+}}(\omega) = e^{-\hbar\omega/k_BT}J_{b_{\mathbf{K}}^{+}b_{\mathbf{K}}}(-\omega).$$

This gives

$$\begin{aligned}\langle H_{int}\rangle &= -2\sum_{\mathbf{K}}[E(\mathbf{K})-2\mu](n_{B,\mathbf{K}}+1/2)+ \\ &\frac{1}{2\pi}\sum_{\mathbf{K}}\int_{-\infty}^{\infty}\omega[J_{b_{\mathbf{K}}^{+}b_{\mathbf{K}}}(\omega)-e^{-\hbar\omega/k_BT}J_{b_{\mathbf{K}}^{+}b_{\mathbf{K}}}(-\omega)]d\omega.\end{aligned} \tag{26}$$

In the last term replacing the argument $-\omega$ by ω we finally have

$$\begin{aligned}\langle H_{int}\rangle &= -2\sum_{\mathbf{K}}[E(\mathbf{K})-2\mu](n_{B,\mathbf{K}}+1/2) \\ &+\frac{1}{2\pi}\sum_{\mathbf{K}}\int_{-\infty}^{\infty}\omega(1+e^{\hbar\omega/k_BT})J_{b_{\mathbf{K}}^{+}b_{\mathbf{K}}}(\omega)d\omega\,.\end{aligned} \tag{27}$$

This result expresses the contribution from H_{int} to the total energy $\mathcal{E} \equiv \langle\mathcal{H}\rangle$ of a BF-mixture. Note the *minus* sign preceding twice the free boson energy $\langle\mathcal{H}_B^o\rangle =$

$\sum_{\mathbf{K}}[E(\mathbf{K})-2\mu](n_{B,\mathbf{K}}+1/2)$. Using (27) in (8) gives

$$\mathcal{E}=\sum_{\mathbf{k},\sigma}(\epsilon_{\mathbf{k}}-\mu)n_{\mathbf{k}\sigma}-\sum_{\mathbf{K}}[E(\mathbf{K})-2\mu](n_{B,\mathbf{K}}+1)+$$
$$\frac{1}{2\pi}\sum_{\mathbf{K}}\int_{-\infty}^{\infty}\omega(1+e^{\hbar\omega/k_BT})J_{b_{\mathbf{K}}^{+}b_{\mathbf{K}}}(\omega)d\omega. \tag{28}$$

We need $J_{b_{\mathbf{K}}^{+}b_{\mathbf{K}}}(\omega)$ to symplify (28). Using Eq. (50) of Ref. 12, namely

$$J_{b_{\mathbf{K}}^{+}b_{\mathbf{K}}}(\omega)=2\pi(e^{\hbar\omega/k_BT}-1)^{-1}\frac{1}{B(\omega,\mathbf{K})}\delta(A(\omega,\mathbf{K})/B(\omega,\mathbf{K}))$$

where

$$A(\omega,\mathbf{K})\equiv\omega-E(\mathbf{K})+2\mu-\frac{f^2}{L^d}\sum_{\mathbf{q}}\frac{F(\mathbf{q},\mathbf{K})}{\omega-\zeta(\mathbf{q},\mathbf{K})}$$

$$B(\omega,\mathbf{K})\equiv 1+\frac{f^2}{L^d}\sum_{\mathbf{q}}\frac{F(\mathbf{q},\mathbf{K})}{[\omega-\zeta(\mathbf{q},\mathbf{K})]^2}$$

where $\zeta(\mathbf{q},\mathbf{K})\equiv\epsilon_{-\mathbf{q}+\mathbf{K}/2}+\epsilon_{\mathbf{q}+\mathbf{K}/2}-2\mu$ and $F(\mathbf{q},\mathbf{K})\equiv 1-n_{\mathbf{q}+\mathbf{K}/2\uparrow}-n_{-\mathbf{q}+\mathbf{K}/2\downarrow}$ with

$$n_{\mathbf{k}}=\frac{1}{2}\left\{1-\frac{\epsilon_{\mathbf{k}}-\mu(V,T)}{\sqrt{[\epsilon_{\mathbf{k}}-\mu(V,T)]^2+f^2n_{B0}(T)}}\tanh\frac{\sqrt{[\epsilon_{\mathbf{k}}-\mu(V,T)]^2+f^2n_{B0}(T)}}{2k_BT}\right\}.$$

Magnitude of the term proportional f^2 in $B(\omega,\mathbf{K})$ is much smaller than the corresponding correction $\sim f^2$ in $A(\omega,\mathbf{K})$. This is because the sign changes in $F(\mathbf{q},\mathbf{K})$ under the summation $\sum_{\mathbf{q}}$ reduce the magnitude of $B(\omega,\mathbf{K})$ due to the positive factor $[\omega-\zeta(\mathbf{q},\mathbf{K})]^2$. As a result of this reduction, $B(\omega,\mathbf{K})$ differs very slightly from 1. Assuming $B(\omega,\mathbf{K})\approx 1$ we immediately get for the total system energy

$$\mathcal{E}=\sum_{\mathbf{k},\sigma}(\epsilon_{\mathbf{k}}-\mu)n_{\mathbf{k}\sigma}-\sum_{\mathbf{K}}[E(\mathbf{K})-2\mu]n_{B,\mathbf{K}}+$$
$$\sum_{\mathbf{K}}\{\hbar\Omega_{\mathbf{K}}\coth(\hbar\Omega_{\mathbf{K}}/2k_BT)-[E(\mathbf{K})-2\mu]\} \tag{29}$$

where $\hbar\Omega_{\mathbf{K}}$ is a root of $A(\hbar\Omega,\mathbf{K})=0$. Regarding the energies $\hbar\Omega_{\mathbf{K}}$ at which the single-particle GF $\left\langle\left\langle b_{\mathbf{Q}}\mid b_{\mathbf{Q}}^{+}\right\rangle\right\rangle_{\omega}$ has a singular behavior, they are essentially boson spectral-energy values renormalized by BF interactions. Physically, $\hbar\Omega_{\mathbf{K}}$ must at least be nonnegative. From (3) and the condition $A(\hbar\Omega,\mathbf{K})=0$ one then has

$$\hbar\Omega_{\mathbf{K}}=2(E_F-\mu)+2\Delta+\frac{f^2}{L^d}\sum_{\mathbf{q}}\frac{F(\mathbf{q},\mathbf{K})}{\hbar\Omega_{\mathbf{K}}-\zeta(\mathbf{q},\mathbf{K})}. \tag{30}$$

Let us now consider temperatures T low enough so that the condition $\hbar\Omega_{\mathbf{K}}/2k_BT \gg 1$ is satisfied. For such temperatures we may put $\coth(\hbar\Omega_{\mathbf{K}}/2k_BT) \simeq 1$ in (29) and obtain

$$\mathcal{E} \simeq \sum_{\mathbf{k},\sigma}(\epsilon_{\mathbf{k}}-\mu)n_{\mathbf{k}\sigma} - \sum_{\mathbf{K}}[E(\mathbf{K})-2\mu]n_{B,\mathbf{K}} + \sum_{\mathbf{K}}\left[\frac{f^2}{L^d}\sum_{\mathbf{q}}\frac{F(\mathbf{q},\mathbf{K})}{\hbar\Omega_{\mathbf{K}}-\zeta(\mathbf{q},\mathbf{K})}\right]. \tag{31}$$

This shows how the total energy of an attractively-interacting N-fermion system changes as it evolves into a BF mixture. The second term in (31) is associated with the appearance of composite bosons and *lowers* the total energy $\mathcal{E}$. This energy lowering is greater for *positive* 2Δ in (3). Replacing the summation in (31) over free-fermion momenta $\mathbf{q}$ by an energy ϵ integration within the interval $\mu - \hbar\omega_D \leq \epsilon \leq \mu + \hbar\omega_D$ we may estimate, at least graphically, the terms within the last square brackets in (31). This was done by graphing $1/[\hbar\Omega_{\mathbf{K}} - \zeta(\epsilon,\mathbf{K})]$ and $F(\epsilon,\mathbf{K})$ vs. ϵ and noting the regions contributing with positive and negative signs to integral $\int F(\epsilon,\mathbf{K})[\hbar\Omega_{\mathbf{K}}-\zeta(\epsilon,\mathbf{K})]^{-1}d\epsilon$. When $\Omega_{\mathbf{K}} \ll \omega_D$ these integrals appear to be negative. Therefore, introducing an attractive interaction between electrons gives rise to continual processes of pair formation with pair energies *above* the Fermi sea, and their subsequent disintegration into two unpaired electrons, all of which leads to a significant lowering in the total energy $\mathcal{E}$ of the system. As a result, and contrary to conventional wisdom, a mixture of unpaired-electron/bosonic-excitations above the Fermi sea turn out to be energetically lower compared with a single-component Fermi system of interacting fermions without explicit bosons. This BF mixture state develops gradually as T is lowered and is different from the BCS state where a phase transition to a coherent state occurs abruptly. Physically, this is clear from the fact that the BF mixture includes nonzero-CMM pairs whereas BCS theory does not.

4. Pseudogap

As a result of pairing, the number of unpaired fermions obviously decreases. The accompanying boson production processes occurring at $T \leq T^*$ lead to changes in the temperature- and coupling-dependent chemical potential $\mu(\lambda, T)$ which shifts *down* from its original value E_F. Physically, the number of bosons just formed is expected to be proportional to the difference $E_F - \mu(\lambda,T)$. Therefore, ignoring the small variation of the density of single-fermion states $N(\epsilon)$ within the interval $[E_F - \hbar\omega_D, E_F + \hbar\omega_D]$, we may take it as constant $N(E_F)$. Assuming that $\hbar\omega_D \gg E_F - \mu(\lambda,T)$ the density of composite bosons may then be written quite accurately as

$$n_B(\lambda,T) \simeq N(E_F)[E_F - \mu(\lambda,T)]. \tag{32}$$

This equation was first applied in Ref. 14. It must be satisfied for any $T \leq T^*$ for which $\mu(\lambda,T)$ begins to differ from E_F; its detailed derivation is given in Ref. 12. This remarkable expression allows us to define T^*, referred as a pseudogap or depairing temperature, as the solution of $n_B(\lambda,T^*) = 0$ or, equivalently, of

$E_F - \mu(\lambda, T^*) = 0$. This defines T^* as the temperature below which a transition occurs from the normal state with no composite bosons to one *with* such bosons, i.e., to the binary BF mixture. To determine $E_F - \mu(\lambda, T)$ in (32) we use the identity

$$\left\langle b_{\mathbf{Q}}^{+}\right\rangle = -\frac{1}{E(\mathbf{Q}) - 2\mu}\frac{f}{L^{d/2}}\sum_{\mathbf{q}}\left\langle a_{-\mathbf{q}+\mathbf{Q}/2\downarrow}^{+} a_{\mathbf{q}+\mathbf{Q}/2\uparrow}^{+}\right\rangle, \tag{33}$$

which follows directly on setting $A = b_{\mathbf{Q}}^{+}$ in the general relation $\langle [A, \mathcal{H}]\rangle \equiv Tr\{(A\mathcal{H} - \mathcal{H}A)\exp(-\beta\mathcal{H})\} = 0$ valid for any operator A and any Hamiltonian $\mathcal{H}$. In particular, (33) relates $\langle b_{\mathbf{0}}^{+}\rangle$ via a linear combination of anomalous averages $\left\langle a_{-\mathbf{q}\downarrow}^{+} a_{\mathbf{q}\uparrow}^{+}\right\rangle$. These averages may be calculated by means of GFs such as $\left\langle\left\langle a_{\mathbf{q}\uparrow}^{+} \mid a_{-\mathbf{q}\downarrow}^{+}\right\rangle\right\rangle_{\omega}$. Choosing $A \equiv a_{\mathbf{q}\uparrow}^{+}$ and $B \equiv a_{-\mathbf{q}\downarrow}^{+}$ in (12) with $\eta = -1$ we write an equation for $\left\langle\left\langle a_{\mathbf{q}\uparrow}^{+} \mid a_{-\mathbf{q}\downarrow}^{+}\right\rangle\right\rangle_{\omega}$ the solution of which, on using (13), leads to the result Ref. 12

$$\left\langle a_{-\mathbf{q}\downarrow}^{+} \mid a_{\mathbf{q}\uparrow}^{+}\right\rangle_{\mathcal{H}} = \frac{f}{2L^{d/2}}\frac{\langle b_{\mathbf{0}}^{+}\rangle}{\sqrt{(\epsilon_{\mathbf{q}} - \mu)^2 + f^2 n_{B0}(T)}}\tanh\frac{\sqrt{(\epsilon_{\mathbf{q}} - \mu)^2 + f^2 n_{B0}(T)}}{2k_B T} \tag{34}$$

with $n_{B0}(T)$ being a number-density of condensed bosons. Combining (33) with $\mathbf{Q} = \mathbf{0}$ and (34), and then replacing the sum $\sum_{\mathbf{q}}$ by integration over free-electron energies ϵ, we arrive at

$$E_F - \mu(\lambda, T) = -\Delta(\lambda, T) + \frac{\lambda\hbar\omega_D}{2}\int_{-\hbar\omega_D}^{\hbar\omega_D}\frac{dx}{\sqrt{x^2 + f^2 n_{B0}(T)}}\tanh\frac{\sqrt{x^2 + f^2 n_{B0}(T)}}{2k_B T} \tag{35}$$

which must be satisfied at any $T \leq T^*$. Because the integral is always positive, (35) applied to define T^* as that temperature below which pairs form (i.e., from the condition $E_F - \mu(\lambda, T^*) = 0$ on the lhs of (35) or $n_B(\lambda, T^*) = 0$), the equation is satisfied only provided $2\Delta(\lambda, T) > 0$, i.e., for resonant pairs which are excited *above* E_F. In fact, if $2\Delta(\lambda, T) < 0$ in (35) then one could get a real nonzero T^* only on condition that $|E_F - \mu(\lambda, T^*)| > |\Delta(\lambda, T^*)| \neq 0$. This implies that transition from the free-fermion system to the new state is not a gradual process (beginning when $n_B(\lambda, T)$ just ceases to be zero) but is associated with finite changes in $E_F - \mu(\lambda, T)$ at T^* when a macroscopic number of fermions simultaneously leave the fermionic system and enter into the bosonic system.

In Fig. 1 the depairing temperature T^* from (35) is plotted for $\Theta_D/T_F = 0.1$ (thick dashed curve) and for $\Theta_D/T_F = 0.05$ (thick full curve) as a function of dimensionless interelectron coupling $\lambda = N(E_F)V$ for the Cooper/BCS model interaction. The BCS expression $\Delta(\lambda) = \hbar\omega_D/\sinh(1/\lambda)$ implied in the $T = 0$ result of Refs. 7,8 was employed. We believe that the figure reflects, at least schematically, the correct behavior of T^* even though a finite-T expression for $\Delta(\lambda, T)$ should have been used in (35). The rapid decrease of T^* with λ is appreciable. It shows

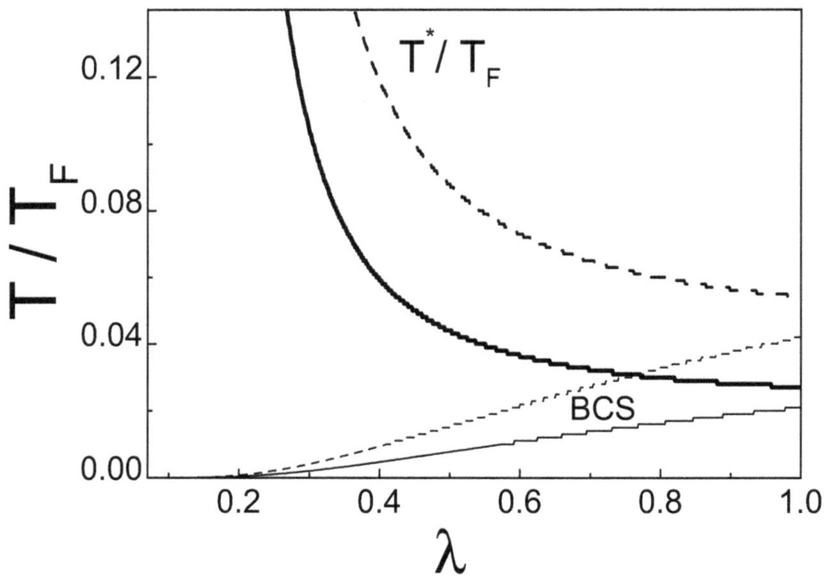

Fig. 1. Depairing temperature phase boundary T^*/T_F as a function of dimensionless coupling parameter λ for $\Theta_D/T_F = 0.1$ (thick dashed curve) and $\Theta_D/T_F = 0.05$ (thick full curve). The normal state lies above T^*/T_F below which the BF mixture state emerges. Thin curves are the BCS values from $T_c/T_F \simeq 1.134(\Theta_D/T_F)\exp(-1/\lambda)$ for $\Theta_D/T_F = 0.1$ (upper curve) and $\Theta_D/T_F = 0.05$ (lower curve).

that bosons energetically closer to Fermi surface, i.e., with a smaller gap 2Δ in (3), turn out to be better "protected" against thermal disruptions and thus survive over a wider temperature range. At first glance, this curious T^* vs. λ behavior may be understood in terms of the gapped dispersion relation (3).

In fact, for breakup to occur fermionic levels that are occupied by electrons resulting from bosonic breakup must be empty. The expectation of *empty* single-fermion **k**-states needed to be occupied as a result of a breakup, however, is smaller if these **k**-states are situated in the sea of densely-populated fermion states or near it. By contrast, bosons propagating with energies above E_F break up into unpaired fermions more easily because there are many admissible states to be occupied above E_F. As to thew magnitude of T^* at very small λ, it is widely believed that it must vanish in this limit, e.g., Ref. 15. This does not occur here. On the contrary, bosons created in the $\lambda \to 0$ limit, (i.e., pairs emerging within the Fermi sea) are stable, at least the zero-CMM ones, up to much higher temperatures than bosons created with larger λ. We believe that this reflects the fact that as soon as λ differs from zero, pair fluctuations in the electron medium become possible. This is similar to what occurs in the BCS scenario where the slightest infinitesimal attractive interaction leads to the creation of pairs below some threshold temperature, except that here T^* and T_c coincide while in the BF case they need not be identical. Of course, thermal fluctuations will destroy these pairs. However, independently of how large T and how small λ are, this formation mode for pairs will always be nonnegligible. This feature has no analogy either in conventional superconductors, as in the BCS model where the "hardness" of pairs is associated with large λ, nor in BF models[1] with a gapless boson dispersion curve. It seems to be an inherent property of a BF mixture with an interaction-dependent gap such as 2Δ in its composite-boson spectrum.[7,8] To understand this result intuitively, note that in BF mixture models pair stability (or CP lifetime) depends not only on how large the interaction parameter λ is but also on the degree to which the corresponding bosonic states are energetically separated from the Fermi surface. The latter is determined by the energy scale 2Δ in the dispersion relation (3) and plays a decisive role in confining two electrons to form a CP with a finite lifetime. Indeed, increasing λ increases the separation of the bosonic states from the Fermi sea, thus making pairs more loosely-bound, e.g., with respect to decay by thermal effects. As stressed above, an expression for Δ used in Fig. 1 stems from the $T = 0$ calculations.[7,8] In fact, at nonzero T the actual attractive interaction necessary for pair formation has to be stronger than it is at $T = 0$, as is needed to overcome thermal disruptions. Therefore, the λ values involved in $\Delta(\lambda, T)$ which would be obtained from a $T > 0$ treatment are expected to be shifted for larger λ's with respect to those in $\Delta(\lambda, 0)$. This circumstance must "tame" the divergence in the $\lambda \to 0$ limit seen in Fig. 1. However, we believe the schematic T^* vs. λ behavior shown in Fig.1 will not change due to finite-T corrections.

5. Conclusions

Using two-time dependent GF techniques it was showed that positive-energy resonant bosonic pairs formed above the Fermi sea are energetically favorable in a ground-state of two coexisting and dynamically interacting many-particle subsystems of unpaired fermions and composite bosons. It is argued that an interaction between free fermions and bosons excited above the Fermi sea in the mixture, namely, the continuous processes of pair formation and disintegration into two unpaired electrons, results in lowering the total system energy considerably. This system is described with the Friedberg-T.D. Lee phenomenological Hamiltonian. Positive-energy composite-bosons appear at temperatures below the depairing T^* and their coupling- and temperature-dependent number density $n_B(\lambda, T)$ increases gradually from zero. The resulting BF mixture with positive-energy resonant pairs very naturally leads to the pseudogap phenomenon observed in high-T_c cuprates.

Acknowledgments

MdeLl thanks UNAM-DGAPA-PAPIIT (Mexico) IN106401 and CONACyT (Mexico) 41302F for partial support. Both MdeLl and TAM thank the Max Planck Institute for the Physics of Complex Systems, Dresden, Germany, for its support and hospitality.

Appendix

Choosing $A \equiv b^+_{\mathbf{Q}}$ and $B \equiv b_{\mathbf{Q}}$ in (12) with $\eta = -1$ we write (12) as $\omega \left\langle\left\langle b^+_{\mathbf{Q}} \mid b_{\mathbf{Q}} \right\rangle\right\rangle_\omega = -1 + \left\langle\left\langle [b^+_{\mathbf{Q}}, \mathcal{H}]_- \mid b_{\mathbf{Q}} \right\rangle\right\rangle_\omega$ where

$$[b^+_{\mathbf{Q}}, \mathcal{H}]_- = -[E(\mathbf{Q}) - 2\mu] b^+_{\mathbf{Q}} - \frac{f}{L^{d/2}} \sum_{\mathbf{q}} a^+_{-\mathbf{q}+\mathbf{Q}/2\downarrow} a^+_{\mathbf{q}+\mathbf{Q}/2\uparrow}$$

since $[b^+_{\mathbf{Q}}, b^+_{\mathbf{K}}]_- = 0$ while $[b^+_{\mathbf{Q}}, b_{\mathbf{K}}]_- = -\delta_{\mathbf{Q},\mathbf{K}}$ and $[b^+_{\mathbf{Q}}, b^+_{\mathbf{K}} b_{\mathbf{K}}]_- = -b^+_{\mathbf{K}} \delta_{\mathbf{Q},\mathbf{K}}$. After simple algebra we get

$$[\omega + E(\mathbf{Q}) - 2\mu] \left\langle\left\langle b^+_{\mathbf{Q}} | b_{\mathbf{Q}} \right\rangle\right\rangle_\omega = -1 - \frac{f}{L^{d/2}} \sum_{\mathbf{q}} \left\langle\left\langle \sum_{\mathbf{q}} a^+_{-\mathbf{q}+\mathbf{Q}/2\downarrow} a^+_{\mathbf{q}+\mathbf{Q}/2\uparrow} \mid b_{\mathbf{Q}} \right\rangle\right\rangle_\omega . \tag{36}$$

Taking first $\omega = \omega + i\varepsilon$ with $\varepsilon \to 0^+$ in (36) gives

$$\begin{aligned} &[\omega + i\varepsilon + E(\mathbf{Q}) - 2\mu] \left\langle\left\langle b^+_{\mathbf{Q}} \mid b_{\mathbf{Q}} \right\rangle\right\rangle_{\omega+i\varepsilon} \\ &= -1 - \frac{f}{L^{d/2}} \sum_{\mathbf{q}} \left\langle\left\langle \sum_{\mathbf{q}} a^+_{-\mathbf{q}+\mathbf{Q}/2\downarrow} a^+_{\mathbf{q}+\mathbf{Q}/2\uparrow} \mid b_{\mathbf{Q}} \right\rangle\right\rangle_{\omega+i\varepsilon} . \end{aligned} \tag{37}$$

Then, taking $\omega = \omega - i\varepsilon$ again with $\varepsilon \to 0^+$ yields

$$\begin{aligned} &[\omega - i\varepsilon + E(\mathbf{Q}) - 2\mu] \left\langle\left\langle b^+_{\mathbf{Q}} \mid b_{\mathbf{Q}} \right\rangle\right\rangle_{\omega-i\varepsilon} \\ &= -1 - \frac{f}{L^{d/2}} \sum_{\mathbf{q}} \left\langle\left\langle \sum_{\mathbf{q}} a^+_{-\mathbf{q}+\mathbf{Q}/2\downarrow} a^+_{\mathbf{q}+\mathbf{Q}/2\uparrow} \mid b_{\mathbf{Q}} \right\rangle\right\rangle_{\omega-i\varepsilon} . \end{aligned} \tag{38}$$

Consider the difference between (37) and (38). For the lhs of this difference we have

$$\begin{aligned} &[\omega + i\varepsilon + E(\mathbf{Q}) - 2\mu] \left\langle\left\langle b^+_{\mathbf{Q}} \mid b_{\mathbf{Q}} \right\rangle\right\rangle_{\omega+i\varepsilon} - [\omega - i\varepsilon + E(\mathbf{Q}) - 2\mu] \left\langle\left\langle b^+_{\mathbf{Q}} \mid b_{\mathbf{Q}} \right\rangle\right\rangle_{\omega-i\varepsilon} \\ &= -i[\exp(\hbar\omega/k_B T) - 1][\omega + E(\mathbf{Q}) - 2\mu] J_{b_{\mathbf{Q}} b^+_{\mathbf{Q}}}(\omega) \end{aligned} \tag{39}$$

where according to (14) $J_{b_{\mathbf{Q}} b^+_{\mathbf{Q}}}(\omega)$ is a spectral density for $\left\langle\left\langle b_{\mathbf{Q}} \mid b^+_{\mathbf{Q}} \right\rangle\right\rangle_\omega$. The difference between the two rhs's (37) and (38), on passing from $\omega + i\varepsilon$ to $\omega - i\varepsilon$, yields

$$i[\exp(\hbar\omega/k_B T) - 1] \frac{f}{L^{d/2}} \sum_{\mathbf{q}} J_{b_{\mathbf{Q}} a^+_{-\mathbf{q}+\mathbf{Q}/2\downarrow} a^+_{\mathbf{q}+\mathbf{Q}/2\uparrow}}(\omega). \tag{40}$$

Equating (18) and (19) we have

$$\frac{f}{L^{d/2}} \sum_{\mathbf{q}} J_{b_{\mathbf{Q}} a^+_{-\mathbf{q}+\mathbf{Q}/2\downarrow} a^+_{\mathbf{q}+\mathbf{Q}/2\uparrow}}(\omega) = -[\omega + E(\mathbf{Q}) - 2\mu] J_{b_{\mathbf{Q}} b^+_{\mathbf{Q}}}(\omega). \tag{41}$$

Summing (20) over $\mathbf{Q}$, integrating over ω and dividing by 2π immediately leads to

$$\left\langle H_{int}^{(2)} \right\rangle = -\sum_{\mathbf{K}} [E(\mathbf{K}) - 2\mu] \left\langle b_{\mathbf{K}} b_{\mathbf{K}}^{+} \right\rangle - \frac{1}{2\pi} \sum_{\mathbf{K}} \int_{-\infty}^{\infty} \omega J_{b_{\mathbf{K}} b_{\mathbf{K}}^{+}}(\omega) d\omega.$$

References

1. R. Friedberg, T.D. Lee, and H.-C. Ren, Phys. Rev. B **45**, 10732 (1992) and refs. therein.
2. V.V. Tolmachev, Phys. Lett. A **266**, 400 (2000).
3. M. de Llano and V.V. Tolmachev, Physica A **317**, 546 (2003).
4. L.N. Cooper, Phys. Rev. **104**, 1189 (1956).
5. M. de Llano, in *Frontiers in Superconductivity Research* (ed. B.P. Martins) (Nova Science Publishers, NY, 2004). Cond-mat/0405071.
6. J. Bardeen, L.N. Cooper, and J.R. Schrieffer, Phys. Rev. **108**, 1175 (1957).
7. M. Fortes, M.A. Solís, M. de Llano, and V.V. Tolmachev, Physica C **364**, 95 (2001).
8. V.C. Aguilera-Navarro, M. Fortes, and M. de Llano, Sol. St. Comm. **129**, 577 (2004).
9. T. Timusk and B. Statt, Rep. Prog. Phys. **62**, 61 (1999).
10. J.L. Tallon and J.W. Loram, Physica C **349**, 53 (2001).
11. T. Mamedov, J. Low Temp. Phys., **131**, 217 (2003).
12. T.A. Mamedov and M. de Llano, *Bose-Einstein-condensation temperatures in a boson-fermion model of superconductors*, to be published.
13. D.N. Zubarev, Sov. Phys. Uspekhi **3**, 320 (1960) [English trans. of Usp. Fiz. Nauk **71**, 71 (1060)].
14. T.A. Mamedov and M. de Llano, in *New Challenges in Superconductivity: Experimental Advances and Emerging Theories*, ed. by J. Ashkenazi *et al.* pp. *135-138* (Kluwer, NY, 2004).
15. Q. Chen, J. Stajic, Sh. Tan, and K. Levin, Phys. Repts. **412**, 1 (2005).

THE CONCEPT OF CORRELATED DENSITY AND ITS APPLICATION

KLAUS MORAWETZ[1,2], PAVEL LIPAVSKÝ[3,4], JAN KOLÁČEK[4],
ERNST HELMUT BRANDT[5] and MICHAEL SCHREIBER

[1]*Institute of Physics, Chemnitz University of Technology, 09107 Chemnitz, Germany*
[2]*Max-Planck-Institute for the Physics of Complex Systems, Noethnitzer Str. 38, 01187 Dresden, Germany*
[3] *Faculty of Mathematics and Physics, Charles University, Ke Karlovu 3, 12116 Prague 2, Czech Republic*
[4] *Institute of Physics, Academy of Sciences, Cukrovarnická 10, 16253 Prague 6, Czech Republic*
[5]*Max Planck Institute for Metals Research, 70506 Stuttgart, Germany*
E-mail: morawetz@physik.tu-chemnitz.de, lipavsky@fzu.cz

The correlated density appears in many physical systems ranging from dense interacting gases up to Fermi liquids which develop a coherent state at low temperatures, the superconductivity. One consequence of the correlated density is the Bernoulli potential in superconductors which compensates forces from dielectric currents. This Bernoulli potential allows to access material parameters. Though within the surface potential these contributions are largely canceled, the bulk measurements with NMR can access this potential. Recent experiments are explained and new ones suggested. The underlying quantum statistical theory in nonequilibrium is the nonlocal kinetic theory developed earlier.

Keywords: Correlated density; nonlocal kinetic theory; Bernoulli potential; superconductivity; NMR; YBCO.

1. Introduction

The theory of dense quantum gases distinguishes the number of free particles and the number of particles in a correlated state, e.g. in bound states. Therefore any kinetic theory should result in a balance equation which in equilibrium leads to the Beth-Uhlenbeck equation of state[1–3] where the density of free and correlated particles are distinguished $n = n_f + n_{\rm corr}(n_f)$. In contrast to that the Landau theory of quasiparticles describes dense interacting Fermi systems only in terms of the quasiparticle density. This is supported by the Luttinger theorem[4] which states that the Fermi momentum for any correlated Fermi system in the ground state is exclusively determined by the total number of particles and there is no shift due to correlations. This apparent contradiction between the two approaches is solved within the nonlocal kinetic theory which unifies both pictures and provides them as limiting cases of a more general treatment.[5] How is this possible? First of all an ideal Fermi liquid obeys the Luttinger theorem and any correlated density has to vanish in the ground state. This is in a sense a tautology.[6] But one has to observe that most Fermi systems do not turn to ideal Fermi liquids at low temperatures. Electrons in metals and nucleons, e.g. develop a coherent state, the superconducting state. There is a small fraction of correlated density in the superconducting state. Such correlated density shifts the Fermi momentum and consequently the chemical potential μ. Since the electrochemical potential has to remain constant there must be an electrostatic potential ϕ compensating this shift according to $\mu + e\varphi = 0$. This electrostatic potential turns out to be of the form of a Bernoulli potential in superconductors. If this potential could be measured we have a proof of the concept of correlated density.

This overview discusses the theoretical treatments of the Bernoulli potential and the experimental realizations to measure it. First we describe shortly how the correlated density appears in superconducting states and how it is related to the electrostatic potential. Then we review the history of this Bernoulli potential and explain why surface measurements have not been successful in the past. We found the reason in the Budd-Vannimenus theorem which explains why large-scale cancellations appear on the surface of superconductors. Then we calculate what potential can be measured deeper in the bulk by calculating the density and magnetic profile of superconductors. This allows us to explain a recent NMR experiment and to predict inhomogeneous fields to be measured above vortices in type-II superconductors. Finally we conclude with a short description of the nonlocal kinetic theory which provides the description of the correlated density in nonequilibrium.

2. The Concept of Correlated Density

2.1. *Correlated density within BCS approach*

In superconductors the Wigner distribution function has a two-part structure[7]

$$f_W = \frac{1}{2}\left(1+\frac{\xi}{E}\right)f(E) + \frac{1}{2}\left(1-\frac{\xi}{E}\right)f(-E) = \frac{1}{2} - \frac{\xi}{2E}\tanh\frac{1}{2}\beta E \qquad (1)$$

where $\xi = \epsilon_p - \mu$ is the free-particle energy ϵ_p minus the chemical potential μ. The quasiparticle energy $E = \sqrt{\xi^2 + \Delta^2}$ describes the influence of the superconducting gap Δ on the excitation spectrum of the superconducting state. The Fermi-Dirac distribution is $f(x) = 1/(e^{\beta x} + 1)$ with the inverse temperature $\beta = 1/k_B T$. There are two branches of energies. For states inside the Fermi sphere, $\xi < 0$, the negative branch given by the second term dominates. For the states above the Fermi sphere the role of the contributions is reversed. Close to the Fermi sphere both contributions are comparable.

The density n is obtained by the momentum integral over (1) and introducing the density of states $h(\xi) = 2\sum_p 2\pi\delta(\xi - \epsilon_p)$, one has

$$n = \int_{-\bar{\mu}}^{\infty} \frac{d\xi}{2\pi} h(\bar{\mu} + \xi)\left(\frac{1}{2} - \frac{\xi}{2E}\tanh\frac{1}{2}\beta E\right). \qquad (2)$$

Here we account for a possible electrostatic potential φ and the velocity v of superconducting electrons by $\bar{\mu} = \mu - e\varphi - mv^2/2$. For a vanishing gap we obtain the corresponding density n_n of normal electrons with the chemical potential $\bar{\mu}$ by $n_n = n(\Delta = 0)$. The difference

$$n_{\rm corr} = n - n_n \tag{3}$$

describes the correlated density. Let us discuss it first in the ground state. There the normal density turns into

$$n_n = 2\sum_p \Theta(\bar{\mu} - \epsilon_p) \approx n_0 - (e\varphi + \frac{m}{2}v^2)\frac{h(\mu)}{2\pi} \tag{4}$$

where we have expanded $\bar{\mu}$ in first order around the Fermi energy and n_0 describes the number of particles with no motion and no electrostatic potential. The correlated density (3) splits into two parts in the zero-temperature limit of (2)

$$n_{\rm corr} = \frac{1}{2}\int\limits_0^\infty \frac{d\xi}{2\pi} h(\bar{\mu}+\xi)\frac{\sqrt{\xi^2+\Delta^2}-\xi}{\sqrt{\xi^2+\Delta^2}} - \frac{1}{2}\int\limits_{-\bar{\mu}}^0 \frac{d\xi}{2\pi} h(\bar{\mu}+\xi)\frac{\sqrt{\xi^2+\Delta^2}+\xi}{\sqrt{\xi^2+\Delta^2}} \tag{5}$$

which vanishes for vanishing gap. Since the gap is only nonzero in the vicinity of the Fermi level given by the Debye frequency ω_D we can restrict the integration to the $\pm\omega_D$-range. Expanding the density of states for $\xi < \omega_D$ we obtain finally

$$n_{\rm corr} = \frac{\partial h}{\partial \mu}\frac{\Delta^2}{4\pi}\left[\ln\left(\frac{\omega_D}{\Delta} + \sqrt{\frac{\omega_D^2}{\Delta^2}+1}\right) - \frac{1}{1+\sqrt{1+\frac{\Delta^2}{\omega_D^2}}}\right] \approx \frac{\partial h}{\partial \mu}\frac{\Delta^2}{4\pi}\ln\left(\frac{2\omega_D}{\sqrt{e}\Delta}\right) \tag{6}$$

for $\omega_D \gg \Delta$ in the last step.

Since the total system should stay neutral we expect $n = n_0$ and the two contributions, $n_n - n_0$ according to (4) and $n_{\rm corr}$ of (6), should cancel. Therefore the required electrostatic potential must read

$$e\varphi = -\frac{m}{2}v^2 + \frac{\partial \ln h}{\partial \mu}\frac{\Delta^2}{2}\ln\left(\frac{2\omega_D}{\sqrt{e}\Delta}\right). \tag{7}$$

We see that it has the form of a Bernoulli potential which compensates the contribution of the supercurrent velocity and the associated inertial and Lorentz forces and has a part directly linked to a material parameter namely the gap. The latter is called thermodynamic corrections. This internally developed Bernoulli potential and field keep the charge neutrality and compensate the inhomogeneities due to diamagnetic currents.

The great hope is now to measure the Bernoulli potential in order to access directly the gap parameter.

2.2. *Bernoulli potential*

A way to discuss the Bernoulli potential closely connected to experimental facts is the London theory. The London condition

$$m\mathbf{v} = -e\mathbf{A} \tag{8}$$

fixes the trajectories of superconducting electrons due to the transverse vector potential. This condition explains the Meissner effect and contains both London equations. One can consider it as the simplest way to account for basic electrodynamic features of superconductors. The time derivative of this London condition gives the force

$$m\dot{\mathbf{v}} = -e\frac{\partial \mathbf{A}}{\partial t} - e(\mathbf{v}\nabla)\mathbf{A} = e(\mathbf{E} + \mathbf{v}\times\mathbf{B}) + \nabla\left(e\varphi + \frac{1}{2}mv^2\right) \tag{9}$$

exerted on the electrons. Comparing this with Newton's equation of motion $m\dot{\mathbf{v}} = e(\mathbf{E}+\mathbf{v}\times\mathbf{B})+\mathbf{F}_s$ where the unknown force on the superconducting pair is denoted with $\mathbf{F}_s$ one obtains

$$\nabla e\varphi = \mathbf{F}_s - \nabla\frac{1}{2}mv^2. \tag{10}$$

The different expressions in literature differ by the different treatment of the unknown force $\mathbf{F}_s$. London neglected this force consistent with the London condition. This corresponds to the picture of hydrodynamics of a charged ideal gas and it results in $e\varphi = -\frac{1}{2}mv^2$. The diamagnetic current is maintained by the electrostatic potential.[8] Sorkin[9] suggested that the free energy responsible for superconductors must contribute, too, which yields $e\varphi = -\frac{1}{2}mv^2 - \frac{\partial f_s}{\partial n_s}$. As a further step van Vijfeiken and Staas[10] considered the force resulting from the interaction between electrons and condensate acting on electrons to keep them at rest $\mathbf{F}_n + e\mathbf{E} = 0$. Taking into account the balance of forces $n_n\mathbf{F}_n + n_s\mathbf{F}_s = 0$ as well as that $e\mathbf{E} = -e\nabla\varphi$ the force becomes $\mathbf{F}_s = -\frac{n_n}{n_s}e\nabla\varphi$. Introducing this into (10) one obtains that the Bernoulli potential is reduced by the fraction of superconducting electrons $e\varphi = -\frac{n_s}{n}\frac{1}{2}mv^2$ called quasiparticle screening or fountain effect. Rickayzen[11] finally started from the kinetic energy density $f_{kin} = n_s\frac{1}{2}mv^2$ of the condensate which determines the change in the chemical potential

$$e\varphi = -\mu = -\frac{\partial}{\partial n}f_{kin} = -\frac{\partial n_s}{\partial n}\frac{1}{2}mv^2 = -\frac{n_s}{n}\frac{1}{2}mv^2 + 4\frac{n_n}{n}\frac{\partial \ln T_c}{\partial \ln n}\frac{1}{2}mv^2. \tag{11}$$

This expression provides, besides the quasiparticle screening, also thermodynamic corrections. This offers the experimental possibility to access directly material parameters similar to the BCS expression (7).

The experimental attempts to measure it, however, have yielded no result. Since Ohmic contact measurements cannot provide a result due to the constant electrochemical potential,[12,13] Bok and Klein[14] and later more precisely Morris and Brown[15] measured the Bernoulli potential at the surface via capacitive pickup. For the investigated Pb at 7K the thermodynamic corrections have been expected to be larger by a factor 30 than the first term in (11). Surprisingly the observed data agree only with the first term of (11).

Why no signal of thermodynamic corrections is seen remained a puzzle for nearly 30 years. Recently we found the solution by a modification[16] of the Budd-Vannimenus theorem[17]

$$\rho_{\rm lat}\left(\varphi_0-\varphi(0)\right)=f_{\rm el}-n\frac{\partial f_{\rm el}}{\partial n}=-n^2\frac{\partial}{\partial n}\left(\frac{f_{\rm el}}{n}\right), \tag{12}$$

which states that the extension of the surface profile towards the surface, $\varphi(0)$ has a step at the surface due to the surface dipoles which is given exclusively in terms of the electronic free energy with no regard to the potential inside. Writing the change of the free energy as Rickayzen, $e\varphi=-\frac{\partial}{\partial n}\delta f$ and $\delta f=n_s\frac{1}{2}mv^2$, the potential at the surface becomes

$$e\varphi_{\rm surf}=e\varphi+n\frac{\partial}{\partial n}\frac{\delta f}{n}=e\varphi+\frac{\partial}{\partial n}\delta f-\frac{\delta f}{n}=-\frac{n_s}{n}\frac{1}{2}mv^2. \tag{13}$$

The surface dipoles compensate the thermodynamic corrections exactly.[16]

This is of course strictly valid only for homogeneous superconductors. Even the appearance of vortices in type-II superconductors requires a more detailed discussion. In order to describe such inhomogeneous situation we use the extension of the Landau-Ginzburg equation towards lower temperatures by Bardeen.[18]

2.3. *Inhomogeneous superconductors within extended Ginzburg-Landau approach*

We consider the free energy as composed of the condensation energy, the kinetic energy, the electrostatic energy as well as the magnetic energy, $\mathcal{F}[\psi,\mathbf{A},n_n]=\int dr f=\mathcal{F}_s+\mathcal{F}_{\rm kin}+\mathcal{F}_C+\mathcal{F}_M$. The condensation energy is constructed according to the two-fluid model of Gorter and Casimir,[19] where one describes two experimental facts, firstly that the fraction of superconducting electrons shows a temperature dependence of nearly $\varpi=1-\frac{T^4}{T_c^4}\approx\frac{n_s}{n_s+n_n}$ and secondly that the condensation energy is an expression containing the specific heat parameter γ and the critical temperature as $\varepsilon_{\rm con}=\frac{1}{4}\gamma T_c^2$. Eliminating the critical temperature one sees that the condensation energy must have the form $\varepsilon_{\rm con}=\frac{\gamma T^2}{4\sqrt{1-\varpi}}$ in equilibrium. This relation should result from a variation of the free energy $\delta\mathcal{F}_s/\delta\varpi=\partial f_s/\partial\varpi=0$ such that

$$f_s=U-\varepsilon_{\rm con}\varpi-\frac{1}{2}\gamma T^2\sqrt{1-\varpi}. \tag{14}$$

The kinetic part is taken as a form proposed by Ginzburg and Landau,[20] who suggested to represent the superconducting density by a pseudo wave function $|\psi|^2=\frac{n_s}{2}$ such that the kinetic energy can be given in Schrödinger form ($m^*=2m_e$, $e^*=2e$)

$$\mathcal{F}_{\rm kin}=\int d\mathbf{r}\frac{1}{2m^*}\left|\left(-i\hbar\nabla-e^*\mathbf{A}\right)\psi\right|^2. \tag{15}$$

Variation of the total free energy with respect to ψ^* yields the Ginzburg-Landau equation (GL) for the wave function

$$\frac{1}{2m^*}(-i\hbar\nabla - e^*\mathbf{A})^2\psi + \chi\psi = 0 \tag{16}$$

with the potential

$$\chi = \frac{\partial f_s}{\partial|\psi|^2} - 2\frac{\partial f_s}{\partial n_n} = -2\frac{\varepsilon_{\rm con}}{n} + \frac{\gamma T^2}{2n}\frac{1}{\sqrt{1-\frac{2|\psi|^2}{n}}}. \tag{17}$$

This extends the GL equation towards lower temperatures. Close to T_c the potential takes the form $\chi \to \alpha + \beta|\psi|^2$ and the original GL equation is recovered.

The GL equation was solved numerically[21] for Nb; the results are seen in figure 1. The vortices arrange themselves in the hexagonal structure of an Abrikosov-type lattice. The superconducting density is zero at the vortex centers and reaches nearly the nonmagnetic value at the borders of the vortex. The fact that n_s is smaller there than its nonmagnetic value is a result of the nonlocal effects. The corresponding magnetic field reaches its maximum $B_{\max}$ at the vortex centers. In fact, it is higher than the applied field in the core which means that the superconductor compresses the magnetic field in vortices.

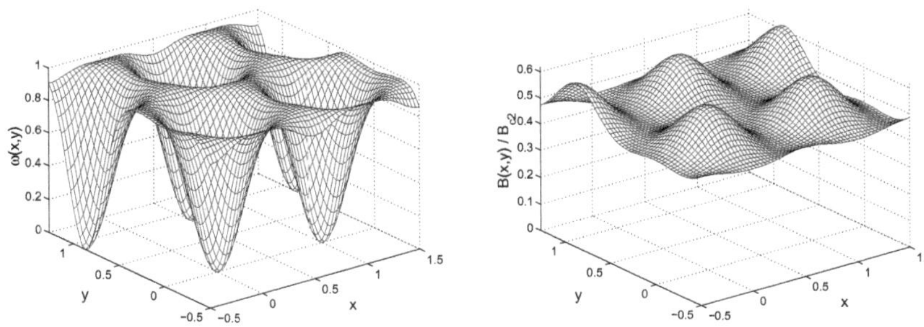

(a) (b)

Fig. 1. (a) Condensate fraction $|\psi/\psi_0|^2$ normalized to the magnetic-field-free value ψ_0 for $T/T_c = 0.5$, the magnetic induction $\bar{B}/B_{c2} = 0.5$, and the GL parameter $\kappa_0 = 1.5$. (b) The magnetic field in units of the upper critical field B_{c2}.[21]

2.4. *Charge profile*

The numerical solution of the GL equation allows us to discuss the detailed charge density and potential profile towards the surface. For magnetic fields perpendicular to the surface we found a linearized solution of the GL equation (16) with respect to low magnetic field $\psi = \sqrt{n(1-(T/T_c)^4)/2} + \delta\psi$ with[22]

$$\delta\psi = \frac{\psi_\infty}{2-\kappa^2}\frac{e^{*2}\lambda^4 B^2}{2\hbar^2}\left(\mathrm{e}^{-2x/\lambda} - \frac{\sqrt{2}}{\kappa}\mathrm{e}^{-\sqrt{2}x/\xi}\right), \tag{18}$$

where $\lambda = m^*/\mu_0 n e^2$ is the London penetration depth indicating how far the magnetic field penetrates the superconductor, $\xi = \hbar^2/2m^*\alpha$ is the GL coherence length in terms of $\alpha = \gamma T_c(T-T_c)/n$, and the ratio between both quantities is the GL parameter κ. One sees that the two length scales determine the profile corresponding to whether κ is larger or smaller than $\sqrt{2}$, i.e. type II or type I superconductors. The resulting charge profile results from the variation of the free energy with respect to the normal density and reads

$$e\delta\varphi = -\frac{1}{2}mv^2\left(\frac{n_s}{n}+4\frac{n_n}{n}\frac{\partial \ln T_c}{\partial \ln n}\right)\frac{1}{1+\frac{\sqrt{2}}{\kappa}}\left(1+\frac{\mathrm{e}^{\left(\frac{2}{\lambda}-\frac{\sqrt{2}}{\xi}\right)x}-1}{1-\frac{\kappa}{\sqrt{2}}}\right). \tag{19}$$

This result remains finite at the limiting case of $\kappa = \sqrt{2}$ and yields Rickayzen's result (11) for $\kappa \to \infty$.

With the help of the Poisson equation the linearized density profile reads[22]

$$\rho = \frac{2e\epsilon_0 B^2}{m\left(1-\frac{2}{\kappa^2}\right)}\left(\frac{n_s}{n}+4\frac{n_n}{n}\frac{\partial \ln T_c}{\partial \ln n}\right)\left(\mathrm{e}^{-2x/\lambda}-\frac{\kappa}{\sqrt{2}}\mathrm{e}^{-\sqrt{2}x/\xi}\right). \tag{20}$$

The surface value, $\rho(x=0)$, is always negative while the bulk value, $\rho(x \gg 0)$, remains positive. This allows us to define the width of the surface by $\rho(w) = 0$ with the result

$$w = \frac{\lambda}{2}\frac{\ln\frac{\kappa}{\sqrt{2}}}{\frac{\kappa}{\sqrt{2}}-1}. \tag{21}$$

For type-II superconductors, $\kappa > \sqrt{2}$, the surface charge is formed by the contribution on the scale of the GL coherence length ξ while for $\kappa \to \infty$ the width $w \to \frac{\xi}{\sqrt{2}}\ln\frac{\kappa}{\sqrt{2}}$ and the bulk charge extends on the scale of λ. For type-I superconductors, $\kappa < \sqrt{2}$, the situation is reversed, the surface charge is formed by the contribution on the scale of the London penetration depth λ and for $\kappa \ll \sqrt{2}$ the width $w \to \frac{\lambda}{2}\ln\frac{\sqrt{2}}{\kappa}$ and the bulk charge extends on the scale of ξ.

3. Application of the Bernoulli Potential

3.1. *Surface potential*

Summarizing, the electrostatic potential can leak out of a superconductor by three types of charges: (i) The bulk charge which describes the transfer of electrons from the inner to the outer regions of vortices creating a Coulomb force. This force has to balance the centrifugal force by the electrons rotating around the vortex center, the outward push of the magnetic field via the Lorentz force and the outward force coming from the fact that the energy of Cooper pairs is lower than the one of free electrons such that unpaired electrons in the vortex core are attracted towards the condensate around the core.[22] (ii) The surface dipole which cancels all contributions of pairing forces[23] resulting in an observable surface potential of

$$e\phi_0 = -\frac{f_{\mathrm{el}}}{n}. \tag{22}$$

(iii) The surface charge[24] distributed on the scale of the Thomas-Fermi screening length $\lambda_{\rm TF}$ with $\xi, \xi_0, \lambda \gg \lambda_{\rm TF} \to 0$.

Within the discussed Bardeen's extension of the GL equation the free electron energy has three components

$$f_{\rm el} = \frac{1}{2}\gamma T^2 + \frac{1}{2m^*}\psi^* \left(-i\hbar\nabla - e^*\mathbf{A}\right)^2 \psi - \varepsilon_{\rm con}\frac{2|\psi|^2}{n} - \frac{1}{2}\gamma T^2\sqrt{1-\frac{2|\psi|^2}{n}} \quad (23)$$

which determines, according to the modified Budd-Vannimenus theorem (22), the surface potential. Near the critical temperature this becomes very simple $e\phi_0 = \frac{1}{2n}\beta|\psi|^4$ in terms of $\beta = \gamma T_c^2/2n^2$. Without surface dipoles, the surface potential equals the internal potential

$$e\phi = -\frac{1}{2m^* n}\psi^* \left(-i\hbar\nabla - e^*\mathbf{A}\right)^2 \psi + \frac{\partial \varepsilon_{\rm con}}{\partial n}\frac{2|\psi|^2}{n} - \frac{T^2}{2}\frac{\partial\gamma}{\partial n}\left(\frac{|\psi|^2}{n} + \frac{|\psi|^4}{2n^2}\right) \quad (24)$$

where the first part are the inertial and Lorentz forces neglecting pairing forces. The approximation by Khomskii and Kusmartsev[25] adopted by Blatter,[26] $e\phi_{\rm Bl} = \frac{\gamma T_c}{n}\frac{\partial T_c}{\partial n}|\psi|^2$, takes only the pairing force into account. Before comparing the different approximations of the literature with our result, it is illustrative to discuss the magnetic field and temperature dependence. For thin layers and a B field close to B_{c2} the spatially averaged values of the superconducting fraction $\omega = \frac{|\psi|^2}{|\psi_\infty|^2}$ can be given by general arguments[27] as $\langle\omega\rangle = \frac{1-b}{\beta_{\rm A}}$ and $\langle\omega^2\rangle = \frac{(1-b)^2}{\beta_{\rm A}}$ with a number β_A characteristic for the vortex geometry and $b = \frac{B}{B_{c2}}$. Knowing our averaged surface potential we compare it now with Blatter's approximation

$$\langle e\phi_0\rangle = \frac{\varepsilon_{\rm con}}{n\beta_{\rm A}}\left(1-t^2\right)^2(1-b)^2 \quad (25)$$

$$\langle e\phi_{\rm Bl}\rangle = \frac{\varepsilon_{\rm con}}{n\beta_{\rm A}}\frac{\partial \ln T_c}{\partial \ln n}2\left(1-t^4\right)(1-b) \quad (26)$$

where $t = T/T_c$. Obviously the surface dipole strongly modifies the magnitude of the potential, in particular when the GL wave function has a small magnitude we have $\phi_0 \propto |\psi|^4$, while without dipoles $\phi_{\rm Bl} \propto |\psi|^2$.

Possible experimental consequences for which the presented theory can be tested are at temperatures close the critical temperature, $t \to 1$, where $|\psi|^2 \propto 1-t$, therefore $\phi_0 \propto (1-t)^2$ while $\phi_{\rm Bl} \propto 1-t$ and at magnetic fields close to the upper critical field, $b \to 1$, where $|\psi|^2 \propto 1-b$ so that $\phi_0 \propto (1-b)^2$ while $\phi_{\rm Bl} \propto 1-b$.

The numerical comparison[24] for Nb is shown in figure 2. The internal potential and Blatter's result are similar. The full theory and the approximations of only inertial/Lorentz forces are much smaller. The surface dipole cancels a major part of the pairing forces. The full theory and the inertial/Lorentz forces even result in different profiles and sign. This illustrates the delicate balance between the three contributions in (23). It is not justified to neglect just one of them. Corresponding results in the literature should be critically reexamined, e.g. the vortex induced deforma-

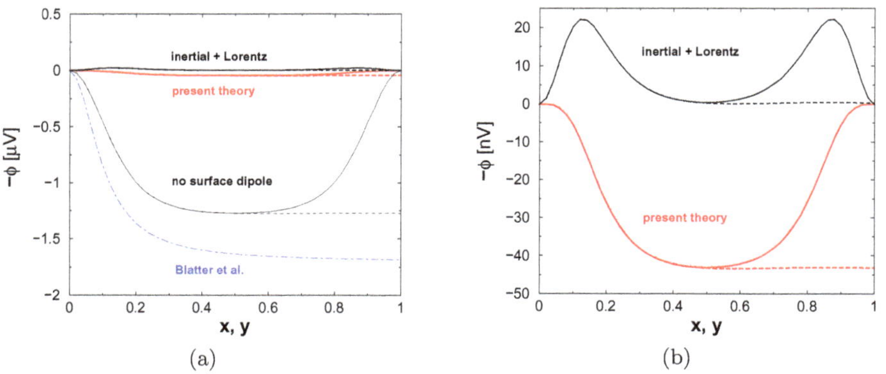

Fig. 2. (a) Surface potential for different approximations at $\frac{T}{T_c} = 0.95$, $\kappa = 0.78$ and $\frac{B}{B_{c2}} = 0.7818$. The broken lines indicate the behavior along y directions different from the x directions due to the hexagonal structure. (b) An enlarged view of the approximation of only inertial and Lorentz forces compared to the present theory.[24]

tion and the experimentally accessible effective mass of vortices show remarkably different results.[28]

3.2. *Charged vortices probed by NMR*

Though the capacitive coupling measurements of the surface potential do not yield an access to thermodynamic corrections and the high precision measurements of the electrostatic potential leaking out of the surface remain to be performed, there is a direct attempt to measure the charge deeper in the bulk by NMR. Kumagai et al.[29] have measured the quadrupole resonance lines in the high-T_c material YBCO. The polarization of the Cu atoms leads to a coupling of spins with the electrical field gradient and to a splitting of the quadrupole resonance

$$\nu_Q^{\rm NQR} = E_{\pm 3/2} - E_{\pm 1/2} = A\rho + C \tag{27}$$

proportional to the charge density ρ. This allows to measure the charge of the vortices. If one advocates the picture that the positive charges are expelled from the vortex to the outer regions one would expect a BCS estimate of the measured charge per length $Q \approx \frac{2ek_F}{\pi^3}\left(\frac{\lambda_{\rm TF}}{\xi}\right)^2\left(\frac{d\ln T_c}{d\ln\mu}\right)$. This means for the underdoped regime $Q > 0$ and for the overdoped $Q < 0$, which is just the opposite to what is observed experimentally. Moreover the BCS estimate gives a charge of $10^{-5}e$ per vortex while the experiment yielded a value three orders of magnitude larger. We solved this apparent contradiction with the help of the Bernoulli potential as follows.

One has to consider that YBCO is built up from Cu chains and planes where in the latter the superconductivity and the vortices are located. The discussed Bernoulli potential therefore appears in the planes. On the other hand the charge neutrality requires that the exceeding charge has to be compensated in the Cu planes such that a charge transfer appears between planes and chains. Taking this

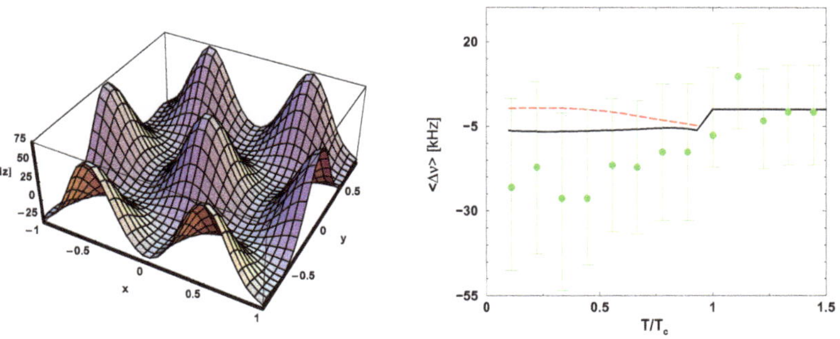

(a) (b)

Fig. 3. (a) Spatial variation of the line shift $\nu_2 = E_{-1/2} - E_{-3/2}$. The units are in vortex sizes. (b) Shift $\langle \Delta\nu \rangle$ of $\Gamma = 140$kHz of a single crystal (dashed line), and after averaging over grain orientation (solid line) compared to the experimental values.[21]

and the two-dimensional geometry into account we have been able to explain the observed quadrupole line shift,[30] see figure 3. The experimental line width is comparable to the spatial variation of the charge and magnetic field profile. Therefore the actual spatial profile of the magnetic field and charge density is important as presented in figure 3.

4. Consequences to the Kinetic Theory

4.1. *Extended quasiparticle approximation*

We have seen that the concept of correlated density leads in a natural way to the Bernoulli potential, which is useful in explaining recent experiments and promising to access a material parameter in superconductors. The underlying nonequilibrium theory has to distinguish between parts of the Wigner distribution described by quasiparticle poles and the rest called off-shell parts. This off-shell motion can be eliminated from the kinetic equation, which requires to introduce an effective distribution (the quasiparticle distribution f) from which the Wigner distribution f_W can be constructed

$$f_W[f] = f + \int \frac{d\omega}{2\pi} \frac{\wp}{\omega - \varepsilon} \frac{\partial}{\partial \omega} \left((1-f)\sigma_\omega^< - f\sigma_\omega^> \right). \tag{28}$$

Here $\sigma^>$ and $\sigma^<$ denote the selfenergies describing all correlations and ε is the quasiparticle energy. This relation represents the extended quasiparticle picture derived for small scattering rates.[5,31–33] The limit of small scattering rates has been first introduced by Craig.[34] An inverse relation $f[f_W]$ has been constructed.[35] For equilibrium non-ideal plasmas this approximation has been employed by[36,37] and has been used under the name of the generalized Beth-Uhlenbeck approach by[38] in nuclear matter for studies of the correlated density. The authors in[39] have used this approximation with the name "extended quasiparticle approximation" for the study of the mean removal energy and high-momenta tails of Wigner's distribution.

The non-equilibrium form has been derived finally as the modified Kadanoff and Baym ansatz.[40] We will call this the extended quasiparticle approximation.

This extended quasiparticle picture leads to balance equations which include explicit correlation parts analogous to the virial corrections. The firmly established concept of the equilibrium virial expansion has been extended to nonequilibrium systems[5] although a number of attempts have been made to modify the Boltzmann equation so that its equilibrium limit would cover at least the second virial coefficient.[41–43] The corrections to the Boltzmann equation have the form of gradients or nonlocal contributions to the scattering integral. The nonlocal quasiparticle kinetic equation for the momentum-, space-, and time-dependent distribution function of particle a, $f_1 \equiv f_a(k,r,t)$, derived within the non-equilibrium Green's function technique[5,31] has the form of a Boltzmann equation with the quasiparticle energy $\varepsilon_1 = \varepsilon(k,r,t)$,

$$\frac{\partial f_1}{\partial t} + \frac{\partial \varepsilon_1}{\partial k}\frac{\partial f_1}{\partial r} - \frac{\partial \varepsilon_1}{\partial r}\frac{\partial f_1}{\partial k} = s\sum_b \int \frac{dpdq}{(2\pi)^5\hbar^7}\mathcal{P}_\pm$$
$$\times\Big[(1-f_1)(1-f_2^-)f_3^- f_4^- - f_1 f_2^\pm(1-f_3^\pm)(1-f_4^\pm)\Big] \tag{29}$$

and the spin-isospin etc. degeneracy s. The superscripts $\pm$ denote the signs of nonlocal corrections: $f_2^\pm \equiv f_b(p, r\pm\Delta_2, t)$, $f_3^\pm \equiv f_a(k-q\pm\Delta_K, r\pm\Delta_3, t\pm\Delta_t)$, and $f_4^\pm \equiv f_b(p+q\pm\Delta_K, r\pm\Delta_4, t\pm\Delta_t)$. For the out-scattering part of (29) either the plus or minus signs can be chosen.[44] The scattering measure is given by the modulus of the scattering T-matrix $\mathcal{P}_\pm = |\mathcal{T}_\pm^R|^2\delta(\varepsilon_1+\varepsilon_2-\varepsilon_3-\varepsilon_4\pm 2\Delta_E)$. All corrections Δ, describing the nonlocal and non-instant collision are given by derivatives of the scattering phase shift $\phi = \text{Im}\,\ln\mathcal{T}^R(\Omega,k,p,q,t,r)$

$$\begin{aligned}
\Delta_t &= \left.\frac{\partial\phi}{\partial\Omega}\right|_{\varepsilon_1+\varepsilon_2}, & \Delta_2 &= \left(\frac{\partial\phi}{\partial p} - \frac{\partial\phi}{\partial q} - \frac{\partial\phi}{\partial k}\right)_{\varepsilon_1+\varepsilon_2}, \\
\Delta_E &= -\left.\frac{1}{2}\frac{\partial\phi}{\partial t}\right|_{\varepsilon_1+\varepsilon_2}, & \Delta_3 &= -\left.\frac{\partial\phi}{\partial k}\right|_{\varepsilon_1+\varepsilon_2}, \\
\Delta_K &= \left.\frac{1}{2}\frac{\partial\phi}{\partial r}\right|_{\varepsilon_1+\varepsilon_2}, & \Delta_4 &= -\left(\frac{\partial\phi}{\partial k} + \frac{\partial\phi}{\partial q}\right)_{\varepsilon_1+\varepsilon_2}.
\end{aligned} \tag{30}$$

The nonlocal kinetic equation (29) covers all quantum virial corrections on the binary level and conserves density, momentum and energy including the corresponding two-particle correlated parts.[5] It requires no more computational power than solving the Boltzmann equation.[45,46]

4.2. *Conservation Laws*

Neglecting all shifts, the time-invariant observables are the mean quasiparticle density $n^{\rm qp}$, the mean momentum $\mathcal{Q}^{\rm qp}$, the mass current $j^{\rm qp}$, the mean energy $\mathcal{E}^{\rm qp}$ and the stress tensor $\mathcal{J}_{ij}^{\rm qp}$ in the form

$$\begin{aligned} n^{\rm qp} &= \sum_k f, & \mathcal{Q}^{\rm qp} &= \sum_k k\, f, & j^{\rm qp} &= \sum_k \frac{\partial \varepsilon}{\partial k} f, \\ \mathcal{E}^{\rm qp} &= \sum_k (\tfrac{k^2}{2m} + \tfrac{1}{2}\sigma_{\rm mf}) f_k, & \mathcal{J}_{ij}^{\rm qp} &= \sum_k \left(k_j \frac{\partial \varepsilon}{\partial k_i} + \delta_{ij}\varepsilon \right) f - \delta_{ij}\mathcal{E}^{\rm qp}, \end{aligned} \tag{31}$$

where we abbreviated the mean-field part $\sigma_{\rm mf} = \int \frac{dp}{(2\pi)^3} t^R(\omega + \varepsilon_p, k, p, 0) f_p$ of the selfenergy.

Taking now into account the binary correlations which we have reformulated into the shifts, we obtain from the non-local kinetic equation the modified balance equations[5]

$$\frac{\partial (n^{\rm free} + n^{\rm corr})}{\partial t} + \frac{\partial (j^{\rm qp} + j^{\rm corr})}{\partial r} = 0 \tag{32}$$

$$\frac{\partial (\mathcal{Q}_j^{\rm qp} + \mathcal{Q}_j^{\rm corr})}{\partial t} + \sum_i \frac{\partial (\mathcal{J}_{ij}^{\rm qp} + \mathcal{J}_{ij}^{\rm corr})}{\partial r_i} = 0 \tag{33}$$

$$\frac{\partial \mathcal{E}}{\partial t} = \frac{\partial (\mathcal{E}^{\rm qp} + \mathcal{E}^{\rm corr})}{\partial t} = 0. \tag{34}$$

This shows that the conserving observables now consist of the sum of the quasiparticle parts (31) and the correlated parts which we can interpret as the parts coming from correlated pairs or molecules,

$$\begin{aligned} n^{\rm corr} &= \int d\mathcal{P}\Delta_t & j^{\rm corr} &= \int d\mathcal{P}\Delta_3 \\ \mathcal{Q}^{\rm corr} &= \int d\mathcal{P}\tfrac{k+p}{2}\Delta_t & \mathcal{E}^{\rm corr} &= \int d\mathcal{P}\tfrac{\epsilon_k+\epsilon_p}{2}\Delta_t \end{aligned}$$

$$\mathcal{J}_{ij}^{\rm corr} = \frac{1}{2}\int d\mathcal{P}\Big\{ k_j\Delta_{3i} + p_j(\Delta_{4i} - \Delta_{2i}) + q_j(\Delta_{4i} - \Delta_{3i}) \Big\}, \tag{35}$$

where $d\mathcal{P}$ is the probability to form a pair per unit of time. This immediate interpretation of $d\mathcal{P}$ is obvious from the expressions (35). The density of pairs is given, if $d\mathcal{P}$ is multiplied by the lifetime of the molecule Δ_t. The energy, the momentum and mass current which are carried by the pairs result analogously. As one can see, the correlated part of the stress tensor (35) takes the form of a virial which is well known for the collision flux in dense gases. The corresponding momenta are connected to the corresponding offsets. The diagonal part of this stress tensor is of course the correlated part of the pressure. The proofs that these conservation laws indeed follow from the nonlocal kinetic equation and are consistent with the extended quasiparticle picture can be found in detail in the appendix G of.[5]

5. Conclusion

The nonlocal kinetic equation unifying the achievements of transport theory in dense gases with the quantum transport of dense Fermi systems is presented. The quasiparticle drift of Landau's equation for the evolution of the quasiparticle distribution is connected with a dissipation governed by a nonlocal and non-instantaneous scattering integral in the spirit of Enskog corrections.[47] In this way quantum transport can be recast into a quasiclassical picture. The balance equations for the density, momentum and energy include quasiparticle contributions and the correlated two-particle contributions beyond the Landau theory.[48]

Compared with the Boltzmann equation, the presented form of virial corrections only slightly increases the numerical demands in implementations. Within INDRA collaboration, we have performed numerical studies and compared them with the experimental data from GANIL. The nonlocal corrections with parameters found from collisions of two isolated nucleons[47] have been implemented into numerical simulations of heavy ion reactions in the non-relativistic regime.[45,46] Keeping all simulation parameters as in the local approximation, the temperature of mono-nucleon products of central reactions increases due to nonlocal effects towards experimentally observed values, while the temperature of more complex clusters remains unchanged.[45] For the proton distribution in very peripheral reactions we have achieved an agreement between theoretical predictions and experimental data.[46]

As a consequence of the nonlocal kinetic theory, the correlated density appears which results in the Bernoulli potential in superconductors. This Bernoulli potential allows to access material parameters. Since at the surface large-scale cancellations appear, the charge of vortices as measured by NMR could be explained. We believe the correlated density to be a fruitful concept to describe strong correlations in interacting systems.

References

1. G. E. Beth and E. Uhlenbeck, *Physica* **3**, p. 729 (1936).
2. G. E. Beth and E. Uhlenbeck, *Physica* **4**, p. 915 (1937).
3. M. Schmidt, G. Röpke and H. Schulz, *Ann. Phys. (NY)* **202**, p. 57 (1990).
4. J. M. Luttinger, *Phys. Rev.* **119**, p. 1153 (1960).
5. P. Lipavský, K. Morawetz and V. Špička, *Kinetic equation for strongly interacting dense Fermi systems*, Annales de Physique, Vol. 26,1 (EDP Sciences, Paris, 2001).
6. B. Farid, *Phil. Mag. B* **79**, p. 1097 (1999).
7. V. Ambegaokar, in *Superconductivity*, (Marcel Dekker, New York, 1969), New York, ch. 5.
8. F. Bopp, *Z. Phys.* **107**, p. 623 (1937).
9. V. S. Sorokin, *JETP* **19**, p. 553 (1949).
10. A. G. van Vijfeijken and F. S. Staas, *Phys. Lett.* **12**, p. 175 (1964).
11. G. Rickayzen, *J. Phys. C* **2**, p. 1334 (1969).
12. H. W. Levis, *Phys. Rev.* **92**, p. 1149 (1953).
13. H. K. Hunt, *Phys. Lett.* **22**, p. 42 (1966).
14. J. Bok and J. Klein, *Phys. Rev. Lett.* **20**, p. 660 (1968).
15. T. D. Morris and J. B. Brown, *Physica* **55**, p. 760 (1971).
16. P. Lipavský, J. Kolacek, J. J. Mares and K. Morawetz, *Phys. Rev. B* **65**, p. 012507 (2001).
17. H. F. Budd and J. Vannimenus, *Phys. Rev. Lett* **31**, p. 1218 (1973).
18. J. Bardeen, *Phys. Rev.* **94**, p. 554 (1954).
19. C. J. Gorter and H. B. G. Casimir, *Phys. Z.* **35**, p. 963 (1934).
20. W. L. Ginzburg and L. D. Landau, *Zh. Eksp. Teor. Fiz* **20**, p. 1064 (1950).

21. P. Lipavský, J. Kolacek, K. Morawetz and E. H. Brandt, *Phys. Rev. B* **65**, p. 144511 (2002).
22. P. Lipavský, K. Morawetz, J. Kolacek, J. J. Mares, E. H. Brandt and M. Schreiber, *Phys. Rev. B* **69**, p. 024524 (2004).
23. P. Lipavský, K. Morawetz, J. Kolacek, J. J. Mares, E. H. Brandt and M. Schreiber, *Phys. Rev. B* **70**, p. 104518 (2004).
24. P. Lipavský, K. Morawetz, J. Kolacek, J. J. Mares, E. H. Brandt and M. Schreiber, *Phys. Rev. B* **71**, p. 024526 (2005).
25. D. I. Khomskii and F. V. Kusmartsev, *Phys. Rev. B* **46**, p. 14245 (1992).
26. G. Blatter, M. Feigel'man, V. Geshkenbein, A. Larkin and A. van Otterlo, *Phys. Rev. Lett.* **77**, p. 566 (1996).
27. P. deGennes, *Superconductivity of Metals and Alloys* (Addison–Wesley, New York, 1989).
28. P. Lipavský, K. Morawetz, J. Kolacek and E. H. Brandt, *Phys. Rev. Lett.* (2006), sub. cond-mat/0609669.
29. K. Kumagai, K. Nozaki and Y. Matsuda, *Phys. Rev. B* **63**, p. 144502 (2001).
30. P. Lipavský, J. Kolacek, K. Morawetz and E. H. Brandt, *Phys. Rev. B* **66**, p. 134525 (2002).
31. V. Špička, P. Lipavský and K. Morawetz, *Phys. Lett. A* **240**, p. 160 (1998).
32. K. Morawetz, P. Lipavský and V. Špička, *Ann. of Phys.* **294**, p. 134 (2001).
33. K. Morawetz (ed.), *Nonequilibrium Physics at Short Time Scales - Formation of Correlations* (Springer, Berlin, 2004).
34. R. A. Craig, *Ann. Phys.* **40**, p. 416 (1966).
35. B. Bezzerides and D. F. DuBois, *Phys. Rev.* **168**, p. 233 (1968).
36. H. Stolz and R. Zimmermann, *phys. stat. sol. (b)* **94**, p. 135 (1979).
37. D. Kremp, W. D. Kraeft and A. D. J. Lambert, *Physica A* **127**, p. 72 (1984).
38. M. Schmidt and G. Röpke, *phys. stat. sol. (b)* **139**, p. 441 (1987).
39. H. S. Köhler and R. Malfliet, *Phys. Rev. C* **48**, p. 1034 (1993).
40. V. Špička and P. Lipavský, *Phys. Rev. B* **52**, p. 14615 (1995).
41. K. Bärwinkel, *Z. Naturforsch. (a)* **24**, p. 38 (1969).
42. K. Bärwinkel, The generalized Boltzmann equation and some of its applications, in *Proceedings of the 14th International Symposium on Rarified Gas Dynamics*, (University of Tokyo Press, Tokyo, 1984).
43. R. F. Snider, *J. Stat. Phys.* **63**, p. 707 (1991).
44. V. Špička, K. Morawetz and P. Lipavský, *Phys. Rev. E* **64**, p. 046107 (2001).
45. K. Morawetz, V. Špička, P. Lipavský, G. Kortemeyer, C. Kuhrts and R. Nebauer, *Phys. Rev. Lett.* **82**, p. 3767 (1999).
46. K. Morawetz, P. Lipavský, J. Normand, D. Cussol, J. Colin and B. Tamain, *Phys. Rev. C* **63**, p. 034619 (2001).
47. K. Morawetz, P. Lipavský, V. Špička and N.-H. Kwong, *Phys. Rev. C* **59**, p. 3052 (1999).
48. P. Lipavský, V. Špička and K. Morawetz, *Phys. Rev. E* **59**, p. R1291 (1999).

COMPETING LOCAL AND NON-LOCAL PHASE CORRELATIONS IN FERMIONIC SYSTEMS WITH RESONANT PAIRING: THE BOSON–FERMION SCENARIO

J. RANNINGER*

Centre de Recherches sur les Très Basses Températures, CNRS, BP 166, 38042 Grenoble Cedex 9, France

* *E-mail: julius.ranninger@grenoble.cnrs.fr*

We consider systems where the amplitude and the phase of a superfluid order parameter are controlled independently by local, respectively non-local phase correlations of the internal degrees of freedom of quasi-particles which characterize the different phases of such fermionic systems. A possible realization of such a situation is when pairing into the bosonic entities occurs *via* resonant scattering of itinerant fermionic particles. Examples for that are the ultracold fermionic gases in optical traps and lattices, the correlated electronic systems and strongly coupled electron-phonon systems, such as seen in the high T_c cuprates. We demonstrate the occurrence of a quantum phase transition from a superfluid to a phase uncorrelated boson liquid state in such systems, which is controlled by the competing effects of long and short range phase correlations of the constituents of composite quasi-particles.

Keywords: Superconductor–Insulator Transition.

While superfluidity in bosonic systems is controlled by a macroscopic coherent quantum state describing a Bose condensed state with phase locked bosons in real space, the situation for fermionic systems is more complex. It is a phase locked state of bosonic electron pairs in momentum space, as initially conjectured by F. and H. London[1] on the basis of the experimentally observed macroscopic screening currents implied by the magnetic flux expulsion in a bulk superconducting material. The phase locked nature of such a state was clearly spelled out by P. W. Anderson[2] for the case of the classical low temperature superconductors in the framework of the BCS theory.[3] There, the pairing amplitude $| < c_{k,\uparrow}^{+} c_{-k,\downarrow}^{+} > |$ builds up in a small region around the Fermi surface, but only when simultaneously phase locking between pairs with different momenta sets in and a macroscopic coherent quantum state —showing superconductivity— is established. In general however that needs not be the case. A finite pairing amplitude can form without a long range phase coherence driving it. In this more challenging problem of an independent onset of a local pairing amplitude and long range phase correlations, such as it can occur in ultra-cold optically trapped fermionic gases and the high T_c cuprates, a finite pairing amplitude results from strong local phase correlations between fermions in different degenerate two-particle states. In fact, such strong local phase correlations, necessary to provide pairing in the first place, must be partly decorrelated before a global phase locked superconducting state can even set in. Such systems are hence controlled by a competitions between local and non-local phase correlations and this will be the issue discussed in this lecture.

The formation of bosonic entities such as Cooper pairs in weak coupling systems happens generally via two-particle scattering processes which involve an exchange of certain bosonic entities such a phonons, excitons, magnons, etc. and which serve as glue. Cooper pairing correlations between two Fermions with two given momenta then happens over a characteristic finite length scale of the order of a few thousands Å's and a typical time scale of 10^{-12} sec which is controlled by the dynamics of those excitations. In the case of strong coupling between the fermions and the bosonic entities, providing the glue at weak coupling, this gives rise to pairing correlations between the fermions which in general is of a very different kind. For the cases we are interested here (pairing of fermionic atoms in optical traps and fermionic charge carriers in the high T_c cuprates) pairing is controlled by kinematic interactions which can be attributed to a degeneracy of localized bound pair states of fermions and uncorrelated itinerant fermion pairs. The essence of such a situation can be described by a simple, but physically very rich and highly non-trivial phenomenological model such as the Boson–Fermion Model (BFM) —representing a mixture of itinerant fermions and localized bound fermion pairs which are exchange coupled with each other.

In the case of ultracold fermion gases, atoms, such as Li and K, exhibit a strong spin orbit coupling with concomitant fluctuations of the electron spin and orbital momentum. Two such atoms in a singlet electron spin configuration tend to be in scattering states, while in a triplet configuration they exhibit a shallow bound state. Applying a magnetic field to such a system permits to shift those two corresponding inter-atomic potentials with respect to each other. For a dilute system of such fermionic atoms, this give rise to *Feshbach* resonant pairing[4] of the fermions at the Fermi level and which can be described by the above mentioned BFM. It involves bosons —representing the bound fermions pairs in the triplet configuration— and itinerant fermions to which they are exchange coupled. A similar situation is encountered in electron-phonon coupled systems in the intermediate coupling regime. There, single electrons are in a superposition of itinerant and localized small polaron states, as was recognized early on.[5–7] For a many-electron system, this picture originally led to the conjecture[8] of an exchange coupled mixture of localized bipolarons (tightly bound electron pairs trapped in their selfinduced lattice deformed environment) and itinerant electrons (essentially uncoupled to the lattice deformations). It ultimately resulted in the proposition of the BFM[8] and the picture of resonating bipolarons inside a Fermi sea, analogous[9] to the Feshbach resonance of atomic gas systems.

The idea behind the proposition of this BFM was to find a description of the cross-over regime between a weak coupling BCS type superconductor and the hypothetical yet to be experimentally verified Bipolaronic Superconductor.[10] The scenario of the BFM was initially intended to capture simultaneously two features: those which could account for the itinerant behaviour of the fermions and those which could describe their tendency to localized pairing in real space and on a finite time scale. This was done by replacing the initial fermionic single-component system by an effective two-component system with two types of commuting fields describing: itinerant fermionic quasi-particles and bosonic quasi-particles describing tightly bound localized fermion pairs, which are coupled to the itinerant fermion subsystem via an exchange term. The two subsystems are in a chemical equilibrium, implying a single chemical potential which is common to both the fermions and the bosons such as to keep the total number of both species constant and equal to the number of initial fermions.

The theoretical study of such a scenario describing a cross-over situation, then has to be done within non-perturbative techniques, such as self-consistent diagrammatic approaches,[11,12] Dynamical mean field analysis,[13] renormalization group techniques,[14] functional integral approaches[15] and schemes using bond operators.[16] All those methods aim to disentangle the dynamics of the fermions from that of the bosons, and by doing that result in interdependent coupling constants characterizing the two formally *decoupled* subsystems of fermionic and bosonic nature. This permits to handle situations where the amplitude of the order parameter forms independently of the onset of a global superfluid or charge ordered state, which is what is required if we want to address our studies to such a systems as atomic gases and the high T_c cuprate superconductors.

The physically relevant variables in the BFM scenario are itinerant fermions $c_{i\sigma}^{(\dagger)}$, itinerant fermion pairs $\tau_i^+ = c_{i\uparrow}^\dagger c_{i\downarrow}^\dagger$, $\tau_i^- = (\tau_i^+)^\dagger$ on a single site i with opposite spins $\sigma = [\uparrow, \downarrow]$ and localized tightly bound fermions pairs $\rho_i^\pm$. The operators describing the pair states have spin-$\frac{1}{2}$ commutation relations, taking into account their hard core nature imposed by the Pauli principle of their internal fermionic degrees of freedom. Within this Hilbert space we have the following Hamiltonian describing the BFM:

$$
\begin{aligned}
H = (D-\mu)\sum_{i\sigma} c_{i\sigma}^\dagger c_{i\sigma} + (\Delta_B - 2\mu)\sum_i (\rho_i^z + \frac{1}{2}) \\
+ \sum_{i\neq j,\,\sigma} t_{ij}(c_{i\sigma}^\dagger c_{j\sigma} + H.c.) + g\sum_i \left(\rho_i^+\tau_i^- + \rho_i^-\tau_i^+\right).
\end{aligned}
\tag{1}
$$

g denotes the strength of the exchange coupling and t the nearest neighbor hopping integral for the itinerant fermions with a band half-width (which will serve as energy unit) $D = zt$, z being the coordination number of the underlying lattice. The energy of the bound fermion pairs is Δ_B. The number of the ensemble $n_{\text{tot}} = n_F + 2n_B$ of n_B bosons and $n_F = n_{F\uparrow} + n_{F\downarrow}$ fermions being conserved, implies a common chemical potential μ for both subsystems. For the purpose of this lecture, we shall restrict ourselves to the analysis of the fully symmetric half-filled band case ($\Delta_B \equiv 0$), which in such a two-component system means $n_{tot} = 2$ and $\mu \equiv 0$. This greatly facilitates the algebra and thus brings out the underlying physics more clearly. This presentation follows closely a very detailed study[16] on the issues we shall discuss here, that is to say: how strong local phase correlations leading to fermion pairing must get partially decorrelated in order to permit a superfluid state of such a system. This local phase correlation is inherent in the atomic limit of this problem for which we have the following set of eigenstates and eigen energies

$$\begin{array}{lcl}
\varepsilon_b = -g & \frac{1}{\sqrt{2}}\left[\rho^+ - \tau^+\right]|0\rangle \Longrightarrow b^\dagger|v\rangle \\
\varepsilon_a = +g & \frac{1}{\sqrt{2}}\left[\rho^+ + \tau^+\right]|0\rangle \Longrightarrow a^\dagger|v\rangle \\
\varepsilon_f = 0 & \rho^+\tau^+|0\rangle \quad \Longrightarrow f^\dagger|v\rangle \\
\varepsilon_d = 0 & |0\rangle \quad \Longrightarrow d^\dagger|v\rangle \\
\varepsilon_h = 0 & c_\sigma^\dagger|0\rangle \quad \Longrightarrow h_\sigma^\dagger|v\rangle \\
\varepsilon_s = 0 & \rho^+ c_\sigma^\dagger|0\rangle \quad \Longrightarrow s_\sigma^\dagger|v\rangle
\end{array} \tag{2}$$

Since it is those local states which will largely determine the physics of the system we are interested in, we introduce a new vacuum state $|v\rangle$ on which we define operators $[b, a, d, f, h, s]$ which create the ensemble of such states, but which then have to be constraint by a local closure relation

$$f^\dagger f + \sum_\sigma s_\sigma^\dagger s_\sigma + b^\dagger b + a^\dagger a + \sum_\sigma h_\sigma^\dagger h_\sigma + d^\dagger d = 1\,. \tag{3}$$

For such a single-site problem we derive the following single-fermion spectral function[17]

$$G_{at}(\omega) = \frac{Z^F}{\omega} + (1 - Z^F)\frac{1}{2}\left[\frac{1}{\omega - g} + \frac{1}{\omega + g}\right], \quad Z^F = \frac{2}{3 + \cosh\beta g} \tag{4}$$

which formally is of a very similar form as that for a BCS superconductor, showing a sharp drop in the single Fermion spectral weight Z_F at $\omega = 0$ as the temperature drops below $T = g$. This feature implies the opening of a kind of gap in the density of states, which, when hopping between nearest neighbors is introduced, results in a pseudogap caused by the increase in spectral weight of the bonding and anti-bonding states $b^\dagger|v\rangle, a^\dagger|v\rangle$. This feature is visible in the strengthening of the local phase coherence between pairs of the itinerant fermions τ^+ and the tightly bound fermions pairs ρ^+ as the corresponding correlator

$$\langle\rho^+\tau^-\rangle = \frac{1}{2}\tanh(\frac{\beta g}{2}) = \frac{1}{2}\frac{\sinh(\beta g)}{1 + \cosh(\beta g)} \tag{5}$$

shows in form of a sharp increase as T decrease below $T = g$. The foreshadowing of the opening of the pseudogap in the density of states and the disappearance of the single-fermion spectral weight is manifest in the building up of strong local fermion pairing, as apparent from the low frequency pair susceptibility

$$\chi_F = FT\langle T\{c_\downarrow(t)c_\uparrow(t)c_\uparrow^\dagger(0)c_\downarrow^\dagger 0)\}\rangle_{\omega\to 0} = \frac{1}{g}\frac{\sinh(\beta g)}{3 + \cosh(\beta g)} \Rightarrow \frac{1}{g}(T = 0). \tag{6}$$

It clearly shows a deviation from the free particle behavior as T is decreases below $T = g$ tending to a finite value $1/g$ rather than diverging as $1/T$ in the absence of any fermion pair correlations.

Having gained this insight into the problem in the extreme strong coupling limit $g/t = \infty$, we shall now examine how the physics changes upon reducing the exchange coupling constant g/t. This is best be done upon using the local states defined in eq. (2), in analogy to the bond operator formalism successfully applied to a variety of different systems in which the evolution of short range pairing correlations and the onset of long range phase coherence was studied.[18] To proceed with that we must re-express the initial fermionic and bosonic variables by the set of bonding operators which results in:

$$\begin{aligned} c^\dagger_\sigma &= p_\sigma f^\dagger s_{-\sigma} + \tfrac{1}{\sqrt{2}} s^\dagger_\sigma\,(b+a) + \tfrac{p_\sigma}{\sqrt{2}}\left(a^\dagger - b^\dagger\right) h_{-\sigma} + h^\dagger_\sigma d \\ \rho^+ &= \tfrac{1}{\sqrt{2}} f^\dagger (a-b) + \sum_\sigma s^\dagger_\sigma h_\sigma + \frac{1}{\sqrt{2}}(b^\dagger + a^\dagger) d\,, \end{aligned} \tag{7}$$

with $p_\sigma = (+1,-1)$ for $\sigma = \uparrow,\downarrow$. Rephrasing the initial Hamiltonian in terms of such bond operators enables one to obtain non-trivial results within a mean field analysis for the various phases in the $[T/t, g/t]$ parameter space. A simple mean field treatment of the Hamiltonian in the initial operators would totally decouple the local exchange term and hence can not account for the local phase correlations which are vital for this problem. Being interested in the stability of a superconducting state in these systems we look for a macroscopic coherent quantum state of the form

$$|\psi_c\rangle = \prod_i (u_v + u_b b^\dagger_i + u_a a^\dagger_i + u_d d^\dagger_i + u_f f^\dagger_i)|v\rangle, \tag{8}$$

for the bosonic entities involving the bosonic bond operators b, a, d, f, restricting ourselves to homogeneous solutions. Denoting the amplitudes of the various bosonic operators by $\bar{b}^\dagger_i, \bar{a}^\dagger_i, \bar{d}^\dagger_i, \bar{f}^\dagger_i$, given by the expectation values of $b^\dagger_i, a^\dagger_i, d^\dagger_i, f^\dagger_i$ with respect to the coherent macroscopic quantum state, eq. (8), the corresponding mean field version of the Hamiltonian is:

$$\begin{aligned} H &= H_t + H_{ex} + H_0, \\ H_t &= \sum_k D^\dagger_k \hat{U}_k D_k + c, \quad H_{ex} = L\, g(\bar{a}^2 - \bar{b}^2) \\ H_0 &= -L\,\mu[4\bar{f}^2 + 2(\bar{b}^2 + \bar{a}^2)] + L\,\lambda]\bar{f}^2 + \bar{b}^2 + \bar{a}^2 + \bar{d}^2 - 1], \end{aligned} \tag{9}$$

where now all the coupling constants of the bosonic part H_0 (including the constraint (eq. (3) controlled by a Lagrange parameter λ), of the exchange term H_{ex} and of the kinetic term H_t for the two types of Fermions h and s depend on the amplitudes of the bosonic bond operators which have to be determined such as to minimize the free energy of the system. L denotes the total number of bond sites. The diagonalization of H_t involves a generalization of the usual Bogoliubov two-component spinors $(c_{k,\uparrow}, c^\dagger_{-k,\downarrow})$ in form of four component spinors:

$$D_k = (s_{k,\uparrow}, s^\dagger_{-k,\downarrow}, h_{k,\uparrow}, h^\dagger_{-k,\downarrow}), \quad D^\dagger_k = (s^\dagger_{k,\uparrow}, s_{-k,\downarrow}, h^\dagger_{k,\uparrow}, h_{-k,\downarrow}). \tag{10}$$

The matrix elements U_k^{ij} linking them together are given by

$$\begin{bmatrix} \gamma_k^s & \Delta_k^s & \gamma_k^{\bar{s}h} & \gamma_k^{\bar{s}\bar{h}} \\ \Delta_k^s & -\gamma_k^s & \gamma_k^{sh} & -\gamma_k^{\bar{s}h} \\ \gamma_k^{\bar{s}h} & \gamma_k^{sh} & \gamma_k^h & \Delta_k^h \\ \gamma_k^{\bar{s}\bar{h}} & -\gamma_k^{\bar{s}h} & \Delta_k^h & -\gamma_k^h \end{bmatrix} \tag{11}$$

and reflect the intricate competition between the local phase correlations, determined by the amplitudes $\bar{b}, \bar{a}$ and long range phase correlations determined by non-zero values of the amplitudes $\bar{d}, \bar{f}$. The competition between the local and global phase coherence in this kinetic terms is apparent in the following features:

- The spectra of the s and h fermions determined by

$$\begin{aligned} \gamma_k^s &= \epsilon_k[0.5 * (\bar{b}+\bar{a})^2 - f^2] - 3\mu + \lambda \\ \gamma_k^h &= \epsilon_k[\bar{d}^2 - 0.5 * (\bar{b}-\bar{a})^2] - \mu + \lambda \end{aligned} \tag{12}$$

  are strongly renormalized by two competing effects: the condensation of the bonding and antibonding a and b bosons and of the d and f bosons whose densities $\bar{a}^2, \bar{b}^2, \bar{d}^2, \bar{f}^2$ enter in form of their differences
- The gap in the spectra of the s and h fermions determined by

$$\Delta_k^s = \epsilon_k\sqrt{2}[\bar{f}(\bar{b}+\bar{a})], \quad \Delta_k^h = \epsilon_k\sqrt{2}[\bar{d}(\bar{a}-\bar{b})] \tag{13}$$

  respectively depend on the products of $\bar{a}$ and $\bar{b}$ with $\bar{d}$ and $\bar{f}$ which is an indication that for a global superfluid state one needs simultaneously both of them to be non zero.
- The particle-hole and particle-particle hybridization between the two types of fermions determined by

$$\begin{aligned} \gamma_k^{\bar{s}h} &= \tfrac{\epsilon_k}{\sqrt{2}}\left\{[(\bar{b}+\bar{a})\bar{d}] - [(\bar{a}-\bar{b})\,\bar{f}]\right\} \\ \gamma^{\bar{s}\bar{h}} &= \epsilon_k\left[(\bar{f}\,\bar{d}) + \tfrac{1}{2}\left(\bar{a}^2-\bar{b}^2\right)\right]. \end{aligned} \tag{14}$$

  The first one depends on products of $\bar{a}$ and $\bar{b}$ and of $\bar{d}$ and $\bar{f}$, similar to the structure of the matrix elements characterizing the gap. The second, on the contrary, depend on exclusively the density of the various bond bosons ie., $\bar{a}^2, \bar{b}^2, \bar{d}^2, \bar{f}^2$.

Let us now sketch the results of these competing local and non-local phase correlations as far as the zero temperature phases are concerned upon varying the strength of the exchange coupling constant g/t. For a full account of these results together with their extension to finite temperatures can be found in Ref. 16. The ground state energy of our system is given by

$$E_0 = H_{\text{exc}} + H_0 + L\sum_k E_k^0\theta(-E_k^0) \tag{15}$$

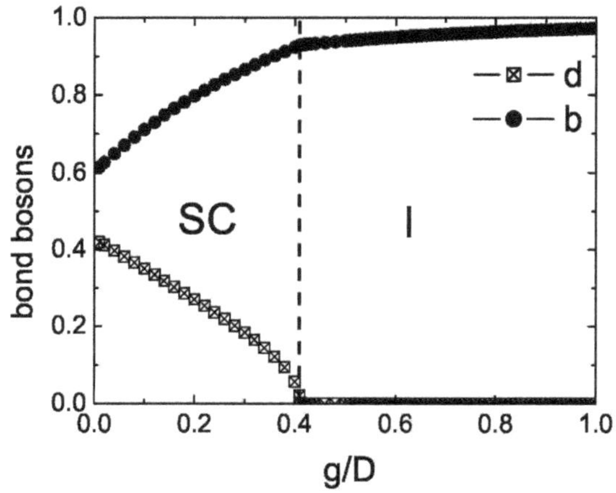

Fig. 1. Evolution of the amplitudes of the bonding bosons b and bond holes d at $T = 0$ as g increases and one passes from the superconducting into the insulating bond liquid phase (after Ref. 16).

where E_k^0 denotes the lowest eigen state of H_t which is obtained by diagonalizing the matrix $U_k^{i,j}$. For the specific case we discuss in this lecture, i.e., $\Delta = 0, n_F = 1, n_B = 0.5$ we have a chemical potential independent on g/t, being fixed at $\mu \equiv 0$. We then find independent on g/t: $\bar{a} = 0$ and $\bar{d} = \bar{f}$. The stable phases are thus determined by minimizing E_0 with respect to $\bar{b}$, $\bar{d}$ and λ. As can be seen from Fig. 1, as the exchange coupling g/t increases towards a critical value g_c/t, $\bar{d}$ decreases to zero. Simultaneously $\bar{b}$ increases steadily up to g_c/t where it shows a kink before it smoothly approaches the value of unity. The opposite trend with increasing g/t of $\bar{b}$ and $\bar{d}$ is a clear signature of the competition between the local and the non-local phase correlation and shows that in order for superfluidity to set, the local amplitude of the bonding bosons must be diminished when at the same time the superconducting order wants to get stronger. For $g/t \geq g_c/t$ superconductivity can not be maintained and the system enters a phase, corresponding to an uncorrelated bond-pair liquid state

$$|\Psi_{\mathrm{BPL}}\rangle \prod_i (u_v + u_b b_i^\dagger)|v\rangle \tag{16}$$

One of the startling features of this BFM scenario is the transition between this phase uncorrelated bond pair liquid state and a superfluid state of bonding pairs. Both states have a gap but their physical implications are quite different. This gap is seen in the single Fermion spectrum for the initial Fermions ($[c_i^\dagger, c_i^\dagger]$ respectively their Bloch representation in momentum space and given by the spectral function

$$A_{k\sigma}^c(\omega) = (-1/\pi)\,\mathrm{Im}[G_{k\sigma}^c(\omega)], \quad G_{k\sigma}^c(\tau) = -\left\langle T\left\{c_k(\tau)c_k^\dagger(0)\right\}\right\rangle \tag{17}$$

The evaluation of that proceedes via the related to it matrix Green's function for

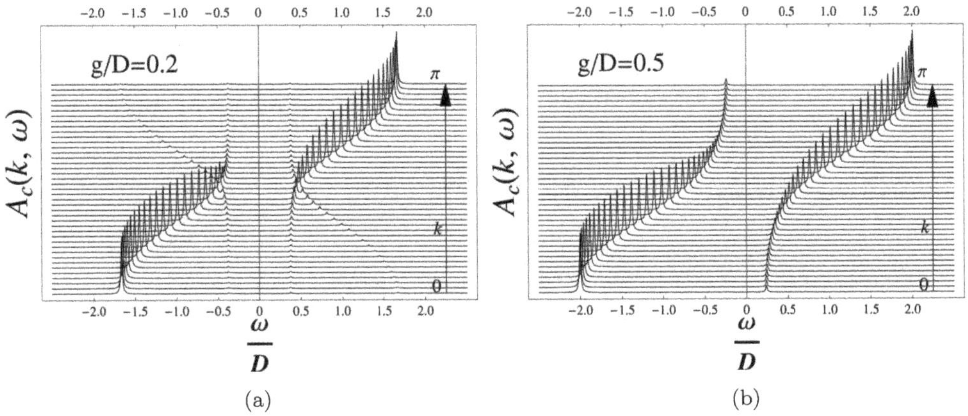

Fig. 2. The single particle fermionic spectral function for a 1D system where the k vectors run (vertical line) from 0 to π and the chemical potential μ is situated at $k_F = \pi/2$. (a) Superconducting phase with $g/t = 0.2$, (b) Insulating bond pair liquid phase with $g/t = 0.5$.

the bond fermions

$$\hat{G}_k^D(\tau) = -\left\langle T\left\{\vec{D}_k(\tau)\vec{D}_k^\dagger(0)\right\}\right\rangle \tag{18}$$

obtained from the diagonalizing of H_t. We illustrate in fig. 2 the single particle fermionic spectral function for two characteristic phases, the superconducting ($g/t = 0.2$) and the bond boson liquid phase ($g/t = 0.5$). Both spectra show a clear two-band structure with the lower band filled and the upper one empty. This feature originates from the bonding, respectively anti-bonding atomic states and hence is visible not only in the phase uncorrelated bond pair liquid state but also in the superconducting state where it has nothing to do with the superconducting gap. The single particle gap, given by the minimal energy between the two bands varies, as illustrated in fig. 3. The gap varies linearly in g as expected from the difference in energy for $g \geq g_c$ between the bonding and antibonding atomic states. This gap extrapolates to zero at $g = 0$, which suggests that, as we enter the superconducting phase from the phase uncorrelated bond pair liquid state, the gap can be imagined as a superposition of a "insulating gap" and a superconducting one. This is because in the superconducting phase we still must have strong local correlations which produce the necessary Cooperon pairing and which is of an energy of the order of g. But since now the Cooper pairs exist in a phase locked state. It gives rise to an additional contribution to the gap, arising from the superconducting phase correlations. These global phase correlations are apparent in form of a Bogoliubov like mode in the region around $k_F = \pi/2$, as can be seen from fig. 2.

In this lecture we focused on the intricate interplay between local phase correlations which drive local pairing among the fermions and global phase correlations which drive superfluidity and by doing so have to decorrelate partly the local phase correlations. The basic building stones of such a system are bonds between pairs of itinerant fermions and tightly bound fermion pairs. As the exchange coupling

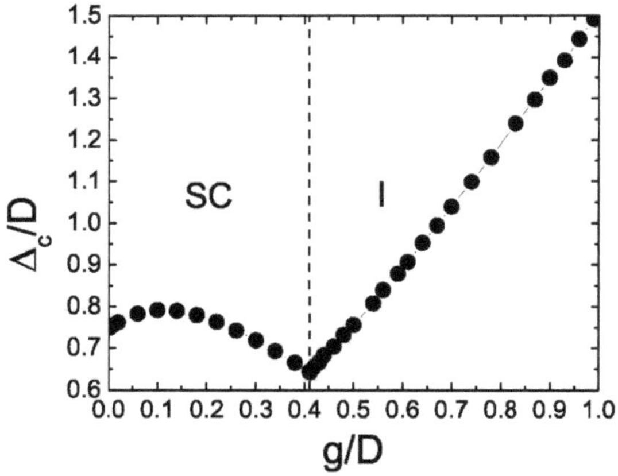

Fig. 3. Evolution of the charge gap, associated with the excitation of a c fermion, as g increases and one passes from the superconducting into the insulating bond liquid phase (after Ref. 16).

between the two, which causes the local phase correlations, decreases, one changes from a phase uncorrelated bond pair liquid to a phase correlated superfluid bond pair liquid. The nature of the uncorrealted bond pair liquid will be discussed in some future work, where the corresponding elementary excitations of this state will be worked out and compared with those of the superfluid state at lower exchange coupling.

I am indebted to Mario Cuoco for his collaboration on such matters and permitting me to publish unpublished data showing the spectral properties of Fermions in fig. 2.

References

1. F. London and H. London, *Proc. Roy. Soc.* (London) **A147**, 71 (1935).
2. P. W. Anderson, *Phys. Rev.* **112**, 1900 (1958).
3. J. Bardeen, L. N. Cooper, and J. R. Schrieffer, *Phys. Rev.* **108**, 1175 (1957).
4. E. Timmermans *et al.*, *Phys. Rep.* **315**, 199 (1999).
5. D. M. Eagles, *Phys. Stat. Solidi B* **48**, 407 (1971).
6. H. B. Shore and L. M. Sander, *Phys. Rev. B* **7**, 4537 (1973).
7. K. Cho and Y. Toyozawa, *J. Phys. Soc. Jap.* **33**, 1555 (1971).
8. J. Ranninger and S. Robaszkiewicz, *Physica B* **135**, 468 (1985).
9. J. Ranninger and A. Romano, *Eur. Phys. Lett.* **75**, 461 (2006).
10. A. S. Alexandrov and J. Ranninger, *Phys. Rev. B* **23**, 1796 (1981).
11. J. Ranninger, J.-M. Robin, and M. Eschrig, *Phys. Rev.Lett.* **74**, 4027 (1985).
12. J. Ranninger and L. Tripodi, *Phys. Rev. B* **67**, 174521 (2003).
13. J.-M. Robin, A. Romano, and J. Ranninger, *Phys. Rev. Lett.* **81**, 2755 (1998).
14. T. Domanski and J. Ranninger, *Phys. Rev. B* **63**, 134505 (2001).
15. M. Cuoco and J. Ranninger, *Phys. Rev. B* **70**, 104509 (2004).
16. M. Cuoco and J. Ranninger, *Phys. Rev. B* **74** 94511 (2006).
17. T. Domanski, J.-M. Robin, and J. Ranninger, *Solid State Commun.* **105**, 473 (1998).
18. P. Phillips and D. Davidovich, Science **302**, 243 (2003).

SUPERCONDUCTING ORDER PARAMETERS IN THE EXTENDED HUBBARD MODEL: A SIMPLE MEAN-FIELD STUDY

J. S. THAKUR

Department of Electrical and Computer Engineering,
Wayne State University, Detroit, Michigan 48202, USA
jagdish@wayne.edu

M. P. DAS

Department of Theoretical Physics, Institute of Advanced Studies,
The Australian National University, Canberra, ACT 0200, Australia
mukunda.das@anu.edu.au

The extended Hubbard model with nearest neighbor hopping and intersite interaction for a 2D square lattice has been studied within the frame work of a simple mean-field formalism. The self-consistent solutions of superconducting order parameters have been obtained for various fillings. The results indicate that superconducting pairing occurs on doping for the repulsive on-site U and attractive nearest neighbor V.

Keywords: Hubbard model; extended Hubbard model; nearest neighbor interaction; mean-field superconducting order parameter; repulsive on-site U; attractive nearest neighbor V.

1. Introduction

Ever since the discovery of high temperature superconductors in layered oxides,[1] two dimensional Hubbard model is being studied very extensively by various analytical and numerical techniques.[2] The Hubbard model, originally designed to describe itinerant magnetism in narrow band systems, has been found to be a successful model for many strongly correlated electronic systems.[3] Mott (metal-insulator) transition, band insulator with antiferromagnetic ordering, non-Fermi liquid behavior etc. are some of the well-investigated topics with the Hubbard model. Although the model is only exactly solvable in one dimension, a number of results have been obtained in the mean field approximation, occasionally including fluctuations. In 2-dimension mean field phase diagrams[4] have been studied implying the richness of various ordered phases[5] (e.g. charge density wave states, anti-ferromagnetism, para and ferromagnetism and singlet pairing states).

In 1987 Anderson[6] made an important suggestion that the Hubbard model is a suitable candidate for high T_c superconductivity. Based on the observation that a Mott insulator when doped, becomes a superconductor, Anderson postulated that upon doping, the loss of antiferromagnetic long range order produces a so-called "resonating valence bond" (RVB) phase — a quantum liquid of singlet pairs and stabilizes the ground state with a short range magnetic order. In the strong coupling limit Anderson and coworkers[7] believed that superconductivity can appear in a mean field theory, where the pair excitations are formed in the strange normal state and these pair can undergo charged boson condensation, when the temperature is lowered. Although this idea has passed through several modifications in the past decade,[8] there have been other serious suggestions even at the weak coupling level (e.g. spin bag,[9] antiferromagnetic Fermi liquid[10] and paramagnon exchange[11] models) that superconductivity can be found in a Hubbard model.

On the other hand, during the past few years there have been great efforts on the elaborate computer simulations on the question as to whether a Hubbard ($U > 0$)-only model can produce superconductivity. Recently lmada[12] showed strong numerical evidence against superconductivity in 2D U-only Hubbard model near half-filling. Consequently an obvious step is to include the next neighbor interaction in the so-called extended Hubbard model. There are now some evidence[13–21] that onsite (U) and the next neighbor (V) interactions can produce pairing and hence the superconductivity. The parameters, U and V can be positive or negative which depend on the coupling of the electrons to phonons or to other bosonic modes of the system. A familiar example is a negative U model. The direct Coulomb interaction can be suppressed by the strong coupling of phonons as has been pointed out by Anderson[22] in the context of amorphous semiconductors.

Apart from pure Hubbard model there are variants of it in various limits, known as t-J model family. Extensive numerical work has been reported on several mysterious anomalies in the normal phases, besides the superconducting states. Bulut[23] reviews effective pairing interaction which is repulsive at (π, π) momentum transfer, and enhanced single-particle spectral weight near the $(\pi, 0)$ and $(0, \pi)$ points of the Brillouin zone, create optimum conditions for $d_{[x]2-[y]2}$-wave pairing. Maier *et al.*[24] examine superconducting instabilities of the doped repulsive 2D Hubbard model, which are studied in the intermediate to strong coupling regime with the help of the dynamical cluster approximation. To solve the effective cluster problem they employ an extended noncrossing approximation, which allows for a transition to the broken symmetry state. Roth's two-pole approximation has been used by Calegari *et al.*[25] to investigate the role of d-p hybridization in the superconducting properties of an extended d-p Hubbard model. Metzner[26] examines the existence of d-wave superconductivity in the two-dimensional Hubbard model close to half-filling, by using a renormalization group analysis at the one-loop level.

As indicated earlier, research with Hubbard model and its variants have been carried out intensely with many diverging point of view, even whether the Hubbard model is the right model for the superconductivity. In the absence of a general consensus and a definite proof, we consider the Hubbard model as a very basic many-body model that takes appropriately electron correlations. In this paper we consider a pedagogic view in terms of basic interactions and attempt a study of the superconducting order parameters in an extended Hubbard Model and treat onsite interaction U and nearest neighbor interaction V as phenomenological parameters. In Sec. 2 we discuss the mean field superconducting order parameters from the extended Hubbard Hamiltonian within a generalized (a simple) Hartree-Fock approximation (GHFA) in the ground state. Self-consistent solutions are obtained by varying the parameters for various fillings in Sec. 3. The effective potential is found to be attractive in certain conditions and the results are discussed with its implication to high T_c superconductivity. We conclude with this simple picture that one can have a qualitative understanding of how the superconductivity is manifested through electron correlation only.

2. Mean-Field Superconducting Order Parameter

We consider an extended Hubbard Hamiltonian given by

$$H = -\sum_{ij\sigma}(t + \mu\delta_{ij})C_{i\sigma}^{+}C_{j\sigma} + \frac{1}{2}\sum_{ij\sigma\sigma'}(S_u U\delta_{ij}\delta_{\sigma-\sigma'} + S_v V)n_{i\sigma}n_{j\sigma'} \tag{1}$$

where $C_{i\sigma}^{+}$, $C_{i\sigma}$ and $n_{i\sigma} = C_{i\sigma}^{+}C_{i\sigma}$ are the creation, destruction and number operators respectively at the site i with spin $\sigma = \uparrow$ or $\downarrow$. j is restricted only to the nearest-neighbor sites of the site i. μ stands for chemical potential. S_u, and S_v take care of the sign of the interactions. The three cases of interactions which we study are; (1) when both U and V are attractive, which mean $S_u = S_v = -1$; (2) when U is attractive and V is repulsive, here $S_u = -1$ and $S_v = +1$ and (3) when U is repulsive and V is attractive, i.e., $S_u = +1$ and $S_v = -1$. The Eq. (1), in Bloch representation becomes

$$H = H_0 + H_1 \,, \tag{2a}$$

where

$$H_0 = -\sum_{ij\sigma}(\varepsilon_k + \mu)C_{k\sigma}^{+}C_{k\sigma} \tag{2b}$$

$$H_1 = \frac{1}{2N}\sum_{kk'q\sigma\sigma'}(S_u U\delta_{ij}\delta_{\sigma-\sigma'} + zS_v V f_q)C_{k\sigma}^{+}C_{k-q\sigma}C_{k'\sigma'}^{+}C_{k'+q\sigma'} \,. \tag{2c}$$

Here N is the total number of sites $\varepsilon_k = ztf_k$ is the Fourier transform of hopping integral and $f_k = \frac{1}{Z}\sum_J \exp(Ik\cdot r_j)$, z is the co-ordination number and r_j is the nearest-neighbor vector. Within the generalized Hartree-Fock approximation (GHFA) the singlet-superconducting order parameter for the square lattice at $T = O$ is given by

$$\delta_k = -\frac{1}{N}\sum_{k'}(S_u U + zS_v V f_k f_{k'})\frac{\Delta_{k'}}{[E_{k'}^2 + \Delta_{k'}^2]^{1/2}} \tag{3}$$

where

$$E_k = (zt + zS_v V\eta)f_k - \mu + \frac{n}{2}(S_u U + 2zS_v V) \tag{4}$$

The mean value of carrier concentration, n, is given by

$$n = 1 - \frac{1}{N}\sum_{k} \frac{E_k}{[E_{k'}^2 + \Delta_{k'}^2]^{1/2}} \tag{5}$$

and the quantity η

$$\eta = \frac{1}{N}\sum_{k} \frac{f_k E_k}{[E_{k'}^2 + \Delta_{k'}^2]^{1/2}} \tag{6}$$

renormalizes the hopping integral. The expression for Δ_k, Eq. (3), is written as $\Delta_k = \Delta_0 + f_k\Delta_1$ for the superconducting order (SC) order parameter. So one gets, from Eq. (3), the coupled equations for the on-site and off-site component of the order parameters as (8), respectively. Now we perform self-consistent numerical calculations of the relevant equations.

$$\Delta_0 = -\frac{US_u}{N}\sum_{k} \frac{(\Delta_0 + f_k\Delta_1)}{[E_{k'}^2 + \Delta_{k'}^2]^{1/2}}, \tag{7}$$

and

$$\Delta_1 = -\frac{zS_vV}{N}\sum_{k} \frac{f_k(\Delta_0 + f_k\Delta_1)}{[E_{k'}^2 + \Delta_{k'}^2]^{1/2}}, \tag{8}$$

respectively. Now we perform self-consistent numerical calculations of the relevant equations.

3. Numerical Results

Rewriting $E_k = t_0 f_k - \mu_0$ where, from Eq. (4) $t_0 = zt + zS_vV\eta$, $\mu_0 = \mu - n(S_uU + 2zS_vV)/2$ and defining band width $W = 2zt\gamma$, here, $\gamma = \max|f_k|$, it is convenient to study Eqs. (5) to (8) in terms of the dimensionless variables of the system. The dimensionless parameters are $\alpha = U/W$ and $\beta = zV/W$. In what follows, we define the dimensionless variables $\Delta_0 = \Delta_0/W$, $\Delta_1 = \Delta_1/W$ and $\mu_0 = \mu_0/W$. The k-space integration of the Eqs. (2.5) to (2.8) is easily done, assuming that f_k is a variable such that $0 \le f_k \le \gamma$. This approximation is equivalent to a square density of states model. With the square density of states model, the pairing is taking place in the whole Brilloun zone, contrary to the BCS model. However, if one includes the van-Hove singularity in the density of states, then the strength of the Cooper pairing increases. After some algebra, the coupled Eqs. (5) to (8) are simplified and we get

$$\mu_0 = \frac{1}{2X_0}[-X_1 - (X_1^2 - 4X_0X_2)^{1/2}] \tag{9}$$

$$X_0 = t_0^2 - (\delta t_0 - \Delta_0\Delta_1 S_u/\alpha)^2; \quad X_1 = -2t\Delta_0\Delta_1 \tag{10}$$

$$X_2 = (\Delta_0\Delta_1)^2 + \gamma^2(\delta t_0 - \Delta_0\Delta_1 S_u/\alpha)^4 - [\Delta_0^2 + \gamma^2 C](\delta t_0 - \Delta_0\Delta_1 S_u/\alpha)^2 \tag{11}$$

Other coupled equations are

$$2\gamma\sqrt{C}(t_0\mu_0 - \Delta_0\Delta_1)[(\delta t_0 - \Delta_0\Delta_1 S_u/\alpha)^2 - C]$$

$$= -(\delta t_0 - \Delta_0\Delta_1 S_u/\alpha)(t_0\Delta_0 + \mu_0\Delta_1)^2 \sinh\left[\frac{2\gamma\sqrt{C}(\delta\Delta_1 + t_0\Delta_0 S_u/\alpha)}{t_0\Delta_0 + \mu_0\Delta}\right] \quad (12)$$

$$-\frac{2\Delta_1}{\beta} CS_v = 2t_0\Delta_0(\delta - \mu_0 S_u/\alpha) + \Delta_1(\delta\mu_0 + \Delta_0^2 S_u/\alpha) - \Delta_1\frac{(t_0\mu_0 - \Delta_0\Delta_1)}{(\delta t_0 - \Delta_0\Delta_1 S_u/\alpha)} \quad (13)$$

and

$$\eta_v = -\frac{1}{2C}[2\Delta_0\Delta_1(\delta - \mu_0 S_u/\alpha) - t_0(\delta\mu_0 + \Delta_0^2 S_u/\alpha)] - \frac{t_0}{2C}\frac{(t_0\mu_0 - \Delta_0\Delta_1)}{(\delta t_0 - \Delta_0\Delta_1 S_u/\alpha)} \quad (14)$$

where $C = t_0^2 + \Delta_0^2$. The above four equations Eq. (9) and Eqs. (12–(14) are solved for the unknown Δ_0, Δ_1 and μ_0 self-consistently as a function of system's parameters α, β and δ. Here $\delta = 1 - n$, is taken positive definite to account for the hole density.

3.1. *The influence of U on Δ_0, Δ_1 and μ_0*

In the absence of an extended Hubbard interaction, $V = 0$, the s-wave SC order parameter (since, Δ_1 comes out to be zero) is given by a simple expression,

$$\Delta_1 = \gamma t_0(1 - \delta^2)^{1/2} \sinh(\gamma t_0/\alpha)^{-1}, \quad (15)$$

while the chemical potential obtained from Eq. (9) is

$$\mu_0 = -\delta\gamma t_0 \coth(\gamma t_0/\alpha). \quad (16)$$

With the inclusion of V, the above simple expression of Δ_0 gets coupled to Δ_1. In all our calculations we have taken, $\gamma = 1$. We discuss here the solutions of Eq. (9) and Eqs. (12)–(14) for $\beta = 0.5$, $\delta = 0.5$ while α is varied from 0 to 1. We choose $\delta = 0.5$ because the order parameters at this value of δ show maximum deviation (see Sec. 3e) in the results from the U Hubbard model's results. Fig. 1(a) shows the results for Δ_0. As expected, Δ_0 gets enhanced with attractive V as compared with the values of the U Hubbard model, while it gets suppressed in the region where $\alpha < \beta$ for repulsive V. For all values of α at fixed β, Δ_0 obeys an inequality $-\Delta_0(-U, +V) \le -\Delta_0(-U, V = 0) \le -\Delta_0(-U, -V)$. From the qualitative behavior of Δ_0, it is clear that, when U is attractive, a repulsive or an attractive V only renormalizes the effect of U in a simple manner, i.e., the V seems to get either added or subtracted to U. However, for a repulsive U, the mean field equations still provide an effective attractive interaction required for SC pairing, though its magnitude is very small as is evident from the values of Δ_0. With repulsive U, the

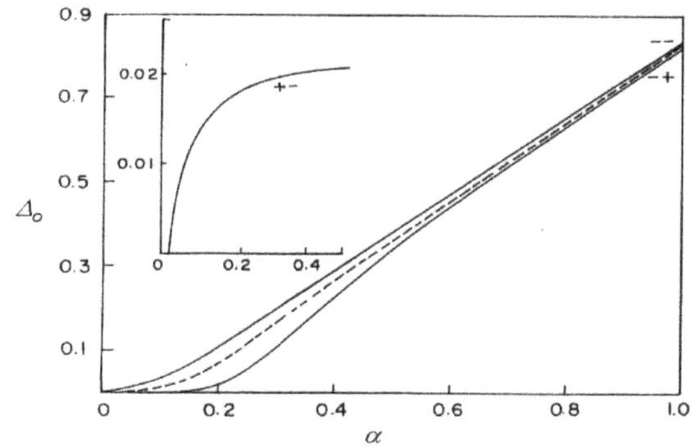

(a)

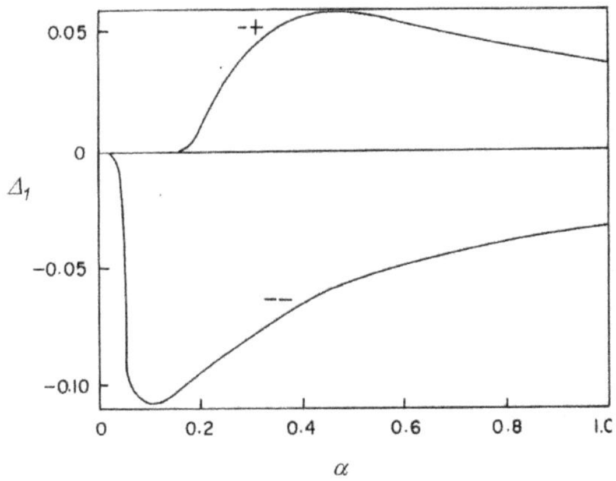

(b)

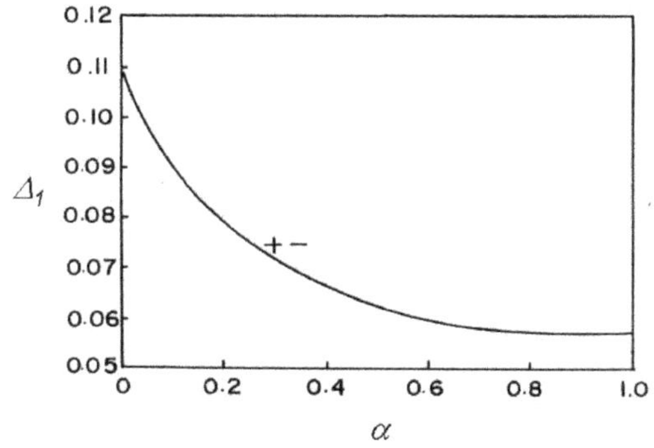

(c)

Fig. 1. (a) Δ_0 versus α at $\beta = 0.5$ and $\delta = 0.5$. The $---$ curve is the U Hubbard model's values for Δ_0. In all the figures, the curves with $--$ mark corresponds to the case when U and V are attractive. The $-+$ curves are the results for attractive U and repulsive V, while $+-$ curves represent the repulsive U and attractive V, respectively. (b) Δ_1 versus α at $\beta = 0.5$ and $\delta = 0.5$. (c) Δ_1 versus α at $\beta = 0.5$ and $\delta = 0.5$. (d) μ_0 versus α at $\beta = 0.5$ and $\delta = 0.5$. The $---$ curve is the U Hubbard model's values for μ_0.

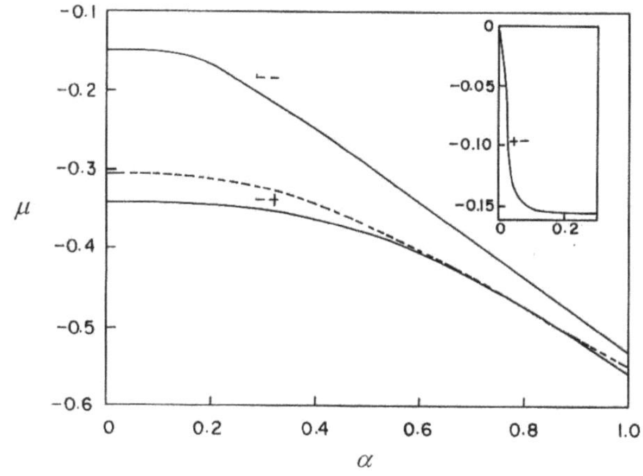

(d)

Fig. 1. (*Continued*).

behavior of Δ_0 changes drastically — it shows a non-BCS kind of take-off for small values of α and goes to a constant for $\alpha > 0.45$.

The extended-wave pairing as a function of α is shown in Figs. 1(b) and 1(c). Notice, that for attractive V, Δ_1 becomes negative, however, $\Delta = \Delta_0 + \gamma\Delta_1$ remains positive definite for all values of α. For the case of repulsive V, no SC pairing exist ($\Delta_0 = 0$, $\Delta_1 = 0$) for small values of α (e.g., for $\alpha < 0.16$). Pairing starts building as α is increased away from 0.16 and acquires a maximum value around $\alpha = 0.46$. Note that, the strong attractive U does not promote the extended-wave pairing, for attractive or repulsive V. In fact, U suppresses it (6.1) monotonically. For repulsive U, Δ_1 decreases exponentially and gets saturated for larger values of α [see Fig. 1(c)]. A glance at the magnitudes of Δ_0 and Δ_1 shows that the SC pairing is dominated by the extended-wave pairing in the case of repulsive U.

The chemical potential μ_0 is plotted in Fig. 1(d). The two points are to be noted about μ_0. First μ shows a jump as V is switched-on from repulsive to attractive interaction. Second, μ_0 remains constant up to certain values of α, when α is small. This region where μ_0 is constant is in proximity with non-SC phase. For large α, μ_0 decreases linearly with α. In the SC phase, μ_0 decreases with α. In the case of repulsive U, μ_0 starts from 0 and decreases with α, and for $\alpha > 0.2$ it acquires a constant values.

3.2. *The influence of V on Δ_0, Δ_1 and μ_0*

Note that it due to the k-dependence of Δ_k which couples Δ_0 to the V. The effect of an extended Hubbard interaction on SC pairing is shown in Figs. 2(a) and 2(b) for various values of β (β varies from 0 to 1) at $\alpha = 0.5$ and $\delta = 0.5$. At the origin of β axis are the results of U Hubbard model, where, the only pairing exhibited is the s-wave pairing. Now if we switch-on an attractive V, then physically one

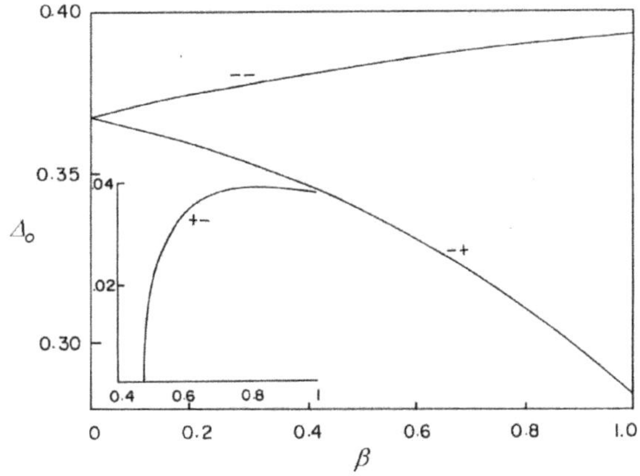

(a)

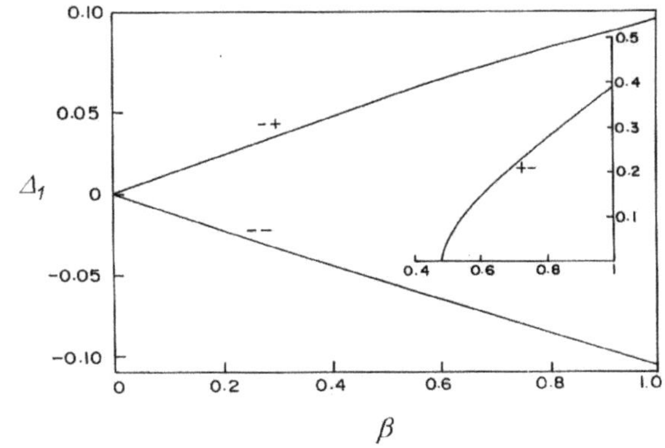

(b)

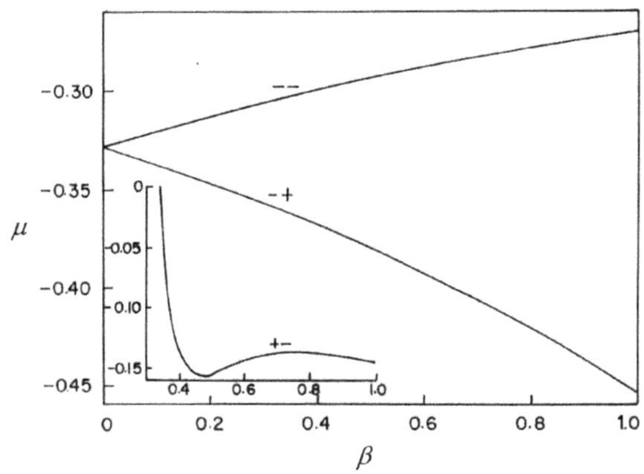

(c)

Fig. 2. (a) Δ_0 versus β at $\alpha = 0.5$ and $\delta = 0.5$. (b) Δ_1 versus β at $\alpha = 0.5$ and $\delta = 0.5$ and (c) μ_0 versus β at $\alpha = 0.5$ and $\delta = 0.5$.

expects, the SC-pairing to get enhanced, while the reverse effect should be observed for the repulsive case. As shown in Fig. 2(a), Δ_0 increases linearly with attractive V almost till $\beta = 1$. While for repulsive V, Δ_0 decreases comparatively faster with β and deviates from linearity also, as we know that, for $+U$ Hubbard model, in the mean field approximation, there is no s-wave pairing. However, one expects SC-pairing with attractive V if its magnitude is greater than some critical value. This critical value depends upon the magnitude of repulsive U. In this case when $\alpha = 0.5$, we find that s-wave pairing starts building when $\beta > 0.5$ and increases rapidly in an non-exponential manner with β and gets saturated for $\beta > 0.8$. In the mean field calculations, we find that there is a qualitative difference in the behavior of s-wave pairing due to an on-site Hubbard interaction and an extended Hubbard interaction.

Figure 2(b) shows the variations of Δ_1 with β. For an attractive V, Δ_1 becomes negative, while it is positive for repulsive V. In both the cases it varies linearly with β. Notice, that the magnitude of Δ_1 is larger than Δ_0 for repulsive U, manifesting the dominance of the extended-wave pairing.

The values of μ_0 [see Fig. 2(c)] increases with attractive V as compared with the U Hubbard model's values. For repulsive V, it decreases. In the case when U is repulsive, μ_0 shows little variation with β for $\beta > 0.48$.

3.3. *The influence of δ on Δ_0, Δ_1 and μ_0*

The dependence of Δ_0 on δ is shown in Figs. 3(a) for $\alpha = 0.2$, and $\beta = 0.5$. We have chosen $\alpha < \beta$ because, as we shall see latter on that for these values of interaction parameters, SC order parameter exhibits an interesting phase diagram. In U Hubbard model, Δ_0 is proportional to $\sqrt{1-\delta^2}$, [see Eq. (15)]. The inclusion of an attractive V produces qualitative changes in Δ_0 it decreases faster than $\sqrt{1-\delta^2}$ for intermediate values of δ. In the above mean field calculations, one finds that, the s-wave pairing is maximum at half-filling and vanishes as expected at no carrier concentration, i.e., at $\delta = 1$. Since we took $\alpha > \beta$, so one excepts a destabilizing effect of pairing with repulsive V at some critical carrier concentration. Which happens for $n \geq 0.42$. However, the s-wave pairing which has been maximum at half-filling for attractive U, got completely suppressed to zero for repulsive U, see Fig. 3(b). This can be seen from Eq. (12), that for repulsive U, the only solution of Δ_0 at half-filling is $\Delta_0 = 0.0$. Departure from half-filling starts building s-wave pairing and it acquires a maximum value at $\delta = 0.75$ then it vanishes in the usual way like $\sqrt{1-\delta^2}$ at $\delta \to 1$.

On the contrary, for attractive U (and for attractive or repulsive V) Δ_1 is zero at half-filling, see Fig. 3(c). The zero solution of Δ_1 at half-filling, $\mu_0 = 0$, can be trivially seen from the GHFA equations for zero *or* non-zero values of Δ_0' As $\delta \to 1$. Δ_1 seems to vanish like $\sqrt{1-\delta^2}$. With attractive U and repulsive V, the variation in Δ_1 is small and it (Δ_1) vanishes at $\delta = 0.58$. For repulsive U, Δ_1 is zero for $\delta \leq 0.05$ and as δ increases further, like Δ_0, Δ_1 passes through a maximum

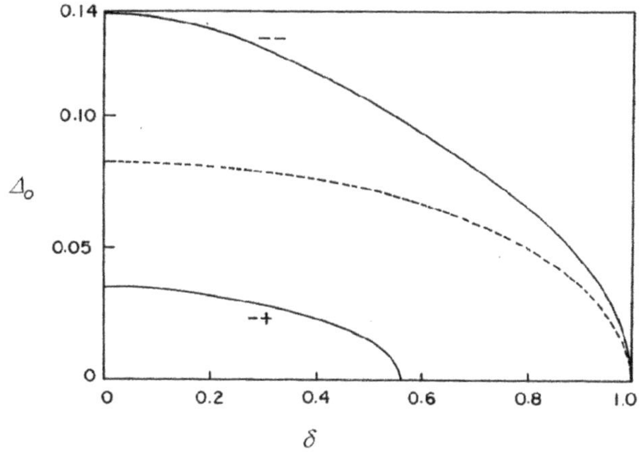

(a)

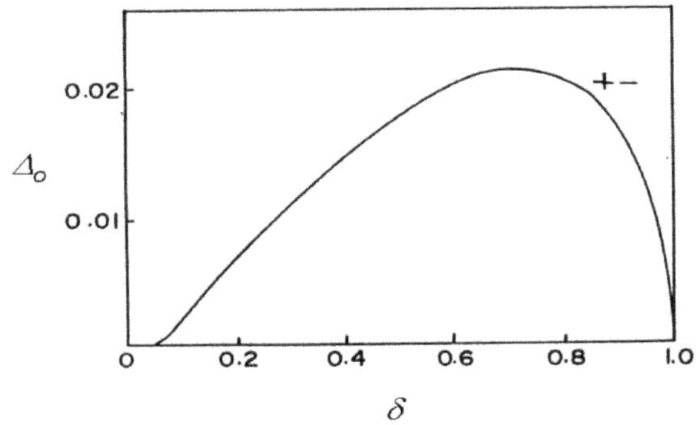

(b)

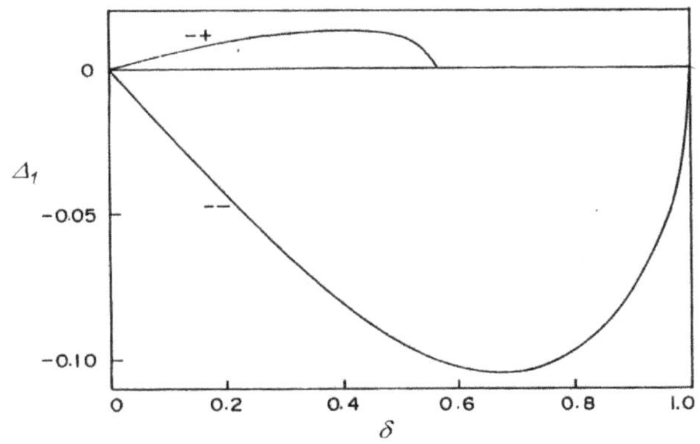

(c)

Fig. 3. (a) Δ_0 versus δ at $\alpha = 0.2$ and $\beta = 0.5$. The – – – curve is the U Hubbard model's values for Δ_0. (b) Δ_0 versus δ at $\alpha = 0.2$ and $\beta = 0.5$. (c) Δ_1 versus δ at $\alpha = 0.2$ and $\beta = 0.5$. (d) Δ_1 versus δ at $\alpha = 0.2$ and $\beta = 0.5$. (e) μ_0 versus δ at $\alpha = 0.2$ and $\beta = 0.5$.

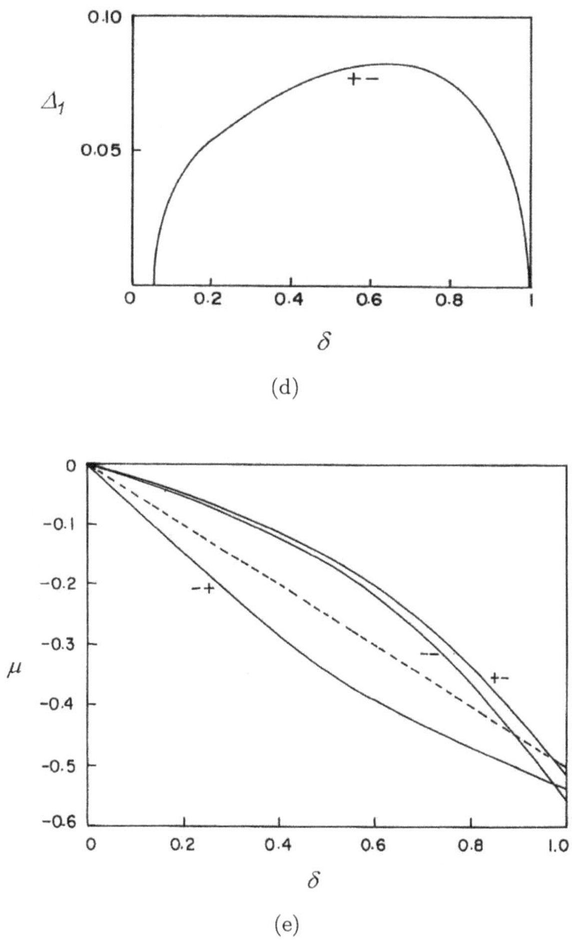

Fig. 3. (*Continued*).

around $\delta = 0.75$, than drops to zero with the usual mean field manner. The results are plotted in Fig. 3(d). The qualitative features of the SC diagram, defined by $\Delta = \Delta_0 + \gamma\Delta_1$ as a function of δ for the case of weak repulsive U and comparatively stronger V.

The chemical potential μ_0 for U Hubbard model (under the mean field approximation) is a linear function of δ, see Eq. (16). With the inclusion of V, μ_0 remains linear for $\delta < 0.2$ and $\delta > 0.8$ for all the cases of interactions. The slope of μ_0 is determined by the case of interaction — for small δ the slope is largest for attractive U and repulsive V, and it is smallest for repulsive U and attractive V. At intermediate values of δ, μ_0 shows maximum departure from the — $\gamma t_0 \coth(\gamma t_0/\alpha)$ which is the slope of μ_0 for U Hubbard model. For non-SC phase which is observed in repulsive V when $\delta > 0.58$, μ_0 varies linearly like — γt_0 (see Eq. (3.1)). At the level of chemical potential, one can identify (within the mean-field approximation) the transition from a SC-phase to non-SC phase from the slope of μ_0 against μ. One notice that μ_0 varies non-linearly in SC phase or linearly with slope $-\gamma t_0 \coth(\gamma t_0/\alpha)$ as in U Hubbard model, while in the non-SC phase ($\Delta_0 = 0.$, $\Delta_1 = 0$), it varies linearly with slope $-\gamma t_0$.

4. Conclusions

The reliability of the present mean-field calculations is restricted to the regions where electron correlations are not too strong ($\alpha < 1$ and $\beta < 1$) and the carrier concentration is not too low. This situation is physically appropriate for superconducting phase of the oxides. One knows that on doping the carriers are delocalized reducing the strong short-range correlations. A doped material is a poor insulator in comparison with an undoped one. For stronger correlations, deviation[27] from the above results are expected. In summary, we found, that the contribution to the pairing from on-site interaction is always stronger than the off-site interaction, except for the case when U is repulsive, where pairing is dominated by extended wave pairing. The effect of V on μ_0 and order parameters, shows linear dependence in the whole range of its variation, again except for the case when U is repulsive. The deviation of Δ_0 and μ from the U Hubbard results due to the inclusion of V is also discussed. The SC-phase diagram versus doping in the repulsive U and attractive V case has to be carefully analysed that would make a connection with the quasi-particle energy[28] for experimental comparison.

In summary, we have investigated possibilities of superconducting order in an extended Hubbard model with onsite U and nearest neighbor interaction V as phenomenological parameters. While we are aware that a huge number of papers have been written on this aspect, our aim has been to provide in this work a simple and a pedagogic generalized mean-field picture. We demonstrate the manifestation of superconductivity arising out of purely electronic interactions over a broad range of parameters in particle density and interaction strengths.

Acknowledgment

One of us (JST) would like to thank the Smart Sensors and Integrated Microsystems at Wayne Sate University for the financial support.

References

1. J. B. Bednorz and K. A. Mueller, Possible high-T_c superconductivity in the Ba–La–Cu–O system, *Z. Phys. B* **64**, 189 (1986).
2. A. Montorsi (ed.), *Hubbard Model* (World Scientific, Singapore, 1990).
3. M. P. Das and J. Mahanty (eds.), *Modern Perspectives in Many-Body Physics* (World Scientific, Singapore, 1994).
4. See e.g., M. W. Long in the *Hubbard Model Recent Results*, M. Rosetti (ed.) (World Scientific, Singapore, 1991).
5. B. den Hertog and M. P. Das, Aspects of the normal-state phase of copper oxide planes in high-T_c superconductors, *Phys. Rev. B* **54**, 6693 (1996).

6. P. W. Anderson, The resonating valence bond state in La_2CuO_4 and superconductivity, *Science* **235**, 1196 (1987).
7. P. W. Anderson, G. Baskaran, Z. Zou and T. Hsu, Resonating–valence-bond theory of phase transitions and superconductivity in La_2CuO_4-based compounds, *Phys. Rev. Lett.* **58**, 2790 (1987).
8. P. W. Anderson, Present status of the theory of high-T_c cuprates, cond-mat. 0510053, pp. 1-9 (2005).
9. J. R. Schrieffer, X. G. Wen and S. C. Zhang, Topology of the resonating valence-bond state: Solitons and high-T_c superconductivity, *Phys. Rev. B* **35**, 8865 (1989).
10. P. Monthoux, A. Balatsky and D. Pines, Weak-coupling theory of high-temperature superconductivity in the antiferromagnetically correlated copper oxides, *Phys. Rev. B* **46**, 14803 (1992).
11. D. J. Scalapino, J. E. Hirsch and E. Y. Loh, *d*-wave pairing near a spin-density-wave instability, *Phys. Rev. B* **34**, 8190 (1986).
12. M. Imada, Superconducting correlation of two-dimensional Hubbard model near half-filling, *J. Phys. Soc. Japan* **60**, 2740 (1992).
13. V. Emery, Theory of high-T_c superconductivity in oxides, *Phys. Rev. Lett.* **58**, 2794 (1987).
14. J. E. Hirsch, Antiferromagnetism and superconductivity: Can a Hubbard *U* do it all by itself? *Physica C* **153–155**, 549 (1987).
15. J. E. Hirsch, S. Tang, E. Loh and D. J. Scalapino, Pairing interaction in two-dimensional CuO_2, *Phys. Rev. Lett.* **60**, 1668 (1988).
16. H. Chen and D. C. Mattis, Toward a rigorous theory of the screened electron-electron interactions in metals, *Int. J. Mod. Phys. B* **5**, 2951 (1991).
17. C. D. Hu, The indirect attraction of holes in the extended Hubbard model, *J. Supercond.* **4**, 481 (1991).
18. B. R. Bulka, Superconductivity in the extended Hubbard model, *Phys. Stat. Solidi B* **163**, 197 (1991).
19. M. S. Laad and D. K. Ghosh, Extended Hubbard model in two dimensions, *J. Phys. Cond. Matt.* **3**, 9723 (1991).
20. T. C. Choy and M. P. Das, Attractive pairing potential from the repulsive Hubbard model, *Phys. Rev. B* **47**, 8942 (1993).
21. R. Micnas, J. Ranninger and S. Robaszkiewicz, Superconductivity in narrow-band systems with local nonretarded attractive interactions, *Rev. Mod. Phys.* **62**, 113 (1990).
22. P. W. Anderson, Model for the electronic structure of amorphous semiconductors, *Phys. Rev. Lett.* **34**, 953 (1975).
23. N. Bulut, $dx2 - y2$ superconductivity and the Hubbard model, *Adv. Phys.* **51**, 1587 (2002).
24. Th. Maier, M. Jarrell, Th. Pruschke and J. Keller, *d*-Wave superconducti-vity in the Hubbard model, *Phys. Rev. Lett.* **85**, 1524 (2000).
25. E. J. Calegari, S. G. Magalhães and A. A. Gomes, Superconductivity in a two dimensional extended Hubbard model, *Eur. Phys. J. B* **45**, 485 (2005).
26. W. Metzner, Superconductivity in the two-dimensional Hubbard model, *Foundations of Physics* **30**, 2101 (2000).
27. R. T. Scalettar, E. Y. Low, J. E. Gubernatis, A. Mores, S. R. White, D. J. Scalapino, R. L. Sugar and E. Dagotto, Phase diagram of the two-dimensional negative-*U* Hubbard model, *Phys. Rev. Lett.* **62**, 1407 (1989).
28. R. Citro and M. Marinaro, Energy scales and quasiparticle properties in an extended Hubbard model for HTC, *Eur. Phys. J. B* **37**, 15 (2004).

Part F

Nuclear systems

DISTRIBUTION OF MAXIMA OF THE ANTISYMMETRIZED WAVE FUNCTION FOR THE NUCLEONS OF A CLOSED-SHELL AND FOR THE NUCLEONS OF ALL CLOSED-SHELLS IN A NUCLEUS

GERASIMOS S. ANAGNOSTATOS

Institute of Nuclear Physics, National Center for Scientific Research Demokritos, Aghia Paraskevi, Attiki, 15310 Greece
anagnos4@otenet,gr

The significant features of exchange symmetry are displayed by simple systems such as two identical, spinless fermions in a one-dimensional well with infinite walls. The conclusion is that the maxima of probability of the antisymmetrized wave function of these two fermions lie at the same positions as if a repulsive force (of unknown nature) was applied between these two fermions. This conclusion is combined with the solution of a mathematical problem dealing with the equilibrium of identical repulsive particles (of one or two kinds) on one or more spheres like neutrons and protons on nuclear shells. Such particles are at equilibrium only for specific numbers of particles and, in addition, if these particles lie on the vertices of regular polyhedra or their derivative polyhedra. Finally, this result leads to a pictorial representation of the structure of all closed shell nuclei. This representation could be used as a laboratory for determining nuclear properties and corresponding wave functions.

Keywords: Many fermion system; equilibrium polyhedra; closed shells; magic numbers.

1. Introduction

In this work we search for the distribution of maxima of the antisymmetric wave function of many fermions, specifically, of the nucleons of a nuclear shell or of the nucleons of all closed shells in a nucleus, without a precise knowledge of the wave function. This effort is materialized by strictly following the consequences of the nature of the particles involved, i.e., the fermionic nature of nucleons. Apparently, this is a many-body problem in its specific case where the particles are fermions and where their quantity is more than a few (i.e., it is not a few-body problem) and where the fermion quantity is not so high as in a solid body (i.e., it is not an excessive number of particles problem). In other words, it is a *mesoscopic* physics problem.

As known, as a many-body problem could be considered any problem with any number of particles more than one. Specifically, in this work we start with the simplest possible problem considering the very specific case of two identical fermions in one-dimensional square-well potential with infinite walls. Afterwards, we extend the research to the number of identical nucleons of a closed shell and to the number of identical nucleons of all closed shells in a nucleus by taking advantage of the conclusions of the above elementary problem of two identical spinless fermions together with the solutions of a well-analyzed relevant mathematical problem.

2. The Nuclear Hamiltonian

As a first step of this work we analyze the nuclear Hamiltonian into partial Hamiltonians one per nuclear shell and different for protons and neutrons as follows

$$H = {}_pH_{1s} + {}_pH_{1p} + {}_pH_{1d2s} + \cdots + {}_nH_{1s} + {}_nH_{1p} + {}_nH_{1d2s} + \cdots \tag{1}$$

In the following first we search for the maxima of probability of the antisymmetric wave function for the identical nucleons for each nuclear shell and then for the nucleons of all shells in a closed-shell nucleus. As a first step we examine the simplest possible problem described below.

After the above search of maxima is completed, we can go backwards to determine explicitly the relevant Hamiltonian in Eq. (1) and the corresponding wave functions as suggested at the end of this work.

3. Two Identical Fermions in a Square Well Potential with Infinite Walls

The total energy for two non-interacting spinless fermions is

$$\begin{aligned} &-(\hbar^2/2m)(\partial^2\Psi/\partial x_1^2) - (\hbar^2/2m)(\partial^2\Psi/\partial x_2^2) + V_1(x_1)\Psi + V_2(x_2)\Psi \\ &\quad = (\hbar/i)\partial\Psi/\partial t \end{aligned} \tag{2}$$

where

$$V_1(x_1) = 0 \text{ for } 0 \le x_1 \le L \text{ and } \infty \text{ elsewhere}$$

$$V_2(x_2) = 0 \text{ for } 0 \le x_2 \le L \text{ and } \infty \text{ elsewhere and}$$

$$\Psi = \Psi(x_1, x_2, t)$$

Equation (2) is separable in its three variables and the two coordinate-dependent equations are

$$d^2\psi_1/dx_1^2 + (2m/\hbar^2)[W_1 - V_1(x_1)]\psi_1 = 0 \tag{3}$$

$$d^2\psi^2/dx_2^2 + (2m/\hbar^2)[W_2 - V_2(x_2)]\psi_1 = 0 \tag{4}$$

where

$$W = W_1 + W_2 = \hbar^2 n^2\pi^2/2mL^2 + \hbar^2 k^2\pi^2/2mL^2 \tag{5}$$

$n = 1, 2, 3, \ldots$ and $k = 1, 2, 3, \ldots$.

The one-particle eigenfunctions are

$$\Psi_1(\chi_1) = \sqrt{2}/L\ \sin(n\pi x_1/L) \text{ and} \tag{6}$$

$$\Psi_2(\chi_2) = \sqrt{2}/L \sin(k\pi x_2/L) \tag{7}$$

and the two-particle eigenfunctions are

$$\psi_1^0(\chi_1, \chi_2) = (2/L)\sin(n\pi x_1/L)\sin(k\pi x_2/L) \text{ and} \tag{8}$$

$$\psi_2^0(\chi_1, \chi_2) = (2/L)\sin(n\pi x_2/L)\sin(k\pi x_1/L) \tag{9}$$

These wave functions are apparently degenerate if $n \neq k$. This degeneracy is due to the symmetry with respect to the interchange of the two identical fermions. If, however, there is some mutual interaction between the two identical fermions in the form of some mutual potential energy H' (x_1, x_2) which is unchanged in sign and magnitude upon the interchange of x_1 and x_2, then there are only the following two possible choices for the zero-order wave function for the two fermions system.
Symmetric wave function:

$$\Psi_s = (1/\sqrt{2})(2/L)[\sin(n\pi x_1/L)\sin(k\pi x_2/L) + \sin(n\pi x_2/L)\sin(k\pi x_1/L)] \tag{10}$$

Antisymmetric wave function:

$$\Psi_a = (1/\sqrt{2})(2/L)[\sin(n\pi x_1/L)\sin(k\pi x_2/L) - \sin(n\pi x_2/L)\sin(k\pi x_1/L)] \tag{11}$$

The above results come from the following secular equation

$$\begin{vmatrix} (H'_{11} - W') & H'_{12} \\ H'_{21} & (H'_{12} - W') \end{vmatrix} = 0 \tag{12}$$

The determinant becomes

$$(H'_{11} - W')^2 = (H'_{12})^2 , \tag{13}$$

since $H'_{12} = H'_{21}$,

$$H'_{11} = H'_{22}, \text{ given that}$$

$$H'_{11} = \int \psi_1^0 * H'\psi_1^0 dx_1 dx_2 \text{ , etc.}$$

In Fig. 1 a plot of Eqs. (10) and (11) is shown, i.e., a plot for both symmetric [part (a) of the figure] and antisymmetric [part (b) of the figure] wave function of the system. The horizontal axis of the plot is used to mark positions inside the potential well from 0 to L for the particle assigned the number 1, likely the vertical axis for the particle assigned the number 2. In both parts of the figure the probability amplitudes are considered perpendicularly to the paper and are presented by counters numbered 1.0, 1.5, 2.0. As apparent from the figure the maxima of probability in the case of symmetric wave function lie on the shown broken line where both particles lie at $x_1 = x_2$, either at the position $x_1 = x_2 = L/4$ or at the position $x_1 = x_2 = 3L/4$, while in the case of antisymmetric wave function the maxima of probability lie out of this line and specifically the one at the position $L/4$ and the other at the position $3L/4$. This means that for the symmetric wave function of the system

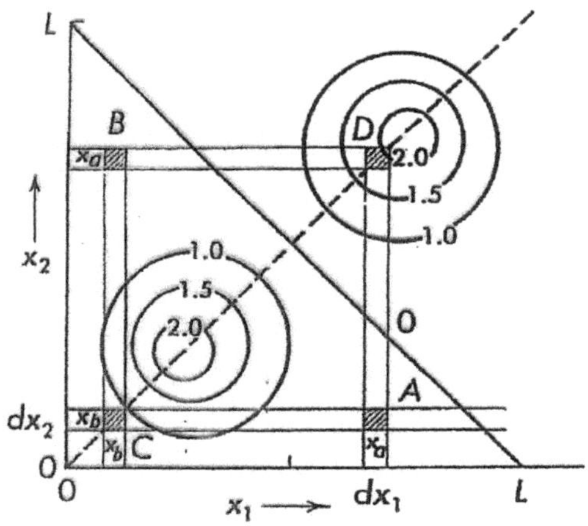

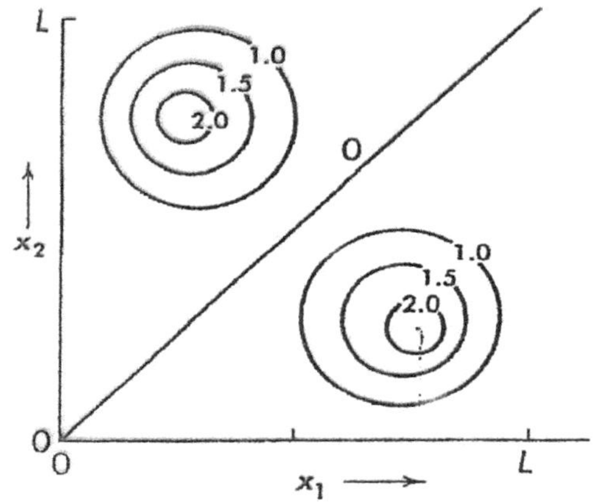

Fig. 1. (a) Symmetric wave function: $\psi_s^*\psi_s$ with $n = 1$ and $k = 2$, according to Eq. (10), (b) antisymmetric wave function: $\psi_a^*\psi_a$ with $n = 1$ and $k = 2$, according to Eq. (11).

the maxima of probability of the two particles coincide and the two particles are found together there, while for the antisymmetric wave function of the system the particles have zero probability of being together. This conclusion is equivalent to saying that among the particles, an attractive force was applied in the case of symmetric wave function and a repulsive force (of unknown nature) was applied in the case of antisymmetric wave function.[1]

Thus, the antisymmetrization requirement of two identical fermions in a potential well results in a situation for the maxima of probability of these identical fermions equivalent to the situation we would have if among these two fermions a repulsive force (of unknown nature) was applied.

4. Nucleons in a Closed Shell and in All Shells of a Closed-Shell Nucleus

In the rest of this work we will take advantage of the above conclusion in order to proceed with the problem of the many nucleons of a nuclear shell or of the many nucleons of all shells in a closed-shell nucleus. In order to do so we extensively use the paper of Leech[2] dealing with the equilibrium of one or two kinds of repulsive nucleons on a sphere resembling the behavior of one or two kinds of identical fermions on a nuclear shell or shells.

According to this paper of Leech, repulsive particles on a sphere are at equilibrium (or at their positions of maximum probability) only for *specific numbers of particles and if these particles lie on the vertices of one of the regular polyhedra or of one of their derivative polyhedra or simultaneously of a regular polyhedron and one or two derivative polyhedra.* We call derivative polyhedra of a regular polyhedron those polyhedra which are derived by considering the middles of faces or the middles of edges of the initial polyhedron.

Fig. 2 represents the Leech polyhedra in three rows. Vertical blocks in the first two rows of the figure represent reciprocal polyhedra, that is, polyhedra where the vertices of the one are the middles of faces of the other. The third row makes more apparent the relative orientation of the above reciprocal pairs for the special case where orthogonal edges of the reciprocal pairs of polyhedra bisect each other. At each block of the figure the number of vertices of the relevant polyhedron is given in parentheses, while at the bottom of the block the official name and/or its mathematical abbreviation of this polyhedron is also given.

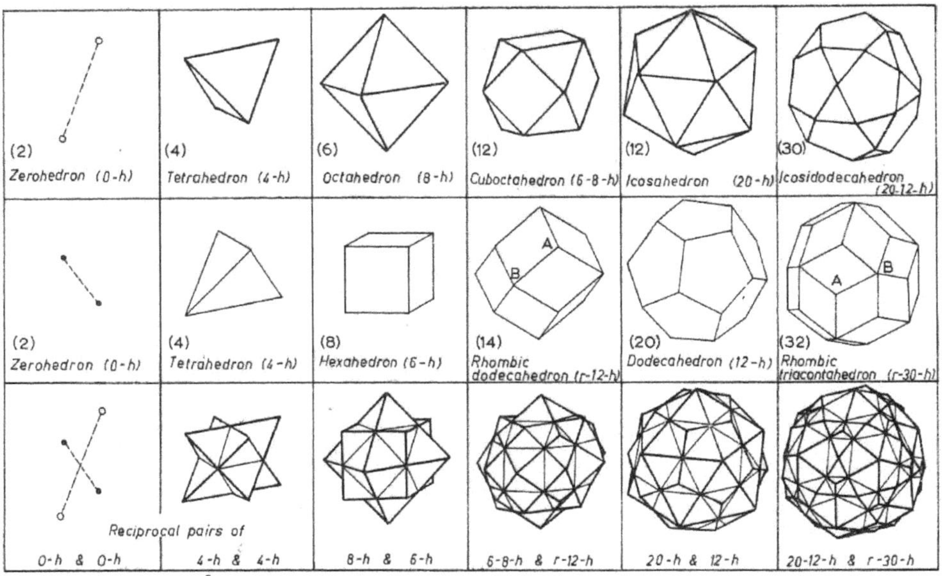

Fig. 2. Leech-type equilibrium polyhedra.

The Leech paper as mentioned earlier, also deals also with the equilibrium of two categories of identical repulsive particles on spheres. The particles of each category are at equilibrium with themselves and in addition at equilibrium with the particles of the other category, which means that the equilibrium polyhedron of the one category should be reciprocal to the equilibrium polyhedron of the other category. Thus, here as two categories of identical repulsive particles, we have two categories of identical fermions as neutrons and protons on neutron and proton shells. The probability that these nucleons lie on the vertices of two reciprocal polyhedra is at a maximum. Thus, these polyhedra represent the most probable forms of a neutron and of a relevant proton shell. Two reciprocal polyhedra (e.g., a cube and an octahedron, or a dodecahedron and an icosahedron) have the same rotational symmetry, as expected for a proton shell and its relevant neutron shell, e.g., an $1p$ proton shell and its relevant $1p$ neutron shell or an $1d2s$ proton shell and its relevant $1d2s$ neutron shell. We will comment on relevant proton and neutron shells again when we deal with orbital-angular-momentum quantization of direction vectors of these shells.

Fig. 3 represents the most probable forms of the $1p$ and the $1d2s$ nuclear shells accommodating 6 and 12 neutrons or protons, respectively. The vectors shown in the figure stand *identically* (that is, precisely to any order of accuracy) for the orbital-angular-momentum quantization of directions with quantum numbers $\ell = 1$ and 2, respectively and $-\ell \leq m \leq \ell$.[3] This should happen any way since the most probable form of each of the $1p$ and $1d2s$ nuclear shells represented in Fig. 3 is also one out of the infinite number of instantaneous nuclear forms these shells could have and as such all nuclear properties should be valid and should have a geometrical representation. Specifically, the quantization of direction vectors of orbital-angular-momentum with quantum numbers ℓ and m should result as property of the symmetries of the relevant polyhedron for each shell. Indeed, these orbital-angular-

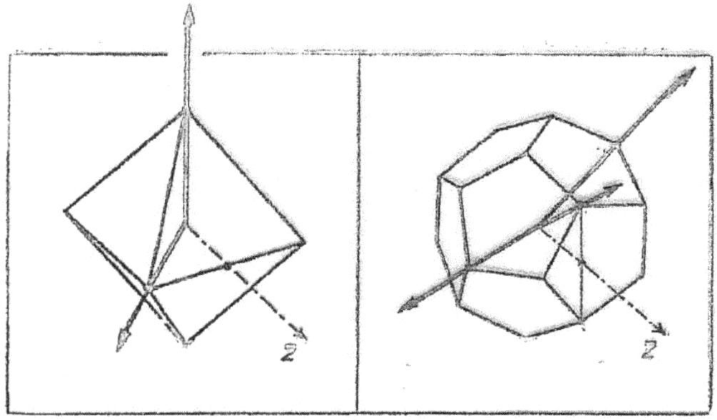

Fig. 3. Most probable forms of the $1p$ and the $1d2s$ nucleon shells, and the corresponding orbital-angular-momentum quantization of direction vectors for $\ell = 1$, $-1 \leq m \leq 1$ and $\ell = 2$, $-2 \leq m \leq 2$.

momentum quantization of direction vectors for the $1p$ shell (which is represented by the octahedron in Fig. 3) pass through the polyhedral vertices (or through the middles of faces of the reciprocal shell which has a cube as its most probable form). Also, the orbital-angular-momentum quantization of direction vectors for the $1d2s$ shell (which is represented by the dodecahedron in Fig. 3) also pass through the polyhedral vertices (or through the middles of faces of the reciprocal shell which has an icosahedron as its most probable form). These orbital-angular-momentum quantization of direction vectors with respect to the shown quantization axis z for each shell (which z axes coincide and become a common quantization axis for both shells when the two relevant polyhedra are superimposed with common center and orientations as shown) form *identically* the angles

$$_n\theta_\ell^m = \cos^{-1} m/[\sqrt{\ell}(\ell+1)]\,. \tag{14}$$

In Fig. 4 it is shown that similar properties to those represented in Fig. 3 are also valid for all orbital angular momenta with quantum numbers $\ell = 1-6$ and $-\ell \leq m \leq \ell$ with respect to the polyhedra in the first three rows of the figure. In each block of the fourth row the relation between the polyhedra in the same column (rows 1–3) is given.[3] The polyhedra represented in Fig. 4 have symmetries in relation to orbital-angular-momentum quantization of direction vectors which are included in the symmetries of the polyhedra standing for the most probable forms

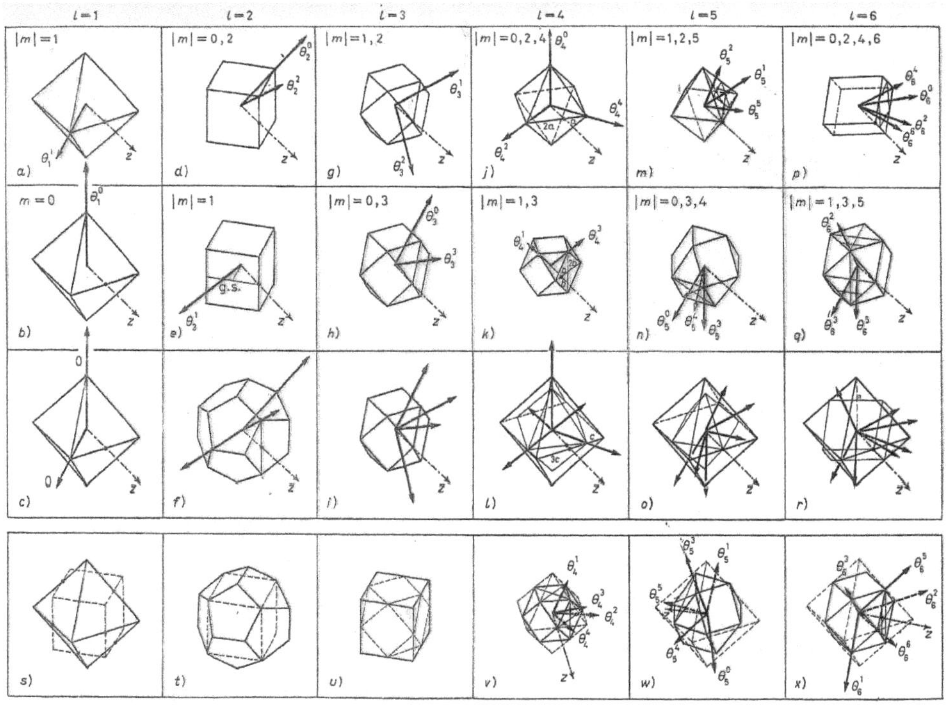

Fig. 4. Angular-momentum quantization of direction vectors in relation to the polyhedra shown.

of nuclear shells shown in the next figure.

In Fig. 5 all polyhedra representing the most probable forms of all nuclear shells involved up to Pb are given and are considered superimposed with a common center and orientations as shown.[4] At the bottom of each block of the figure the order of a proton (Z) or of a neutron (N) shell is given (i.e., Z_1, N_2 etc). In the next row inside parentheses the number of vertices for the relevant polyhedron, inside brackets the cumulative number of vertices of all previous proton or neutron polyhedra and that polyhedron, and the radius of that polyhedron (obtained by packing the shells themselves, and by considering the average size of a proton and of a neutron to be 0.860 fm and 0.974 fm, respectively) are also given. More details are not given here for this figure, but they are extensively included in Ref. 4, where also an explanation

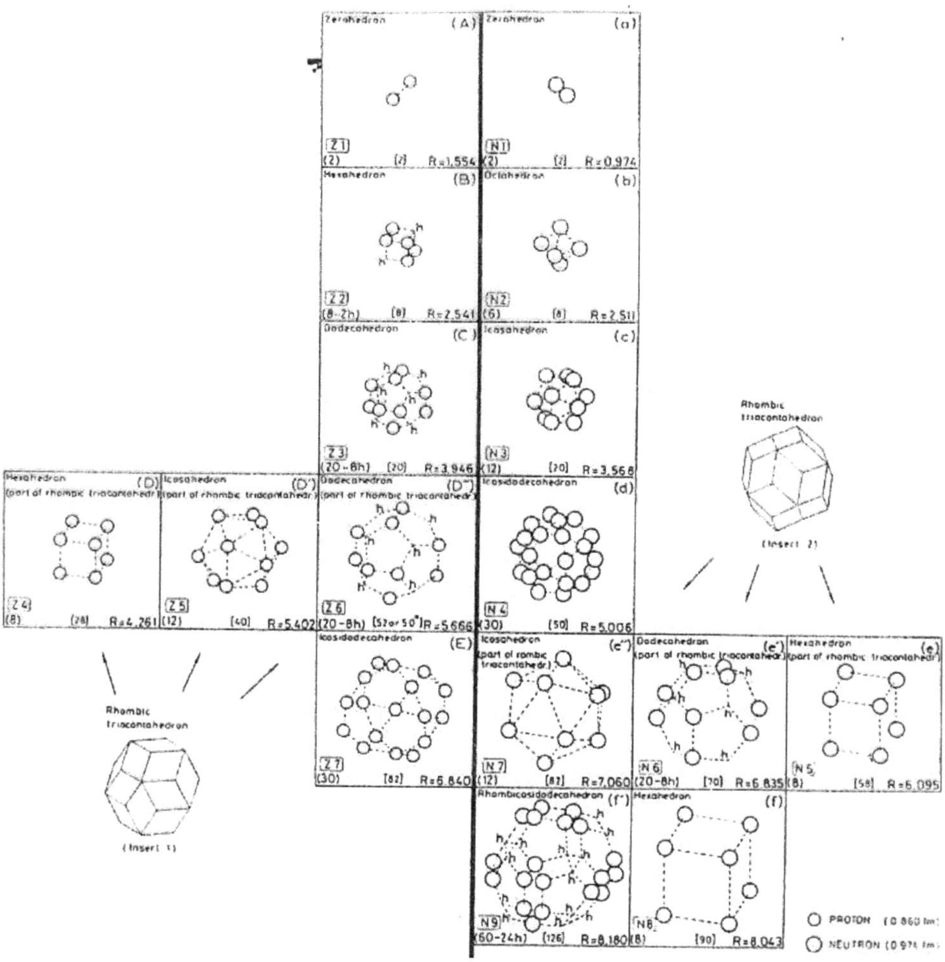

Fig. 5. Pictorial average nuclear structure of all closed shells up to ^{208}Pb. The numbers in brackets given at the bottom of the blocks of the figure coincide with the experimentally known magic and semi-magic numbers.

for the reason why some of Leech-polyhedra are taken as most probable forms of proton shells and some of them as most probable forms. of neutron shells.

In this figure we consider that all nucleons *simultaneously* reside at their most probable positions, which coincide with the vertices of the known polyhedral shells. Such a structure resembles a crystal structure, but it is not, since it is not permanent. It represents only a momentary (the most probable) structure which, of course, reappears *periodically* due to the fact that the orbital-angular-momentum quantization of direction vectors are axes of symmetry of the relevant polyhedra. The same property makes Fig. 5 as a whole to have a meaning as a pictorial representation of the shell structure of a nucleus. It is very interesting to notice that the cumulative numbers in brackets of Fig. 5 (mentioned earlier) are identical to the nuclear magic and semi-magic numbers 2, 8, 20, 28, 40, 50, 70, 82, and 126. Thus, the magic numbers here are strictly the result of the fermionic nature of identical nucleons and not the result of a strong spin orbit coupling as assumed in the conventional shell model. It is in support of the present explanation of nuclear magic numbers the fact that identical magic numbers appear in certain cases of atomic clusters[5–7] where a strong spin orbit coupling certainly does not exist. There, apparent fermions are the decoupled electrons, but also neutral atoms in the cluster if the number of the electrons in an atom is an odd number, or the positive ions in the core of the cluster after an odd number of electrons has been decoupled from each neutral atom with an even number of electrons. The neutral atoms and the positive ions previously described behave like heavy fermions and obey all equations of Sec. 2 here. Thus, it is expected this type of fermions to form polyhedral shells of Leech type and the decoupled electrons to behave like electrons in a giant atom where the polyhedra of positive ions play the role of the core in this giant atom. The common ground of these seemingly irrelevant structures is that they refer to *mesoscopic* physics of fermions.

Now, having this momentary (most probable) shell structure of a nucleus, we can proceed either in a semiclassical way by employing a two-body potential, e.g.[8]

$$V(r_{jj}) = V_{\mathrm{R}} e^{-\mu_{\mathrm{R}}r_{\mathrm{ij}}}/r_{ij} - V_{\mathrm{A}} e^{-\mu_{\mathrm{A}}r_{\mathrm{ij}}}/r_{ij} \,, \tag{15}$$

or we can proceed in a purely quantum mechanical way by deriving the proper $\hbar\omega$ value for each nuclear shell by employing Eq. (16), where the value of $\langle r_i^2 \rangle^{1/2}$ for each shell is taken[9] from Fig. 5, as explained earlier

$$\hbar\omega_i = \hbar^2(n + 3/2)/[m\langle r_i^2 \rangle] \,. \tag{16}$$

Thus, due to the antisymmetrization requirement of the wave function for the many nucleons in a closed shell nucleus, we first arrive at the distribution of maxima of this wave function and the size of the corresponding shell, then going backwards we can write the wave function itself.

5. Conclusions

Starting with the simplest possible problem of two identical fermions in a square-well potential with infinite walls, we arrive at the conclusion that identical fermions in a common potential behave as if a repulsive force (of unknown nature) is acting among the particles.[1] Furthermore, we take advantage of this conclusion together with a mathematical paper[2] dealing with identical repulsive particles on a sphere like protons and neutrons in a nuclear shell, and we come to the conclusion that only a certain number of particles can be accommodated in a shell which has as its most probable form that of a regular polyhedron or its derivative polyhedra.[2] The same paper is used to examine the case where two kinds of fermions, like protons and neutrons in a nucleus, are in a common potential. In this case the proton and the relevant neutron shells have as their most probable forms reciprocal polyhedra (i.e., polyhedra where the vertices of the one are the middle of faces of the other).[2]

A proper superposition of such polyhedral shells leads to a pictorial shell-structure of a close-shell nucleus.[4] Each time a polyhedral shell (of neutrons or protons) is completed, a magic or semi-magic number appears, i.e., the sum of vertices of all proton or neutron previous polyhedra and that polyhedron is equal to one of the numbers 2, 8, 20, 28, 40, 50, 70, 82, and 126. In addition, by assuming that the average size of a proton and of a neutron has the value 0.860 fm and 0.974 fm, respectively, and by further considering that the above polyhedral shells are superimposed with common center and in contact with each other, we arrive at the average structure and the size of all closed-shell nuclei.[4]

By utilizing this structure we can proceed in calculations of nuclear observables either in a semi-classical way taking a two-fermion potential like that of Eq. (15) or in a purely quantum mechanical way estimating the $\hbar\omega$ for each shell from Eq. (16).

Finally, making a long story short, we can say that the fermionic nature of protons and neutrons alone leads to the distribution of maxima of probability of the antisymmetrized wave function of protons and neutrons on the vertices of equilibrium polyhedra[2] and to the appearance of magic numbers.[4] The average structure thus derived is that represented in Fig. 5 and could serve as a nuclear laboratory for calculations and a pictorial explanation of nuclear observables and nuclear phenomena.

References

1. C. W. Sherwin, *Introduction to Quantum Mechanics* (Holt, Rinehart and Winston, New York, 1959), p. 205.
2. J. Leech, *Mathematical Gazette* **41**, 81 (1957).
3. G. S. Anagnostatos, *Lettere al Nuovo Cimento* **29**, 188 (1980).
4. G. S. Anagnostatos, *Int. J. Theor. Phys.* **24**, 579 (1985).
5. G. S. Anagnostatos, *Phys. Lett. A* **128**, 266 (1988).
6. G. S. Anagnostatos, *Phys. Lett. A* **154**, 169 (1991).
7. G. S. Anagnostatos, *Phys. Lett. A* **157**, 65 (1991).
8. G. S. Anagnostatos and C.N. Panos, *Phys, Rev. C* **26**, 260 (1982).
9. G. S. Anagnostatos, *Can. J. Phys.* **70**, 361 (1992).

PAIRING OF STRONGLY CORRELATED NUCLEONS

W. H. DICKHOFF

Department of Physics, Washington University
St. Louis, Missouri 63130, USA
E-mail: wimd@wuphys.wustl.edu

Pairing properties of infinite matter are surveyed that are obtained from self-consistent Green's function calculations. A complete treatment of off-shell propagation that incorporates the effects of short-range correlations is included. Mean-field calculations based on the BCS approach are superseded by the present results, which resolve a long-standing puzzle associated with $^3S_1$-$^3D_1$ (proton-neutron) pairing in symmetric nuclear matter. Results for $^1S_0$ pairing in pure neutron matter are in agreement with recent Monte-Carlo calculations for this system. The possibility of proton pairing in finite nuclei, as a result of increasing proton correlations with increasing nucleon asymmetry, is pointed out.

Keywords: Pairing correlations; nuclear matter; neutron matter; proton pairing in nuclei.

1. Introduction

The description of pairing in infinite nuclear systems has a long history.[1] In the development of a detailed understanding of pairing phenomena involving nucleons in symmetric or pure neutron matter several obstacles have been encountered. Some of the associated puzzles will be addressed in the present discussion. Studies with realistic nucleon-nucleon (NN) interactions that describe experimental NN scattering data up to pion production threshold, have been mostly confined to the solution of the gap equation that employs a description of nucleons as mean-field particles. Such calculations for symmetric nuclear matter lead to gaps for the $^3S_1$-$^3D_1$ coupled channel of the order of 10 MeV at densities around saturation.[2–6] The puzzle associated with these results is the absence of empirical information in finite nuclei that can be reconciled with such huge gaps for the proton-neutron pairing in the $^3S_1$-$^3D_1$ channels. For the $^1S_0$ channel the situation is distinctly different. Typical gap sizes around normal density have quite similar magnitudes to those encountered in nuclei for like-nucleon pairing, *i.e.* with $^1S_0$ quantum numbers.

Three main issues need to be addressed before these results can be considered definitive. The first is to clarify in what way results from nuclear and isospin polarized matter can be related to pairing phenomena in finite nuclei. Secondly, and related to the first, it is necessary to assess the relevance and the importance of the influence of polarization effects. We will employ the convention to use this term to represent the presence of additional contributions to the irreducible interaction in the gap equation beyond the bare NN interaction. For neutron matter such effects must be accurately incorporated before final conclusions about pairing phenomena in that system can be drawn. For nuclear matter this is less obvious, since the relevance of volume polarization phenomena for the understanding of pairing phenomena in finite nuclei is limited, whereas surface polarization effects in finite systems must be deemed different and of course more relevant. A third critical issue is related to the treatment of modifications of nucleon propagation in the nuclear medium when pairing correlations are studied.

We start in Sec. 2 with a brief review of both experimental and theoretical evidence that clarifies the influence of long-range (LRC) and short-range correlations (SRC) on the propagation properties of nucleons in nuclei. The emerging picture suggests that both correlation effects are substantial and play a significant role for particles considered as candidates to participate in pairing phenomena. Since only SRC can be comfortably studied in infinite matter with the confidence that such studies are relevant for finite systems, we proceed in Sec. 3 to describe how such correlations can be incorporated in nuclear and neutron matter. In particular, we emphasize the change in the single-particle (sp) strength distribution when such effects are included. The extension of this description to the solution of the gap equation is also discussed together with the corresponding results. In Sec. 4 a brief discussion is presented to illustrate an unexpected form a proton-proton pairing that may occur in nuclei where the number of neutrons far exceeds that of the number of protons. Concluding remarks are presented in Sec. 5.

2. Role of Correlations in Low-Energy Nucleon Propagation

Advances in the understanding of the sp properties of nucleons in nuclei and nuclear matter[7] demonstrate the dominant influence of short-range and tensor correlations in generating the distribution of the spectral strength associated with the addition or removal of protons from the ground states of closed-shell nuclei. Short-range repulsion in the NN interaction leads to a suppression of the relative wave function that in turn requires the admixture of high-momentum components. The latter lead to the presence of faster nucleons in the ground state of a many-nucleon system than can be expected on the basis of the Pauli principle alone, *i.e.* in an infinite system it will lead to the occupation of states with momentum above k_F. A similar occupation of high-momentum states beyond the typical values associated with confining nucleons to a finite volume, is predicted for finite nuclei.[8] Associated with this occupation of high-momentum states is the depletion of states that are occupied in the noninteracting system. A depletion of states with momentum less than k_F is therefore inevitable for the case of the infinite Fermi system.

Similar depletion effects have been predicted for finite nuclei.[9] Such a global depletion of Fermi sea due to these correlations. has recently been observed in an $(e, e'p)$ experiment from NIKHEF. The analysis of this experiment puts the depletion of the proton Fermi sea in ^{208}Pb at a little less than 20%[10] in accordance with earlier nuclear matter calculations.[9] The experimental occupation numbers of the various proton shells that would be completely filled by protons in the independent-particle model relevant for this nucleus, is strikingly similar to the one for nuclear matter and exhibits the effect of SRC as a global depletion, essentially independent of the nature of sp orbit. Superimposed on this global depletion is an enhanced depletion for those orbits that reside in the vicinity of the Fermi surface. The latter depletion is associated with the coupling of such particles with low-energy (mostly) surface excitations[11] and can therefore be regarded as due to LRC.

As discussed above, the other predicted consequence of the presence of SRC is the appearance of high-momentum components in the ground state to compensate for the depleted strength of the mean field. Recent JLab experiments[12] indicate that the amount and location of this strength is consistent with earlier predictions for finite nuclei[8] and studies of infinite nuclear matter[13] Of particular relevance for the discussion of pairing in this chapter is the observed and calculated energy distribution of the sp strength of the high-momentum nucleons that is located at energies far removed from the Fermi energy. The situation is similar to the case of nuclear matter[14] and even more fragmented distributions are obtained above the Fermi energy. Such sp strength distributions lead to substantial modifications of the calculated properties of low-energy phenomena, like pairing, as compared to a those generated by a traditional mean-field treatment.

As indicated above, details of the nucleon sp strength distribution have become well established experimentally and understood theoretically. In particular, the crucial effects of short-range and tensor correlations on the distribution of the sp strength have recently been treated at a very sophisticated level. Since these correlations have demonstrated relevance for understanding finite nuclei, it is important that their consequences for pairing are studied in detail.

3. Short-Range Correlations and Pairing in Infinite Matter

The effects of short-range and tensor correlations on the distribution of the sp strength has been studied at a very sophisticated level by different groups. Techniques have been developed to evaluate the sp strength from realistic NN interactions, like the Argonne V18[15] or the CDBonn[16] interaction, within the self-consistent Green's function (SCGF) approach.[17–25] A recent reference for the Green's function method in bookform is also available.[26] Such calculations reproduce the energy- and momentum-distribution of the sp strength corresponding to high-momentum nucleons as observed in experiments[12] and account for the depletion of the mean-field strength obtained in Ref. 10.

In these calculations the scattering equations for two nucleons in the medium are solved, employing dressed sp propagators that contain the complete information about the energy- and momentum-distribution of the sp strength. The resulting scattering matrix T is then used to calculate the nucleon self-energy. The solution of the Dyson equation, employing this complex and energy dependent self-energy, yields the sp propagator that enters the T matrix equation to close the self-consistency cycle. In determining the two-nucleon Green's function or the corresponding scattering matrix T, one has to face the problem of the so-called pairing instabilities that reflect the existence of NN bound state solutions in the scattering equation. The presence of such solutions provides a major numerical obstacle for a self-consistent evaluation of one- and two-body propagator within the normal Green's function approach at densities where such instabilities can occur. For this reason, recent SCGF calculations have been performed for temperatures above the critical temperature for a possible phase transition to pairing condensation.[17,27]

In order to proceed to a study of the density and temperature dependence of pairing, some results for fully self-consistent calculations above the critical temperature T_c for pairing are presented first. The detailed procedure is explained in Refs. 27,28. The results discussed below have been reported in Ref. 29. The imaginary part of the nucleon self-energy is displayed in the upper panel of Fig. 1 for a temperature of 5 MeV. The purpose of this figure is to visualize some differences between various models of the NN interaction and between symmetric nuclear matter and pure neutron matter. Therefore we consider a large interval for the energy variable ω. The imaginary parts of the self-energy derived from the CDBonn interaction[16] and the Argonne V18 (ArV18) interaction[15] are very similar at energies around $\omega = 0$. At those energies the ArV18 yields a slightly weaker imaginary part than CDBonn. The differences get larger at positive values of ω, where the imaginary part of the self-energy derived from ArV18 reaches a minimum of around -100 MeV at an energy ω around 1.7 GeV. The minimum for the CDBonn interaction is only about -35 MeV and occurs at energies ω around 0.5 GeV.

Figure 1 also displays results for the real and imaginary part of the self-energy for neutrons with the same momentum ($k = 225$ MeV) in pure neutron matter. The density of neutron matter considered in this figure is one half of the empirical saturation density of nuclear matter, which implies that these systems have the same Fermi momentum. The imaginary part of the self-energy in neutron matter is weaker than the corresponding one for symmetric nuclear matter, reflecting the dominance of proton-neutron correlations. For both interactions a minimum is obtained around 1.7 GeV. At these high energies the absolute value for the imaginary part of the self-energy is about a factor three larger for ArV18 than for CDBonn. This means that the distribution of sp strength to high energies due to central short-range correlations is much stronger for the local ArV18 interaction than for the non-local meson-exchange model CDBonn. The results for the lower energies are closer to each other. The differences in the amount of NN correlations is also reflected in the occupation probability $n(k)$. For symmetric nuclear matter at saturation density we obtain for $n(k = 0)$ the values 0.89 and 0.87 for CDBonn and ArV18, respectively. The corresponding value for neutron matter are $n(k = 0) = 0.968$ and 0.963. These numbers for nuclear matter together with the result of 0.85 for the Reid potential, suggest that there is a direct correlation with the strength of the core of the interaction and the depletion at $k = 0$. Since this depletion can now be

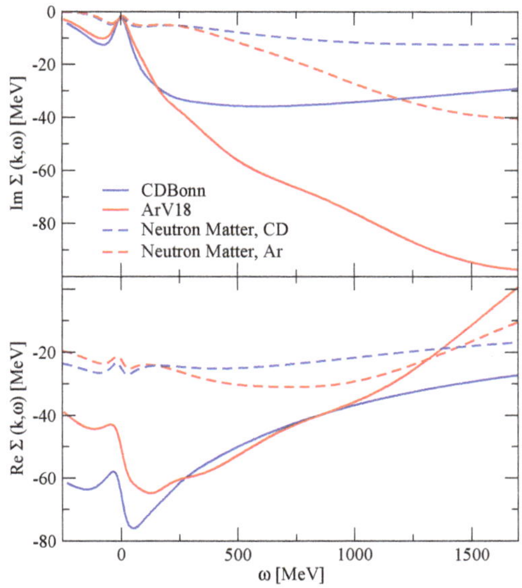

Fig. 1. Imaginary (upper panel) and real part (lower panel) of the retarded self-energy for nucleons with momentum $k = 225$ MeV/c in symmetric nuclear matter at the empirical saturation density ($\rho = 0.16$ fm^{-3}). Results obtained for CDBonn interaction are compared to those resulting from SCGF calculations using ArV18. Also included are results for neutrons with the same momentum in neutron matter at $\rho = 0.08$ fm^{-3}. The temperature in all these calculations was fixed at T=5 MeV.

experimentally scrutinized and is found to be substantial, it is clear that a much softer interaction is ruled out by such data.

At temperatures below the critical temperature for a transition to a superfluid one has to supplement the evaluation of the normal Green's function with the anomalous Green's function. While the self-consistent inclusion of ladder diagrams has reached quite a sophistication, it remains to fully account for the possibility of a pairing solution in such calculations for realistic NN interactions. A first step towards such a complete scheme is to include the full self-consistent dressing due to normal self-energy terms generated by ladder diagrams, in the calculation of the anomalous self-energy and the solution of the corresponding generalized gap equation. The expression for the gap function Δ can be written[30] in a partial wave expansion

$$\Delta_l^{JST}(p) = \sum_{l'} \frac{2}{\pi} \int_0^\infty \mathrm{d}k\, k^2 \int_{-\infty}^{+\infty} \frac{\mathrm{d}\omega}{2\pi} \int_{-\infty}^{+\infty} \frac{\mathrm{d}\omega'}{2\pi} \langle p|V_{ll'}^{JST}|k\rangle \times A(k,\omega) A_s(k,\omega') \frac{1 - f(\omega) - f(\omega')}{-\omega - \omega'} \Delta_{l'}^{JST}(k)\,. \tag{1}$$

If we ignore for a moment the difference between the spectral functions A and A_s, we see that this equation for the self-energy Δ corresponds to the homogeneous scattering equation for the T-matrix at energy $\Omega = 0$ and center-of-mass momentum $P = 0$. This means that a non-trivial solution of Eq. (1) is obtained if and only if the scattering matrix T generates a pole at energy $\Omega = 0$, which reflects a bound two-particle state. This is precisely the condition for the pairing instability, demonstrating that this treatment of pairing correlations is compatible with the T-matrix approximation in the non-superfluid regime.

One may also consider Eq. (1) in the limit in which the spectral functions $A(k,\omega)$ and $A_s(k,\omega)$ are approximated by the corresponding mean-field and BCS approximation. The BCS approximation for the spectral function yields

$$A_s(k,\omega) = 2\pi \left(\frac{E_k + \chi_k}{2E_k} \delta(\omega - E_k) + \frac{E_k - \chi_k}{2E_k} \delta(\omega + E_k) \right), \tag{2}$$

with the quasi-particle energy

$$E_k = \sqrt{\chi_k^2 + \Delta^2(k)}\,. \tag{3}$$

Inserting these approximations for the spectral function into Eq. (1) and taking the limit $T = 0$ reduces to the usual BCS gap equation

$$\Delta_l^{JST}(p) = \sum_{l'} \frac{2}{\pi} \int_0^\infty \mathrm{d}k\, k^2 \left\langle p \right| V_{ll'}^{JST} \left| k \right\rangle \frac{1}{-2E_k} \Delta_{l'}^{JST}(k)\,. \tag{4}$$

Therefore we can consider Eq. (1) as a generalization of the usual gap equation. It accounts for the spreading of sp strength leading to a generalization of the form

$$\frac{1}{-2E_k} \quad \to \quad \int_{-\infty}^{+\infty} \frac{\mathrm{d}\omega}{2\pi} \int_{-\infty}^{+\infty} \frac{\mathrm{d}\omega'}{2\pi} A(k,\omega) A_s(k,\omega') \frac{1 - f(\omega) - f(\omega')}{-\omega - \omega'}\,. \tag{5}$$

The next step involves to consider the effects of temperature and short-range correlations in the solution of the gap equation. For that purpose one may reconsider the two-particle propagator of Eq. (5) but replace the spectral function of the superfluid phase, $A_s(k,\omega')$, by the corresponding one for the normal phase, $A(k,\omega')$. This leads to a definition of an average energy denominator $\widetilde{\chi}_k$ of the form

$$\frac{1}{-2\widetilde{\chi}_k} =: \int_{-\infty}^{+\infty} \frac{\mathrm{d}\omega}{2\pi} \int_{-\infty}^{+\infty} \frac{\mathrm{d}\omega'}{2\pi} A(k,\omega) A(k,\omega') \frac{1 - f(\omega) - f(\omega')}{-\omega - \omega'}\,. \tag{6}$$

If we consider this propagator in the limit of the mean-field approximation ($A(k,\omega) = \delta(\omega - \varepsilon_k)$) at $T = 0$, it reduces to an energy denominator of the form

$$\frac{1}{-2\widetilde{\chi}_k} \quad \overset{\mathrm{mf},T=0}{\longrightarrow} \quad \frac{1}{-2|\chi_k|}\,. \tag{7}$$

This means that the energy $\widetilde{\chi}_k$ has been defined in this equation to exhibit the effects of finite temperature and correlations on the two-particle propagator. Figure 2 displays results for this quantity $\widetilde{\chi}_k$, the inverse of this propagator multiplied by -2, and compares it with $|\chi_k|$ using the corresponding quasi-particle energies. The dashed-dotted line represents the effects of temperature, *i.e.* the propagator has been calculated using the quasi-particle approach for the spectral function and the Fermi function for the temperature under consideration, while the solid line accounts for

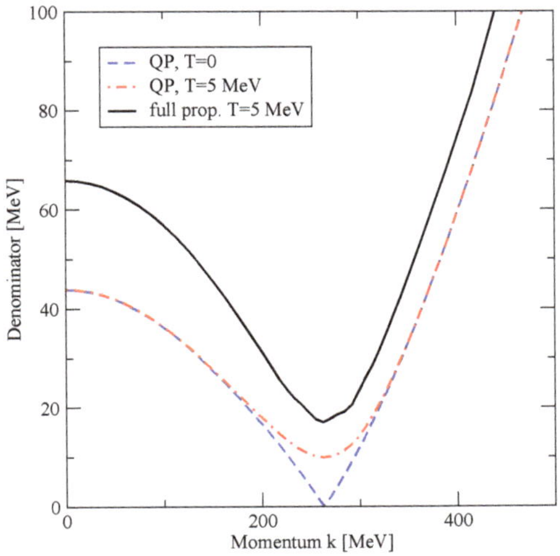

Fig. 2. The quantity $\tilde{\chi}_k$ defined in Eq. (6), which represents the energy denominator for the propagator of two nucleons in the medium. Results are displayed for the quasi-particle approximation in the limit $T = 0$ (QP, T=0), the quasi-particle approximation for finite temperature $T = 5$ MeV (QP, T=5 MeV) and the dressed propagator resulting from SCGF calculations (full prop., T=5 MeV). All results displayed in this figure were obtained for symmetric nuclear matter at density $\rho = 0.16$ fm^{-3}, using CDBonn interaction.

finite temperature and correlation effects. The finite temperature yields an enhancement of the effective sp energy $\tilde{\chi}_k$ for momenta around the Fermi-momentum, only. Including in addition the effects of correlations in the propagator, we obtain larger values for $\tilde{\chi}_k$ for all momenta. This corresponds to the well known feature that a finite temperature yields a depletion of the occupation probability of sp states only for momenta just below the Fermi momentum, while strong short-range correlation provide such a depletion for all momenta of the Fermi sea.

Results for the gap parameter $\Delta(k_F)$ in symmetric nuclear matter of various densities are presented in Fig. 3 as a function of the temperature T. We will first discuss the results obtained within the usual BCS approximation in the $^3S_1 -^3 D_1$ partial wave. At the empirical saturation density the CDBonn interaction yields a gap parameter $\Delta(k_F)$ at temperature $T = 0$ of 8.6 MeV, which decreases with increasing temperature until it vanishes at $T = 5.2$ MeV. At $\rho = 0.08$ fm^{-3}, which is about half the empirical density, the value of the gap parameter at $T = 0$ is even larger ($\Delta(k_F) = 10.6$ MeV) and the gap calculated within the usual BCS approach disappears only at a temperature of 5.9 MeV. This increase of the pairing gap with decreasing density can be related to the momentum-dependence of the pairing gap $\Delta(k)$: The gap function increases with decreasing momentum. Therefore, as the Fermi momentum decreases with density, the value $\Delta(k_F)$ tends to decrease with density. At even lower densities, however, this effect is more than compensated by the feature, that the phase-space of two-hole configurations decreases with density, so that ultimately the gap parameter will approach the binding energy of the

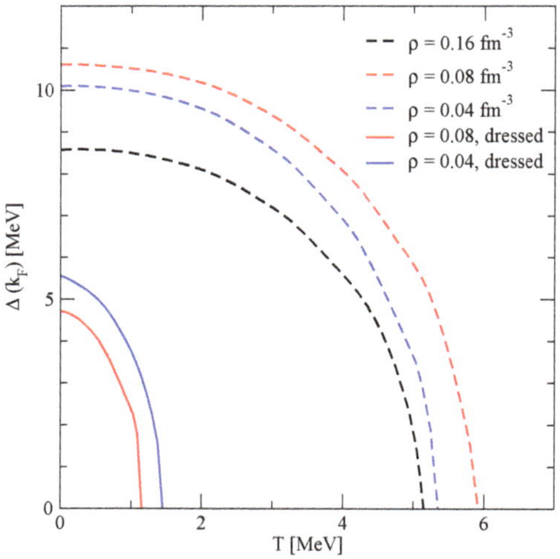

Fig. 3. Gap parameter $\Delta(k_F)$ in symmetric nuclear matter as a function of temperature T. Results are presented for various densities, with and without taking into account the dressing of the sp propagator due to short-range correlations. The pairing gap disappears at ρ=0.16 fm^{-3}, if dressed propagators are considered. All results displayed in this figure were obtained using the CDBonn interaction.

deuteron in the limit of $\rho \to 0$. This explains the decrease of the gap parameter going from $\rho = 0.08$ fm^{-3} to $\rho = 0.04$ fm^{-3}.

If we take the effects of short-range correlations into account, the generalized gap equation of Eq. (1) does not give a non-trivial solution for symmetrical nuclear matter at $\rho = 0.16$ fm^{-3}. This means that a proper treatment of correlation effects in nuclear matter at normal density yields a disappearance of the proton-neutron pairing predicted by the usual BCS approach. The effects of short-range correlations tend to decrease with density. As a consequence, one obtains non-vanishing gaps for proton-neutron pairing at lower densities (see solid lines in Fig. 3). This also leads to an increase of the critical temperature and the value of $\Delta(k_F)$ at $T = 0$ going from $\rho = 0.08$ fm^{-3} to $\rho = 0.04$ fm^{-3}. Note, that these functions are qualitatively different from the corresponding BCS predictions.

Finally, consider the case of neutron-neutron pairing in pure neutron matter. We will focus the attention to densities, where the pairing correlations in the $^1S_0$ partial wave are dominating. Results for the gap parameter $\Delta(k_F)$ as a function of temperature are displayed in Fig. 4. Using the BCS approximation with sp energies derived from the quasi-particle energies of SCGF calculations we obtain a gap at $T = 0$, which, for the range of densities considered, increases with decreasing density. This is in agreement with results of similar calculations, which are summarized *e.g.* in Ref. 1.

The effects of short-range correlations are weaker in neutron matter than in nuclear matter. Nevertheless, also these weaker effects of short-range correlations in neutron matter are sufficient to suppress the formation of a pairing gap in neutron

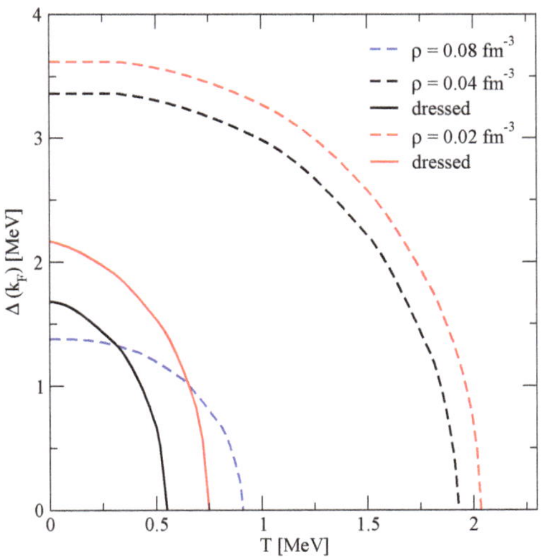

Fig. 4. Gap parameter $\Delta(k_F)$ in neutron matter as a function of temperature T. Results are presented for the usual BCS approximation (dashed lines) and the solution of the generalized gap equation (1) in the $^1S_0$ partial wave using the CDBonn interaction.

matter at $\rho = 0.08$ fm^{-3}. Such a suppression of pairing correlations is also observed at smaller densities. In this case, however, the inclusion of the correlation effects just leads to a reduction of the gap parameter at a given temperature and a reduction of the critical temperature (see Fig. 4). Recent results for neutron matter obtained with state of the art Monte Carlo and CBF techniques[31] indicate that in the density range where our results overlap, there is complete agreement for the gap at $T = 0$ with the Monte Carlo results that are also discussed by Schmidt in another contribution to this volume.

4. Proton Correlations at Large Asymmetry

Although there are numerous studies of the effect of correlations on the properties of sp levels for nuclei near stability, there are only a few studies for very neutron or proton-rich nuclei. From neutron knock-out reactions, Gade *et al.*[32] infer the occupancy of the $0d_{5/2}$ neutron hole state in the proton-rich ^{32}Ar nucleus is considerably reduced relative to those for stable nuclei. An alternate method to study sp strength is through the use of the dispersive optical model (DOM) developed by Mahaux and Sartor.[33] This description employs the Kramers-Kronig dispersion relation that links the imaginary and real parts of the nucleon self-energy.[26] This procedure links optical-model (OM) analyses of reaction data at positive energies to structural information at negative energies. In Ref. 34, the properties of proton levels in Ca nuclei as a function of asymmetry $\delta = \frac{N-Z}{A}$ have been investigated with the DOM. Previously measured elastic-scattering and reaction-cross-section data for protons on ^{40}Ca and ^{48}Ca as well as level properties of hole states in these

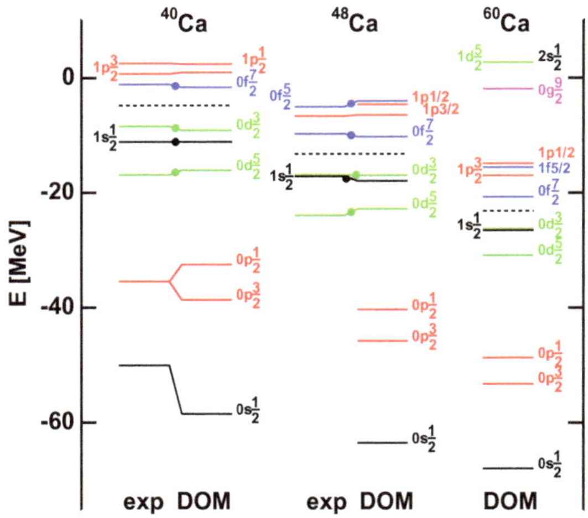

Fig. 5. Comparison of experimental proton single-particle levels to DOM calculations. For the levels indicated with the solid dots, their energies were included in the fits. The dashed lines indicate the Fermi energy.

nuclei, inferred from $(e, e'p)$ reactions, were simultaneously fit. The dependence on δ is extracted and used to predict level properties of ^{60}Ca. Result of this procedure yield good agreement for all considered scattering data and structure information in a large energy window.[34] A fit to all these data requires a stronger imaginary surface potential near the Fermi energy leading to deeper binding of proton levels in ^{48}Ca as compared to ^{40}Ca. The effect on the level structure around the Fermi energy that is obtained for ^{60}Ca after the extrapolating the potential is shown in Fig. 5. The levels in the immediate vicinity of E_F are focused closer to E_F, increasing the density of sp levels. A reduced gap between the particle and hole valence levels implies that the closed-shell nature of this nucleus has diminished and proton pairing may be important. This possibility of proton pairing at large nucleon asymmetry therefore has as its origin the larger strength of p-n relative to the p-p and n-n interactions. Hence, protons (neutrons) experience larger (weaker) correlations in neutron-rich matter. The reversed is true for proton-rich matter.

5. Conclusions

An attempt has been made to treat the effects of short-range and pairing correlations in a consistent way within the T-matrix approach of the self-consistent Green's function (SCGF) method. The pairing effects are determined from a generalized gap equation that employs sp propagators fully dressed by short-range and tensor correlations. While short-range and tensor correlations yield a redistribution of sp strength over a wide range of energies, the effects of pairing correlations on the spectral function are limited in nuclear matter to a relatively small interval in energy and momentum around the Fermi surface. The formation of a pairing gap is very sensitive to the quasi-particle energies and strength distribution at the Fermi surface and can be suppressed by moderate temperatures. From this sensitivity to different areas in momentum and energy one may conclude that the features of short-range correlations should be rather similar in studies of nuclear matter and finite nuclei. The investigation of pairing phenomena, however, is rather sensitive *e.g.* to the energy spectrum around the Fermi energy. Therefore the shell effects of finite nuclei may lead to quite different results for pairing properties than corresponding studies in infinite matter.

The redistribution of sp strength due to the short-range correlations has a significant effect on the formation of a pairing gap. While the usual BCS approach predicts a gap for proton-neutron pairing in nuclear matter at saturation density as large as 8 MeV, the inclusion of short-range correlations suppresses this gap completely. Correlation effects are weaker at smaller densities, but still lead to a significant quenching of the proton-neutron pairing gap and to a reduction of the critical temperature for the phase transition. Compared to symmetric nuclear matter correlation effects are weaker in neutron matter. Nevertheless, the inclusion of correlations suppresses the formation of a gap for neutron-neutron pairing at $\rho = 0.08$ fm^{-3} completely and yields a significant quenching at lower densities. While the results now appear consistent with empirical information on pairing of both isoscalar (pn) and isovector (pp or nn) kind, we note that our calculations do not include polarization contributions to the pairing interaction.

A final observation concerns the possibility of observing pairing phenomena in nuclei far from stability that result from the increase of the correlations of one kind of particle when a large excess of the other kind is present.

Acknowledgments

The work reported in this contribution is based on research performed in collaboration with Herbert Müther, Bob Charity, and Lee Sobotka.

References

1. D. Dean and M. Hjorth-Jensen, *Rev. Mod. Phys.* **75**, 607 (2003).
2. T. Alm, G. Röpke and M. Schmidt, *Z. Phys. A* **337**, 355 (1990).
3. B. E. Vonderfecht, C. C. Gearhart, W. H. Dickhoff, A. Polls and A. Ramos, *Phys. Lett. B* **253**, 1 (1991).
4. M. Baldo, I. Bombaci and U. Lombardo, *Phys. Lett. B* **283**, 8 (1992).
5. T. T. and R. Tamagaki, *Suppl. Prog. Theor. Phys.* **112**, 27 (1993).
6. M. Baldo, U. Lombardo and P. Schuck, *Phys. Rev. C* **52**, 975 (1995).
7. W. H. Dickhoff and C. Barbieri, *Prog. Part. Nucl. Phys.* **52**, 377 (2004).
8. H. Müther and W. H. Dickhoff, *Phys. Rev. C* **49**, R17 (1994).
9. B. E. Vonderfecht, W. H. Dickhoff, A. Polls and A. Ramos, *Phys. Rev. C* **44**, R1265 (1991).

10. M. F. van Batenburg, Deeply bound protons in ^{208}Pb, PhD thesis, University of Utrecht (2001).
11. M. G. E. Brand, G. A. Rijsdijk, F. A. Muller, K. Allaart and W. H. Dickhoff, *Nucl. Phys. A* **531**, 253 (1991).
12. D. Rohe *et al.*, *Phys. Rev. Lett.* **93**, 182501 (2004).
13. T. Frick, K. S. A. Hassaneen, D. Rohe and H. Müther, *Phys. Rev. C* **70**, 054308 (2004).
14. B. E. Vonderfecht, W. H. Dickhoff, A. Polls and A. Ramos, *Nucl. Phys. A* **555**, p. 1= (1993).
15. R. B. Wiringa, V. G. J. Stoks and R. Schiavilla, *Phys. Rev. C* **51**, 38 (1995).
16. R. Machleidt, F. Sammarruca and Y. Song, *Phys. Rev. C* **53**, R1483 (1996).
17. W. H. Dickhoff and E. P. Roth, *Acta Phys. Pol. B* **33**, 65 (2002).
18. E. P. Roth, Self-consistent green's functions in nuclear matter, PhD thesis, Washington University in St. Louis (2000).
19. Y. Dewulf, D. V. Neck and M. Waroquier, *Phys. Rev. C* **65**, 054316 (2002).
20. P. Bożek and P. Czerski, *Eur. Phys. J. A* **11**, 271 (2001).
21. P. Bożek, *Phys. Rev. C* **65**, 054306 (2002).
22. P. Bożek, *Eur. Phys. J. A* **15**, 325 (2002).
23. Y. Dewulf, W. H. Dickhoff, D. V. Neck, E. R. Stoddard and M. Waroquier, *Phys. Rev. Lett.* **90**, 152501 (2003).
24. A. E. L. Dieperink, Y. Dewulf, D. V. Neck and M. Waroquier, *Phys. Rev. C* **68**, 064307 (2003).
25. T. Frick, H. Müther, A. Rios, A. Polls and A. Ramos, *Phys. Rev. C* **71**, 014313 (2005).
26. W. Dickhoff and D. V. Neck, *Many-Body Theory Exposed!* (World Scientific, Singapore, 2005).
27. T. Frick and H. Müther, *Phys. Rev. C* **68**, 034310 (2003).
28. T. Frick, Self-consistent green's functions in nuclear matter at finite temperature, PhD thesis, University of Tübingen (2004).
29. H. Müther and W. H. Dickhoff, *Phys. Rev. C* **72**, 054313 (2005).
30. P. Bożek, *Nucl. Phys. A* **657**, 187 (1999).
31. A. Fabrocin, S. Fantoni, A. Y. Illarionov and K. E. Schmidt, *Phys. Rev. Lett.* **95**, 192501 (2005).
32. A. Gade *et al.*, *Phys. Rev. Lett* **93**, 042501 (2004).
33. C. Mahaux and R. Sartor, *Adv. Nucl. Phys.* **20**, 1 (1991).
34. R. J. Charity, L. G. Sobotka and W. H. Dickhoff, *Phys. Rev. Lett.* **97**, in press (2006).

SHORT RANGE CORRELATIONS IN RELATIVISTIC NUCLEAR MODELS

PRAFULLA K. PANDA

Institute of Physics, Bhubaneswar, India – 751005
E-mail:prafulla_panda@rediffmail.com

C. PROVIDÊNCIA and J. DA PROVIDÊNCIA

Departamento de Física, Universidade de Coimbra, P-3004-516 Coimbra, Portugal
E-Mails: cp@teor.fis.uc.pt and providencia@teor.fis.uc.pt

Short range correlations are introduced using the unitary operator method in a relativistic theory of nuclear matter based on the Hartree-Fock approximation. Pion exchange, including the tensor contribution, is taken into account. It is found that both the tensor contribution of pion exchange and short range correlations soften the equation of state. The neutron matter equation of state is also discussed.

Keywords: Short range correlations; relativistic models; equation of state.

1. Introduction

We discuss the role of short range correlations in a relativistic approach to the description of nuclear matter. Short range correlations may be introduced into the model wave function by several procedures which have something in common with the so called e^S method or coupled-cluster expansion, introduced by Coester and Kümmel,[1] and extensively developed by R. Bishop.[2]

In this short note, we consider the unitary operator method as proposed by Villars,[3] which automatically guarantees that the correlated state is normalized. The general idea of introducing short range correlations in systems with short range interactions exists for a long time[4,5] but has not been pursued for the relativistic case.

Non-relativistic calculations based on realistic NN potentials predict equilibrium points which do not reproduce simultaneously the binding energy and saturation density. Either the saturation density is reproduced but the binding energy is too small, or the binding energy is reproduced at too high a density.[6] In order to solve this problem, the existence of a repulsive potential or density-dependent repulsive mechanism[7] is usually assumed. Due to Lorentz covariance and self-consistency, relativistic mean field theories[8] include automatically contributions which are equivalent to n-body repulsive potentials in non-relativistic approaches. The relativistic quenching of the scalar field provides a mechanism for saturation, though, by itself it may lead to too small an effective mass and too large incompressibility of nuclear matter, a situation which is encountered in the Walecka model.[8]

In the non-relativistic case, we find a wound in the relative wave function if the vector interaction is stronger than the scalar interaction. However, this may not be the case if the interaction is strong enough for short distances. Then, the effective potential can actually turn attractive at short distances and the wave function may well have a node, as proposed by V. Neudatchin[9] and advocated by S. Moszkowski,[10] on the basis of the the quark structure of the nucleon. The situation with short range correlations may be more subtle than might be thought from a simple non-relativistic model. We show that a short range node in the relative wave function may be encountered in relativistic models, although it remains to be seen to what extent the relativistic description simulates the quark structure.

In non-relativistic models the saturation arises from the interplay between a long range attraction and a short range repulsion, so strong that it is indispensable to take short range correlations into account. In relativistic mean field models, the parameters are phenomenologically fitted to the saturation properties of nuclear matter. Although in this approach short range correlation effects may be accounted for, to some extent, by the model parameters, it is our aim to study explicitly the consequences of actual short range correlations. In a previous publication[11] we have discussed the effect of the correlations in the ground state properties of nuclear matter in the framework of the Hartree-Fock approximation using an effective Hamiltonian derived from the $\sigma - \omega$ Walecka model. We have shown, for interactions mediated only by sigma and omega mesons, that the equation of state (EOS) becomes considerably softer when correlations are taken into account, provided the correlation function is treated variationally, always paying careful attention to the constraint imposed by the "healing distance" requirement. In the present note we will work within the same approach and will include also the exchange of pions and ρ-mesons. Preliminary results of the present work have been presented in refs.[12,13].

2. Effective Relativistic Model Hamiltonian

We consider the effective Hamiltonian

$$\begin{aligned} H = &\int \psi_\alpha^\dagger(\vec{x})(-i\vec{\alpha}\cdot\vec{\nabla} + \beta M)_{\alpha\beta}\psi_\beta(\vec{x})\, d\vec{x} \\ &+\frac{1}{2}\int \psi_\alpha^\dagger(\vec{x})\psi_\gamma^\dagger(\vec{y})V_{\alpha\beta,\gamma\delta}(|\vec{x}-\vec{y}|)\psi_\delta(\vec{y})\psi_\beta(\vec{x})\, d\vec{x}\, d\vec{y}, \end{aligned} \tag{1}$$

where the exchange of σ, ω and ρ mesons is taken into account through the interaction

$$V_{\alpha\beta,\gamma\delta}(r) = \sum_{i=\sigma,\omega,\rho} V^i_{\alpha\beta,\gamma\delta}(r) \tag{2}$$

with

$$V^\sigma_{\alpha\beta,\gamma\delta}(r) = -\frac{g_\sigma^2}{4\pi}(\beta)_{\alpha\beta}(\beta)_{\gamma\delta}\frac{e^{-m_\sigma r}}{r},$$

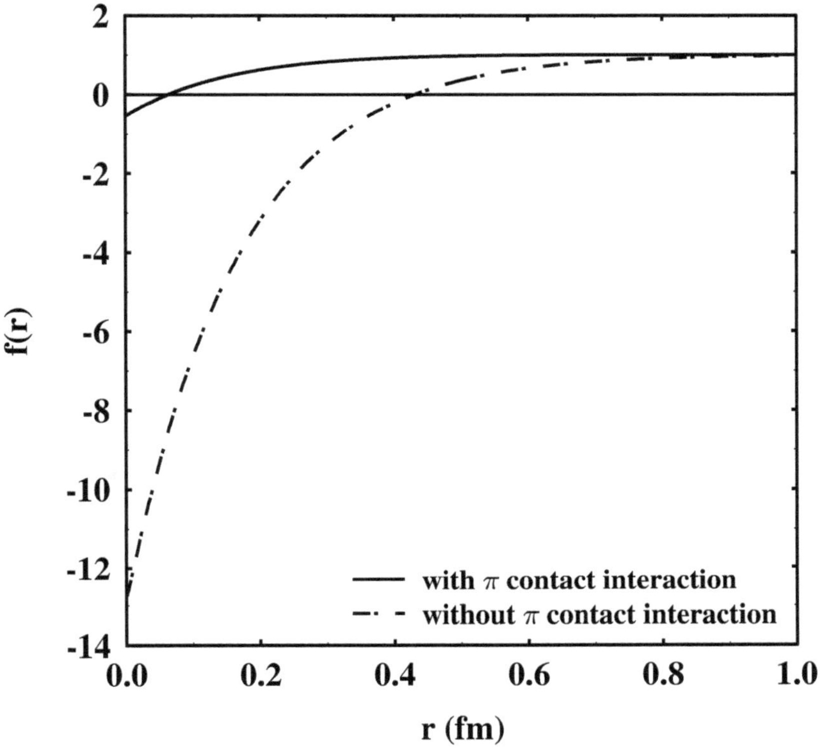

Fig. 1. Influence of the contact term of the pion exchange interaction on the correlation function, suggesting the need for a more flexible form of the Jastrow factor.

$$V^{\omega}_{\alpha\beta,\gamma\delta}(r) = \frac{g^2_\omega}{4\pi}\left(\delta_{\alpha\beta}\delta_{\gamma\delta} - \vec{\alpha}_{\alpha\beta}\cdot\vec{\alpha}_{\gamma\delta}\right)\frac{e^{-m_\omega r}}{r},$$

$$V^{\rho}_{\alpha\beta,\gamma\delta}(r) = \frac{g^2_\rho}{4\pi}\left(\delta_{\alpha\beta}\delta_{\gamma\delta} - \vec{\alpha}_{\alpha\beta}\cdot\vec{\alpha}_{\gamma\delta}\right)\vec{\tau}_1\cdot\vec{\tau}_2\frac{e^{-m_\rho r}}{r}.$$

The exchange of π mesons is described by an interaction of the form[14]

$$V^{\pi}_{\alpha\beta,\gamma\delta}(\vec{r}) = \frac{G^2_{\pi NN}}{4\pi 4M^2}(\vec{\tau}_1\cdot\vec{\tau}_2)(\vec{\sigma}_1\cdot\vec{\nabla})_{\alpha\beta}(\vec{\sigma}_2\cdot\vec{\nabla})_{\gamma\delta}\frac{e^{-m_\pi r}}{r}, \tag{3}$$

where $\vec{\sigma}_i = \vec{\alpha}_i\gamma_5$ and M is the nucleon mass. This can be rewritten in momentum space as

$$V^{\pi}_{\alpha\beta\gamma\delta}(\vec{q}) = \frac{G^2_{\pi NN}}{4\pi 4M^2}(\vec{\tau}_1\cdot\vec{\tau}_2)(\vec{\sigma}_1\cdot\vec{q})_{\alpha\beta}(\vec{\sigma}_2\cdot\vec{q})_{\gamma\delta}\frac{1}{q^2+m^2_\pi}. \tag{4}$$

As explained in the paper by Bouyssy *et al*,[14] eq. (3) may be approximated as

$$V^{\pi}_{\alpha\beta,\gamma\delta}(\vec{r}) \approx \frac{G^2_{\pi NN}}{4\pi 4M^2}(\vec{\tau}_1\cdot\vec{\tau}_2)((\vec{\sigma})_{\alpha\beta}\cdot(\vec{\sigma})_{\gamma\delta})\frac{m^2_\pi}{3}\left[\frac{4\pi}{m^2_\pi}\delta(\vec{r}) - \frac{e^{-m_\pi r}}{r}\right]. \tag{5}$$

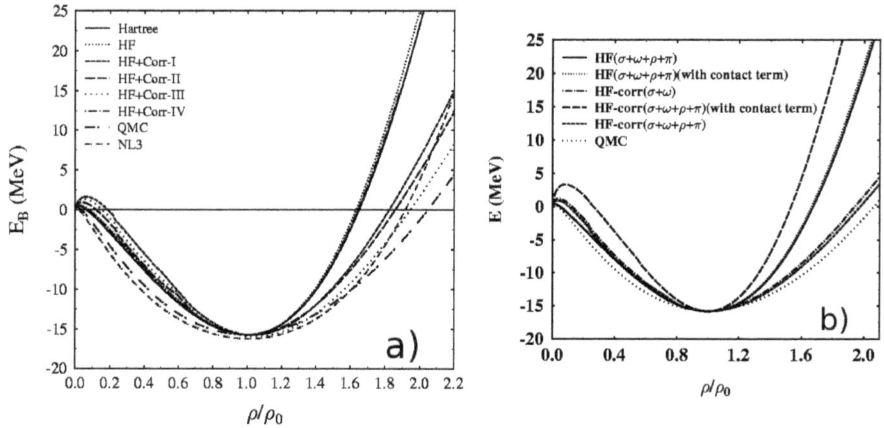

Fig. 2. Influence of ρ and π exchange on the EOS. a) EOS when the exchange of ρ and π mesons is not considered, for several choices of the correlation function. In the curve labeled HF+Corr-IV, which almost coincides with the curve labeled Hartree, the Jastrow factor was not determined variationally. Instead, it was chosen so as to reproduce familiar correlations in non-relativistic calculations. For comparison, the EOS obtained within the QMC[20] and the NL3[21] parametrizations of the $\sigma - \omega$ model are also shown. b) EOS when rho and pion exchange is taken into account. The stiffness of the long-dashed curve is due to the insufficient flexibility of the correlation factor. For comparison, a typical EOS in the absence of π and ρ mesons exchange is also shown.

The first term in the above is the repulsive contact interaction and the second term is an attractive Yukawa potential. In equation (1), ψ is the nucleon field and $\vec{\alpha}$, β are the Dirac-matrices. The equal time quantization conditions

$$[\psi_\alpha(\vec{x},t), \psi_\beta(\vec{y},t)^\dagger]_+ = \delta_{\alpha\beta}\delta(\vec{x}-\vec{y})$$

are satisfied. At time $t=0$, the field ψ may be expanded as[15]

$$\psi(\vec{x}) = \frac{1}{\sqrt{V}} \sum_{r,k} \left[U_r(\vec{k}) c_{r,\vec{k}} \; + \; V_r(-\vec{k}) \tilde{c}^\dagger_{r,-\vec{k}} \right] e^{i\vec{k}\cdot\vec{x}},$$

where U_r and V_r are given by

$$U_r(\vec{k}) = \begin{pmatrix} \cos\frac{\chi(\vec{k})}{2} \\ \vec{\sigma}\cdot\hat{k}\sin\frac{\chi(\vec{k})}{2} \end{pmatrix} u_r; \quad V_r(-\vec{k}) = \begin{pmatrix} -\vec{\sigma}\cdot\hat{k}\sin\frac{\chi(\vec{k})}{2} \\ \cos\frac{\chi(\vec{k})}{2} \end{pmatrix} v_r.$$

For free fields, we have $\cos\chi(\vec{k}) = M/\epsilon(\vec{k})$, $\sin\chi(\vec{k}) = |\vec{k}|/\epsilon(\vec{k})$ with $\epsilon(\vec{k}) = \sqrt{\vec{k}^2+M^2}$, while, for interacting fields, $\cos\chi(\vec{k}) = M^*(\vec{k})/\epsilon^*(\vec{k})$, $\sin\chi(\vec{k}) = |\vec{k}^*|/\epsilon^*(\vec{k})$, with $\epsilon^*(\vec{k}) = \sqrt{\vec{k^*}^2 + M^{*2}(\vec{k})}$, where $\vec{k}^*$ and $M^*(\vec{k})$ denote, respectively, the effective momentum and mass, determined self-consistently by the Hartree-Fock (HF) prescription. The equal time anti-commutation conditions $[c_{r,\vec{k}}, c^\dagger_{s,\vec{k}'}]_+ = \delta_{rs}\delta_{\vec{k},\vec{k}'} = [\tilde{c}_{r,\vec{k}}, \tilde{c}^\dagger_{s,\vec{k}'}]_+$ are satisfied. We wish to describe hadronic matter in terms of the occupation of positive energy states, leaving empty the negative energy states, so that the vacuum denoted by $\mid 0\rangle$ is defined by $c_{r,\vec{k}} \mid 0\rangle =$

$\tilde{c}^{\dagger}_{r,\vec{k}} \mid 0\rangle = 0$. One-particle states are written as $|\vec{k}, r\rangle = c^{\dagger}_{r,\vec{k}} \mid 0\rangle$. Two-particle and three-particle uncorrelated states are written as $|\vec{k}, r; \vec{k}', r'\rangle = c^{\dagger}_{r,\vec{k}}\, c^{\dagger}_{r',\vec{k}'} \mid 0\rangle$, and $|\vec{k}, r; \vec{k}', r'; \vec{k}'', r''\rangle = c^{\dagger}_{r,\vec{k}}\, c^{\dagger}_{r',\vec{k}'}\, c^{\dagger}_{r'',\vec{k}''} \mid 0\rangle$, respectively; and so on. By minimizing the energy expectation value

$$E = \frac{\langle\Phi|H|\Phi\rangle}{\langle\Phi|\Phi\rangle},$$

the HF method determines the Slater determinant $|\Phi\rangle$, which is obtained by filling up all positive energy states up to the Fermi momentum.

We take into account the effect of short range correlations through the unitary operator method. The correlated wave function[16] is $|\Psi\rangle = e^{i\Omega}|\Phi\rangle$ where Ω is, in general, a n-body Hermitian operator, splitting into a 2-body part, a 3-body part, etc.. The expectation value of H becomes

$$E = \frac{\langle\Psi|H|\Psi\rangle}{\langle\Psi|\Psi\rangle} = \frac{\langle\Phi|e^{-i\Omega}\ H\ e^{i\Omega}|\Phi\rangle}{\langle\Phi|\Phi\rangle}. \tag{6}$$

In the present calculation, we only take into account two-body correlations. Let us denote the two-body correlated wave function by $\overline{|\vec{k}, r; \vec{k}', r'\rangle} = e^{i\Omega}|\vec{k}, r; \vec{k}', r'\rangle \approx f_{12}|\vec{k}, r; \vec{k}', r'\rangle$ where f_{12} is the short range correlation factor, the so-called Jastrow factor.[17] For simplicity, we consider $f_{12} = f(|\vec{r}_{12}|)$, $\vec{r}_{12} = \vec{r}_1 - \vec{r}_2$, and $f(r) = 1 - (\alpha + \beta r)\, e^{-\gamma r}$ where α, β and γ are parameters. The choice of a real function to describe the effect of the unitary operator has to be supplemented by a normalization condition

$$\int \left(f^2(r) - 1\right)\, d^3r = 0, \tag{7}$$

which assures unitarity to leading cluster order. As we will discuss later, in the present case the function $f(r)$ introduces a node in the wave function as shown in Fig. 1.

The important effect of the short range correlations is the expression for the correlated ground-state energy. Here, in the leading order of the cluster expansion, the interaction matrix element $\langle\vec{k}, r; \vec{k}', r'|V_{12}|\vec{k}, r; \vec{k}', r'\rangle$ of the HF expression is replaced by $\overline{\langle\vec{k}, r; \vec{k}', r'|}V_{12} + t_1 + t_2\overline{|\vec{k}, r; \vec{k}', r'\rangle} - \langle\vec{k}, r; \vec{k}', r'|t_1 + t_2|\vec{k}, r; \vec{k}', r'\rangle$, where t_i is the kinetic energy operator of particle i. As argued by Moszkowski[18] and Bethe,[19] it is expected that the true ground-state wave function of the nucleus, containing correlations, coincides with the independent particle, or HF wave function, for inter particle distances $r \geq r_h$, where $r_h \approx 1$ fm is the so-called "healing distance". This behavior is a consequence of the restrictions imposed by the Pauli Principle. A natural consequence of having the correlations introduced by a unitary operator is the normalization constraint on $f(r)$ expressed by eq. (7). The correlated ground state energy of symmetric nuclear matter with σ, ω and ρ potentials reads

$$\mathcal{E}_{\sigma,\omega,\rho} = \frac{\nu}{\pi^2}\int_0^{k_F} k^2\, dk\ [|k|\sin\chi(k)\ +\ M\cos\chi(k)] - \frac{4}{(2\pi)^4}\int_0^{k_F} k^2\, dk\ k'^2\, dk'$$
$$\times\Big\{\Big[|k|\sin\chi(k) + 2\ M\cos\chi(k)\Big]I(k,k') + |k|\ \sin\chi(k')\ J(k,k')\Big\}$$
$$+\frac{1}{(2\pi)^4}\int_0^{k_F} k\ dk\ k'\ dk'\Bigg[\sum_{i=\sigma,\omega,\rho} A_i(k,k') + \cos\chi(k)\cos\chi(k')\sum_{i=\sigma,\omega,\rho} B_i(k,k')$$
$$+\sin\chi(k)\sin\chi(k')\sum_{i=\sigma,\omega,\rho} C_i(k,k')\Bigg]\ +\ \frac{\tilde{F}_\sigma(0)}{2}\rho_s^2 + \frac{\tilde{F}_\omega(0)}{2}\rho_B^2 \tag{8}$$

where A_i, B_i, C_i , I and J are exchange integrals defined in the Appendix. In the above equation, the first term comes from the kinetic contribution, the last two terms come from the σ and ω direct contributions to the correlated potential energy, the remaining terms arise from the exchange correlation contribution to the kinetic energy, and from the meson exchange contributions to the correlated potential energy. The direct term of the correlation contribution to the kinetic energy vanishes due to (7). The baryon and the scalar densities are, respectively, ρ_B and ρ_s.

2.1. *Inclusion of pion correlations*

We next describe in more detail the treatment of pionic exchange. Starting with eq. (3), the one pion exchange potential may also be expressed as

$$V^\pi(r) = \frac{G_{\pi NN}^2}{16\pi M^2}(\vec{\tau}_1\cdot\vec{\tau}_2)\left[(\vec{\sigma}_1\cdot\hat{r})(\vec{\sigma}_2\cdot\hat{r})(\mu^2r^2+3\mu r+3) - \vec{\sigma}_1\cdot\vec{\sigma}_2(\mu r+1)\right]\frac{e^{-\mu r}}{r^3}. \tag{9}$$

In the above, $\mu = m_\pi$ is the pion mass, $\vec{r} = \vec{x} - \vec{y}$ and $r = |\vec{r}|$. Thus, the pion contribution to H is

$$\mathcal{H}_\pi = \frac{G_{\pi NN}^2\mu^2}{48\pi M^2}\int dy\Big[\psi(\vec{x})^\dagger\sigma_i\tau_a\psi(\vec{x})\psi(\vec{y})^\dagger\sigma_i\tau_a\psi(\vec{y})\ (y_0(r) - y_2(r))$$
$$+3y_2(r)\ \psi(\vec{x})^\dagger(\sigma\cdot\hat{r})\tau_a\psi(\vec{x})\psi(\vec{y})^\dagger(\sigma\cdot\hat{r})\tau_a\psi(\vec{y}).\Big] \tag{10}$$

where $y_0(r)$ and $y_2(r)$ are given by

$$y_0(r) = -\frac{e^{-\mu r}}{r},\quad y_2(r) = \left(1+\frac{3}{\mu r}+\frac{3}{(\mu r)^2}\right)\frac{e^{-\mu r}}{r}. \tag{11}$$

In order to introduce correlations we are led to replace the potential $g(r) = \frac{e^{-\mu r}}{r}$ by the potential $g_{corr}(r)$, solution of the differential equation:

$$\left(\frac{1}{r}\frac{d^2}{dr^2} - \frac{1}{r^2}\frac{d}{dr}\right)g_{corr}(r) = \frac{1}{r}U(r)f^2(r), \tag{12}$$

where $U(r)$ is given by

$$\frac{1}{r}U(r) = \left(\frac{1}{r}\frac{d^2}{dr^2} - \frac{1}{r^2}\frac{d}{dr}\right) g(r) \tag{13}$$

and $f(r)$ is the Jastrow factor. There remains one extra correction term, which may be easily determined. From (9) we have

$$f^2 V^\pi(r) = \frac{G^2_{\pi NN}}{16\pi M^2}(\vec{\tau}_1 \cdot \vec{\tau}_2)\left[V_1(r) + V_2(r)\right], \tag{14}$$

where

$$\begin{aligned} V_1(r) &= (\vec{\sigma}_1 \cdot \hat{r})(\vec{\sigma}_2 \cdot \hat{r})\left(\frac{1}{r}\frac{d^2}{dr^2} - \frac{1}{r^2}\frac{d}{dr}\right) g_{corr}(r) + \vec{\sigma}_1 \cdot \vec{\sigma}_2 \left(\frac{1}{r}\frac{d}{dr}\right) g_{corr}(r), \\ &= (\vec{\sigma}_1 \cdot \vec{\nabla})(\vec{\sigma}_2 \cdot \vec{\nabla}) g_{corr}(r) \end{aligned} \tag{15}$$

and

$$V_2(r) = \vec{\sigma}_1 \cdot \vec{\sigma}_2\, h(r), \quad h(r) = f^2\left(\frac{1}{r}\frac{d}{dr}\right) g(r) - \left(\frac{1}{r}\frac{d}{dr}\right) g_{corr}(r). \tag{16}$$

In momentum space:

$$V_1(q) = (\vec{\sigma}_1 \cdot \vec{q})(\vec{\sigma}_2 \cdot \vec{q})\tilde{g}_{corr}(q), \quad V_2(q) = \vec{\sigma}_1 \cdot \vec{\sigma}_2 \tilde{h}(q). \tag{17}$$

The function $g_{corr}(r)$ is the solution of the differential equation

$$\left(\frac{d^2}{dr^2} - \frac{1}{r}\frac{d}{dr}\right) g_{corr}(r) = f^2(3 + 3\mu r + \mu^2 r^2)\frac{e^{-\mu r}}{r^3}. \tag{18}$$

The contribution of the pion exchange potential to the correlated Hamiltonian becomes:

$$\begin{aligned} \widetilde{\mathcal{H}}_\pi = \frac{G^2_{\pi NN}}{16\pi M^2}\int dy \Big[&\psi(\vec{x})^\dagger\ (\vec{\sigma}\cdot\vec{\nabla})\tau_a\ \psi(\vec{x})\ \psi(\vec{y})^\dagger (\vec{\sigma}\cdot\vec{\nabla})\tau_a \psi(\vec{y})\ g_{corr}(r) \\ &+ \psi(\vec{x})^\dagger\ \sigma_i\ \tau_a \psi(\vec{x})\ \psi(\vec{y})^\dagger \sigma_i \tau_a \psi(\vec{y})\ h(r).\Big] \end{aligned} \tag{19}$$

The direct term of the expectation value of $\widetilde{\mathcal{H}}_\pi$ vanishes. The exchange term splits into to terms: $I^e_\pi = I^{e1}_\pi + I^{e2}_\pi$. The first one reads

$$\begin{aligned} I^{e1}_\pi = \frac{G^2_{\pi NN}}{4M^2}\frac{3\nu}{2\,V^2}\sum_{\vec{k},\vec{k}'} \tilde{g}_{corr}(\vec{k}-\vec{k}')\Big[&(\vec{k}-\vec{k}')^2\{1 + \cos\chi(\vec{k})\cos\chi(\vec{k}')\} \\ &-(2kk' - \hat{k}\cdot\hat{k}'(k^2 + k'^2))\sin\chi(\vec{k})\sin\chi(\vec{k}')\Big]. \end{aligned} \tag{20}$$

In the above the factor 3 comes from isospin. The second term reads

$$I^{e2}_\pi = \frac{G^2_{\pi NN}}{4M^2}\frac{3\nu}{2\,V^2}\sum_{\vec{k},\vec{k}'} \tilde{h}(\vec{k}-\vec{k}')\left[1 + \cos\chi(\vec{k})\cos\chi(\vec{k}') - \hat{k}\cdot\hat{k}'\ \sin\chi(\vec{k})\sin\chi(\vec{k}')\right]. \tag{21}$$

Finally the total energy contribution is

$$\mathcal{E}_{total} = \mathcal{E}_{\sigma,\omega,\rho} + I_\pi^e. \qquad (22)$$

The reported calculation involves the following approximation. The interaction $V_1(q)$, eq. (17), may be separated into a tensor term and a central term,

$$V_1(q) = \left((\vec{\sigma}_1\cdot\vec{q})(\vec{\sigma}_2\cdot\vec{q})\tilde{g}_{corr}(q) - \frac{1}{3}(\vec{\sigma}_1\cdot\vec{\sigma}_2)q^2\tilde{g}_{corr}(q)\right) + \frac{1}{3}(\vec{\sigma}_1\cdot\vec{\sigma}_2)q^2\tilde{g}_{corr}(q).$$

We have replaced $V_1(q)$ by its central part, $\frac{1}{3}(\vec{\sigma}_1\cdot\vec{\sigma}_2)q^2\tilde{g}_{corr}(q)$. This means replacing I_π^e by

$$I_\pi^e = \frac{G_{\pi NN}^2}{4M^2}\frac{3\nu}{2\,V^2}\sum_{\vec{k},\vec{k}'}(\tilde{h}(\vec{k}-\vec{k}') + \frac{1}{3}(\vec{k}-\vec{k}')^2\tilde{g}_{corr}(\vec{k}-\vec{k}'))$$
$$\times\left[1+\cos\chi(\vec{k})\cos\chi(\vec{k}') - \hat{k}\cdot\hat{k}'\,\sin\chi(\vec{k})\sin\chi(\vec{k}')\right].$$

Moreover, we have treated this term within two distinct schemes: leaving the term as it stands, and replacing in it $(\vec{k}-\vec{k}')^2\tilde{g}(\vec{k}-\vec{k}')$ by $((\vec{k}-\vec{k}')^2\tilde{g}(\vec{k}-\vec{k}') - \lim_{q\to 0} q^2\tilde{g}(q))$. The second scheme amounts to neglecting the contact term in the pion interaction.

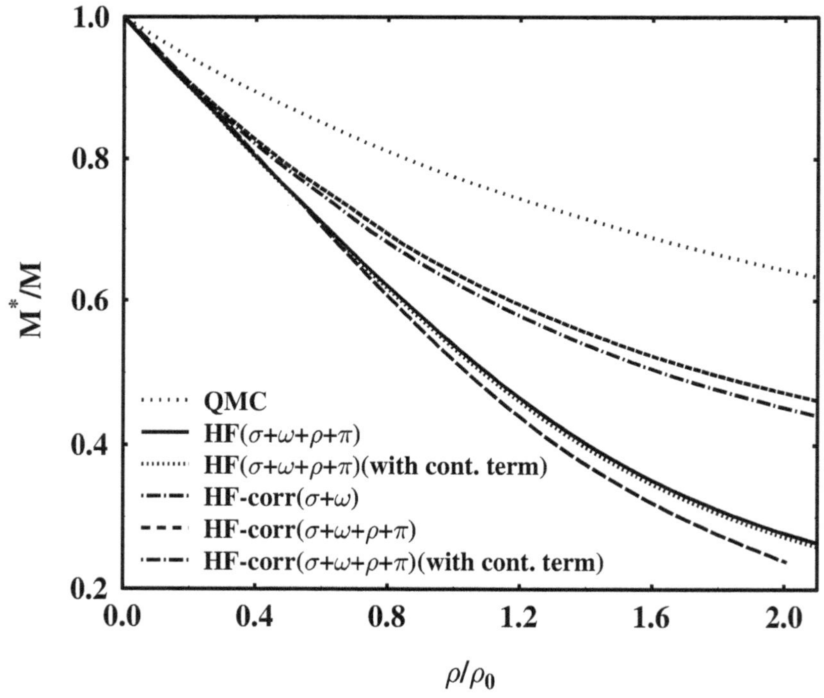

Fig. 3. Behavior of the effective mass. When the contact term (eq. (5)) is omitted, the decrease is less pronounced than for the corresponding HF result. The lowest curve is the one labeled **HF-corr** $(\sigma+\omega+\rho+\pi)$ **(with cont. term)**, but we believe that the presence of the contact term may obscure the results and requires a special interpretation, as explained in the text.

3. Results and Discussions

The models contain the following parameters: the couplings g_σ, g_ω, g_ρ, g_π, the meson masses, m_i, $i = \sigma, \omega, \rho, \pi$ and the three parameters specifying the short range correlation function, α, β and γ. The couplings g_σ and g_ω are chosen so as to reproduce the ground state properties of nuclear matter. For the ρ and π-meson couplings we take the usual values $g_\rho^2/4\pi = 0.55$ and $f_\pi^2/4\pi = 0.08$, where $f_\pi = G_{\pi NN} m_\pi/(2M)$. We choose $m_\sigma = 550$ MeV, $m_\omega = 783$ MeV, $m_\rho = 770$ MeV and $m_\pi = 138$ MeV. The normalization condition (7) determines β. We fix α variationally, by minimizing the energy. The parameter γ is such that it reproduces a reasonable healing distance, assuming that this quantity decreases as k_F increases. Therefore, we assume that γ depends on k_F according to $\gamma = a_1 + a_2\, k_F/k_{F0}$, where the parameters a_1 and a_2 are conveniently chosen. The behavior of the EOS when only the exchange of sigma and omega mesons is taken into account is shown in Fig. 2 (a) and the corresponding model parameters are displayed in Table 1. In Fig. 2 (a), the curves labeled I, II, III, correspond to different choices of the parameters a_1, a_2. The curve labeled IV refers to a model in which α was not determined variationally, but was kept fixed so as to generate a hole in the relative wave function.

Table 1. Model parameters and ground state properties of nuclear matter in the absence of π, ρ exchange at saturation density are given. These results were obtained with fixed: $M = 939$ MeV $m_\sigma = 550$ MeV, $m_\omega = 783$ MeV at $k_{f0} = 1.3$ fm^{-1} with binding energy $E_B = \varepsilon/\rho - M = -15.75$ MeV. We have used 4 choices of the parameters for the correlation. HF,corr-I corresponds to $600 + 200 k_f/k_{f0}$; HF,corr-II corresponds to $750 + 50 k_f/k_{f0}$; HF,corr-III corresponds to $700 + 200 k_f/k_{f0}$; HF,corr-IV corresponds to $\alpha = 1$ and γ fixed. Here, $\mathcal{T}'$ stands for $\mathcal{T} - M\rho_B$.

	g_σ	g_ω	α	β (MeV)	γ (MeV)	M^*/M	$\mathcal{T}'/\rho_B$ (MeV)	$\mathcal{V}_d/\rho_B$ (MeV)	$\mathcal{V}_e/\rho_B$ (MeV)
Hartree	11.08	13.81				0.540	8.11	-23.86	
HF	10.43	12.22				0.515	5.87	-37.45	15.83
HF, corr-I	4.326	2.431	13.47	-1867.	800	0.568	14.96	-96.81	26.77
HF, corr-II	4.326	2.431	13.47	-1867.	800	0.568	14.96	-96.81	26.77
HF, corr-III	4.359	2.489	13.66	-2068.	900	0.608	15.08	-83.01	23.34
HF, corr-IV	11.84	13.61	1.0	-244.7	750	0.515	7.16	-45.17	17.94

The behavior of the EOS when the exchange of all the mesons, σ, ω, ρ and π, is taken into account, is shown, in Fig. 2 (b) and the corresponding model parameters are displayed in Tables 2 and 3 . We found that short-range correlations have important consequences and should not be neglected. From Figs. 2 (a) and 2 (b), it may be seen that their main effect is the tendency to considerably soften the EOS. Moreover, as seen in Tables 1 and 2, the phenomenological values of the couplings g_σ and g_ω are strongly dependent on the correlations, being considerably reduced by their presence. When the contact term (c.t.) in the pion exchange interaction is included, the EOS appears to show a stiffer behavior, even when correlations are taken into account, but this is due to the very simplified parametrization of the adopted correlation function which was not flexible enough to respond to both repulsive and

Table 2. Parameters of nuclear matter. We have used a density dependent parameter (HF,corr) $\gamma = 600 + 400\ k_F/k_{F0}$ MeV for the correlation. These parameters were obtained with fixed: M = 939 MeV, m_σ = 550 MeV, m_ω = 783 MeV, m_ρ = 770 MeV, m_π = 138 MeV, $g_\rho^2/4\pi$ = 0.55 and $f_\pi^2/4\pi$ = 0.08 at k_{F0} = 1.3 fm^{-1} with binding energy $E_B = \varepsilon/\rho - M = -15.75$ MeV. The value of γ refers to saturation density.

	g_σ	g_ω	α	β (MeV)	γ (MeV)
Hartree	11.079	13.806			
HF	10.432	12.223			
HF, corr, σ, ω	4.4559	2.6098	13.855	-2252.448	1000
HF, corr, σ, ω, ρ, π	3.1925	2.199	13.822	-2258.037	1000
HF, corr, σ, ω, ρ, π and c. t.	11.662	12.930	1.5435	-497.391	1000

Table 3. Ground state properties of nuclear matter at saturation density.

	M^*/M	$\mathcal{T}/\rho_B - M$ (MeV)	$\mathcal{V}_d/\rho_B$ (MeV)	$\mathcal{V}_e/\rho_B$ (MeV)	$\mathcal{T}^C/\rho_B$ (MeV)	c. t. (MeV)
Hartree	0.540	8.11	-23.86			
HF	0.515	5.87	-37.45	15.83		
HF, corr, σ, ω	0.625	15.95	-73.12	20.46	19.96	
HF, corr, σ, ω, ρ, π	0.645	16.41	-17.76	-34.57	20.16	
HF, corr, σ, ω, ρ, π and c. t.	0.517	6.67	-42.30	11.11	2.53	6.21

attractive components of the interactions. In Fig. 1, we plot the correlation function when the c.t. is taken into account and when it is omitted. Short range correlations give rise to an extra node in the dependence of the ground state wave-function on the relative coordinate, contrary to what generally happens in non-relativistic calculations with hard core, when the wave function acquires a wound. The curves shown in Fig. 1 suggest that the present parametrization of $f(r)$ is not able to respond adequately to the very short range repulsion associated with the contact term. A richer parametrization, such as $f(r) = 1 - (1 + \alpha r + \beta r^2)e^{-\gamma r}$, seems to be required when the c.t. is included.

As shown in Fig. 3, if short-range two-body correlations are included and the c. t. in the pion exchange interaction is neglected, the effective mass does not decrease so fast with the increase of density as in the Hartree and HF calculations, which explains the softer behavior of the EOS with correlations. The faster decrease of the mass found when the c.t. is not neglected, may be attributed to an insufficient flexibility of the correlation factor. Correlations increase the kinetic energy to the order of magnitude one expects from non-relativistic approaches. This is an important effect since the Hartree and, even more dramatically, the HF approximation, predict an unrealistically low value of the kinetic contribution to the total energy at saturation, as shown in Table 2.

The EOS of neutron matter, which requires the inclusion of σ, ω, ρ and π mesons for its description, is strongly affected by correlations. The presence of the c.t. makes

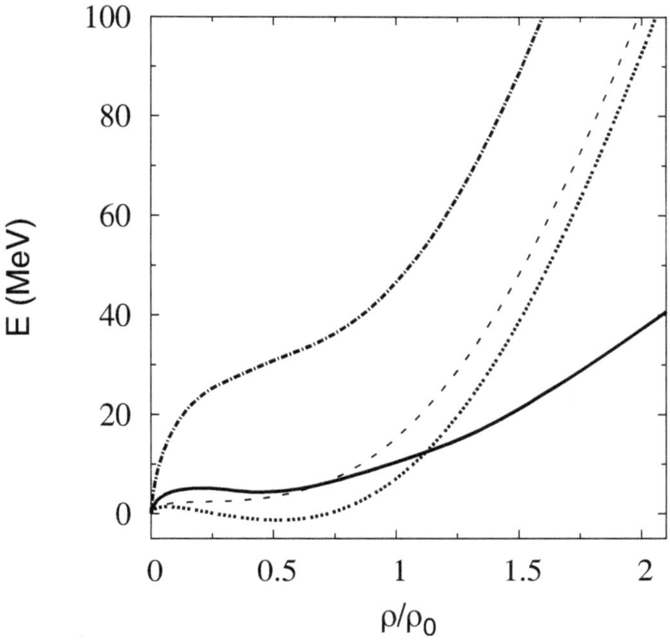

Fig. 4. Equation of state of neutron matter (solid line: $\sigma+\omega+\rho+\pi+$correlation (without delta), dash-dotted line: $\sigma+\omega+\rho+\pi+$ correlation (with delta), dotted line: $\sigma+\omega+\rho+\pi$ (without delta), dashed line: $\sigma+\omega+\rho+\pi$ (with delta).

the EOS of neutron matter very hard, but this is because a correlation function has been used which was not flexible enough. As shown in Fig. 4, the EOS of neutron matter without the c.t. shows binding in the HF calculation and a shallow minimum in HF plus correlations, which is not in consonance with neutron star observation data requiring zero density at surface. This is not possible if a minimum at finite density occurs in neutron matter EOS. Correlations almost lift this undesirable behavior, typical of the Walecka type models, which seems to require a proper treatment of the pion contribution, for its solution.

Appendix

The angular integrals in (8) are given by

$$A_i(k,k') = B_i(k,k') = 2\pi\,\frac{g_i^2}{4\pi}\int_0^{\pi} d\cos\theta\;\tilde{F}_i(k,k',\cos\theta),$$

$$C_i(k,k') = 2\pi\,\frac{g_i^2}{4\pi}\int_0^{\pi}\cos\theta\;d\cos\theta\;\tilde{F}_i(k,k',\cos\theta),$$

$$I(k,k') = 2\pi\int_0^{\pi} d\cos\theta\;\tilde{C}_1(k,k',\cos\theta),$$

$$J(k,k') = 2\pi \int_0^\pi \cos\theta \; d\cos\theta \; \tilde{C}_1(k,k',\cos\theta),$$

$$\tilde{F}_i(\vec{k},\vec{k}') = \int [f(r)V_i(r)f(r)] \; e^{i(\vec{k}-\vec{k}')\cdot\vec{r}} \; d\vec{r},$$

$$\tilde{C}_1(\vec{k},\vec{k}') = \int (f^2(r)-1) \; e^{i(\vec{k}-\vec{k}')\cdot\vec{r}} \; d\vec{r}.$$

Acknowledgments

Valuable discussions with S. Moszkowski are gratefully acknowledged. This work was partially supported by FCT (Portugal) under the projects POCI/FP/63918/2005 and POCI/FP/63419/2005.

References

1. F. Coester, Nucl. Phys. **7**, 421 (1958); F. Coester and H. Kümmel, Nucl. Phys. **17**, 477 (1960).
2. R. F. Bishop and H. Kümmel, Phys. Today **40(3)**, 95 (1991); R. F. Bishop, Theor. Chem. Acta **80**, 95 (1991).
3. F. Villars, "Proceedings of the International School of Physics, 'Enrico Fermi'-Course 23, (1961)." Academic Press, New York, 1963; J.S. Bell, "Lectures on the Many-Body Problem, First Bergen International School of Physics." Benjamin, New York, (1962).
4. J. da Providência and C.M. Shakin, Ann. Phys.(NY) **30**, 95 (1964).
5. H. Feldmeier, T. Neff, R. Roth, and J. Schnack, Nucl. Phys. **A 632**, 61 (1998); T. Neff and H. Feldmeier, Nucl. Phys. **713**, 311 (2003).
6. F. Coester, S. Cohen, B.D. Day, and C.M. Vincent, Phys. Rev. **C 1**, 769 (1970); R. Brockmann and R. Machleidt, Phys. Rev. **C 42**, 1965 (1990).
7. R.B. Wiringa, V. Fiks and A. Fabrocini, Phys. Rev. **C 38**, 1010 (1988); W. Zuo, A. Lejeune, U. Lombardo and J.-F. Mathiot, Nucl. Phys. **A 706**, 418 (2002).
8. B.D. Serot, J.D. Walecka, Int. J. Mod. Phys. **E6**, 515 (1997).
9. V.G. Neudatchin, I.T. Obukhovsky, V.I. Kukulin and N.F. Golovanova, Phys. Rec. **C 11**, 128 (1975).
10. S. A. Moszkowski, Proceedings online of the Conference on Microscopic Approaches to Many-Body Theory (MAMBT), in honor of Ray Bishop, Manchester, UK, http://www.qmbt.org/MAMBT/pdf/Moszkowski.pdf, 2005.
11. P.K. Panda, D.P. Menezes, C. Providência and J. da Providência, Phys. Rev. **C 71**, 015801 (2005); P.K. Panda, D.P. Menezes, C. Providência and J. da Providência, Braz. J. Phys **35**, 873 (2005).
12. P.K. Panda, J. da Providência, C. Providência and D.P. Menezes, Proceedings online of the Conference on Microscopic Approaches to Many-Body Theory (MAMBT), in honor of Ray Bishop, Manchester, UK, http://www.qmbt.org/MAMBT/pdf/Providencia.pdf, 2005.
13. P.K. Panda, C. Providência and J. da Providência, Phys. Rev. **C 73**, 035805 (2006).
14. A. Bouyssy, J.-F. Mathiot, N.Van Giai and S. Marcos, Phys. Rev. **C 36**, 380 (1987).
15. A. Mishra, P.K. Panda, S. Schramm, J. Reinhardt and W. Greiner, Phys. Rev. **C 56**, 1380 (1997).
16. J. da Providencia and C. M. Shakin, Phys. Rev **C 4**, 1560 (1971); C. M. Shakin, Phys. Rev. **C 4**, 684 (1971).

17. R. Jastrow, Phys. Rev. **98**, 1479 (1955).
18. S.A. Moszkowski and B.L. Scott, Ann. Phys. (N.Y.) **11**, 65 (1960).
19. H. Bethe, Ann. Rev. Nucl. Sci. **21**, 93 (1971).
20. P. A. M. Guichon, Phys. Lett. **B 200**, 235 (1988); K. Saito and A.W. Thomas, Phys. Lett. B **327**, 9 (1994); P.K. Panda, A. Mishra, J.M. Eisenberg, W. Greiner, Phys. Rev. **C 56**, 3134 (1997).
21. G.A. Lalazissis, J. König, P. Ring, Phys. Rev. **C55**, 540 (1997).

QUARTETTING IN ATTRACTIVE FERMI-SYSTEMS AND ALPHA PARTICLE CONDENSATION IN NUCLEAR SYSTEMS

P. SCHUCK

Institut de Physique Nucléaire, 91406 Orsay Cedex, France

Y. FUNAKI

The Institute of Physical and Chemical Research (RIKEN), Wako, Saitama 351-0198, Japan

H. HORIUCHI

Research Center for Nuclear Physics (RCNP), Osaka University, Ibaraki, Osaka 567-0047, Japan

G. RÖPKE

Institut für Physik, Universität Rostock, D-18051 Rostock, Germany

A. TOHSAKI

Research Center for Nuclear Physics (RCNP), Osaka University, Ibaraki, Osaka 567-0047, Japan

T. YAMADA

Laboratory of Physics, Kanto Gakuin University, Yokohama 236-8501, Japan

The famous Hoyle state ($0_2{}^+$ at 7.654 MeV in ^{12}C) is identified as being an almost ideal condensate of three α-particles, hold together only by the Coulomb barrier. It, therefore, has a ^{8}Be-α structure of low density. Transition probability and inelastic form factor together with position and other physical quantities are correctly reproduced without any adjustable parameter from our two parameter wave function of α-particle condensate type. The possibility of the existence of α-particle condensed states in heavier $n\alpha$ nuclei is also discussed.

Keywords: Alpha particle condensation; Hoyle state; quartetting.

1. Introduction

Quantum condensation of particles is one of the most amazing phenomena of many body systems. Striking well known examples are superconducting metals and superfluid ^{4}He. Also nuclei are superfluid. However, in nuclei the most tightly bound cluster is not a pair but a quartet. Therefore, what about α-particle condensation in nuclei? The only nucleus which in its ground state has a pronounced α-cluster structure is ^{8}Be. In Fig. 1(a) we show the result of an exact calculation with a realistic N-N interaction for the density distribution in the laboratory frame, whereas in Fig. 1(b) we see the same in the intrinsic, deformed frame where in addition the question has been asked where to find the second α-particle when the first is placed at a given position. So we see that the two α's are $\sim$ 4 fm apart giving raise to a very low average density $\rho \sim \rho_0/3$ as seen on Fig. 1(a) where ρ_0 is the nuclear saturation density. ^{8}Be is also a very large object with an rms radius of $\sim$ 3.7 fm to be compared with the nuclear systematics of $R = r_0 A^{1/3} \sim 2.44$ fm. Definitely ^{8}Be is a rather unusual nucleus. One may ask the question what happens when one brings a third α-particle alongside of ^{8}Be. We know the answer: the 3-α system collapses to the ground state of ^{12}C which is much denser than ^{8}Be and cannot accommodate with its small rms radius of 2.4 fm three α-particles barely touching one another. One nevertheless may ask the question whether the dilute three α configuration ^{8}Be-α may not form an isomeric or excited state of ^{12}C. This will be the main subject of our considerations.

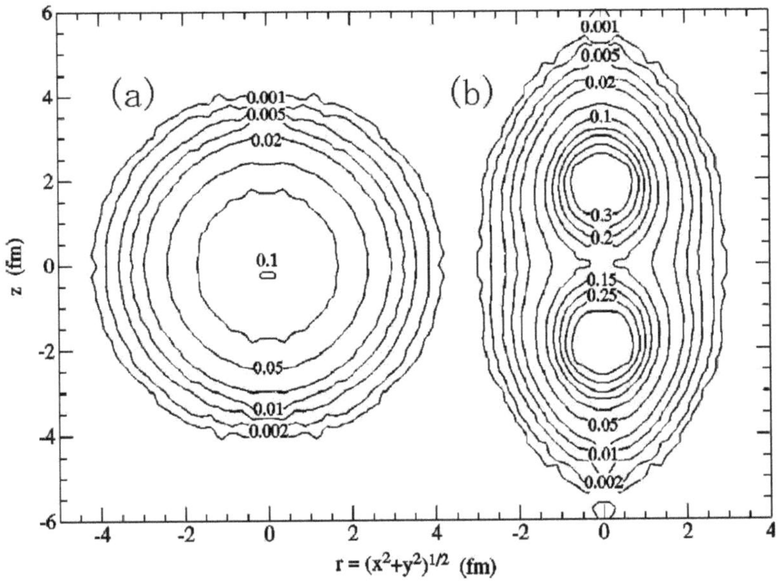

Fig. 1. Contours of constant density (taken from[1]), plotted in cylindrical coordinates, for ^{8}Be(0^+). The left side (a) is in the "laboratory" frame while the right side (b) is in the intrinsic frame.

2. α Condensate States in Self-Conjugate $4n$ Nuclei

We now will show that α-particle condensation most likely exists in $n\alpha$ nuclei around energies of α-particle break up thresholds. We want to give the demonstration here that, due to the existence of ample experimental data, we have identified at least one nucleus where such an α-particle condensed state exists and then we discuss the indications and the likelyhood that such states very naturally also are present in other nuclei and that it may be a quite general phenomenon in nuclear systems. The nucleus we want to draw our attention to is ^{12}C. We, indeed, will give strong arguments that the 0_2^+ state at 7.654 MeV in ^{12}C is a state of α-particle condensate nature.

First, it should be noticed that the 0_2^+ state in ^{12}C is actually, as ^{8}Be, unstable and situated about 300 keV above the three α-break up threshold.
This state only is stabilised by the Coulomb barrier. It has a width of 8.7 eV and a corresponding lifetime of 7.6×10^{-17} s. As well known, this state is of extreme astrophysical importance concerning the synthesis of ^{12}C in the universe and its existence was predicted in 1953 by the astrophysicist F. Hoyle [2] and shortly after discovered by W. A. Fowler in 1957 .[3] It is also well known that this Hoyle state, as it is called now, is a notoriously difficult state for any nuclear theory and for example the most modern no-core shell model calculations predict the 0_2^+ state in ^{12}C to occur at around 17 MeV, that is more than two times its actual value.[4] This fact alone may tell us that the Hoyle state must have a very unusual structure and it is easy to understand that, should it indeed have a loosely bound three α-particle structure, a shell model type of calculation would have great difficulties to reproduce its properties. However, about 30 years ago, two Japanese physicists, M. Kamimura[5] and K. Uegaki,[6] with their collaborators, almost simultaneously achieved to reproduce the Hoyle state from a microscopic theory, i.e. employing a twelve nucleon wavefunction together with a Hamiltonian containing an effective nucleon-nucleon interaction. Their works, at that time, probably did not attract the wide spread attention they desserved and the true importance of their achievement is eventually only recognised now. Both authors started from practically the same ansatz for the ^{12}C wavefunction, having the following three α-cluster structure: $\langle \vec{r}_1 ... \vec{r}_{12} | ^{12}\mathrm{C} \rangle = \mathcal{A}[\chi(\vec{\mathcal{R}}, \vec{s}) \phi_1 \phi_2 \phi_3]$ where ϕ_i is an intrinsic α-particle wavefunction of prescribed Gaussian form, i.e. $\phi = \exp([(\vec{r}_1 - \vec{r}_2)^2 + (\vec{r}_1 - \vec{r}_3)^2 + ...]/b^2)$ where the size parameter is adjusted to get the rms value of the free α-particle radius right. $\chi(\vec{\mathcal{R}}, \vec{s})$ is the yet to be determined three body wavefunction for the center of mass motion of the three α's with $\vec{\mathcal{R}}$ and $\vec{s}$ the corresponding Jakobi coordinates. The unknown function χ was then determined via a GCM[6] and RGM[5] calculation using the Volkov I and Volkov II nucleon-nucleon forces which well reproduce α-α phase shifts. The precise solution of this complicated three body problem was, 30 years back, a truely pioneering work. The results were up to expectation. The position of the Hoyle state as well as other properties like inelastic form factor and transition probability successfully reproduced the experimental data. Other states of ^{12}C below and around the energy of the Hoyle state were also successfully described. It also was already recognised that the three α's in the Hoyle state form sort of gas like state, a feature which had already been pointed out by H. Horiuchi[7] prior to the works of[5] and[6] using the orthogonality condition model (OCM).[8] All of these authors also

concluded from their studies that the linear chain state of three α-particles, as this was postulated by Morinaga many years back,[9] had to be excluded.

Though, as already said, the afore mentioned authors all had already stressed the somewhat α-gas like nature of the Hoyle state, eventually two important aspects were missed at that time. First comes the fact that, because all three α's move in identical S-wave orbits, this forms an α-condensate state, albeit not in the macroscopic sense. Second is that the complicated three body wave function can be replaced by a structurally and conceptually very simple microscopic three α wave function of the condensate type which has practically 100 percent overlap with the previous ones.[10] We now shortly want to describe this condensate wave function.

For this we make an analogy to the Cooper pair BCS wave function of ordinary pairing. The latter wave function can be written in position space as

$$\langle \vec{r}_1 ... \vec{r}_N | \text{BCS} \rangle = \mathcal{A}[\phi(\vec{r}_1, \vec{r}_2)\phi(\vec{r}_3, \vec{r}_4)...\phi(\vec{r}_{N-1}, \vec{r}_N)] \tag{1}$$

where $\phi(\vec{r}_1, \vec{r}_2)$ is the Cooper pair wave function , including spin and isospin, which is being determined variationally by the well known BCS equations. As before $\mathcal{A}$ is the antisymmetriser. The condensate character of the BCS ansatz is born out by the fact that we have a product of $N/2$ times the same pair wave function ϕ. Formally it now is a simple matter to generalise (1) to α-particle condensation. We write

$$\langle \vec{r}_1 ... \vec{r}_N | \Phi_{n\alpha} \rangle = \mathcal{A}[\phi_\alpha(\vec{r}_1, \vec{r}_2, \vec{r}_3, \vec{r}_4)\phi_\alpha(\vec{r}_5, .., \vec{r}_8)...\phi_\alpha(\vec{r}_{N-3}, .., \vec{r}_N)] \tag{2}$$

where ϕ_α is the wave function common to all condensed α-particles. Of course, in general, the variational solution for $\phi_\alpha(\vec{r}_1, .., \vec{r}_4)$ from (2) is extraordinarily more complicated than to find the Cooper pair wave function of (1). However, in the case of the α-particle and for relatively light nuclei, the complexity of the problem can be reduced dramatically. This stems from the fact that, as was already recognised by the authors in,[5,6] an intrinsic wave function of the α-particle of Gaussian form with only the size parameter b to be determined , is an excellent variational ansatz (see above). In addition, and here resides the essential and crucial novelty of our wave function, even the center of mass motion of the various α-particles can very well be described by a Gaussian wave function with, this time, a size parameter $B \gg b$ to account for the motion over the whole nucleus. We therefore write

$$\phi_\alpha(\vec{r}_1, \vec{r}_2, \vec{r}_3, \vec{r}_4) = e^{-\vec{R}^2/B^2}\phi(\vec{r}_1 - \vec{r}_2, \vec{r}_1 - \vec{r}_3, ...) \tag{3}$$

where $\vec{R} = (\vec{r}_1 + \vec{r}_2 + \vec{r}_3 + \vec{r}_4)/4$ is the c.m. coordinate of one α-particle and $\phi(\vec{r}_1 - \vec{r}_2, ...)$ is the same intrinsic α-particle wave function of Gaussian form as already used in[5,6] and written out above. Of course, in (2) the center of mass $\vec{X}_{cm}$ of the three α's, i.e. of the whole nucleus, should also be eliminated what is easily achieved by replacing $\vec{R}$ by $\vec{R} - \vec{X}_{cm}$ in each of the α wave functions in (2). The α-particle condensate wave function (2) with (3), proposed in[11] and henceforth called THSR-wavefunction, now depends only on two parameters, B and b. The expectation value of the microscopic Hamiltonian

$$\mathcal{H}(B,b) = \langle \Phi_{n\alpha}(B,b)|H|\Phi_{n\alpha}(B,b)\rangle / \langle \Phi_{n\alpha}|\Phi_{n\alpha}\rangle \tag{4}$$

can be evaluated and the corresponding two dimensional energy surface quantised in using the two parameters B and b as Hill-Wheeler coordinates.

Before coming to the results, let us discuss the THSR- wave function a little more. The inocuously looking ansatz (2) with (3) is actually more subtle as it might seem. One should realise that it contains two limits exactly: if $B = b$, then (2) boils down to a standard Slater determinant with harmonic oscillator wave functions with oscillator length b as the single variational parameter. This holds because (3), with $B = b$, becomes a product of four identical Gaussians and the antisymmetrisation creates all the necessary P, D, etc. harmonic oscillator wave functions automatically.[11] On the contrary, when $B \gg b$, the density of α-particles is very low and in the limit $B \to \infty$, the average distance between α-particles is so large that the antisymmetrisation between α's can be neglected, i.e. the operator '$\mathcal{A}$' in front of (2) can be taken off. Our wave function then becomes an ideal gas of independent condensed α-particles, i.e. a pure product state of α's! On the other hand, in realistic cases, the antisymmetriser $\mathcal{A}$ can not be neglected and the evaluation of the expectation values in (4) becomes an analytical (but non-trivial) task. For the Hamiltonian in (4) we took the one of[12] with an effective nucleon-nucleon force of the Gogny type whose parameters have been adjusted to α-α scattering phase shifts about 15 years back. It also leads to very reasonable properties of nuclear matter. Our theory is therefore free of any adjustable parameter. The energy land scapes $\mathcal{H}(B,b)$ for various $n\alpha$ nuclei are interesting by themselves[13] but for brevity not shown here.

3. Results for Finite Nuclei

As already mentioned above, the wave function constructed from the Hill-Wheeler equation based on (2),(3),(4), has practically 100 percent overlap with the ones in,[5,6] once the same Volkov force is used.[10] It is, therefore, not astonishing that we also get very similar results to theirs. For ^{12}C we obtain two eigenvalues: the ground state and the Hoyle state. Theoretical values for positions, rms values, transition probabilities, compared to the data, are given in Table 1. From the comparison of the rms radii we see that the volume of the Hoyle state is a factor 3 to 4 larger than the one of the ground state of ^{12}C. This is the aspect of dilute gas state we were talking about above. Constructing an α-particle density matrix $\rho(\vec{R},\vec{R}')$, in integrating out of the total density matrix all intrinsic α-particle coordinates, we find in diagonalising this density matrix that the corresponding $0S$ α-particle orbit is occupied to 70 percent by the three α-particles.[14,15] This is a huge percentage, underlining the almost ideal α-particle condensate aspect of the Hoyle state. In this regard one should remember that superfluid ^{4}He has only 10 percent of the particles in the condensate! Let us also mention that for the ground state of ^{12}C the α-particle occupations are equally shared between $0S$, $0D$, and $0G$ orbits, thus invalidating a condensate picture for the ground state. Please notice that also the ground state energy of ^{12}C is reasonably reproduced by our theory.

Table 1. Comparison of the binding energies, rms radii ($R_{\rm r.m.s.}$), and monopole matrix elements ($M(0_2^+ \to 0_1^+)$) for ^{12}C given by solving Hill-Wheeler equation based on (2) and by RGM.[5] Volkov No. 2 force as the effective two-nucleon force is adopted in the two cases, for which the 3α threshold energy is calculated to be -82.04 MeV.

		condensate w. f. (H. W.)	RGM[5]	Exp.
E (MeV)	0_1^+	-89.52	-89.4	-92.2
	0_2^+	-81.79	-81.7	-84.6
$R_{\rm r.m.s.}$ (fm)	0_1^+	2.40	2.40	2.44
	0_2^+	3.83	3.47	
$M(0_2^+ \to 0_1^+)$ (fm^2)		6.45	6.7	5.4

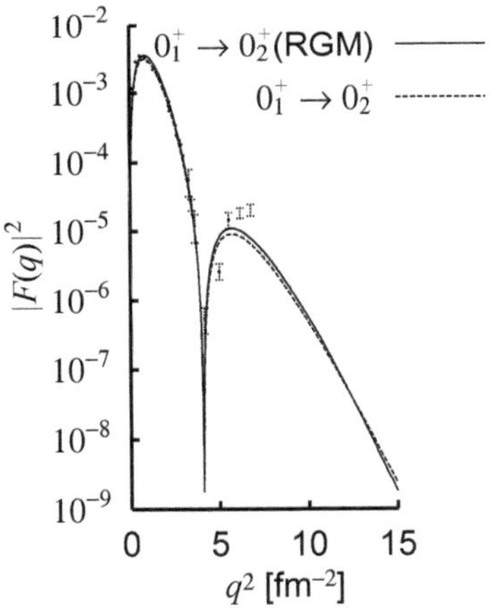

Fig. 2. Experimental values of inelastic form factor in ^{12}C to the Hoyle state[16] are compared with our values and those given by Kamimura et al. in[5] (RGM). In our result, the Hoyle state wave function is obtained by solving the Hill-Wheeler equation based on (2).

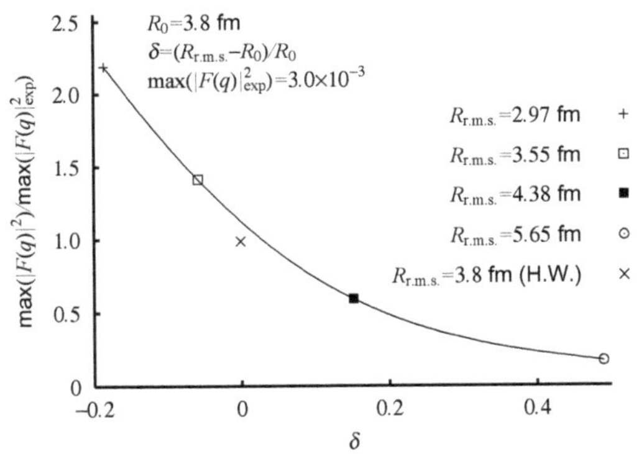

Fig. 3. The ratio of the value of maximum height, theory versus experiment, for the inelastic form factor, i.e. $\max|F(q)|^2$ / $\max|F(q)|^2_{\rm exp}$, is plotted as a function of δ, which is defined as δ=$(R_{\rm r.m.s.} - R_0)/R_0$. $R_{\rm r.m.s.}$ and R_0 are the rms radii corresponding to the wave function (2) and the one obtained by solving the Hill-Wheeler equation based on (2), respectively.

Let us now discuss the, to our mind, most convincing feature that our description of the Hoyle state is the correct one. As the authors of,[5] we reproduce very accurately the inelastic form factor $0_1^+ \to 0_2^+$ of ^{12}C. This is shown in Fig. 2. The agreement with experiment is as such already quite impressive. Additionally, however, we made the following study shown in Fig. 3. We artificially varied the extension of the Hoyle state and studied the influence on the form factor. We found that the overall shape of the form factor only varies little, for instance in what concerns the position of the minimum. On the contrary, we found a strong dependence on the absolute magnitude of the form factor and in Fig. 3 we plot the variation of the height of the first maximum of the inelastic form factor as a function of the percentage change of the rms radius of the Hoyle state.[17] It can be seen that a 20 percent increase of the rms radius decreases the maximum by a factor of two! This strong dependence of the magnitude of the form factor makes us firmly believe that the agreement with the actual measurement is practically a proof that our calculated wide extension of the Hoyle state corresponds to reality.

The Hoyle state can be considered as the ground state of the α-particle condensate. Exciting one α-particle out of the condensate and putting it into the $0D$ orbit reproduces the experimentally measured position of the 2_2^+ state in ^{12}C. Without going into details, we also state that the width of this state is correctly reproduced.[18] It is tempting to imagine that the 0_3^+ state which experimentally is almost degenerate with the 2_2^+ state is obtained by lifting one α-particle into the $1S$ orbit. First theoretical studies[19] indicate that this view might indeed be true. However, its width ($\sim$ 3MeV) is very broad what makes a theoreticla treatment rather delicate and further investigations are necessary to validate this picture. At any rate, it would be very satisfying, if the the triplet of states, i.e. $0_2{}^+, 2_2{}^+, 0_3{}^+$, could all be explained from the α-particle point of view, since those three states are precisely the ones which can not be explained within a (no core) shell model approach.[4]

In conclusion, in what concerns ^{12}C, we think we have accumualted enough facts to become convinced that the Hoyle state is, indeed, what one can call an α-particle condensate state, being aware of the fact that 'condensate' for only three particles constitutes a certain abuse of the word. We, however, should remember in this context that also in the case of nuclear Cooper pairing, only a few Cooper pairs are sufficient to obtain clear signatures of superfluidity in nuclei!

What about α-particle condensation in heavier nuclei? Of course, once one accepts the idea that the Hoyle state is essentially a state of three free α-particles hold together only by the Coulomb barrier, it is hard to believe that analogous states should not also exist in heavier $n\alpha$ nuclei like ^{16}O, ^{20}Ne, ^{24}Mg, ... At least our calculations systematically always show a 0^+-state close to the α-particle disintegration threshold. For example in ^{16}O we obtain three 0^+-states: the ground state at $E_0 = -124.8$ MeV (experimental value: -127.62 MeV), a second state at $E_{0_2{}^+} = 8.8$ MeV and a third one at $0_3{}^+ = 14.1$ MeV. The threshold in ^{16}O is at 14.4 MeV. Unfortunately the experimental situation in ^{16}O is by far not so complete as the one in ^{12}C. For example no transition probability measurement of 0^+-states around threshold in ^{16}O nor inelastic form factors do exist. Recently Wakasa[20] identified a new 0^+-state at 13.5 MeV in ^{16}O which is the 5-th 0^+-state. There are indications that it might be the α-condensate state.[21]

An intersting question is how many α's can maximally be in a self bound α-gas state. For answering this question, a schematic investigation using an effective α-α interaction in an α-gas mean field calculation of the Gross-Pitaevsky type was performed. Our estimate yields[22] a maximum of ten α-particles which can be held together in a condensate. However, a couple of extra neutrons may stabilise larger condensates.

Another interesting idea concerning α-particle condensates was put forward by von Oertzen and collaborators.[23,24] Adding more and more α-particles to e.g. the ^{40}Ca core, one sooner or later will arrive at the α-particle drip. Therefore it may need little further excitation energy to shake loose further α-particles, so that an $n\alpha$-condensate could be created on top of an inert ^{40}Ca core. Similar ideas also have been advanced by Ogloblin[25] who imagines a three α-particle condensate on top of ^{100}Sn and earlier by Brenner and Gridnev who think having detected gaseous α-particles in ^{28}Si and ^{32}S on top of an inert ^{16}O core.[26]

4. Conclusion

In conclusion, we see that the idea of α-particle condensation in nuclei has already triggered a lot of new works and ideas inspite of the fact that so far strong identification of such a state only exists in ^{12}C. However, the possibility of the existence of a completely new nuclear phase where α-particles play the role of quasi-elementary constituents is surely fascinating and hopefully many more α-particle states will be detected in the near future.

References

1. R. B. Wringa, S. C. Pieper, J. Carlson, and V. R. Pandharipande, Phys. Rev. **C 62**, 014001 (2000).
2. F. Hoyle, D. N. F. Dunbar, W. A. Wenzel, W. Whaling, Phys. Rev. 92, 1095 (1953).
3. C. W. Cook, W. A. Fowler, C. C. Lauritsen, T. B. Lauritesen, Phys. Rev. 107, 508 (1957).
4. P. Navrátil, J. P. Vary, and B. R. Barrett, Phys. Rev. Lett. **84**, 5728 (2000); Phys. Rev. C **62**, 054311 (2000); B. R. Barrett, B. Mihaila, S. C. Pieper, and R. B. Wiringa, Nucl. Phys. News, **13**, 17 (2003).
5. Y. Fukushima and M. Kamimura, *Proc. Int. Conf. on Nuclear Structure*, Tokyo, 1977, ed. T. Marumori (Suppl. of J. Phys. Soc. Japan, **44**, 225 (1978)); M. Kamimura, Nucl. Phys. **A 351**, 456 (1981).
6. E. Uegaki, S. Okabe, Y. Abe, and H. Tanaka, Prog. Theor. Phys. **57**, 1262 (1977); E. Uegaki, Y. Abe, S. Okabe, and H. Tanaka, Prog. Theor. Phys. **59**, 1031 (1978); **62**, 1621 (1979).
7. H. Horiuchi, Prog. Theor. Phys. **51**, 1266 (1974); **53**, 447 (1975).
8. S. Saito, Prog. Theor. Phys. **40** (1968); **41**, 705 (1969); Prog. Theor. Phys. Suppl. **62**, 11 (1977).
9. H. Morinaga, Phys. Rev. **101**, 254 (1956); Phys. Lett. **21**, 78 (1966).
10. Y. Funaki, A. Tohsaki, H. Horiuchi, P. Schuck, and G. Röpke, Phys. Rev. C **67**, 051306(R) (2003).
11. A. Tohsaki, H. Horiuchi, P. Schuck, and G. Röpke, Phys. Rev. Lett. **87**, 192501 (2001).
12. A. Tohsaki, Phys. Rev. C 49, 1814 (1994).
13. A. Tosaki, H. Horiuchi, P. Schuck, and G. Röpke, *Proc. of the 8th Int. Conf. on Clustering Aspects of Nuclear Structure and Dynamics*, Nara, Japan, 2003, ed. K. Ikeda, I. Tanihata and H. Horiuchi (Nucl. Phys. **A 738**, 259 (2004)).

14. T. Yamada, P. Schuck, Eur. Phys. J. A, 26, 185 (2005).
15. H. Matsumura and Y. Suzuki, Nucl. Phys. **A 739**, 238 (2004).
16. I. Sick and J. S. McCarthy, Nucl. Phys. **A 150**, 631 (1970); A. Nakada, Y. Torizuka and Y. Horikawa, Phys. Rev. Lett. **27**, 745 (1971); and 1102 (Erratum); P. Strehl and Th. H. Schucan, Phys. Lett. **27B**, 641 (1968).
17. Y. Funaki, A. Tohsaki, H. Horiuchi, P. Schuck and G. Röpke, Eur. Phys. J. A **28**, 259 (2006).
18. Y. Funaki, H. Horiuchi, A. Tohsaki, P. Schuck and G. Röpke, Eur. Phys. J. A, 24, 321 (2005).
19. C. Kurakowa, K. Kato, Phys. Rev. C 71, 021301 (2005).
20. T. Wakasa, private communication.
See also http://www.rcnp.osaka-u.ac.jp/ annurep/2002/sec1/wakasa2.pdf
21. Y. Funaki et al, RCNP workshop, April, 2006, Osaka, to appear in Mod. Phys. Lett. A, World Scientific.
22. T. Yamada, P. Schuck, Phys. Rev. C **69**, 024309 (2004).
23. Tz. Kokalova, N. Itagaki, W. von Oertzen, and C. Wheldon, Phys. Rev. Lett. **96**, 192502 (2006).
24. W. von Oertzen et al, to appear in Eur. Phys. J. A.
25. A. A. Ogloblin et al, Proceedings of the International Nuclear Physics Conference, Peterhof, Russia, June 28-July 2, 2005.
26. M. W. Brenner et al, Proceedings of the International Conference " Clustering Phenomena in Nuclear Physics", St. Petersburg, published in 'Physics of Atomic Nuclei (Yadernaya Fizika), 2000.

ALPHA-ALPHA AND ALPHA-NUCLEUS POTENTIALS: AN ENERGY-DENSITY FUNCTIONAL APPROACH

Z. F. SHEHADEH*

Physics Department, Taif Teachers College,
Taif, P.O.Box 1070, Saudi Arabia
**E-mail: z_shehadeh@yahoo.com*
www.ttc.edu.sa

A. K. BASAK, M. N. A. ABDULLAH and M. A. UDDIN

Physics Department, Rajshahi University,
Rajshahi 6205, Bangladesh

I. REICHSTEIN

School of computer Science, Carleton University,
Ottawa, Ontario K1S 5B6, Canada

M. S. SABRA and F. B. MALIK

Physics Department, Southern Illinois University,
Carbondale, Illinois 62901, USA

The real parts of the alpha-alpha and alpha-nucleus potential, particularly for ^{40,44,48}Ca and ^{58}Ni targets are determined from a realistic two-nucleon potential using an energy-density functional approach and are found to be non-monotonic with a short range repulsion in their functional forms that are similar to those needed in accounting for elastic scattering data. The deduced alpha-alpha potential has no bound states and the calculated decay width of ^{8}Be of about 6.4 eV at about 200 keV energy is in agreement with the measurement.

Keywords: Alpha-nucleus potential.

1. Introduction

Although alpha-ray was discovered over a century ago and identified to be doubly charged helium atom for nearly a hundred years[1,2], the actual nature of alpha-alpha and alpha-nucleus potential, particularly for light and medium-light nuclear targets remains incomplete. Recently, however, significant progress has been made in this direction both phenomenologically and from the fundamental aspect, relating these potentials to basic two-nucleon interaction using energy-density functional, noted henceforth as EDF, approach, by the groups of Basak in Bangladesh, Reichstein in Canada and ours in the U.S.A. In this lecture, dedicated to the celebration of the first centennial of the discovery of alpha particles, I present a gist of this endeavor.

The puzzle of alpha-nucleus potential started with the experiment of Corelli[3] *et al.* who measured unexpected large cross section, termed ALAS, of elastically scattered alpha particles by a number of nuclei, e.g. ^{28}Si, ^{32}S, which could not be analyzed by the standard optical model. This feature has subsequently been detected for many light nuclear targets as noted in ref. 4. Phenomenologically, two groups of local potentials, one proposed by Michel and his collaborators[5,6] which has a deep attractive real part having functional form of the squared Woods-Saxon potential, noted henceforth as the Michel potential, and the other by Malik and his group, a complex molecular potential with a non-monotonic real part, termed henceforth as NMS, have been reasonably successful in accounting for the data in many cases. Attempts to relate Michel potentials to two-nucleon potential in terms of folding model (FM) have met with a limited success. Quite often the potentials derived from FM need renormalization and the density distribution of alpha-particle used in folding procedure may not reproduce adequately its observed binding energy. Furthermore, in many particle-transfer reactions, e.g. ^{28,29,30}Si(α,d)^{30,31,32}P and ^{28}Si(α,p)^{31}P, the data are better accounted for by NMS potentials rather than Michel potentials in the entrance channel[7–9]. The general forms of these NMS potentials can be derived from realistic two-nucleon interactions within the context of the EDF approach outlined in the next section. In section 3, we present the results for a number of targets with a particular emphasis on alpha-alpha interaction at low energies.

2. Energy-Density Functional Theory

The EDF theory for nuclear reaction is based on the Kohn-Hohenberg theorem[10] which states that the total energy, E, of a system of interacting fermions can always be expressed as a functional of its energy-density, $\varepsilon[\rho(r)]$,

$$E = \int \varepsilon[\rho(r)]\rho(r)d^3r. \tag{1}$$

The energy density of nucleonic matter may be written as [11,12]

$$\varepsilon[\rho(r)] = T(\rho) + V_m(\rho) + V_g(\rho) + V_c(\rho). \tag{2}$$

In (2), the contribution of the kinetic energy to the energy-density, $T(\rho)$ for a nucleonic matter with Z protons and N neutrons can be calculated using the statistical theory and is given by

$$T(\rho) = (\frac{3}{5})(\frac{\hbar^2}{2M})(\frac{3\pi^2}{2})^{\frac{2}{3}}(\frac{1}{2})[(1-\alpha)^{\frac{5}{3}} + (1+\alpha)^{\frac{5}{3}}]\rho^{\frac{2}{3}}. \tag{3}$$

In (3),α is the neutron excess $(N - Z)/A$.

The mean potential,$V_m(\rho)$, generated by the mutual interaction among nucleons, i.e. the Hartree-Fock potential, can not be calculated directly from the two-nucleon interaction due to the strong short range repulsion but can be approximately obtained by evaluating the K-matrix elements using plane wave basis and is given by

$$V_n(k) = \sum_{q<k_{nf}} [(kq|K_{nn}|kq) - \text{Exchange}] + \sum_{q<k_{pf}} (kq|K_{np}|kq). \quad (4)$$

$$V_p(k) = \sum_{q<k_{pf}} [(kq|K_{pp}|kq) - \text{Exchange}] + \sum_{q<k_{nf}} (kq|K_{np}|kq). \quad (5)$$

In (4) and (5), n and p stand for neutron and proton, respectively, and k and q are wave number and momentum transfer, respectively, and f stands for Fermi-momentum. Since k is related to density, the mean nuclear potential, $V_m(\rho)$, is density dependent which can be represented as

$$V_m(\rho) = b_1(1+a_1\alpha^2)\rho + b_2(1+a_2\alpha^2)\rho^{\frac{4}{3}} + b_3(1+a_3\alpha^2)\rho^{\frac{5}{3}}\,, \quad (6)$$

where a_i and b_i (i=1,2,3) are parameters determined by fitting the calculated energy per nucleon versus the density of nucleonic matter curve.

$V_g(\rho)$ in (3) represents inhomogenity correction to $T(\rho)$ and the nucleonic correlation energy not included in the mean field and is taken to be

$$V_g(\rho) = \eta(\frac{\hbar^2}{8M})\frac{(\nabla\rho)^2}{\rho}. \quad (7)$$

In (7), M is the nucleonic mass and the parameter η is determined by fitting the known nuclear masses with observed density distribution functions[11]. The Coulomb energy among protons with the correction due to the Pauli principle is given by

$$V_c(\rho) = 0.359(1-\alpha)\int d\acute{r}\frac{\rho(\acute{r})}{|r-\acute{r}|} - 1.0636[\frac{(1-\alpha)}{2}]\rho^{\frac{1}{3}}. \quad (8)$$

The observed nuclear masses calculated using (1) are very well reproduced with observed density distributions through proper root mean squared radii[11].

The potential between an alpha-particle and a target nucleus, $V(R)$, in the sudden approximation is given by[12]

$$V(R) = E[\rho(r,R)] - E[\rho_\alpha(r,R=\infty)] - E[\rho_T(r,R=\infty)]\,, \quad (9)$$

where the first term in the right hand side, RHS, represents the energy of the composite overlapped system of an alpha-particle and a target nucleus at a separation distance, R. The second and the third terms on RHS are, respectively, masses of alpha particle and target nucleus which can be taken from the observed masses or calculated using the EDF theory[11]. In the sudden approximation employed here, the density of the composite system, $\rho(r,R)$, is the sum of alpha-particle and target nuclear densities[12]

$$\rho(r) = \rho_\alpha(r) + \rho_T(r). \quad (10)$$

For target nuclei, other than alpha-particle, considered here, $\rho_T(r)$ are taken from ref. 13. The density distribution function of alpha particle, $\rho_\alpha(r)$, is reasonably represented by the following Gaussian function

$$\rho_\alpha(r) = 4(\frac{\gamma}{\pi})^{\frac{3}{2}}\exp(-\gamma r^2)\,, \quad (11)$$

with η=7.56 fm^3 and γ=0.5, the binding energy (BE) and root mean squared radius (RMSR) of alpha-particle are, respectively, given by 19.01 MeV and 1.732 fm that are to be compared to the observed values of BE=28.3 MeV, and RMSR between 1.67 and 1.70 fm, respectively. The central density of 0.254 fm^{-3} of alpha-particle is slightly lower than 0.32 fm^{-3} needed to fit electron scattering data.

3. Results and Discussion

The main features of alpha-nucleus potentials calculated for various target nuclei up to Zr in the context of EDF theory, are that its functional form is non-monotonic and the magnitude of the attractive part is shallow, i.e., the potential is non-monotonic shallow, as the acronym NMS signifies. In fact, by slightly varying the parameters characterizing the calculated NMS potential with an imaginary part, one can usually fit the elastic scattering data well, as noted for ^{28}Si target for the first time[14]. Empirical complex NMS potentials describe very well low energy alpha-scattering data, for elastic and non-elastic processes by ^{24}Mg and ^{28}Si (ref. 8, 15). The standard form used for this empirical complex NMS potential, $V(R)$, is the following:

$$\mathrm{Re}V(r) = -V_0[1+\exp(\frac{r-R_0}{a})]^{-1} + V_1 \exp[-(\frac{r-D_1}{R_1})^2] + V_C(r). \tag{12}$$

$$\mathrm{Im}V(r) = -W_0 \exp(-\frac{r^2}{R_W^2}) - W_S \exp[-(\frac{r-D_S}{R_S})^2]. \tag{13}$$

With

$$V_C(r) = \{ \begin{matrix} (Z_1Z_2/2R_C)(3-r^2/R_C^2) & \text{for} \quad r \leq R_C \\ (Z_1Z_2e^2/r) & \text{for} \quad r > R_C \end{matrix} \tag{14}$$

In the above, Z_1, Z_2, and R_C are the charges of projectile, target, and Coulomb radius, respectively. In the following, we discuss the extents to which the real parts of the empirical NMS potential in (12) (a) conform to those calculated using the EDF theory and (b) describe the elastic scattering data of alpha-particle by ^{4}He, ^{40,44,48}Ca and ^{58}Ni, noting that these have also been successful for other targets. We note here that the EDF calculations, presented here, are strictly valid for zero kinetic energy or, at energies equal to Coulomb barrier where two colliding particles have almost no kinetic energy. At much higher energies, the formalism may only serve as a guide to the general feature of the functional nature of the interaction.

3.1. $\alpha - \alpha$ *Potential*

The calculated potential of the $\alpha - \alpha$ system, shown by solid triangles in the upper insert of Fig. 1, can be well described by V_{EDF} of the functional form (12), with V_0=15.56 MeV, R_0=3.25 fm, a=0.492 fm, V_1=49.24 MeV, R_1=1.510 fm and D_1=0 fm, as well as a functional form where the first term is replaced by a Gaussian function[16]. Since non-elastic processes start well above 20 MeV excitation energies, there is no need to add an imaginary part to this potential. Except for the 5.26 MeV (lab) data, the EDF potential can generally account for the shape and magnitudes of the data at 2.0, 3.0, 8.87, 11.88 and 15.2 MeV (lab), as can be seen by comparing the calculations, noted as dashed lines, to the experimental data, shown as solid dots, in Fig. 1.

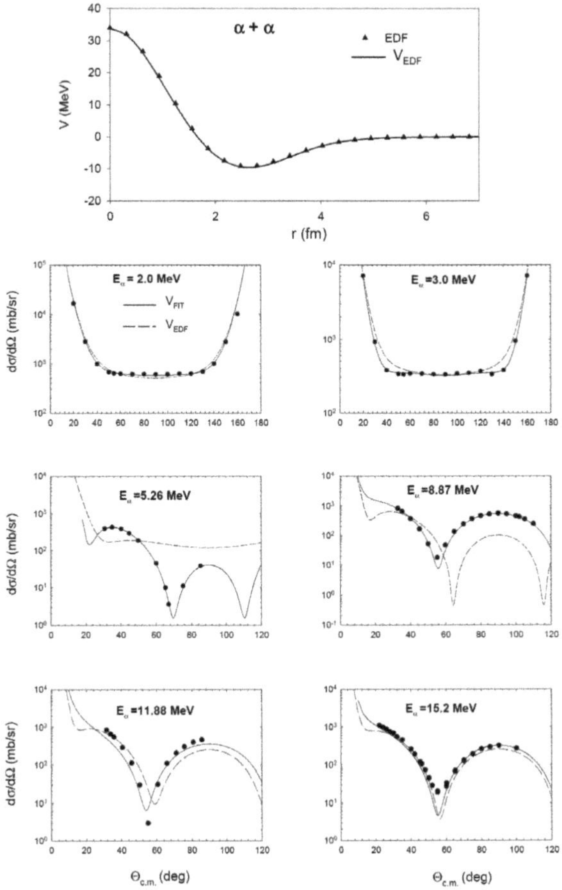

Fig. 1. EDF calculations (solid triangles) for $\alpha + \alpha$ elastic scattering are fitted, in solid line, with V_{EDF} comprising the WS attractive part and Gaussian repulsive core. The predicted cross sections using V_{EDF} (dashed lines) and those (solid lines) with the NMS potential, V_{FIT}, in Table 1, are compared to the experimental data (solid dots) at 2.0, 3.0, 5.26, 8.87, 11.88 and 15.2 MeV (lab).

The inability of the EDF potential in explaining the 5.26 MeV (lab) data points towards its failure to account for the onset of the threshold of d-wave to elastic scattering. One can, however, describe the data at all six energies by slightly changing the potential parameters, V_0 to 15.40 MeV, R_0 to 3.95 fm, a to 0.40 fm,V_1 to 150.0 MeV, and R_1 to 1.68 fm, leading to the new NMS set, V_{FIT} in Table 1.

Table 1. Comparison of the parameters of the EDF generated potential, V_{EDF}, and those of the real part of the NMS potential, V_{FIT}, needed to fit the experimental data.

Target	Set	V_0	R_0	a	V_1	R_1	D_1	R_C
^{4}He	V_{EDF}	15.56	3.25	0.492	49.24	1.51	0.00	4.00
	V_{FIT}	15.40	3.95	0.40	150.00	1.68	0.00	5.80
^{40}Ca	V_{EDF}	29.85	5.38	0.64	16.00	2.15	1.75	4.48
	V_{FIT}	31.80	5.38	0.54	18.00	2.15	1.75	6.00
^{44}Ca	V_{EDF}	29.20	5.52	0.64	15.80	2.10	1.75	4.62
	V_{FIT}	32.40	5.52	0.67	14.00	2.10	1.75	6.00
^{48}Ca	V_{EDF}	41.20	5.32	0.70	32.00	2.54	1.68	4.50
	V_{FIT}	44.00	5.32	0.70	31.70	2.54	1.68	6.00
^{58}Ni	V_{EDF}	28.50	5.92	0.668	20.00	2.30	2.15	8.50
	V_{FIT}	36.00	5.92	0.668	15.00	2.30	2.15	8.70

The fits to the data using V_{FIT} are shown as solid lines in the lower inserts of Fig. 1 and are very good. In the tail region, the two potentials are very similar, whereas they differ from each other in the interior region, mainly in the magnitudes of the repulsive part, which is primarily determined by the density function of alpha-particle near the origin. The Gaussian function (11) used in describing the alpha density does not determine the magnitude of its central part very well and hence, this repulsive part of the potential is not well described by the EDF calculation. Nevertheless, the non-monotonic nature of the $\alpha - \alpha$ potential is established and the theory determines the key part of the potential responsible in describing the low energy elastic scattering, namely, the tail region, from a realistic two-nucleon interaction, rather well. The data in this region are also well described by a simple deep monotonic potential[16,17]. However, such potentials bind two alpha particles forming ^{8}Be in s- and d-states. On the other hand, ^{8}Be is, experimentally, unbound with a decay width of (6.8±1.7) eV (ref. 18). The NMS potential deduced here does not bind ^{8}Be and has a resonance at about the right energy with a decay width of 6.36 eV, thus solving an old problem.

3.2. $\alpha-^{40,44,48}$*Ca Potential*

The detailed analyses[19] of elastic scattering data by ^{40,44,48}Ca indicate that the observed data are well explained by the empirical deep monotonic Michel[5,6,20] potentials as well as NMS potentials derived from the EDF calculations. The observed density distributions of all three isotopes are described by three parameters Fermi function[13]. The BE calculated using EDF theory are, respectively, 340.44, 381.95 and 419.02 MeV for ^{40}Ca, ^{44}Ca and ^{48}Ca which are very close to the observed values of 342.06, 380.97, and 416.0 MeV, respectively. For alpha particle, γ in (11) for this case is taken to be 0.45 fm leading to a BE of 20.96 MeV, RMSR of 1.825 fm and central density of 0.217 fm^{-3}. The calculated potentials for all three targets are shown in the left insert of Fig. 2 with solid triangles. The functional form of the potential, while maintaining a NMS character, dips down near r=0 which is the

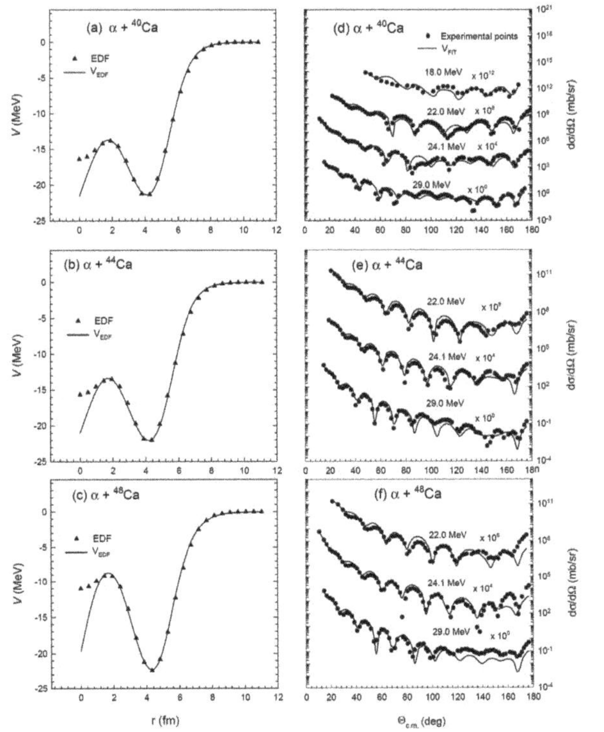

Fig. 2. Same as in Fig. 1 with the shifted Gaussian repulsive core for (a) α+^{40}Ca, (b) α+^{44}Ca and (c) α+^{48}Ca. The corresponding fits (solid lines) to the data using the V_{FIT} parameters are shown in (d), (e) and (f), respectively.

consequence of the wine-bottle shape of the density distribution functions of these nuclei.

Calculated potentials are reasonably given by the parameter set, marked as V_{EDF} in Table 1, and shown as solid lines in Fig. 2. The data are reasonably reproduced by using the parameter sets, noted as V_{FIT}, in Table 1 using the imaginary components of the potential described in ref. 19. The closeness of V_{EDF} and V_{FIT} parameters in Table 1 exemplifies once again that EDF method provides a reasonable way to ascertain alpha-nucleus potential from a realistic two-nucleon interaction.

3.3. $\alpha-^{58}$*Ni Potential*

The real part of α-Ni potentials derived from the EDF theory is also reasonably close to those needed to fit the data using NMS potentials[4]. For example, the calculated potential of the alpha-^{58}Ni system, which also has a wine-bottle density distribution, is shown in the upper insert of Fig. 3, as solid triangles. Except for the interior part, this potential can reasonably be expressed by the parameters, noted in Table 1 as V_{EDF}, and is shown as solid line in the upper insert of Fig. 3. A very similar potential, the parameters of which are termed as V_{FIT} in Table 1, can reasonably account for the data between 18.0 and 50.2 MeV, as exhibited in the lower part of Fig. 3. The data for ^{60,62,64}Ni targets can similarly be described by NMS potentials, whose real parts are close to the ones derived from the EDF calculations.

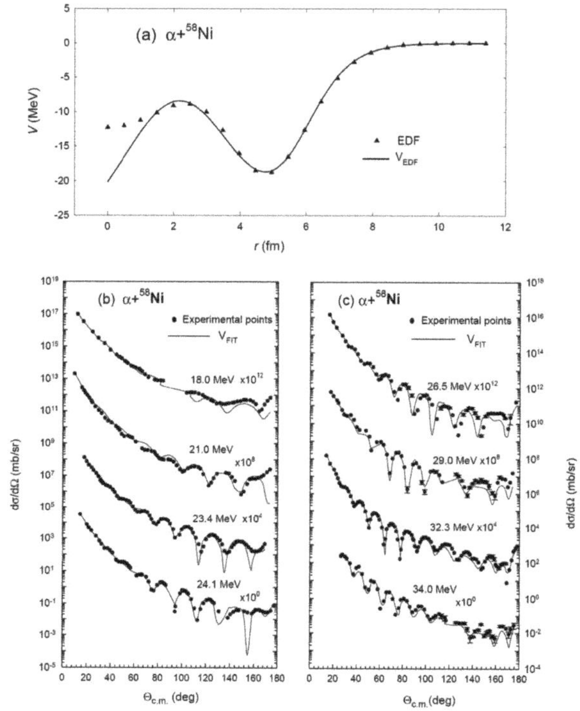

Fig. 3. Same as in Fig. 2 for α+^{58}Ni. The EDF potential (solid triangles) is fitted with the V_{EDF} set in (a). The fits to the data using the V_{FIT} set (Table 1) are shown in (b) and (c).

4. Conclusion

The examples and calculations for a number of medium light targets indicate that the general forms of real parts of alpha-nucleus potential can reasonably be derived from the realistic two-nucleon potential using the EDF method.

Acknowledgments

The authors are pleased to acknowledge the U.S. National Science Foundation grant INT-0209583 and INT-0209584 and a grant from the University Grants Commission of Bangladesh. One of the authors, FBM, is further thankful to the U.S. Army Research Office for a travel grant.

References

1. E. Rutherford, *Phil. Mag.* **11**, 166 (1906); *ibid* **12**, 134 (1906).
2. E. Rutherford and T. Royds, *Phil Mag.* **17**, 281 (1909).
3. J. C. Corelli, E. Bleuler and D. J. Tandem, *Phys. Rev.* **116**, 1184 (1958).
4. M. M. Billah, M. N. A. Abdullah, S. K. Das, M. A. Uddin, A. K. Basak, I. Reichstein, H. M. Sen Gupta and F. B. Malik, *Nucl. Phys. A* **762**, 50 (2005).
5. F. Michel, *Phys. Rev. C* **13**, 1446 (1976).
6. F. Michel and R. Vanderpoorten, *Phys. Lett. B* **82**, 183 (1978).
7. S. K. Das, A. S. B. Tariq, M. A. Uddin, A. S. Mondal, A. K. Basak, K. M. Rashid, H. M. Sen Gupta and F. B. Malik, *Phys. Rev. C* **62**, 054605 (2000).
8. S. K. Das, A. K. Basak, K. Banu, A. S. Mondal, A. S. B. Tariq, A. F. M. M. Rahman, H. M. Sen Gupta and F. B. Malik, *Phys. Rev. C* **62**, 054606 (2000).
9. S. K. Das, A. S. B. Tariq, A. F. M. M. Rahman, S. Hossain, A. K. Basak, H. M. Sen Gupta and F. B. Malik, *Phys. Rev. C* **64**, 034605 (2001).
10. P. Hohenberg and W. Kohn, *Phys. Rev.* **136**, B864 (1964).
11. M. A. Hooshyar, I. Reichstein and F. B. Malik, *Nuclear Fission and Cluster Radioactivity* (Springer Verlag 2005).
12. I. Reichstein and F. B. Malik, *Phys. Lett. B* **67**, 344 (1971).
13. H. de Vries, C. W. de Jager and C. de Vries, *Atomic Data and Nucl. Data* **36**, 495 (1987).
14. P. Manngård, M. Brenner, M. M. Alam, I. Reichstein and F.B. Malik, *Nucl. Phys. A* **504**, 130 (1989).
15. A. S. B. Tariq, A. F. M. M. Rahman, S. K. Das, A. S. Mondal, M. A. Uddin, A. K. Basak, H. M. Sen Gupta and F. B. Malik, *Phys. Rev. C* **59**, 2558 (1999).
16. M. N. A. Abdullah, M. S. Sabra, M. M. Rashid, Z. F. Shehadeh, M. M. Billah, S. K. Das, M. A. Uddin, A. K. Basak,I. Reichstein, H. M. Sen Gupta and F. B. Malik, *Nucl. Phys. A* **775**, 1 (2006).
17. B. Buck, H. Friedrich and C. Wheatley, *Nucl. Phys. A* **275**, 246 (1971).
18. J. Benn, E. B. Dally, H. H. Muller, R. E. Pixley, H. H. Staub and H. Winkler, *Phys. Lett.* **20**, 43 (1966).
19. M. N. A. Abdullah, A. B. Idris, A. S. B. Tariq, M. S. Islam, S. K. Das, M. A. Uddin, A. S. Mondal, A. K. Basak, I. Reichstein, H. M. Sen Gupta and F. B. Malik, *Nucl. Phys. A* **700**, 40 (2005).
20. F. Michel, J. Albinski, P. Belery, Th. Delbar, Gh. Grégoire, B. Tasiaux and G. Reidemeister, *Phys. Rev. C* **28**, 1904 (1983).

Part G

Density Functional Theory and MD Simulations

DYNAMICS OF METAL CLUSTERS IN RARE GAS CLUSTERS

M. BAER

Institut für Theoretische Physik, Universität Erlangen, Staudtstrasse 7, D-91058 Erlangen, Germany

G. BOUSQUET and P. M. DINH*

Laboratoire de Physique Théorique, UMR 5152, Université P. Sabatier 118 Rte de Narbonne, F-31062 Toulouse cedex, France
**E-mail: dinh@irsamc.ups-tlse.fr*

F. FEHRER and P.-G. REINHARD**

Institut für Theoretische Physik, Universität Erlangen, Staudtstrasse 7, D-91058 Erlangen, Germany ,
***E-mail: mpt218@theorie2.physik.uni-erlangen.de*

E. SURAUD

Laboratoire de Physique Théorique, Université P. Sabatier 118 Rte de Narbonne, F-31062 Toulouse cedex, France

We investigate the dynamics of Na clusters embedded in Ar matrices. We use a hierarchical approach, accounting microscopically for the cluster's degrees of freedom and more coarsely for the matrix. The dynamical polarizability of the Ar atoms and the strong Pauli-repulsion exerted by the Ar-electrons are taken into account. We discuss the impact of the matrix on the cluster gross properties and on its optical response. We then consider a realistic case of irradiation by a moderately intense laser and discuss the impact of the matrix on the hindrance of the explosion, as well as a possible pump probe scenario for analyzing dynamical responses.

Keywords: Metal cluster; rare gas matrix; laser irradiation; Density Functional Theory.

1. Introduction

Structural and dynamical properties of clusters, especially in the case of simple alkali clusters, have focused many experimental as well as theoretical investigations.[1–5] The case of clusters embedded in a matrix or deposited on a surface requires specific treatments, because the environment influences cluster properties and has thus to be incorporated in the analysis. It should nevertheless be noted that the presence of an environment (surface or matrix) may simplify the experimental handling: it usually implies localization, fixed geometry, well controlled temperature, and higher yields. Deposited and/or embedded clusters thus often represent a preferred alternative for cluster studies. But the presence of an environment complicates the theoretical modeling. It requires a handling of the cluster-matrix interface and the huge number of atoms in the matrix or substrate may become prohibitive. But because many interesting experiments have been, or can only be, done with clusters in contact with a carrier material such theoretical investigations, although challenging, become unavoidable. Let us mention as examples of experimental motivations the systematics of optical response in large noble-metal clusters[6,7] and its dependence on the environment.[8]

While static properties of embedded/deposited clusters have already been studied since long,[2,3] dynamical scenarios still remain little explored[5] but promise challenging questions. As in free clusters, the optical response represents a key tool of analysis, both of cluster structure, in particular in metal clusters[1] and in most dynamical situations. Let us mention the case of moderate perturbations, as, *e.g.*, in the case of laser irradiation in which resonant coupling between laser and cluster eigenfrequencies represents a basic doorway mechanism for many dynamical scenarios.[5] The study of the optical response thus constitutes a key "entry point" for the cluster response to electromagnetic probes.

Among the various possible combinations of clusters with substrate/matrix, the case of inert or moderately active environments is especially interesting. It implies only moderate perturbations of cluster properties and one can thus still benefit from the well defined geometry of the system and access predominantly the cluster properties themselves. From the theoretical side, the environment can be included at a lower level of description (hierarchical method), which simplifies the handling. This is the type of situations we shall focus on in the following, namely the case of simple metal clusters (sodium) inside a moderately active environment (rare gas matrix).

The term "hierarchical method", used above, requires some comments. It refers to the fact that the interface may be treated at a simpler level than the cluster's degrees of freedom, because of the moderate interactions between cluster and environment. Such approaches are well known in many complex systems, as for example the shell model in surface chemistry[9,10] or the coupled quantum-mechanical with molecular-mechanical method (QM/MM) often used in bio-chemistry.[11–13] Hierarchical modeling is nevertheless not free from difficulties, especially at the technical side, because the interface interaction requires a very careful adjustment. And this may become very tough when one is dealing with a huge range of energies (cluster *vs.* matrix), as in the case of an alkaline cluster embedded in a rare gas (RG) matrix, where one has to deal with eV (Na) as well as meV (RG-RG and RG-Na). But once properly tuned the model becomes able to simulate rather easily realistic situations, especially in terms of system's sizes.

The paper is organized as follows. A brief presentation of the hierarchical approach is given in section 2. We then discuss the impact of the matrix on gross properties of a small metal cluster. We then pursue the analysis by considering the impact of the matrix on the optical response of the embedded cluster. We finally illustrate possible violent irradiation scenarios on a realistic example.

2. Model

The model has been presented in detail in[14] and we just remind here the basic ingredients. The Na cluster is described in the TDLDA-MD approach, an approach which has been validated for linear and non-linear dynamics of free metal clusters.[5,15] It combines a (time-dependent) density-functional theory at the level of the local-density approximation (TDLDA) for the electrons, to a classical molecular (MD) dynamics for the ions. The electron-ion interaction is described by soft, local pseudo-potentials.[16] Two classical degrees-of-freedom are associated with each Ar atom : its center-of-mass and its electrical dipole moment. The dipoles allow to explicitely treat the dynamical polarizability of the atoms with help of polarization potentials.[17] The atom-atom interactions are described by a standard Lennard-Jones potential, and for the Ar-Na^+ subsystem we employ effective potentials from the literature.[18] The pseudo-potential for the electron-Ar core repulsion has been modeled in the form proposed by,[19] and we slightly readjust it by a final fine-tuning to the NaAr molecule as benchmark (bond length, binding energy, and optical excitation). A Van der Waals interaction is also added and computed *via* the variance of dipole operators.[14,19,20]

The starting quantity is the total energy which is constructed from the various pieces as outlined above. From the total energy, the corresponding equations of motion can be derived by standard variation. This leads to the time-dependent Kohn-Sham equations for the cluster electrons. And one obtains Hamiltonian equations of motion for the classical degrees of freedom (namely Na^+ ions and Ar atom positions and dipoles). The initial condition is obtained from the corresponding stationary equations. The Ar-sodium configuration is produced as follows. One cuts a finite piece from a fcc Ar crystal and optimizes the structure. One then carves a central cavity to place the Na cluster inside. The total configuration is then re-optimized by means of cooled molecular dynamics for the ions and atoms coupled to stationary Kohn-Sham solution for the cluster electrons.

We compute several observables to analyze the statics and dynamics of the system. The ionic and electronic cluster structure is characterized in terms of r.m.s. radii

$$r_{I,e} = \sqrt{< (x^2 + y^2 + z^2) >_{I,e}} \tag{1}$$

where $< . >_I = \sum_I . / N_I$ and $< . >_e = \int d^3r . \rho_e(\mathbf{r})/N_e$. We also analyze the shape of the system by computing the reduced quadrupole moment β_2 which reads

$$\beta_2 = \frac{4\pi}{5} \frac{\sqrt{\frac{5}{16\pi}} < (2z_I^2 - x_I^2 - y_I^2) >_{I,e}}{< r^2 >_{I,e}} \tag{2}$$

with obvious notations. The optical response is obtained from the dipole momemt $D_{\rm el} = \langle \sum_n z \rangle$ as quantum mechanical expectation value over the Kohn-Sham state. The dynamics is initiated by an instantaneous dipole boost of the cluster electrons. The dipole signal is followed in the time domain $D_{\rm el}(t)$ and then Fourier transformed to the frequency domain to obtain the dipole strength as imaginary part of the Fourier transformed dipole moment.[21–23]

The numerical solution proceeds with standard methods as described in detail in.[15] The (time-dependent) Kohn-Sham equations for the cluster electrons are solved using real space grid techniques. The time propagation proceeds using a time-splitting method, and the stationary solution is attained by accelerated gradient iterations. We furthermore employ the cylindrically-averaged pseudo-potential scheme (CAPS) as introduced in,[24,25] which is justified for the chosen test cases. The Na^+ ions as well as the Ar atoms are nevertheless still treated in full 3D.

3. Gross Properties of Embedded Clusters

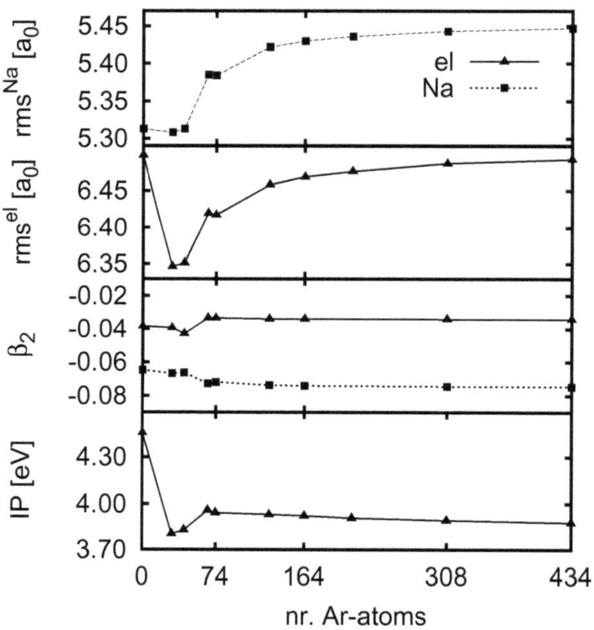

Fig. 1. Trend of basic observables with matrix size: ionic and electronic r.m.s. radius, ionic and electronic quadrupole deformation β_2, electronic ionization potential (IP). The matrix is optimized anew for each configuration. From.[14]

We take as test case a Na_8 cluster embedded into an Ar matrix of various sizes. Fig. 1 summarizes a few basic properties of the embedded cluster as a function of increasing matrix size. For moderate size matrices the electronic radius is first reduced (as compared to the free cluster), because of the strong repulsive Ar-core potentials. But with increasing matrix size both the electronic and the ionic radii expand in a similar way because of the monopole-dipole interaction between Na^+ and the Ar-atoms. The global deformation (third panel from top in Fig. 1) is nearly independent of matrix size, although one can spot a slight trend for small sizes. The ionization potential (IP) is, as usual, defined as the energy difference between Na_8 and Na_8^+ for ionic and atomic positions frozen. The IP globally exhibits a decrease mostly due to the short range compression. There is indeed a big jump from the free to the clusters inside small matrices and very little changes amongst all embedded systems, in accordance with the short-range nature of the Ar core potentials. Generally speaking one can identify from the figure two regimes, with a threshold at 164 Ar atoms. Below that limit one observes specific size effects, which tend to be smoothened out for larger sizes for which one observes a monotonous trend towards the asymptotic value. This again reflects the balance between the two competing effects, namely the short range repulsion due to compression effects and the long range attraction due to polarization effects.

4. Linear Optical Response

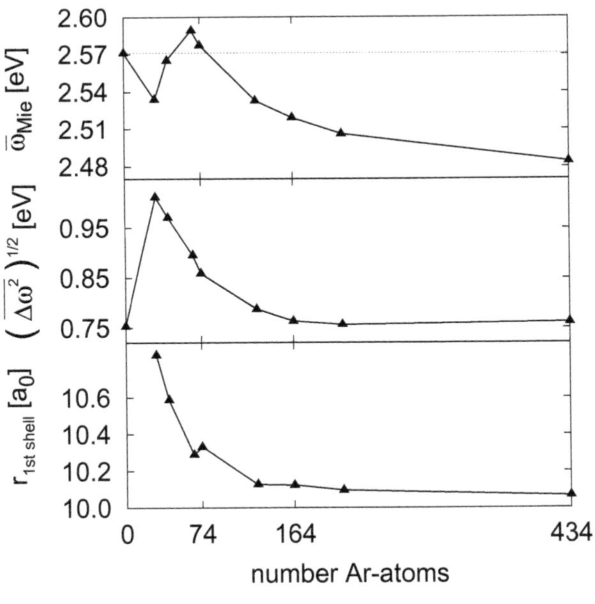

Fig. 2. Average peak position, width and radius of the innermost Ar shell, as a function of matrix size. From.[14]

Fig. 2 presents the trend of the average plasmon peak position $\bar{\omega}_{Mie}$ and width $(\overline{\Delta\omega^2})^{1/2}$ with matrix size, as well as the radius of the innermost Ar shell. The value $\bar{\omega}_{Mie}$ is the averaged dipole transition over the interval [1.9 eV; 3.2 eV]. The width is the variance of the frequency computed in the same averaging interval. Once again, visible structures are seen only below 164 Ar atoms. While the Na_8 exhibits a clean Mie plasmon peak, spectral structures of the clusters embedded in small matrices are strongly fragmented, corresponding to a larger width (see middle panel). This effect is clearly due to Landau fragmentation, since the surrounding Ar atoms modify the spectral density of one-electron-one-hole states near the resonance. Indeed the latter emerges from a subtle interplay of core repulsion and dipole attraction. For matrices larger than 164 Ar atoms, the variance decreases rapidly and saturates at the same value as that of the free Na_8, leading to a fairly clean plasmon peak again.

Note however that the changes in the average resonance position are very little or even insignificant, that is, less than 1/10 of an eV. This is the result of an almost complete cancellation of core and polarization potentials.[14] Indeed we have computed the dipole spectrum for Na_8Ar_{164} while switching off the polarization potentials from the Ar atoms. Thus the cluster electrons feel only the repulsion of the Ar core potentials. The effect is a strong blue-shift of the resonance by about 0.8 eV. Switching on the polarization potentials produces an equally strong red-shift. At the end, the peak position comes out almost unchanged as compared to the free cluster. It is obvious that the final resonance position, resulting from two large and counteracting effects, is extremely sensitive to details of the model. One should thus be careful not to overinterpret the trends at the meV scale.

Looking back to Fig. 2 with that mechanism in mind, the radius of the innermost Ar shell shrinks strongly up to about 74 atoms (bottom panel). This corresponds to the regime where core repulsion increases. The dipole attraction increases as well and we see practically constant average peak position. For larger systems, whereas the core effect stabilizes, the dipole attraction still increases, although on a lower rate. This explains why a small but steady red-shift is seen for further increasing system sizes. We have estimated the effects for further shells towards the full crystal and we found about 14 meV more red-shift asymptotically. It is a systematic effect, but a very small one. Thus we will ignore it for the now following explorations of violent excitations.

5. Strongly Excited Na Cluster in an Ar Environment

In this section, we present a realistic intense laser excitation of the system $Na_8@Ar_{434}$. The laser features are the following: polarization along the symmetry axis of the system (and denoted below as the z axis), frequency of 1.9 eV, intensity of 2×10^{12} W.cm^{-2} and FWHM of 50 fs. The system reacts mainly by a strong dipole moment which leads to a direct emission of 3 electrons which escape instantaneously (i.e. within 3-10 fs). One may wonder whether the strong external or effective Coulomb field at Ar sites would also trigger electron emission from Ar. This was checked by TDLDA calculations for an Ar atom and we found a critical field strength of about 0.1 Ry/a_0 for that process. In our calculations, we record the actual field strengths at all Ar sites. They stay safely below this critical value all times. Electron emission thus comes exclusively from the Na cluster ($Na_8 \rightarrow Na_8^{3+}$).

In Fig. 3, showing the time evolution of the Na ions and Ar atoms coordinates, we observe a rearrangement of the whole system at ionic/atomic time scale, due to

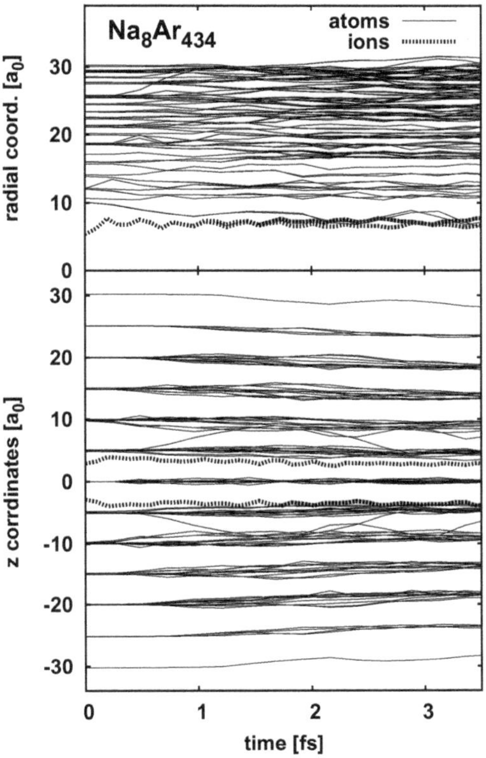

Fig. 3. Time evolution of the the atomic (full lines) and ionic (dotted lines) z-coordinates (lower panel) and radial distances $r = \sqrt{x^2 + y^2 + z^2}$ (upper panel) for Na_8Ar_{434} excited with a laser (see text for details).

the thus produced large Coulomb pressure. We note however different interesting time scales. First the Na ions start, up to about 200 fs, a Coulomb explosion, almost identical to the similar case of a free Na_8 cluster lifted to charge 3^+. Second when the ions hit the repulsive core of the first shell of Ar, the "explosion" is abruptly stopped. Then the ionic motion turns to damped oscillations around a (r.m.s.) radius of about 7 a_0. Note that the shape oscillations of the cluster have the typical cycle of about 250 fs, well known for free Na clusters[26] and re-established for embedded ones.[27]

Two stages in the matrix are also seen, although on longer time scales. We first observe a "diffusion" of the perturbation (due to rearrangement of Na ions) into the various Ar shells. This is especially visible along the z axis which allows to read off the propagation speed of this perturbation as 20-30 a_0/ps. We have estimated the sound velocity in the corresponding large pure Ar cluster (Ar_{447}) by computing its vibrational spectrum. The radial compression mode corresponds to the sound mode in bulk material. We found a frequency of $\omega_{\rm vib} = 1.8$ meV. The momentum of the radial wave is $q = \pi/R$ where $R = 30$ a_0 is the cluster radius. The sound velocity is then estimated as $v_{\text{sound}} = \omega_{\text{vib}}/q \approx 30\ \text{a}_0/\text{ps}$ which is very close to the propagation speed as observed in the figure. One may thus interpretate the perturbation as a sound wave sent by the initial bounce of the Na ions.

Then the perturbation generates oscillations combined with some diffusion which, after about 1.5 ps, has spread over all shells. Even the outermost shell is perturbed and the matrix seems to get an oblate deformation, as the Na_8 does. However whereas the Na cluster exhibits a global relaxed trend to oblate shapes, the relaxation of the Ar atoms seems much longer than that of the Na ions and far beyong the time scale computed here. These long time scales for full relaxation and evaporation of Ar atoms are well known from experiments of dimer molecules embedded in Ar clusters, see *e.g.*[28]

Let us now briefly comment on the finite net charge of our system. It emerges because the electrons propagate almost unhindered through the surrounding Ar cluster and eventually escape to infinity. We may wonder whether, in a macroscopically large matrix, the electrons would be stuck somewhere in the range of their mean free path, drift very slowly back towards the now attractive cluster well, and eventually recombine there. This has been checked by using reflecting boundary conditions rather than absorbing ones. Up to 4 ps, no recombination has been observed. Furthermore the ionic/atomic rearrangements following the irradiation turn out to be qualitatively very similar whatever the boundary conditions. Thus the present scenario should provide a pertinent picture for a few ps, the time window studied here.

Finally one may wonder how to analyze the key pattern of the embedded cluster dynamics experimentally. The energy transfer to the Ar part could be measured from the Ar evaporation spectra as long as one deals with metal clusters embedded in finite Ar drops. But the most interesting effect is the hindered explosion of the imprisoned Na cluster and its subsequent shape oscillations in the Ar cavity. Here we can exploit the pronounced features of the Mie plasmon resonance of the embedded metal cluster. Indeed, it couples strongly to light, the coupling is highly frequency selective, and the plasmon frequencies are uniquely related to the cluster shape. This suggests to track the shape oscillations by pump and probe analysis. In free clusters, a proper setup allows to map in a unique fashion the evolution of radial shape[29] and of quadrupole deformations,[30] through the time evolution of the Mie plasmon resonance. We have computed that for the case considered here. We take the instantaneous configuration at a given time and compute the optical response in all three spatial directions for the actual charge state of the system. The strong oblate deformation leads to a splitting of the resonance peak where the shorter extension along z is associated to a blue-shift of the mode and the larger extension in orthogonal direction to a red-shift. Our findings are in qualitative agreement with experimental results obtained for Ag in glass,[31] using a similar pump probe scenario. Our model thus provides a microscopic interpretation, the first one, to the best of our knowledge, of these experimental investigations.

6. Conclusion

We have presented a robust model for the description of simple Na clusters embedded in an Ar matrix. The dynamics of the Na cluster is treated in the TDLDA-MD *ab initio* framework, allowing any dynamical regime (linear and non-linear) at the side of the Na cluster. The matrix is treated at the simplest level of modeling still accommodating polarization and full MD motion at the side of Ar atoms. Both sodium and argon are properly coupled following previous detailed quantum chemistry calculations. On the basis of that model, we have investigated the structure and dynamics (low and high energy) of the embedded Na cluster.

We have considered in this paper different systems, namely $Na_8@Ar_n$ with n varying from 0 to 434. The optical response provides an experimental access to the cluster's structure. When the cluster is embedded, the peak position results from a subtle cancellation between two large effects : matrix compression of the ionic configuration, which tends to blue-shift the response, and electronic polarization effects which tend to red-shift the response. All in all the net result, in large matrices, comes very close to the free response. This, by the way, also holds true in the non-linear domain of the plasmon response where the same cancellation effect is found to persist. These results are actually quite compatible with related experiments on Ag in rare gas matrices where it was found that the plasmon peak is very little modified as a function of matrix size. The effect of matrix size was also explored in more detail and some results presented. Again the effects are relatively small although one observes specific behaviors for small matrices. Indeed below 164 Ar atoms we are mostly facing an Ar cluster with an inclusion of a Na cluster, with properties strongly depending on the system size. For larger numbers of Ar atoms, we clearly see more regular patterns, hinting to a convergence of the system's properties towards a "bulk-like" limit.

In the last part of the paper, we have presented results specifically for the case of non-linear processes, especially in relation to intense laser irradiation. The surrounding Ar stops a Coulomb explosion which would else-wise have taken place due to the strong laser ionization of the cluster. Steady shape oscillations of the cluster are then observed, while the absorbed momentum spreads in a radial sound wave propagating through the atoms and triggers global radial oscillations thereof. The relaxation process can be nicely seen in oscillations of cluster radius and deformation. Both these observables lead to unique signatures in the plasmon spectrum of the cluster which, in turn, provides an ideal handle for pump and probe analysis. More investigations are in progress to quantify these effects.

In this paper, we have focused mainly on the actual response of the Na cluster itself, not discussing deeply the induced impact on the matrix. It should thus be highly interesting to investigate the impact of this emission and/or energy deposit on the matrix itself. Our model was originally tailored for this purpose and it is indeed well suited to it. We shall present these results in a forthcoming publication. Moreover we have started some research program in the case of more realistic environments such as MgO or SiO_2. Another line of development, which we also investigate, concerns clusters and molecules of biological interest, which again have to be treated in relation to an environment, often water. Work along that line is also in progress with aim to consider the effects of irradiation on such systems.

References

1. U. Kreibig and M. Vollmer, *Optical Properties of Metal Clusters* (Springer Series in Materials Science, 1993).
2. H. Haberland (ed.), *Clusters of Atoms and Molecules 1- Theory, Experiment, and Clusters of Atoms* (Springer Series in Chemical Physics, Berlin, 1994).
3. H. Haberland (ed.), *Clusters of Atoms and Molecules 2- Solvation and Chemistry of Free Clusters, and Embedded, Supported and Compressed Clusters* (Springer Series in Chemical Physics, Berlin, 1994).
4. W. Ekardt (ed.), *Metal Clusters* (Wiley, New York, 1999).
5. P.-G. Reinhard and E. Suraud, *Introduction to Cluster Dynamics* (Wiley, New York, 2003).
6. N. Nilius, N. Ernst and H.-J. Freund, *Phys. Rev. Lett.* **84**, p. 3994 (2000).
7. M. Gaudry, J. Lermé, E. Cottancin, M. Pellarin, J.-L. Vialle, M. Broyer, B. Prével, M. Treilleux and P. Mélinon, *Phys. Rev. B* **64**, p. 085407 (2001).
8. T. Diederich, J. Tiggesbäumker and K. H. Meiwes-Broer, *J. Chem. Phys.* **116**, p. 3263 (2002).
9. P. J. Mitchell and D. Fincham, *J. Phys.: Condensed Matter* **5**, p. 1031 (1993).
10. A. Nasluzov et al. , *J. Chem. Phys.* **115**, p. 17 (2001).
11. M. J. Field, P. A. Bash and M. Karplus, *J. Comp. Chem.* **11**, p. 700 (1990).
12. J. Gao, *Acc. Chem. Res.* **29**, p. 298 (1996).
13. N. Gresh and D. R. Garmer, *J. Comp. Chem.* **17**, p. 1481 (1996).
14. F. Fehrer, P.-G. Reinhard, E. Suraud, E. Giglio, B. Gervais and A. Ipatov, *Appl. Phys. A* **82**, p. 152 (2005).
15. F. Calvayrac, P.-G. Reinhard, E. Suraud and C. A. Ullrich, *Phys. Rep.* **337**, p. 493 (2000).
16. S. Kümmel, M. Brack and P.-G. Reinhard, *Euro. Phys. J. D* **9**, p. 149 (1999).
17. B. G. Dick and A. W. Overhauser, *Phys. Rev.* **112**, p. 90 (1958).
18. G. R. Ahmadi, J. Almlöf and J. Roegen, *Chem. Phys.* **199**, p. 33 (1995).
19. F. Duplàe and F. Spiegelmann, *J. Chem. Phys.* **105**, p. 1492 (1996).
20. B. Gervais, E. Giglio, E. Jaquet, A. Ipatov, P.-G. Reinhard and E. Suraud, *J. Chem. Phys.* **121**, p. 8466 (2004).
21. F. Calvayrac, P.-G. Reinhard and E. Suraud, *Phys. Rev. B* **52**, p. R17056 (1995).
22. F. Calvayrac, P.-G. Reinhard and E. Suraud, *Ann. Phys. (NY)* **255**, p. 125 (1997).
23. K. Yabana and G. F. Bertsch, *Phys. Rev. B* **54**, p. 4484 (1996).
24. B. Montag and P.-G. Reinhard, *Phys. Lett. A* **193**, p. 380 (1994).
25. B. Montag and P.-G. Reinhard, *Z. f. Physik D* **33**, p. 265 (1995).
26. P.-G. Reinhard and E. Suraud, *Euro. Phys. J. D* **21**, p. 315 (2002).
27. F. Fehrer, P.-G. Reinhard and E. Suraud, *Appl. Phys. A* **82**, p. 145 (2005).
28. V. Vorsa et al., *J. Chem. Phys.* **105**, p. 2298 (1996).
29. K. Andrae, P.-G. Reinhard and E. Suraud, *J. Phys. B* **35**, p. 1 (2002).
30. K. Andrae, P.-G. Reinhard and E. Suraud, *Phys. Rev. Lett.* **92**, p. 173402 (2004).
31. G. Seifert, M. Kaempfe, K.-J. Berg and H. Graener, *Appl. Phys. B* **71**, p. 795 (2000).

KOHN-SHAM CALCULATIONS COMBINED WITH AN AVERAGE PAIR-DENSITY FUNCTIONAL THEORY

P. GORI-GIORGI* and A. SAVIN

Laboratoire de Chimie Théorique, CNRS UMR7616, Université Pierre et Marie Curie, 4 Place Jussieu, F-75252 Paris, France
** E-mail: paola.gori-giorgi@lct.jussieu.fr*
www.lct.jussieu.fr/pagesequipe/DFT/gori

A recently developed formalism in which Kohn-Sham calculations are combined with an "average pair density functional theory" is reviewed, and some new properties of the effective electron-electron interaction entering in this formalism are derived. A preliminary construction of a fully self-consitent scheme is also presented in this framework.

Keywords: Electronic structure; density functional theory; pair density.

1. Introduction

Kohn-Sham (KS) Density Functional Theory[1–3] (DFT) is nowadays one of the most popular methods for electronic structure calculations both in chemistry and solid-state physics, thanks to its combination of low computational cost and reasonable performances. The accuracy of a KS-DFT result is limited by the approximate nature of the exchange-correlation energy density functional $E_{xc}[n]$. Typical cases in which present-day DFT fails are strongly correlated systems, the description of van der Waals forces, the handling of near degeneracy. Much effort is put nowadays in trying to improve the DFT performances via the construction of better approximations for the KS $E_{xc}[n]$ (for recent reviews, see, e.g., Refs. 2–4), or via alternative routes, like, e.g., the use of non-KS options.[5] A popular trend in the development of new KS $E_{xc}[n]$ is the use of the exact exchange functional $E_x[n]$ (in terms of the KS orbitals), and thus the search for an approximate, compatible, correlation functional $E_c[n]$.

In this work we review the basis of a theoretical framework[6,7] in which KS-DFT is combined with an "average pair density functional theory" (APDFT) that provides an explicit construction for $E_c[n]$, transfering the work of finding an approximate functional to the search of an effective particle-particle interaction. A self-consitent scheme for this approach is presented, and some new properties of the effective interaction that enters in this combined formalism are derived. Very preliminary applications are discussed.

2. Definitions

Our target problem is finding the ground-state energy of the standard N-electron hamiltonian in the Born-Oppenheimer approximation (in Hartree atomic units, $\hbar = m = a_0 = e = 1$, used throughout),

$$H = T + V_{ee} + V_{ne}, \tag{1}$$

$$T = -\frac{1}{2}\sum_{i=1}^{N} \nabla_i^2, \tag{2}$$

$$V_{ee} = \frac{1}{2}\sum_{i\neq j}^{N} \frac{1}{|\mathbf{r}_i - \mathbf{r}_j|}, \tag{3}$$

$$V_{ne} = \sum_{i=1}^{N} v_{ne}(\mathbf{r}_i), \tag{4}$$

where v_{ne} is the external potential due to nuclei. Given Ψ, the exact ground-state wavefunction of H, we consider two reduced quantities that fully determine, respectively, the expectation values $\langle\Psi|V_{ne}|\Psi\rangle$ and $\langle\Psi|V_{ee}|\Psi\rangle$, i.e., the electronic density,

$$n(\mathbf{r}) = N \sum_{\sigma_1 \ldots \sigma_N} \int |\Psi(\mathbf{r}\sigma_1, \mathbf{r}_2\sigma_2, \ldots, \mathbf{r}_N\sigma_N)|^2 d\mathbf{r}_2 \ldots d\mathbf{r}_N, \tag{5}$$

and the spherically and system-averaged pair density $f(r_{12})$ (APD), which is obtained by first considering the pair density $P_2(\mathbf{r}_1, \mathbf{r}_2)$,

$$P_2(\mathbf{r}_1, \mathbf{r}_2) = N(N-1) \sum_{\sigma_1 \ldots \sigma_N} \int |\Psi(\mathbf{r}_1\sigma_1, \mathbf{r}_2\sigma_2, \mathbf{r}_3\sigma_3, \ldots, \mathbf{r}_N\sigma_N)|^2 d\mathbf{r}_3 \ldots d\mathbf{r}_N, \tag{6}$$

and then by integrating it over all variables except $r_{12} = |\mathbf{r}_1 - \mathbf{r}_2|$,

$$f(r_{12}) = \frac{1}{2}\int P_2(\mathbf{r}_1, \mathbf{r}_2) \frac{d\Omega_{\mathbf{r}_{12}}}{4\pi} d\mathbf{R}, \tag{7}$$

where $\mathbf{R} = \frac{1}{2}(\mathbf{r}_1 + \mathbf{r}_2)$, $\mathbf{r}_{12} = \mathbf{r}_2 - \mathbf{r}_1$. The function $f(r_{12})$ is also known in chemistry as intracule density,[8–14] and, in the special case of a an electron liquid of uniform density n, is related to the radial pair-distribution function $g(r_{12})$ by $g(r) = 2f(r)/(nN)$. We thus have

$$\langle\Psi|V_{ne}|\Psi\rangle = \int n(\mathbf{r}) v_{ne}(\mathbf{r}) d\mathbf{r} \tag{8}$$

$$\langle\Psi|V_{ee}|\Psi\rangle = \int \frac{f(r_{12})}{r_{12}} d\mathbf{r}_{12} = \int_0^\infty \frac{f(r_{12})}{r_{12}} 4\pi r_{12}^2 dr_{12}. \tag{9}$$

In the following text we will also deal with modified systems in which the external potential and/or the electron-electron interaction is changed. Thus, the notation V_{ne} and V_{ee} will be used to characterize the physical system, while the modified systems will be defined by $V = \sum_{i=1}^{N} v(\mathbf{r}_i)$ and $W = \frac{1}{2}\sum_{i\neq j}^{N} w(|\mathbf{r}_i - \mathbf{r}_j|)$, where the pairwise interaction w always depends only on $|\mathbf{r}_i - \mathbf{r}_j|$.

3. The Exchange-Correlation Functional of KS-DFT

In standard DFT one defines a universal functional of the one-electron density $n(\mathbf{r})$ as resulting from a constrained search over all the antisymmetric wavefunctions Ψ that yield n[15]

$$\tilde{F}[n; V_{ee}] = \min_{\Psi \to n} \langle \Psi | T + V_{ee} | \Psi \rangle, \tag{10}$$

or, more completely, as a Legendre transform[16]

$$F[n; V_{ee}] = \sup_{v} \left\{ \min_{\Psi} \langle \Psi | T + V_{ee} + V | \Psi \rangle - \int n(\mathbf{r}) v(\mathbf{r}) d\mathbf{r} \right\}. \tag{11}$$

In both Eqs. (10) and (11), the dependence on the electron-electron interaction has been explictly shown in the functional. The universality of the functional F stems exactly from the fact that the e-e interaction is always $1/r_{12}$. The ground-state energy E_0 of the system can then be obtained by minimizing the energy with respect to n,

$$E_0 = \min_{n} \left\{ F[n; V_{ee}] + \int n(\mathbf{r}) v_{ne}(\mathbf{r}) d\mathbf{r} \right\}. \tag{12}$$

A possible way to derive the Kohn-Sham equations in DFT is to define a set of hamiltonians depending on a real parameter λ,[17–19]

$$H^{\lambda} = T + W^{\lambda} + V^{\lambda}, \tag{13}$$

having all the same one-electron density, equal to the one of the physical system

$$n^{\lambda}(\mathbf{r}) = n(\mathbf{r}) \qquad \forall \lambda. \tag{14}$$

If $W^{\lambda_{\rm phys}} = V_{ee}$ and $W^{\lambda=0} = 0$ (e.g., $W^{\lambda} = \lambda V_{ee}$), one can slowly switch off the electron-electron interaction, while keeping the density fixed via a suitable external potential V^{λ}. Obviously, the APD $f(r_{12})$ changes with λ. By the Hellmann-Feynmann theorem,

$$\frac{\partial E_0^{\lambda}}{\partial \lambda} = \langle \Psi^{\lambda} | \frac{\partial W^{\lambda}}{\partial \lambda} + \frac{\partial V^{\lambda}}{\partial \lambda} | \Psi^{\lambda} \rangle = \int f^{\lambda}(r_{12}) \frac{\partial w^{\lambda}(r_{12})}{\partial \lambda} d\mathbf{r}_{12} + \int n(\mathbf{r}) \frac{\partial v^{\lambda}(\mathbf{r})}{\partial \lambda} d\mathbf{r}, \tag{15}$$

so that by directly integrating Eq. (15), and by combining it with Eq. (12), one obtains

$$F[n; V_{ee}] = T_s[n] + \int_0^{\lambda_{\rm phys}} d\lambda \int d\mathbf{r}_{12} f^{\lambda}(r_{12}) \frac{\partial w^{\lambda}(r_{12})}{\partial \lambda}, \tag{16}$$

where $T_s[n] = F[n; 0]$ is the kinetic energy of a noninteracting system of N spin-$\frac{1}{2}$ fermions with density $n(\mathbf{r})$. The adiabatic connection in DFT thus naturally defines the Kohn-Sham non-interacting kinetic energy functional $T_s[n]$. The second term in the right-hand-side of Eq. (16) is an exact expression, in terms of the APD $f^{\lambda}(r_{12})$, for the Hartree and the exchange-correlation functional, $E_H[n] + E_{xc}[n]$. The one-body potential at $\lambda = 0$ is the familiar Kohn-Sham potential, $v^{\lambda=0}(\mathbf{r}) = v_{\rm KS}(\mathbf{r})$.

The traditional approach of DFT to construct approximations for $E_{xc}[n]$ is based on the idea of universality. For example, the familiar local-density approximation (LDA) consists in transfering, in each point of space, the pair density from the uniform electron gas to obtain an approximation for $f^\lambda(r_{12})$ in Eq. (16). Our aim is to develop an alternative strategy in which realistic APD $f^\lambda(r_{12})$ along the DFT adiabatic connection are constructed via a formally exact theory that must be combined with the KS equations in a self-consitent way. The formal justification for this "average pair-density functional theory" (APDFT) is the object of the next Sec. 4.

4. Average Pair Density Functional Theory

As shown by Eqs. (8) and (9), the APD $f(r_{12})$ couples to the operator V_{ee} in the same way as the electronic density $n(\mathbf{r})$ couples to V_{ne}. In order to derive an "average pair density functional theory" (APDFT) we thus simply repeat the steps of the previous Sec. 3 by switching the roles of f and n, and of V_{ee} and V_{ne}.[7]

We thus define a system-dependent functional (i.e., a functional depending on the external potential V_{ne}, and thus on the specific system) of the APD $f(r_{12})$ as

$$\tilde{G}[f; V_{ne}] = \min_{\Psi\to f} \langle\Psi|T + V_{ne}|\Psi\rangle, \tag{17}$$

where, again the minimum is over all antisymmetric wavefunction Ψ that yield a given $f(r_{12})$. We can also define the system-dependent functional G as

$$G[f; V_{ne}] = \sup_{w} \left\{ \min_{\Psi} \langle\Psi|T + W + V_{ne}|\Psi\rangle - \int f(r_{12}) w(r_{12}) d\mathbf{r}_{12} \right\}. \tag{18}$$

The ground-state energy could then be obtained as

$$E_0 = \min_{f\in\mathcal{N}_f} \left\{ G[f; V_{ne}] + \int \frac{f(r_{12})}{r_{12}} d\mathbf{r}_{12} \right\}, \tag{19}$$

where $\mathcal{N}_f$ is the space of all N-representable APD (i.e., coming from the contraction of an N-particle antisymmetric wavefunction). The definition of the space $\mathcal{N}_f$ is evidently related to the N-representability conditions for the pair density, for which recent interesting progresses have been made.[20] In our case, however, we combine APDFT with DFT so that the minimization of Eq. (19) is never directly carried on.

In order to find the analog of the KS system for APDFT, we define an adiabatic connection similar to the one of Eq. (13) in which, this time, we switch off the external potential. We thus introduce a set of hamiltonians depending on a real parameter ξ,

$$H^\xi = T + W^\xi + V^\xi, \tag{20}$$

in which the function $f(r_{12})$ is kept fixed, equal to the one of the physical system,

$$f^\xi(r_{12}) = f(r_{12}) \qquad \forall\xi. \tag{21}$$

If $V^{\xi_{\rm phys}} = V_{ne}$ and $V^{\xi=0} = 0$ (e.g., $V^{\xi} = \xi V_{ne}$), we are switching continuously from the physical system, to a system of N free electrons interacting with a modified potential $w^{\xi=0}(r_{12})$. That is, $f(r_{12})$ is kept fixed as ξ varies by means of a suitable electron-electron interaction W^{ξ} while the one-electron density $n(\mathbf{r})$ changes with ξ. Again, by the Hellmann-Feynman theorem, we find

$$\frac{\partial E_0^{\xi}}{\partial \xi} = \langle \Psi^{\xi} | \frac{\partial W^{\xi}}{\partial \xi} + \frac{\partial V^{\xi}}{\partial \xi} | \Psi^{\xi} \rangle = \int f(r_{12}) \frac{\partial w^{\xi}(r_{12})}{\partial \xi} d\mathbf{r}_{12} + \int n^{\xi}(\mathbf{r}) \frac{\partial v^{\xi}(\mathbf{r})}{\partial \xi} d\mathbf{r}, \quad (22)$$

so that

$$G[f; V_{ne}] = T_{\rm f}[f] + \int_0^{\xi_{\rm phys}} d\xi \int d\mathbf{r}\, n^{\xi}(\mathbf{r}) \frac{\partial v^{\xi}(\mathbf{r})}{\partial \xi}, \quad (23)$$

where $T_{\rm f}[f]$ is the kinetic energy of a system of N free (zero external potential) interacting spin-$\frac{1}{2}$ fermions having the same $f(r_{12})$ of the physical system. In the case of a confined system (atoms, molecules) the effective interaction $w^{\xi=0}(r_{12})$ must have an attractive tail: the hamiltonian corresponding to $\xi = 0$ in Eq. (20) describes a cluster of fermions whose center of mass is translationally invariant. The functional $T_{\rm f}[f]$ is the internal kinetic energy of this cluster.

To fix the ideas, consider the simple case of two electrons, e.g. the He atom. When $\xi = 0$, we have two fermions in a relative bound state (similar to the case of positronium, but with a different interaction). This relative bound state is such that the square of the wavefunction for the relative coordinate r_{12} is equal to $f(r_{12})$ of the starting physical system. The corresponding effective interaction $w^{\xi=0}(r_{12})$, obtained[7] by inversion from a very accurate wavefunction,[21] is shown in Fig. 1, for the case of the He atom.

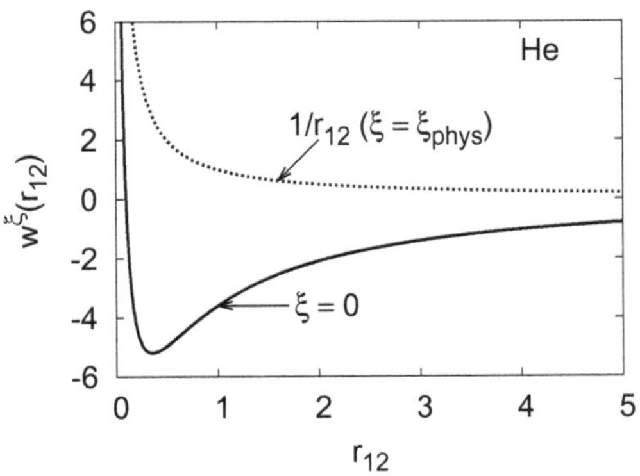

Fig. 1. The electron-electron interaction at the two ends of the APDFT adiabatic connection of Eqs. (20)-(21) in the case of the He atom.

For more than two electrons, $T_{\rm f}[f]$ is still a complicated many-body object. Moreover, since the corresponding $w^{\xi=0}(r_{12})$ can have an attractive tail (as in the case

$N = 2$), we may have an "exotic" true ground-state for the cluster, i.e., the cluster state with the same $f(r_{12})$ of the physical system can be an excited state. However, we have to keep in mind that our aim is not to solve the many-electron problem by means of APDFT alone: we want to use APDFT to produce realistic $f(r_{12})$ along the DFT adiabatic connection of Eqs. (13)-(14). To this end, we proposed[6,7] an approximation for the functional $T_{\rm f}[f]$ based on a geminal decomposition,

$$T_g[f] = \min_{\{\psi_i\}\to f} \sum_i \vartheta_i \, \langle \psi_i | -\nabla^2_{r_{12}} | \psi_i \rangle, \tag{24}$$

where $\psi_i(r_{12})$ are some effective geminals (orbitals for two electrons, but only depending on the relative distance r_{12}), and ϑ_i some occupancy numbers to be chosen. For example, one can always make a "bosonic" choice, by occupying only one geminal,[22] equal to $\sqrt{f(r_{12})}$. The geminals $\psi_i(r_{12})$ then satisfy the equations

$$\begin{cases} [-\nabla^2_{r_{12}} + w_{\rm eff}(r_{12})]\psi_i(r_{12}) = \epsilon_i \, \psi_i(r_{12}) \\ \sum_i \vartheta_i |\psi_i(r_{12})|^2 = f(r_{12}), \end{cases} \tag{25}$$

with

$$w_{\rm eff}(r_{12}) = \frac{1}{r_{12}} + \frac{\delta G[f]}{\delta f(r_{12})} - \frac{\delta T_g[f]}{\delta f(r_{12})}. \tag{26}$$

The approximation of Eq. (24) is mainly motivated by the need of having simple equations for $f(r_{12})$ (the one-dimensional character of Eqs. (25) is of course very appealing). Notice that only in the case $N = 2$ we have $T_g[f] = T_{\rm f}[f]$, and thus the effective interaction $w_{\rm eff}(r_{12})$ of Eqs. (25) becomes equal to $w^{\xi=0}(r_{12})$.

5. Combining DFT and APDFT in a Self-Consistent Way

As explained in the previous sections, to compute the expectation value of the physical hamiltonian of Eq. (1) we only need $f(r_{12})$ and $n(\mathbf{r})$ for V_{ee} and V_{ne}, and the non-interacting KS kinetic energy $T_s[n]$ plus the APD $f^\lambda(r_{12})$ along the DFT adiabatic connection (13) for the expectation value of T; schematically:

$$\langle \Psi | H | \Psi \rangle = \underbrace{\langle \Psi | T | \Psi \rangle}_{T_s[n]+f^\lambda(r_{12})} + \underbrace{\langle \Psi | V_{ee} | \Psi \rangle}_{f(r_{12})} + \underbrace{\langle \Psi | V_{ne} | \Psi \rangle}_{n(\mathbf{r})}. \tag{27}$$

Our aim is to obtain $n(\mathbf{r})$ and $T_s[n]$ via the KS equations, and $f^\lambda(r_{12})$ via Eqs. (25), which can be generalized to any hamiltonian along the DFT adiabatic connection, by simply replacing the physical hamiltonian H with H^λ of Eqs. (13)-(14) in the steps of Sec. 4.

A self-consitent scheme for this construction reads

$$(T + V_{\rm KS})\, \Phi_{\rm KS} = E_{\rm KS} \Phi_{\rm KS} \qquad \Rightarrow \qquad n(\mathbf{r}),\ T_s[n] \tag{28}$$

$$\begin{cases} [-\nabla^2_{r_{12}} + w^\lambda_{\rm eff}(r_{12};[n])]\psi^\lambda_i(r_{12}) = \epsilon^\lambda_i \, \psi^\lambda_i(r_{12}) \\ \sum_i \vartheta_i |\psi^\lambda_i(r_{12})|^2 = f^\lambda(r_{12}), \end{cases} \qquad \Rightarrow \qquad f^\lambda(r_{12}) \tag{29}$$

$$E_0 = \min_{v_{\rm KS}} \left\{ T_s[n] + \int_0^{\lambda_{\rm phys}} d\lambda \int d\mathbf{r}_{12} f^\lambda(r_{12}) \frac{\partial w^\lambda(r_{12})}{\partial \lambda} + \int d\mathbf{r}\, n(\mathbf{r}) v_{ne}(\mathbf{r}) \right\}. \tag{30}$$

The computation starts with a trial $v_{\rm KS}(\mathbf{r})$ in the KS equations, schematically represented by Eq. (28), where $\Phi_{\rm KS}$ is the Slater determinant of KS orbitals. From the KS equations we thus get a first approximation for the density $n(\mathbf{r})$ and the non-interacting kinetic energy $T_s[n]$. Provided that we have a prescription to build an approximate $w_{\rm eff}^{\lambda}$ for a given density $n(\mathbf{r})$ (see next Sec. 6), we can obtain $f^{\lambda}(r_{12})$ along the DFT adiabatic connection from Eqs. (29). In general, this step is not expensive: Eqs. (29) are unidimensional, and if the dependence of $w^{\lambda}(r_{12})$ on λ is smooth, few λ values (5-20) are enough to provide a good estimate of the coupling-constant average. The physical ground-state energy E_0 is then evaluated via Eq. (30). The procedure should then be repeated by optimizing the KS potential $v_{\rm KS}$, so that E_0 is minimum. The N-representability problem of the KS exchange-correlation functional is clearly shifted to the N-representability problem for $f^{\lambda}(r_{12})$. In view of the new conditions derived for the pair density,[20] this seems to leave space for improvements.

6. Properties of the Effective Electron-Electron Interaction

So far, we have only replaced the problem of finding an approximation for $E_{xc}[n]$ with the problem of constructing $w_{\rm eff}^{\lambda}(r_{12};[n])$. In order to proceed further, we study here the properties of $w_{\rm eff}^{\lambda}(r_{12};[n])$.

If we want our Eqs. (28)-(30) to be fully self-consitent, we should impose that for $\lambda=0$ Eqs. (29) yield $f^{\lambda=0}(r_{12})=f_{\rm KS}(r_{12})$, i.e., the same APD we would obtain by inserting the KS Slater determinant of Eq. (28) in Eqs. (6)-(7). This corresponds, in the usual DFT language, to treat exchange exactly. The first property we should thus impose to $w_{\rm eff}^{\lambda}(r_{12})$ is

$$w_{\rm eff}^{\lambda=0}(r_{12})=w_{\rm eff}^{\rm KS}(r_{12}). \tag{31}$$

If we use only one geminal to define $T_g[f]$ in Eq. (24), the property (31) corresponds to $w_{\rm eff}^{\lambda=0}(r_{12})=\nabla^2\sqrt{f_{\rm KS}(r_{12})}/\sqrt{f_{\rm KS}(r_{12})}$. For more than one geminal we need more sophisticated constructions, mathematically equivalent to those used to construct the KS potential $v_{\rm KS}(r)$ for a given spherical density $n(r)$.[23] Equation (31) also provides a very good starting point to build $w_{\rm eff}^{\lambda}(r_{12})$: the KS system already takes into account the fermionic structure and part of the effect of the external potential in the physical problem. What is then left, that needs to be approximated, is the effect of turning on the electron-electron interaction without changing the one-electron density $n(\mathbf{r})$, and the difference between $T_{\rm f}[f]$ and $T_g[f]$.

For confined systems (atoms, molecules) another property to be imposed on $w_{\rm eff}^{\lambda}(r_{12})$ concerns the eigenvalue $\epsilon_{\rm max}^{\lambda}$ corresponding to the highest occupied geminal in Eqs. (29). In fact, the asymptotic behavior of the pair density $P_2(\mathbf{r}_1,\mathbf{r}_2)$ of Eq. (6) for $|\mathbf{r}_1|\to\infty$ (or $|\mathbf{r}_2|\to\infty$) is, in this case,[24]

$$\lim_{|\mathbf{r}_1|\to\infty}P_2(\mathbf{r}_1,\mathbf{r}_2)=n(\mathbf{r}_1)n_{N-1}(\mathbf{r}_2)\{\hat{\mathbf{r}}_1\}, \tag{32}$$

where $n_{N-1}(\mathbf{r})$ is one of the degenerate ground-state densities of the $(N-1)$-electron system (in the same external potential V_{ne}), with the choice depending parametrically upon the direction $\hat{\mathbf{r}}_1 = \mathbf{r}_1/|\mathbf{r}_1|$. A similar expansion holds for the KS pair density, obtained from the KS Slater determinant of Eq. (28),

$$\lim_{|\mathbf{r}_1|\to\infty} P_2^{\rm KS}(\mathbf{r}_1,\mathbf{r}_2) = n(\mathbf{r}_1)n_{N-1}^{\rm KS}(\mathbf{r}_2)\{\hat{\mathbf{r}}_1\}, \tag{33}$$

where we have used the fact that, by construction, the N-electron density is the same for the physical system and for the KS one (while, of course, the corresponding $(N-1)$-electron densities are in general different). For a given attractive (atomic, molecular) external potential vanishing at large distances, the N-electron density is in general more diffuse (decaying slower at large distances) than the $(N-1)$-electron density, so that the asymptotic behavior of the APD $f(r_{12})$ is, for large r_{12}, dominated by the N-electron density decay at large distances. We thus see, from Eqs. (32) and (33), that the corresponding APD's, $f(r_{12})$ and $f_{\rm KS}(r_{12})$, will have the same large-r_{12} decay, $\propto e^{\sqrt{-2\epsilon_{\max}}\,r_{12}}$, with a different prefactor (which can also include a polynomial function of r_{12}), depending on the difference between $n_{N-1}(\mathbf{r})$ and $n_{N-1}^{\rm KS}(\mathbf{r})$. Since the same expansion holds for any $P_2^\lambda(\mathbf{r}_1,\mathbf{r}_2)$ along the DFT adiabatic connection,

$$\lim_{|\mathbf{r}_1|\to\infty} P_2^{\lambda}(\mathbf{r}_1,\mathbf{r}_2) = n(\mathbf{r}_1)n_{N-1}^{\lambda}(\mathbf{r}_2)\{\hat{\mathbf{r}}_1\}, \tag{34}$$

the highest eigenvalue $\epsilon_{\max}^\lambda$ in Eqs. (29) must be independent of λ and equal to the one for the KS APD,

$$\epsilon_{\max}^\lambda = \epsilon_{\max}^{\lambda=0} = \epsilon_{\max}^{\rm KS}. \tag{35}$$

In particular, if we choose only one geminal[22] for the definition of $T_g[f]$, there is only one eigenvalue, which must be the same for every λ.

For an extended system we have scattering states in Eqs. (29). For the special case of the uniform electron gas, Eqs. (29) become equivalent to an approach that was first proposed by Overhauser,[25] and further developed by other authors in the past five years.[26–28] In this approach, the geminal occupancy numbers ϑ_i are the same as the ones for a Slater determinant: occupancy 1 for singlet states (even relative angular momentum ℓ), and occupancy 3 for triplet states (odd relative angular momentum ℓ), up to $N(N-1)/2$ geminals. Rather simple approximations for the effective potential $w_{\rm eff}^\lambda(r_{12})$ gave good results[26–28] for the radial distribution function $g(r_{12})$, when compared with quantum Monte Carlo data. The long-range asymptotic behavior, in this case, corresponds to a phase-shift sum rule for the interaction $w_{\rm eff}^\lambda(r_{12})$.[29] The choice of one geminal for the uniform electron gas has been explored, with remarkable success, in Ref. 30. In this case, the formal similarity with the Fermi-hypernetted-chain approach[31] (FHCN) was exploited to build a good approximation for $w_{\rm eff}^\lambda(r_{12})$ (which was split into $\uparrow\uparrow$ and $\uparrow\downarrow$ contributions).

Finally, the small-r_{12} behavior of the effective potential $w_{\rm eff}^\lambda(r_{12})$ is determined by the choice of the adiabatic connection path. For instance, if we choose $w^\lambda(r_{12}) = \lambda/r_{12}$ in Eqs. (13)-(16), then the APD $f^\lambda(r_{12})$ displays the electron-electron cusp $f^\lambda(r_{12}\to 0) = f^\lambda(0)(1+\lambda r_{12}+...)$, which implies, in turn, that also $w_{\rm eff}^\lambda$ must behave, *for small* r_{12}, as $w_{\text{eff}}^{\lambda}(r_{12} \to 0) = \lambda/r_{12} + ...$ In particular, for $\lambda = \lambda_{\text{phys}}$, we always have $w_{\text{eff}}^{\lambda_{\text{phys}}}(r_{12} \to 0) = 1/r_{12} + ...$, as shown, for the case of the He atom in Fig. 1. If we choose a cuspless nonlinear path, like $w^{\lambda}(r_{12}) = \text{erf}(\lambda r_{12})/r_{12}$, then the small r_{12} behavior of $w_{\text{eff}}^{\lambda}(r_{12})$ is known only when we are approaching the physical interaction.[32]

7. Preliminary Applications

The construction of an approximate $w_{\text{eff}}^{\lambda}(r_{12};[n])$ can thus start with the decomposition

$$w_{\text{eff}}^{\lambda}(r_{12};[n]) = w_{\text{eff}}^{\text{KS}}(r_{12}) + w^{\lambda}(r_{12}) + \Delta w_{\text{eff}}^{\lambda}(r_{12};[n]), \tag{36}$$

where the term $\Delta w_{\text{eff}}^{\lambda}(r_{12};[n])$ should take care of the fact that, when the electron-electron interaction is turned on, the one-electron density $n(\mathbf{r})$ and (for confined systems) the highest eigenvalue $\epsilon_{\text{max}}^{\lambda}$ do not change.

As a starting point, we applied the method of Eqs. (28)-(30) to the He isoelectronic series. In this simple (yet not trivial) 2-electron case, we have the advantage that we can treat $T_f[f]$ exactly. We developed an approximation[6] for $\Delta w_{\text{eff}}^{\lambda}(r_{12};[n])$ based on the one used for the uniform electron gas.[26] This approximation is designed to mimic the conservation of $n(\mathbf{r})$ along the DFT adiabatic connection, but does not take into account the eigenvalue conservation. It works remarkably well when combined with a nonlinear adiabatic connection path $w^{\lambda}(r_{12}) = \text{erf}(\lambda r_{12})/r_{12}$ that separates short- and long-range correlation effects, and is reported in the Appendix of Ref. 6. Preliminary implementations of the self-consitent procedure of Eqs. (28)-(30) yield ground-state energies within 1 mH with respect to full configuration interaction (CI) calculations in the same basis set. However, the way we carried out these first tests was simply based on a direct minimization of few variables parametrizing $v_{\text{KS}}(\mathbf{r})$. This rather inefficient way to implement Eqs. (28)-(30) needs further improvment. Besides, the Kohn-Sham potentials we obtain in this way are very unstable (like the ones of Ref. 33), although the corresponding total energies and electronic densities are stable, and do not display the variational collapse of perturbation-theory-based approximate $E_c[n]$.[34]

8. Perspectives

The generalization to many-electron systems of nonuniform density of the approximation built in Ref. 6 for $w_{\text{eff}}^{\lambda}(r_{12})$ is not straightforward. If we simply apply it to the Be atom case (by using only one geminal), we obtain energy errors of 300 mH. Adding the eigenvalue conservation can improve the results, but, of course, there is not a unique way to impose it. So far, we found that the final outcome strongly depends on how we impose the eigenvalue conservation (i.e., if we use a functional form with one parameter adjusted to keep the eigenvalue independent of λ, the results drastically depend on the chosen functional form). It seems thus necessary to switch to more than one geminal, and/or to find better constructions for $w_{\text{eff}}^{\lambda}(r_{12})$.

In particular, it may be promising to explore the possibility to construct approximations inspired to the FHNC,[30,31,35] and to try to include some of the new results on N-representability conditions for the pair density.[20] Different approximations with respect to the one of Eq. (24) for the functional $T_{\rm f}[f]$, and the use of Eq. (23) also deserve further investigation.

Acknowledgments

We thank E.K.U. Gross for useful discussions and suggestions. One of the author (P.G.G.) gratefully aknowledges the *30th International Worksohop on Condensed Matter Theory* organizers for supporting her participiation to the meeting.

References

1. W. Kohn, Rev. Mod. Phys. **71**, 1253 (1999).
2. A.E. Mattsson, Science **298**, 759 (2002).
3. C. Fiolhais, F. Nogueira, and M. Marques (eds.), *A Primer in Density Functional Theory* (Springer-Verlag, Berlin, 2003).
4. J. P. Perdew, A. Ruzsinszky, J. Tao, V. N. Staroverov, G. E. Scuseria, and G. I. Csonka, J. Chem. Phys. **123**, 062201 (2005).
5. see, e.g., T. Leininger, H. Stoll, H.-J. Werner, and A. Savin, Chem. Phys. Lett. **275**, 151 (1997); R. Pollet, A. Savin, T. Leininger, and H. Stoll, J. Chem. Phys. **116**, 1250 (2002); J. Toulouse, F. Colonna, and A. Savin, Phys. Rev. A **70**, 062505 (2004); J. Toulouse, F. Colonna, and A. Savin, J. Chem. Phys. **122**, 014110 (2005); J. G. Ángyán, I.C. Gerber, A. Savin, and J. Toulouse, Phys. Rev. A **72**, 012510 (2005); E. Goll, H.-J. Werner and H. Stoll, Phys. Chem. Chem. Phys. **7** , 3917 (2005); E. Goll, H.-J. Werner, H. Stoll, T. Leininger, P. Gori-Giorgi, and A. Savin, Chem. Phys. **329**, 276 (2006).
6. P. Gori-Giorgi and A. Savin, Phys. Rev. A **71**, 032513 (2005).
7. P. Gori-Giorgi and A. Savin, Philos. Mag. **86**, 2643 (2006).
8. A.J. Thakkar, A.N. Tripathi, and V.H. Smith, Jr., Int. J. Quantum Chem. **26**, 157 (1984), and references therein.
9. C.A. Coulson and A.H. Neilson, Proc. Phys. Soc. London **78**, 831 (1961).
10. J. Cioslowski, B.B. Stefanov, A. Tan, and C.J. Umrigar, J. Chem. Phys. **103**, 6093 (1995).
11. J. Cioslowski and G. Liu, J. Chem. Phys. **109**, 8225 (1998).
12. E. Valderrama, J.M. Ugalde, and R.J. Boyd, in *Many-electron densities and reduced density matrices*, edited by J. Cioslowski (Kluwer Academic/Plenum Publishers, New York, 2000).
13. E.R. Davidson, *Reduced Density Matrices in Quantum Chemistry* (Academic Press, New York, 1976).
14. A.J. Coleman and V.I. Yukalov, *Reduced Density Matrices: Coulson's Challenge* (Springer-Verlag, New York, 2000).
15. M. Levy, Proc. Natl. Acad. Sci. U.S.A. **76**, 6062 (1979).
16. E. Lieb, Int. J. Quantum Chem. **24**, 243 (1983).
17. J. Harris and R. Jones, J. Phys. F **4**, 1170 (1974); D.C. Langreth and J.P. Perdew, Solid State Commun. **17**, 1425 (1975); O. Gunnarsson and B.I. Lundqvist, Phys. Rev. B **13**, 4274 (1976).
18. W. Yang, J. Chem. Phys. **109**, 10107 (1998).

19. A. Savin, F. Colonna, and R. Pollet, Int. J. Quantum Chem. **93**, 166 (2003).
20. P. W. Ayers and E. R. Davidson, Int. J. Quantum Chem. **106**, 1487 (2006); P. W. Ayers, Phys. Rev. A **74**, 042502 (2006); P. W. Ayers and E. R. Davidson, Adv. Chem. Phys., in press.
21. D.E. Freund, B.D. Huxtable, and J.D. Morgan III, Phys. Rev. A **29**, 980 (1984). We used an improved version (provided to us by C. Umrigar) of the accurate variational wavefunctions described in this work. See also C.J. Umrigar and X. Gonze, Phys. Rev. A **50**, 3827 (1994), and Ref. 10.
22. A. Nagy, preprint.
23. F. Colonna and A. Savin, J. Chem. Phys. **110**, 2828 (1999); Q. Zhao, R. C. Morrison, and R. G. Parr, Phys. Rev. A **50**, 2138 (1994); R. van Leeuwen and E. J. Baerends Phys. Rev. A **49**, 2421 (1994).
24. M. Levy, J.P. Perdew, and V. Sahni, Phys. Rev. A **30**, 2745 (1984); M. Ernzerhof, K. Burke, and J.P. Perdew, J. Chem. Phys. **105**, 2798 (1996).
25. A.W. Overhauser, Can. J. Phys. **73**, 683 (1995).
26. P. Gori-Giorgi and J.P. Perdew, Phys. Rev. B **64**, 155102 (2001).
27. B. Davoudi, M. Polini, R. Asgari, and M.P. Tosi, Phys. Rev. B **66**, 075110 (2002).
28. M. Corona, P. Gori-Giorgi, and J.P. Perdew, Phys. Rev. B **69**, 045108 (2004); I. Nagy, R. Diez Muiño, J.I. Juaristi, and P.M. Echenique, Phys. Rev. B **69**, 233105 (2004).
29. P. Ziesche, Phys. Rev. B **67**, 233102 (2003); physica status solidi (b) **242**, 2051 (2005).
30. B. Davoudi, R. Asgari, M. Polini, and M. P. Tosi Phys. Rev. B **68**, 155112 (2003).
31. see, e.g., E. Krotscheck and M. Saarela, Phys. Rep. **232**, 1 (1993).
32. P. Gori-Giorgi and A. Savin, Phys. Rev. A **73**, 032506 (2006).
33. V. N. Staroverov, G. E. Scuseria, and E. R. Davidson, J. Chem. Phys. **124**, 141103 (2006).
34. P. Mori-Snchez, Q. Wu, and W. Yang, J. Chem. Phys. **123**, 062204 (2005); D. R. Rohr, O. V. Gritsenko, and E. J. Baerends, Chem. Phys. Lett., in press.
35. E. Krotscheck, Phys. Lett. A **190**, 201 (1994).

CORRELATIONS, COLLISION FREQUENCY AND OPTICAL PROPERTIES IN LASER EXCITED CLUSTERS

H. REINHOLZ*, T. RAITZA and G. RÖPKE

Universität Rostock, Institüt für Physik, D-18051 Rostock, Germany

**E-mail: heidi.reinholz@uni-rostock.de*

Within linear response theory, the dielectric and optical properties of a charged particle system are related to equilibrium correlation functions. In particular, the dynamical conductivity and the dynamical collision frequency are expressed in terms of the current-current or force-force correlation function, which can be evaluated analytically using perturbation expansions or numerically by MD simulations. Results are given for bulk material. Furthermore, finite systems such as laser excited clusters are considered. It is shown that the collision frequency is reduced in finite systems. The interaction with the laser field is discussed with respect to the current-current correlation function which changes with time due to the expansion of the laser-irradiated cluster.

Keywords: Laser excited clusters; dielectric function; molecular dynamics simulation; absorption; dynamical conductivity.

1. Introduction

The response of a charged particle system to an external field contains many different phenomena, in particular transport phenomena as well as absorption and emission of radiation. As long as the external fields are weak the reaction of the system can be considered within linear response theory. For systems near to thermal equilibrium, the corresponding theory has been elaborated since 1957 when the Kubo formula[1] was invented which relates current-current correlation functions in equilibrium to the electrical conductivity. The relation between fluctuations in equilibrium and friction, however, had already been given a long time ago in the famous Einstein paper on Brownian motion in 1905[2] and is a special version of the more general fluctuation-dissipation theorem. A general treatment of linear response can be given starting from non-equilibrium statistical mechanics, see Zubarev *et al.*[3] These general relations expressing frequency and wave number dependent transport properties via equilibrium correlation functions are valid for interacting systems at arbitrary interaction strength and degeneracy. They may serve as starting point for analytical calculations within perturbative expansions of equilibrium correlation functions, e.g. using Green function techniques, but alternatively also for simulations.

In particular, MD simulations have been elaborated which treat the system in quasi-classical approximation replacing the Coulomb interaction by a pseudopotential.[4] A strict quantum treatment can be achieved by applying, e.g., wave packet molecular dynamics,[5] quantum molecular dynamics[6] or path integral Monte Carlo techniques.[7,8]

A recent description of linear response theory and applications to optical properties of dense plasmas has been given by Reinholz.[9] Starting from an extended set of relevant fluctuating observables, the approach has been developed further in two directions, the inclusion of strong external fields leading to a non-linear response of the system[11,12] but also to non-equilibrium situations such as a two-temperature plasma.[13] We are interested here in finite systems, in particular nano-plasmas produced from laser-excited atomic clusters, and in non-equilibrium situations during the expansion of the highly excited cluster.

Plasma properties such as reflectivity and absorption of electromagnetic waves are related to the correlation functions, in particular the dipole-dipole correlation function. For bulk plasmas in equilibrium, correlation functions in relation to dielectric function and optical properties have been considered by Reinholz *et al.*[9,10] The investigation of highly excited finite systems is a challenging problem from experimental[14,15] as well as from theoretical[16–20] point of view.

Within the present paper we start from the description of dielectric response in bulk material in Sec. 2, present some first exploratory results for MD simulations of clusters in strong laser fields in Sec. 3 and discuss the results in Sec. 4.

2. Dielectric Response in Bulk Material

The response of a charged particle system to an external, frequency and wave-vector dependent electrical field is given by the dielectric function $\epsilon(\vec{k},\omega)$ which, for isotropic media, can be decomposed into transverse and longitudinal components. Restricting ourselves to the long-wavelength limit $k \to 0$ what is justified in the optical region discussed here, both components become identical so that we focus here only on the longitudinal response.

The dielectric function is related to the polarization function $\Pi(\vec{k},\omega)$ and the dynamical conductivity $\sigma(\vec{k},\omega)$ according to

$$\epsilon(\vec{k},\omega) = 1 - V(k)\Pi(\vec{k},\omega) = 1 + i\frac{\sigma(\vec{k},\omega)}{\epsilon_0\omega} \tag{1}$$

where $V(k) = (\epsilon_0 \Omega k^2)^{-1}$ is the Coulomb interaction, with Ω as renormalization volume. According to Kubo[1] the dynamical conductivity can be related to the current-current correlation function

$$\sigma(\vec{k},\omega) = \beta\Omega\langle j_k; j_k\rangle_{\omega+i\eta}. \tag{2}$$

The electrical current density is due to all charge carriers c,

$$\vec{j}_k = \frac{1}{\Omega}\sum_{c,i} e^{i\vec{k}\cdot\vec{r}_{c,i}}\frac{e_c}{m_c}\vec{p}_{c,i} = \frac{1}{\Omega}\sum_{p,i}\frac{e_c}{m_c}\hbar\vec{p}a^{\dagger}_{c,p+k/2}a_{c,p-k/2}\,. \tag{3}$$

The averaging is over the equilibrium distribution $\varrho = \exp(-\beta H + \beta\mu N)/Z$,

$$\langle j_k; j_k\rangle_{\omega+i\eta} = \frac{1}{\beta}\int_0^\infty dt \int_0^\beta d\tau e^{i(\omega+i\eta)t}\mathrm{Tr}\left\{\varrho j_k^\dagger(t - i\hbar\tau) j_k\right\} . \tag{4}$$

The time evolution $\vec{j}(t)$ is determined by the Hamiltonian which, in the case of a Coulomb system, contains the Coulomb interaction $V_{cd}(k) = e_c e_d V(k)$. In the classical limit, the τ integral reduces to a factor β.

2.1. *Dynamical collision frequency*

The calculation of the current-current correlation function (4) has been performed for quasiclassical systems. In order to take into account quantum effects the Coulomb interaction was replaced by the Kelbg interaction. Details of the interaction potential and the MD simulation technique are given by Reinholz *et al.*[21] In the long-wavelength limit $k \to 0$, an oscillatory behavior is obtained with a damping rate, similar to the behavior obtained from a simple Drude model. The phenomenological Drude ansatz assumes a static collision frequency $\nu(0)$ which is related to the dc conductivity via $\sigma_{\rm dc} = \epsilon_0\,\omega_{\rm pl}^2/\nu(0)$.

For the analysis of the MD data, however, we introduce a generalized Drude formula which contains a frequency dependent collision frequency $\nu(\omega)$,

$$\epsilon(\omega) = 1 + \frac{\omega_{\rm pl}^2}{\omega(\omega + i\nu(\omega))} , \tag{5}$$

where $\omega_{\rm pl} = \sqrt{e^2 n_{\rm e}/(\epsilon_0 m_{\rm e})}$ is the plasma frequency, with $n_{\rm e}$ and $m_{\rm e}$ the electron density and mass, respectively. As was shown by Reinholz *et al.*[22] for a description consistent with known limiting cases and generally valid sum rules for the dielectric function, the collision frequency has to be a complex frequency dependent quantity. This can be achieved following from a generalized linear response theory.[3,22] The collision frequency is expressed in terms of equilibrium correlation functions which is a realization of the fluctuation dissipation theorem. Results for special plasma parameter values[21] are shown in Fig. 1.

The analytical evaluation of the correlation functions can be performed within perturbation theory. For the current-current correlation function, a direct perturbative treatment is not possible since the conductivity of a non-interacting plasma is diverging already in the zeroth order term. In contrast, for the collision frequency a perturbative expansion is possible starting with zero for the non-interacting plasma.

Relating $\nu(\omega)$ to a force-force correlation function, we find in Born approximation

$$\nu^{\rm Born}(\omega) = -\frac{i\epsilon_0 n_{\rm i}\,\Omega^2}{6\pi^2 e^2\, n_{\rm e} m_{\rm e}}\int_0^\infty dq\, q^6 V_{\rm ei}^2(q) S_{\rm i}(q)\frac{1}{\omega}[\epsilon_{\rm RPA}(q,\omega) - \epsilon_{\rm RPA}(q,0)]. \tag{6}$$

For the electronic part of the dielectric function in random phase approximation (RPA), $\epsilon_{\rm RPA}(q,\omega)$, there are analytical expressions available.[24] The ionic distribution is taken into account via the ionic static structure factor $S_{\rm i}(q)$ assuming that

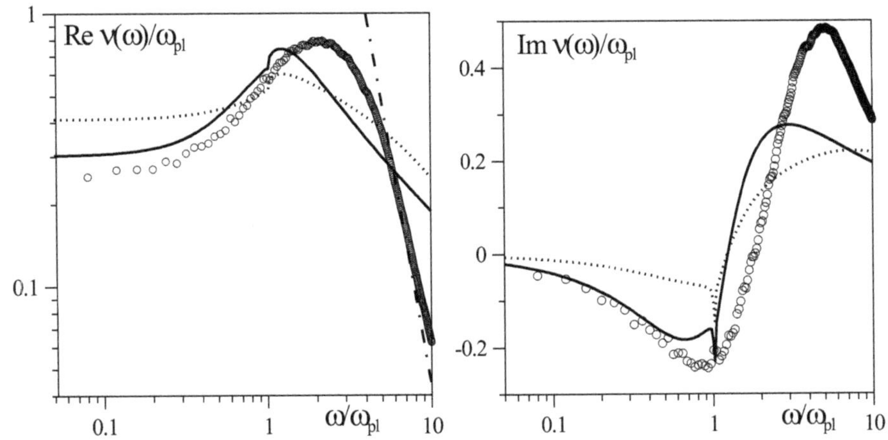

Fig. 1. Comparison between MD data for the Kelbg pseudopotential (○) and the quantum statistical treatment for the Coulomb potential.[23] The Gould–DeWitt scheme, see text, is used. Full line – including renormalization factor $r(\omega)$ in Eq. (9), dotted line – without renormalization factor. The dot-dashed line gives the analytical result for the high-frequency behavior of Re $\nu(\omega)$ for the Born approximation Eq. (6) with Kelbg potential. $\Gamma = 1.28$, $T = 33\,000$K.

the electrons are scattered from ions at fixed positions. $V_{\rm ei}(q)$ is the Coulomb potential. The Born approximation is calculated for a statically screened potential (Debye potential, screening only by electrons, $S_{\rm i}(q) = 1$) as

$$\nu^{\rm D}(\omega) = ig \int_0^\infty dy\, \frac{y^4 e^{-y^2}}{(\bar{n} + y^2)^2} \int_{-\infty}^\infty dx\, \frac{e^{-2xy} - e^{2xy}}{xy} \, \frac{e^{-x^2}}{xy - (\bar{\omega} + i\bar{\eta})} \tag{7}$$

with

$$g = \frac{1}{24\sqrt{2}\pi^{5/2}} \frac{n_{\rm e} e^4 \beta^{3/2}}{m_{\rm e}^{1/2} \epsilon_0^2}, \qquad \bar{\omega} = \frac{\hbar\omega}{4k_{\rm B}T}, \qquad \bar{n} = \frac{\hbar^2 e^2 n_{\rm e}}{8\epsilon_0 m_{\rm e} (k_{\rm B}T)^2}.$$

In an attempt to improve the Born approximation (6) to higher order perturbation theory, several aspects have been discussed and evaluated in the literature, see, e.g., Reinholz[9] and references given therein. Considering screening processes, the Lenard-Balescu collision term (LB) is obtained where the bare Coulomb potential is replaced by a dynamically screened one. The behavior at small electron–ion distances is determined by strong binary collisions which are accounted for by a higher order perturbation expansion, replacing the interaction potential $V_{\rm ei}$ by a t-matrix (TM). A dynamically screened TM approximation gives a consistent description of collisions combining dynamical screening and strong collisions. It can be approximated by the following so-called Gould-DeWitt ansatz

$$\nu^{\rm GD}(\omega) = \nu^{\rm TM}(\omega) - \nu^{\rm Born}(\omega) + \nu^{\rm LB}(\omega) \tag{8}$$

which is equivalent to the dynamically screened binary collision approximation for the collision frequency. For the TM approximation a static potential such as the

Debye potential is taken. To avoid double counting, the collision frequency in Born approximation (6) with respect to the static potential has to be subtracted since the Born approximation is also taken into account in $\nu^{\rm LB}(\omega)$.

In context with Eq. (5), the frequency-dependent conductivity will be rewritten as follows[9,22]

$$\sigma(\omega) = \frac{\epsilon_0\,\omega_{\rm pl}^2}{-i\,\omega + r(\omega)\,\nu^{\rm GD}(\omega)}\,, \tag{9}$$

where $\nu^{\rm GD}(\omega)$ is determined according to Eq. (8). The renormalization factor $r(\omega)$ takes into account corrections due to higher moments of the distribution function, in particular electron-electron correlations which is equivalent to, e.g., the Chapman-Enskog method[25] in solving kinetic equations. In the special case of zero-frequency limit (the dc conductivity $\sigma_{\rm dc} = \epsilon_0\,\omega_{\rm pl}^2/\nu(0)$), the renormalization factor $r(0)$ gives the correct prefactor known as Spitzer result given below, see also Reinholz *et al.*[22]

2.2. *DC Conductivity*

We give a short review of the electrical dc conductivity. For a neutral system of free charged particles, e.g. electrons and singly charged ions with respective densities $n_{\rm e} = n_{\rm i}$, interacting via Coulombic forces, the dc conductivity $\sigma_{\rm dc}^{\rm Coulomb}$ depends on the density $n_{\rm e}$, the temperature T and the mass ratio $m_{\rm i}/m_{\rm e}$. However, we consider here only the adiabatic limit of infinite mass ratio $m_{\rm i}/m_{\rm e} \to \infty$. Instead of $n_{\rm e}$ and T, we can also introduce dimensionless parameters, the coupling parameter

$$\Gamma = \frac{(4\pi n_{\rm e}/3)^{1/3}\;e^2}{4\pi\epsilon_0 k_{\rm B}T} \tag{10}$$

and the degeneracy parameter $\Theta = (3\pi^2 n_{\rm e})^{-2/3}\;2m_{\rm e}k_{\rm B}T/\hbar^2$.

The Coulomb conductivity for a fully ionized plasma in the classical ($\Theta \ll 1$), low-density ($\Gamma \ll 1$) limit has been given by Spitzer[26]

$$\sigma_{\rm Sp}(\Gamma) = \lim_{\Gamma\to 0}\lim_{\Theta\to\infty}\sigma_{\rm dc}^{\rm Coulomb}(\Gamma,\Theta) = 0.591\frac{2(4\pi\epsilon_0)^2(k_BT)^{3/2}}{e^2 m_{\rm e}^{1/2}\ln(\frac{3}{2}\Gamma^{-3})}\,. \tag{11}$$

Analytical expressions applicable to a larger region of temperatures and densities have since been derived within the relaxation time approximation[27,28] and linear response theory.[10,21,29] In the latter approach, transport coefficients are expressed in terms of equilibrium correlation functions. In the weak coupling regime $\Gamma \leq 1$ where perturbation expansions are applicable, a systematic treatment of many-particle effects at arbitrary degeneracy is possible.[29,30] At higher coupling strengths, simulation techniques can be applied.[23,31–33] A good coincidence between both approaches, the analytical results of perturbative expansions in the low-coupling limit and molecular dynamics simulations using pseudo-potentials, has been demonstrated recently.[23] On the other hand, comparison with experiments is of interest.

The following interpolation formula for the static case has been constructed based on correct analytical behavior in limiting cases and simulation data for intermediate regions,[30]

$$\sigma_{\mathrm{dc}}^{\mathrm{ERR}}(\Gamma,\Theta) = a_0 T^{3/2}\left(1+\frac{b_1}{\Theta^{3/2}}\right)\left[D\ln(1+A+B) - C - \frac{b_2}{b_2+\Gamma\Theta}\right]^{-1} \tag{12}$$

where the temperature T in K, σ in $(\Omega\mathrm{m})^{-1}$, and with the functions

$$A = \Gamma^{-3}\frac{1+a_4/\Gamma^2\Theta}{1+a_2/\Gamma^2\Theta + a_3/\Gamma^4\Theta^2}\left[a_1 + c_1\ln(c_2\Gamma^{3/2}+1)\right]^2,$$

$$B = \frac{b_3(1+c_3\Theta)}{\Gamma\Theta(1+c_3\Theta^{4/5})},$$

$$C = \frac{c_4}{\ln(1+\Gamma^{-1}) + c_5\Gamma^2\Theta},$$

$$D = \frac{\Gamma^3 + a_5(1+a_6\Gamma^{3/2})}{\Gamma^3 + a_5}.$$

The set of parameters is presented in Tab. 1. The validity was checked to be extended to about $\Gamma \le 2$, but covers any degree of degeneration.

a_0	a_1	a_2	a_3	a_4	a_5	a_6
0.03064	1.1590	0.698	0.4876	0.1748	0.1	0.258

b_1	b_2	b_3	c_1	c_2	c_3	c_4	c_5
1.95	2.88	3.6	1.5	6.2	0.3	0.35	0.1

In Fig. 1, we compare MD data with the evaluation of analytical expressions for the real and imaginary part of the collision frequency.[23] For the classical MD simulations, quantum effects were taken into account by constructing pseudopotentials using thermodynamic properties. Therefore, the static and low frequency behavior shows good agreement. In the high-frequency limit, quantum effects for Coulomb systems are not reflected adequately in the classical MD simulation results using the Kelbg pseudopotential.

From MD simulations for the current auto-correlation function, the dc conductivity σ_{dc} is obtained. Calculations have been performed for a fixed temperature of $T = 33\,000$ K and varying coupling parameter Γ. Results for the static collision frequency and the static conductivity are shown in Fig. 2.[23] Considering the MD simulations for $\Gamma \le 1$, the systematic behavior agrees very well with the analytical results obtained from the interpolation formula (12) shown for comparison. Discrepancies arise for higher values of Γ. However, the principal behavior of the MD simulations can be reproduced with the interpolation formula.

The theoretical description, based on analytical expressions and MD simulations, leads to a good understanding of experimental results which are shown in the left figure as well. Experimentally, the dc conductivity is determined for partially

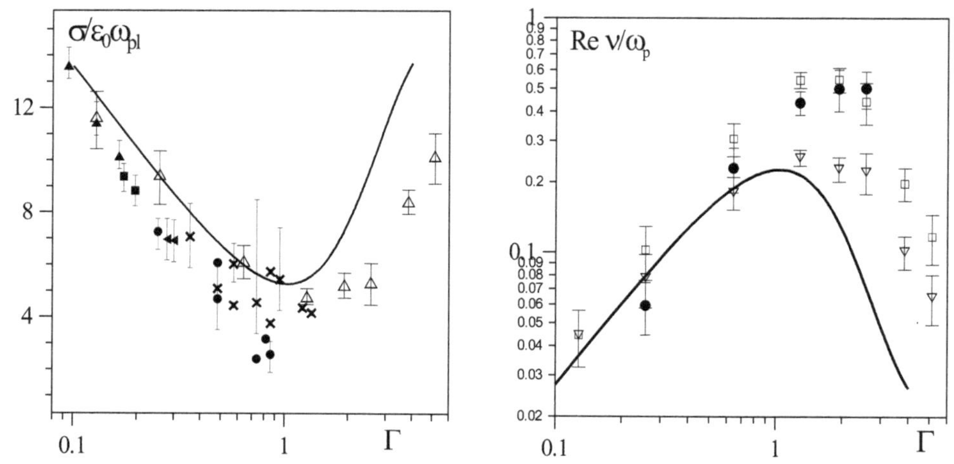

Fig. 2. Static conductivity (**left**) and dynamical conductivity (**right**) in dependence on the coupling parameter Γ. MD results for the Kelbg pseudopotential for $T = 33\,000$ K: Δ, ∇ – $\omega = 0$, □ – $\omega = \omega_{\rm pl}$ (plasma frequency), • – from collisional damping of the Langmuir waves, full line – interpolation formula (12) for $T = 33\,000$ K, experimental data: ●[34], ×[35], ■[36], ▲[37] and ◀.[38]

ionized plasma. In order to compare with a fully ionized plasma model we have to separate the contribution of free charge carriers.[39] However, the plasma parameters like temperature and density are not directly accessible within the shock wave experiments

The dielectric function is a quantity which is of relevance in various situations where the coupling of a charged particle system with a radiation field is considered. In a first approximation, often the RPA is applied to describe the dielectric response of the Coulomb system. Screening effects are treated on a self-consistent mean field level, but collisions are neglected. It provides closed analytical expressions[24] and can be implemented in an easy way, see textbooks, e.g. Mahan.[40] Collective excitations such as plasmons are due to the polarization of free charged particles, leading to self-consistent treatment of screening. Despite the fact that collisions are not considered, the RPA has been proven to be successful in describing the plasma response in many applications. However, different phenomena demand for a description that goes beyond RPA in the sense that collisions will be relevant. Exemplarily, we mention bremsstrahlung,[41] reflectivity[42] and Thomson scattering.[43] In each case, there is a very close link to experiments which have been either done or are intended. Of interest are also further applications to physical situations where the interaction of charged particle systems with radiation, in particular high-intensity lasers, is relevant.

3. MD Simulations of Laser Excited Clusters

In the previous section, we evaluated current correlation functions in bulk material and derived the dynamical collision frequency, comparing analytical calculations with MD simulations. Now, we consider finite systems. The behavior of clusters excited by intense laser beams is subject of current research, see Sec. 1. An important feature is the dipole resonance (Mie plasmon) at the frequency $\omega_{\rm Mie} = \omega_{\rm pl}/\sqrt{3}$ which is essential for the dielectric behavior. To describe the damping of these oscillations, collisions have to be considered. The investigation of the dynamical collision frequency for clusters has been performed recently using MD simulations.[20] We will describe the behavior of relatively small clusters after excitation with intense short-pulse laser beams.

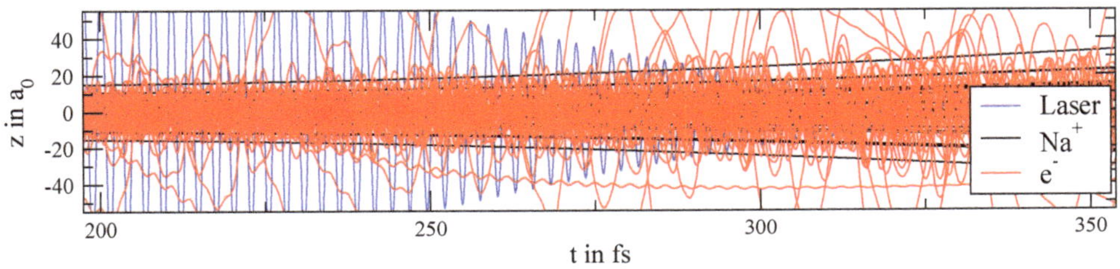

Fig. 3. Time evolution of laser field (periodic oscillations, arb. units), electron (disordered trajectories) and ion (straight trajectories) motion from MD simulations. (color online)

We consider a nano cluster consisting of 55 sodium atoms (one electron per singly charged ion). For the classical MD simulation, the electron-ion interaction is approximated by an error-function-potential $V_{\rm ei}^{\rm erf}(r) = -\frac{Z_{\rm i}e^2}{r}{\rm erf}(\frac{r}{\sigma_{\rm ei}})$. For single ionization, $Z_{\rm i} = 1$ and $\sigma_{\rm ei} = 6.03\ {\rm a_B}$ were chosen in such a way that the ionization energy is reproduced correctly. The cluster is irradiated by a laser field which exerts an additional force on the electrons and ions. Initially, the electrons are positioned on top of the ions for which a three-shell configuration is assumed.

The time evolution of the cluster under the influence of the external laser field is calculated using a MD code from Suraud *et al.*[19] The amplitude of the laser field of 800 nm wavelength and 100 fs pulse width (Gaussian shaped, intensity 10^{12} W $\rm cm^{-2}$) as well as the electron and ion trajectories are shown in Fig 3. Absorbing energy from the electromagnetic field, electrons become unbound within the cluster, forming a nanoplasma. On the time scale of 100 fs hot electrons escape from the cluster. Ion motion becomes essential on the timescale of ps, leading to the expansion of the cluster.

We evaluated the electrical current due to all particles within the cluster which is given as function of time in Fig. 4. We have separated the emitted electrons from the ones still left within the cluster which leads to an overall 20 fold positive charge of the cluster. The current in the cluster is fluctuating even after the pulse has ceased. The internal structure of fluctuations will become more obvious by considering the correlation functions. For the charged particles within the cluster we determine the

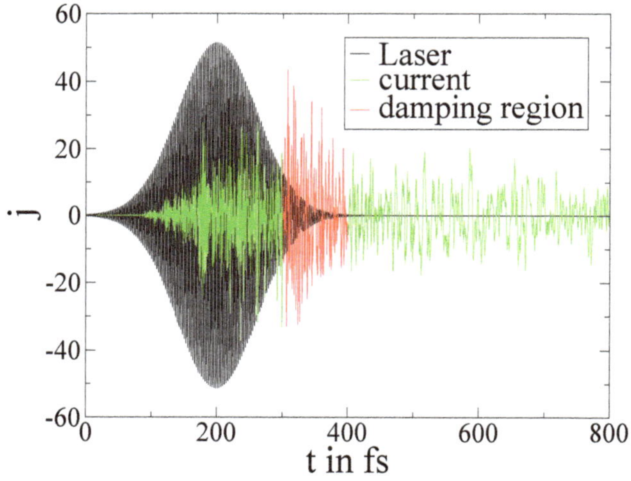

Fig. 4. Electrical current in the laser excited cluster as function of time. (color online)

classical normalized current-current correlation function (4) at $k = 0$

$$K(t) = \frac{1}{J^2(t)\Delta t} \int_0^{\Delta t} \mathrm{d}\tau J(t-\tau)J(\tau), \tag{13}$$

see Fig. 5. In addition, the real part of the Laplace transform $K(\omega)$,

$$K(\omega) = \frac{\beta\Omega}{\epsilon_0 \omega_{\rm pl}^2} \langle J^z; J^z \rangle_{\omega+i\eta} = \lim_{\epsilon\to 0} \int_0^\infty e^{i(\omega+i\epsilon)t} K(t)\, \mathrm{d}t \tag{14}$$

is shown on the right. $K(\omega)$ is related to the frequency dependent collision frequency

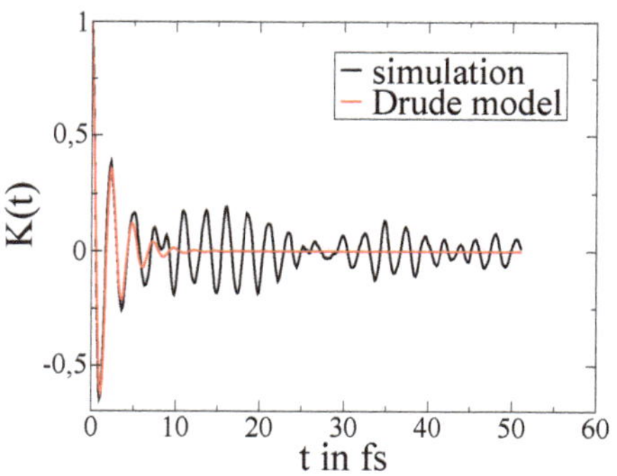

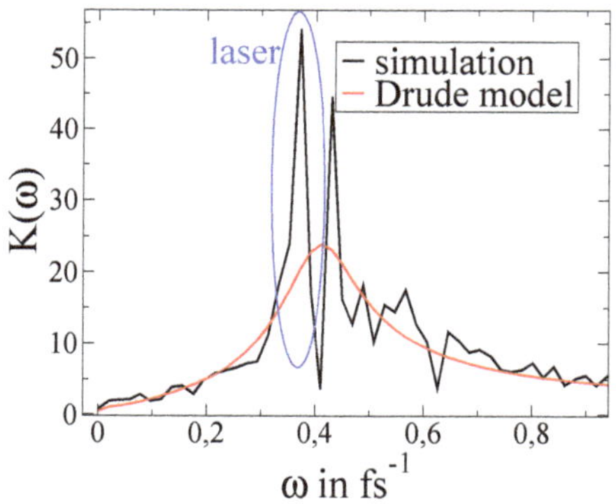

Fig. 5. Normalized current-current correlation function during laser irradiation of cluster (**left**) and its Laplace transfoms (**right**) from MD simulations (black) and Drude model (grey). (color online)

$\nu(\omega)$ via

$$\frac{\nu(\omega)}{\omega_{\rm pl}} = \frac{\epsilon_0 \omega_{\rm pl}}{K(\omega)} + i\left(\frac{\omega}{\omega_{\rm pl}} - \frac{\omega_{\rm pl}}{\omega}\right) . \tag{15}$$

In addition to the oscillatory behavior, a beat frequency is observed due to the overlap of irradiation field and plasma oscillations. This is seen as a double peak in the frequency space. For comparison, a simple Drude behavior[21]

$$K^{\rm Drude}(t) = \exp\left\{-\frac{\nu}{2}t\right\}\left[-\frac{\nu}{2z}\sin(zt) + \cos(zt)\right] \quad z = \sqrt{\omega_{\rm Mie}^2 - \frac{\nu^2}{4}} \tag{16}$$

is assumed and a resonance frequency $\omega_{\rm Mie} = 2.95\ {\rm fs}^{-1}$ and a collision frequency of $\nu = 1.39\ {\rm fs}^{-1}$ can be deduced. We found that $\omega_{\rm Mie}$ agrees well with the Mie frequency found directly from the free electron density in the cluster obtained during the simulations.

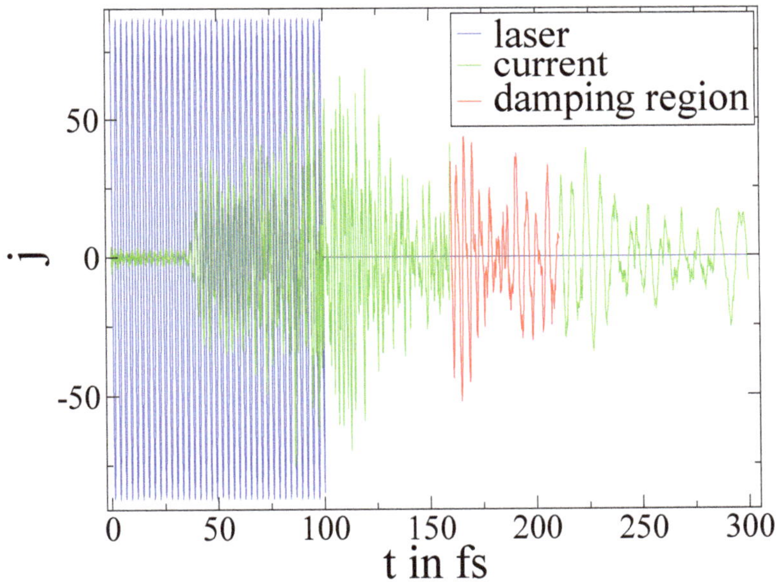

Fig. 6. Electrical current fluctuations after rectangular pulse shape for the time interval from 160 till 210 fs is analyzed. (color online)

In order to avoid the interference of plasma oscillations with the laser field we now consider a square shaped laser pulse and calculate the current–current correlation functions 60 fs after the pulse ended, in an additional simulation run. A typical trajectory for the fluctuation of the current is shown in Fig. 6. Averaging over 10 runs, the current-current correlation function is shown in Fig. 7. From this, the plasma frequency $\omega_{\rm Mie} = 1.37\ {\rm fs}^{-1}$ and the collision frequency $\nu = 0.6\ {\rm fs}^{-1}$ has been obtained. The change in the plasma frequency corresponds to the change of density of the expanding cluster.

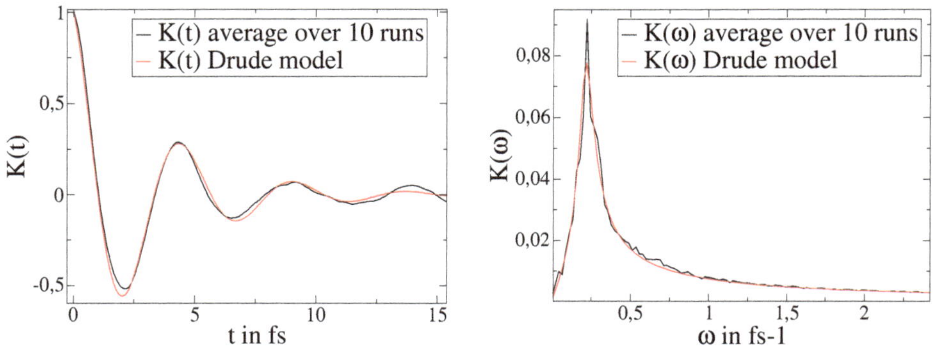

Fig. 7. As in Fig. 5 considered 60 fs after irradiation with a rectangular shaped laser pulse. (color online)

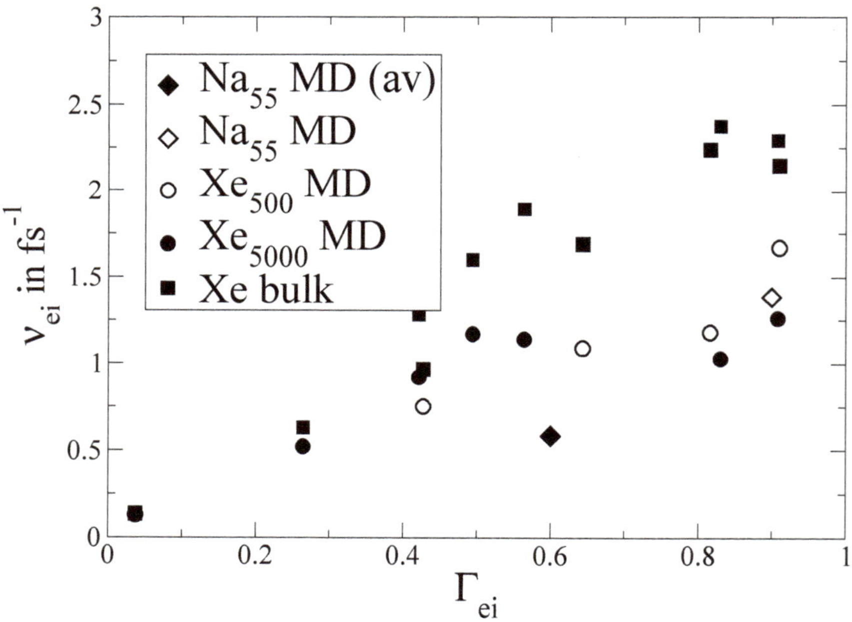

Fig. 8. Collision frequency for plasmas with different coupling strength Γ for clusters of different size and bulk material. (color online)

We can attribute the density and temperature values obtained from the MD simulations to a value of the coupling parameter Γ, Eq. (10). In Fig. 8, the collision frequency is shown as function of Γ. The fit to a static collision frequency $\nu(0)$, see Fig. 2, is compared with calculations for the bulk material[9] and simulations for large clusters.[20] The collision frequency depends on the cluster size and decreases with decreasing cluster size. This may be related to the increasing proportion of particles near the surface.

4. Discussion

Our focus is the dynamical collision frequency which gives the possibility to include the effects of collisions and correlations into the theory of the dielectric function. The results obtained in the long-wavelength limit $k \to 0$ may be extended to the entire $\vec{k}, \omega$ plane using the Mermin approach.[44,45] In general, the dynamical collision frequency is a complex quantity obeying the Kramers-Kronig relations.

For *bulk* materials, a reasonable agreement between MD simulations and analytical calculations using quantum statistical methods has been achieved. In particular, in the dc limit a good coincidence of calculated results for the conductivity is observed, and comparison with experiments in high dense plasmas is satisfactorily. The results are of relevance for other optical properties such as reflectivity, inverse bremsstrahlung or Thomson scattering. An open problem, however, is the application to partially ionized plasmas where in addition to the free electrons also bound electrons arise. First results are obtained within the chemical picture, but further work is needed to get a unified picture to describe correlations in strongly coupled Coulomb systems.

The extension of the optical response to *finite* systems is an actual topic of the present investigation. Using a MD simulation code, the properties of laser-irradiated clusters are calculated, including a quasiclassical description based on a pseudopotential. Averaging over an adequate ensemble of realizations is needed. Deviations from the bulk behavior are strong for small cluster sizes as considered here. Similar calculations for comparatively large Xe clusters (5000 atoms) have been performed by Brabec *et al.*[20] We found that the collision frequency depends on the cluster size and becomes smaller with decreasing number of atoms. The deviation from bulk behavior may be related to the increasing contribution of particles which are located near the surface.

From the current-current correlation function we can derive optical properties such as the emission, absorption and scattering of light by the excited clusters. Such results would be of high interest with future experiments planned at new facilities such as X-ray FELs. The proper treatment of collisions in dense Coulomb systems is mandatory to describe the properties of these highly excited states of condensed matter, in particular for diagnostic tools.

In contrast to near-equilibrium situations in bulk materials, laser excited clusters follow a time evolution where the density and temperature changes. For large clusters such as liquid droplets, hydrodynamical calculations can be used to describe the expansion of excited droplets. Time dependent density and temperature profiles can also be obtained from MD simulations. The correlation functions are explicitly depending on time reflecting a nonequilibrium situation. Performing the evaluation of the correlation function at different times, we observed variation of the plasma frequency as well as of the collision frequency. In further calculations, the variation in space would also be of interest. The evaluation of correlation functions could be performed at different positions such as the central or peripheral part of the cluster.

As already mentioned for the bulk systems, for a more strict description of the nanoplasma in laser excited clusters we have to consider quantum effects. In particular, ionization and recombination are of importance to describe the initial process of energy deposition, the charging of the cluster by emission of hot electrons, and the recombination of electrons with ions in particular in the late stages of cluster expansion.

Acknowledgment

The authors are indebted to Eric Suraud for help with the MD code. This work was supported by the DFG funded SFB 652.

References

1. R. Kubo, *J. Phys. Soc. Jpn.* **12**, 570 (1957).
2. A. Einstein, *Ann. Physik (Leipzig)* **17**, 132 (1905).
3. D. Zubarev, V. Morozov, and G. Röpke, *Statistical Mechanics of Nonequlibrium Processes* (Akademie-Verlag, Berlin, 1997), Vol. 1,2.
4. I. Morozov, G. Norman and A. Valuev, *Phys. Rev. E* **63**, 036405 (2001); I. Morozov and G. Norman, *JETP* **100**, 370 (2005).
5. M. Knaup, P.-G. Reinhard and C. Toepffer, *Contrib. Plasma Phys.* **41**, 159 (2001).
6. M. P. Desjarlais, *Contrib. Plasma Phys.* **45**, 300 (2005).
7. B. Militzer and D. M. Ceperley, *Phys. Rev. Lett.* **85**, 1890 (2000).
8. V. S. Filinov, M. Bonitz, P. Levashov, V. E. Fortov, W. Ebeling, M. Schlanges and S. W. Koch, *J. Phys. A* **36**, 6069 (2003).
9. H. Reinholz, *Annales de Physique* **30**, 1 (2005).
10. H. Reinholz, R. Redmer, G. Röpke and A. Wierling, *Phys. Rev. E* **62** 5648 (2000).
11. V. Silin, *Zh. Eksp. Teor. Fiz.* **47**, 2254 (1964).
12. P. Hilse, M. Schlanges, T. Bornath and D. Kremp, *Phys. Rev. E* **43**, 056408 (2005).
13. M. Schlanges, T. Bornath, P. Hilse and D. Kremp, *Contrib. Plasma Phys.* **43**, 360 (2003).
14. T. Döppner, S. Teuber, J. Tiggesbäumker and K. H. Meiwes-Broer, *Appl. Phys. B* **71**, 357 (2000).
15. K. Y. Kim, I. Alexeev, E. Parra and H. Milchberg, *Phys. Rev. Lett.*, **90**, 023401 (2003).
16. H. M. Milchberg, S. J. McNaught and E. Parra, *Phys. Rev. E* **64**, 056402 (2001).
17. T. Ditmire, T. Donnelly, A. M. Rubenchik, R. W. Falcone and M. D. Perry, *Phys. Rev. A*, **53**, 3379 (1996).
18. T. Fennel, G. F. Bertsch and K.-H. Meiwes-Broer, *Eur. Phys. J. D*, **29**, 367 (2004).
19. M. Belkacem, F. Megi and E. Suraud, *Eur. Phys. J. D*, **40**, 247 (2006).
20. L. Rammuno, C. Jungreuthmayer, H. Reinholz and T. Brabec, *J. Phys. B* **39**, 4923 (2006).
21. H. Reinholz, I. Morozov, G. Röpke, and T. Millat, *Phys. Rev. E* **69** 066412 (2004).
22. H. Reinholz, R. Redmer, G. Röpke and A. Wierling, *Phys. Rev. E* **62**, 5648 (2000).
23. I. Morozov, H. Reinholz, G. Röpke, A. Wierling and G. Zwicknagel, *Phys. Rev. E* **71**, 066408 (2005).
24. N. Arista and W. Brandt, *Phys. Rev. A* **29**, 1471 (1984).
25. S. Chapman and T. Cowling, *The Mathematical Theory of Non-Uniform Gases* (University Press, London, 1952).
26. L. Spitzer and R. Härm, *Phys. Rev.* **89**, 977 (1953).

27. Y. T. Lee and R. M. More, *Phys. Fluids* **27**, 1273 (1984).
28. W. A. Stygar, G. A. Gerdin and D. L. Fehl, *Phys. Rev. E* **66**, 046417 (2002).
29. G. Röpke, *Physica* **121**A 92 (1983); *Phys. Rev. A* **38** 3001 (1988).
30. A. Esser and G. Röpke, *Phys. Rev. E* **58** 2446 (1998); A. Esser, R. Redmer and G. Röpke, *Contrib. Plasma Phys.* **43** 39 (2003).
31. J. P. Hansen and I. R. McDonald, *Theory of Simple Liquids* (Academic Press, London, 1986).
32. J. P. Hansen and I. R. McDonald, *Phys. Rev. A* **23** 2041 (1981).
33. J. P. Hansen in *Strongly Coupled Plasma Physics*, Ed. by F. Rogers and H. DeWitt (Plenum Press, New York, 1987) p.111.
34. V. Sechenov, E. Son and O. Shchekotov, *Sov. Phys.-TVT* **15**, 415 (1977).
35. Y. Ivanov, V. Mintsev, V. Fortov and A. Dremin, *Sov. Phys.-TVT* **71**, 216 (1976).
36. A. Bakeev and R. Rovinskii, *Sov. Phys.-TVT* **8**, 1121 (1970).
37. V. Batenin and P. Minaev, *Sov. Phys.-TVT* **92**, 676 (1971).
38. S. Andreev and T. Gavrilova, Sov. Phys.-TVT **13**, 176 (1975).
39. J. Adams, H. Reinholz, R. Redmer and G. Röpke, *Phys. Plasmas* subm. (2007).
40. G. D. Mahan, *Many-Particle Physics*, 2nd ed. (Plenum Press, New York, 1990).
41. C. Fortmann, R. Redmer, H. Reinholz, G. Röpke, A. Wierling and W. Rozmus, *High Energy Density Phys.* **2**, 57 (2006).
42. T. Raitza, H. Reinholz, G. Röpke, V. Mintsev and A. Wierling, *J. Phys. A: Math. Gen.* **39**, 4393 (2006).
43. R. Thiele, R. Redmer, G. Röpke and H. Reinholz, *J. Phys. A: Math. Gen.*, **39**, 4365 (2006).
44. A. Selchow, G. Röpke, A. Wierling, H. Reinholz, T. Pschiwul and G. Zwicknagel, *Phys. Rev. E* **64**, 056410 (2001).
45. A. Selchow, G. Röpke and A. Wierling, *Contrib. Plasma Phys.* **42**, 43 (2002).

Part H

Biophysics

CONDENSED MATTER PHYSICS OF BIOMOLECULE SYSTEMS IN A DIFFERENTIAL GEOMETRIC FRAMEWORK

HENRIK BOHR

Department of Physics, Quantum Protein Centre, The Technical University of Denmark, Building 309, DK-2800 Lyngby, Denmark

JOHN H. IPSEN

Memphys, Department of physics, South Danish University, SDU, Campusvej 55, DK-5230, Odense M, Denmark

STEEN MARKVORSEN

Department of Mathematics, The Technical University of Denmark, Building 303S, DK-2800 Lyngby, Denmark

In this contribution biomolecular systems are analyzed in a framework of differential geometry in order to derive important condensed matter physics information. In the first section lipid bi-layer membranes are examined with respect to statistical properties and topology, e.g. a relation between vesicle formation and the proliferation of genus number. In the second section differential geometric methods are used for analyzing the surface structure of proteins and thereby understanding catalytic properties of larger proteins.

Keywords: Biomolecular systems; differential geometry; protein; surface topology.

SECTION I: Thermodynamics and Topology of Closed Membranes

In this section we shall describe the lipid bilayer membrane system in differential geometrical terms by a curvature elasticity Hamiltonian (Canham-Helfrich) and analyse it especially concerning it's topological features. Phase transitions can be studied when varying the coefficients $\kappa, \bar{\kappa}$ representing respectively the bending rigidity and the Gaussian curvature. It is found that the Hamiltonian has no lower bounds for $\bar{\kappa}$ positive, indicating that handles and tunnels will be formed in large numbers spontaneously in the phase region of $(\kappa, \bar{\kappa} \geq 0)$. For $\bar{\kappa} \leq 0$ apparently there are two phases where either small vesicles are formed in large numbers or large, single vesicles appear, with a second order phase transition in between.

1. Introduction

Aggregates formed by small amphiphillic molecules in water display a remarkable structural richness, e.g. micellar, hexagonal and bilayer structures.[1] Furthermore monolayer structures can be formed in water-air or water-hydrocarbon interfaces.

Amphiphiles constitute a very extensive class of compounds, which count common substances like soaps, alcohols and lipids. The lipid bilayer structures are of particular interest because they play an essential role in the organization of biological cells. Hydrated lipid bilayer systems display a wealth of polymorphic transitions, however their possible significance in biological membranes are still unrevealed. Although the discussion in this section is restricted to lipid bilayer structures it may be applicable to a range of other amphiphillic surfactant systems.

Theory has been of limited help in the characterization at the large-scale structural transition properties of lipid bi-layers in excess water. The stabilisation of the lipids in a bilayer structure is understood in the framework of a thermodynamics theory describing the interplay between molecular interaction free energy, molecular geometry and entropy.[2] This analysis has been supplemented by thermodynamics considerations based on electrolyte double-layer theory which can give a description of the stability of simple bilayer shapes like cylinders and spheres.

A popular phenomenological theory for the description of shapes of individual lipid bi-layers is the Canham-Helfrich model.[3] This model has even proven to be successfull in the description of shape transformations of biological membranes like erythocytes.[4] Furthermore this model has served as the basis for recent studies on the effects of thermal undulations
undulations?
on the forces between bi-layers[5] and the possibility of order-disorder transitions in macroscopic conformations of membranes.[6–8] However, a full statistical mechanical treatment of this model is not possible before the partition function over all shapes and topologies can be handled analytically or numerically. This is the issue the present paper is concerned with.

In this contribution we demonstrate that it is possible, within a mean-field approach, to obtain important information about the phase behaviour of the theory when only closed membranes are considered. The information is concerned with the membrane stability against changes in surface connectivity.

1.1. *Topology in membrane phenomenology*

A range of experimental techniques have been applied in the characterization of the phase behaviour and morphologies of lipid-water systems. The existence of the lamellar bilayer phases and a number of bilayer phases with bicontinous structures exhibiting cubic symmetries have been complimented and confirmed with NMR[9] and freeze-fracture electron microscopy.[10] In the more diluted regimes of the phase diagram the characterization of the phases is complicated due to the absence of long range order in the phases and coexistence of a large number of bilayer structures. Direct visualization by microscope tecniques probably gives the best insight in the nature of the phases in this regime.[11,12]

A considerable effort has been directed toward characterization of the phases in terms of surface geometries.[13] The multi-lamellar and cubic phases have been characterized in terms of infinite periodic minimal surfaces (IPMS), e.g. the surfaces having zero mean curvatures and separating the ambient space into periodic subspaces. An IPMS has thus an associated point group symmetry and characteristic dimensions of its unit cell. The topology of an IPMS can be very complicated, e.g. represented by the number of genus per unit cell. Properties of minimal surfaces can be derived from complex analysis through their representation by Weierstrass-polynomials. However, IPMS is still not a fully determined group of surfaces, and the non-periodic minimal surfaces with non-trivial topology has only recently been explored. It is thus evident that surfaces are difficult to treat in a statistical mechanical frame if they are assumed to be minimal. Another difficulty in dealing with minimal surfaces in membrane physics is that a physical principle, which dictates the crystalline properties of the IPMS, is not known. No packing condition or internal symmetri property of the constituents can guide us, as in the case of molecular crystals. With these difficulties we find that minimal surfaces at present do not provide a good starting point for the description of membranes undergoing phase transitions involving topology. We shall in this section restrict ourselves to closed membranes, which actually never can be described as minimal surfaces in R^3.

Our original motivation for studying the differential geometrical model of membranes has been to extend the previously found solutions to simple non-trivial topological aspects, but it turned out that our solutions would predict much more complex topological structures than first anticipated. One might therefore ask if such topological complexity, i.e. a large conglomerate of handles and tubuli, has ever been observed in biological membranes. In Fig. 1 examples from intra-cellular membrane structures are shown, e.g. the Golgi apparatus, the Endoplasmic Reticulum, etc., where complicated topological structures of tunnels and pipes are interwoven in each other. Such structures are difficult to observe experimentally but can within a certain accuracy be derived from e.g. x-ray diffraction data.

1.2. *Model*

In this subsection the Canham-Helfrich model of membrane elasticity will be briefly described. This model considers only fluid membranes which at length-scales much larger than the molecular distances and the bilayer thickness can be modeled as a mathematical surface without any internal structure. The lipid bi-layers exhibit a number of low-temperature solid-like phases with in-plane order of the lipid molecules, but they do no demonstrate the deluge of large-scale structural transitions displayed by the fluid membranes. The model is based on the following Hamiltonian that takes the form

$$\mathcal{H} = \mu \int dA + \frac{\kappa}{2} \int dA(\frac{1}{r_1} + \frac{1}{r_2} - \frac{2}{r_0})^2 + \bar{\kappa} \int dA \frac{1}{r_1 r_2} \quad , \tag{1}$$

where the integrations are performed over the surface area, $dA = dx_1 dx_2$, x_1, x_2 being the embedded coordinates of the membrane surface. Here r_1 and r_2 are the

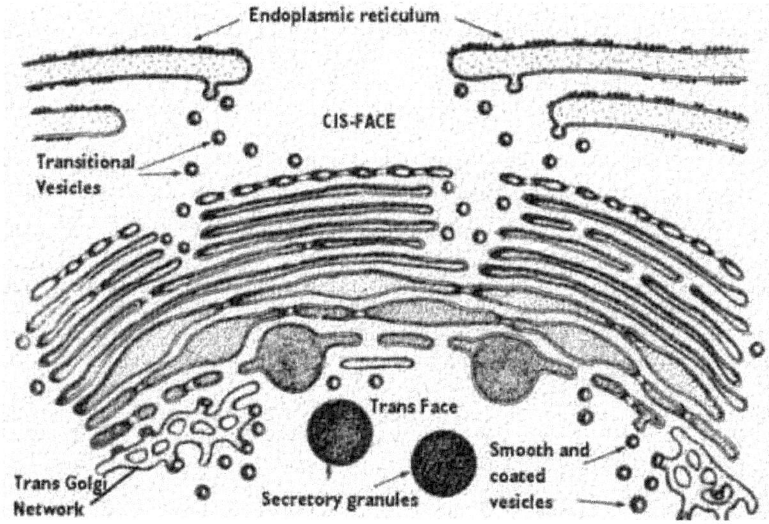

Fig. 1. Schematic picture of a typical part of the endoplasmatic reticulum in animal cells.[17]

local principal curvatures of the surface and r_0 is the spontaneous curvature, which can arise in bi-layers with an intrinsic asymmetry between the mono-layers of the bilayer. In this work we consider full symmetry between the two bilayer halves, which is the case when the bilayer is composed of a single molecular constituent. The notion of spontaneous curvature will thus be omitted in the following. The mean curvature $\frac{1}{r_1} + \frac{1}{r_2}$ and the gaussian curvature $\frac{1}{r_1 r_2}$ are surface invariants, i.e. independent of the chosen parametrisation of the surface. The surface tension μ, which couples to the surface area, must be considered as a chemical potential for the lipids in the membrane due to the fixed cross-sectional areas of the lipids. In most thermodynamic problems involving interfaces the chemical potential controls the interface.

When all the lipids in the lipid-water mixture form bilayer structures with a fixed cross-sectional area per lipid molecule, μ is negligible since the free lipid molecules are insoluble in water. μ may be non-negligible in realistic cases when a lipid bilayer phase is coexisting with a non-bilayer phase, when different bilayer phases are coexisting or in the case of black films. Other terms may be included in Eq. (1) . If the membrane has boundaries a line tension term

$$\mu_L \int_{\text{boundary}} d\ell \tag{2}$$

must be added. However the line tension μ_L is so large that even the presence of small boundaries are suppressed for free membranes. Boundaries can occur if the membrane can be attached to hydrophobic or hydrophillic elements of the experimental setup. Here, we do not consider these cases and just assume that the membranes are without boundaries. Furthermore anharmonic terms are neglegted in Eq.(1). The model parameters κ and $\bar{\kappa}$ are difficult to obtain experimentally.

However, some concensus have been reached regarding the value of κ for artificial membranes.

1.3. *The Willmore functional*

In this subsection we discuss some results from the mathematical literature concerning the properties of the functional which appears as the second term in Eq.(1), the Willmore functional. The Willmore functional is written as

$$W(\Sigma) = \frac{1}{2} \int_\Sigma H^2 dA \tag{3}$$

where $H = \frac{1}{r_1} + \frac{1}{r_2}$ is the mean curvature and dA is the area element of a surface Σ. Here Σ is *any* compact surface in $\mathbf{R}^3$ and we assume that it has no boundaries and no self-intersections. The functional W is invariant under conformal mappings of the ambient 3-space. Thus, if $\tilde{\Sigma}$ is the image of Σ under a Möbius transformation (an isometry, a scaling or an inversion in a sphere with center not in Σ), then $W(\tilde{\Sigma}) = W(\Sigma)$.[14] Recently, considerable effort have been directed toward a solution of the Willmore problem for surfaces of any genus (concerning the infimum of W and the related variational problem). A few results relevant for our purpose will be given.

Following L. Simon[16] we write $\beta_g = \inf W(\Sigma)$, where the infimum is taken over compact genus g surfaces without self-intersections. The following inequality is fundamental:

$$8\pi \leq \beta_g < 16\pi. \tag{4}$$

Equality holds on the left if and only if $g = 0$ and Σ is a round sphere.[14] The right hand side inequality was observed independently by U. Pinkall and R. Kusner, see.[15] Simon then showed,[16] that if we consider $e_g = \beta_g - 8\pi$, then

$$e_g \leq \sum_{j=1}^{q} e_{\ell_j} \tag{5}$$

for any integers $q \geq 2$ and $\ell_1, ..., \ell_q$ with $\sum_{j=1}^{q} \ell_j = g$. Further he proved the *existence* of W-minimizers in the following sense: For any genus g there exists a genus g surface Σ with $W(\Sigma) = \beta_g$, unless equality holds in Eq. (5) in which case there exists a sequence Σ_k of genus g surfaces and a genus $g_\circ$ surface $\Sigma_\circ$ (with $g_\circ \leq g$) such that $W(\Sigma_k) \to W(\Sigma_\circ) = \beta_{g_\circ}$ for $k \to \infty$. Thus for a given surface the minimization of W may cause a drop in genus number. However, it is not known whether equality actually can occur in Eq. (5).

1.4. *Thermodynamics of surface topology*

Going back to the original Hamiltonian Eq.(1) we have in the last chapter given bounds on the second term, the Willmore functional. The third term is easily evaluated by using the Gauss-Bonnet theorem:

$$\int_{\Sigma} dA \frac{1}{r_1 r_2} = 2\pi\chi \tag{6}$$

When the surface is without boundaries the Euler characteristic χ is simply related to the genus number by $\chi = 2\pi(2-2g)$.

From the previous section it is clear that the Willmore functional restricted to compact embedded surfaces Σ without boundaries is related to $\mathcal{H}$: $W(\Sigma) = \frac{1}{\kappa}(\mathcal{H} - \bar{\kappa}\chi(\Sigma))$. In particular $\inf_g \mathcal{H}(\Sigma_g) = \kappa \inf_g W(\Sigma_g) - 4\pi\bar{\kappa}(1-g)$, where $\inf_g W$ represents the infimum of $W(\Sigma_g)$ for all boundary-less compact, embedded surfaces Σ_g with genus g.

The next step is to sum up the contribution of the Willmore functional from each topological genus class. We shall be taking the extremum of the Willmore functional from each genus class. It's measure contributes to a number summed up to either almost 0 or to a finite number that could become infinite dependent on the degeneracy of the extremum. The sum would be correspondingly 0 or infinite. Hence, in order to obtain a meaningful result of the genus sum we shall assume each Willmore contribution in each class to be finite or that there is a finite degeneracy of the functional. We have reasons to believe that the Willmore functional is defined up to homotopy in finite exemplar. The assumption of finiteness is also rendered probable by computer simulation where surfaces can be triangularized and thus make a computer estimate of the Willmore functional possible.

We are now in the position to set up the partition function for a single closed membrane made out of A lipids. Note, by introduction of a fixed membrane area we break the conformal invariance explictly.

$$\begin{aligned} Z(A) &= \sum_{g=0}^{G_A} \mathrm{Tr}_g(\exp(-\beta\mathcal{H})) \\ &= \sum_{g=0}^{G_A} \exp(-4\pi\bar{\kappa}(1-g))\mathrm{Tr}_g(\exp(-\beta\frac{\kappa}{2}W)) \end{aligned} \tag{7}$$

The trace Tr_g represents the summation of all physical distinct surfaces Σ_g with fixed area corresponding to A. Surfaces which are identical apart from a reparametrisation of the surface are not physically distinct. G_A represents a cut-off in the number of genus for a membrane of size A. G_A exists if the diameter and the sectional curvatures have upper bounds. This is indeed the case for a closed membrane due limited number of lipids A involved and material parameters determined by size and physical properties of the molecular constituents. A first approximation to $Z(A)$ can be obtained by restricting Tr_g to surfaces which realize the minimum of W. This is - according the previous considerations - the case for one surface for each g. The structure of this surface is not known for general g at present, which makes it impossible to go beyond this simple thermodymanical level of description. We will further reduce the number of degrees of freedom and assume that the membrane is motionally hindered, so g is the only available degree of freedom and the minima of $W(\Sigma)$ are non-degenerate up to homotopy.

Case 1.:

$$Z(A) \approx Z^{\text{SP}}(A) = \sum_{g=0}^{G_A} \exp(-4\pi\beta\bar{\kappa}(1-g)) \exp(-\beta\frac{\kappa}{2}\inf_g(W)) \tag{8}$$

By use of Eq.(4) and Eq.(7)

$$Z_1(A) \leq Z^{\text{SP}}(A) \leq Z_2(A) \quad , \tag{9}$$

where

$$Z_c(A) = \exp(-4\pi\beta(2\kappa+\bar{\kappa})) + \exp(-8\pi\beta c\kappa)\frac{1-\exp(4\pi\bar{\kappa}\beta G_A)}{1-\exp(4\pi\beta\bar{\kappa})} \qquad c=1,2 \tag{10}$$

For $G_A \to \infty$, Z_c and Z^{SP} are analytic for $\bar{\kappa} < 0$, and Z_c is solely controlled by the second term in Eq.(9) for $\bar{\kappa} \to 0^-$. From Eq.(8) it is evident that Z^{SP} is also governed by a singularity of this nature as $\bar{\kappa} \to 0^-$. Z_c can thus be considered as a good approximation for Z^{SP} under these conditions. Note that Z_1 in this limit is essentially the partition function for a quantum harmonic oscillator. However, $1 \ll G_A < \infty$ is the regime of interest in the description of the physical system. Here Z_c and Z^{SP} are analytic for $\bar{\kappa} \neq 0$. The inequality in Eq.(8) holds term by term in an expansion of Z_c and Z^{SP} in $\exp(4\pi\bar{\kappa}(1-g))$ like Eq.(7). It is then trivial to show that the inequality holds term by term in an expansion in $\bar{\kappa}$. The expansion coefficients are thermal expectation values which will be considered in the following.

The free energy derived from Z_c, $F_c = -\beta^{-1}\ln(Z_c(A))$, and its thermal behaviour may be analyzed. The obvious orderparameter in this problem is the averaged genus $< g >$ or the average Euler-characteristics $< \chi >=4\pi(1- < g >)$.

$$\begin{aligned} < \chi > &= -\frac{\partial F_c}{\partial(\beta\bar{\kappa})} \\ &= \beta^{-1}\frac{1}{Z_c(A)}\frac{\partial Z_c(A)}{\partial(\beta\bar{\kappa})} && (11) \\ &= 4\pi\beta^{-1}\frac{1}{Z(A)}(-\frac{a}{x} + b\frac{-G_A x^{G_A} + (G_A-1)x^{G_A+1} + x}{(1-x)^2}) && (12) \end{aligned}$$

We have here used the notation $x = \exp(4\pi\beta\bar{\kappa})$, $a = \exp(-8\pi\kappa\beta)$ and $b = \exp(-8\pi\kappa\beta c)$. The fluctuations in $< \chi >$ can be expressed as

$$\begin{aligned} \sigma(\chi) &= < \chi^2 > - < \chi >^2 \\ &= -\frac{\partial^2 F_c}{\partial(\beta\bar{\kappa})^2} \\ &= \beta^{-2}(\frac{1}{Z_c(A)^2}\frac{\partial^2 Z_c(A)}{\partial^2(\beta\bar{\kappa})} - (\frac{1}{Z_c(A)}\frac{\partial Z_c(A)}{\partial(\beta\bar{\kappa})})^2) && (13) \end{aligned}$$

$< \chi >$ display a sudden, but continous change at $\bar{\kappa} = 0$ from $< \chi >\approx -4\pi$ corresponding to a sphere for $\bar{\kappa} < 0$ to $< \chi >\approx -4\pi G_A$ for $\bar{\kappa} > 0$, see Fig. 2. This transition is governed by strong fluctuations in g, which is manifested by a peak in $\sigma(\chi)(\approx (2\pi G_A)^2)$ at $\bar{\kappa} = 0$.

Case 2.

The simplest extension of *case 1* is to consider a collection of immobile closed membranes which only interact through exchange of lipids giving rise to a vesicle size distribution. The partition function thus takes the form:

$$Z = \sum_{N_A} \frac{e^{-\nu \sum_A N_A A}}{\prod_A (N_A A)!} \prod_A Z(A)^{N_A} \tag{14}$$

where the sum runs over all possible vesicle size distributions $\{N_A\}$. The total amount of lipid in the system is controlled by the chemical potential ν. A has a lower cut-off which is determined by the molecular details.[2]

Case 3.

This case represents a collection of closed membranes, which can exchange lipids and are free to move in space. Only the translational degrees of freedom will be taken into account, because handling of the rotational degrees of freedom require detailed information about the structures of the Willmore surfaces for all g, which is not available.

In Fig. 2 we summarise these thermodynamical results by showing the various phases of case 1, 2 and 3 in phase diagrams with the parameters κ and $\bar{\kappa}$, the coefficients of the bending rigidity and the Gaussian curvature respectively. A phase transition line along $\bar{\kappa} = 0$ is separating a topological phase, in the upper region of the diagram, with infinitely many handles (high genus number g, case 1) and a phase ($g = 0$) with many vesicles, case 2 and 3, in the lower region of the phase diagram. This region is furthermore divided by a phase transition line, approximately along $\kappa = -2\bar{\kappa}$, into a phase containing many smaller vesicles and a phase with one single large vesicle. In Fig. 3 we have focused on the case 2, 3 and shown the average order parameter $< -1/2\chi >$ and the fluctuation of χ as a function of $\bar{\kappa}$. Furthermore, the size distribution of vesicles at different $\bar{\kappa}$ and fixed κ is presented in Fig. 4. These distributions are exactly what we expect of the theory in case 3.

1.5. *Discussion and Conclusion*

In the previous section the biophysical problem on the thermodynamic properties of closed membranes is related to a variational problem on Willmore surfaces. This mathematical problem is still unresolved but the results obtained so far gives sufficient information to provide valuable results on the phase behaviour of membranes. Three simple cases was considered: *case* 1 an ensemble of independent and motionally restricted membranes. In this case the system display an abrupt, continous change in $< g >$ at $\bar{\kappa} = 0$. Although it is accompanied by strong fluctuations in $< g >$, the transition is neither 1. order or 2. order, but rather ∞ order in the sense that $\frac{\partial^n F}{\partial(\beta\bar{\kappa})} \propto (G_A)^n \to \infty$ for $n \to \infty$ at $\bar{\kappa} = 0$. In an ensemle of weakly interacting membranes this may be changed to a 1. order or a 2. order transition.

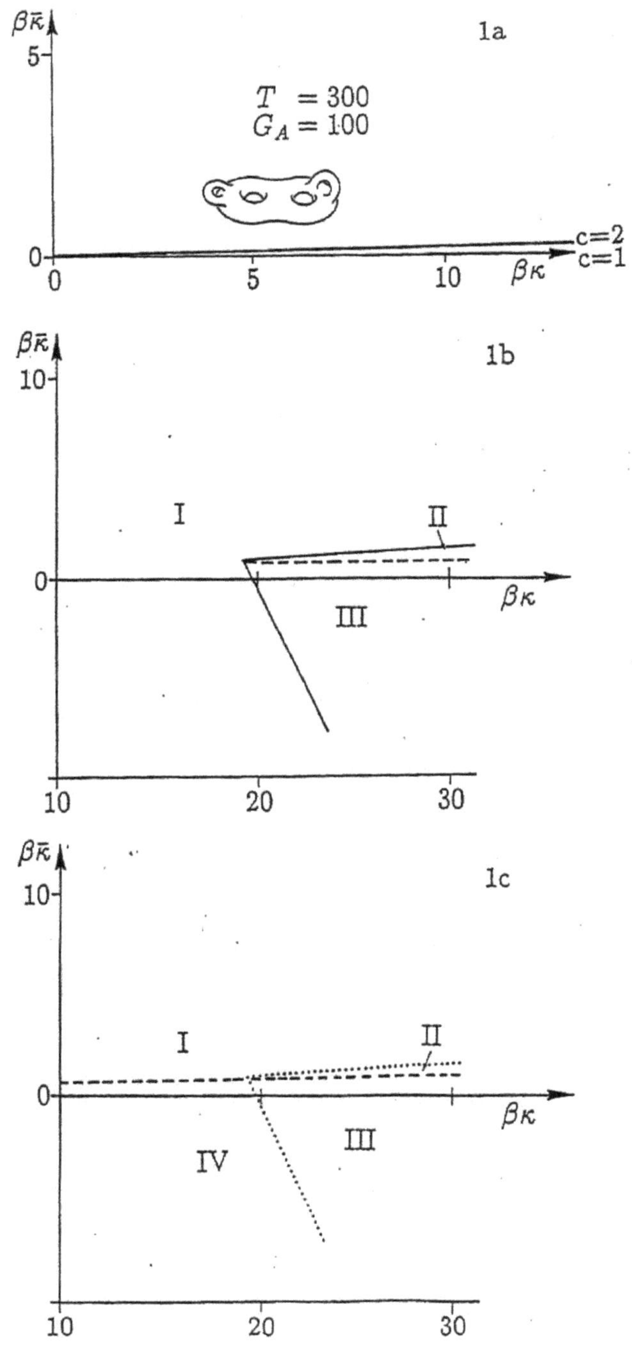

Fig. 2. The figure shows 3 phase diagrams in the parametres $\beta\kappa$ and $\beta\bar{\kappa}$ of the partition function for the lipid system. Fig. 1a, 1b and 1c correspond respectively to case 1, case 2 and case 3.

The principles behind the interplay between three dimensional structure and the structural transitions of proteins and their biological function are to a large extent understood. A similar relationship for the biological membranes is still considered at a hypothetical level.[19] Whether the extended lipid polymorphism has any significance in biological system is still unclear.

Of great interest in this contribution are the topological properties of membranes. It is observed[22] that large-scale structures of lipid membranes exhibit great topological complexity which is realized in biological membranes, for example in the Golgi-apparatus, see Fig. 1, inside the cell. Here are lots of handles and tubuli connecting different compartments between layers. The purpose of this complex topological structure is a need for filtration of proteins in the cellular liquid. An interesting question is then how one could explain dynamically the formation of handles or tunnel structure and division into small vesicles. Such an explanation requires a theory of topology dynamics.

The topological complexity appearing in biological membrane structures, e.g. the intra-cellular Golgi apparatus, can largely be explaned by the phase diagram of Figs. 2 and 3, where an "explosion" of handles and tubuli occur and only limited by material constraints such as a finite lipid size. The phase transition around $\kappa = -2\bar{\kappa}$, between single vesicle structure and many small vesicles can also be observed in experiments.[23] It seems crucial for cell growth, in an early state of life, that the membrane structure is balancing between the two situations as if life tends to stay in non-equilibrium.

2. SECTION II: The Topology of 2-Dimensional Protein Surfaces

Previously we considered smooth, biomolecular systems at the typical size of hundreds to thousands of nanometers. We shall now zoom down to particular biomolecules, proteins typically of the size of five to ten nanometers, that also have a well-defined differential geometric structure that explains aspects of their biofunctionality and material science features.

2.1. *Construction of surfaces for proteins from their coordinates*

It is not a trivial task to construct a smooth surface from a 3-dimensional protein structure, especially if the surface has to contain all the protein backbone. In order to see that, one should visualize the protein as a cluster of atoms connected to each other by an unbranched chain that winds in and out of the center of the cluster.

In X-ray crystallography as well as in NMR the problem is to construct the 3-dimensional protein structure from spectroscopic data that provide distances between some of the atoms in the protein. Such construction is an expansion of 2-dimensional information in the form of distance maps to 3-dimensional structural information which requires large processing capacity and some extra chiral information or chain topology. Nevertheless it has been one of the most important processing tasks of this century since it gave science all the known 3-dimensional structures

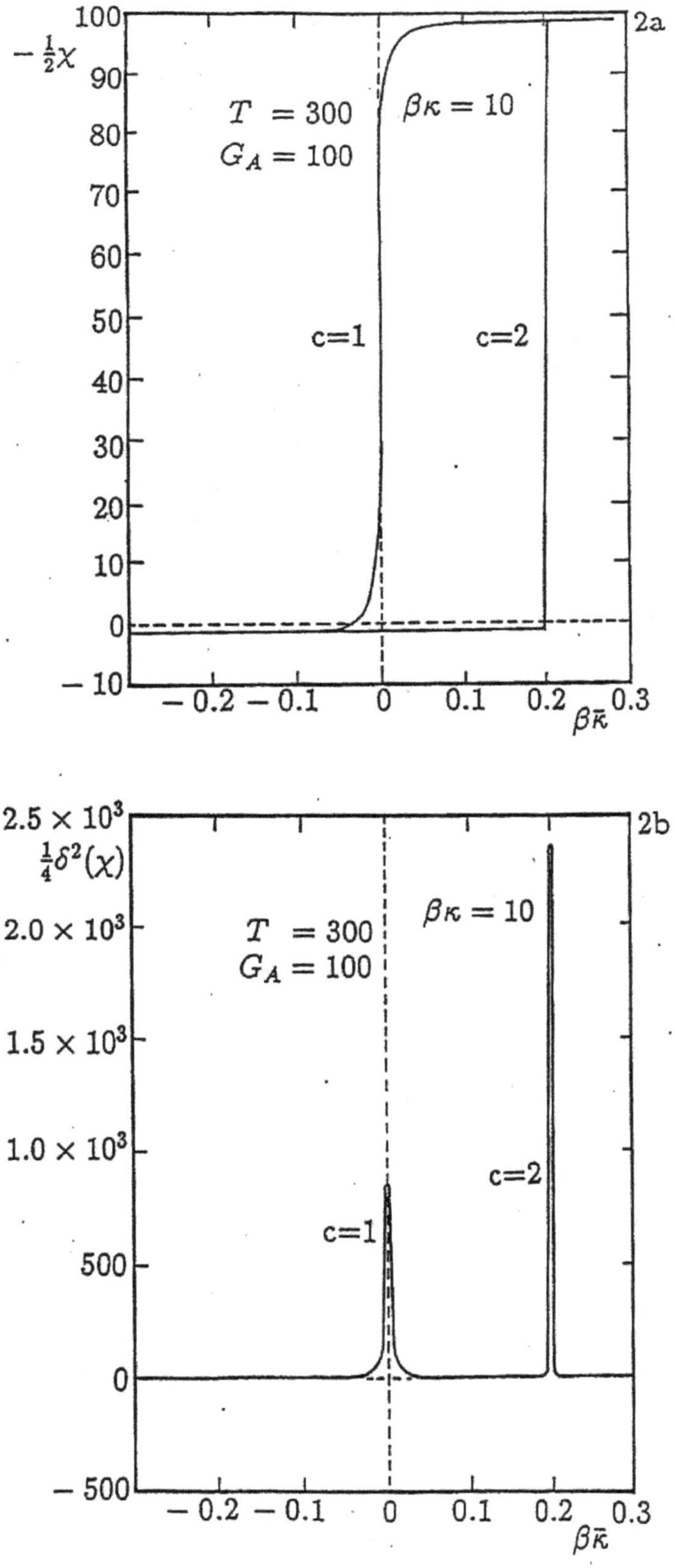

Fig. 3. Profiles of genus g, or actually $-1/2\chi$, are showns along line $\kappa = 3.0$ which has disconti-

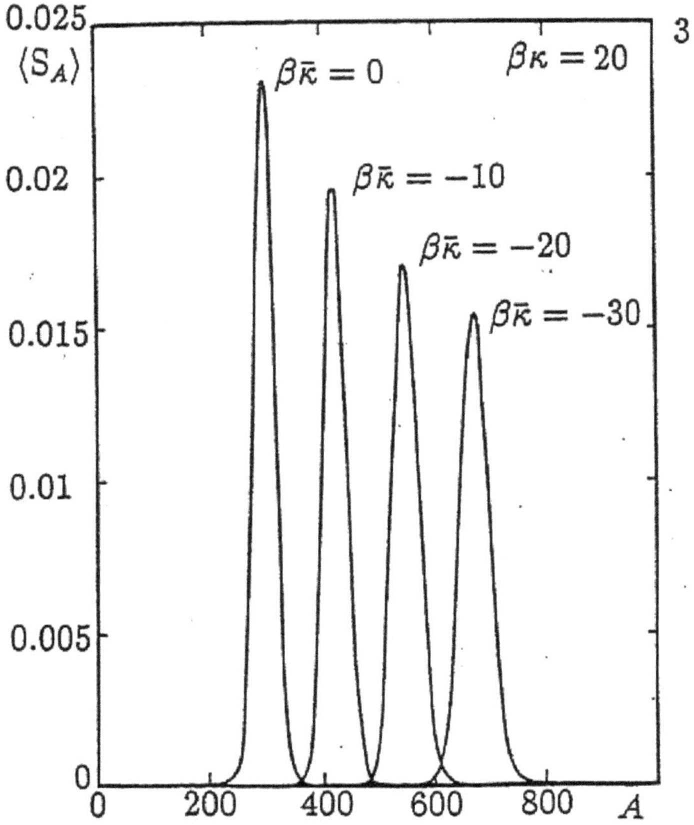

Fig. 4. Several size distributions of vesicles with different values of $\bar{\kappa}$ and fixed $\mu\kappa$.

of bio-molecules, as well as other molecules, gathered together in huge databases. There has now been devised various short-cuts to the problem of 2d to 3d expansion.[24]

Before discussing how to construct surfaces one should decide what kind of surface one wants. The most obvious one is the "outer" surface or also called the *solvent accessible surface* for a protein.

This solvent accessible surface is most commonly obtained by rolling a ball on the protein structure given by coordinates thereby obtaining a subset of the protein coordinate set that could be in contact to a solvent with the size of the rolling ball. This is of course all done in a simulation by a computer. The ball size can then be varied in order to get into grooves and cavities or, oppositely, in order to avoid them by making it smaller, respectively, larger. There are many standard programs for this procedure. In this review we have produced some pictures with the help of the Connolly packet.[25] Once the outer solvent accessible surface has been made one can then construct (a triangulation of) the *electrostatic potential surface* or the *Van der Vaal contact surface* as shown in Fig. 5, and hence find the curvature tensors.

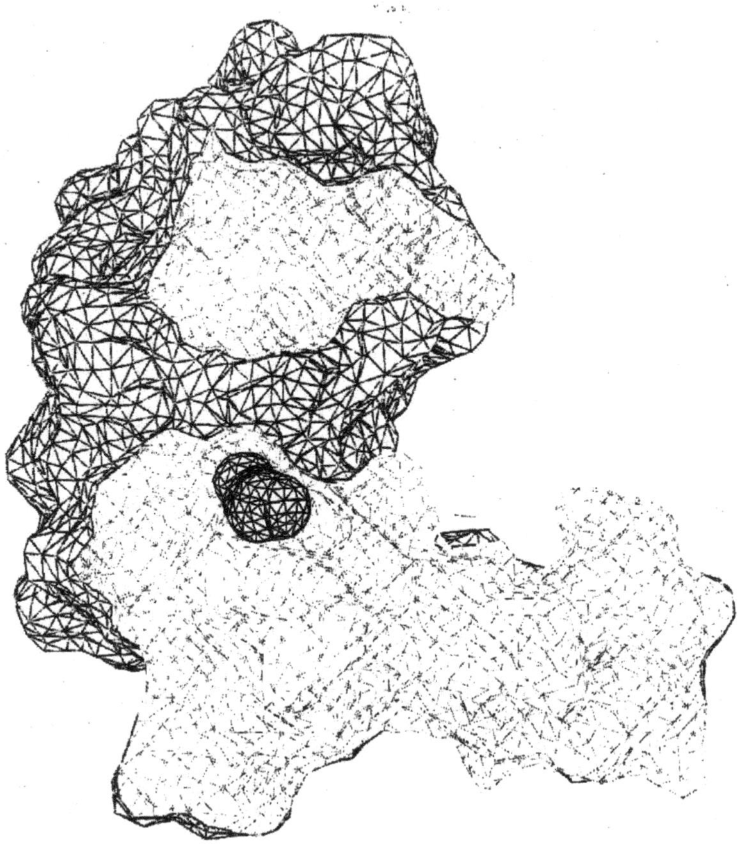

Fig. 5. A construction of the solvent accessible surface[25] for the protein, lysozyme. The darker color represents the outer surface while the lighter color represents the inner surface. A ligand is embedded in the surface.

2.2. *Differential geometry of the protein surface*

We start again by considering the constructed 2-dimensional protein surface embedded in a 3-dimensional Euclidean space. We shall only be considering embedded surfaces with no intersections. The surface $\{\bar{x}(\sigma)\}$, which can be described by a parametrization $\sigma(\sigma_1, \sigma_2)$, is completely determined, up to Euclidean transformations, by the two tensors, the metrical tensor g_{ij} and the extrinsic curvature tensor (the second fundamental form) K_{ij}. The *induced* metric tensor g_{ij} is of course:

$$g_{ij} = \partial_i \bar{x} \partial_j \bar{x} \tag{15}$$

where $\partial_i = \frac{\partial}{\partial \sigma_i}$. The extrinsic curvature tensor, K_{ij} is:

$$K_{ij} = \bar{n}\partial_i\partial_j\bar{x} \tag{16}$$

where $\bar{n}$ is the unit normal perpendicular to the tangential surface, $\bar{n} \cdot \partial_i\bar{x} = 0$ for $i = 1, 2$. $\bar{x}, g_{ij}$ and K_{ij} are related through the Gauss-Weingarten equations:

$$\partial_i\partial_j\bar{x} = \Gamma^{\alpha}_{ij}\partial_\alpha\bar{x} + K^{\alpha}_{ij}\bar{n} \quad \text{where} \quad \Gamma^{\alpha}_{ij}g_{\alpha\kappa} = \partial_i\partial_j\bar{x}\partial_\kappa\bar{x} \quad . \tag{17}$$

If we want to construct a surface model with an energy function it must be invariant with respect to changes of the coordinate system. Therefore the terms of an eventual energy function must contain surface invariants and they can only be constructed from the the curvature, metrical and torsional tensors up to third order in the derivatives. The first invariants to be included in an expansion of the free energy of the surface are:

$$1, \; g^{ij}K_{ij} \quad \text{and} \quad K^i_j K^i_j \quad , \tag{18}$$

where K^i_i is the mean (exterior) curvature and $G = K^i_i K^j_j - K^j_i K^i_j$ is the Gaussian curvature. These quantities can also be expressed in term of the local principal radii of curvature r_1 and r_2:

$$K^i_i = (\frac{1}{r_1} + \frac{1}{r_2}), \quad G = \frac{1}{r_1 r_2} \quad . \tag{19}$$

The next invariants will contain 2-forms of higher order derivatives.

One can now write the Hamiltonian function as in section 1 but this time for the protein surface that contains these surface invariants:

$$H = \mu \int d^2\sigma\sqrt{det(g)} + k_1 \int d^2\sigma\sqrt{det(g)}\,(K^i_i)^2 + k_2 \int d^2\sigma\sqrt{det(g)}\,G \tag{20}$$
$$+ \text{...higher order terms} \quad .$$

2.3. *Minimal surfaces of proteins*

We shall in this chapter discuss the importance of minimal surfaces for protein functionality and especially enzymatic reaction kinetics. In the last chapter we saw how surfaces for proteins could be constructed from the set of atomic coordinates by triangularization. As it turns out several important proteins have a solvent accessible surface that resembles that of minimal surfaces and their enzymatic functioning is very dependent on that.

It is not surprising that protein surfaces often appear as minimal surfaces since, as e.g. in the case of beta-sheets, a network of hydrogen bonds is forcing the surface to maintain constant zero mean curvature such that a bend in one direction will cause an opposite bend in another direction. This can be understood by applying the Poisson equation that seek to maintain constant average density on the protein surface.

First, some words about minimal surfaces arising from infinite systems. Minimal surfaces have their mean curvature overall equal to zero: $K^i_i = 0$. In general

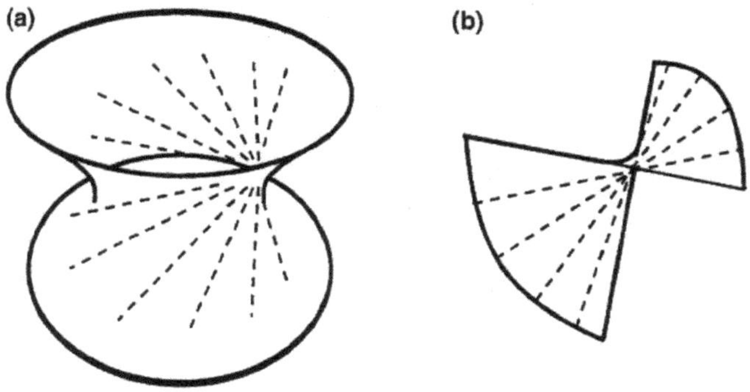

Fig. 6. A picture of the two most well-known minimal surfaces (shown cut-out with boundary), (a) the catenoid and (b) the helicoid.[26]

the minimal surfaces are infinite but they are often considered and drawn with a boundary, such as for the catenoid or helicoid, see Fig. 6. One of the most famous minimal surfaces is the helicoid, which can be generated as a ruled surface. The simplest minimal surface is the plane. A minimal surface generally has a Weierstrass representation where the coordinates of a point in 3D Euclidean space on the surface is given by the function $\mathcal{F}$ of complex variables $\omega = \sigma + i\tau$ in the plane produced from the minimal surface by a conformal parametrization:

$$\begin{aligned} x &= \mathrm{Re}\int (1-\omega^2)\mathcal{F}(\omega)\,d\omega, \\ y &= \mathrm{Im}\int (1+\omega^2)\mathcal{F}(\omega)\,d\omega, \\ z &= \mathrm{Re}\int 2\omega\mathcal{F}(\omega)\,d\omega. \end{aligned} \tag{21}$$

3. The Surface Geometry of Protein Reaction Kinetics

When looking at protein functionality in general, and especially for enzymatic reactions, it is not surprising that the geometry of the surface of the protein plays an important role. By the surface we of course again mean the solvent accessible surface, SAS. This dependence on the surface geometry is especially seen when the functionality is about binding to a substrate, antibody-antigen interaction, enzymatic hydrolysis of peptide bonds, charge transfer etc. We shall in the following consider the particular process of substrate (ligand) binding or ligand transferring and relate that to the minimal surface property.

3.1. *Minimal surfaces and protein-ligand interaction*

Coming back to protein surfaces there are obvious cases where the SAS of a protein seems to resemble a minimal surface, at least partly, and which seems to have profound reasons for the functioning of that particular protein. Such cases are often involving particular topological patterns of secondary structures. The word *topology* is here used in the meaning of a graph theoretical sense where the various secondary structure elements are connected to each other in a particular way. Typically these patterns consist of an alternating, symmetric occurrence of alpha helices and beta strands, α/β. This is seen in the α/β barrel proteins, such as the triose phosphate isomerase where the beta strands, see Fig. 7, intertwine the alpha helices and form a particular core structure with a barrel structure and which can be described by the center portion of the catenoid minimal surface. The open end of the catenoid is then supposed to contain the active site while the tunnel of the catenoid is formed by the hydrophobic residues of the beta strands which are tilted relative to the axis of the tunnel. The beta sheets are straight meaning that the normal curvature is zero and their relative twist is equal to the geodesic torsion. The value of the twist gives the magnitude of the Gaussian curvature, G. The less symmetrical α/β protein, the redox favodoxin, has similarly a beta sheet core that is well described by the helicoid minimal surface. Thus, we base our analysis of enzymatic processes on the curvature tensors K and G calculated from triangularization of the SAS.

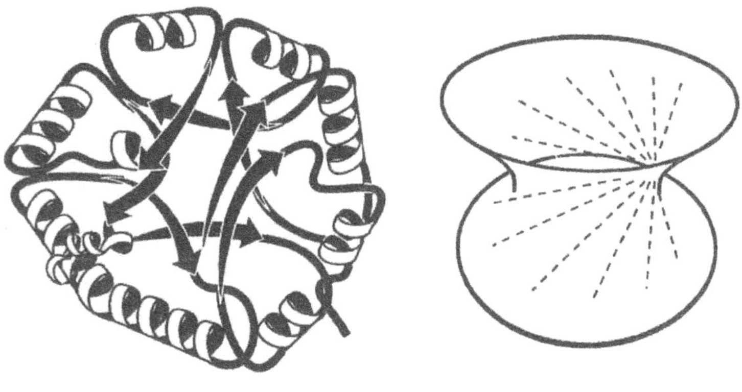

Fig. 7. A constructed picture of the protein, Triose Phosphate Isomerase, schematically embedded into a catenoid surface.[27]

We shall now, following the argument in the book of Hyde et al.,[26] see how these minimal surfaces can be important for the enzymatic processes the two mentioned proteins are involved with. Most commonly the enzymatic processes of a protein involve hydrolysis of some sort or binding to a substrate, e.g. a particular ligand. In the latter case it is important for the understanding of the process and its effectiveness that one can explain how the diffusion of the ligand is steered through the protein rather than being a matter of "random encounter chemistry" which never would be able to explain the extremely high yield of protein catalysed reaction kinetics. The driving mechanism in the so-called "forced diffusion" of the catenoid modelled beta-barrel protein is the guiding gradient of the Gaussian curvature which attain its most negative value in the waist of the catenoid. Thus a ligand approaching the protein from infinity will be smoothly guided through the catenoid tunnel of the protein towards the waist and the active site. The minimal surface will guarantee smooth and swift diffusion of the ligand towards and through the protein. Indeed, the negative Gaussian curvature of any minimal surface locally speeds up the Brownian diffusion process in comparison with the corresponding diffusion on a flat surface, see.[29] However, a spherically cut off part of any minimal surface also has larger area than a similarly cut off part (disk) in the plane. It turns out, as a consequence of minimality, that the larger area balances precisely the faster Brownian motion, so that the Brownian motion mean exit times are exactly identical on the two domains: the sperically cut off part of the minimal surface and the disk, respectively. The specificity of the reaction is given by the particular combination of the geometry and side-chain composition of the ligand and protein. Concerning construction of minimal surfaces we can refer to.[29,30]

3.2. *Conclusion*

In section 2 we have analyzed constructed protein surfaces that can be triangularized in order to provide differential geometric information. Usually the protein folding problem is seen as a matter of going from 1 dimensional representations to 2 dimensional (secondary) structures and then going to the full 3-dimensional (tertiary) structures. However, when going to 2-dimensional surface structures one can by means of minimal surfaces understand enzymatic activities.

Acknowledgment

We wish to thank the employees at department of physics at DTU and the group at Memphys of SDU for support. Especially Dr Gerhard Besold is acknowledged.

References

1. V. Luzzati, A. Tardieu, *Annu. Rev. Phys. Chem.* **25**, 79-92 (1974).
2. D. Israelachvili, J. Mitchell, B. W. Ninham, *Biochim. Biophys. Acta* **470**, 185-201 (1977).
3. P. B. Canham, *J. Theoret. Biol.* **26**, 61-81 (1970); W. Helfrich, *Z. Naturforsch.* **28c**, 693-703 (1973).
4. H. J. Deurling, W. Helfrich, *Biophys. J.* **13**, 941(1980).
5. W. Helfrich, *Z. Naturforsch.* **33c**, 305 (1978); R. Lipowsky, S. Leibler, *Phys. Rev. Lett.* **56**, 2541 (1986).
6. W. Helfrich, *J. Physique* **46**, 1263 (1985).
7. L. Peliti, S. Leibler, *Phys. Rev. Lett.* **56**, 1690 (1985) .
8. Y. Kantor, M. Kardar, D. R. Nelson, *Phys. Rev. Lett.* **60**, 238 (1988).
9. G. Lindblom, H. Wennerström, *Biophys. Chem.* **6**, 167-171 (1977).
10. T. Gulik-Krzywicki, L. P. Aggerbeck, K. Larsson, in *Surfactants in Solutions* (Eds.: K. L. Mittel, B. Lindman) Vol. 1 pp 237-257, Plenum, New York 1984.
11. D. D. Miller, J. R. Bellore, D. F. Evans, Y. Talmon, B. W. Ninham, *J. Phys. Chem.* **91**, 674-685 (1987).

12. R. M. Servus, *Chemistry and Physics of Lipids* **50**, 87-97 (1989).
13. S. Anderson, S. T. Hyde, K. Larsson, S. Lidin, *Chem. Rev.* **88**, 221-242 (1988).
14. J. Weiner, *Indiana Univ. Math. J.* **27**,19-35 (1978).
15. R. Kusner, *Pacific J. Math.* **138**, 317-345 (1989).
16. L. Simon, *Proc. Cont. Math. Anal. Aust. Natl. Univ.* **10**, 187-216 (1986).
17. University of Texas Medical Branch: www.cytochemistry.net/Cell-biology/golgi.htm
18. T. J. Willmore, *Total Curvature in Riemanian Geometry*, Chichester, Wiley (1982).
19. P.R. Cullis, M. J. Hope, B. de Kruijff, A. T. Verkleij, C. P. S. Tilcock, *Phospholipids and Cellular Regulation* (Ed.: J. F.Kou) CRC Press, Boca Raton, Florida (1985) Vol. 1.
20. H. Bohr and J. H. Ipsen, in *Characterizing Complex Systems* (Ed: H. Bohr), pp. 74-79, Singapore, World Scientific Publishing Co. (1989).
21. W. Helfrich, W. Harbich in *Physics of Amphiphilic layers* (Eds.: J. Meunier, D. Langevin, N. Boccara) pp 58, Springer, Berlin, (1987).
22. R. E. Goldstein and S. Leibler, *Phys. Rev. Lett.*, **61**, 2213-2216 (1988).
23. B. J. Litman, *L. Biochemistry* **12**, 2545 (1973); F. Szoka and D. Papahadjopoulos, *Annual Review of Biophys. and Bioeng.* **9**, 467 (1980).
24. J. Bohr, H. Bohr, S. Brunak, R. M. J. Cotterill, B. Lautrup and S. B. Petersen, *J. Mol. Biol.* **231**, 861-871 (1993).
25. M. L. Connolly, *Molecular Surface Package Manual*, Version 2.6, Menlo Park, CA (1994).
26. S. Hyde, S. Andersen, K. Larson Z. Blum, T. Landh, S. Ledin and B. W. Ninham, *The Language of Shape*, Elsevier, Amsterdam (1997).
27. S. Andersen, S. T Hyde, K. Larson and S. Ledin, *Chem. Rev.* **88**, 221 (1988).
28. H. A. Schwartz, in *Gesammelte Mathematische Abhandlungen*, Springer, Berlin (1890).
29. S. Markvorsen and M. Min-Oo, *Global Riemannian Geometry: Curvature and Topology*, Birkhäuser Verlag, Basel - Boston - Berlin (2003).
30. T. H. Colding and W. P. Minicozzi II, PNAS **103**, 11106-11111 (2006).

THE BRAIN'S VIEW OF THE NATURAL WORLD IN MOTION: COMPUTING STRUCTURE FROM FUNCTION USING DIRECTIONAL FOURIER TRANSFORMATIONS

B. K. DELLEN

Bernstein Center for Computational Neuroscience,
D-37018 Göttingen, Germany
E-mail: bkdellen@bccn-goettingen.de

J. W. CLARK and R. WESSEL

Department of Physics, Washington University
St. Louis, MO 63130 USA
E-mail: jwc@wuphys.wustl.edu

Attractive possibilities exist for the brain to solve important problems of visual processing in Fourier space. First, we describe a wave-interference model for computing motion contrast directly from the moving intensity distribution, without the need for directionally selective motion sensors. We then propose a global method for motion-based image segmentation based on unsupervised clustering in a three-dimensional Fourier space. The Fourier components of coherently moving entities are segregated from the remainder by means of a simple velocity proximity measure. This is accomplished without altering the spatial frequency components encoding the object, thereby ensuring that the spatiotemporal features of the segregated object can be reconstructed.

Keywords: Visual motion; biological motion processing; motion contrast; Fourier analysis; image segmentation.

1. Introduction

In biological and machine vision, information about the environment is inferred from time-dependent, two-dimensional (2D) distributions of intensity, i.e., the image sequence. Movements of the observer or of objects in the outside world produce characteristic changes in the image that carry information on the visual world or on the state of the observer. For example, when objects move against a background, groups of image points are displaced coherently, providing evidence for the segmentation of the image into figure and ground. In human vision, this principle is known as the Gestalt law of "common fate"; image points moving with the same velocity are bound together to give the percept of a coherently moving entity.[1,2] How this can be achieved in neural circuitry or electronic hardware is one of the most basic questions confronted in computational neuroscience or artificial vision research.

The answer or answers to this question remain elusive. However, we suggest that a partial answer lies in the following observation:

There are advantages to be gained by machine vision systems, or by visual systems in the brain, from generating and processing Fourier-related transforms of local information about motion in the visual scene, as contained in its time-dependent 2D intensity distribution $I(\mathbf{x}, t)$.

This observation or assertion reflects the idea that motion is as fundamental as spatial location in time, not merely a derived quantity, $\mathbf{v} = d\mathbf{x}/dt$. As physicists, we appreciate the truth and value of this perspective in the Hamiltonian formulation of classical mechanics and, of course, in quantum theory. Is it not possible that the duality of the global concept of state of motion and the local concept of space-time point has been naturally embodied by evolution in the inner workings of the visual systems of living creatures?

The above assertion has recently been exemplified in a schematic model of how vertebrate visual systems might perform the tasks of detecting and estimating motion contrast, i.e., relative motion of two objects, or of an object moving against a background. Self-evidently, these are tasks of fundamental importance to visual processing in vertebrates, since an animal must be able to distinguish salient motion in its environment from irrelevant whole-field motion due to body, head, and eye movements. Sec. 2 describes a wave-interference algorithm that we have proposed for solution of the motion-contrast problem,[5] based quite explicitly on global features of the changing intensity pattern, as revealed by Fourier analysis.

The principal aim of the present contribution is to offer a further exemplification of the above assertion. We shall demonstrate that, in certain cases, the problem of image segmentation also becomes more accessible to solution when the image is transformed into Fourier space. Motion induces an oscillation of the phase of the spatial-frequency components of the moving part of the image. Accordingly, the components that we wish to segregate from the remainder are conveniently labeled by their respective temporal frequencies. This key observation leads to an algorithm for motion binding based on temporal filtering in Fourier space. In Sec. 3, the essence of the proposed segmentation algorithm is spelled out analytically. In Sec. 4, its operation is illustrated by three sets of computer simulations. There follows (Sec. 5) a discussion of the merits (and drawbacks) of global methods for performing image segmentation, in comparison to local methods.

2. A Wave-Interference Model of Motion Contrast Detection

We have recently proposed and tested an algorithm for motion-contrast detection and estimation that does not require the participation of directionally selective motion sensors.[5] Considering first the case of one-dimensional (1D) motion, the intensity distribution $I(x, t)$ is convoluted with a cosine function over an interval L:

$$J(X, t) = \int_L I(x, t) \cos\left[k(X - x)\right] . \tag{1}$$

It is easily shown that this transformed distribution has the alternative expression

$$J(X, t) = |\mathcal{F}_k(t)| \cos\left[kX - \phi(t)\right] \tag{2}$$

in terms of the Fourier transform

$$\mathcal{F}_k(t) = \int_L I(x,t)\exp(ikx)dx \tag{3}$$

of the intensity distribution, with $\phi(t) = \arctan(\mathrm{Im}\mathcal{F}_k(t)/\mathrm{Re}\mathcal{F}_k(t))$. For coherent motion (two objects moving in the same direction at the same speed), the transform $J(X,t)$ reduces to a traveling wave of fixed amplitude, while incoherent motion causes the amplitude to oscillate. The frequency of the amplitude oscillation provides a measure of the relative velocity of the moving objects.

To make this behavior more explicit, let the visual scene be composed of the sum of two intensity distributions $I_a(x,t)$ and $I_b(x)$ that move rigidly with velocities v_a and v_b, respectively, i.e., let the total visual stimulus be

$$I(x,t) = I_a(x - v_at) + I_b(x - v_bt)\,. \tag{4}$$

Straightforward manipulations[5] based on trigonometric identities permit one to write the transformed intensity for this case as

$$J(X,t) = A(t)\cos\left[kX - \Phi(t)\right], \tag{5}$$

where

$$A(t) = |\mathcal{F}_k[I(x,t)]| = \left\{(a-b)^2 + 4ab\cos^2\left[\frac{k(v_b - v_a)t}{2} + \frac{\beta - \alpha}{2}\right]\right\}^{1/2}, \tag{6}$$

$$\Phi(t) = \phi(t) = kv_at + \alpha + \arctan\left[\frac{b\sin(k(v_b - v_a)t + \beta - \alpha)}{a + b\sin(k(v_b - v_a)t + \beta - \alpha)}\right]. \tag{7}$$

The amplitude of the transformed intensity $J(X,t)$ is seen to oscillate with a frequency $f = k(v_b - v_a)/4\pi$, while the factor $\cos\left[kX - \phi(t)\right]$ describes a carrier wave moving at constant velocity. Motion contrast within the visual field, as represented by the input (4), can therefore be directly determined from the frequency of the amplitude oscillation, which thereby provides a suitable measure of motion contrast. In the absence of contrast, i.e., if the motion of the two intensity components is coherent and $v_a = v_b$, the amplitude $A(t)$ is constant in time and the frequency is zero.

As an instructive and important example, we consider two identical but independently moving test spots. In this case $a = b$ and $\alpha = \beta$, and the transformed intensity distribution (5)-(7) collapses to

$$J(X,t) = 2a\cos\left(k\frac{v_b - v_a}{2}t\right)\cos\left(kX - k\frac{v_b + v_a}{2}t - \alpha\right). \tag{8}$$

We see immediately that for coherent motion ($v_a = v_b$), the transformed intensity distribution is a *traveling wave* with constant amplitude that propagates with the velocity v_a of the test spots. On the other hand, for incoherent (opposing) motion ($v_a = -v_b$), the transformed intensity distribution, $J(X,t) = 2a\cos(kv_at)\cos(kX - \alpha)$, is a *standing wave* with an oscillating amplitude. The frequency f of the amplitude oscillation is given by $f = k(v_b - v_a)/4\pi = kv_a/2\pi$.

The efficacy of the algorithm is illustrated in two computer simulations, the frequency of the amplitude oscillations being determined as follows. First, the absolute value of $J(X,t)$ is integrated over an integral multiple m of the transformation wavelength $\lambda = 2\pi/k$ to extract a characteristic time-dependent carrier wave

$$\tilde{J}(t) = |A(t)| \int_0^{m\frac{2\pi}{k}} |\cos(kX - \Phi(t))| dX \qquad (9)$$

that is independent of the transformed spatial variable X. Next, we introduce an amplitude modulation function $M(t) = d\tilde{J}(t)/dt$. For coherent motion $M(t)$ is identically zero, whereas for incoherent motion it oscillates about zero with a frequency proportional to the motion contrast of the visual scene. Lastly, the frequency of oscillation of the function $M(t)$ is determined with high accuracy from the power spectrum of its temporal Fourier transform. [Note that this frequency differs from that of $A(t)$ by a factor two, because the absolute value is taken in Eq. (9).]

In part A of Fig. 1, the algorithm is applied to a visual scene consisting of two test spots of the same intensity, each moving with a speed 8 deg/s through the visual field of the motion detector. For $t \in [0, 3.8]$s, both stimuli are moving in the same direction, and the amplitude modulation $M(t)$ is zero. At $t = 3.8$s, the direction of motion of one of the stimuli is reversed, and the stimuli undergo opposing motion. As a result, $M(t)$ oscillates with a frequency $f_M = 8$ Hz. Thus, the algorithm correctly determines the relative velocity to be $\Delta v = 2\pi f_M/k = 16$ deg/s.

In part B of Fig. 1, the algorithm is applied to a solid moving object of uniform intensity that moves in front of a random-dot background. This visual scene is chosen to correspond with experiments performed for neurons in the avian optic tectum.[3] The visual field of the detector spans a visual angle of 80 deg, and the object size is 5 deg. The random dots have size 0.1 deg and density 1 dot/deg. For $t \in [0, 3.8]$s, the object and background are moving in the same direction with the same speed 8 deg/s. This exemplifies whole-field motion. The amplitude modulation $M(t)$ remains predominantly zero, with fluctuations coming from edge effects. (Dots moving in and out of the visual field disturb the convolution process in an unpredictable way.) At $t = 3.8$s, the motion of the background is reversed and $M(t)$ oscillates with a frequency $f_M = 8Hz$. Again the algorithm determines the relative velocity to be 16 deg/s, the correct result.

Motion in the 2D visual field is dealt with by projecting the velocity vectors of the objects along an arbitrary direction. The algorithm is then found to be successful in reproducing tuning curves derived from measurements of motion-contrast sensitivity in the avian (pigeon) tectum[3,4] and in the primate (monkey) middle temporal area.[6] Such curves, which represent the normalized response of motion-contrast-sensitive neurons as a function of the angle between the directions of two moving stimuli, have a cosine shape, as predicted by the model.

A plausible biological implementation[5] of the wave-interference algorithm, sketched in Fig. 2, features a layer of spatially distributed subunits (Layer 1) that

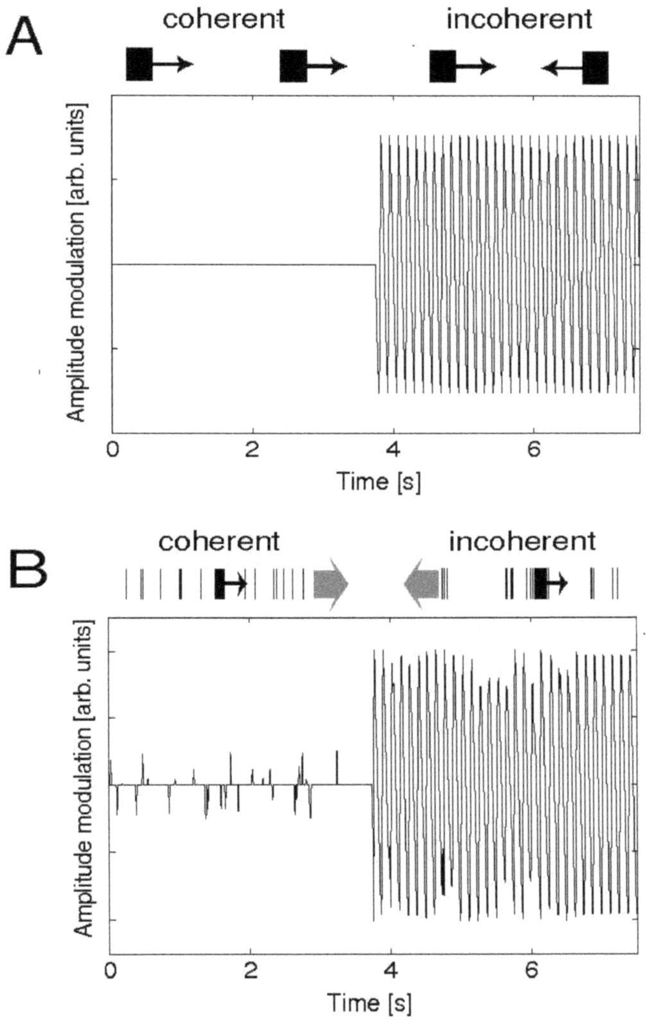

Fig. 1. Amplitude modulation $M(t)$ as a function of time for different stimuli in one dimension. (A) Two test spots (black squares) move with the same speed through the visual field, first in the same direction and later in opposite directions. (B) A solid object of finite size moves in front of a random-dot background, object and background having the same speed. (For the sake of clarity, every point of the 1D stimulus is represented by a vertical line in the image icon.) First, the object and background move in the same direction, and later the direction of motion of the background is reversed.

provide input to a deeper layer (Layer 2) of linear summing units in a periodic fashion. Each subunit (or "neuron") in Layer 1 responds to motion within its receptive field (upper gray line). For a small moving object, this creates a wave of activity (lower gray line) spanning the neural population of Layer 2. When the object moves

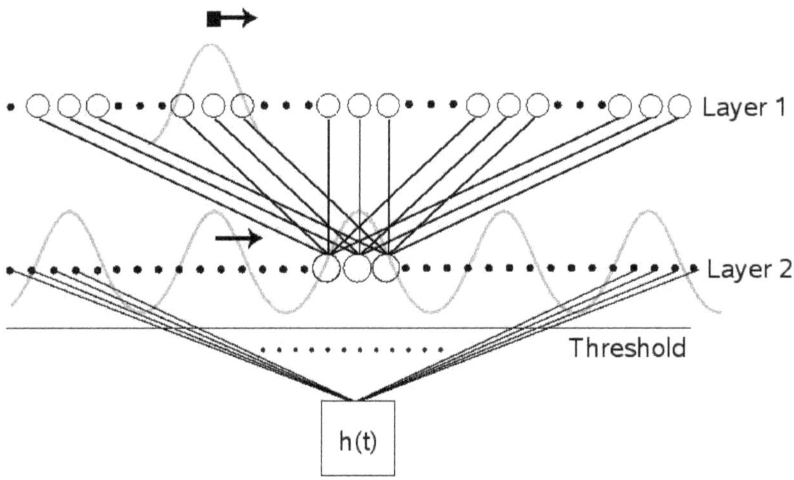

Fig. 2. Schematic view of a plausible biological implementation of the wave-interference model of motion-contrast computation. See text for explanation.

along the dimension of Layer 1, the wave of activity will move with the point along the neural population in Layer 2. As a result, the amplitude of the wave will oscillate in response to motion contrast. The population response of the second layer is then rectified, integrated, and presented to an appropriate temporal filter $h(t)$ to read out the oscillation frequency and hence the relative speed of the object moving against a background.

3. Formulation of the Segmentation Algorithm

Let us describe the image by a 2D time-dependent intensity function $I(\mathbf{x}, t)$. Its 2D spatial Fourier transform is

$$F(\mathbf{p}, t) = \int_{\mathcal{X}} (\mathbf{x}, t) \exp(2\pi i \mathbf{p} \cdot \mathbf{x}) d\mathbf{x} \,, \tag{10}$$

where $\mathbf{p} = (p_x, p_y)$ is the 2D spatial wave vector of the transformation and $\mathbf{x} = (x, y)$ is a point vector in real space $\mathcal{X}$. The inverse transformation

$$I(\mathbf{x}, t) = \int_{\mathcal{P}} F(\mathbf{p}, t) \exp(-2\pi i \mathbf{p} \cdot \mathbf{x}) d\mathbf{p} \tag{11}$$

reconstructs the original image in $\mathcal{X}$ space. Thus, $F(\mathbf{p}, t)$ and $I(\mathbf{x}, t)$ are equivalent representations of the image.

Coherent motion of image points generally indicates the presence of a moving object. Considering the case of two independently moving objects, denoted o and b, the image is composed of two coherently moving groups of points. Thus the intensity may be written as

$$I(\mathbf{x}, t) = I_o(\mathbf{x} - \mathbf{v_o} t) + I_b(\mathbf{x} - \mathbf{v_b} t) \,, \tag{12}$$

where all points of object o have the same velocity $\mathbf{v_o} = (v_x^o, v_y^o)$, and likewise for object (or background) b. Transforming to frequency space, we have

$$\begin{aligned}
F(\mathbf{p}, t) &= F_o(\mathbf{p}, t) + F_b(\mathbf{p}, t) \\
&= \int_{\mathcal{X}} I_o(\mathbf{x} - \mathbf{v_o}t) \exp(2\pi i \mathbf{p} \cdot \mathbf{x}) d\mathbf{x} \\
&\quad + \int_{\mathcal{X}} I_b(\mathbf{x} - \mathbf{v_b}t) \exp(2\pi i \mathbf{p} \cdot \mathbf{x}) d\mathbf{x} \\
&= \exp(2\pi i \mathbf{p} \cdot \mathbf{v_o}t) \int_{\mathcal{X}} I_o(\mathbf{x}) \exp(2\pi i \mathbf{p} \cdot \mathbf{x}) d\mathbf{x} \\
&\quad + \exp(2\pi i \mathbf{p} \cdot \mathbf{v_b}t) \int_{\mathcal{X}} I_b(\mathbf{x}) \exp(2\pi i \mathbf{p} \cdot \mathbf{x}) d\mathbf{x} .
\end{aligned} \tag{13}$$

As seen in Eq. (13), image motion results in characteristic temporal oscillations at frequency $2\pi \mathbf{p} \cdot \mathbf{v_o}$ and $2\pi \mathbf{p} \cdot \mathbf{v_b}$, corresponding to a given spatial wave vector $\mathbf{p}$. We perform a temporal Fourier analysis by multiplying every point of the Fourier transform $F(\mathbf{p}, t)$ with a complex time-dependent function $T(\omega, t) = \exp(2\pi i \omega t)$ and integrating over time, obtaining

$$\begin{aligned}
F_\omega(\mathbf{p}, t) &= \int_{\mathcal{T}} T(\omega, t) F(\mathbf{p}, t) dt \\
&= \int_{\mathcal{T}} \exp(2\pi i \omega t) F(\mathbf{p}, t) dt \\
&= \int_{\mathcal{T}} \exp[2\pi i (\mathbf{p} \cdot \mathbf{v_o} - \omega) t] dt \int_{\mathcal{X}} I_o(\mathbf{x}) \exp(2\pi i \mathbf{p} \cdot \mathbf{x}) d\mathbf{x} \\
&\quad + \int_{\mathcal{T}} \exp[2\pi i (\mathbf{p} \cdot \mathbf{v_b} - \omega) t] dt \int_{\mathcal{X}} I_b(\mathbf{x}) \exp(2\pi i \mathbf{p} \cdot \mathbf{x}) d\mathbf{x} .
\end{aligned} \tag{14}$$

Now let both objects move along the x-axis but in opposite directions, such that $\mathbf{v_o} = (v_o, 0)$ and $\mathbf{v_b} = (-v_b, 0)$ with $v_o > 0$ and $v_b > 0$. Eq. (14) then becomes

$$F_\omega(\mathbf{p}, t) = \begin{cases} \displaystyle\int_{\mathcal{T}} e^{[2\pi i (\mathbf{p} \cdot \mathbf{v_o} - \omega) t]} dt \int_{\mathcal{X}} I_o(\mathbf{x}) e^{(2\pi i \mathbf{p} \cdot \mathbf{x})} d\mathbf{x} & \text{if } \operatorname{sgn}(\omega / p_x) = 1 , \\ \displaystyle\int_{\mathcal{T}} e^{[2\pi i (\mathbf{p} \cdot \mathbf{v_b} - \omega) t]} dt \int_{\mathcal{X}} I_b(\mathbf{x}) e^{(2\pi i \mathbf{p} \cdot \mathbf{x})} d\mathbf{x} & \text{if } \operatorname{sgn}(\omega / p_x) = -1 . \end{cases} \tag{15}$$

Thus, both objects occupy distinct, non-overlapping areas of the three-dimensional (3D) Fourier space defined by (p_x, p_y, ω); they are linearly separable. In this case, the objects can be segregated from one another by applying "directional" Fourier transformations if we restrict the 3D Fourier transformation to points of the 3D Fourier space that are occupied by one of the two objects but not the other. Upon segregation of the object, we can return to real space by applying the inverse 3D Fourier transformation.

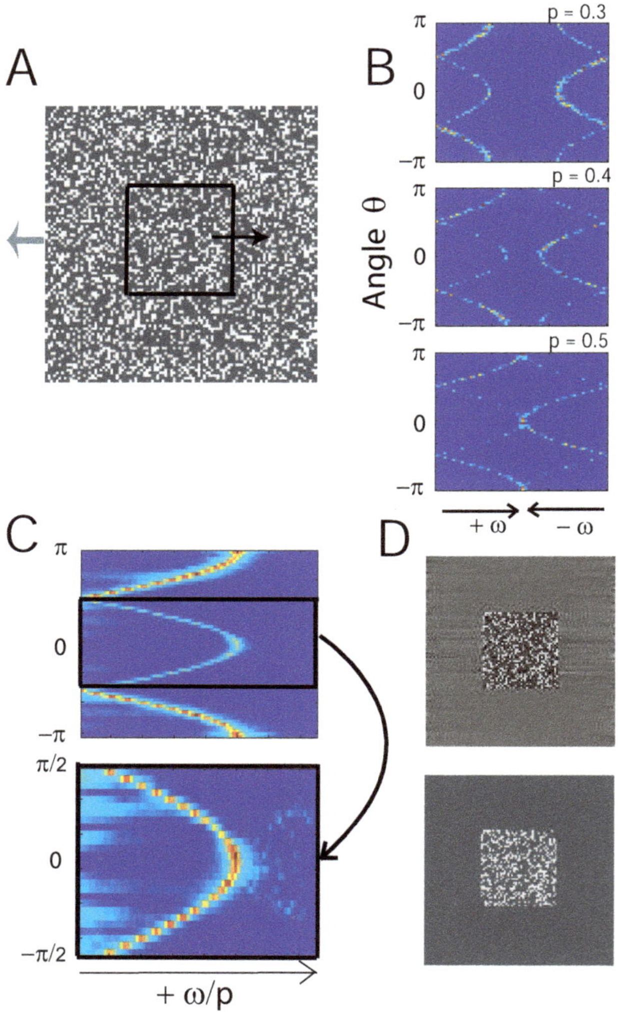

Fig. 3. Image segmentation in Fourier space: object and background moving with constant velocities.

4. Computer Demonstrations

In this section, we present the results from computer simulations that reveal the strengths and weaknesses of the proposed algorithm.

First we demonstrate the algorithm for an image sequence which consists of a random-dot square 40×40 pixels^2 moving in the positive x-direction and a random-dot background of size 100×100 pixels^2 in moving in front of the background in the negative x-direction (Fig. 3A), both with a constant absolute speed of 1 pixel/frame. The Fourier coefficients of the image sequence are computed as a function of the temporal frequency ω and the angle $\theta = \arctan(p_y/p_x)$ of the 3D Fourier transform, while keeping the value of $p = |\mathbf{p}|$ fixed. In Fig. 3B, the intensity plots are shown for $p = 0.3$, $p = 0.4$, and $p = 0.5$ pixel^{-1}, from top to bottom. There emerges a characteristic pattern, indicating the presence of two translating objects. The intensity plots for different values of p are scaled and summed to allow easier identification of the moving objects. Here we only considered positive temporal frequencies ω. The resulting intensity plot is shown in Fig. 3C (top panel). The traces of both the random-dot square object and the background are clearly visible. For illustration, we have surrounded the area occupied by the random-dot square by a black frame. The inner part of the black frame is shown in the lower panel of Fig. 3C. By visual inspection of the intensity plot, we conclude that it is sufficient to restrict the 3D Fourier transformation to parameters $\omega > 0$ for $p_x > 0$ and $\omega < 0$ for $p_x < 0$, in order to segregate the random-dot square from the background. A snapshot of the reconstructed segregated scene is depicted in Fig. 3D (upper panel). For comparison, the original random-dot square is plotted in the lower panel of Fig. 3D. We computed the correlation coefficient of the original object and the reconstructed object for every time point and calculated the mean to be $c = 0.776$ with a standard deviation of $\sigma = \pm 0.067$. For the mean correlation coefficient of the reconstructed object and the original background we have $c = 0.081$ and $\sigma = \pm 0.062$.

In most image-segmentation problems, the objects do not have a constant velocities, and the characteristic Fourier patterns of the moving objects will show a greater variety and complexity. Again, let us consider an image sequence in which moving random-dot square 40×40 pixels^2 moves in front of a moving random-dot background spanning 100×100 pixels^2. The background is moving to the left with a constant speed of $|\mathbf{v_b}| = 1$ pixel/frame. The random-dot square is moving to the right with a mean speed of $|\mathbf{v_o}| = 0.5$ pixels/frame, but its speed is not constant. Instead, its speed randomly alternates between 1 and 0 pixels/time-step, resulting in a jerky motion. A snapshot of the image sequence is shown in Fig. 4A (left panel). In the right panel, the position of the random-dot square is plotted as a function of the number of frames. We compute the Fourier coefficients of the image sequences as a function of the temporal frequency ω and the angle $\theta = \arctan(p_y/p_x)$ of the 3D Fourier transform while keeping the value of $p = |\mathbf{p}|$ fixed; then, we scale and sum the resulting intensity plots for positive values of ω. In Fig. 4B (upper panel), the resulting intensity plot is shown. Although the background objects are cleary visible, the Fourier coefficients belonging to the random-dot square are much fainter and spread out over a larger area. However, upon zooming into the area occupied by the random-dot square, the pattern formed by the Fourier coefficients still allows an unambiguous identification of the object. As in the previous example, we restrict the 3D Fourier transformation to parameters $\omega > 0$ for $p_x > 0$ and $\omega < 0$ for $p_x < 0$ to segregate the random-dot square from the background. The reconstructed random-dot square is depicted in Fig. 4C (upper panel). For comparison, the original object is shown below. The mean correlation coefficient of the reconstructed object and the original objects is $c = 0.7284$ with a standard deviation of $\sigma = \pm 0.0589$.

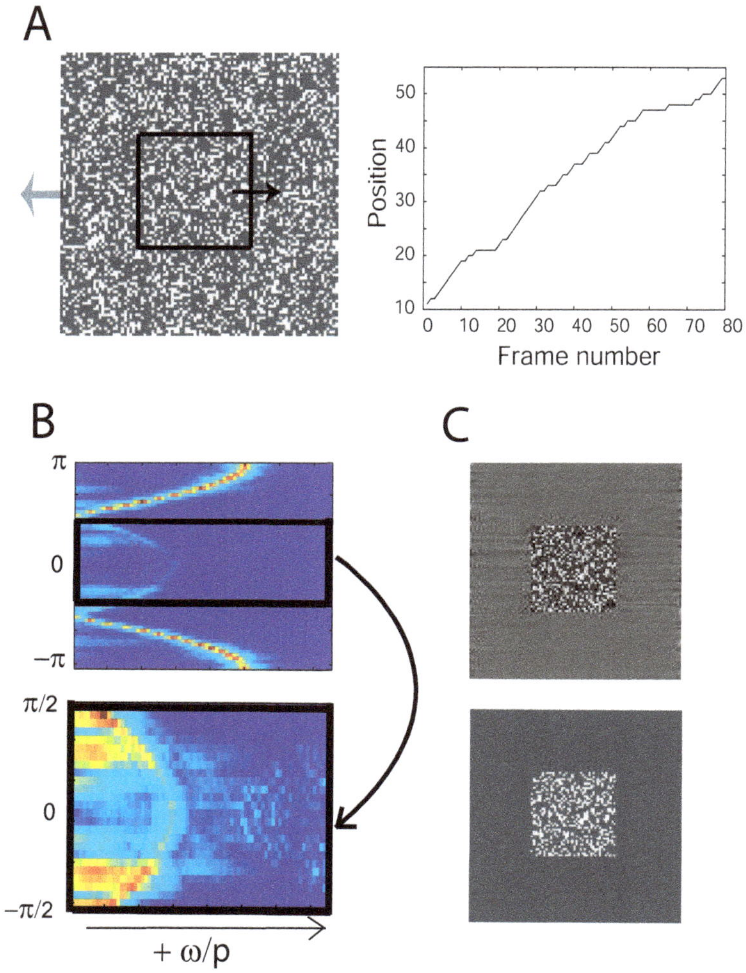

Fig. 4. Image segmentation in Fourier space: as in Fig. 4 but with accelerated object.

For other image sequences, notably those in which two objects move at an angle relative to each other smaller than 180 degrees, the objects are no longer linearly separable. The smaller the angle, the more ambiguous the segregation becomes, and other cues need to be employed to help segregate the objects. This ambiguity is inherent to the problem of motion-based image-segmentation; it is not a flaw of the algorithm. In Fig. 5A, the intensity plot for an image sequence containing an random-dot square moving in the positive x-direction in front of random-dot background which moving in the negative y-direction is displayed. Both object and background move with constant speed. We observe that the characteristic patterns overlap in Fourier space. To segregate object and background from each other, it is necessary to restrict the directional Fourier transform to only those points of

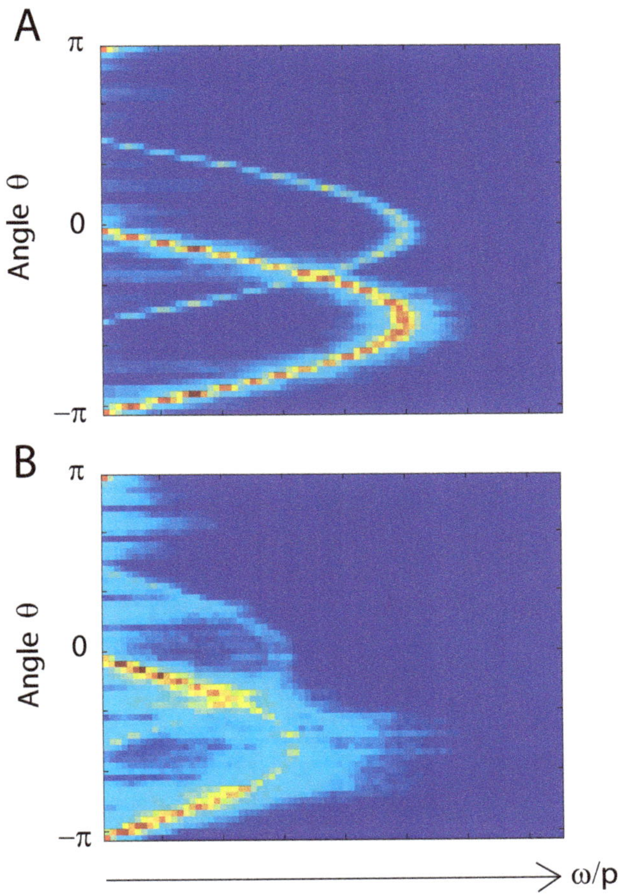

Fig. 5. Fourier analysis of image sequences containing object and background that overlap in Fourier space.

the Fourier space which lie exactly on the arch-shaped characteristic pattern of the targeted object (results not shown). When the velocity of both the background and the object alternates between 1 pixel/frame and 0 pixel/frame (jerky motion), the characteristic patterns of the object and background are only partly separable (Fig. 5B), reducing the quality of the segmentation.

5. Discussion

Currently, most image-segmentation algorithms are based on the extraction of local measurements such as the computation of normal components of local velocity or local correlation surfaces, resulting in a two-dimensional optic flow field.[7] In contrast, the algorithm proposed in this paper employs only global methods to perform image segmentation. Local algorithms suffer from the so-called aperture problem – local measurements produce ambiguous information when the image is observed through a small opening. This problem is absent in the global method. Furthermore, in the locally driven algorithms, image regions of zero contrast (or intensity) are not recognized by local measurements and thus are not bound to the object, while in the algorithm presented a moving object can be segmented in its entirety. The locally-driven algorithm might be severely challenged when presented with objects which overlap in real space, which is the case for transparent motion; in this particular case, the global algorithm presented here is expected to perform better than the local method. However, in cases where the moving objects of an image sequence are only poorly separable in Fourier space (see example Fig. 5B), locally-driven methods have the advantage that objects do not necessarily overlap in real space and thus might outperform the global algorithm.

The segmentation algorithm developed above offers a new perspective on computational processes executed in the brain. When we examine the basic computations that are performed in the visual cortex, neurons can be described as two-dimensional filters, tuned to different spatial frequencies and orientations.[8–10] It has been suggested that the ensemble of such cells could encode, with some precision, the two-dimensional Fourier spectrum of a patch of visual space.[10] The algorithm presented here gives new ideas how two objects which lie partly inside the same patch could be segregated, and how spatial information could be retained and processed by neural populations.

Acknowledgments

We thank Charles Anderson for useful comments. The work has received support from the Whitehall Foundation, the McDonnell Center for Higher Brain Function, and the National Science Foundation under Grant No. PHY-0140316. JWC acknowledges partial support from FCT POCTI, FEDER in Portugal and the hospitality of the Centro de Ciências Mathemáticas at the University of Madiera.

References

1. G. Wertheimer, *Zeitung für Psychologie* **6**, 161 (1912).
2. K. Nakayama, *Vision Research* **25**, 625 (1985).
3. B. J. Frost and K. Nakayama, *Science* **220**, 744 (1983).
4. H. J. Sun, J. Zhao, T. Southhall, and B. Xu, *Visual Neurosci.* **19**, 133 (2002).
5. B. K. Dellen, J. W. Clark, and R. Wessel, *Phys. Rev. E* **70**, 031907 (2004).
6. R. M. Davidson and D. B. Bender, *J. Neurophysiol.* **65**, 1115 (1991).
7. J. L. Barron, D. J. Fleet, and S. S. Beauchemin, *Int. J. Comp. Vision* **12**, 43 (1994).
8. J. A. Movshon, I. Thompson, and D. Tolhurst, *Journal of Physiology* **283**, 79 (1978).
9. J. S. Movshon, I. D. Thompson, and D. J. Tolhurst, *Journal of Physiology* **283**, 101 (1978).
10. K. K. De Valois, R. L. De Valois, and E. W. Yund, *J. Physiol.* **291**, 438 (1979).

Part I

Quantum Information

CONTROL AND ERROR PREVENTION IN CONDENSED MATTER QUANTUM COMPUTING DEVICES

M. S. BYRD

Department of Physics and Computer Science, Southern Illinois University, Carbondale, Illinois 62901, USA
E-mail: mbyrd@physics.siu.edu

L.-A. WU

Chemical Physics Theory Group, University of Toronto
Toronto, Ontario M5S 3H6

Proposals for scalable quantum computing devices suffer not only from decoherence due to their interaction with the environment, but also from severe engineering constraints. For example, our ability to implement quantum gates is determined, in part, by the experimentally available interactions with which quantum information may be processed. Here we review a practical solution to some of the major concerns, control and error prevention, addressing solid state proposals for quantum computing devices. Some noise is eliminated by encoding a logical qubit into two qubits, other noise is reduced by an efficient set of decoupling pulse sequences. The same encoding removes the need for single-qubit operations which pose a difficult design constraint. We also discuss several generalizations which follow from this work.

Keywords: Quantum computing; quantum error correction; decoherence.

1. Introduction

There are many different physical systems which are being explored as likely candidates from which to build a quantum computing (QC) device. Many of these are quite promising. However, given the success of our present-day computers, and the associated solid-state technologies, many researchers believe that when a scalable quantum computer is built, it will use similar condensed matter systems for its operation. One of the main problems with these, as well as other proposals, is the noise present in the experimental system. The noise is caused by both imprecise control as well as environmental interactions. The noise causes errors which must be reduced or eliminated if we are to build a reliable quantum computing device.

Quantum error correcting codes (QECC) are viable provided the error rate is below a certain "threshold" which is estimated to be about 10^{-4} depending on the assumptions used for the estimation. (See Refs.[1] and references therein for the threshold as well as fault-tolerant recovery requirements.) While these codes are very promising, they do not take into account particular design constraints associated with building a quantum computing device from a given system. In addition, the threshold has been quite difficult to achieve.

To overcome problems of design for the purposes of control and noise reduction, several combinations different methods of error correction, error avoidance, and error reduction techniques have been proposed. (See Ref.[2] for a review and summary recent work.) These methods are known as, respectively, the aforementioned quantum error correcting codes[3,4] used to detect and correct errors, decoherence-free subsystems (DFS)[5–9] used to avoid errors, and dynamical decoupling controls (DD)[10–21] used to reduce or eliminate errors.

Specifically addressing the problems of control and noise for condensed matter quantum computing proposals, we have proposed several possible approaches, first presented in Refs.,[22–24] which use a decoherence-free subspace encoding with recoupling operations (which "turn on and off" certain interactions) and decoupling operations (which "turn off" certain interactions). The encoding enables recoupling and decoupling operations to reduce noise in the system while enabling the control of interactions between various parts of the system. In this paper we will explain this specific proposed solution to the problems of noise and control, as well as several generalizations of the methods used in this particular proposal.

2. Quantum Computing in Condensed Matter Systems

Almost all promising solid-state quantum computing proposals are based on direct or effective exchange interactions between qubits and can be described by a Hamiltonian of the following form:

$$H_{\rm ex} = \sum_{i<j}(J_{ij}^x X_i X_j + J_{ij}^y Y_i Y_j + J_{ij}^z Z_i Z_j), \tag{1}$$

where X_i represents the Pauli matrix σ_x acting on the $i^{\rm th}$ qubit, Y_i represents the Pauli matrix σ_y acting on the $i^{\rm th}$ qubit, and Z_i represents the Pauli matrix σ_z acting on the $i^{\rm th}$ qubit. Examples are quantum dots,[25,26] donor atoms in silicon,[27] and quantum Hall systems.[28] These proposals would combine potential scalability with controllability of qubit interactions in the form of tunable exchange couplings J_{ij}^α and single qubit operations (e.g., Zeeman). There are two separate problems with these proposals. Problem I is the noise in the system which introduces computational errors. As with all QC proposals, the experiment inevitably couples to the environment (lattice, impurities, and other degrees of freedom). Problem II is that single-qubit gates can be problematic and this is true for many condensed matter systems. As discussed more thoroughly in Refs.[29–31] single qubit gates can require extreme g-factor engineering, highly localized magnetic fields, or have excess heating due to high-intensity RF fields. *Here, we discuss a solution to both problems, noise and control, by making use of a simple encoding and efficient pulse sequences.* The encoding uses a decoherence-free subspace and dynamical decoupling pulses[10,13–15,32] combined with selective recoupling.[30] We will show that a DD pulse, in the limit of strong, fast pulses (often referred to as the bang-bang

(BB) limit), eliminates all leakage errors (i.e., errors that mix states inside the DFS subspace with states outside the subspace) *with a single pair of pulses per cycle using the experimentally available exchange interaction.* We first discuss the use of this encoding for solid-state quantum dots and then provide some generalizations which include a set of compatibility conditions for the use of DD with quantum error correcting codes.

2.1. *Encoding*

To begin, we use a well-known code[33] of blocks of two qubits encoding a single logical qubit:

$$|0_L\rangle_i \equiv |0\rangle_{2i-1} \otimes |1\rangle_{2i}\,, \qquad |1_L\rangle_i \equiv |1\rangle_{2i-1} \otimes |0\rangle_{2i}\,. \tag{2}$$

Here $i = 1, ..., N/2$ indexes logical qubits, and N is the number of physical qubits. Logical operations (denoted by a bar) act on the encoded qubits as unencoded operations act on physical qubits. For example, $\bar{X}_i |0_L\rangle_i = |1_L\rangle_i$, $\bar{X}_i |1_L\rangle_i = |0_L\rangle_i$. Then, the single-encoded-qubit logical operations, $\bar{X}_i = (X_{2i-1}X_{2i} + Y_{2i-1}Y_{2i})/2$ and $\bar{Z}_i = (Z_{2i-1} - Z_{2i})/2$, generate all encoded-qubit transformations ($SU(2)$), through time-evolution. A logical operation between two logical qubits which can produce entanglement is given by $\bar{Z}_i\bar{Z}_{i+1} = Z_{2i}Z_{2i+1}$ which couples physical qubits in neighboring blocks. These operations form a *universal set of Hamiltonians*[30] meaning that by selectively turning on/off $\{\bar{X}_i, \bar{Z}_i, \bar{Z}_i\bar{Z}_{i+1}\}$ it is possible to generate a dense subgroup of the unitary group $U(2^{N/2})$. This set forms a complete set of logical operations on the $N/2$ logical qubits if we assume that the single-qubit spectrum is non-degenerate (e.g., due to Zeeman splitting), but the $\bar{Z}_i$ are not necessarily controllable. Then, controlling *only* $\bar{X}_i$ is sufficient to achieve (encoded) universality in the Heisenberg ($J^x_{ij} = J^y_{ij} = J^z_{ij}$), XXZ ($J^x_{ij} = J^y_{ij} \neq J^z_{ij}$), and XY ($J^x_{ij} = J^y_{ij}$, $J^z_{ij} = 0$) instances of the general exchange Hamiltonian, Eq. (1). This eliminates the need for controlling single, physical qubit interactions in exchange-based quantum computer architectures and thus solves Problem II.

Another advantage of the encoding given in Eq. (2) is that it is immune to collective dephasing.[6,33,34] Suppose the system interacts with a bath through the Hamiltonian $H_I = S_z \otimes B_z$, where $S_z = \sum_i Z_i$ and B_z is an operator which acts on the bath. Logical qubit states are superpositions of the logical zero and logical one states: $|\psi_L\rangle = a\,|0_L\rangle + b\,|1_L\rangle$. For such states $S_z|\psi_L\rangle = 0$, so H_I does not affect the code. This immunity to the system-bath interaction, is the reason that the encode state, or code word, $|\psi_L\rangle$ is decoherence-free. One might expect collective errors to be particularly relevant for solid-state systems at low temperatures and dephasing is one of the main problems in the corresponding class of QC devices. However, in reality there are other sources of errors and our goal is to show how the aforementioned methods can be extended to treat these as well. We will do this by introducing BB pulses as an added layer of protection, except that, using our encoding, we apply the pulses to *encoded* qubits. The DFS encoding together with the decoupling operations will serve to counter decoherence, thus solving Problem I, while the encoded states allow for universal quantum computing. We note that an interesting alternative proposal for dealing with Problems I and II is to combine the DFS method with energetic suppression of decoherence.[35] However, this perturbative result requires an encoding into at least four two-state systems. Here we use only two.

2.2. *Dynamical decoupling operations*

Dynamically decoupling operations are controls which are implemented in order to decouple the system from its environment. These are a generalization of the decoupling controls used in NMR. (See for example Ref.[36]) Here we assume the availability of BB controls which are decoupling controls in the limit of strong and fast pulses.[10] This assumption is unnecessary, however, as explained in Ref.[37] The simplest example of a BB pulse is the "parity-kick"[10,15] or π-pulse. Suppose that an error acts on the system which is represented by the operator E, and that a pulse U (a unitary operator) anticommutes with E. This unitary transformation inverts the sign of this error:

$$\{E, U\} = 0, \ \Rightarrow \ U^\dagger E U = -E. \tag{3}$$

Repeatedly implementing the cycle: {apply U, evolve freely under E (for time Δt), apply U^{-1}, evolve freely}, averages out the errors (or "symmetrizes"[13,14,19,32] the evolution), thus *decoupling* system and bath. The time for a complete cycle, T_c, must be shorter than τ_c, the inverse of the bath spectral density high-frequency cutoff:

$$\Delta t \leq T_c \ll \tau_c. \tag{4}$$

If the time scales are close, one can still achieve noise reduction.[10,15] Knowledge of τ_c is clearly useful for the success of the procedure. However, it is not always available. Therefore, we also outline an alternative method for the determination and empirical evaluation of the BB operations. Let us now show how BB can be applied to the encoded qubits, enabling the comprehensive solution to the problems posed above.

2.3. *Applying bang-bang operations on a decoherence-free subspace*

Logical qubits encoded as in Eq. (2) are immune to collective dephasing errors $Z_{2i-1} + Z_{2i}$. When considering other errors on the system, it is useful to classify the possible errors using a basis of Hermitian operators. We will use a basis composed of the 2^4 different tensor products of Pauli matrices (including the identity I) acting on two qubits. Four types of operations can be identified: (i) two operations which leave the DFS unchanged – $(I, Z_1 + Z_2)$; (ii) three would act to interchange states outside the DFS. (iii) three logical operations – $[\bar{X}, \bar{Y}, \bar{Z}]$ which represent operations on the *logical* subspace. (iv) eight operations which mix DFS states with states out of the DFS [Eq. (5)]. These cause leakage from, and to, the DFS. Both (i) and (ii) have no effect on the DFS while sets (iii) and (iv) damage the encoding. Both (iii) and (iv) can cause errors by entangling the encoded information with the bath. A basis for leakage errors (iv) on the first logical qubit is represented by the following set:

$$\{X_1, X_2, Y_1, Y_2, X_1Z_2, Z_1X_2, Y_1Z_2, Z_1Y_2\}. \tag{5}$$

These can be seen to take the encoded states of Eq. (2) out of the DFS (and vice versa) since they involve single bit flips, or bit and phase flips on individual physical qubits. We now come to a key observation.

Theorem 1: Let $U_{\bar{X}_i}(\phi) \equiv \exp(-i\phi\bar{X}_i)$. Then cycles of a *single* pair of BB pulses $U_{\bar{X}_i}(\pi)$ and $U_{\bar{X}_i}(\pi)^\dagger$, where

$$U_{\bar{X}_i}(\pi) = e^{-i\pi(X_{2i-1}X_{2i}+Y_{2i-1}Y_{2i})/2} = -Z_{2i-1}Z_{2i}, \tag{6}$$

can eliminate all type (iv) leakage errors on the ith logical qubit.

Proof: $U_{\bar{X}_1}(\pi)$ anticommutes with all errors in Eq. (5) so it satisfies the parity-kick condition (3). QED

The fact that a *single pair of pulses* will eliminate all leakage errors is extremely important given the time constraints of Eq. (4). In order to implement $U_{\bar{X}_1}(\pi)$ one must switch on $J(X_1X_2 + Y_1Y_2)$ for a time $t = \pi/2J$. This (XY) Hamiltonian is available in a number of QC proposals (quantum dots/atoms in cavities[26,38] and quantum Hall systems[28]). Systems governed by Heisenberg or XXZ Hamiltonians can be made to simulate the XY type using encoded recoupling.[30] The Heisenberg case applies to the spin-coupled quantum dots[25] and donor-spin proposals;[27] and the XY and Heisenberg if symmetry breaking mechanisms are present.[30] Spin-orbit coupling induces corrections to Eq. (1),[39] which can be overcome by a number of methods.[31,40]

Compared with the previous proposal of concatenation of a DFS and QECC,[34] eliminating all leakage errors on a DFS encoding a logical qubit by cycles of a single pair of BB pulses is very different and quite an advantage since it saves qubit resources. However, if logical errors (iii) are present, which the DFS-QECC method can correct at no extra cost[34] we would require another decoupling pulse, $U_{\bar{X}_i}(\pi/2) = -i\bar{X}_i$, which anticommutes with both $\bar{Y}_i$ and $\bar{Z}_i$ (this also implies that this pulse can be used to *create* the conditions of collective dephasing[41,42]). Hence *all but one error* ($\bar{X}_i$ itself) *can be eliminated using only the single BB control Hamiltonian* $\bar{X}_i$, which, as discussed above, is experimentally viable in many QC proposals. To eliminate $\bar{X}_i$ itself without destroying the previous step of eliminating $\bar{Y}_i$ and $\bar{Z}_i$, we must introduce two more BB controls, e.g., $\bar{Z}_i$ and $\bar{Y}_i$. The encoded recoupling method[30] can be used to switch on/off $\bar{Z}_i$ *by controlling only* $\bar{X}_i$, and $e^{-i\theta\bar{Y}_i} = e^{-i\frac{\pi}{4}\bar{Z}_i}e^{-i\theta\bar{X}_i}e^{i\frac{\pi}{4}\bar{Z}_i}$. This procedure already involves several pulses and may not need meet the strict BB time-constraints. We note that cycles of three parity-kick pulses (+ the identity operation) can suppress all single-qubit errors *without* encoding.[13,32] In contrast, the advantages of our method, which is compatible with encoded recoupling, are: (i) the elimination of *leakage* errors on a logical qubit with a *single* parity kick sequence depending on a controllable XY Hamiltonian, (ii) the elimination of all other errors using control of the *same* Hamiltonian.

3. Empirical Decoupling

Accurate models of complicated systems, such as condensed matter systems, can be difficult to develop. An alternative to model-based approaches of determining the bath cutoff and dynamical decoupling controls is provided by "empirical decoupling". This requires neither a microscopic understanding nor a detailed experimental analysis of each of the decoherence processes in the system. Instead, it requires quantum process tomography measurements (QPT)[43] to determine the *types* of errors. It is then possible, at least in principle, to empirically determine the *required* corrective pulses and the effectiveness of an *experimentally available* set.[16] While quantum process tomography can be experimentally taxing, once the appropriate measurements have been made on a repeatable experiment, one can use the data to determine, through repeated trials, a sufficient set of decoupling operations.

This empirical method is based on the evolution of an open quantum system, described by a density matrix ρ, that satisfies

$$\rho(t) = \sum_{\alpha,\beta} \chi_{\alpha\beta}(t) K_\alpha \rho(0) K_\beta^\dagger, \tag{7}$$

where the matrix $\chi_{\alpha\beta}(t)$ is Hermitian and $\{K_\alpha\}$ is a system operator basis.[43,44] The χ matrix is the output of QPT,[43] i.e., it is *measurable.* For BB operations, we should ideally employ a short-time expansion of Eq. (7). In this case, choosing a Hermitian operator basis $\{K_\alpha\}$ ($K_0 = I$), to first order in τ (where $\tau \ll \tau_c$): $\rho(\tau) = i[S(\tau), \rho(0)]$. Here $S(\tau) = \sum_{\alpha \geq 1} \mathrm{Im}(\chi^{(1)}_{\alpha 0}(\tau)) K_\alpha$, and $\chi^{(1)}_{\alpha 0}(\tau) = \tau (d(\chi^{(1)}_{\alpha 0})/dt)_{t=0}$.[44] $S(\tau)$ plays the role of a Hamiltonian. Under the action of a group $\mathcal{G} = \{U_k\}_{k=1}^N$ of unitary BB controls, the operator basis transforms as $K_\alpha \to \frac{1}{N}\sum_k U_k^\dagger K_\alpha U_k$. This implies a transformation of χ under the adjoint representation of $\mathcal{G}$, defined by $\sum_\beta R^{(k)}_{\alpha\beta} K_\beta = U_k^\dagger K_\alpha U_k$ (e.g., $R \in SO(3)$ for $U \in SU(2)$,[16,19]). Specifically, using the abbreviation $\bar{\chi}_\alpha \equiv \mathrm{Im}(\chi^{(1)}_{\alpha 0}(\tau))$, we have under BB that $\sum_{\alpha \geq 1} \bar{\chi}_\alpha K_\alpha \to \sum_{\beta \geq 1} \tilde{\chi}_\beta K_\beta$, where

$$\tilde{\chi}_\beta = \frac{1}{N} \sum_k \sum_{\alpha \geq 1} \bar{\chi}_\alpha R^{(k)}_{\alpha\beta}. \tag{8}$$

The coefficients $\tilde{\chi}_\beta$ are the expansion coefficients of a "desired" Hamiltonian. For example, for *storing* information we would require the decoupling pulses to eliminate all errors due to $S(\tau)$, so that all $\tilde{\chi}_\beta$ vanish. For *computation* we would have non-vanishing $\tilde{\chi}_\beta$ describing the Hamiltonian we wish to implement.[16] *The key idea of empirical BB is to use the experimentally determined* $\bar{\chi}_\alpha$, *together with a specified set of* $\tilde{\chi}_\beta$ *(corresponding to a desired evolution), to solve Eq.* (8) *for the rotation matrices* $R^{(k)}_{\alpha\beta}$. These determine a set of BB operations. Thus, using empirical BB, *one may determine the required BB operations directly from experimental data.* Repeatedly performing such a procedure (measure χ, apply BB) determines the optimal BB process, given the available controls and accounting for constraints, through a learning loop.[16,45] In this manner only the experimentally relevant errors are addressed, thus potentially reducing the set of BB operations. In addition, a reduction in decoherence, as manifested in χ, is a direct indication that condition in Eq. (4) is satisfied. For example, the use of empirical BB allows for a direct test of the feasibility of our encoded parity-kick scheme.

We may also use this formalism to provide a set of decoupling operations which is sufficient to remove noise from a DFS. To do this, we may assume that a DFS is unaffected by some given set of algebraic operations S. In the example above, Eq. (2) is unaffected by any operator proportional to S_z. The set of such operations can be used to define a group called the stabilizer $\mathcal{S}$ of the code (the code being the set of encoded, or logical, qubit states).[8] The stabilizer leaves code words invariant. To eliminate errors on the DFS encoded subspace, it is sufficient to ensure, not that the set $\tilde{\chi}_\beta$ vanish, but only that they are contained in the stabilizer subspace. Similarly, for computation, errors eliminated, up to a stabilizer element are acceptable for reliable quantum computing. We thus interpret the condition

$$\sum_{\alpha\geq 1} \bar{\chi}_\alpha K_\alpha \in \mathcal{S} \tag{9}$$

in two different ways. 1) Given a code which we wish to protect from noise, this indicates the objective of a set of decoupling operations. 2) Given an experimental set of available decoupling operations, a code with the given, achievable stabilizer can be protected against errors. We can think of this as projecting into the code subspace. This condition was first derived in Ref.[16] It bears a striking resemblance to the so-called operator quantum error correction recently developed in Ref.[46] There, the idea is to correct errors in a QECC only up to an element of the stabilizer of a DFS, similar to the proposal of Ref.[16] In either case, 1) or 2), *we do not require that a particular noise-producing interaction be eliminated, only that it be modified such that it is contained in the stabilizer of the code.*

4. Generalizations

Let us now discuss generalizations which will shed further light and suggest additional applications.

4.1. *Dynamically decoupling logical states*

Let $\bar{\mathcal{G}}$ denote the group of logical operations (e.g., for $SU(2)$, generated by $\{\bar{X}, \bar{Y}, \bar{Z}\}$), acting as *gates* on encoded qubits. In analogy to standard BB theory[10,13–15,32] we define "symmetrization of a Hamiltonian H with respect to $\bar{\mathcal{G}}$" as: $H \mapsto (1/|\bar{\mathcal{G}}|)\sum_{U\in\bar{\mathcal{G}}} U^\dagger H U$. We then have:

Theorem 2: Symmetrization with respect to $\bar{\mathcal{G}}$ suffices to completely decouple the dynamics of the encoded subspace from the bath.

Proof: Symmetrization takes any system-bath Hamiltonian and projects it onto the centralizer of the group $\bar{\mathcal{G}}$ (i.e., the set of elements that commute with all elements of this group). By irreducibility of the representation of $\bar{\mathcal{G}}$, it follows, from Shur's Lemma, that the BB-modified system-bath Hamiltonian is proportional to identity on the code space. In other words, the code space dynamics will be decoupled. QED.

The theorem shows that *any encoding* may be combined with decoupling operations using logical operations and the generalization to many qubits is immediate. The relevance of this is that if one chooses an particular code subspace on which to perform quantum computing, a set of logical operations must be given. In addition, it motivates us to consider combining encoded BB with QECCs. Such codes can often be described by a stabilizer group $\mathcal{S} = \{S_i\}$: a group that has all codewords as eigenstates with eigenvalue 1.[4] The set of errors $\mathcal{E} = \{E_j\}$ a stabilizer code can detect is the same as the set of operators that *anticommute* with some element of $\mathcal{S}$. This naturally leads to an encoded parity kick scheme: *to suppress* $\mathcal{E}$, *apply the generators of* $\mathcal{S}$ *as a set of BB operations.* Consider as a simple, but important example, the case of protecting an encoded qubit against all single physical qubit errors. The smallest QECC uses five physical qubits per logical qubit.[47] Instead, we could start by encoding one logical qubit into three: $|0\rangle_L = |000\rangle$, $|1\rangle_L = |111\rangle$. This protects against independent bit flip errors $\mathcal{E}_X = \{X_1, X_2, X_3\}$, but leaves us with the independent phase flip errors $\mathcal{E}_Z = \{Z_1, Z_2, Z_3\}$. The phase errors can be suppressed using a decoupling pulse on the encoded qubits. The stabilizer for this code is generated by $\mathcal{S}_X = \{X_1X_2, X_2X_3\}$, which anticommutes with $\mathcal{E}_Z$. Therefore, frequent application of X_1X_2 and X_2X_3 as parity kicks will suppress the $\mathcal{E}_Z$ errors. Note that it is possible to implement these using simultaneous application of single-body Hamiltonians X_i and X_j, i.e., there is an inherent parallelism. The advantage of this compared to the five-qubit code is that this code uses fewer physical qubits. The drawback is the additional gate operations that must be included in the QECC circuitry. However, this may well be a worthwhile tradeoff when qubits are scarce, as will be the case with near-future experiments. Another possibility, that illustrates *Theorem 2* directly, is to use the normalizer elements of the QECC.[4] These are logical operations which commute with stabilizer elements and preserve the code subspace. Therefore they can be applied at any time during a QECC circuit (with the exception of during the recovery operations), as BB pulses. As an example, let us choose a subset of logical operations $\bar{g}_i \in \bar{\mathcal{G}} \neq \mathcal{S}$ which satisfy the parity-kick condition: $\{\bar{g}_i, E_j\} = 0$. For the three-qubit code for phase flips, $\bar{Z} = X_1X_2X_3$, and $\{\bar{Z}, \mathcal{E}_Z\} = 0$. Thus the theorem identifies many different combinations of error prevention methods and is illustrated in these examples.

4.2. *Eliminating leakage*

As stated before, leakage operators are those operators which couple states inside an encoded subspace to states outside of that subspace. If an encoded qubit is to be effective preventing errors and/or enabling universal quantum computing, then the quantum information must properly be confined to the encoded subspace. The analysis of the two qubit code given in Eq. (2) led, not only to the identification of the leakage operators in Eq. (5) for that code, but also to more general results. Leakage elimination operators and the decomposition of the code subspace, as explained below, can be identified for many other codes and experimental proposals and this has been done in Refs.[23,24]

We now present a general argument for the a existence of leakage elimination operator (LEO), an operator which would eliminate all leakage errors on encoded or physical qubits. Suppose that n two-level systems (e.g., electron spins in quantum dots[25]) are used to encode one logical qubit, or that a N-level Hilbert space $\mathcal{H}_N$ supports a two-dimensional physical qubit subspace (e.g., hyperfine energy levels of an ion[48]). Let us arrange the basis vectors $\{|n\rangle\}_{n=0}^{N-1}$ so that $|0\rangle$ and $|1\rangle$ represent the (physical or encoded) qubit states. We refer to this as the "ordered basis". All system operators are of one of the following three types:

$$E = \begin{pmatrix} B & 0 \\ 0 & 0 \end{pmatrix} \quad E^{\perp} = \begin{pmatrix} 0 & 0 \\ 0 & C \end{pmatrix} \quad L = \begin{pmatrix} 0 & D \\ F & 0 \end{pmatrix}, \tag{10}$$

where B is 2×2 (for a single qubit, logical or physical), C is an $(N-2) \times (N-2)$ block and D, F are $2 \times (N-2)$, $(N-2) \times 2$ blocks. Operators of type E represent logical operations, i.e., they act entirely within the qubit subspace. $E^{\perp}$ operators, on the other hand, do not effect the qubit since they act entirely outside the qubit subspace. Finally, L represents possible leakage operators. The total system-bath Hamiltonian can be written as $H_{SB} = H_E + H_{E^{\perp}} + H_L$, where H_E ($H_{E^{\perp}}, H_L$) is a linear combination of tensor products of bath operators with elements of the set E ($E^{\perp}, L$). Now consider

$$R_L = e^{i\phi} \begin{pmatrix} -I & 0 \\ 0 & I \end{pmatrix}, \tag{11}$$

where the blocks have the same dimensions as in Eq. (10) and ϕ is an overall phase. This operator satisfies $\{R_L, L\} = 0$, while $[R_L, E] = [R_L, E^{\perp}] = 0$. Using a BB parity-kick sequence[10,15,49] it follows that R_L is a *leakage-elimination operator* (LEO):

$$\lim_{n\to\infty} (e^{-iH_{SB}t/n} R_L^{\dagger} e^{-iH_{SB}t/n} R_L)^n = e^{-iH_E t} e^{-iH_{E^{\perp}} t}. \tag{12}$$

In practice one takes $n = 1$. Eq. (12) then holds to order t^2 so that one intersperses periods of free evolution for time t with $R_L, R_L^{\dagger}$ pulses which strong enough that H_{SB} is negligible during the pulses. Transformations of the form $e^{-iH_{E^{\perp}} t}$ in Eq. (12) has no effect on the qubit subspace. Transformations of the form $e^{-iH_E t}$ may result in logical errors, which will have to be treated by other methods, e.g., concatenation with a QECC,[34,50–52] or additional BB pulses.[13,22] Note that R_L commutes with the logical operations, so they can be implemented at the same time, i.e., *our leakage elimination procedure is fully compatible with universal quantum computation.*

We can also provide a procedure for generating LEOs from a controllable system Hamiltonian H_S acting for a time τ, i.e., $R_L = \exp(-iH_S\tau)$. From Eq. (11) it follows that H_S acts (mathematically) as a projection operator P onto the qubit subspace. (However, it is not necessary that it act in this way.[24]) The time, τ must be chosen so that R_L acts as $-I$ in the qubit subspace. A general choice is

$$R_L^{\mathrm{E}(1)} = \exp\left(\pm i\pi\widehat{n}\cdot\vec{\sigma}P\right) \tag{13}$$

where $\vec{\sigma}$ denotes the vector of Pauli matrices, which we refer to as logical X, Y, Z operations, and $\widehat{n}$ is a real unit vector. This is a valid LEO since $\exp\left(-i\pi\widehat{n}\cdot\vec{\sigma}\right)$ expresses a 2π rotation about the axis $\hat{n}$ on the qubit Bloch sphere, upon which the qubit state acquires a minus sign. A useful example is $\tau = \pi$ and $H_S = |0\rangle\langle 0| + |1\rangle\langle 1|$, which is a projector onto the qubit subspace and acts as identity there. This can be generalized to logical operations by replacing X, Y, Z with $\bar{X}, \bar{Y}, \bar{Z}$.

Several examples of the usefulness of this method have been given in Ref.[23] They include electron spin qubits in quantum dots and a nuclear spin qubits in donor atoms in Silicon. In addition, Ref.[24] shows the possible use of three-to-one and four-to-one DFS encodings which will protect a logical qubit from arbitrary collective errors on the physical qubits while enabling exchange-only universal computation.

5. Conclusions

In conjunction with the elimination of the need for difficult-to-implement single qubit operations by the encoded recoupling method,[30] we believe that we offer a realistic and comprehensive solution to some of the major difficulties related to decoherence in, and the design of, quantum dot, and other exchange-based quantum computers. Our solution combines ideas from the theory of decoherence-free subspaces, decoupling controls, and encoded recoupling. By encoding physical qubits into logical qubits, a first level of protection against collective dephasing is obtained. Control of an exchange Hamiltonian suffices to implement universal quantum computation on this code. Cycles of pairs of hard, fast, decoupling pulses generated from the *same exchange Hamiltonian*, can be used to eliminate all *leakage* errors on a logical qubit, and to further suppress *all* other errors, two more pulses can be used. We note also, that the same Hamiltonian can be used to *create* the conditions of collective dephasing, using decoupling pulses.[14,16,41,42]

We emphasize that these methods are not only useful for condensed matter systems. For example, it has recently been shown[53] these methods also apply to the ion trap quantum computing proposal.[54] Similar to the systems discussed here, this encoding enables universal quantum computing while satisfying the experimental constraints of the physical system. In addition, we have presented several generalizations of these methods. First, we provided an empirical means for determining a set of decoupling operations which is compatible with any logical encoding and which can be described using a stabilizer formalism. Second we showed that logical operations can provide a sufficient set of decoupling pulses for a set of encoded qubits; and thrid we extended the methods of eliminating leakage errors, which are a particularly devastating class of errors. The examples show that our results apply to a wide range of experiments and codes. We believe this work will help to design a robust quantum computing device.

Acknowledgments

MSB would like to thank the Max Plank Institute for Complex Systems (MPIPKS) in Dresden, Germany, Southern Illinois University, the National Science Foundation NSF-CCF0545798, and the Army for support.

References

1. J. Preskill, Fault-Tolerant Quantum Computation, in *Introduction to Quantum Computation and Information*, ed. H.K. Lo, S. Popescu and T.P. Spiller (World Scientific, Singapore, 1999) A.M. Steane, *Phys. Rev. A* **68**, p. 042322 (2003).
2. M. S. Byrd, L.-A. Wu and D. A. Lidar, *J. Mod. Optics* **51**, p. 2449 (2004).
3. P.W. Shor, *Phys. Rev. A* **52**, p. 2493 (1995). A.R. Calderbank and P.W. Shor, *Phys. Rev. A* **54**, p. 1098 (1996). A. Steane, *Rep. Prog. Phys.* **61**, p. 117 (1998).
4. D. Gottesman, *Phys. Rev. A* **54**, p. 1862 (1996).
5. P. Zanardi and M. Rasetti, *Phys. Rev. Lett.* **79**, p. 3306 (1997). D.A. Lidar, I.L. Chuang and K.B. Whaley, *Phys. Rev. Lett.* **81**, p. 2594 (1998).
6. L.-M Duan and G.-C. Guo, *Phys. Rev. A* **57**, p. 737 (1998).
7. E. Knill, R. Laflamme and L. Viola, *Phys. Rev. Lett.* **84**, p. 2525 (2000).
8. J. Kempe, D. Bacon, D.A. Lidar, and K.B. Whaley, *Phys. Rev. A* **63**, p. 042307 (2001).
9. D.A. Lidar, D. Bacon, J. Kempe, and K.B. Whaley, *Phys. Rev. A* **63**, p. 022306 (2001).
10. L. Viola and S. Lloyd, *Phys. Rev. A* **58**, p. 2733 (1998).
11. M. Ban, *J. Mod. Opt.* **45**, p. 2315 (1998).
12. L.-M. Duan and G. Guo, *Phys. Lett. A* **261**, p. 139 (1999).
13. L. Viola, E. Knill and S. Lloyd, *Phys. Rev. Lett.* **82**, p. 2417 (1999).
14. P. Zanardi, *Phys. Lett. A* **258**, p. 77 (1999).
15. D. Vitali and P. Tombesi, *Phys. Rev. A* **59**, p. 4178 (1999).
16. M.S. Byrd and D.A. Lidar, *Phys. Rev. A* **67**, p. 012324 (2003).
17. L. Viola, S. Lloyd and E. Knill, *Phys. Rev. Lett.* **83**, p. 4888 (1999).
18. L. Viola, E. Knill, and S. Lloyd, *Phys. Rev. Lett.* **85**, p. 3520 (2000).
19. M. S. Byrd and D.A. Lidar, *Qu. Inf. Proc.* **1**, p. 19 (2001).
20. D.G. Cory, R. Laflamme, E. Knill, L. Viola, T.F. Havel, N. Boulant, G. Boutis, E. Fortunato, S. Lloyd, R. Martinez, C. Negrevergne, M. Pravia, Y. Sharf, G. Teklemariam, Y.S. Weinstein, W.H. Zurek, *Fortschritte der Physik* **48**, p. 875 (2000). D. Vitali and P. Tombesi, *Phys. Rev. A* **65**, p. 012305 (2002).
21. G.S.Agarwal, M.O.Scully and H.Walther, *Phys. Rev. Lett.* **86**, p. 4271 (2001). C. Uchiyama and M. Aihara, *Phys. Rev. A* **66**, p. 032313 (2002). C. Uchiyama and M. Aihara, *Phys. Rev. A* **68**, p. 052302 (2003).
22. M. S. Byrd and D. A. Lidar, *Phys. Rev. Lett.* **89**, p. 047901 (2002).
23. L.-A. Wu, M.S. Byrd and D.A. Lidar, *Phys. Rev. Lett.* **89**, p. 127901 (2002).
24. M.S. Byrd, D.A. Lidar, L.-A. Wu and P. Zanardi, *Phys. Rev. A* **71**, p. 052301 (2005).
25. D. Loss and D.P. DiVincenzo, *Phys. Rev. A* **57**, p. 120 (1998). J. Levy, *Phys. Rev. A* **64** (2001).
26. A. Imamoḡlu, D.D. Awschalom, G. Burkard, D.P. DiVincenzo, D. Loss, M. Sherwin and A. Small, *Phys. Rev. Lett.* **83**, p. 4204 (1999).

27. B.E. Kane, *Nature* **393**, p. 133 (1998). R. Vrijen, E. Yablonovitch, K. Wang, H.W. Jiang, A. Balandin, V. Roychowdhury, T. Mor, and D. DiVincenzo, *Phys. Rev. A* **62**, p. 012306 (2000).
28. D. Mozyrsky, V. Privman and M. L. Glasser, *Phys. Rev. Lett.* **86**, p. 5112 (2001).
29. D.P. DiVincenzo, D. Bacon, J. Kempe, G. Burkard, and K. B. Whaley, *Nature* **408**, p. 339 (2000).
30. D.A. Lidar and L.-A. Wu, *Phys. Rev. Lett.* **88**, p. 017905 (2002).
31. L.-A. Wu, D.A. Lidar, *Phys. Rev. A* **66**, p. 062314 (2002).
32. P. Zanardi, *Phys. Rev. A* **60**, p. R729 (1999).
33. G.M. Palma, K.-A. Suominen and A.K. Ekert, *Proc. Roy. Soc. London Ser. A* **452**, p. 567 (1996).
34. D.A. Lidar, D. Bacon and K.B. Whaley, *Phys. Rev. Lett.* **82**, p. 4556 (1999).
35. D. Bacon, K.R. Brown, and K.B. Whaley, *Phys. Rev. Lett.* **87**, p. 247902 (2000).
36. R.R. Ernst, G. Bodenhausen and A. Wokaun, *Principles of nuclear magnetic resonance in one and two dimensions* (Clarendon Press, Oxford, 1987).
37. L. Viola, *J. Mod. Optics* **51**, p. 2357 (2004).
38. S.-B. Zheng and G.-C. Guo, *Phys. Rev. Lett.* **85**, p. 2392 (2000).
39. K.V. Kavokin, *Phys. Rev. B* **64**, p. 075305 (2001).
40. N.E. Bonesteel, D. Stepanenko and D.P. DiVincenzo, *Phys. Rev. Lett.* **87**, p. 207901 (2001). G. Burkard and D. Loss, *Phys. Rev. Lett.* **88**, p. 047903 (2002).
41. L.-A. Wu and D.A. Lidar, *Phys. Rev. A* **65** (2002).
42. L. Viola, *Phys. Rev. A* **66**, p. 012307 (2002).
43. I.L. Chuang and M.A. Nielsen, *J. Mod. Optics* **44**, p. 2455 (1997).
44. D.A. Lidar, Z. Bihary, and K.B. Whaley, *Chem. Phys.* **68**, p. 35 (2001).
45. C. Brif, H. Rabitz, S. Wallentowitz, and I.A. Walmsley, *Phys. Rev. A* **63**, p. 063404 (2001).
46. D. Kribs, R. Laflamme, and D. Poulin, *Phys. Rev. Lett.* **94**, p. 180501 (2005).
47. R. Laflamme, C. Miquel, J.P. Paz and W.H. Zurek, *Phys. Rev. Lett.* **77**, p. 198 (1996).
48. J.I. Cirac and P. Zoller, *Phys. Rev. Lett.* **74**, p. 4091 (1995). M. Plenio and P. Knight, *Proc. Roy. Soc. London Ser. A* **453**, p. 2017 (1997).
49. P. Zanardi, *Phys. Rev. A* **63**, p. 012301 (2001), eprint quant-ph/9910016.
50. J. Preskill, *Proc. Roy. Soc. London Ser. A* **454**, p. 385 (1998).
51. E. Knill, R. Laflamme and W. Zurek, *Science* **279**, p. 342 (1998).
52. D.A. Lidar, D. Bacon, J. Kempe, and K.B. Whaley, *Phys. Rev. A* **63**, p. 022307 (2001).
53. D.A. Lidar and L.-A Wu, *Phys. Rev. A* **67**, p. 032313 (2003).
54. D. Kielpinski, V. Meyer, M. A. Rowe, C. A. Sackett, W. M. Itano, C. Monroe, and D. J. Wineland, *Science* **291**, p. 1013 (2001).

MAXENT APPROACH TO QUBITS

C. M. SARRIS and A. N. PROTO*

Laboratorio de Sistemas Complejos, Facultad de Ingenieria
Universidad de Buenos Aires
Av. Paseo Colon 850 4th F (1063)
Buenos Aires, Argentina
E-mail: aproto@fi.uba.ar

F. B. MALIK

Physics Department, Southern Illinois University, Carbondale, IL 62901 USA
**E-mail: fbmalik@physics.siu.edu*

We consider the Hubbard-like Hamiltonian used in Physica B 352 (2004) 269-279, using the MaxEnt approach and a CSCO, as the set of relevant operators which, obviously, close a partial Lie algebra with the Hamiltonian. So, it is possible to define a positive definite metric on the space spanned by the mean values of the CSCO. Particularly, the selected operators represent the occupation number, the aligned and antialigned pairs per site, respectively. The interesting question to be examined here is whether the above mentioned Hamiltonian have the property that the expectation values of a suitable operator are indeed zero and one (or a constant) qubits, and zero and +1 and -1, i.e., qutrits.

Keywords: Generating qubits.

1. Introduction

According to the mathematical theory of information, the amount of information needed to characterize an element of a set of power n is n. This follows from the observation that one needs elements to characterize the information containing " n" characters of binary digits. The corresponding lack of information corresponds to Shannon entropy, $S(sh)$, (also independently noted by Wiener),[1,2] given by:

$$S(sh) = -k \sum_{m=0}^{N} \Phi_m \ln \Phi_m \tag{1}$$

where Φ_m is the average information gained from a particular element m and k is an arbitrary positive constant. m runs over all messages. The quantum mechanical expression of the Shannon entropy can be obtained by introducing the concept of density matrix, $\hat{\rho}$, introduced by von Neuman.[3] The representation of "words" or information by bits or binary digits, zero and one, in classical physics, when translated into quantum physics, reads that any word can be represented by the expectation values of appropriate operators with respect to the density matrix $\hat{\rho}$ that are zero or one. The interesting question to be examined here is whether there could be physical systems that could have the property that expectation values of suitable operator representing "words" are indeed zero and one (or a constant) i.e., qubits. To search for such systems, we adopt the maximum entropy principle (MEP) of Jaynes.[4,5] In the next section, we first present the key features of MEP necessary to search for a suitable system and then discuss possible systems characterized by a Hamiltonian. The results and discussion are presented in the last section

2. A Brief Overview of MEP

2.1. *Closure condition and density matrix*

Within the context of the density matrix theory, the Shannon entropy can be written as:[4,5]

$$S = -kTr\left(\hat{\rho}\ln\hat{\rho}\right) \tag{2}$$

where the density matrix $\hat{\rho}$ is defined in terms of $N+1$ relevant operators $\hat{O}_r$ ($\hat{O}_0 = \hat{I}$) and their Lagrange multipliers λ's associated to them:

$$\hat{\rho} = \exp\left(-\lambda_0\hat{I} - \sum_{j=1}^{N}\lambda_j\hat{O}_j\right), \tag{3}$$

subject to the constraints,

$$Tr\left(\hat{\rho}\right) = 1, \tag{4}$$

(where λ_0 is associated to the identity operator, $Tr\left(\hat{\rho}\right) = 1$), and:

$$\langle\hat{O}_r\rangle = Tr(\hat{\rho}\hat{O}_r)\,. \tag{5}$$

The set of $N+1$ relevant operators $(RO) = \left\{\hat{I}, \hat{O}_1, ..., \hat{O}_N\right\}$ are those which close a partial Lie algebra under commutation with the Hamiltonian of the system through the well known prescription:[6]

$$[\hat{H}, \hat{O}_j] = i\hbar\sum_{r=0}^{N} g_{rj}\hat{O}_r, \quad j = 1, ..., N \tag{6}$$

where the g_{rj} are elements of a $N\times N$ matrix G which contains the complete dynamics of the system. The closure condition (6) enables us to find a Complete Set of Non-Commuting Operators ($CSNCO$) in general, and a Complete Set of Commuting Observables ($CSCO$), in particular (when $g_{rj} = 0\ \forall_{rj}$). In both cases, the RO defined trough (6) are linearly independent.[6] The density matrix (3) follows the Liouville equation: $[\hat{H}, \hat{\rho}] = i\hbar\frac{\partial\hat{\rho}}{\partial t}$, so the entropy S remains as a constant of the motion and can be expressed in the fashion:

$$S = \lambda_0 + \sum_{r=1}^{N}\lambda_r\left\langle\hat{O}_r\right\rangle, \tag{7}$$

where the $\left\langle \hat{O}_r \right\rangle$'s are obtained through Eq.(5), or equivalently by the following:

$$\left\langle \hat{O}_r \right\rangle = -\frac{\partial \lambda_0(\lambda_1,....,\lambda_N)}{\partial t}, \quad r = 1, ..., N. \tag{8}$$

2.2. *Metric and uncertainty principle*

The closure condition (6) enables us to define a generalized phase space $\mathbb{V} = span\left\{\left\langle \hat{O}_1 \right\rangle, ..., \left\langle \hat{O}_q \right\rangle\right\}$ which is $N-$dimensional and in which the dynamics of the system takes place. Let's define a real scalar-valued function $\bullet$ of ordered pairs of elements belonging to the $\mathbb{V} - space$ such that $\bullet : \mathbb{V} \times \mathbb{V} \longrightarrow \mathbb{R}$

$$\left\langle \hat{O}_i \right\rangle_t \bullet \left\langle \hat{O}_j \right\rangle_t = \frac{1}{2} Tr\left(\hat{\rho}\left[\hat{O}_i, \hat{O}_j\right]_+\right)_t - Tr\left(\hat{\rho}\hat{O}_i\right)_t Tr\left(\hat{\rho}\hat{O}_i\right)_t, \tag{9}$$

where $[\hat{O}_i, \hat{O}_j]_+$ indicates anti-commutation and $\mathbb{R}$ is the set of real numbers (the field associated with the $\mathbb{V} - space$) which can be rewritten as:

$$\left\langle \hat{O}_i \right\rangle_t \bullet \left\langle \hat{O}_j \right\rangle_t = \frac{1}{2}\left\langle \left[\hat{O}_i, \hat{O}_j\right]_+ \right\rangle_t - \left\langle \hat{O}_i \right\rangle_t \left\langle \hat{O}_i \right\rangle_t = K_{ij}(t) = K_{ji}(t). \tag{10}$$

This inner product $\langle \hat{O}_i \rangle_t \bullet \langle \hat{O}_j \rangle_t$ defines a positive definite metric on $\mathbb{V} - space$[7] and also the components of the second-rank covariant metric tensor, $K(t)$, $K_{ij}(t) = K_{ji}(t)$. As the metric is a positive definite one, it follows that: a) the Schwarz inequality holds, b) all the principal minors of order $s \leq N$ belonging to the metric tensor are also positive, and then, c) the summation over these principal minors is also positive (called S_r in[8]). It should be noticed that the components $K_{ij}(t)$ of the covariant metric tensor $K(t)$ are the quantum correlation coefficients between the operators $\hat{O}_i$ and $\hat{O}_j$ belonging to the *CSNCO*. The metric defined by Eqs.(9) and (10) is closely related to the uncertainty principle. In fact, because of the positive definitness of the metric the Schwarz inequality can be written as, :

$$\left|\left\langle \hat{O}_i \right\rangle_t \bullet \left\langle \hat{O}_j \right\rangle_t\right|^2 \leq \left(\left\langle \hat{O}_i \right\rangle_t \bullet \left\langle \hat{O}_i \right\rangle_t\right)\left(\left\langle \hat{O}_j \right\rangle_t \bullet \left\langle \hat{O}_j \right\rangle_t\right), \tag{11}$$

which, in fact, is:

$$\left(\Delta\hat{O}_i\right)^2 \left(\Delta\hat{O}_j\right)^2 \geq \left[\frac{1}{2}\left\langle \hat{O}_i\hat{O}_j + \hat{O}_j\hat{O}_i \right\rangle - \left\langle \hat{O}_i \right\rangle \left\langle \hat{O}_j \right\rangle\right]^2. \tag{12}$$

As the operators belonging to the *CSNCO* are non-commuting ones they obey the relation:

$$[\hat{O}_m, \hat{O}_k] = i\hat{O}_l\,, \tag{13}$$

so, because of the uncertainty relation between them we can write the following:

$$\left(\Delta\hat{O}_m\right)^2\left(\Delta\hat{O}_k\right)^2-\left[\frac{1}{2}\left\langle\hat{O}_m\hat{O}_k+\hat{O}_k\hat{O}m\right\rangle-\left\langle\hat{O}_m\right\rangle\left\langle\hat{O}_k\right\rangle\right]^2$$
$$\geq -\frac{1}{4}\left\langle\left[\hat{O}_m,\hat{O}_k\right]\right\rangle^2. \tag{14}$$

Schwarz's inequality (12) and the uncertainty relation (14), establish a connection between the metric and the uncertainty principle as said previously. Finally, we want to emphazise that the generalized uncertainty principle (GUP) is written as:[9]

$$\sum_{m,k=1}^{q}\left\{\left(\Delta\hat{O}_m\right)^2\left(\Delta\hat{O}_k\right)^2-\left[\left\langle\hat{L}_{mk}\right\rangle-\left\langle\hat{O}_m\right\rangle\left\langle\hat{O}_k\right\rangle\right]^2\right\}$$
$$\geq -\frac{1}{4}\sum_{m,k=1}^{q}\left\langle\left[\hat{O}_m,\hat{O}_k\right]\right\rangle^2, \tag{15}$$

(with $\left\langle\hat{L}_{mk}\right\rangle=\frac{1}{2}\left\langle\hat{O}_m\hat{O}_k+\hat{O}_k\hat{O}_m\right\rangle$ and $m<k$). This results in the summation over the principal minors of order 2 belonging to the covariant metric tensor $K(t)$. Summarizing, we have shown the connection between the closure condition (6) and the existence of the metric $\mathbb{V}-$ *space*.

3. A Model for Qubits

The quantum analogue of the classical bits would be to find an appropriate relevant operator whose expectation value with respect to the MaxEnt density are zero and one. To this end, we consider two fermions in general and electrons in particular. They could be paired both in singlet and triplet states, the expectation value of z-projection of which are zero and one, respectively (We note that in quantum mechanics, one has the option of having z-projection 0, 1, and −1. However, we shall be referring here primarily to the absolute value of z-projection, which are 0 and 1). Since a free-fermion pair could be both in singlet and triplet states, it would be difficult to confine them in one or the other state alone without an external field, which could be an inherent internal magnetic field of the system or external magnetic field, or both. To characterize information requires an array or a chain or qubits. One, therefore, is to consider a large system of fermions interacting pair-wise at each site, where a site is defined in the following sense: Let a complete set of commuting observables, possessing well-defined values, be associated to each spin-carrying "particle". Then any set of such values less spin degrees of freedom constitutes in the present context a site.

In metallic superconductors, electrons pair off their spin to 0 in such sites below critical temperature; however, one needs materials where a pair of electrons could also be aligned to a triplet state at each site. As noted in[10] strongly correlated systems under suitable conditions could exhibit both singlet and triplet states at every site. The equation of a pair of fermions in a finite medium has been derived by a variational method in[11] for a single pair treating other fermions in Hartree-Fock approximation and in,[12,13] in pair-pair approximation for a finite system in the absence of magnetic field using again a variational method. The equation of pair represents an effective Hamiltonian of a pair which contains a kinetic energy, a mean field, an interaction between two members of a pair and the interaction of the pair with the rest of the medium terms. The incorporation of an internal and an external magnetic field would merely add additional terms representing interaction between spin degree of freedom and magnetic field at each site. In the presence of an internal and external magnetic field, B and $\mathcal{H}$, respectively, the following form of a Hamiltonian essentially incorporate the above salient features:

$$\hat{H} = \sum_i E(i)\hat{n}_i + (B\Delta + \mathcal{H})\hat{q}_i + U(i)\hat{r}_i + t\Delta^2\hat{I} \equiv \sum_i \hat{h}(i), \tag{16}$$

where, $\hat{n}_i = \hat{n}_{i\uparrow} + \hat{n}_{i\downarrow}$; $\hat{q}_i = \hat{n}_{i\uparrow} - \hat{n}_{i\downarrow}$ and $\hat{r}_i = \hat{n}_{i_\uparrow}\hat{n}_{i\downarrow}$. In (16) $\hat{n}_{i_\uparrow}$ and $\hat{n}_{i\downarrow}$ are related to fermion creation and annihilation operators of two-spin states as follows: $\hat{n}_{i_\uparrow} = \hat{c}^+_{i\uparrow}\hat{c}_{i\uparrow}$ and $\hat{n}_{i\downarrow} = \hat{c}^+_{i\downarrow}\hat{c}_{i\downarrow}$. The index i runs over all sites. The first term in (16) represents the motion of a fermion in a mean-field. $E(i)$ is the sum of one-body energy which may or may not contain a chemical potential. Δ in (16) represents average interaction of a particle in the site 'i' with the rest of the sites. As noted in,[14] by defining:

$$\Delta = \left\langle \hat{n}^+_{i\uparrow} + \hat{n}^+_{j\uparrow} \right\rangle + \left\langle \hat{n}^+_{i\downarrow} + \hat{n}^+_{j\downarrow} \right\rangle, \tag{17}$$

the Hamiltonian (16) is equivalent to the t-J Hamiltonian,[15,16] used widely to describe physical properties of strongly correlated systems in the absence of B and $\mathcal{H}$. The second term in (16) represents interaction of fermions with internal and applied magnetic field which act on two-spin states of fermion at the site $'i'$. The third term is the effective attraction between a pair at site $'i'$ caused in the presence of a medium and the last terms reflects the interaction of the site "i" with the rest of the sites. This Hamiltonian has been used to explain colossal magnetoresistance exhibited by suitably doped manganites.[17] The question to be investigated now is whether a system represented by Hamiltonian (16) could generate expectation values of z-component of spin per site that are 1 or 0, in terms of their absolute magnitudes, and if so, determine the conditions for that. The relevant operators for this purpose is chosen to be $RO = \{\hat{I}, \hat{h}(i), \hat{n}_i, \hat{r}_i, \hat{q}_i\}$ with: $\hat{h}(i)$ energy per site, $\hat{n}_i$: number of particles per site, which is relevant in determining the distribution function per site, $\hat{r}_i$: the number of anti-aligned pairs per site, $\hat{q}_i$: the number of aligned pairs per site. Aside from these operators, one has the identity operator, $\hat{I}$ in order to fulfill the condition (4). The inclusion of the per site Hamiltonian $\hat{h}(i)$ as a relevant operator does not modify the dynamics of the system (see Eq.(6)) but transforms it in a thermodynamical one, given that now it is possible to define the temperature $\beta = 1/K_BT$ of the system. So, the generalized phase space is

3–dimensional and it is defined :$\mathbb{V}$ =*span* $\{\langle\hat{n}_i\rangle, \langle\hat{r}_i\rangle, \langle\hat{q}_i\rangle\}$. It is easy to see that, because of the fermion and creation annhihilation operators of two-spin states obey the anticommutation relations:

$$\left\{\hat{c}_{i\uparrow}, \hat{c}^{+}_{j\uparrow}\right\} = \left\{\hat{c}_{i\downarrow}, \hat{c}^{+}_{j\downarrow}\right\} = \delta_{ij}, \tag{18}$$

the $RO = \left\{\hat{I}, \hat{h}(i), \hat{n}_i, \hat{r}_i, \hat{q}_i\right\}$ constitutes a $CSCO$ and thet fulfill the following: $\hat{n}^2 = \hat{n} + 2\,\hat{r}$, $\hat{q}^2 = \hat{n} - 2\,\hat{r}$, $\hat{r}^2 = \hat{r}$, $\hat{n}\hat{q} = \hat{q}\hat{n} = \hat{q}$, $\hat{n}\hat{r} = \hat{r}\hat{n} = 2\hat{r}$, $\hat{q}\hat{r} = \hat{r}\hat{q} = 0$.

The density matrix is given by:

$$\hat{\rho}(i) = \exp\left(-\lambda_0\hat{I} - \lambda_1\hat{h}(i) - \lambda_2\hat{n}_i - \lambda_3\hat{r}_i - \lambda_4\hat{q}_i\right). \tag{19}$$

To evaluate the thermodynamic quantities at each site, one now considers only the spin degrees of freedom explicitly which, for a pair of fermion, are the following: $\hat{c}_\uparrow|0\rangle = 0$ and $\hat{c}_\downarrow|0\rangle = 0$; $|3\rangle = \hat{c}^+_\uparrow\hat{c}^+_\downarrow|0\rangle$; $|2\rangle = \frac{1}{\sqrt{2}}\left(\hat{c}^+_\uparrow + \hat{c}^+_\downarrow\right)|0\rangle$; $|1\rangle = \frac{1}{\sqrt{2}}\left(\hat{c}^+_\uparrow - \hat{c}^+_\downarrow\right)|0\rangle$, $|0\rangle$.

The condition (4) leads to:

$$\begin{aligned}\lambda_0 = &-\lambda_1 t\Delta^2 + \\ &\ln\{1 + \exp(\Lambda_1) + 2\exp(\Lambda_2)\cosh[\lambda_1(B\Delta + \mathcal{H}) + \lambda_4]\}\end{aligned} \tag{20}$$

where, $\Lambda_1 = -2\,\lambda_2 - \lambda_3 - \lambda_1(U + 2E)$, $\Lambda_2 = -\lambda_3 - \lambda_1\,E$. Clearly $\hat{h}(i)$ is the energy per site and hence, to preserve the proper relation between free energy F and entropy S, λ_1 is to be identified with $\beta = 1/K_BT$. The least biased mean value of other operators can be obtained by partially differentiating (20) with respect to β, λ_2, λ_3 and λ_4. For the present case, we also set $\lambda_2 = \lambda_3 = \lambda_4 = 0$. This leads to the following expressions for the expectation values of the chosen relevant operators:

$$\langle\hat{h}\rangle = E\,\langle\hat{n}\rangle + (B\Delta + \mathcal{H})\,\langle\hat{q}\rangle + U\,\langle\hat{r}\rangle + t\Delta^2, \tag{21}$$

$$\langle\hat{n}\rangle = \frac{2}{D}\{\exp[-\beta(U + 2E)] + \exp(-\beta E)\cosh[\beta(B\Delta + \mathcal{H})]\}, \tag{22}$$

$$\langle\hat{r}\rangle = \frac{1}{D}\exp[-\beta(U + 2E)], \tag{23}$$

$$\langle\hat{q}\rangle = -\frac{2}{D}\exp[-\beta E]\sinh[\beta(B\Delta + \mathcal{H})], \tag{24}$$

with D given by:

$$D = 1 + \exp[-\beta(U + 2E)] + 2\exp(-\beta E)\cosh[\beta(B\Delta + \mathcal{H})], \tag{25}$$

Free energy, F, is now given by:

$$F = t\Delta^2 - \frac{1}{\beta}\ln D. \tag{26}$$

As noted in the previous section, MaxEnt entropy S corresponds to a minimum in free-energy F. This is achieved by minimizing the expression (26) with respect to Δ and leads to the following relation:

$$\Delta = -\frac{B}{2t} \langle \hat{q} \rangle \,. \tag{27}$$

This result is often a postulate in Weiss mean filed theory but derived in.[18] Eq. (27) provides the important information that for alignment to occur, Δ, *the interaction of the site 'i' with the rest of the medium must be non-zero.* Thus, aligned states with spin projection can only be generated in strongly correlated systems such as semi-conductor or insulator having an inherent internal magnetic field or a system exhibiting spontaneous magnetization. The presence of an applied field $\mathcal{H}$ is not critical but the system must have an internal magnetic field. However, for such systems, as noted in,[17] the minimum of F could also occur at $\Delta = 0$. In those cases, the fermions are paired in such that the z-projection per site is zero.

Now, we are in a position to calculate the covariant metric tensor´s components:

$$K_{nn} = (\Delta \hat{n})^2 = \left\langle \hat{n}^2 \right\rangle - \langle \hat{n} \rangle^2 = \langle \hat{n} \rangle + 2 \langle \hat{r} \rangle - \langle \hat{n} \rangle^2 \,, \tag{28}$$

$$K_{qq} = (\Delta \hat{q})^2 = \left\langle \hat{q}^2 \right\rangle - \langle \hat{q} \rangle^2 = \langle \hat{n} \rangle - 2 \langle \hat{r} \rangle - \langle \hat{q} \rangle^2 \,, \tag{29}$$

$$K_{rr} = (\Delta \hat{r})^2 = \left\langle \hat{r}^2 \right\rangle - \langle \hat{r} \rangle^2 = \langle \hat{r} \rangle - \langle \hat{r} \rangle^2 \,, \tag{30}$$

$$K_{nq} = K_{qn} = \langle \hat{q} \rangle - \langle \hat{n} \rangle \langle \hat{q} \rangle \,, \tag{31}$$

$$K_{nr} = K_{rn} = 2 \langle \hat{r} \rangle - \langle \hat{n} \rangle \langle \hat{r} \rangle \,, \tag{32}$$

$$K_{qr} = K_{rq} = - \langle \hat{q} \rangle \langle \hat{r} \rangle \,, \tag{33}$$

and the principal minors of order 2 belonging to the covariant metric tensor K (which are always positive due to the metric properties):

$$I_{nq} = (\Delta \hat{n})^2 (\Delta \hat{q})^2 - \left[\langle \hat{n}\hat{q} \rangle - \langle \hat{n} \rangle \langle \hat{q} \rangle \right]^2 > 0, \tag{34}$$

$$I_{nr} = (\Delta \hat{n})^2 (\Delta \hat{r})^2 - \left[\langle \hat{n}\hat{r} \rangle - \langle \hat{n} \rangle \langle \hat{r} \rangle \right]^2 > 0, \tag{35}$$

$$\begin{aligned} I_{qr} &= (\Delta \hat{q})^2 (\Delta \hat{r})^2 - \langle \hat{q} \rangle^2 \langle \hat{r} \rangle^2 > 0 \\ &= \langle \hat{n} \rangle - \langle \hat{n} \rangle \langle \hat{r} \rangle - 2 \langle \hat{r} \rangle + 2 \langle \hat{r} \rangle^2 - \langle \hat{q} \rangle^2 > 0 \,. \end{aligned} \tag{36}$$

The generalized uncertainty principle (GUP) (see Eq.(15)) is given by:

$$I^H = I_{nq} + I_{nr} + I_{qr} > 0, \tag{37}$$

which results the summation over the principal minors of order 2 given by: (34), (35) and (36).

The entropy of the system is the following:

$$S = \frac{\beta(U+2E)}{D} e^{-\beta(U+2E)} + \ln(D) + \\ + \frac{2\beta e^{-\beta E}}{D} \{E \cosh[\beta(B\Delta + \mathcal{H})] - (B\Delta + \mathcal{H}) \sinh[\beta(B\Delta + \mathcal{H})]\} \quad (38)$$

and, due to the fact that we have well defined te temperature (because of the inclusion of the Hamiltonian as a relevant operator) we can calculate the specific heat C of the system:

$$C = -\beta \frac{\partial S}{\partial \beta}, \quad (39)$$

which, in light of the former defined metric, can be expressed not only in terms of the system´s parameters but also in terms of the covariant metric tensor´s components, enhancing a topological nature of the specific heat:

$$C = \beta^2 \left[E^2 K_{nn} + (B\Delta + \mathcal{H})^2 K_{qq} + U^2 K_{rr}\right] + \\ +2\beta^2 \left[(B\Delta + \mathcal{H}) E K_{nq} + U E K_{nr} + U (B\Delta + \mathcal{H}) K_{nr}\right]. \quad (40)$$

4. Results

In this section we will show how qubits could be obtained. Coming back to Eq. (27), we can see that the mean value of aligned pairs per site, $\langle \hat{q} \rangle$, and Δ (the average interaction of a particle in the site "i" with the rest of the sites) are related by $-\frac{B}{2t}$. On the other side, in,[17] it was established, that, through the minimization

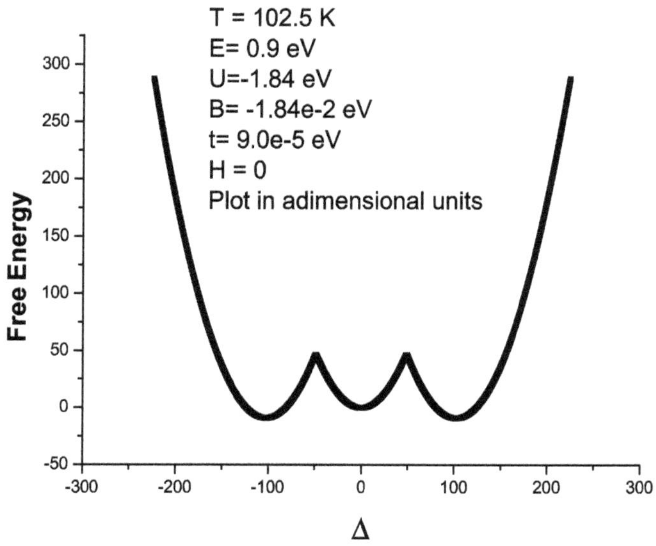

Fig. 1. Free energy vs Δ, in adimensional units.

of the free energy F (26) with respect to Δ, the critical temperature can be found, searching at which temperature three equal minima appear in F. Particularly, for $\mathcal{H} = 0$, $E = 0.9eV$, $U = -1.84eV$, $B = 1.84 \times 10^{-2}eV$ and $t = 9 \times 10^{-5}eV$, the critical temperature results, $T_c = 102.6K^o$. Three absolute minima of equal magnitude occur in F for $\Delta = \pm B/2t \simeq \pm 100$, and $\Delta = 0$ as it is shown in Fig. 1. (for more details see[17])

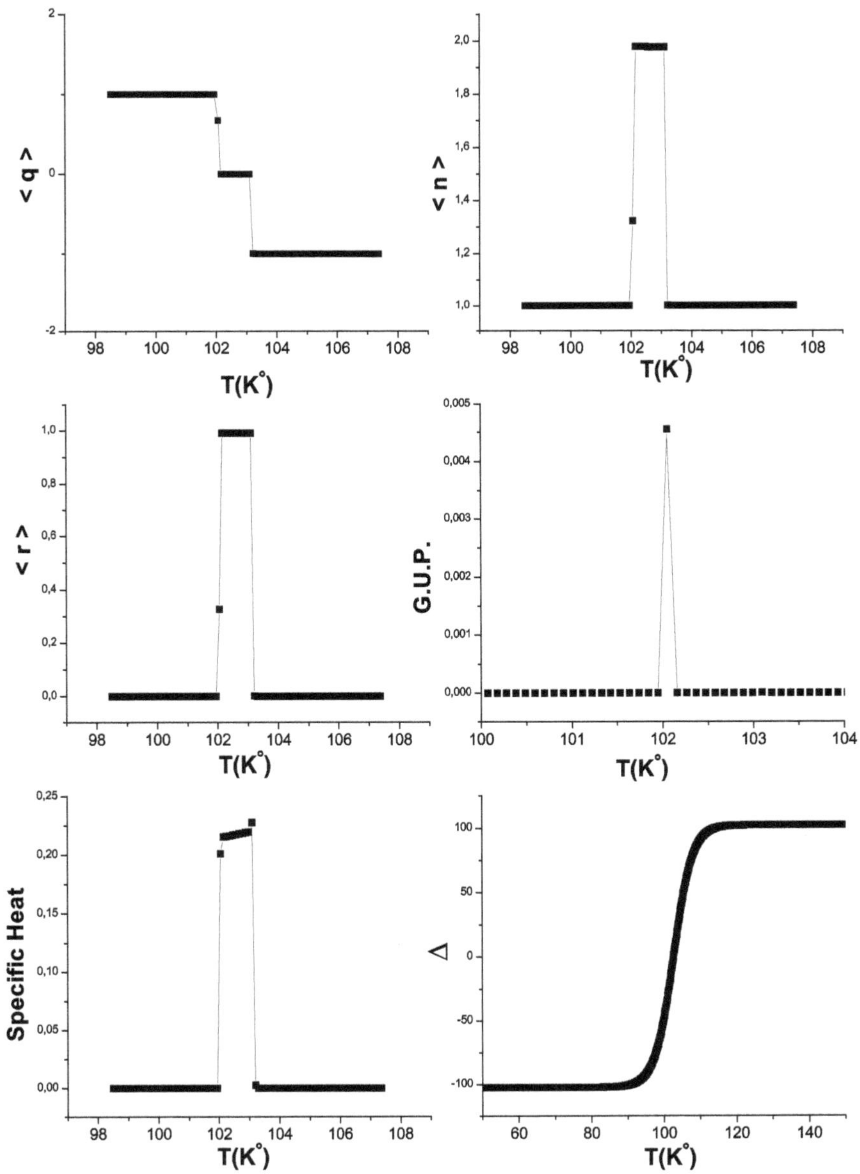

Fig. 2. Temperature dependence of $\langle\hat{q}\rangle$ (aligned-pairs), $\langle\hat{n}\rangle$ (number of pairs per site), $\langle\hat{r}\rangle$, (antialigned pairs), GUP, specific heat, and Δ, respectively.

As the set of relevant operators we have chosen is a $CSCO$, using Eqs. (34) (35), (36) and (37), it is possible to find $\langle\hat{n}\rangle, \langle\hat{q}\rangle, \langle\hat{r}\rangle$ values which do not violate the uncertainty principle (37). These mean values do not evolve any more. Only the Δ parameter is allowed to change with temperature, because the other parameters correspond to a given material. To change the Δ value, means to change the energy of the system, *using information which is not included into neither the Hamiltonian nor the density matrix.*

Notice that the value of F for $(+\Delta = \pm B/2t \cong \pm 100$ are the same, since it depends on Δ^2 and $cosh\,(\beta\, B\, \Delta)$, but the values of $< \hat{q} >$ are different on account of Eq.(27). In Fig. 2 we show the temperature dependence of $\langle\hat{q}\rangle$ (aligned-pairs), $\langle\hat{n}\rangle$ (number of pairs per site), $\langle\hat{r}\rangle$, (antialigned pairs), the uncertainty principle (37) (GUP), the specific heat (40) and Δ, respectively, where variation of Δ the temperature, we assume, it is shown. Notice that this variation is a new information we are including in our analysis, and it is not derived from Fig.1. In this figure, we examine closely the values of Δ and $< \hat{q} >$ near the critical temperature T_c.= $102.6K^o$. The Δ values pass through all the three minima allowed by the free energy minimization (see[17]). In that case, the mean value of the aligned pairs are able to move from 1 to 0 to -1. Below T_c , the absolute minimum of F occurs at $\Delta = -100$ and the system is in state $< \hat{q} > = +1$. Above T_c, the absolute minimum stays at $\Delta = 0$ but there is a local minimum $\Delta = 100$ corresponding to the state $< \hat{q} > = -1$. Thus, the system forms stable qubits, if it is kept at different temperatures, and *could also exhibit a tribits behavior, as the temperature moves from lower to upper values around T_c.*

As we can see from the plots, neither the specific heat nor the uncertainty principle values depend on the signature of $< \hat{q} >$, the specific heat detects the jumps from $< \hat{q} >= 1$ to $< \hat{q} >= 0$ to $< \hat{q} >= -1$.

5. Conclusions

The method proposed in[14,17] basically means to reduce an $n-$dimensional spin chain system to a mean field Hamiltonian through the minimization of the free energy, which defines the allowed values of Δ for a Hamiltonian defined by a set of parameters. This fact makes possible to fulfill the closure relationship (6) for both a $CSNCO$, as considered in ref.,[14] or a $CSCO$, as it was noted in[17] and in the present contribution here. Eq. (6) not only allows one to find a $CSCO$, but also it defines a 3− dimensional generalized phase space, which is spanned by, the mean values $\langle\hat{n}\rangle, \langle\hat{q}\rangle, \langle\hat{r}\rangle$. Taking advantage of the metric's properties of this space, one can get the expression of a Generalized Uncertainty Principle. Finally, making the numerical simulations wich corresponds to a variation of Δ with the temperature, around the T_c founded in,[17] we demonstrate that the aligned pairs of the proposed Hamiltonian jump from 1 to 0, and to -1 for the chosen parameters values. In other words, *qubits can be found using the same material, simply varying the temperature.* Also we show: a) the *jump is done without violating the UP*, which in fact reflects the jump. *b*) the specific heat of the system can be expressed not only in terms of the temperature and the Hamiltonian parameters, but also in terms of the metric tensor's components demonstrating a topological nature of the specific heat, and *c*) the specific heat detects the jumps in $\langle \hat{q} \rangle$.

Acknowledgment

Discussion with Professor M. Byrd and Mr. A.A. Shams are much appreciated and thankfully acknowledged. One of us, FBM, is pleased to acknowledge a grant from U.S. Army Research Office for travel support. The views, opinions, and/or findings contained in this article are those of the authors and should not be construed as an official position, policy or decision of the department of the army, unless so designated by other documentation.

References

1. C. E. Shannon, Bell System Tech. Jour. 27, 379, 623 (1948).
2. C. E. Shannon and W. Weaver, the Mathematical Theory of Communication University of Illinois, 1949).
3. J. von Neumann, Mathematical *Foundations of Quantum Mechanics* (Princeton University Press, 1955); Gött. Nachr., 273 (1929); Zeit. Physik 57, 30 (1929).
4. E. T. Jaynes, Phys. Rev. 106, 620 (1957).
5. E. T. Jaynes, Phys. Rev. 108, 171 (1957).
6. Y. Alhassid and R.D. Levine, J. Chem. Phys., **67** (1977) 4321; Phys. Rev. A **18** (1978) 89.
7. C.M. Sarris, F. Caram and A.N. Proto, Phys. A, **331** (2004) 125-139.
8. C.M. Sarris and A.N. Proto, Phys. A, **348** (2005) 97-109.
9. C.M. Sarris, F. Caram and A.N. Proto, Phys. Lett. A, **324** (2004) 1-8.
10. S. Picozzi, A. N. Proto, and F. B. Malik, Cond. Matter Theories 12, 319 (1997).
11. F. B. Malik, Ann. Phys. (N.Y.) 36, 86 (1966).
12. J. Shapiro, R. Richardson, and F. B. Malik, J. Phys. G: Nucl. Phys. 14, 535 (1988).
13. J. Shapiro and F. B. Malik, Cond. Matter Theories, 3, 11 (1988).
14. J. Aliaga, A. N. Proto, and V. Zunino, Cond. Matter Theories 7, 335 (1992).
15. J. Hubbard, Proc. Roy. Soc. (Loud) A 276, 238 (1965).
16. P. W. Anderson, Science 235, 1196 (1987).
17. A. A. Shams, S. Picozzi, and F. B. Malik, Physica 352, 269 (2004).
18. S. Picozzi, Doctoral Dissertation (1997) Southern Illinois University Carbondale.

Part J

New Formalisms

THERMAL COHERENT STATES, A BROADER CLASS OF MIXED COHERENT STATES, AND GENERALIZED THERMO-FIELD DYNAMICS

RAYMOND F. BISHOP

School of Physics and Astronomy, The University of Manchester, Sackville Street Building, P.O. Box 88, Manchester, M60 1QD, UK (email: raymond.bishop@manchester.ac.uk)

APOSTOLOS VOURDAS

Department of Computing, University of Bradford, Richmond Road, Bradford, West Yorkshire, BD7 1DP, UK

We introduce coherent mixed states (or thermal coherent states) associated with the displaced harmonic oscillator at nonzero temperature ($T \neq 0$), as a "random" (or "thermal" or "noisy") basis in Hilbert space. A resolution of the identity for these states is proven and is used to generalize the usual pure ($T = 0$) coherent state formalism to the mixed ($T \neq 0$) case. This new formalism for thermal coherent states is then further generalized to a broader class of so-called negative-binomial mixed states. It is known that the negative-binomial distribution is itself intimately related to the discrete series of SU(1,1) representations. We consider the pure SU(1,1) coherent states in the two-mode harmonic oscillator space, and show how our negative-binomial mixed states arise from taking the partial trace with respect to one of these two modes. This observation is then used to show how the formalism of thermo-field dynamics may be generalized to a correspondingly much broader negative-binomial-field dynamics, which we expect to have many uses.

Keywords: Glauber coherent states; SU(1,1) coherent states; mixed coherent states; thermal coherent states; negative-binomial mixed coherent states; generalized therm-field dynamics; negative-binomial-field dynamics; coherence and decoherence.

1. Introduction

The pure Glauber (harmonic oscillator) coherent states provide a very useful basis for many purposes. They are complete since an arbitrary state in the Hilbert space may be expanded in terms of them. Some of their main properties are discussed in Sec. 2. Our main aim in the present paper is to generalize the Glauber coherent states in several directions. The first involves an extension to the idea of mixed coherent states, and we show in particular in Sec. 3 that we may introduce a set of thermal mixed coherent states that completely generalize the usual pure Glauber formalism to the mixed case where the oscillator is in contact with a heat bath at a nonzero temperature ($T = 0$). We show, for example how the thermal coherent states permit a generalization to the $T \neq 0$ case of the usual P- and Q-representations of operators.

A second generalization in Sec. 4 is to consider various (one- and two-mode) realizations of the pure (squeezed) coherent states associated with the group SU(1,1). One of these is associated with the so-called k-squeezed states of a single boson (or mode), and this introduces us to the pure SU(1,1) negative-binomial coherent states. These are then further generalized in Sec. 5 to the so-called negative-binomial mixed coherent states, which are associated with the discrete series of representations of SU(1,1) characterized by the quantum number $k = \frac{1}{2}, 1, \frac{3}{2}, \ldots$. We show that the thermal coherent states are the particular limiting case $k = \frac{1}{2}$ of this much larger class.

In Sec. 4 we also consider squeezed states associated with two different modes (or bosons). The Hilbert space of such a two-mode oscillator can be expressed as a direct sum of an infinite number of subspaces, each of which is a particular k-representation of SU(1,1). In Sec. 6 we reconsider the pure SU(1,1) coherent states in the two-boson Hilbert space, and show that the partial trace over one of the two boson Hilbert spaces leads, rather miraculously, to our previous negative-binomial mixed states of Sec. 5. We hence show how thermo-field dynamics may be generalized to a correspondingly much broader "negative-binomial-field dynamics," which we expect to have many uses. The work is summarized in Sec. 7, where we also briefly consider potential applications.

2. Single-Boson Pure (Glauber) Coherent States

Consider a single-mode harmonic oscillator with Hamiltonian,

$$\hat{H}_0 = \frac{\hat{p}^2}{2m} + \tfrac{1}{2}m\omega^2\hat{x}^2 = \hbar\omega\left(\hat{a}_0^\dagger\hat{a}_0 + \tfrac{1}{2}\right) \equiv \hbar\omega\left(\hat{n}_0 + \tfrac{1}{2}\right) , \tag{1}$$

in terms of the usual boson creation and destruction operators, $\hat{a}_0^\dagger$ and $\hat{a}_0$ respectively,

$$\hat{a}_0 \equiv \sqrt{\frac{m\omega}{2\hbar}}\left(\hat{x} + \frac{\mathrm{i}}{m\omega}\hat{p}\right) , \tag{2}$$

and which hence obey the usual canonical commutation relations, $[\hat{a}_0, \hat{a}_0^\dagger] = \hat{\mathbb{1}}$. The set of orthonormal number eigenstates,

$$|n\rangle \equiv (n!)^{-1/2}\left(\hat{a}_0^\dagger\right)^n |0\rangle \; ; \; n = 0, 1, 2, \ldots , \tag{3}$$

is defined in terms of the normalized vacuum state $|0\rangle$, $\hat{a}_0|0\rangle = 0$. They obey the completeness relation,

$$\sum_{n=0}^{\infty} |n\rangle\langle n| = \hat{\mathbb{1}} . \tag{4}$$

The normalized state $|z\rangle \equiv |x,p\rangle$ is a Glauber coherent state if it is a right eigenstate of $\hat{a}_0$,

$$\hat{a}_0|z\rangle = z|z\rangle \iff \left(\hat{x} + \frac{\mathrm{i}}{m\omega}\hat{p}\right)|x,p\rangle = \left(x + \frac{\mathrm{i}}{m\omega}p\right)|x,p\rangle \; ; \; x,p \in \mathbb{R} \, , \tag{5}$$

with

$$z \equiv \sqrt{\frac{m\omega}{2\hbar}}\left(x + \frac{\mathrm{i}}{m\omega}p\right) \; ; \; x,p \in \mathbb{R} \, . \tag{6}$$

Expanding $|z\rangle$ in terms of number eigenstates, we easily find

$$|z\rangle = \exp\left(-\frac{1}{2}|z|^2\right)\sum_{n=0}^{\infty}\frac{z^n}{\sqrt{n!}}|n\rangle = \exp\left(-\frac{1}{2}|z|^2\right)\exp\left(z\hat{a}_0^\dagger\right)|0\rangle \, . \tag{7}$$

Alternatively, by making use of the disentangling relation for the Heisenberg-Weyl group,

$$\exp\left(z\hat{a}_0^\dagger - z^*\hat{a}_0\right) = \exp\left(-\tfrac{1}{2}|z|^2\right)\exp\left(z\hat{a}_0^\dagger\right)\exp\left(-z^*\hat{a}_0\right) \, , \tag{8}$$

we may write $|z\rangle$ in terms of the unitary Weyl displacement operator,

$$\hat{D}(z) \equiv \exp\left(z\hat{a}_0^\dagger - z^*\hat{a}_0\right) \iff \hat{D}(x,p) \equiv \exp[\mathrm{i}(p\hat{x} - x\hat{p})/\hbar] \, , \tag{9}$$

as

$$|z\rangle = \hat{D}(z)|0\rangle \iff |x,p\rangle = \hat{D}(x,p)|0\rangle \, . \tag{10}$$

We may readily check that the mode of action of the Weyl displacement operator on the elementary group operators is as follows,

$$\hat{D}(z)\hat{a}_0\hat{D}^\dagger(z) = \hat{a}_0 - z\hat{\mathbb{1}} \equiv \hat{a}_z \; ; \; \hat{D}(z)\hat{a}_0^\dagger\hat{D}^\dagger(z) = \hat{a}_0^\dagger - z^*\hat{\mathbb{1}} \equiv \hat{a}_z^\dagger \, , \tag{11}$$

$$\hat{D}(x_0,p_0)\hat{x}\hat{D}^\dagger(x_0,p_0) = \hat{x} - x_0\hat{\mathbb{1}} \; ; \; \hat{D}(x_0,p_0)\hat{p}\hat{D}^\dagger(x_0,p_0) = \hat{p} - p_0\hat{\mathbb{1}} \, . \tag{12}$$

Very importantly, the Glauber coherent states provide a resolution of the identity operator,

$$\iint_C \frac{\mathrm{d}^2 z}{\pi}|z\rangle\langle z| = \hat{\mathbb{1}} \iff \int_{-\infty}^{\infty}\int_{-\infty}^{\infty}\frac{\mathrm{d}x\,\mathrm{d}p}{2\pi\hbar}|x,p\rangle\langle x,p| = \hat{\mathbb{1}} \, , \tag{13}$$

despite being massively overcomplete, where the integration over z is over the entire complex plane, C, and where the integration element is $\mathrm{d}^2 z \equiv \mathrm{d}z_{\mathrm{R}}\,\mathrm{d}z_{\mathrm{I}} = \mathrm{d}z\,\mathrm{d}z^*/2\mathrm{i}$.

More general coherent states $|\ell\rangle$ are required to satisfy the two conditions that they are strongly continuous functions of the label $\ell \in L$, where the (often multi-dimensional) label space L is continuous, and that they provide a resolution of the identity operator,

$$\int_L \delta\ell\,|\ell\rangle\langle\ell| = \hat{\mathbb{1}} \, , \tag{14}$$

with a suitable choice of positive measure $\delta\ell$. We note that Eq, (14) implies that all admissible wave functions $\langle \ell|\Psi\rangle$ satisfy the nontrivial homogeneous integral equation,

$$\langle \ell|\Psi\rangle = \int_L \delta\ell' \,\langle \ell|\ell'\rangle\langle \ell'|\Psi\rangle \,, \tag{15}$$

involving the reproducing kernel $K(\ell,\ell') \equiv \langle \ell|\ell'\rangle = \langle \ell'|\ell\rangle^*$. For the Glauber coherent states we may use Eq. (7) to show that

$$K(z,z') \equiv \langle z|z'\rangle = \exp\left(-\tfrac{1}{2}|z|^2 - \tfrac{1}{2}|z'|^2 + z^* z'\right) \,. \tag{16}$$

2.1. *Glauber coherent states and phase-space distribution functions*

It is well-known in quantum optics that all operators belonging to the trace class (i.e., all operators $\hat{\Theta}$ for which $\mathrm{Tr}\,\hat{\Theta} < \infty$) are uniquely specified by either their Q- or their (Glauber-Sudarshan) P-representation,

$$Q(\hat{\Theta};z) \equiv \langle z|\hat{\Theta}|z\rangle \iff Q(\hat{\Theta};x,p) \equiv \langle x,p|\hat{\Theta}|x,p\rangle \,, \tag{17}$$

$$\hat{\Theta} \equiv \iint_C \frac{\mathrm{d}^2 z}{\pi}\, P(\hat{\Theta};z)|z\rangle\langle z| \iff \hat{\Theta} \equiv \int_{-\infty}^{\infty}\int_{-\infty}^{\infty} \frac{\mathrm{d}x\,\mathrm{d}p}{2\pi\hbar}\, P(\hat{\Theta};x,p)|x,p\rangle\langle x,p| \,. \tag{18}$$

Equations (16)–(18) readily yield a relationship between these two representations,

$$Q(\hat{\Theta};z) = \iint_C \frac{\mathrm{d}^2 z'}{\pi} P(\hat{\Theta};z')\mathrm{e}^{-(z-z')^2} = \exp\left(\tfrac{1}{4}\Delta_z\right) P(\hat{\Theta};z) \,, \tag{19}$$

in terms of the two-dimensional Laplacian operator,

$$\Delta_z \equiv 4\frac{\partial^2}{\partial z\,\partial z^*} = \frac{\partial^2}{\partial z_{\mathrm{R}}^2} + \frac{\partial^2}{\partial z_{\mathrm{I}}^2} = 2\hbar\left(\frac{1}{m\omega}\frac{\partial^2}{\partial x^2} + m\omega\frac{\partial^2}{\partial p^2}\right) \,. \tag{20}$$

For an arbitrary function $f(z) = f(z,z^*)$ we now introduce its two-dimensional Fourier transform,

$$\tilde{f}(w) \equiv \iint_C \frac{\mathrm{d}^2 z}{\pi}\, \mathrm{e}^{\mathrm{i}z\cdot w} f(z) \iff f(z) = \iint_C \frac{\mathrm{d}^2 w}{4\pi}\, \mathrm{e}^{-\mathrm{i}z\cdot w}\tilde{f}(w) \,, \tag{21}$$

where $z\cdot w \equiv \frac{1}{2}(zw^* + z^*w) = z_{\mathrm{R}}w_{\mathrm{R}} + z_{\mathrm{I}}w_{\mathrm{I}}$. Alternatively, if we express w as

$$w \equiv \sqrt{\frac{2}{m\hbar\omega}}(-P + \mathrm{i}m\omega X) \,;\;\; X,P \in \mathbb{R} \,, \tag{22}$$

as the analogue of Eq. (6), then Eq. (21) may be expressed in the equivalent forms,

$$\tilde{f}(X,P) \equiv \int_{-\infty}^{\infty}\int_{-\infty}^{\infty} \frac{\mathrm{d}x\,\mathrm{d}p}{2\pi\hbar}\, \mathrm{e}^{\mathrm{i}(Xp-Px)/\hbar} f(x,p) \iff$$
$$f(x,p) = \int_{-\infty}^{\infty}\int_{-\infty}^{\infty} \frac{\mathrm{d}X\,\mathrm{d}P}{2\pi\hbar}\, \mathrm{e}^{\mathrm{i}(xP-pX)/\hbar}\tilde{f}(X,P) \,. \tag{23}$$

An equivalent form of Eq. (19) may thus be given as

$$\tilde{Q}(\hat{\Theta}; w) = \exp\left(-\tfrac{1}{4}|w|^2\right) \tilde{P}(\hat{\Theta}; w) \,. \tag{24}$$

Finally, we define the Weyl representation of $\hat{\Theta}$ as follows,

$$\tilde{W}(\hat{\Theta}; w) \equiv \mathrm{Tr}\left\{\hat{\Theta}\hat{D}\left(\tfrac{1}{2}\mathrm{i}w\right)\right\} \,. \tag{25}$$

Its Fourier transform, $W(\hat{\Theta}; z)$, is the Wigner or W-representation of $\hat{\Theta}$. In terms of position eigenstates $|x\rangle_x$, where $\hat{x}|x\rangle_x = x|x\rangle_x$, we may readily derive the more familiar forms,

$$\begin{aligned} W(\hat{\Theta}; z) &= \int_{-\infty}^{\infty} \mathrm{d}X \; {}_x\langle \sqrt{2}z_{\mathrm{R}} + \tfrac{1}{2}X|\hat{\Theta}|\sqrt{2}z_{\mathrm{R}} - \tfrac{1}{2}X\rangle_x \exp\left(-\mathrm{i}\sqrt{2}z_{\mathrm{I}}X\right) \,, \\ \tilde{W}(\hat{\Theta}; w) &= \int_{-\infty}^{\infty} \mathrm{d}x \; {}_x\langle x + 2^{-3/2}w_{\mathrm{I}}|\hat{\Theta}|x - 2^{-3/2}w_{\mathrm{I}}\rangle_x \exp\left(\mathrm{i}w_{\mathrm{R}}x/\sqrt{2}\right) \,, \end{aligned} \tag{26}$$

in units where we have now put $m = \omega = \hbar = 1$, or equivalently,

$$\begin{aligned} W(\hat{\Theta}; x, p) &= \int_{-\infty}^{\infty} \mathrm{d}X \; {}_x\langle x + \tfrac{1}{2}X|\hat{\Theta}|x - \tfrac{1}{2}X\rangle_x \exp(-\mathrm{i}pX/\hbar) \,, \\ \tilde{W}(\hat{\Theta}; X, P) &= \int_{-\infty}^{\infty} \mathrm{d}x \; {}_x\langle x + \tfrac{1}{2}X|\hat{\Theta}|x - \tfrac{1}{2}X\rangle_x \exp(-\mathrm{i}Px/\hbar) \,. \end{aligned} \tag{27}$$

We also easily find the following relations between the P-, Q-, and W-representations,

$$\begin{aligned} Q(\hat{\Theta}; z) = \exp\left(+\tfrac{1}{8}\Delta_z\right) W(\hat{\Theta}; z) &\iff \tilde{Q}(\hat{\Theta}; w) = \exp\left(-\tfrac{1}{8}|w|^2\right) \tilde{W}(\hat{\Theta}; w) \,, \\ P(\hat{\Theta}; z) = \exp\left(-\tfrac{1}{8}\Delta_z\right) W(\hat{\Theta}; z) &\iff \tilde{P}(\hat{\Theta}; w) = \exp\left(+\tfrac{1}{8}|w|^2\right) \tilde{W}(\hat{\Theta}; w) \,. \end{aligned} \tag{28}$$

Equation (28) also implies our previous Eq. (19), which may be rewritten in the form,

$$Q(\hat{\Theta}; z) = \exp\left(+\tfrac{1}{4}\Delta_z\right) P(\hat{\Theta}; z) \iff \tilde{Q}(\hat{\Theta}; w) = \exp\left(-\tfrac{1}{4}|w|^2\right) \tilde{P}(\hat{\Theta}; w) \,. \tag{29}$$

3. Mixed Thermal Coherent States

Our basic idea[1] to extend the pure-state formalism outlined in Sec. 2 above is now to use a set of mixed states as a basis in Hilbert space. Previously we used the pure coherent states $|z\rangle$ with projectors $\hat{\Pi}(z) \equiv |z\rangle\langle z|$, $\hat{\Pi}^2(z) = \hat{\Pi}(z)$. The resolution of the identity operator given by Eq. (13) then allowed us to resolve an arbitrary state $|f\rangle$ as

$$|f\rangle = \iint_C \frac{\mathrm{d}^2 z}{\pi} \langle z|f\rangle |z\rangle \,. \tag{30}$$

A mixed state is described by a density operator $\hat{\rho}$ with eigenstates $|e^{(n)}\rangle$ and corresponding eigenvalues p_n,

$$\hat{\rho} = \sum_{n=0}^{\infty} p_n |e^{(n)}\rangle\langle e^{(n)}| \,; \quad \sum_{n=0}^{\infty} p_n = 1 \,; \quad 0 \le p_i \le 1 \,, \forall i \,. \tag{31}$$

It clearly represents a set of states $\{|e^{(n)}\rangle\}$ with a probability distribution $\{p_n\}$. Thus, the idea of using mixed states as a basis replaces the "fixed" vectors usually used with "noisy" or "random" vectors.

In order to make these ideas concrete we first consider the thermodynamic density operator for a canonical ensemble of single-frequency (ω) harmonic oscillators, with Hamiltonian $\hat{H}_0$ given by Eq. (1), in thermal equilibrium at a temperature $T = (k_B\beta)^{-1}$,

$$\hat{\rho}_0(\beta) \equiv \exp(-\beta\hat{H}_0)/\mathrm{Tr}\{\exp(-\beta\hat{H}_0)\} = (1 - \mathrm{e}^{-\beta\hbar\omega})\exp(-\beta\hbar\omega\hat{a}_0^\dagger\hat{a}_0)\,. \tag{32}$$

Of particular importance to us, as we shall see below, is the displaced linear harmonic oscillator with Hamiltonian given by

$$\hat{H}_z \equiv \hat{D}(z)\hat{H}_0\hat{D}^\dagger(z) = \hbar\omega\left[(\hat{a}_0^\dagger - z^*\hat{\mathbb{1}})(\hat{a}_0 - z\hat{\mathbb{1}}) + \tfrac{1}{2}\right] \equiv \hbar\omega(\hat{a}_z^\dagger\hat{a}_z + \tfrac{1}{2})\,, \tag{33}$$

with the various operators as defined previously in Sec. 1. Its density operator, which represents the Boltzmann distribution for the monochromatic oscillator, is given by

$$\hat{\rho}(z;\beta) \equiv \hat{D}(z)\hat{\rho}_0(\beta)\hat{D}^\dagger(z) = (1 - \mathrm{e}^{-\beta\hbar\omega})\exp(-\beta\hbar\omega\hat{a}_z^\dagger\hat{a}_z)\,, \tag{34}$$

and hence where $\hat{\rho}(0;\beta) \equiv \hat{\rho}_0(\beta)$, or equivalently, as

$$\hat{\rho}(z;\beta) = \sum_{n=0}^{\infty} p_n(\beta)|n;z\rangle\langle n;z|\,; \tag{35}$$

where

$$|n;z\rangle \equiv \hat{D}(z)|n\rangle\,;\quad p_n(\beta) \equiv (1 - \mathrm{e}^{-\beta\hbar\omega})\,\mathrm{e}^{-n\beta\hbar\omega}\,. \tag{36}$$

Its zero-temperature limit just represents the above coherent-state projector $\hat{\Pi}(z)$,

$$\hat{\rho}(z;\beta) \xrightarrow[\beta\to\infty]{} |z\rangle\langle z| \equiv \hat{\Pi}(z)\,. \tag{37}$$

Using the definitions of Eqs. (17) and (18) we may easily calculate its Q- and P-representations,

$$Q(\hat{\rho}(\zeta;\beta);z) = (1 - \mathrm{e}^{-\beta\hbar\omega})\exp[-(1 - \mathrm{e}^{-\beta\hbar\omega})|z-\zeta|^2]\,, \tag{38}$$

$$P(\hat{\rho}(\zeta;\beta);z) = (\mathrm{e}^{\beta\hbar\omega} - 1)\exp[-(\mathrm{e}^{\beta\hbar\omega} - 1)|z-\zeta|^2]\,. \tag{39}$$

Using Eqs. (38) and (39) we can trivially prove the relations

$$\iint_C \frac{\mathrm{d}^2\zeta}{\pi}\,Q(\hat{\rho}(\zeta;\beta);z) = 1 = Q(\hat{\mathbb{1}};z)\,;\quad \iint_C \frac{\mathrm{d}^2\zeta}{\pi}\,P(\hat{\rho}(\zeta;\beta);z) = 1 = P(\hat{\mathbb{1}};z)\,; \tag{40}$$

From Eq. (40) follows immediately the extremely important resolution of the identity,

$$\iint_C \frac{\mathrm{d}^2 z}{\pi}\,\hat{\rho}(z;\beta) = \hat{\mathbb{1}}\,, \tag{41}$$

which is the generalization to nonzero temperatures of Eq. (13). Eq. (41) allows us to resolve an arbitrary state $|f\rangle$ as

$$|f\rangle = \sum_{n=0}^{\infty} p_n \iint_C \frac{\mathrm{d}^2 z}{\pi} \langle n; z|f\rangle |n; z\rangle \,, \tag{42}$$

which is a generalization of the pure coherent-state expansion of Eq. (30) to the "noisy" thermal basis.

3.1. *Relationship to the Glauber-Lachs formalism in quantum optics*

Before proceeding further it is worthwhile to pause for a moment to point out the connection between our density operator $\hat{\rho}(z;\beta)$ and the Glauber-Lachs formalism in quantum optics for mixtures of incoherent (or, loosely, thermal) and coherent radiation. Thus, it has been shown by Glauber[2] that the P-representation for a field which is a superposition of two separate fields with individual P-representations $P_1(z)$ and $P_2(z)$ is given by their convolution,

$$P(z) = \iint_C \frac{\mathrm{d}^2 z'}{\pi} P_1(z') P_2(z - z') \,. \tag{43}$$

In the case of a single-frequency (ω) mode, the P-representation of an oscillator of the same frequency in the coherent state $|\zeta\rangle$ is clearly given from Eq. (18) as

$$P_1(z) = \pi \delta^{(2)}(z - \zeta) \,. \tag{44}$$

Furthermore, Glauber has shown[2] that the P-representation of the corresponding mode of a random incoherent field, with average number of photons $\bar{n}$, is given by

$$P_2(z) = (1/\bar{n}) \exp\left(-|z|^2/\bar{n}\right) \,. \tag{45}$$

Equations (43), (44), and (45) hence yield that the P-representation for the admixture of an incoherent and coherent field is

$$P(z) = (1/\bar{n}) \exp\left(-|z - \zeta|^2/\bar{n}\right) \,, \tag{46}$$

which provides the basis for the Glauber-Lachs formalism[2,3] in quantum optics.

In our case we have essentially, for a single-frequency (ω) mode, a superposition of a coherent field in a state $|\zeta\rangle$ with a thermal (incoherent) black-body radiation field at temperature T. The mean number of thermal photons, n_T, in the thermal field itself,

$$n_T = \langle \hat{n}_0 \rangle_0 \equiv \mathrm{Tr}\big(\hat{\rho}_0(\beta)\hat{n}_0\big) = \left(\mathrm{e}^{\beta\hbar\omega} - 1\right)^{-1} \,, \tag{47}$$

which is simply the usual Planck distribution. Equations (44)–(47) then show the expected agreement with our previously proven P-representation for $\hat{\rho}(z;\beta)$ given by Eq. (39). We can similarly show rather easily that the mean number of bosons in our mixed coherent state $\hat{\rho}(\zeta;\beta)$ is given by the sum of the number n_T from the incoherent thermal field and the number $|\zeta|^2$ from the pure coherent source,

$$n_T(z) = \langle \hat{n}_0 \rangle_\zeta \equiv \mathrm{Tr}\big(\hat{\rho}(\zeta;\beta)\hat{n}_0\big) = n_T + |\zeta|^2 \,. \tag{48}$$

3.2. Generalized P- and Q-representations for nonzero temperatures

As we have already seen in Sec. 1, the formalism of P- and Q-representations is based on the (pure) Glauber coherent states. Although the latter form an overcomplete basis they are of practical usefulness due to the existence of a resolution of the identity operator in terms of them, as given by Eq. (13). Thus, since our new mixed coherent states $\hat{\rho}(z;\beta)$ generalize the pure coherent states to nonzero temperatures, $T \neq 0$, and also provide a resolution of the identity operator given by Eq. (41), we are naturally led to enquire whether we can generalize the P- and Q-representations of an arbitrary operator $\hat{\Theta}$ to the $T \neq 0$ case, in terms of them.

In the first place an obvious generalization to the $T \neq 0$ case of the $(T = 0)$ Q-representation of Eq. (17) is

$$Q(\hat{\Theta}; z; \beta) \equiv \mathrm{Tr}\{\hat{\rho}(z;\beta)\hat{\Theta}\} , \tag{49}$$

and as the corresponding generalization of the P-representation we hypothesize that an arbitrary trace-class operator permits the expansion

$$\hat{\Theta} \equiv \iint_C \frac{\mathrm{d}^2 z}{\pi} P(\hat{\Theta}; z; \beta)\hat{\rho}(z;\beta) . \tag{50}$$

It is clear that both $Q(\hat{\Theta}; z; \beta)$ and $P(\hat{\Theta}; z; \beta)$ reduce to their previous pure-state counterparts $Q(\hat{\Theta}; z)$ and $P(\hat{\Theta}; z)$ respectively in the zero-temperature limit, $\beta \to \infty$. Equations (49) and (50) yield immediately the following relation between the generalized P- and Q-representations of an arbitrary operator $\hat{\Theta}$,

$$Q(\hat{\Theta}; z; \beta) = \iint_C \frac{\mathrm{d}^2 z'}{\pi} \mathrm{Tr}\{\hat{\rho}(z;\beta)\hat{\rho}(z';\beta')\} P(\hat{\Theta}; z'; \beta') . \tag{51}$$

By making use of the general relation

$$\mathrm{Tr}\{\hat{\Theta}_1\hat{\Theta}_2\} = \iint_C \frac{\mathrm{d}^2 z}{\pi} P(\hat{\Theta}_1; z) Q(\hat{\Theta}_2; z) , \tag{52}$$

which follows immediately from Eqs. (17) and (18), we may use Eqs. (38) and (39) to derive the result,

$$\mathrm{Tr}\{\hat{\rho}(z;\beta)\hat{\rho}(z';\beta')\} = \frac{2}{g(\beta,\beta')} \exp\left\{-\frac{2|z-z'|^2}{g(\beta,\beta')}\right\} , \tag{53}$$

$$g(\beta,\beta') \equiv \frac{\sinh\left[\frac{1}{2}(\beta+\beta')\hbar\omega\right]}{\sinh\left(\frac{1}{2}\beta\hbar\omega\right)\sinh\left(\frac{1}{2}\beta'\hbar\omega\right)} . \tag{54}$$

Equations (51) and (53) then yield, as the $T \neq 0$ generalization of Eq. (29),

$$\begin{aligned} Q(\hat{\Theta}; z; \beta) &= \exp\left[+\tfrac{1}{8}g(\beta,\beta')\Delta_z\right] P(\hat{\Theta}; z; \beta') \iff \\ \tilde{Q}(\hat{\Theta}; w; \beta) &= \exp\left[-\tfrac{1}{8}g(\beta,\beta')|w|^2\right] \tilde{P}(\hat{\Theta}; w; \beta') . \end{aligned} \tag{55}$$

By taking the limit $\beta' \to \infty$ in Eq. (55), and by making use of Eq. (29), we find

$$\tilde{Q}(\hat{\Theta}; w; \beta) = \exp\left[-\tfrac{1}{4}\left(\mathrm{e}^{\beta\hbar\omega} - 1\right)^{-1} |w|^2\right] \tilde{Q}(\hat{\Theta}; w) . \tag{56}$$

By making repeated use of Eqs. (55) and (56) we may also prove the relations

$$\tilde{Q}(\hat{\Theta}; w; \beta) = \exp\left[-\tfrac{1}{8}g(\beta, -\beta')|w|^2\right] \tilde{Q}(\hat{\Theta}; w; \beta') \; ; \tag{57}$$

$$\tilde{P}(\hat{\Theta}; w; \beta) = \exp\left[+\tfrac{1}{8}g(\beta, -\beta')|w|^2\right] \tilde{P}(\hat{\Theta}; w; \beta') \; . \tag{58}$$

Finally, by using the above relations together with Eq. (28), we also find the very important relations

$$\tilde{Q}(\hat{\Theta}; w; \beta) = \exp\left[-\tfrac{1}{8}\coth\left(\tfrac{1}{2}\beta\hbar\omega\right)|w|^2\right] \tilde{W}(\hat{\Theta}; w) \; ; \tag{59}$$

$$\tilde{P}(\hat{\Theta}; w; \beta) = \exp\left[+\tfrac{1}{8}\coth\left(\tfrac{1}{2}\beta\hbar\omega\right)|w|^2\right] \tilde{W}(\hat{\Theta}; w) \; . \tag{60}$$

Equations (59) and (60) show the formal result that the generalized ($T \neq 0$) P-representation may be viewed as the analytic continuation to negative temperatures of the corresponding generalized Q-representation, and vice versa. We find this unexpected result both fascinating and suggestive. It surely merits further investigation.

4. (Squeezed) SU(1,1) Pure Coherent States

Having introduced our thermal mixed coherent states in Sec. 3, it is natural to consider next whether it might be possible to generalize them to a broader class. To this end we first discuss a separate class of pure coherent states built on the SU(1,1) group rather than the Heisenberg-Weyl group that underpins the Glauber coherent states. The SU(1,1) Lie algebra is based on the three operators $\{\hat{K}_x, \hat{K}_y, \hat{K}_z\}$ obeying the commutation relations,

$$[\hat{K}_x, \hat{K}_y] = -\mathrm{i}\hat{K}_z \; ; \; [\hat{K}_y, \hat{K}_z] = +\mathrm{i}\hat{K}_x \; ; \; [\hat{K}_z, \hat{K}_x] = +\mathrm{i}\hat{K}_y \; , \tag{61}$$

or, equivalently, on the three operators $\{\hat{K}_\pm \equiv \hat{K}_x \pm \mathrm{i}\hat{K}_y, \; \hat{K}_0 \equiv \hat{K}_z\}$ obeying

$$[\hat{K}_0, \hat{K}_\pm] = \pm\hat{K}_\pm \; ; \; [\hat{K}_+, \hat{K}_-] = -2\hat{K}_0 \; . \tag{62}$$

The SU(1,1) Casimir operator is given by

$$\hat{K}^2 \equiv \hat{K}_z^2 - \hat{K}_x^2 - \hat{K}_y^2 = \hat{K}_0^2 - \tfrac{1}{2}(\hat{K}_+\hat{K}_- + \hat{K}_-\hat{K}_+) \; . \tag{63}$$

The standard SU(1,1) basis (for $k > \frac{1}{2}$ is then $\{|k;\mu\rangle; \; \mu = k + n; \; n = 0, 1, 2, \ldots\}$ with

$$\hat{K}^2|k;\mu\rangle = k(k-1)|k;\mu\rangle \; ; \; \hat{K}_0|k;\mu\rangle = \mu|k;\mu\rangle \; . \tag{64}$$

The group SU(1,1) is non-compact, and consequently there is no further restriction in general on the value of $k \in \mathbb{R}$. As a consequence the group has several series of irreducible unitary representations. In particular, there exist two so-called discrete series of representations, T^+ and T^-, of which it is sufficient to consider only one, say T^+, since all the results clearly carry over to the other case under the substitution $k \to 1 - k$. The representations of T^+ are characterized by the values $k = \frac{1}{2}, 1, \frac{3}{2}, \ldots$.

The SU(1,1) group generators now involve the so-called squeezing operators. By making use of various disentangling relations for the group they may be written in a number of equivalent ways as follows,

$$\begin{aligned}\hat{S}(z) \equiv \hat{S}(t,\phi) &\equiv \exp\big(\mathrm{i}\phi\hat{K}_z\big)\exp\big(\mathrm{i}t\hat{K}_y\big)\exp\big(-\mathrm{i}\phi\hat{K}_z\big)\\ &= \exp\big[\tfrac{1}{2}t(\mathrm{e}^{\mathrm{i}\phi}\hat{K}_+ - \mathrm{e}^{-\mathrm{i}\phi}\hat{K}_-)\big] \equiv \exp\big(\mathrm{i}t\mathbf{n}\cdot\hat{\mathbf{K}}\big)\\ &= \exp\big(z\hat{K}_+\big)\exp\big[\ln(1-|z|^2)\hat{K}_0\big]\exp\big(-z^*\hat{K}_-\big)\,, \end{aligned} \tag{65}$$

where

$$\begin{aligned} &z \equiv \tanh(\tfrac{1}{2}t)\exp(\mathrm{i}\phi)\ ;\quad t,\phi\in\mathbb{R}\,,\ 0\le t<\infty\,,\ 0\le\phi<2\pi\ ;\\ &\mathbf{n}\equiv(\sin\phi,\cos\phi,0)\,. \end{aligned} \tag{66}$$

So far our discussion has been completely abstract and general. However, for further progress we must consider specific realizations of the SU(1,1) algebra. In the remainder of this Section we consider three such realizations, each involving harmonic oscillators.

4.1. *Squeezed states of the one-mode oscillator*

A common realization of the SU(1,1) algebra arises in the context of single-mode squeezed-pair states, where it is easy to check that the three operators

$$\hat{K}_+ \to \tfrac{1}{2}\big(\hat{a}_0^\dagger\big)^2\,;\hat{K}_- \to \tfrac{1}{2}\big(\hat{a}_0\big)^2\,;\hat{K}_0 \to \tfrac{1}{2}\big(\hat{a}_0^\dagger\hat{a}_0+\tfrac{1}{2}\big)\,, \tag{67}$$

obey the commutation relations of Eq. (62), with the Casimir operator of Eq. (63) now given by $\hat{K}^2 = -\frac{3}{16}\hat{\mathbb{1}}$, such that this becomes the $k = (\frac{1}{4},\frac{3}{4})$ representation of SU(1,1). The mode of action of the group operators of Eq. (65) on the underlying boson operators is the expected homogeneous Bogoliubov transformation to squeezed bosons,

$$\hat{S}(z)\hat{a}_0\hat{S}^\dagger(z) = \big(1-|z|^2\big)^{-1/2}\big(\hat{a}_0 - z\hat{a}_0^\dagger\big) \equiv \hat{b}_z\,. \tag{68}$$

4.2. *k-squeezed states of the one-mode oscillator*

The SU(1,1) algebra can also be realized in the context of the above single-mode oscillator in a different way,

$$\hat{K}_+ \to \hat{a}_0^\dagger\big(\hat{n}_0+2k\hat{\mathbb{1}}\big)^{1/2}\,;\quad \hat{K}_- \to \big(\hat{n}_0+2k\hat{\mathbb{1}}\big)^{1/2}\hat{a}_0\,;\quad \hat{K}_0 \to \hat{n}_0 + k\hat{\mathbb{1}}\,, \tag{69}$$

which is based on the SU(1,1) analogue of Holstein-Primakoff theory[4] for spin systems obeying the SU(2) algebra. These operators give a general k-realization of SU(1,1), as is readily verified from explicit evaluation of their commutation relations. If we let the group squeezing operators of Eq. (65) in this k-realization of

Eq. (69), $\hat{S}_k(z)$, act on the oscillator (or bosonic) vacuum state we generate a new class of coherent states, now associated with the group SU(1,1),

$$|z;k\rangle \equiv \hat{S}_k(z)|0\rangle = \left(1-|z|^2\right)^k \sum_{n=0}^{\infty} [p(k,n)]^{1/2}\, z^n |n\rangle \; ; \quad |z|<1\,, \tag{70}$$

where

$$p(k,n) \equiv \frac{\Gamma(n+2k)}{\Gamma(n+1)\Gamma(2k)}\,, \tag{71}$$

which were first introduced by Perelomov[5] in a general context, and later studied by one of us[6] in the present harmonic oscillator context. They are defined in the unit disk $D \equiv \{z; |z|<1\}$, and they play a similar role for the SU(1,1) group as do the Glauber coherent states for the Heisenberg-Weyl group. The boson-number distribution of the states $|z;k\rangle$ is the negative-binomial distribution,

$$p_n\left(|z|^2;k\right) \equiv |\langle n|z;k\rangle|^2 = \left(1-|z|^2\right)^{2k} |z|^{2n} p(k,n)\,, \tag{72}$$

and for this reason the states $|z;k\rangle$ have been called[6] negative-binomial states. We note that for the special case $k=\frac{1}{2}$ we get the thermal (Planck) distribution of Eqs. (35,36),

$$p_n\left(|z|^2;\tfrac{1}{2}\right) = p_n(\beta)\; ; \quad |z|^2 = \mathrm{e}^{-\beta\hbar\omega}\,. \tag{73}$$

The states $|z;\frac{1}{2}\rangle$ are the so-called harmonic oscillator phase (or angle) states.[6] One can also simply show that in the limit that $k\to\infty$, $z\to 0$, $\sqrt{2k}z=\zeta=$ const., the negative-binomial coherent state $|z;k\rangle$ becomes equal to the Glauber coherent state $|\zeta\rangle$.

Very importantly for our purposes, these pure negative-binomial states are also coherent states, because they provide an easily proven resolution of the identity operator,

$$\frac{(2k-1)}{\pi} \iint_D \mathrm{d}\mu(z)\, |z;k\rangle\langle z;k| = \hat{1}\; ; \quad k>\tfrac{1}{2}\,, \tag{74}$$

in terms of the invariant measure on D,

$$\mathrm{d}\mu(z) = \left(1-|z|^2\right)^{-2} \mathrm{d}^2 z\,, \tag{75}$$

which is just the expected Lobachewsky metric for the hyperbolic geometry of SU(1,1). The case $k=\frac{1}{2}$ is handled as the limiting case $k=\frac{1}{2}+\epsilon$, in the limit as $\epsilon\to 0^+$. The analogue of Eq. (16), which holds for the Glauber coherent states, for the overlap of two negative-binomial coherent states is

$$\langle z;k|z';k\rangle = \left(1-|z|^2\right)^k \left(1-|z'|^2\right)^k \left(1-z^*z'\right)^{-2k}\,. \tag{76}$$

4.3. *Squeezed states of the two-mode oscillator*

We now consider a two-mode harmonic oscillator Hilbert space $H_1 \otimes H_2$, in which $\hat{a}_1^\dagger \in H_1$ and $\hat{a}_2^\dagger \in H_2$ are the creation operators of the two modes (or bosons). We also consider the joint number eigenstates, $|m, n\rangle \equiv |m\rangle_1 |n\rangle_2$, and the following subspaces $H_{[k]}$ and $H'_{[k]}$ spanned by them,

$$\begin{aligned} H_{[k]} &\equiv \{|n + 2k - 1, n\rangle\, ;\ k \text{ fixed}\, , n = 0, 1, 2, \ldots\}\, , \\ H'_{[k]} &\equiv \{|n, n + 2k - 1\rangle\, ;\ k \text{ fixed}\, , n = 0, 1, 2, \ldots\}\, , \\ k &= \tfrac{1}{2}, 1, \tfrac{3}{2}, \ldots\, , \\ H_1 \otimes H_2 &= H_{[1/2]} \oplus H_{[1]} \oplus H'_{[1]} \oplus H_{[3/2]} \oplus H'_{[3/2]} \oplus \cdots\, . \end{aligned} \tag{77}$$

We may now readily check that the three operators

$$\hat{K}_+ \to \hat{a}_1^\dagger \hat{a}_2^\dagger\, ;\ \ \hat{K}_- \to \hat{a}_1 \hat{a}_2\, ;\ \ \hat{K}_0 \to \tfrac{1}{2}\left(\hat{a}_1^\dagger \hat{a}_1 + \hat{a}_2^\dagger \hat{a}_2 + \tfrac{1}{2}\right)\, , \tag{78}$$

obey the commutation relations of Eq. (62), with the Casimir operator of Eq. (63) now given by

$$\hat{K}^2 \to \tfrac{1}{4}\left(\hat{a}_1^\dagger \hat{a}_1 - \hat{a}_2^\dagger \hat{a}_2\right)^2 - \tfrac{1}{4}\, . \tag{79}$$

In the above Hilbert subspaces $H_{[k]}$ and $H'_{[k]}$ with $k = \frac{1}{2}, 1, \frac{3}{2}, \ldots$, the standard SU(1,1) basis vectors of Eq. (64) are respectively given by

$$|k; \mu \equiv k + n\rangle \to |n + 2k - 1, n\rangle \in H_{[k]}\, ,\ |n, n + 2k - 1\rangle \in H'_{[k]}\, . \tag{80}$$

It is easily seen that the corresponding two-mode squeezing operators of Eq. (65) in this realization of the SU(1,1) algebra commute with the projection operators onto the subspaces $H_{[k]}$ and $H'_{[k]}$, and thus leave these Hilbert subspaces invariant.

The mode of action of the group operators of Eq. (65) on the underlying two boson creation operators is now the Bogoliubov transformation to squeezed two-mode bosons,

$$\begin{aligned} \hat{S}(z) \hat{a}_1 \hat{S}^\dagger(z) &= \left(1 - |z|^2\right)^{-1/2} \left(\hat{a}_1 - z \hat{a}_2^\dagger\right) \equiv \hat{b}_z^{(1)}\, , \\ \hat{S}(z) \hat{a}_2 \hat{S}^\dagger(z) &= \left(1 - |z|^2\right)^{-1/2} \left(\hat{a}_2 - z \hat{a}_1^\dagger\right) \equiv \hat{b}_z^{(2)}\, . \end{aligned} \tag{81}$$

Finally, if we let the squeezing operator in this realization act on either the lowest state $|2k - 1, 0\rangle \in H_{[k]}$ or the analogous state $|0, 2k - 1\rangle \in H'_{[k]}$, we generate the corresponding two-mode SU(1,1) coherent states. In the former case, for example,

$$\begin{aligned} |z; l, 0\rangle &\equiv \hat{S}(z) |l, 0\rangle\, ;\ \ l = 0, 1, 2, \ldots \\ &= \left(1 - |z|^2\right)^{(1+l)/2} \sum_{n=0}^{\infty} \left[\frac{(n + l)!}{n!\, l!}\right]^{1/2} z^n\, |n + l, n\rangle\, ;\ \ |z| < 1\, , \end{aligned} \tag{82}$$

which should be compared with Eqs. (70,71) for the corresponding one-mode k-squeezed state, when $l = 2k - 1$. These states clearly obey the same completeness relation of Eqs. (74,75) with $|z; \frac{1}{2}(1 + l)\rangle \to |z; l, 0\rangle$.

5. Displaced Negative-Binomial Mixed Coherent States

We now return to our second realization of the SU(1,1) algebra from Sec. 4.2 above and the associated two-boson k-squeezed states of the one-mode oscillator. Since these provide a generalization of our earlier Glauber coherent states, we now attempt[7] a similar generalization to a new class of mixed coherent states which, if they exist, will generalize our earlier thermal mixed coherent states. To that end we first introduce the density operator for the negative-binomial mixed states,

$$\hat{\rho}(k,r) \equiv (1-r)^{2k} \sum_{n=0}^{\infty} p(k,n)\, r^n \,|n\rangle\langle n| \; ; \quad 0 \le r \le 1 \; ; \quad k = \tfrac{1}{2}, 1, \tfrac{3}{2}, \ldots \,, \tag{83}$$

where $p(k,n)$ is as given previously in Eq. (71). The special case $k = \frac{1}{2}$ of Eq. (83) simply reduces to our earlier thermal oscillator mixed state of Eq. (32),

$$\begin{aligned} \hat{\rho}_{\rm th}(r) \equiv \hat{\rho}(k = \tfrac{1}{2}, r) &= (1-r) \sum_{n=0}^{\infty} r^n \,|n\rangle\langle n| \\ &= \left(1 - \mathrm{e}^{-\beta\hbar\omega}\right) \exp\left(-\beta\hbar\omega \hat{a}_0^{\dagger}\hat{a}_0\right) \; ; \quad \beta\hbar\omega = -\ln r \,. \end{aligned} \tag{84}$$

The average number of bosons in the negative-binomial mixed state is given by

$$n(k,r) = \langle \hat{n}_0 \rangle_{k,r} \equiv \mathrm{Tr}\left(\hat{\rho}(k,r)\hat{n}_0\right) = 2k\left(\tfrac{1}{r} - 1\right)^{-1} , \tag{85}$$

which reduces to the thermal Planck distribution of Eq. (47) in the special case $k = \frac{1}{2}$.

We note that $\hat{\rho}(k,r)$ may also be written in terms of the thermal density operator,

$$\hat{\rho}(k,r) = [(2k-1)!]^{-1} \left(\tfrac{1}{r} - 1\right)^{2k-1} \left(\hat{a}_0\right)^{2k-1} \hat{\rho}_{\rm th}(r) \left(\hat{a}_0^{\dagger}\right)^{2k-1} . \tag{86}$$

It is also of interest to note that one can easily show from Eq. (70) that the negative-binomial mixed state $\hat{\rho}(k,r)$ can be obtained from the corresponding pure-state projector by randomizing the phase of the argument z,

$$\hat{\rho}(k,r) = \int_0^{2\pi} \frac{\mathrm{d}\theta}{2\pi} \,|z;k\rangle\langle z;k| \; ; \quad z \equiv \sqrt{r} \exp \mathrm{i}\theta \,. \tag{87}$$

Finally, by analogy with our procedure for generating the mixed thermal coherent states in Sec. 3, we now displace $\hat{\rho}(k,r)$, using the Weyl displacement operator of Eq. (9), to yield the displaced negative-binomial mixed (coherent) states,

$$\begin{aligned} \hat{\rho}(z;k,r) &\equiv \hat{D}(z)\hat{\rho}(k,r)\hat{D}^{\dagger}(z) \\ &= (1-r)^{2k} \sum_{n=0}^{\infty} p(k,n)\, r^n \,|n;z\rangle\langle n;z| \; ; \quad 0 \le r \le 1 \,, \end{aligned} \tag{88}$$

which reduces to Eq. (35), as expected in the special case $k = \frac{1}{2}$. In the special case $r = 0$, Eq. (88) reduces to the pure Glauber coherent state,

$$\hat{\rho}(z; k=0, r) = |z\rangle\langle z| \,. \tag{89}$$

We can also show rather easily that the mean number of bosons in our displaced negative-binomial mixed coherent state $\hat{\rho}(\zeta;k,r)$ is given by

$$n(\zeta;k,r) = \langle\hat{n}_0\rangle_{\zeta;k,r} \equiv \mathrm{Tr}\big(\hat{\rho}(\zeta;k,r)\hat{n}_0\big) = 2k\big(\tfrac{1}{r}-1\big)^{-1} + |\zeta|^2\,, \tag{90}$$

which may be compared with its thermal counterpart of Eq. (48). Just as in Eq. (41) for the mixed thermal coherent states, so we can prove[7] the analogous very important relation that the negative-binomial mixed states also provide a resolution of the identity operator,

$$\iint_C \frac{\mathrm{d}^2 z}{\pi}\,\hat{\rho}(z;k,r) = \hat{\mathbb{1}}\,, \tag{91}$$

which shows that they may also be regarded as mixed coherent states, and thus provide the desired generalization of their thermal counterparts. There are many ways in which Eq. (91) may be proved. Perhaps the most direct way is to take matrix elements in the number-state basis, and use the definition of Eq. (88) together with the well-known and easily proven form of the matrix elements of the displacement operator in terms of the associated Laguerre polynomials $L_n^{l-n}(x)$.[7]

In view of the existence of the completeness relation of Eq. (91) it is clear that everything we did in Sec. 3 for the thermal coherent states ($k=\frac{1}{2}$) can now be generalized to the present case $k=\frac{1}{2},1,\frac{3}{2},\ldots$. For example, we can give further (k,r)-extensions of the ordinary P- and Q-representations as the obvious generalizations of Eqs. (49,50),

$$\begin{aligned} Q(\hat{\Theta};z;k,r) &\equiv \mathrm{Tr}\big(\hat{\rho}(z;k,r)\hat{\Theta}\big)\,,\\ \hat{\Theta} &\equiv \iint_C \frac{\mathrm{d}^2 z}{\pi}\,P(\hat{\Theta};z;k,r)\hat{\rho}(z;k,r)\,. \end{aligned} \tag{92}$$

Clearly the case $r=0$ of Eq. (92) just reproduces the ordinary P- and Q-representations of Eqs. (17,18), while the case $\{k=\frac{1}{2},\ \beta\hbar\omega=-\ln r\}$ gives our previous thermal results of Eqs. (49,50). Typical of the many interesting relations that we find are the following,

$$\tilde{Q}(\hat{\Theta};w;k,r) = \exp\left[-\frac{|w|^2}{8}\frac{(1+r)}{(1-r)}\right] L_{2k-1}\left(\frac{|w|^2 r}{4(1-r)}\right)\tilde{W}(\hat{\Theta};w)\,, \tag{93}$$

where $L_n(x)$ is a Laguerre polynomial, and

$$\tilde{P}(\hat{\Theta};w;k,r) = \exp\left[+\frac{|w|^2}{8}\frac{(1+r)}{(1-r)}\right]\left[L_{2k-1}\left(\frac{|w|^2 r}{4(1-r)}\right)\right]^{-1}\tilde{W}(\hat{\Theta};w)\,, \tag{94}$$

which generalize Eqs. (59,60), and

$$\begin{aligned} \tilde{Q}(\hat{\Theta};w;k,r_1) &= \exp\left[-\frac{|w|^2}{4}\frac{(1-r_1r_2)}{(1-r_1)(1-r_2)}\right] L_{2k-1}\left(\frac{|w|^2 r_1}{4(1-r_1)}\right)\\ &\quad\times L_{2k-1}\left(\frac{|w|^2 r_2}{4(1-r_2)}\right)\tilde{P}(\hat{\Theta};w;k,r_2)\,, \end{aligned} \tag{95}$$

which is a generalization of the pure (Glauber) coherent-state result of Eq. (29). A number of other similar results are discussed in more detail in Ref. [7].

Finally, we also note the existence of another resolution of the identity operator over the variable r, for fixed values of z and k,

$$(2k-1)\int_0^1 \mathrm{d}r\,(1-r)^{-2}\hat{\rho}(z;k,r) = \hat{\mathbb{1}}\,;\quad k>\tfrac{1}{2}\,, \tag{96}$$

where the case $k=\frac{1}{2}$ is again handled as the limiting case $k=\frac{1}{2}+\epsilon$, in the limit as $\epsilon \to 0^+$. Eq. (96) may be proven by making use of Eqs. (87) and (88) together with Eqs. (74) and (75). Although Eq. (96) seems not to have been exploited in the present context up till now, we believe that it should be profitable to give it further consideration.

6. Generalized Thermo-Field Dynamics

We turn finally to another application of the thermal and negative-binomial mixed states that we have discussed here. We recall the third realization of SU(1,1) that we introduced in Sec. 4.3 and the corresponding SU(1,1) two-mode pure coherent state of Eq. (82), with respective pure-state density operator in the two-boson Hilbert space $H_1 \otimes H_2$ given by

$$\hat{\rho}(l,0;z) \equiv |z;l,0\rangle\langle z;l,0|\,;\quad l=0,1,2,\dots\,;\quad |z|<1\,. \tag{97}$$

If we take the partial trace of this density operator with respect to the first mode we find,

$$\begin{aligned}\hat{\rho}_2(l,0;z) &\equiv \mathrm{Tr}_1\,\hat{\rho}(l,0;z)\\ &= \hat{\rho}\big(k=\tfrac{1}{2}(1+l), r=|z|^2\big)\,,\end{aligned} \tag{98}$$

which is simply our previous negative-binomial mixed state of Eq. (83). Similarly, by taking the partial trace with respect to the second mode we find,

$$\begin{aligned}\hat{\rho}_1(l,0;z) &\equiv \mathrm{Tr}_2\,\hat{\rho}(l,0;z)\\ &= \big(1-|z|^2\big)^{1+l}\sum_{n=0}^{\infty}\frac{(n+l)!}{n!\,l!}\,|z|^{2n}\,|n+l\rangle\langle n+l|\,,\end{aligned} \tag{99}$$

which is simply a "shifted" negative-binomial mixed state, where the term "shifted" refers here to the fact that the series in Eq. (99) begins with the term proportional to $|l\rangle\langle l|$ rather than with the term proportional to $|0\rangle\langle 0|$. (Note also that this shifted negative-binomial mixed state should not be confused with the quite different displaced negative-binomial mixed state of Eq. (88.) Only in the special case $l=0$, i.e., $k=\frac{1}{2}$, do both partial traces agree with each other and yield the thermal density operator,

$$\hat{\rho}_1(0,0;z) = \hat{\rho}_2(0,0;z) = \hat{\rho}\big(k=\tfrac{1}{2}, r=|z|^2\big) \equiv \hat{\rho}_{\mathrm{th}}\big(r=|z|^2\big)\,;\quad |z|^2 \equiv \mathrm{e}^{-\beta\hbar\omega}\,. \tag{100}$$

We now make use of the above formalism by letting $\hat{\Theta}_2$ be any operator acting in the Hilbert space H_2 and $\hat{\mathbb{1}}_1$ be the unit operator acting in H_1. Then we find,

$$\begin{aligned}\langle z;l,0|\hat{\mathbb{1}}_1\otimes\hat{\Theta}_2|z;l,0\rangle &= \mathrm{Tr}\{(\hat{\mathbb{1}}_1\otimes\hat{\Theta}_2)\hat{\rho}(l,0;z)\} = \mathrm{Tr}_2\{\hat{\Theta}_2\hat{\rho}_2(l,0;z)\}\\ &= \mathrm{Tr}\{\hat{\Theta}\hat{\rho}(k=\tfrac{1}{2}(1+l), r=|z|^2)\}\,. \end{aligned}\tag{101}$$

The special case $k=\frac{1}{2}$, (i.e., $l=0$) of Eq. (101) is the basic relation of thermo-field dynamics,[8]

$$\langle z;0,0|\hat{\mathbb{1}}_1\otimes\hat{\Theta}_2|z;0,0\rangle = \mathrm{Tr}\{\hat{\Theta}\hat{\rho}_{\mathrm{th}}(|z|^2)\}\ ;\ \ |z|^2 = \mathrm{e}^{-\beta\hbar\omega}\,. \tag{102}$$

In thermo-field dynamics[8] the first mode is the so-called "fictitious mode," and the two-mode SU(1,1) pure coherent state $|z;0,0\rangle \in H_{[1/2]}$ is the so-called "thermo-field vacuum." Essentially all of the results of thermo-field dynamics now follow from this last relation and the two-mode squeezing formalism presented here.

What has now been achieved by us in this context is to generalize the basic equation (102) of thermo-field dynamics to the much more general equation (101) of what we now call "negative-binomial-field dynamics." Whereas thermo-field dynamics is a formalism within the two-boson Hilbert subspace $H_{[1/2]}$, negative-binomial-field dynamics exists within the more general two-boson subspace $H_{[k]}$, where $k=\frac{1}{2}(1+l)$, associated with the k-representation of SU(1,1). We discuss possible uses of this formalism below.

7. Discussion

We have shown elsewhere[9] how the thermal ($T\neq 0$) coherent states of Sec. 3 may very profitably be used in the Bargmann representation, thereby extending the usual ($T=0$) formalism. Clearly, the mixed thermal coherent-state formalism should be applicable in any situation where coherent signals co-exist with thermal noise. Examples abound in fields as diverse as quantum optics and quantum information theory. Other potential applications include signal and image processing, where the usual representations of a signal in the time domain or in the frequency domain are often less convenient than one in the joint time-frequency "phase space." Indeed there are many papers in the signal processing context that use the Wigner function method and allied techniques inspired by the Glauber coherent-state formalism. Wavelet approaches are just one such by-product.

We have also shown previously[10] how the pure two-mode SU(1,1) coherent states of Sec. 4.3 arise in the context of a damped linear harmonic oscillator, and how they can be used for general such processes involving fluctuation-dissipation phenomena. The extension to the mixed case and negative-binomial-field dynamics will be interesting. In many practical applications the usefulness of our result of Eq. (101) will arise from the fact that the right-hand sides will often be easier to calculate than the corresponding left-hand sides. Moreover, in many physical situations the so-called "fictitious" mode will actually be quite real. A particular example is the so-called "accelerated observer formalism" investigated by us previously,[11] where the two modes correspond to observers moving in opposite directions, each of whom only has a partial knowledge of the other due to being able to observe only a restricted region of space-time. This formalism provides a physically appealing account of the Hawking radiation from a black hole. There exist many other examples where we only have partial knowledge about the system, and where the "real" and "fictitious" modes represent simply the known and unknown parameters respectively. The unknown parameters are typically traced over, and what was originally a pure state in the "big" Hilbert space $H_1 \otimes H_2$ becomes a mixed state in the "small" space H_1. In all such cases there will be a need for a basis of mixed states in the "small" space. Furthermore, such admixtures of coherence (pure states) with general noise (thermal or more general, such as negative-binomial) or decoherence is often essentially unavoidable in practice, and in all such situations the formalism presented here is potentially useful.

References

1. R.F. Bishop and A. Vourdas, *J. Phys. A: Math. Gen.* **20**, 3743 (1987).
2. R.J. Glauber, *Phys. Rev.* **131**, 2766 (1963).
3. G. Lachs, *Phys. Rev.* **138**, B1012 (1965).
4. T. Holstein and H. Primakoff, *Phys. Rev.* **58**, 1098 (1940).
5. A.M. Perelomov, *Commun. Math. Phys.* **26**, 222 (1972).
6. A. Vourdas, *Phys. Rev. A* **45**, 1943 (1992).
7. A. Vourdas and R.F. Bishop, *Phys. Rev. A* **51**, 2353 (1995).
8. H. Umezawa, H. Matsumoto, and M. Tachiki, *Thermo-Field Dynamics and Condensed States*, (North-Holland, Amsterdam, 1987); S.M. Barnett and P.L. Knight, *J. Opt. Soc. Am. B* **2**, 467 (1985).
9. A. Vourdas and R.F. Bishop, *Phys. Rev. A* **50**, 3331 (1994).
10. R.F. Bishop and A. Vourdas, *J. Phys. A: Math. Gen.* **20**, 3727 (1987).
11. R.F. Bishop and A. Vourdas, *J. Phys. A: Math. Gen.* **19**, 2525 (1986).

ERGODIC CONDITION AND MAGNETIC MODELS

M. HOWARD LEE

Department of Physics and Astronomy,
University of Georgia, Athens, Georgia 30602, USA
(email: MHLee@uga.edu.au)
and
Korea Institute for Advanced Study, Seoul 130-012, Korea

A condition for ergodicity is derived, applicable to a Hermitian many body model in both the classical and quantum domains. Using this ergodic condition, the validity of the ergodic hypothesis is examined in certain solvable 1d magnetic models. A simple but general picture has emerged which shows why the hypothesis can be valid and why it can also fail.

Keywords: Ergodicity; magnets; response function.

1. Introduction

Boltzmann's ergodic hypothesis is a foundation for statistical mechanics. Yet there has been no proof that the ergodic hypothesis (*EH*) is universally valid. In several standard textbooks on statistical mechanics this lack of proof has been noted [1]. In spite of this unsatisfactory state of affairs most people seem to either believe in it or go about blithely in ignorance. It seems to be one instance that physicists in statistical mechanics as a whole act seemingly irrationally. This is all the more surprising since it has already been shown that in several models *EH* definitely does not hold [2].

One reason for not taking this issue more seriously may perhaps be that proving *EH* seems an unduly difficult mathematical task. Indeed this task seems to have been taken over by mathematicians, turning it into a mathematical field known as the ergodic theory or general dynamical theory [3]. A physicist who happens to wander into this mathematical field would soon realize that there appears little or none of the physics from which it had original arisen.

Another reason may also be that in those models where *EH* is shown to be invalid the analysis seems overly technical and possibly ad hoc [4]. No physical insight has emerged as to why *EH* should fail in them. What is really needed is a technique that is both sufficiently general to analyze any model and physically based to provide some underpinnings as to why *EH* may be valid or why it may fail. What is needed is an ergodic condition, a kind of meter with which one can not only measure whether a model is ergodic but also tell why so. If such an ergodic condition can be found, it might also be able to illuminate some theorems of the ergodic theory said to be relevant to *EH* as we understand it in statistical mechanics. To this end we aim at obtaining an ergodic condition.

2. Dynamical Response Function

The ergodic theory in mathematics studies a general function. For *EH* in statistical mechanics it is essential that a dynamical function should be specific. A dynamical function implies a motion or a time evolution. A system will evolve in time if perturbed as by an inelastic scattering process. The perturbed energy will become delocalized. If a dynamical function, which reflects the energy delocalization, is time averaged, *EH* can be tied to this physical process. The time dependent susceptibility is best suited for it.

Let A be the dynamical variable e.g. single or total spin, so that $\langle A(t)\rangle = \mathrm{Tr}\{A\rho(t)\}$ where ρ is the density matrix. Since we will study some magnetic models, let $\langle A(t)\rangle = M(t)$, the magnetization. If $h(t)$ is an external magnetic field, $M(t)$ is expressible in terms of the time dependent susceptibility as follows [5]:

$$M(t) = \int_0^t h(t')\chi(t-t')\mathrm{d}t'. \tag{1}$$

We have already assumed that the response function $\chi(t-t')$ is causal and stationary. If the external field $h(t)$ is sufficiently weak, linear response theory [6] gives

$$\chi(t-t') = \begin{cases} \frac{i}{\beta}\langle[A(t), A(t')]\rangle & t > t' \\ 0, & \text{otherwise} \end{cases} \tag{2}$$

where the bracket now is by the density matrix independent of the external field. The time average over $M(t)$ is thus the time average over the dynamic response function $\chi(t-t')$. According to *EH* [7] then,

$$\lim_{T\to\infty} \frac{1}{T}\int_0^T \int_0^t \chi(t-t')\mathrm{d}t'\,\mathrm{d}t = \chi\ ?, \tag{3}$$

where χ is the static response function, i.e. $M = \chi h$, where h is a constant field. We do not know whether the two sides of (3) are equal, hence the question mark. To test whether they are equal, we shall simplify the *lhs* of (3), first suppressing $\lim_{T\to\infty}$, but always implied when T appears.

To reduce the double integral to a single one, we introduce a new function $R(t)$, $t \geq 0$, having the following properties among others: $\dot{R}(t) = -\chi(t)$, $t > 0$. With this function we can partially integrate the *lhs* of (3) to obtain

$$\tilde{\chi}(0) + R(\infty) - \frac{1}{T}\tilde{R}(0) = \chi\ ?, \tag{4}$$

where we have used the Laplace transformed expression $\tilde{\chi}(z) = \int_0^\infty e^{-zt}\chi(t)\mathrm{d}t$, $\mathrm{Re}(z) > 0$, and similarly for $\tilde{R}(z)$. Of course, there is nothing really new gained here except for the appearance of simplicity. We still do not know whether the two sides are equal. For this we obviously need to introduce an independent element, prefereably an exact relation with which (4) may be compared. From (4) we realize that the new function $R(t)$ seems to play a central role. We look into it to see whether there is an independent element in it.

3. Relaxation Function and Inner Product Space

In linear response theory the function $R(t)$ is known as the relaxation function, having the following definition:

$$R(t) = (A(t), A),\ t \geq 0, \tag{5}$$

where $A = A(t = 0)$. It is an inner product, known in statistical mechanics as the Kubo scalar product [5]. Let A and B be two dynamical variables, assumed to be Hermitian for simplicity. Then, the inner product of A and B is:

$$(A, B) = \frac{1}{\beta}\int_0^\beta \langle e^{\lambda H} A e^{-\lambda H} B\rangle \mathrm{d}\lambda - \langle A\rangle\langle B\rangle, \tag{6}$$

where β is the inverse temperature. If $[H, B] = 0$ or if we are in the classical domain, it becomes

$$(A, B) = \langle AB\rangle - \langle A\rangle\langle B\rangle. \tag{7}$$

From the definition (6), one can show that indeed $\dot{R}(t) = (i/\beta)\langle[A(t), A]\rangle$, $t > 0$, see [2]. In addition, $R(0) = (A, A) = \chi$, the static response function. Taking the Laplace transform on $\dot{R}(t) = \chi(t)$, we obtain at once by setting $z = 0$

$$\tilde{\chi}(0) + z\tilde{R}(z)\Big|_{z=0} = \chi. \tag{8}$$

Evidently (8) is an independent element, having no relationship to *EH*. Evidently it is an exact relation since it follows from a definition. Now we can compare (4) with (8) to see whether (4) is correct. Clearly they do not agree. Since (4) does not agree with an exact relation, *EH* cannot be valid! This conclusion may perhaps be somewhat hasty. Granted that (4) does not look congruent to the "truth", but could it possibly intersect (8) in the space of R? If so, at least at this point of intersection, *EH* could just be valid, but only at this point. There seems no other way that *EH* could be broadly valid.

4. Point of Intersection and Ergodic Condition

Let us compare (4) and (8) again. One may think that each equation delineates a line. As it appears, the two lines go about in separate ways. If, however, they intersect in some region of the space of R, the point of intersection is a congruent point. It is at this intersection, and only at this point of intersection that *EH* is true. The point of intersection is given by $\tilde{R}(0)$ =*finite*. If it is *finite*, the action of $1/T$ before it in (4) will eliminate the third term on the *lhs* of (4). What is more, if it is *finite*, $R(\infty) = 0$. Also $\tilde{R}(0)/T = 0$ as $T \to \infty$ is equivalent to $\lim_{z\to 0} z\tilde{R}(0) = 0$.

At this point of intersection, suddenly the two equations become equal. Since (8) is true, so must (4) be: *EH* is valid at the point of intersection. The finiteness of $\tilde{R}(0)$ must also include $\tilde{R}(0) \neq 0$. If otherwise, it is not removed by the action of $1/T$, $T \to \infty$. That is, it is not being time averaged over a long time. Thus we arrive at a remarkable ergodic condition

$$0 < \tilde{R}(0) < \infty, \tag{9}$$

where $\tilde{R}(z=0) = \int_0^\infty R(t)\mathrm{d}t$. If $R(0) = \chi$ and is a finite constant, we can introduce $r(t) = R(t)/R(0)$ and restate (9) as

$$0 < \tilde{r}(0) < \infty. \tag{10}$$

5. Recurrence Relations Method and $r(t)$

The normalized relaxation function $r(t)$ can be calculated by the recurrence relations method if a system is Hermitian. We shall sketch here the method [8], which is geometrical. An inner product space S realized by the Kubo scalar product, see (6), is not a Euclidean space. It has the following salient properties:

a. $A(t)$ is a vector in S.
b. $\|A(t)\| = \|A\|$. The norm of A is invariant with time.
c. As t evolves, $A(t)$ changes only its directions. The change of direction is described by the equation of motion of $\dot{A}(t) = i[H, A(t)]$. Hence, it is model dependent.
d. The motion of $A(t)$ delineates a line on the surface of S. Thus, the shape of the surface of S gives the time evolution of A.
e. The surface shape of S is given by a set of H dependent d basis vectors that span S: $(f_0, f_1, \ldots, f_{d-1})$ such that $(f_m, f_{m'}) = 0$ if $m \neq m'$, $(m, m' = 0, 1, 2, \ldots, d-1)$.
f. This set of basis vectors can be calculated by a recurrence relation known as *RR1* for a given H.
g. If $d < \infty$, $r(t)$ is always periodic. That is, $r(\infty) \neq 0$. If $d \to \infty$, it is possible that $r(\infty) = 0$.
h. If $N < \infty$, $d < \infty$. If $N \to \infty$, $d < \infty$ or $\to \infty$. For a system to be ergodic, it is thus necessary, but not sufficient, that $N \to \infty$.
i.

$$\begin{aligned}\tilde{r}(z) &= \int_0^\infty r(t)e^{-zt}\mathrm{d}t, \ \mathrm{Re}(z) > 0 \\ &= 1/z + \Delta_1/z + \ldots + \Delta_{d-1}/z,\end{aligned} \tag{11}$$

where $\Delta_1 = \|f_1\|/\|f_0\|$ etc. are called the recurrants. The surface shape of S is denoted by a hypersurface factor σ

$$\sigma = (\Delta_1, \Delta_2, \ldots, \Delta_{d-1}). \tag{12}$$

The continued fraction (11) being given in terms of the recurrants carries a geometrical meaning. Now we can apply the ergodic condition (10) on (11). Since $r(\infty) = 0$, implying an irreversibility [9–11], to be considered ergodic, $d \to \infty$. Hence

$$\begin{aligned} W &\equiv \lim_{z\to 0}\lim_{d\to\infty} \tilde{r}(z; \Delta_1, \Delta_2, \ldots, \Delta_{d-1}) \\ &= \Delta_2.\Delta_4.\Delta_6 \ldots /(\Delta_1.\Delta_3.\Delta_5 \ldots). \end{aligned} \tag{13}$$

It is an infinite product first studied by J. Wallis in 1655, known by Wallis' number $W_{\text{Wallis}} = \pi/2$. We now state the ergodic condition as

$$0 < W < \infty, \tag{14}$$

where W may be given by (13) or by $W = \tilde{r}(0) = \int_0^\infty r(t)dt$.

We shall give an example. Suppose a particular S has $d = \infty$ and has a hypersurface $\sigma = (1^2, 2^2, 3^2, \ldots)$. Then by (13)

$$2^2 4^2 6^2 \ldots /(1^2 3^2 5^2 \ldots) = W_{\text{Wallis}}. \tag{15}$$

It is known that for this shape of S, $r(t) = \sec(ht)$ in some dimensionless units [12]. Thus, $\int_0^\infty \sec(ht)dt = \pi/2$, in agreement with Wallis' number. The recurrence relation method has produced many remarkable dynamical results since its development in 1982. For representative examples, see [13–27]

6. Magnetic Models

In this work, we will apply the ergodic condition W to several well-known spin models to see whether they are ergodic. If their W's are 0 or ∞, the ergodicity fails. $W = 0$ is termed the *localization* limit and $W = \infty$ the *ballistic* limit. If W for a particular model lies in between the two limits, it is deemed ergodic.

6.1. *Single spin model*

It is a trivial model, but it shows that $d \nrightarrow \infty$ since N is finite [28]. Hence *EH* is ruled out at the outset. A typical single spin model ($N = 1$) is given by

$$H = aS_z + bS_x, \tag{16}$$

where a and b are constants, S_x and S_z are the x and z components of the spin-1/2 operator S. For this model, there are two independent choices for A: $A = S_y$ and $A = S_z$ or S_x. If $A = S_y$, $d = 2$, and if $A = S_z$, $d = 3$. We shall look at the simpler case with $d = 2$. For this choice, $\sigma = (\Delta_1)$ only, where $\Delta_1 = a^2 + b^2$. It immediately follows that

$$r(t) = \cos(\omega t), \tag{17}$$

where $\omega = \sqrt{\Delta_1}$. Evidently $r(t = \infty) \neq 0$ being periodic. Thus, this model is neither irreversible or ergodic. For $d = 3$, one obtains a similar result.

6.2. *Ising model*

Now we consider the N spin Ising model in D dimensions [29],

$$H = -J \sum_{(ij)} S_i^z S_j^z, \tag{18}$$

where (ij) denotes nn pairs and we assume $J > 0$ although not absolutely required. We also impose periodic boundary conditions, so that all spins are equivalent. One could not choose $A = S_1^z$, since for this choice $[H, S_1^z] = 0$. For this spin (or any other similar spin at other sites), $d = 1$, implying that there is no time evolution since S_1^z is always in a stationary state of H.

But, if $A = S_1^x$ or S_1^y, $[H, A] \neq 0$. Hence there can be a dynamical development for it. In terms of scattering, it is equivalent to turning an Ising spin into the x or y component by an external field in the transverse direction. An overturned spin, now not at a stationary state of H, tries to restore itself by giving up its energy to its neighbours, to which it is tied by a strength J. A neighbouring Ising spin accomodates by overturning itself, but finds itself in the same position as the one which initiated the process. As a result, the excess energy is being bounced around between the neighbours, and never gets to propagate. The oscillation of the perturbation energy is thus a function of the coordination number q. For simplicity, we consider $q = 2$.

If $q = 2$, $d = 3$. That is, $\sigma = (\Delta_1, \Delta_2)$, where Δ_1 and Δ_2 are some static properties. Their detailed forms are not needed here except for the relation $\Delta_1 + \Delta_2 = \omega^2$, where $\omega = J/\hbar$. Even though $N \to \infty$, the inner product space remains $3d$. We obtain

$$r(t) = \Delta_2/\omega^2 + \Delta_1/\left(\omega^2 \cos(\omega t)\right). \tag{19}$$

Observe that $r(t = 0) = 1$ as required, but $r(t = \infty) \neq 0$. It fails the necessary condition for ergodicity. The Ising model is not ergodic with respect to $A = S_1^x$ or S_1^y. A comparison between sections 6.1 and 6.2 makes it clear that is is not the thermodynamic limit that is sufficient to make a model ergodic. The thermodynamic limit is merely a necessary condition for ergodicity, not a sufficient one.

6.3. *XY chains*

Now we look at a more complex model, defined by

$$H = -J \sum_{j=1}^{N} S_i^x S_j^x + S_i^y S_j^y, \tag{20}$$

where J is a constant assumed to be positive and periodic boundary conditions are implied e.g. $S_{N+1}^x = S_1^x$. For this model, two independent choices for A are also possible: $A = S_1^x$ or S_1^y and $A = S_1^z$. Spins at different sites are all equivalent, so that our choice at site 1 is arbitrary. For both choices $r(t)$ can be evaluated at $T = \infty$.

6.3.1. $A = S_1^x$

For this dynamical variable $d \to \infty$ if $N \to \infty$. If $T = \infty$, $\sigma = (1, 2, 3, \ldots)$ in units of $2J^2 = 1$ [30]. By (13)

$$\begin{aligned} W &= 2.4.6\ldots/(1.3.5\ldots) \\ &= \left[2^2 4^4 6^2/(1^2 3^2 5^2)\right]^{1/2} \;=\; W_{\text{Wallis}}^{1/2} \\ &= \sqrt{\pi/2} \end{aligned} \tag{21}$$

Here we have assumed that the two operations squaring and taking square roots are cummutative. In infinite products, many of the standard operations governing finite products do not apply. Infinite products are generally very delicate. In this case it appears to be justified: this hypersurface yields

$$r(t) = e^{-t^2/2}. \tag{22}$$

Clearly $r(\infty) = 0$, indicating an irreversibility. We obtain $W = \int_0^\infty e^{-t^2/2}\mathrm{d}t = \sqrt{\pi/2}$. Evidently the model is ergodc with respect to $A = S_1^x$ or S_1^y.

6.3.2. $A = S_1^z$

If $A = S_z^z$, $r(t) = J_0(t)^2$, now in units of $J = 1$, where J_0 is the Bessel function of order 0 [31]. It is irreversible since $r(\infty) = 0$, recalling that if $t \to \infty$, $J_0(t)$ behaves as $t^{-1/2}$ with an oscillation. We obtain

$$\tilde{r}(z) = \frac{1}{z} F(1/2, 1/2, 1, -4/z^2), \tag{23}$$

where F is the hypergeometric function. W is obtained by

$$W = \tilde{r}(z \to 0) = -\frac{1}{\pi}\log(z), \tag{24}$$

thus $W \to \infty$ as $z \to 0$, attaining the ballistic limit [32]. The model is not ergodic with respect to $A = S_z^z$.

It is interesting that a model can be both ergodic and nonergodic depending on dynamical variables. This example illustrates that ergodicity is not absolute. It can be said only with respect to a dynamical variable of interest. In this XY spin chain, when $A = S_1^x$ or S_1^y, the excitation energy becomes delocalized through the chain, giving rise to ergodic behaviour. This is quite different from the behaviour in section 6.2. There the excited energy is localized between the *nn*'s because an Ising spin is in a stationary state. In the XY chain, the excited energy is not trapped between neighbours since S_1^x is not in a stationary state of H. As a result, the excited energy is not prevented from being delocalized.

When $A = S_1^z$, the excited energy makes the chain behave like a vortex motion along the longitudinal direction in spin space. The motion is independent of the transverse plane, which is just being carried along. It is very similar to exciting a Brownian particle. When a Brownian mass is excited, its translational motion is independent of the internal degrees of freedom. Both are at the ballistic limit. The main difference is that the spins act irreversibly while a Brownian particle does not. This difference again points to the fact that irreversibility is not a sufficient condition for ergodic behaviour.

6.4. *Transverse Ising model*

We now consider the $1d$ transverse Ising model of the following form:

$$H = -2J_1 \sum_j S_j^x S_{j+1}^x - J_2 \sum_j S_j^z, \tag{25}$$

where J_1 and J_2 are assumed to be positive. There are two dynamical solutions known for special values of J_2/J_1. We shall study their ergodic behaviour.

6.4.1. $J_2 = J_1$ *at infinite temperature*

We choose $A = S_1^x$, the site 1 is arbitrary since all spins are equivalent owing to periodic boundary conditions imposed. In this case the dynamics turns out to be very similar to the dynamics given in section 6.3.1. In fact its hypersurface has the same structure. As a result [30],

$$r(t) = e^{-t^2/2}. \tag{26}$$

It is irreversible and $W = \sqrt{\pi/2}$. This model is thus ergodic with respect to $A = S_1^x$. If a spin S_1^x is excited by an external probe, the excited energy is kept in the transverse plane. Unlike in section 6.3.2, it does not generate a vortex motion along the x or z direction in spin space. Thus, it is unable to reach the ballistic limit, resulting in a finite W. That is, the excited energy is being delocalized completely.

6.4.2. $J_2 = \frac{1}{2}J_1$ *at the ground state*

If we choose $A = S_1^z$, the dynamics resemble those in section 6.3.2. For this model at the ground state it is known that there is a quantum phase transition at $J_2 = J_1/2$. We choose J_2 to be in the neighbourhood of $J_c \equiv J_1/2$, such that $\varepsilon = (J_2 - J_c)/J_c$ and $\varepsilon \to 0$. The hypersurface can be expressed to order ε. The full solution for $\tilde{r}(z)$ is expressible in terms of the hypergeometric functions, not presented here being too involved. What we need are the asymptotic forms only, and they are simple enough [33]

$$\tilde{r}(z \to 0) \sim z^{-1+2s} \tag{27}$$

and

$$R(t \to \infty) \sim t^{-2s}, \tag{28}$$

where $s = -1/\log \varepsilon$. It is irreversible, but $W \to \infty$. The behaviour is typical of a system encountering an onset of critical fluctuations. If a system enters a critical neighbourhood, it behaves ballistically. Thus, it cease to be ergodic.

6.5. *XY chain at a surface*

We now study the XY model (20), but without periodic boundary conditions. Now each spin is not equivalent. In particular, the spin at site 1 has only one *nn* at site 2. As a result, the spin at site 1 appears as a boundary or surface point, and all other spins are embedded in the bulk. For this problem we choose $A = S_1^x$, the same as in section 6.3.1.

If $N \to \infty$ [34], the hypersurface has a strikingly simple form at $T = \infty$. It is that of a hypersphere: in units of $J = 2$, $\sigma = (1, 1, 1, \ldots)$. It implies that

$$r(t) = J_1(2t)/t. \tag{29}$$

It is irreversible $r(t = \infty) = 0$. From the hypersurface it follows at once that

$$W = 1.1.1\ldots/(1.1.1\ldots) = 1. \tag{30}$$

This result is borne out by

$$\tilde{r}(z = 0) = \frac{1}{2}\left(\sqrt{z^2 + 4} - z\right)_{z=0} = 1. \tag{31}$$

Thus model is ergodic with respect to S_1^x or S_1^y. If the spin at site 1 is excited, the excited energy becomes delocalized starting from the spin at site 1 all the way down to site N. As N grows, the process is irreversible. What is perhaps more remarkable is in the comparison of these dynamics (without periodic boundary conditions) to those of section 6.3.1 (with periodic boundary conditions). Although ergodic in both cases, their dynamics are of different character. Compare (22) with (29).

7. Concluding Remarks and Outlook

Our analysis of these magnetic models illustrates how our ergodic condition works to determine whether a model is ergodic with respect to a given dynamical variable. These results show an unexpected feature that ergodicity is not absolute. A model can be both ergodic and nonergodic depending on what the dynamical variable of interest happens to be. Using our ergodic condition we can also gain insight into the physical nature of the workings of *EH* through the energy delocalizability. This is made possible by our choice of a response function with which the condition for *EH* has been formulated.

We have applied our ergodic condition to other models like the electron gas, both classical and quantum, and other systems like harmonic oscillator chains. The picture for the physical mechanisms that underlie *EH* seems simple but general. With this ergodic condition as a tool or a meter, we are now in the position to ask whether some of the mathematical theorems found in the ergodic theory could be made physically sensible. In particular there is an important theorem due to Birkhoff, proved in 1931, for a classical conservative many-body system, using the measure theory [3]. Birkhoff's theorem proves *EH* but the proof is not valid everywhere. It is circumscribed by certain mathematical conditions e.g. metrical transitivity, sets of measure zero etc.

What do they mean physically? How large is the domain of validity of this theorem? To our knowledge no one knows the answer. Obviously this answer is important to statistical mechanics since its very foundations rest in part on the validity of *EH*. As an analogy, one may think of Birkhoff's theorem in the following picture: The theorem says that there is an island on which one can walk. But it says nothing about how small or large the island is. An islet or a continent. If it is as large as a continent, the foundations of statistical mechanics are secure. If it is as small as an islet, we do have a problem. Our ergodic condition being a meter can map this island to determine how large or small it is. We are beckoned to move on to this direction.

Acknowledgments

I thank Prof. H. Park and Prof. M.W. Kim, respectively, chairman of the School of Physics and director of the Korea Institute for Advanced Study, Seoul for their kind hospitality where a portion of this work was accomplished. I also thank Ms. Sherri H. McElroy for helping with the preparation of this manuscript.

References

1. PT Landsberg, *Thermodynamics and statistical mechanics* (Dover, New York, 1990) p. 139.
2. P Mazur, Physica **43**, 533 (1969).
3. IE Farquhar, *Ergodic theory in statistical mechanics* (Wiley, New York, 1964) See Section 3.4 and Appendix 2.
4. M Suzuki, Physica **51**, 277 (1971).
5. W Marshall and RD Loude, Rep. Prog. Phys. **31**, 705 (1968).
6. R Kubo, Rep. Prog. Phys. **29**, 235 (1966); H Nakano, Int. J. Mod. Phys. B **7**, 2397 (1993).
7. MH Lee, Phys. Rev. Lett. **87**, 250601 (2001).
8. U Balucani, MH Lee and V Tognetti, Phys. Rep. **373**, 409 (2003).
9. T Prosen, in *Open problems in strongly correlated electron systems*, eds. J Bonca *et al.* (Kluwer Academic, Amsterdam, 2001) p. 317.
10. MH Vainstein, IVL Costa and FA Oliveira, in *Lecture notes in physics* (Springer-Verlag, Berlin, 2006).
11. P Grigolini, *Quantum mechanical irreversibility and measurement* (World Scientific, Singapore, 1993) p. 129.
12. K Tankeswar and KN Pathak, J. Phys. CM **6**, 591 (1994).
13. J Kim and I Sawada, Phys. Rev. E **61**, R2172 (2000); J Kim, J. Phys. A **35**, 7305 (2002).
14. I Sawada, Phys. Rev. Lett. **83**, 1668 (1999), Physica A **315**, 14 (2002); J. Phys. and Chem. Solids **62**, 373 (2001).
15. ZX Cai, S Sen and SD Mahanti, Phys. Rev. Lett. **68**, 1637 (1992); S Sen, Physica A **360**, 304 (2006).
16. J Florencio and FC Sa Baretto, Phys. Rev. B **60**, 9555 (1999); M Silva Nunez and J Florencio Jr., Phys. Rev. B **68**, 014406 (2003).
17. SG Jo, KH Lee, SC Kim and SD Choi, Phys. Rev. E **55**, 3676 (1997).
18. NL Sharma, Phys. Rev. B **45**, 3552 (1992).

19. AST Pires, Helv. Phys. Acta. **61**, 988 (1988).
20. VS Viswanath and G Mueller, *Recursion Method* (Springer-Verlag, Berlin, 1994).
21. J Dauligault and MS Murillo, Phys. Rev. E **68**, 015401 (2003), J. Phys. A **36**, 6265 (2003).
22. VV Belyi, Phys. Rev. E **69**, 017104 (2004).
23. RM Yulmatyev, AV Moshkin and P Hanggi, Phys. Rev. E **68**, 051201 (2003); AV Moshkin, RM Yulmatyev and P Hanggi, Phys. Rev. Lett. **95**, 200601 (2005); AV Moshkin and RM Yulmatyev, J. Chem. Phys. **121**, 7341 (2004).
24. J Hong and HY Kee, Phys. Rev. B **52**, 2415 (1995).
25. IM Kim and BY Ha, Can. J. Phys. **67**, 31 (1989).
26. M Bohm and H Leschke, Physica A **199**, 116 (1993).
27. ZQ Liu, XM Kong and XS Chen, Phys. Rev. B **73**, 224412 (2006).
28. MH Lee, Can. J. Phys. **61**, 420 (1983).
29. J Florencio, S Sen and MH Lee, Braz. J. Phys. **30**, 725 (2000).
30. J Florencio and MH Lee, Phys. Rev. B **35**, 1835 (1987).
31. Th. Niemeijer, Physica **36**, 377 (1967).
32. R Morgando, FA Oliveira, GG Bartrouni and A Hansen, Phys. Rev. Lett. **89**, 100601 (2002); MH Vainstein, IVL Costa, R Morgando and FA Oliveira, Europhys. Lett. **73**, 726 (2006).
33. C Lee and SI Kobayashi, Phys. Rev. Lett. **62**, 1061 (1989); C Lee, J. Phys. Soc. Jpn. **58**, 3910 (1989).
34. S Sen, Phys. Rev. B **44**, 7444 (1991); S Sen, SD Mahanti and ZX Cai, Phy. Rev. B **43** 10990 (1991).

FROM THERMODYNAMICS TO MAXENT

A. PLASTINO
IFLP-CONICET
Exact Sciences Faculty, National University, C. C. 727, 1900 La Plata, Argentina
E-mail: plastino@fisica.unlp.edu.ar

E. M. F. CURADO
Centro Brasileiro de Pesquisas Fisicas (CBPF)
Rua Xavier Sigaud 150 - Urca - Rio de Janeiro - Brasil

The maximum entropy principle (MaxEnt) is one of the most powerful approaches in the theorist arsenal. It has been used in many areas of science, and, in particular, in both statistical mechanics and the quantum many body problem. Here we show how to derive MaxEnt from the laws of thermodynamics.

Keywords: MaxEnt; thermodynamics.

1. Introduction

Jaynes' maximum entropy principle (MaxEnt) is one of the most powerful approaches in the theorist arsenal.[1] It has fruitfully applied in many areas of scientific endeavor, and, in particular, in both statistical mechanics and the quantum many body problem. A few references regarding this many body aspect are those of, for instance,[2–5]

The laws of thermodynamics (TL) constitute one of physics' most important set of statements, being strong pillars of our understanding of Nature. These laws (or several equivalent sets of statements) can be regarded as axioms, from which the whole of thermodynamics can be derived.[6] In statistical mechanics (SM) an underlying microscopic substratum is added to the picture that is able to explain these laws, and, a posteriori, the edifice of thermodynamics.[1,7–9] One of SM basic ingredients is a microscopic probability distribution (PD) that controls the population of microstates of the system under consideration.[7] We will be concerned here with changes that affect exclusively this microstate-population. The way these changes are related to changes in a system's *extensive quantities* provides the essential ingredient in explaining thermodynamics' laws from a microscopic viewpoint.[8] What happens if we consider the reverse path? We show here that constraining microscopic PD-changes so as to enforce TL-fulfillment is tantamount to using MaxEnt to derive that very PD, i.e., thermodynamics naturally leads to the maximum entropy principle. In other words

- HYPOTHESIS: for
- a given a concave entropic form (or information measure) S together with
- (i) a mean internal energy U, (ii) mean values $A_\nu \equiv \langle \mathcal{A}_\nu \rangle$; $(\nu = 1, \ldots, M)$ of M additional extensive quantities $\mathcal{A}_\nu$, and 3) a temperature T,
- THESIS: then, for any system described by
  (1) a microscopic probability distribution (PD) $\{p_i\}$, and
  (2) assuming a reversible process via $p_i \rightarrow p_i + dp_i$,
  (3) one can verify that, if:

$$dU = TdS - \sum_{\nu=1}^{M} \gamma_\nu \, dA_\nu, \tag{1}$$

  this predetermines a unique PD that maximizes S,

where the last item incorporates contents from thermodynamics' first and second laws.

2. Top Production and Decay

At the Tevatron the top quark is produced mainly in pairs through the process $qq, gg \rightarrow t\bar{t}$ with a cross section of about 6.7 pbfor $m_t = 175\,\mathrm{GeV/c^2}$ and $\sqrt{s} = 1.96\,\mathrm{TeV}$. Within the Standard Model framework the top decays almost 100% of the time via $t \rightarrow Wb$. Therefore it is customary to classify the $t\bar{t}$ final states based on the W decays modes: dileptons ($\ell = e, \mu$), lepton+jets ($\ell = e, \mu$), all-jets and inclusive τ (hadronically decaying) final-state events.

Even if its production cross section is much smaller compared to other Standard Model background processes, top events have very distinctive signatures that guide the overall analysis strategy: given the large top mass the decay products (leptons, jets) have large p_T's and the event topology is central and spherical, and there are always two b jets in the final state. Excellent lepton identification, energy resolution and b identification capabilities are essential for a successful Run II top physics program. The full kinematical and heavy flavor characterization of top enriched data sets in terms of all the known Standard Model processes is important not only for precision measurements but also to test for any new phenomena.

3. Proving the Thesis

We start with a *general* information measure of the form

$$S = k \sum_i p_i \, f(p_i), \tag{2}$$

where, for simplicity's sake, Boltzmann's constant is denoted just by k. The sum runs over a set of quantum numbers, collectively denoted by i (characterizing levels of energy ϵ_i), that specify an appropriate basis in Hilbert's space and $\mathcal{P} = \{p_i\}$

is an (as yet unknown) un-normalized probability distribution such that $\sum_i p_i =$ constant, to be eventually set equal to unity (often it is preferably, for practical purposes, to postpone normalization until the end of the pertinent computation process). Ordinarily one takes in (2)

$$f(p) = -\ln p, \tag{3}$$

but we do not need this extra-assumption for our present purposes.

Let f be an arbitrary smooth function of the p_i, in such a way $p_i f(p_i)$ is a concave function. Further, consider M quantities A_ν that represent mean values of extensive physical quantities $\mathcal{A}_\nu$. These take, for the state i, the value a_i^ν with probability p_i. We assume as well that g is another arbitrary smooth, monotonic function of the p_i such that $g(0) = 0$ and $g(1) = 1$. We do not need to require the condition $\sum_i g(p_i) = 1$. Of course, customarily one just has

$$g(p) \equiv p. \tag{4}$$

In many instances, however, the general forms employing the functions f and p have been used in recent literature (see, for instance,[10,11]). Our discussion will encompass these extraordinary circumstances as well. The mean energy U and the A_ν are given by

$$U = \sum_i \epsilon_i \, g(p_i); \;\; A_\nu = \sum_i a_i^\nu \, g(p_i). \tag{5}$$

Suppose now that the probability-set $\mathcal{P}$ suffers a change of the type

$$p_i \rightarrow p_i + dp_i, \text{ with } \sum_i dp_i = 0 \text{ (normalization!)}, \tag{6}$$

which in turn generates corresponding changes dS, dA_ν, and dU in, respectively, S, the A_ν, and U.

As stated in the Introduction, we want *to make sure that, in the above described circumstances, the following condition, related to the first two laws, is obeyed*

$$dU - TdS + \sum_{\nu=1}^{M} dA_\nu \lambda_\nu = 0, \tag{7}$$

with T the temperature. As a consequence of the constraint (7), a little algebra yields, up to first order in the dp_i, the requirements

$$\begin{aligned} \textstyle\sum_i [C_i^{(1)} + C_i^{(2)}] dp_i &\equiv \textstyle\sum K_i dp_i = 0; \\ C_i^{(1)} &= [\textstyle\sum_{\nu=1}^{M} \lambda_\nu \, a_i^\nu + \epsilon_i] \, g'(p_i); \\ C_i^{(2)} &= -kT[f(p_i) + p_i \, f'(p_i)], \end{aligned} \tag{8}$$

where the primes indicate derivative with respect to p_i. Eqs. (8) should hopefully yield one and just one expression for the p_i.

3.1. All the K_i in Eq. (8) are equal

We proceed to show now the above statement. As the dp_i are linked by the relation

$$\sum_i K_i \, dp_i = 0, \tag{9}$$

we can write, if we are dealing with N micro-states,

$$dp_N = -\sum_{i=1}^{N-1} dp_i. \tag{10}$$

Substituting (10) in (9) we obtain

$$\sum_{i=1}^{N-1} (K_i - K_N) \, dp_i = 0. \tag{11}$$

Now, since the $N-1$ "population-variations" dp_i are independent, this entails that each term in (11) should vanish by itself, which implies that

$$K_i = K_N; \text{ for all } i = 1, \ldots, N-1. \tag{12}$$

We will see below that K_N gets fixed by the normalization condition as $K_N = K$, say. Thus, $K_i =$ constant (in the sense of being independent of i) $= K$ for all i.

3.2. The role of K

Interestingly enough, we do not need to give a specific value to K for our present purposes, although, as just stated, it will become clear below that it is related to the probabilities-normalization constant. We only need to ascertain the $K-$role, in the following sense. We have, on account of (12),

$$\begin{aligned} K &= C_i^{(1)} + C_i^{(2)}; \;\; (for \;\; any \;\; i), \\ C_i^{(1)} &= [\textstyle\sum_{\nu=1}^{M} \lambda_\nu \, a_i^\nu + \epsilon_i] \, g'(p_i), \\ C_i^{(2)} &= -kT[f(p_i) + p_i \, f'(p_i)], \end{aligned} \tag{13}$$

so that, if we redefine things in the fashion

$$\begin{aligned} T_i^{(1)} &= f(p_i) + p_i \, f'(p_i); \;\; -(T_i^{(2)}/\beta) = \\ &= [(\textstyle\sum_{\nu=1}^{M} \lambda_\nu \, a_i^\nu + \epsilon_i) \, g'(p_i) - K], \;\; (\beta \equiv 1/kT), \end{aligned} \tag{14}$$

we can recast (13) as

$$T_i^{(1)} + T_i^{(2)} = 0; \;\; (for \;\; any \;\; i), \tag{15}$$

a relation whose importance will be presently become manifest.

3.3. *Revisiting the MaxEnt path*

Suppose now that you are a MaxEnt practitioner and wish to determine a microscopic PD by extremizing S subject to the constraints of fixed values for

- U and
- the M values A_ν.

This is, of course, achieved via Lagrange multipliers to be called, respectively, (1) β and (2) M γ_ν. As usual, we need also a normalization Lagrange multiplier ξ.

$$\delta_{\{\ p_i\}}\left[S - \beta U - \sum_{\nu=1}^{M} \gamma_\nu A_\nu - \xi \sum_i p_i\right] = 0, \tag{16}$$

immediately leading, with $\gamma_\nu = \beta\lambda_\nu$, to

$$\begin{gathered} 0 = \delta_{p_m} \textstyle\sum_i ()_i\,, \\ ()_i = -[\sum_i \beta g(p_i)(\sum_{\nu=1}^{M} \lambda_\nu\, a_i^\nu + \epsilon_i) + \xi p_i], \end{gathered} \tag{17}$$

so that

$$\begin{gathered} 0 = f(p_i) + p_i f'(p_i) \\ -[\beta g'(p_i)(\textstyle\sum_{\nu=1}^{M} \lambda_\nu\, a_i^\nu + \epsilon_i) + \xi] \Rightarrow \text{ if } \xi \equiv \beta K, \Rightarrow \end{gathered}$$

$$0 = f(p_i) + p_i f'(p_i) - \beta[g'(p_i)(\textstyle\sum_{\nu=1}^{M} \lambda_\nu a_i^\nu + \epsilon_i) + K] \Rightarrow$$

$$0 = T_i^{(1)} + T_i^{(2)}. \tag{18}$$

Clearly, (15) on the one hand, and the last equality of (18), obtained by the MaxEnt practitioner so as to get the desired microscopic PD are one and the same equation! Thus, using thermodynamics' first two laws as constraints in order to obtain the PD of statistical mechanics leads to that PD that the MaxEnt practitioner would have obtained following the original route paved by Jaynes'.[1]

4. Alternative Axiomatics

We discuss now an alternative axiomatics for thermodynamics that helps clarifying the above procedure. The following set of postulates is entirely equivalent to the celebrated three laws (in the sense that either of the two sets can be derived from the other). The alternative axioms read:[6]

(1) For every system there exists a quantity U (the internal energy), with the usual energy-properties.
(2) The equilibrium state is uniquely determined by the value of U together with the values of a set of M relevant extensive variables A_ν.
(3) For every system there is a state function S that is a function of U and of the M quantities A_ν. A corollary of 1-3 is Eq. (1), where the γ_ν. are called generalized pressures.[6]

(4) If internal constraints are removed from an isolated composite system $\mathcal{S}$, consisting of n subsystems $1, 2, \ldots, n$, and $\mathcal{S}$ is allowed to come to an equilibrium state, then the values assumed by the extensive variables of the subsystems, consistent with the remaining internal and external constraints, will be those that*maximize* S.

It is now clear that, in the previous Section, we have deduced, *in microscopic fashion*, the MaxEnt probability distribution from the first three postulates above. An original result of the present discussion can then be seen in the following light. Assuming:

- the existence of a microscopic probability distribution á la Boltzmann,[12] together with
- Eqs. (2)-(5),

we have microscopically deduced axiom 4 from the preceding ones.

5. Discussion

We have here endeavored to show that appropriate manipulation of the first three axioms of thermodynamics, as enumerated in the preceding Section, can be used to generate the equilibrium microscopic probability distribution (PD) that describes a system within the framework of a generalized ensemble,[7] and that such an approach is an alternative to the MaxEnt-one.

We were exclusively concerned with changes that affect exclusively microstate-population and, more specifically, with the way these modifications are related to internal variation of the system's extensive quantities.[8] We started with (1) a given a concave entropic form (or an information measure (IM)) S, (2) a mean internal energy U, and M mean values $A_\nu \equiv \langle \mathcal{A}_\nu \rangle$; $(\nu = 1, \ldots, M)$ of M extensive quantities $\mathcal{A}_\nu$, (3) a temperature T, and demonstrated that, for any system described by a microscopic probability distribution (PD) $\{p_i\}$, assuming a reversible process via $p_i \to p_i + dp_i$ that is forced to verify the relation $dU = TdS - \sum_{\nu=1}^{M} \gamma_\nu \, dA_\nu$, we got an equation that yields a unique PD that maximizes S. By way of contrast, MaxEnt starts from S and, extremizing it with appropriate constraints, allow one to find the system's PD.

In other words,

$$
\begin{gathered}
dU = TdS - \textstyle\sum_{\nu=1}^{M} \gamma_\nu \, dA_\nu \\
\Updownarrow \\
\text{MaxEnt prob. distr. } \{\mathrm{p_i}\}.
\end{gathered}
\tag{19}
$$

Given the unique status of thermodynamics, as a model free theory in the physics edifice,[13] our results should be regarded as further validating, from an epistemological viewpoint, the MaxEnt approach.

References

1. E. T. Jaynes, Phys. Rev. Lett. **106**, 620 (1957); *Papers on probability, statistics and statistical physics*, edited by R. D. Rosenkrantz (Reidel, Dordrecht, Boston, 1987); A. Katz, *Principles of Statistical Mechanics, The Information Theory Approach* (Freeman and Co., San Francisco, 1967).
2. N. Canosa *et al.*, Nucl. Phys. A **453**, 417 (1986).
3. N. Canosa *et al.*, Nucl. Phys. A **512**, 492 (1990).
4. R. Rossignoli and A. Plastino, Phys. Rev. A **42**, 2065 (1990).
5. L. Arrachea *et al.*, Phys. Lett. A **176**, 353 (1993).
6. E. A. Desloge, *Thermal physics* (Holt, Rhinehart and Winston, NY, 1968).
7. R.K. Pathria, *Statistical Mechanics* (Pergamon Press, Exeter, 1993); H. B. Callen, *Thermodynamics* (J. Wiley, NY, 1960).
8. F. Reif, *Statistical and thermal physics* (McGraw-Hill, NY, 1965).
9. J. J.Sakurai, *Modern quantum mechanics* (Benjamin, Menlo Park, Ca., 1985).
10. M. Gell-Mann and C. Tsallis, Eds. *Nonextensive Entropy: Interdisciplinary applications*, Oxford University Press, Oxford (2004).
11. J.P. Boon and C. Tsallis, Eds. Europhysics News (Special Issue) **36**, November/December 2005.
12. D. Lindley, *Boltzmann's atom* (The free press, NY, 2001).
13. M. S. Longhair, *Theoretical concepts in physics* (Cambridge University Press, Cambridge, UK, 1984).

RECENT PROGRESS IN THE DENSITY-MATRIX RENORMALIZATION GROUP

ULRICH SCHOLLWÖCK*

Institute for Theoretical Physics C, RWTH Aachen University,
52056 Aachen, Germany
**E-mail: scholl@physik.rwth-aachen.de*
www.theorie.physik.uni-muenchen.de/ scholl

Over the last decade, the density-matrix renormalization group (DMRG) has emerged as the most powerful method for the simulation of strongly correlated one-dimensional (1D) quantum systems. Input from quantum information has allowed to trace the method's performance to the entanglement properties of quantum states, revealing why it works so well in 1D and not so well in 2D; it has allowed to devise algorithms for time-dependent quantum systems and, by clarifying the link between DMRG and Wilson's numerical renormalization group (NRG), for quantum impurity systems.

Keywords: Density-matrix renormalization; numerical renormalization group.

1. Introduction

The physics of strongly correlated quantum systems continues to pose major challenges in physics. In this context, the *density-matrix renormalization group (DMRG)*[1,2] has emerged as the most powerful method for static and linear-response properties of 1D strongly correlated systems. Recent progress has allowed to get a profound understanding of why the method works so well, when it fails, and most importantly, to devise new modifications of the method that allow to address the physics of time-dependent[3–6] and quantum impurity systems.[7]

The former modification is particularly relevant due to the progress in preparing dilute ultracold atomic gases. In an optical lattice, these gases are the purest realization of the typical model Hamiltonians of strong correlation physics, such as the Hubbard model, while interactions can be tuned in a controlled fashion.[8,9] The latter leads us directly to Kondo physics and impurity solvers for dynamical mean-field theory (DMFT) that are currently of high practical interest.

In this paper, I want to provide a point of view of DMRG which is an alternative to the one usually presented in the literature, but gives faster access to the method.

2. Density-Matrix Renormalization Group

2.1. *Decimation of state spaces*

Key to the simulation of quantum systems is the compression of information that is exponentially diverging with system size: considering N spins-1/2 on a lattice, Hilbert space dimension diverges as 2^N. An important class of methods attempts a systematic choice of a Hilbert subspace which is hoped to contain the physically most relevant states. This implies a physically driven process of discarding states (decimation). All variational and renormalization group techniques are in this group; can we identify a best decimation strategy?

Let me focus on one particular, rather general decimation setup, "one-dimensional" decimation. Let us assume that the degrees of freedom can be arranged on a 1D axis, e.g. an energy axis, on which we place all levels, either empty or occupied, or a real-space axis, on which we place lattice sites.

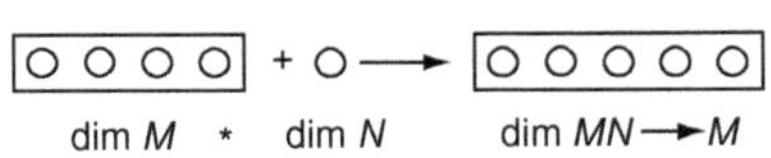

Fig. 1. Growing a system: a local state space of dimension N is added to a block state space of dimension M. Open circles correspond to a site or energy level.

Imagine we grow the system iteratively, adding site by site (whatever the physical interpretation of such a site will be) - Fig. 1. The original system, which I refer to as a block, is assumed to be effectively described within a space $\{|\alpha\rangle\}$ of dimension M, the new site within a space $\{|\sigma\rangle\}$ of dimension N. The state space $\{|\beta\rangle\}$ of the new block will have dimension MN, and will be decimated down to dimension M. The states of the new block will be a linear combination of the old states,

$$|\beta\rangle = \sum_\alpha \sum_\sigma \langle\alpha\sigma|\beta\rangle\,|\alpha\rangle\,|\sigma\rangle \equiv \sum_\alpha \sum_\sigma A_{\alpha\beta}[\sigma]\,|\alpha\rangle\,|\sigma\rangle\,. \tag{1}$$

As a notation, N matrices A of dimension $M \times M$ have been introduced, one for each $|\sigma\rangle$, with $A_{\alpha\beta}[\sigma] = \langle\alpha\sigma|\beta\rangle$. Orthonormality imposes

$$\sum_\sigma A^\dagger[\sigma]A[\sigma] = 1. \tag{2}$$

The states of the block of length, say, $\ell-1$ serving as "input" to produce as "output" the states of the block ℓ via the matrices $A^\ell[\sigma_\ell]$, are themselves the "output" of the construction of the block $\ell-1$ from block $\ell-2$. By recursion, $|\beta\rangle$ for block ℓ is

$$|\beta\rangle = \sum_{\sigma_1\ldots\sigma_\ell} [A^1[\sigma_1]A^2[\sigma_2]\ldots A^\ell[\sigma_\ell]]_\beta\,|\sigma_1\ldots\sigma_\ell\rangle\,. \tag{3}$$

For very short blocks, there are less than M states; the dimensions of the first A-matrices will be $1\times N$, $N\times N^2,\ldots$ until M is exceeded. This turns the matrix product into a vector.

If we consider growing a chain from its right end (site L), similar expressions emerge. We may now choose a "left" and a "right" block to join them to the generic description of a quantum state of the system of length L, matching them directly,

$$|\psi\rangle = \sum_{\sigma_1 \ldots \sigma_L} A^1[\sigma_1] \ldots A^\ell[\sigma_\ell] \times \Psi \times \tilde{A}^{\ell+1}[\sigma_{\ell+1}] \ldots \tilde{A}^L[\sigma_L] \, |\sigma_1 \ldots \sigma_L\rangle \,, \qquad (4)$$

where Ψ is an $M \times M$ matrix. The matrices of the "right" block are denoted by a tilde, as they obey a slightly modified normalization condition (this is conceptually irrelevant and we will ignore this in the following). Alternatively, we can insert one or two sites explicitly between the blocks, leading to ansatzes

$$|\psi\rangle = \sum_{\sigma_1 \ldots \sigma_L} A^1[\sigma_1] \ldots A^\ell[\sigma_\ell] \times \Psi[\sigma_{\ell+1}] \times \tilde{A}^{\ell+2}[\sigma_{\ell+2}] \ldots \tilde{A}^L[\sigma_L] \, |\sigma_1 \ldots \sigma_L\rangle \,, \qquad (5)$$

where there is a Ψ-matrix for each $|\sigma_{\ell+1}\rangle$, and

$$|\psi\rangle = \sum_{\sigma_1 \ldots \sigma_L} A^1[\sigma_1] \ldots A^\ell[\sigma_\ell] \times \Psi[\sigma_{\ell+1}\sigma_{\ell+2}] \times \tilde{A}^{\ell+3}[\sigma_{\ell+3}] \ldots \tilde{A}^L[\sigma_L] \, |\sigma_1 \ldots \sigma_L\rangle \,, \qquad (6)$$

respectively. As it turns out, Eqns. (5) and (6) are the useful ones, leading to what is called single-site and (more conventional) two-site DMRG.

Considering Eq. (4), obviously the junction can be shifted freely, even fictiously to the right end of the chain, with all $\tilde{A}$-matrices disappearing. The generic description of a quantum state obtained by this decimation is therefore given by

$$|\psi\rangle = \sum_{\sigma_1 \ldots \sigma_L} A^1[\sigma_1] A^2[\sigma_2] \ldots A^L[\sigma_L] \, |\sigma_1 \ldots \sigma_L\rangle \,. \qquad (7)$$

In fact, there is no need to worry about this product of matrices yielding a scalar factor because of the dimension of the matrices at the right end: manipulations introduced further down (namely the Schmidt decomposition) can be used to show that without any loss the A-matrices shrink to dimensions mirroring those at the left end, e.g. $N^2 \times N$ and $N \times 1$ for the last two. The generic decimation state encoded as in Eq. (7) is referred to as a *matrix product state*; such states have been studied for a long time.[10–12] Without a limit on M, any finite quantum system can be described exactly, showing the existence of an exact limit to the ansatz. The optimal decimation for finding the *ground state* of a Hamiltonian is to find those A that minimize $\langle\psi| \hat{H} |\psi\rangle$ assuming normalization. This leads to a highly non-linear expression for the energy in A, which is numerically close on useless.

Can we find a method that turns this into a linear (hence solvable) problem in A? In fact this is done by the density-matrix renormalization group.[1,2] Its link to matrix product states has been pointed out by various authors.[13–16]

The problem turns linear by providing some starting set of A-matrices in a warm-up procedure, preferably close to the true solution, and then to proceed iteratively: keeping all elements (i.e. matrices) in $|\psi\rangle$ fixed, with one exception, $\langle\psi| \hat{H} |\psi\rangle$ turns quadratic in the free matrix, and extremizing this expression with respect to the free matrix is a linear procedure. Varying one element after another repeatedly, one may hope to get a very good, even optimal, approximation of the ground state within the state class expressable as above. In the case of Eqns. (5) and (6) we can formulate algorithms where we keep all A-matrices fixed, extremize with respect to the Ψ-matrices, and shift the position of the Ψ matrices forth and back through the entire chain. The remaining problem then is, how do we deduce A-matrices from the newly found Ψ matrices to be able to shift? Can we maintain all the information provided by the Ψ matrices? The answer is the finite-system DMRG algorithm.

2.2. *Finite-system DMRG procedures*

Two-site DMRG. Assume we have our quantum state in the representation of Eq. (6). In that case, the left block of length ℓ formed from A-matrices would be described by a M-dimensional Hilbert space with states $\{|m_\ell^S\rangle\}$ defined as in Eq. (3); similarly the right block of length $L-\ell-2$ by states $\{|m_{L-\ell-2}^E\rangle\}$:

$$|\psi\rangle = \sum_{m^S=1}^{M}\sum_{\sigma^S=1}^{N}\sum_{\sigma^E=1}^{N}\sum_{m^E=1}^{M} \Psi_{m^S m^E}[\sigma^S\sigma^E]\,|m^S\sigma^S\rangle\,|m^E\sigma^E\rangle \equiv \sum_i^{N^S}\sum_j^{N^E}\Psi_{ij}\,|i\rangle\,|j\rangle\,. \tag{8}$$

$\langle\psi|\psi\rangle = 1$, $\{|m^S\sigma^S\rangle\} \equiv \{|i\rangle\}$ and $\{|m^E\sigma^E\rangle\} \equiv \{|j\rangle\}$. The state spaces $\{|i\rangle\}$ and $\{|j\rangle\}$ have dimensions $N^S = MN$ and $N^E = MN$. S and E label system and environment. Using a matrix solver one minimizes $\langle\psi|\hat{H}|\psi\rangle$ with respect to Ψ_{ij}.

In order to make progress, we now have to shift the position of the two active sites, to improve our wave function everywhere. If we shift it by one site to the right, the shrinking right block may easily be constructed from $\tilde{A}$-matrices; for the growing left block, we have to provide A-matrices for the left of the two sites (states $|\sigma^S\rangle$). If one has a suitable truncation of the basis $\{|i\rangle\}$ down to M states, their expansion in $|m^S\rangle$ and $|\sigma^S\rangle$ will just define the desired A-matrices.

To find that truncation, we use a *Schmidt decomposition:* Consider $|\psi\rangle = \sum_{ij}\psi_{ij}\,|i\rangle\,|j\rangle$ as above. Assuming without loss of generality $N^S \geq N^E$, we form the $(N^S \times N^E)$ matrix A with $A_{ij} = \psi_{ij}$. Singular value decomposition guarantees $A = UDV^T$, where U is $(N^S \times N^E)$-dimensional with orthonormal columns, D is a $(N^E \times N^E)$ diagonal matrix with non-negative entries $D_{\alpha\alpha} \equiv \sqrt{w_\alpha}$, and V^T is a $(N^E \times N^E)$ unitary matrix. Then

$$|\psi\rangle = \sum_{i=1}^{N^S}\sum_{\alpha=1}^{N^E}\sum_{j=1}^{N^E} U_{i\alpha}\sqrt{w_\alpha}V_{\alpha j}^T\,|i\rangle\,|j\rangle = \sum_{\alpha=1}^{N^E}\sqrt{w_\alpha}\left(\sum_{i=1}^{N^S}U_{i\alpha}\,|i\rangle\right)\left(\sum_{j=1}^{N^E}V_{j\alpha}\,|j\rangle\right). \tag{9}$$

The properties of U and V^T ensure that $|w_\alpha^S\rangle = \sum_i U_{i\alpha}\,|i\rangle$ and $|w_\alpha^E\rangle = \sum_j V_{j\alpha}\,|j\rangle$ form orthonormal bases of S and E, in which the Schmidt decomposition

$$|\psi\rangle = \sum_{\alpha=1}^{N_{\text{Schmidt}}}\sqrt{w_\alpha}\,|w_\alpha^S\rangle\,|w_\alpha^E\rangle \tag{10}$$

holds. N^SN^E coefficients ψ_{ij} are reduced to $N_{\text{Schmidt}} \leq N^E$ non-zero coefficients $\sqrt{w_\alpha}$, $w_1 \geq w_2 \geq w_3 \geq \ldots$. Relaxing the assumption $N^S \geq N^E$, one has $N_{\text{Schmidt}} \leq$

$\min(N^S, N^E)$. A partial trace gives the reduced density matrices for S and E,

$$\hat{\rho}_S = \sum_{\alpha}^{N_{\text{Schmidt}}} w_\alpha \left|w_\alpha^S\right\rangle \left\langle w_\alpha^S\right|; \quad \hat{\rho}_E = \sum_{\alpha}^{N_{\text{Schmidt}}} w_\alpha \left|w_\alpha^E\right\rangle \left\langle w_\alpha^E\right|. \tag{11}$$

If we demand for a truncation procedure that the approximative wave function $\left|\tilde{\psi}\right\rangle$, where the system space has been truncated to be spanned by only M orthonormal states, minimizes the distance to $|\psi\rangle$ in the quadratic norm, one finds using Eq. (10) that this distance is minimized if one *retains the* M *eigenstates of* $\hat{\rho}_S$ *with the largest eigenvalues* w_α. This is the key truncation prescription. In fact, this prescription maximizes the retained *bipartite entanglement* between system and environment $S = -\sum_\alpha w_\alpha \log_2 w_\alpha$, and the method preserves entanglement or the non-classical nature of the state as well as it can, in contrast to many approximations focused on product states. For local quantities, such as energy, magnetization or density, one can show errors to be of the order of the *truncated weight* $\epsilon_\rho = 1 - \sum_{\alpha=1}^{M} w_\alpha$, which emerges as the key error estimate.

Single site DMRG. In applications one observes that for translationally invariant systems with periodic boundary conditions the position dependency of the matrix-product state (or of observables) does not disappear completely. Also, the method cannot be strictly variational: the ansatz (6) generates (after the Schmidt decomposition and before truncation) ansatz matrices of dimension up to MN at the two local sites, whereas they are of dimension M at all other sites; as the position of these two sites is shifting, this anomaly changes place such that one is not really optimizing within one fixed variational ansatz. Several authors[14,15] have pointed out and numerically demonstrated that DMRG results can be improved by switching, after convergence is reached in the two-site scheme, to a single-site scheme as in ansatz (5) and to carry out some more sweeps. This is now truly variational:[14–16]

In the new scheme the ansatz matrix Ψ can be partially traced to find M states for defining the A-matrix *without* truncation or loss of information or change in energy, because $N_{\text{Schmidt}} \leq M = \min(N^S, N^E)$: $N^E = M$ for the single-site setup. Shifting the "active" site then does not change the energy, and the next minimization can only decrease the energy (or keep it constant). This setup is therefore truly variational in the space of the states generated by the matrices A, $\tilde{A}$ and reaches a minimum of energy within that space. It is important to note that in this setup there is therefore no truncation error as diagnostic tool.

There is of course no guarantee to reach the global minimum: in fact, it turns out that this can usually be achieved by suitable modifications to the state selection scheme, taking the state out of false minima.[17]

2.3. *When and why does DMRG work?*

Obviously, the ordered eigenvalue spectrum w_α of the reduced density-matrix $\hat{\rho}$ should decay as quickly as possible to minimize the truncated weight $\epsilon_\rho = 1 - \sum_{\alpha=1}^{M} w_\alpha$ for optimal DMRG performance. Empirically, in 1D, density-matrix spectra of gapped quantum systems exhibit roughly system-size independent exponential decay of w_α; at criticality, this decay slows down with increasing system size, leading to DMRG failure for $L \to \infty$. In 2D, the number of eigenvalues to be retained to keep a fixed truncation error grows exponentially with system size, restricting DMRG to very small system sizes (typically, the 2D lattice is mapped to a 1D snake with long-ranged interactions).

These empirical observations can be understood from examining the growth of bipartite entanglement between system and environment for various dimensions. Consider the entanglement S_L for 1D systems of length L embedded in a thermodynamic limit universe. One finds[18] that $S_L \to \infty$ for $L \to \infty$ at criticality, but saturates as $S_L \to S_L^*$ for $L \approx \xi$ in the non-critical regime. At criticality the entanglement can be linked[18,19] to the geometric entropy[20] of associated conformal field theories, $S_L^{geo} = (c + \bar{c})/6 \log_2 L$, where c ($\bar{c}$) are the central charges.

Geometric entropy arguments for $(d+1)$-dimensional field theories use a bipartition of d-dimensional space by a $(d-1)$-dimensional hypersurface, which is shared by system S and environment E. By the Schmidt decomposition, S and E share the same reduced density-matrix spectrum, hence entanglement entropy, which is now argued to reside essentially on the shared hypersurface.[21] Taking the thermodynamic (infrared) limit, entropy scales as the hypersurface area, $S_L \propto L^{d-1}$; critical systems will have further logarithmic corrections.[22]

S_L is the number of qubits corresponding to the entanglement information. For a Hilbert space of size M, $S \leq \log_2 M$; alternatively, we need $M \geq 2^{S_L}$ for simulations. We may therefore expect, in perfect agreement with empirical results, that in 1D quantum systems away from criticality, DMRG yields very precise results for the TD limit for some finite number of states kept, $M \sim 2^{S_L^*}$. At criticality, the number of states that has to be kept, will diverge as $M(L) \sim L^k$, with $k = (c + \bar{c})/6$. This explains the failure of DMRG for critical 1D systems as $L \to \infty$. As k is small, DMRG still works for rather large finite systems.

In higher-dimensional quantum systems, however, the number of states to be kept will diverge as $M(L) \sim 2^{L^{d-1}}$, rendering the understanding of thermodynamic limit behavior by conventional DMRG quite impossible. Recent proposals[23] also show that it is possible to formulate generalized DMRG ansatz states in such a way that entropy shows correct size dependency in two-dimensional systems, but their feasability in relevant problems is still unclear.

3. Adaptive Time-Dependent DMRG

The difficulty of quantum dynamics shows if we consider the evolution of $|\psi(0)\rangle$ under a time-independent Hamiltonian $\hat{H}|\psi_n\rangle = E_n |\psi_n\rangle$. We usually have no good knowledge of the eigenstates but use some orthonormal basis with unknown eigenbasis expansion, $|k\rangle = \sum_n a_{kn} |\psi_n\rangle$. The evolution of $|\psi(0)\rangle = \sum_k d_k(0)|k\rangle$ is

$$|\psi(t)\rangle = \sum_k \left(\sum_n d_k(0) a_{kn} e^{-\mathrm{i}E_n t} \right) |k\rangle \equiv \sum_k d_k(t) |k\rangle , \qquad (12)$$

where the modulus of the expansion coefficients $d_k(t)$ is *time-dependent.* For a general basis, Hilbert space truncation at one fixed time (i.e. $t = 0$) as discussed above will therefore not ensure a reliable approximation of the time evolution.

Decisive progress came from quantum information theory, when Vidal proposed an algorithm for simulating quantum time evolutions of one-dimensional systems efficiently on a classical computer.[3,24] His algorithm, known as TEBD [time-evolving block decimation] algorithm, is based on matrix product states;[11,12] because of the strong link between these and DMRG, his ideas could be implemented easily into DMRG, leading to an *adaptive* time-dependent DMRG,[4,5] where the DMRG state space adapts itself in time to the time-evolving quantum state.

If we consider a nearest-neighbor Hamiltonian, such as the conventional Hubbard Hamiltonian, we can split the infinitesimal global time evolution operator into a product of infinitesimal local time evolution operators:[25]

$$e^{-i\hat{H}\Delta t} = e^{-ih_1\Delta t}e^{-ih_2\Delta t}e^{-ih_3\Delta t}\dots e^{-ih_{L-1}\Delta t} + O(\Delta t^2). \tag{13}$$

The h_i are the local Hamiltonians on bonds i; in general neighboring bond Hamiltonians will not commute, giving rise to a decomposition error. The idea is now simply to use finite-system DMRG in the two-site setup: at each step, one carries out the local infinitesimal time evolution exactly on the two adjacent local sites. This will lead to a new state, for which a new Schmidt decomposition is carried out, in which the chain is cut between the two local sites, as before, leading to a new truncation and new reduced basis transformations (2 matrices A adjacent to this bond), which are the choice optimally representing the new state. By doing this for all bonds, one infinitesimal time step is completed.

The performance of this method has been tested in various applications in the context of ultracold atom physics,[4,26–30] but also for far-from-equilibrium dynamics,[31] transport[32] and for spectral functions.[5] It should be mentioned that this method, while very fast, is limited to systems with nearest neighbor interactions on a single chain unless clumsy big unit cells are introduced. A better approach in such cases is to use the so-called time-step targetted method[6] which, at quite some algorithmic cost in time, does not have these limitations.

Let us now move to an application. Following the seminal work of Haldane,[33] it has been understood that the low-energy behaviour of 1D quantum liquids is universally described by the Luttinger liquid picture.[34] A remarkable prediction is spin-charge separation for Fermions: at low energy the excitations of charge and spin completely decouple and propagate with different velocities. Spin-charge separation is an ideal testbed for real-time DMRG: numerical results of the real-time dynamics of a 1D Hubbard model of up to 128 sites can be obtained easily.[27]

We start from the standard Hubbard model

$$H = -J\sum_{j,\sigma}\left(c^\dagger_{j+1,\sigma}c_{j,\sigma} + h.c.\right) + U\sum_j n_{j,\uparrow}n_{j,\downarrow} \tag{14}$$

for Fermions in 1D, where we call the hopping matrix element J. Setting $J = 1$ and $\hbar = 1$ time is measured in units of $\hbar/J = 1$.

In order to study the behaviour of a single particle excitation, the time evolution of the system with one additional particle with spin up added at time $t = 0$ on site j to the ground state, was calculated numerically. In Fig. 2 a snapshot of the resulting evolution of the densities is shown for time $t = 7.4$. Remarkably, even after a such a short time separate wave packets in spin and charge can be seen.

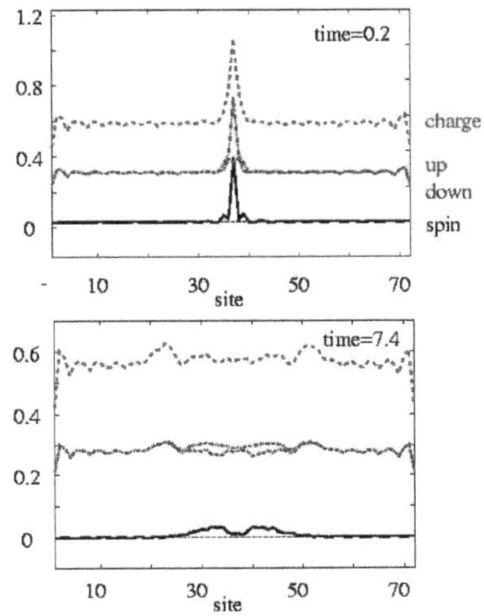

Fig. 2. A snapshot of the time-evolution of the charge and spin densities of a single particle excitation created at time $t = 0$ at site $j = 37$ is shown at $t = 0.2$ and $t = 7.4$. Several hundred DMRG states were kept. DMRG error analysis reveals controlled errors of the order of $O(10^{-3})$. Adapted from.[27]

The previous result indicates (together with the many other applications) that adaptive time-dependent DMRG is a highly performing method, able to answer complicated questions of dynamics in strongly correlated systems. Also note that for *unequal-time* n-point correlators such as the (2-point) real-time Green's function $G_{ij}(t) = \langle\psi| c_i^\dagger(t)c_j(0)|\psi\rangle = \langle\psi| e^{+i\hat{H}t}c_i^\dagger e^{-i\hat{H}t}c_j|\psi\rangle$ one can introduce $|\phi\rangle = c_j|\psi\rangle$ such that the desired correlator is then simply given as an equal-time matrix element between two time-evolved states, $G_{ij}(t) = \langle\psi(t)| c_i^\dagger |\phi(t)\rangle$. As we can calculate both $|\psi(t)\rangle$ and $|\phi(t)\rangle$, a very appealing feature of time-dependent DMRG is that $G_{ij}(t)$ can be evaluated in *a single calculation* for all i and t as time proceeds. Frequency-momentum space is then reached by a double Fourier transformation. This approach might emerge as a very attractive alternative to the current very time-consuming calculations of $G(k,\omega)$ using the dynamical DMRG.[35,36]

4. DMRG Meets NRG

A decisive breakthrough in renormalization techniques was achieved by Wilson's numerical renormalization group (NRG),[37] now the method of choice to address quantum impurity problems such as contained in the Anderson Hamiltonian, where a localized impurity level d with an on-site interaction U hybridizes with a non-interacting conduction band of width D:

$$\hat{H} = U n_{d\uparrow} n_{d\downarrow} + \epsilon_d n_d + \Gamma \sum_\sigma \int dk\, (d^\dagger_\sigma c_{k\sigma} + h.c.) + \sum_\sigma \int dk\, \epsilon(k) c^\dagger_{k\sigma} c_{k\sigma}. \tag{15}$$

The essential and difficult feature of the spectral function of the impurity site for this Hamiltonian is the appearance of a very sharp resonance at the Fermi edge which controls the essential physics here and reflects a quasi-bound many-body state between one electron on the impurity and the conduction band. To simplify this problem, Wilson introduced a logarithmic discretization of the conduction band into intervals $[-1, -\Lambda^{-1}], \ldots, [-\Lambda^{-n}, -\Lambda^{-(n+1)}], \ldots, [\Lambda^{-(n+1)}, \Lambda^{-n}], \ldots, [\Lambda^{-1}, 1]$, assuming band width $D = 2$, where $\Lambda > 1$. Each interval is then approximated by a single electronic state, corresponding to an accumulation point of states at the Fermi edge. Together with a set of other physically motivated approximations that focus on the low-energy physics in the vicinity of the impurity, one arrives as effective model at a semi-infinite electronic chain which formally resembles an electronic Hubbard chain:

$$\hat{H} = U n_{0\uparrow} n_{0\downarrow} + \sum_{n=0}^{\infty} \sum_\sigma t_n (c^\dagger_{n\sigma} c_{n+1,\sigma} + h.c.) \tag{16}$$

However, in the semi-infinite chain only site 0, which is the original impurity site, is interacting and nearest-neighbour inter-site hoppings decay exponentially along the chain: $t_n \sim \Lambda^{-n/2}$. Physically, the first sites correspond to the high-energy physics of the impurity problem (leading to so-called Hubbard sidebands), whereas the later sites encode the low-energy physics of the Kondo-Suhl resonance. NRG sets out by exactly diagonalizing a high-energy fragment of the chain of, say, M states, and then adding site after site, of 4 local states each. As the Hilbert space grows exponentially, a decimation prescription has to be found: in NRG it is to diagonalize in the $4M$-dimensional state space and to keep the M eigenstates of lowest energy. Physically, this focuses on the low-energy physics of the problem. It is now crucial to note that this decimation procedure can be encoded within the matrix product state framework introduced previously, yielding exactly the same ansatz structure as used in DMRG.[7] However, there are two major differences to DMRG: (i) the decimation prescription is different; (ii) as sites are added in a sequence of decreasing energy, information about high-energy states influences the decimation at the low energies, but there is no feedback from the low to the high energy physics. This implies that the low-energy-oriented decimation prescription of NRG can only work if one has sites of differing energy scales and if they are treated sequentially.

As we have shown, within the class of matrix product ansatzes, to which both methods belong, the DMRG approach is optimal (in the sense of yielding the lowest-energy approximation to the ground state). It should therefore be possible to improve on NRG on the level of a more efficient analysis of the semi-infinite chain.[7] If one naively applies DMRG to the semi-infinite chain, improvements in the relevant

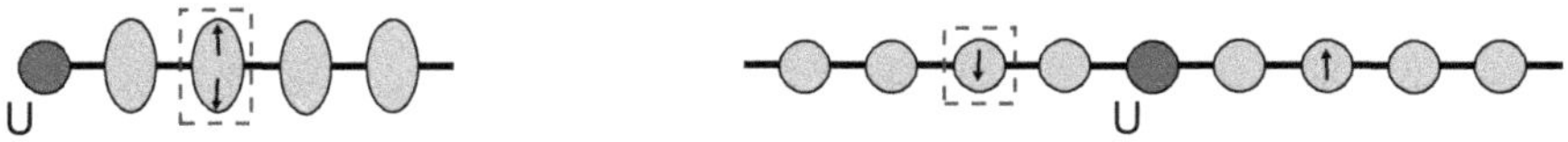

Fig. 3. Original semi-infinite NRG chain of electronic sites (left) and unfolded infinite DMRG chain (right) of two-state sites with spin-↓ on the left and spin-↑ on the right.

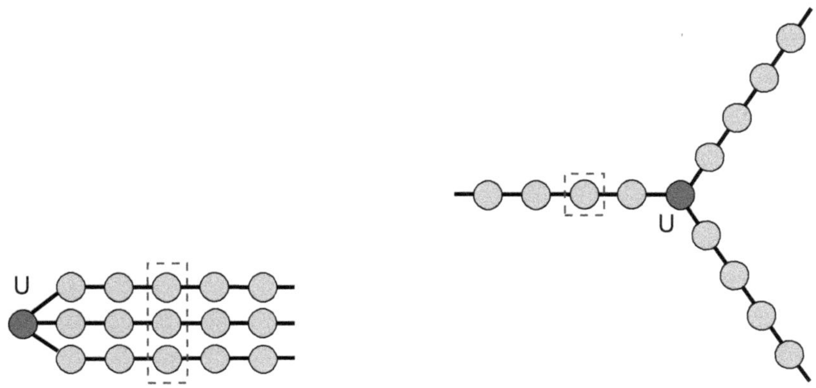

Fig. 4. A single impurity coupling to multiple bands (leads) in the Anderson model as seen in NRG (left) with large effective sites and the unfolded DMRG setup (right) exploiting the non-interacting nature of the bands.

low-energy regions are minor as NRG is (almost) optimal there due to its construction. The big improvements come from relaxing the conceptual assumptions NRG has to make.

As the spin-↑ and spin-↓ electrons on the chain interact only via site 0, one may expect that to a first approximation the ground state will be a product state of a spin-↑ and a spin-↓ part: $|\psi\rangle \approx |\psi_{\uparrow}\rangle |\psi_{\downarrow}\rangle$. Assuming this factorization to be exact, the M states kept would factorize themselves, such that $\sqrt{M}$ states each describe the spin-↑ and spin-↓ part respectively, drastically reducing the number of states effectively available. As same energy scales have to be added to the chain at the same time, NRG cannot avoid this problem. In DMRG, one can unfold the semi-infinite chain into an infinite one, with the interacting site at the center and sites of two possible states to the left and right (Fig. 3): on the left, they are either empty or contain a spin-↓ electron, on the right, they are either empty or contain a spin-↑ electron. This unfolding reflects the state factorization, and for the same precision as in NRG, it is now sufficient to consider $\sqrt{M}$ states only. This is a drastic improvement as computation time scales as the third power in the number of states in both methods. Obviously, states of similar energy are now considered separately, but the DMRG sweeps ensure the flow of information between all energy scales. In practice, it turns out that this approach works extremely well, with the *caveat* that the ground state of course does not factorize completely such that we need somewhat more than $\sqrt{M}$ states: e.g., for reproducing the quality of $M = 1024$

state NRG, we need roughly $M = 64$ states in DMRG (instead of $M = 32$), which is still an improvement of $16^3 = 4096$, which is again improved by the smaller local state space, but slowed down by the need for sweeps (of the order of 10 to 20 or so) and the doubling of the chain length. Overall, a gain of several orders of magnitude remains. At the same time, all analyses typically done in NRG can still be carried out: for example, after the ground state is converged, one can still extract the RG flows typically analyzed in NRG by projecting on parts of the chain.[7]

Conceptually, this unfolding technique can be extended to an impurity coupling to several bands (Fig. 4): while in NRG similar energies in each band have to be considered at the same time (leading to an exponential growth in the local state space), in DMRG we obtain a star-shaped structure, where a generalized state factorization argument shows that for k bands one should get the same precision with of the order of $M^{1/k}$ instead of M states (ignoring the additional help from the spin unfolding). This result might be highly relevant for cluster impurity solvers or general multiband problems.

Another advantage of the DMRG formulation of NRG is that logarithmic discretization can be relaxed: in NRG, it has to remain sufficiently strong (i.e. $\Lambda \sim 2$) to separate energy scales, and there can be only one accumulation point at the Fermi edge. Here, we can choose much smaller Λ, but also to some extent make energy discretization finer at any other point of interest in the spectrum, e.g. the region of the Hubbard side bands, which would be important for uses of NRG-DMRG as an impurity solver for dynamical mean-field theory[38] to have a good spectral function for all frequencies. On the other hand, in impurity problems with an external magnetic field leading to a split of the Kondo resonance, one can introduce two suitable accumulation points to describe both resonance peaks at high accuracy. This is quite beyond the possibilites of NRG.

All these issues are currently being exploited and far form fully understood, but from preliminary results I expect that this will give important additional insights to problems so far studied by NRG.

References

1. S.R. White, *Phys. Rev. Lett.* **69**, 2863 (1992); *Phys. Rev. B* **48**, 10345 (1993).
2. U. Schollwöck, *Rev. Mod. Phys.* **77**, 259 (2005).
3. G. Vidal, *Phys. Rev. Lett.* **93**, 040502 (2004).
4. A. J. Daley, C. Kollath, U. Schollwöck and G. Vidal, *J. Stat. Mech.: Theor. Exp.* (2004) P04005.
5. S. R. White and A. Feiguin, *Phys. Rev. Lett.* **93**, 076401 (2004).
6. A. Feiguin and S. R. White, *Phys. Rev. B* **72**, 020404 (2005).
7. F. Verstraete, A. Weichselbaum, U. Schollwöck, J. von Delft and I. Cirac, *cond-mat/0504305.*
8. M. Greiner *et al.*, *Nature (London)* **415**, 39 (2002).
9. M. Köhl *et al.*, *Phys. Rev. Lett.* **94**, 080403 (2005).
10. M. Fannes, B. Nachtergaele and R. F. Werner, *Europhys. Lett.* **10**, 633 (1989).
11. M. Fannes, B. Nachtergaele and R. F. Werner, *Comm. Math. Phys.* **144**, 3 (1992).

12. A. Klümper and A. Schadschneider and J. Zittartz, *Europhys. Lett.* **24**, 293 (1993).
13. S. Östlund and S. Rommer, *Phys. Rev. Lett.* **75**, 3537 (1995).
14. J. Dukelsky, M.A. Martin-Delgado, T. Nishino and G. Sierra, *Europhys. Lett.* **43**, 457 (1998).
15. H. Takasaki, T. Hikihara and T. Nishino, *J. Phys. Soc. Jpn.* **68**, 1537 (1999).
16. F. Verstraete, D. Porras and J. I. Cirac, *Phys. Rev. Lett.* **93**, 227205 (2004).
17. S. R. White, *cond-mat/0508709.*
18. J. I. Latorre, E. Rico and G. Vidal, *Quantum Inf. Comput.* **4**, 48 (2004).
19. J. Gaite, *quant-ph/0301120.*
20. C. Callan and F. Wilczek, *Phys. Lett.* **333**, 55 (1994).
21. J. Gaite, *Mod. Phys. Lett. A* **16**, 1109 (2001).
22. T. Barthel, M. C. Chung and U. Schollwöck, *Phys. Rev. A* **74**, 022329 (2006).
23. F. Verstraete and J. I. Cirac, *cond-mat/0407066.*
24. G. Vidal, *Phys. Rev. Lett.* **91**, 147902 (2003).
25. M. Suzuki, *Prog. Theor. Phys.* **56**, 1454 (1976).
26. C. Kollath, U. Schollwöck, J. von Delft and W. Zwerger, *Phys. Rev. A* **71**, 053606 (2005).
27. C. Kollath, U. Schollwöck and W. Zwerger, *Phys. Rev. Lett.* **95**, 176401 (2005).
28. S. Trebst, U. Schollwöck, M. Troyer and P. Zoller, *Phys. Rev. Lett.* **96**, 250402 (2006).
29. C. Kollath *et al.*, *Phys. Rev. Lett.* **97**, 050402 (2006).
30. K. Rodriguez *et al.*, *cond-mat/0606166.*
31. D. Gobert, C. Kollath, U. Schollwöck and G. Schütz, *Phys. Rev. E* **71**, 036102 (2005).
32. K. A. Al-Hassanieh *et al.*, *cond-mat/0601411.*
33. F. D. M. Haldane, *J. Phys. C: Solid State Phys. 14*, 2585 (1981); *Phys. Rev. Lett.* **47**, 1840 (1981).
34. J. Voit, *Rep. Prog. Phys.* **58**, 977 (1995); T. Giamarchi, *Quantum physics in one dimension*, Oxford University Press, Oxford (2004).
35. T. Kühner and S.R. White, *Phys. Rev. B* **60**, 335 (1999).
36. E. Jeckelmann, *Phys. Rev. B* **66**, 045114 (2002).
37. K. G. Wilson, *Rev. Mod. Phys.* **47**, 773 (1975).
38. A. Georges, G. Kotliar, W. Krauth and M. J. Rozenberg, *Rev. Mod. Phys.* **68**, 13 (1996).

AUTHOR INDEX

www.ingramcontent.com/pod-product-compliance
Lightning Source LLC
Chambersburg PA
CBHW041754180926
51764CB00011B/132
* 9 7 8 9 8 1 2 7 0 9 5 8 5 *